PRINCIPLES OF ORGANIC CHEMISTRY

PRINCIPLES OF ORGANIC CHEMISTRY

FOURTH EDITION

JAMES ENGLISH, JR.
Yale University

HAROLD G. CASSIDY
Yale University

RICHARD L. BAIRD
E. I. duPont deNemours & Co.

McGRAW-HILL BOOK COMPANY

New York St. Louis San Francisco Düsseldorf Johannesburg
Kuala Lumpur London Mexico Montreal New Delhi Panama
Rio de Janeiro Singapore Sydney Toronto

COVER ILLUSTRATION:

A photomicrograph of an organic reaction—
the formation of a polyester,
printed in reverse.
The micrograph was made by Dr. Roman Vishniac.

PRINCIPLES OF ORGANIC CHEMISTRY

Library of Congress Catalog Card Number 76-137125
07-019520-X

2 3 4 5 6 7 8 9 0 M A M M 7 9 8 7 6 5 4 3 2 1

This book was set in News Gothic by York Graphic Services, Inc., and printed on permanent paper and bound by The Maple Press Company. The designer was Paula Tuerk; the drawings were done by John Cordes, J. & R. Technical Services, Inc. The editors were James L. Smith and Anne Marie Horowitz. John A. Sabella supervised production.

CONTENTS

PREFACE

This text has been designed for use in a full-year course in organic chemistry—a course designed to stress the fundamental principles of the science for students planning to go on to advanced scientific work in medicine, chemistry, chemical engineering, and allied fields.

It is the authors' conviction that the object of a first course in organic chemistry should be to present to the student a selected body of factual and theoretical material and show how this carefully chosen material, the result of experiment, is used in the development and practice of the science. To this end we have chosen those reactions and concepts which seem to us to be of fundamental significance and applicable to a variety of different situations. We have left out of our discussion a great many reactions and compounds often found in organic chemistry texts, recognizing that only a reasonably restricted amount of material can be *fully* mastered by the student in the time available.

In consonance with this conviction, we have omitted the customary "special topics" chapters that adorn the later pages of many textbooks. Such chapters on dyes, alkaloids, drugs, etc., although they furnish useful reference material, have not been found by us to be of importance or value in the elementary course. With the increasing development of organic theory, the existence of specialists in organic chemistry—the alkaloid chemist, the terpene chemist, etc.—is threatened with extinction. With a thorough grounding in the principal reactions and theories of the subject, the student today can safely venture into the literature of these special fields without fear of encountering mysterious new kinds of behavior of carbon compounds. Those who are for one reason or another interested in examining the structures and reactions of specific examples in these fields find it necessary in any event to consult more specialized articles than can properly be included in an elementary text.

This policy of including only selected reactions implies a more detailed study of the underlying principles involved than is usually found in elementary texts. This we have attempted to accomplish by introducing modern ideas on the mechanism of reactions in all cases where they have served to unify the treatment. An attempt has been made to describe in some detail the mechanisms of the principal reactions involved.

A rather large number of exercises and problems is included, and these are assigned liberally. It has been our experience that only from the working of problems such as those found at the end of each chapter can a student acquire facility in handling the reactions discussed. The application of memorized facts and theories to the solution of actual problems has been found to be of great importance to student understanding and appreciation of organic chemistry. The exercises have been so devised that the answers are seldom given by direct quotations from the text; compounds used in the exercises are varied so as to emphasize the essential principles rather than specific isolated facts. Answers to selected exercises (marked with an asterisk) are included as an aid to study.

A course based on this text may be varied from a one-term course for non-science students to a full-year course for chemistry majors, depending on the extent to which the exercises are used. The instructor may introduce additional material at his discretion, and, indeed, is urged to do so.

Aromatic and aliphatic compounds are treated together after Chap. 5. Since the differences in reactions of these two classes are often of a quantitative rather than a qualitative nature, where they exist at all, it is felt that the advantages of this treatment are very real. This order of presentation has been used before by other authors, and the writers have used it for a number of years in their own courses with gratifying results.

In the fourth edition we have tried to retain the essence of the previous editions, that is, to present a core of facts and principles by which the student will be equipped to tackle new problems of greater complexity. In so doing we have tried to approach the subject as a modern organic chemist would, rather than as it has been approached in times past.

Problems of identification, characterization, and detection of both old and new compounds were once solved by laborious and often ambiguous chemical tests. Organic chemists have combined the knowledge gained from these older methods with information provided by spectroscopic techniques and have evolved correlations and analogies that prove extremely useful in unraveling the structures of new compounds, as well as in confirming the identity and presence of known materials. We have greatly increased the space devoted to these newer methods, cutting back and even deleting coverage of the now less-important historical methods.

Certain sections included in the third edition, such as "Historical Notes" and "Industrial Notes," have been deleted, as has been the chapter on heterocycles.

Changes in each of these areas have been so great as to make a respectable treatment too long for a text of this nature. On the other hand, a new chapter on the interaction and competition between various functional groups has been added so as to prepare the student to better understand these phenomena when he encounters them in new situations.

This book owes a great deal to influences exerted by teachers, students, and the "literature"—influences which it would be most difficult to acknowledge in detail, but which we do recognize and acknowledge as best we can. Many of the exercises are, we think, our own. We are indebted to Prof. Carl R. Noller and the editor of the *Journal of Chemical Education* for permission to use several figures.

JAMES ENGLISH, JR.
HAROLD G. CASSIDY
RICHARD L. BAIRD

REMARKS FOR THE STUDENT

It is usually taken for granted that the student who takes up the study of organic chemistry has a thorough knowledge of first-year college chemistry. This means that he is familiar with the symbols for atoms and with simple formulas; that he knows enough about the electronic structures of atoms and the ways that bonds are formed between atoms to be able to write down the structures of simple molecules; that he has a general notion of the properties of solids, liquids, and gases; that he knows how valence numbers and molecular weights are determined; that he knows what ions are, and how their presence is recognized; that he is able to use the concepts embodied in the principle of Le Châtelier; and that he can look at the periodic table with an educated eye. It is quite likely, of course, that many of the finer details of first-year chemistry have been forgotten, and the student of organic chemistry will do well to refer to his first-year text immediately if he feels vague about some point. Every attempt will be made to relate the new materials in this course to the earlier work which the student has had and to suggest its relation to work which he will take in other courses, for example, to work in physical, biological, or technical chemistry.

But there will be a great deal of material which will be quite new to the student. This will be made up partly of familiar materials which will be brought together in new ways, for example, in the use of familiar symbols which are constructed into two- or three-dimensional molecular formulas, and partly of new concepts and ideas which will have to be embodied in additional symbols and special language. The student must, therefore, be prepared for memory work. Since this is the only convenient way to master some of the materials of the course, he might just as well resign himself to it and get on with it. However, he should find that the amount of memory work is no greater than that involved in learning a foreign language, or in learning comparative anatomy. Premedical students regularly pass, and even enjoy, organic chemistry.

Experience has shown that the first half of the first term of the organic chemistry course (about the first eight weeks) is the most crucial part. If the student masters this first work, the rest of the course becomes relatively easy. This is because the subject matter of organic chemistry is very closely bound together internally. The material learned at the beginning of the course will be used at least implicitly, and usually explicitly, throughout it. The course develops by a process of continual enrichment; it is not a journey in which the road is traveled only once. Much of this early material in the course looks easy, and so the student may have a tendency to slight it. In our experience this is one of the chief causes of the difficulty which some students have with organic chemistry.

One of the objectives of science is to gain understanding of the perceptual world through the manipulation of symbols and concepts. In the branch of science called organic chemistry, not only are the methods of manipulating symbols and concepts very clearly illustrated, but the gains in understanding natural phenomena which have resulted are set forth on every hand to be seen. Almost from the very beginning of the course, the student will be able to appreciate the unfolding methods of this science. It is an important quality of this course that throughout it are illustrated in action the elements of scientific philosophy—the elements of part of the culture of the modern educated individual.

ON SOLVING PROBLEMS

A few words need to be said about the exercises at the ends of the chapters and their solution. These problems vary: A few ask purely routine questions. Others ask for the use of something more than memory. In most cases we have tried to provide problems which probe below the surface of the text material. The student should try to approach these problems in a friendly spirit. They are the best means he has for gauging how well he is mastering the textual material.

Although there is no substitute for actual thinking, there is a technique for dealing with and solving most problems—not only the types of problems found in this book, but all problems. This technique (which is an old one) can only be sketched here, but the student who is interested can find a more complete treatment in a book by Polya.[1] This technique consists in an orderly approach to the problem, using a scheme like the following:

First step: State the problem. This means to write down briefly but clearly and in your own words what you want to do and any related information which clarifies your objectives.

[1] G. Polya, "How to Solve It. A New Aspect of Mathematical Method," Princeton University Press, Princeton, N.J., 1945.

Second step: Look for similarities. See if there is any feature of the problem which is like that of a problem you solved once before. If so, perhaps this is like that other problem, only with a different twist. This step is a kind of partial diagnosis. If you find a similarity, go ahead and solve the problem. Analogy is one of your most powerful tools, and is, in fact, used by most investigators to suggest directions for research. If you suspect a similarity but can't think of it, go back through the text and check for the elusive resemblance. Always use the index freely. It is there to be used. Remember that these problems all have solutions. If there is no similarity to a problem you have seen or worked out before, take the third step.

Third step: Devise an attack on the problem. With organic chemistry problems this almost always means starting to work the problem backwards, in a stepwise manner. This is a reasonable thing to do. For example, suppose you are asked to convert compound Q to compound Z. There are a great many reactions which you can apply to Q, only a few of which can ultimately lead to Z. Why not start with Z and work backwards, keeping Q in mind? If at a given point you are "stuck"—say you can't think of any way of making Z—use the index. Look up ways of making Z-like substances. Also, as you go through the course, keep tables of useful reactions, such as those for shortening and for lengthening carbon chains, etc.

Fourth step: Carry out the plan of attack which you devised. This is usually done along with the third step.

Fifth step: Check your solution. Find out if it meets the conditions set down in the original problem.

Next time you get stuck on a problem try out this systematic technique.

The following references relate to studying and to solving problems:

H. M. Dadourian, "How to Study, How to Solve Arithmetic through Calculus," Addison-Wesley Publishing Company, Reading, Mass., 1951.

W. Jepson, "How to Think Clearly," Longmans, Green & Co., Inc., New York, 1937.

C. T. Morgan and J. Deese, "How to Study," McGraw-Hill Book Company, Inc., New York, 1957.

G. Polya, "How to Solve It. A New Aspect of Mathematical Method," Princeton University Press, Princeton, N.J., 1945.

M. Wertheimer, "Productive Thinking," Harper & Brothers, New York, 1945.

1

INTRODUCTION

1.1 Definitions Chemistry is the science that deals with the structures and properties of substances and with the reactions by means of which these structures and properties may be changed. Organic chemistry is the chemistry of those substances containing carbon. Some compounds which, according to this definition, would be called organic have been studied in the inorganic part of first-year chemistry. The carbonates, acetates, cyanides, and complex cyanides may be listed here, together with carbon dioxide, as examples. Some of these compounds, for example, calcium carbonate and potassium cyanide, are easily thought of as being inorganic because their chemistry is so much like that of inorganic substances of various kinds, and because substances like marble are, after all, minerals. But coal and petroleum are also minerals, and no one would argue that they are not organic substances.

Besides, many inorganic substances display properties which are like those of organic substances, so that the difference between organic and inorganic chemistry (which has to be made for very practical reasons, apart from any historical justification) cannot rest on degree of likeness of reactions but is much better made arbitrarily, sharply dependent on the presence of carbon.

1.2 The Carbon Atom Since all organic molecules contain carbon, it is important to know certain facts about the atom itself and the substances formed from it. Carbon has an atomic weight of 12, an atomic number of 6, and is classified in group IV of the second period of the Periodic Classification of the elements.

The first shell of electrons outside the nucleus of the carbon atom contains two electrons and is, therefore, a complete shell; the second, the valence, shell contains four electrons. In atomic carbon the electrons within different shells are not all located identically with respect to the nucleus. They have different energies. The details of this arrangement will be discussed later in this chapter (Sec. 1.5). Since the valence shell is completed only when eight electrons occupy it, it is apparent that the carbon atom requires four additional electrons for completion of this shell.

It is energetically very favorable for atoms to form arrangements in which the valence shell is completely filled or completely empty. This may be accomplished by the gain or loss of electrons, or by a process known as sharing. In either case the process of completing the valence shell is intimately associated with that of chemical bond formation.

1.3 Chemical Bonds Four types of chemical bonds are usually distinguished: electrovalent, covalent, coordinate, and metallic. Metallic bonds will not be considered here.

Electrovalent, or *ionic*, bonds are those in which the bonded atoms carry explicit charges. The familiar case is Na^+Cl^-. Here the two atoms are held together (and to other charged atoms) by the electrostatic (Coulomb) attraction between the positive and negative charges. Na^+ has lost its only valence electron, and so shows the effect of an uncompensated nuclear proton; the chloride ion

$$:\ddot{\underset{\cdot\cdot}{Cl}}:^-$$

has a completed valence shell, which means that it has one more electron than needed to compensate for its nuclear charge and, therefore, is a negative ion of charge -1. An electron has been transferred, in effect, from the valence shell of one atom, Na, to that of another, Cl. Note that the bond between a positive ion and a negative ion does not usually have a favored direction in space.

In crystalline NaCl each Na^+ ion is surrounded by six Cl^- ions, which are in turn surrounded by six Na^+ ions so that the whole mass is bound together into a giant network. This arrangement represents the best compromise between each ion being surrounded by as many unlike ions as possible and the limitations imposed by the finite sizes of the various ions. Such ionic networks, or *lattices*, usually are difficult to break down. For this reason, ionic substances generally exhibit high melting and boiling points.

A *covalent bond* is formed by the sharing of two electrons by the bonded atoms, one electron being, in effect, contributed by each of the bonded atoms (no matter how the bond is actually produced). There is no permanent transfer of electrons. Both bonding electrons contribute to the valence shells of each bonded atom—they are shared.

The actual bonding force is sometimes described as the electrostatic interaction between the nuclei of the bonded atoms and the two bonding electrons, the latter spending most of their time *between* the two nuclei. In the extreme (although not uncommon) case of two essentially identical atoms, the sharing will be equal and there will be no net electrostatic charge separation between the two atoms. In many cases, however, there will be an unequal sharing of electrons due to a greater affinity of one of the bonded atoms for electrons; i.e., it has greater electronegativity. See Sec. 1.18. In such cases there will be a partial charge separation, and the bond is said to be *polarized*. Examples of the above types of bonds are:

$$:\ddot{\underset{..}{Cl}}:\ddot{\underset{..}{Cl}}: \qquad :\overset{\delta^-}{\ddot{\underset{..}{Cl}}}:\overset{\delta^+}{\ddot{\underset{..}{Br}}}: \qquad :\overset{\delta^-}{\ddot{\underset{..}{Cl}}}:\overset{\delta^+}{\ddot{\underset{..}{I}}}:$$

In the first compound, Cl_2, the sharing of the two bonding electrons is equal. In the second and third compounds the bonding electrons spend more time in the vicinity of the chlorine atom than in that of the bromine or iodine atoms. Due to the greater electronegativity of chlorine, the bonds in BrCl and ICl are polarized—with a small excess of negative charge on the chlorine atom and a corresponding excess of positive charge on the bromine and iodine atoms. For convenience we will often use δ^+ and δ^- to indicate this polarity in covalent bonds. δ may be interpreted to mean "relatively a little more than."

Factors such as polarization (and other as yet unmentioned factors) can give covalent bonds a certain degree of *ionic character*. It is, in fact, possible to find bonds ranging from essentially purely covalent ones with little ionic character, all the way to essentially 100 percent ionic ones. There is, thus, no sharp line of demarcation between covalent and ionic bonds.

The vast majority of the bonds that are classified as covalent retain certain characteristics which serve to distinguish them from ionic bonds. These properties are *saturation* and *directional* character. Saturation means that when two atoms are bonded by a covalent bond, the shared electrons contribute essentially only to this one bond; the electrons do not create large interactions between the two bonded atoms and other atoms in the vicinity. The directional characteristic means that covalent bonds about a given atom are formed at certain well-defined angles to one another, as in the case of H_2S, where the H—S—H bond angle is known to be 90°. As the degree of ionic character in a bond is increased, the greater separation of charge leads to increased attractive (electrostatic) forces between atoms in one bond and oppositely charged atoms in other bonds. Ultimately, a situation is reached where these forces of attraction between parts of different molecules rival those forces between bonded atoms in the same molecule. The properties of saturation and directional character are lost and the bond is then best described as ionic. Fortunately, cases in which it is difficult to make a distinction between ionic and covalent bonds are relatively few in organic chemistry; most carbon compounds are made up largely of covalent bonds.

Simple compounds containing only covalent bonds exist in the form of units called *molecules*. Molecules may vary in size from those containing only two atoms to those containing hundreds or thousands of atoms. The atoms within a molecule are covalently bonded to other atoms within the same molecule, but the atoms of one molecule interact only weakly with those of another molecule. Because the forces between molecules are weak, it is comparatively easy to pull them apart from one another. This gives rise to the low melting and boiling points observed for many covalent compounds. When molecules contain bonds tending more toward ionic character, intermolecular forces increase, as do the observed boiling and melting points.

The *coordinate* bond is a type of covalent bond in which there is a "built in" polarization. The bond itself is an ordinary covalent one, but the bonded atoms usually have positive or negative charges due to the manner in which electrons are distributed relative to the nuclear charges. In order to show this clearly it is useful to discuss the "bookkeeping" by which chemists keep track of electrons in complex molecules.

First of all, isolated atoms of a given element are found to possess a nuclear charge which is balanced by the sum of the negative charges of the electrons in the outer shells, so that the whole atom is electrically neutral. It is customary to ignore electrons in shells closer to the nucleus than the valence shell, treating the inner electrons plus the nucleus as the *kernel* of the atom. The kernel will have a net positive charge equal in magnitude to the negative charge provided by the valence electrons, the remainder of the nuclear charge being accounted for by the inner shells of electrons. When an atom is involved in a chemical bond, the number of electrons in the valence shell may not correspond to the number of positive charges in the kernel, and the atom will possess a *formal charge*. The formal charge will be negative if there are more electrons in the valence shell than in the neutral atom; it will be positive if there are fewer electrons than in the neutral atom. For the purpose of calculating formal charges, one counts all the unshared electrons possessed by an atom outside the kernel and one-half all shared electrons. If these add up to be greater than the number of plus charges on the kernel, the difference is the formal negative charge. Most of the atoms in organic compounds turn out to have zero formal charge according to this scheme, but a significant number of atoms do not.

Our bookkeeping scheme is illustrated pictorially by indicating the unshared valence electrons as dots and the shared pairs as two dots or a dash between atoms. Thus the compound dimethylsulfoxide may be represented as:

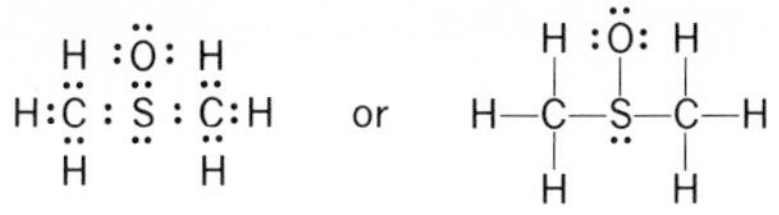

In this compound it can be seen that the hydrogens with a kernel charge of plus one and the carbons with a kernel charge of plus four come out with zero formal charge. However, both oxygen and sulfur have kernel charges of plus six. Our bookkeeping scheme indicates that oxygen is in possession of all six unshared electrons plus half of the two shared electrons for a total of seven electrons. Therefore, oxygen has a formal charge of minus one. Correspondingly, sulfur has a formal charge of plus one. The molecule is, therefore, customarily written:

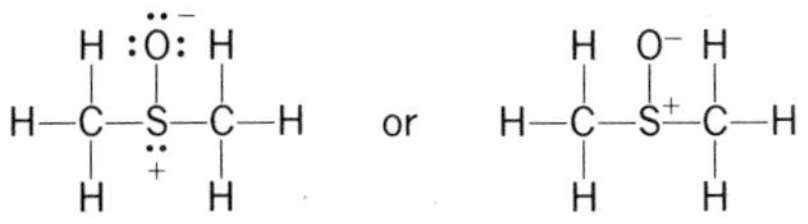

The bond between sulfur and oxygen in this case is in most respects a covalent bond, but it is referred to as a coordinate bond. In coordinate bonds the involved atoms commonly carry formal charges.

1.4 Structure of Carbon in Compounds It has been known for a long time that carbon is capable of forming four covalent bonds with a variety of elements. The organic chemist has been concerned not only with the composition of carbon compounds, but with the *structures* of their molecules. In representing the structure of molecules on paper, several systems of symbols have had to be devised. The most common system is to use the symbol for the element to represent the kernel of the element, i.e., the nucleus and all electrons except the valence electrons. Thus, methane may be represented as CH_4, or as

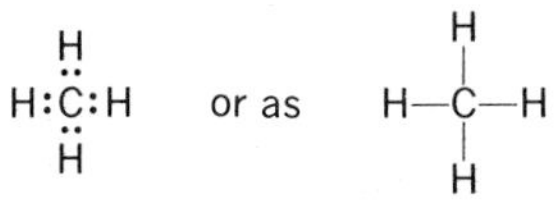

where C and H represent the kernels of carbon and hydrogen atoms respectively, and the dashes or pairs of dots represent the covalent bonds comprising a pair of electrons. In many cases, however, even these detailed diagrams do not adequately represent organic compounds, since the atoms and molecules are three-dimensional.

A large body of evidence exists to support the fact that the carbon atom is *tetrahedral* in many organic compounds. Some of the most convincing evidence is beyond the scope of this book, but sufficient evidence can be described below and in later sections to support this conclusion.

If one or more of the bonds of tetravalent carbon were different from the rest, then it would be expected that there might be more than one compound with the

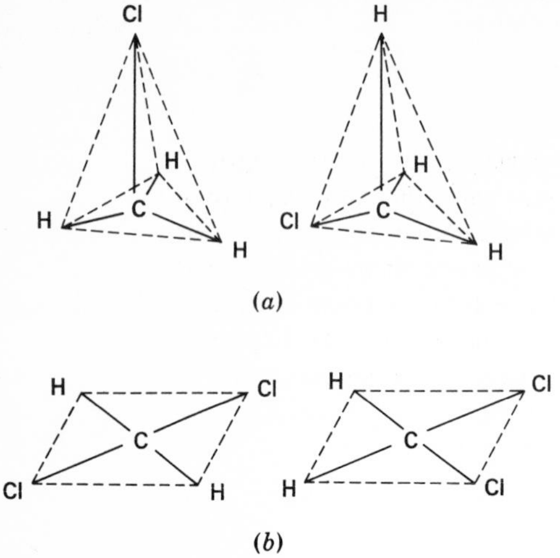

figure 1-1
Conceivable structures. In (*a*), a hypothetical pyramidal structure, there would be two molecules with the formula CH_3Cl, one with the Cl at the apex of the pyramid, and one with the Cl at the base. These would not be identical because they are not superimposable (in imagination, of course). In (*b*), a hypothetical planar structure, there would be two different arrangements possible for a compound of the formula CH_2Cl_2.

formula CH_3Cl. See, for example, Fig. 1-1*a*. However, all pure samples of CH_3Cl, no matter by what different reactions they have been made, have been found to be identical. There is no evidence to indicate that one of the bonds of tetravalent carbon is different from the others.

If carbon compounds were planar (two-dimensional) or pyramidal, it should be possible to find two compounds of the type CH_2Cl_2, namely, those indicated in Fig. 1-1*b*. Again, all pure samples of CH_2Cl_2 which have been prepared have been found to be identical in chemical and physical properties; that is, they are indistinguishable from one another. By comparison, in the inorganic compound $PtCl_2(NH_3)_2$, which is known to exist in the square planar configuration, two different compounds of this formula have been found corresponding to the structures shown in Fig. 1-2.

The validity of these arguments has recently been confirmed by use of physical methods such as microwave and x-ray spectroscopy which yield precise information

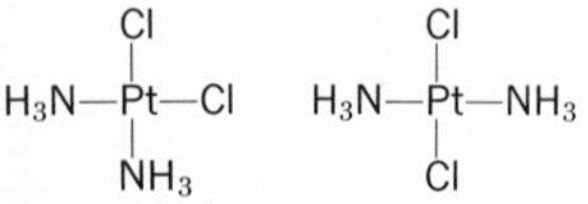

figure 1-2
Square planar platinum complex, $PtCl_2(NH_3)_2$.

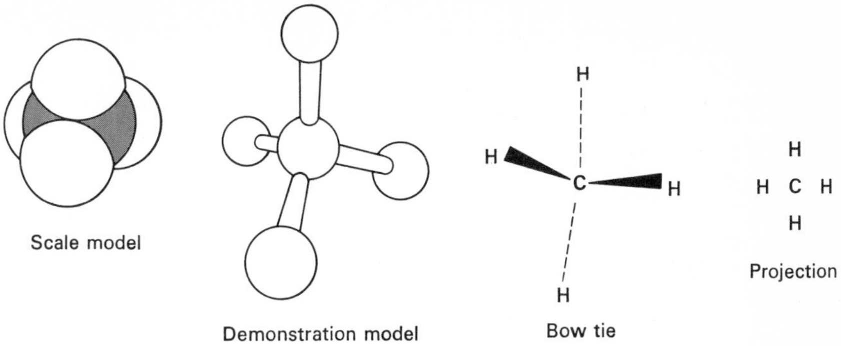
Scale model
Demonstration model
H
H
C
H
H
Bow tie
H
H C H
H
Projection

figure 1-3
Models of the methane molecule, CH_4.

about actual bond lengths and angles, and accordingly, provide a direct demonstration of the tetrahedral nature of carbon.

Tetrahedral carbon may be represented in a variety of ways, as illustrated in Fig. 1-3. The scale model shows CH_4 with the atoms drawn approximately to scale. The demonstration model emphasizes the tetrahedral bond angles and more clearly illustrates the bonds. The "bow-tie" model represents a convenient method for drawing the three-dimensional carbon atom on paper. The wings of the bow tie represent bonds extending above the plane of the paper; the dashed lines represent those extending below. Where no ambiguity is possible it is customary to use the projection model, in which the tetrahedral nature is understood.

1.5 Some Principles of Orbital Theory Since all modern elementary textbooks go into the theories of atomic structure in some detail in terms of shells and orbitals, only a brief review will be given here.

An atom may be visualized as a positively charged nucleus surrounded by negatively charged electrons, which are located in shells at certain average distances from the nucleus. The electrons within a given shell possess roughly comparable average energies, but in shells farther removed from the nucleus, the electrons within a given shell can be further grouped into subshells or *orbitals*. These orbitals differ both in shape (regions of highest average electron density) and in the energy possessed by electrons occupying them. The orbitals of lowest energy within a given shell are called *s* orbitals and are spherical in shape; the ones having the next highest energy are called *p* orbitals and are roughly dumbbell-shaped (Fig. 1-4). Following the *p* orbitals are *d* and *f* orbitals. These latter two orbitals are of more complex shapes than the two preceding orbitals and will not be discussed here as they are not often encountered in carbon compounds.

As one proceeds out from the nucleus, the first shell, with the smallest average energy and distance from the nucleus, is the 1 shell. It contains only one orbital, an *s* orbital, or more specifically, a 1*s* orbital. The next shell is the 2 shell, which

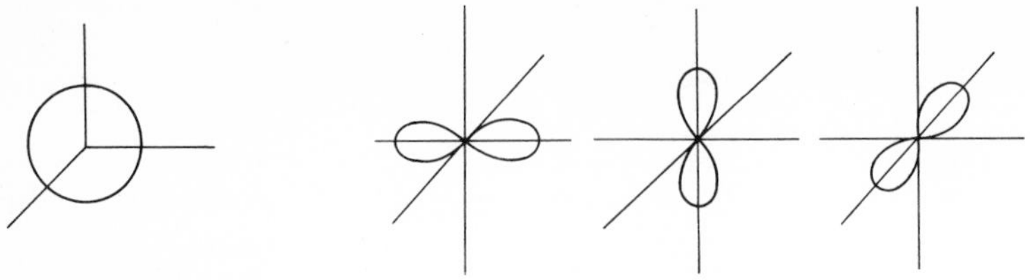

s-Orbital (average electron density)

p-Orbitals (shown along *x*, *y*, and *z* axes)

figure 1-4
Shapes (regions of maximum electron density) for *s* and *p* orbitals.

contains one 2*s* orbital and three 2*p* orbitals. The *p* orbitals differ from one another in that they are located at right angles to one another (usually shown as along the *x*, *y*, and *z* axes). The third shell contains one 3*s* *s* orbital, three 3*p* *p* orbitals, and five 3*d* *d* orbitals. Only in the fourth shell are *f* orbitals found; the chemistry of these (and for the most part of the *d* orbitals) is classified by the lower half of the periodic table and will be of less interest to the student of organic chemistry.

The electronic structures of stable *atoms* can be built up by supplying electrons in numbers equal to the nuclear charge and placing them in the orbitals of lowest energy. The *Pauli exclusion principle* tells us that *any given orbital may contain no more than two electrons* and *these must be of opposite spin*. The spins are conventionally indicated by arrows pointing up or down (↑ or ↓). When more than one orbital of equal energy occurs, e.g., the three *p* orbitals, the electrons are placed according to *Hundt's rule*. This tells us that *it is energetically more favorable to place one electron with the same spin in each of the equivalent orbitals before placing the second electron (with opposite spin) in any of these orbitals*. By following these principles, Table 1.1 can be built up. Table 1.1 shows that the first shell is filled when it contains two electrons, the second when it contains eight electrons, etc.

The concept of *atomic orbitals* may be extended to molecules by means of the concept of *molecular orbitals* (MO). According to the molecular orbital theory, when two atoms share a pair of electrons and become joined by a bond, the process

TABLE 1.1
SHELL STRUCTURE OF STABLE ATOMS

	H	He	Li	Be	B	C	N	O	F	Ne
$n = 1$	(First shell)									
$1s$	↑	↑↓	↑↓	↑↓	↑↓	↑↓	↑↓	↑↓	↑↓	↑↓
$n = 2$	(Second shell)									
$2s$			↑	↑↓	↑↓	↑↓	↑↓	↑↓	↑↓	↑↓
$2p_x$					↑	↑	↑	↑↓	↑↓	↑↓
$2p_y$						↑	↑	↑	↑↓	↑↓
$2p_z$							↑	↑	↑	↑↓

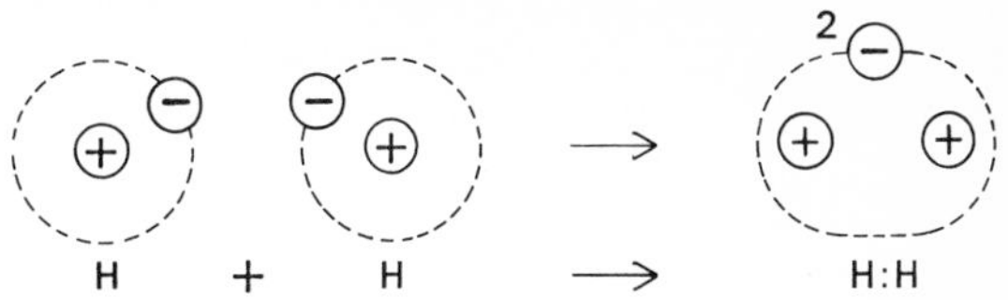
2
H
+
H
H:H

figure 1-5
Union of two hydrogen atoms to form H_2 according to the molecular-orbital picture. In the two separated atoms the electrons are in the $1s\uparrow$ and $1s\downarrow$ states. In the molecule they form a molecular orbital about the nuclei. The bond is called a σ (sigma) bond.

involves the overlapping of two atomic orbitals wherein the separate electrons previously resided. From this overlapping there results a new orbital which includes both kernels, i.e., a molecular orbital is formed (Fig. 1-5). When such a bond is formed, energy is released (by definition of the term *bond*) and the resulting structure is more stable than the system of separated atoms from which it was derived. The most common type of bond is formed by overlap of orbitals along the line of centers of the two atoms involved, and may utilize the overlap of two s orbitals, s and p orbitals, two p orbitals, or others. Such a bond has a high average electron density between the two nuclei and is called a "sigma bond" (σ bond). The *two* electrons in such a bond are now located in a molecular orbital (which takes the place of the *two* atomic orbitals from which it was formed) and must, according to the Pauli exclusion principle, be paired.

1.6 Hybridization of Carbon Bonds The electronic structure of the elementary carbon atom (Table 1.1) comprises a filled shell of two $1s$ electrons and a partially filled second shell containing two $2s$ and two $2p$ electrons. The $1s$ electrons do not become involved in chemical combination and, therefore, belong to the kernel of the atom. Two of the remaining electrons are located in the $2s$ orbital and the other two electrons are in two of the three available $2p$ orbitals. The permitted p orbitals are located with their axes at right angles to one another, as indicated in Fig. 1-4.

As we have seen, carbon in most of its compounds is tetravalent, and the four bonds formed are often equivalent. This situation is interpreted in the following way. Although the most stable arrangement for atomic carbon is that indicated above, when bonds are formed (releasing energy), the electrons are redistributed in such a way that all four orbitals are utilized. These orbitals have comparable energies and the four carbon electrons are distributed between all four of them. The simplest way of utilizing all these orbitals would be to form one bond with the $2s$ orbital and one with each of the three p orbitals. The formation of a covalent bond, however, involves a compromise between the attractive overlap of atomic orbitals and the electrostatic repulsion of the two nuclei as they approach one another. The average distance between two bonded nuclei is thus determined by the point at which these repulsive forces just balance the attractive forces of overlap.

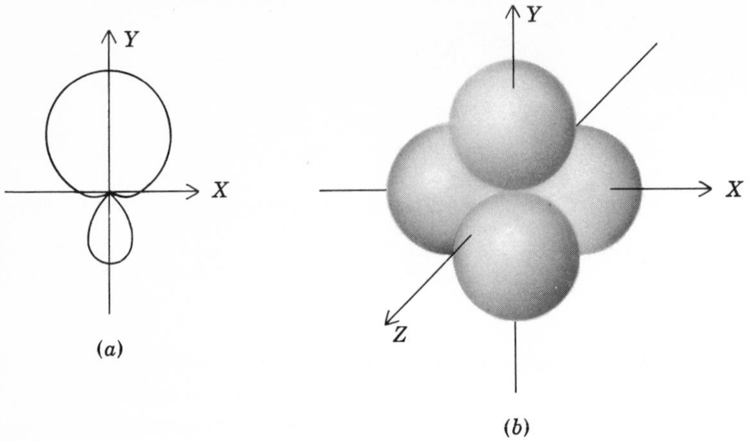
Y
X
(a)
Y
X
Z
(b)

figure 1-6
(*a*) Cross section of a single sp^3 hybrid atomic orbital and (*b*) perspective of four sp^3 hybrid orbitals. [*From C. R. Noller, J. Chem. Ed.,* **27**:*505 (1950).*]

It follows that stronger bonds can be obtained if it is possible to utilize orbitals which normally extend farther from the nucleus, since significant overlap can be obtained before the nuclei approach too closely. In this connection, *s* orbitals are extremely poor at overlap, and *p* orbitals are much better. It is possible to "reshuffle" the *s* and *p* orbitals and, while keeping the same overall electron density about an atom, to construct new *hybrid* orbitals which extend even farther from the nucleus than *s* or *p* orbitals. In the case of carbon new hybrid orbitals can be formed from the 2*s* orbital plus the three 2*p* orbitals. Four hybrid orbitals, called sp^3 orbitals to indicate their origin, result and have the shapes indicated in Fig. 1-6. They are directed at tetrahedral angles. These hybrid orbitals represent an energetically unfavorable arrangement for atomic carbon, but the additional energy gained by bond formation involving overlap with these orbitals more than compensates for this in *compounds* of carbon.

This phenomenon of hybridization is very common and is involved in many ways with both carbon and other elements. *However, it is usually only involved when bond formation can compensate for the orbitals' higher inherent energy.* In cases where four groups are not available to form bonds with carbon, hybrid orbitals are formed which involve the *s* and only one or two of the *p* orbitals (the remaining *p* orbitals being left untouched). These are designated *sp* and sp^2 orbitals, respectively. See Fig. 1-7. They will be of more interest to us in subsequent chapters.

1.7 Resonance The term resonance is used in organic chemistry to describe situations in which certain molecules are found to be much more stable than one would expect by simple analogy with other molecules. It has been possible to formulate a set of rules by which such situations can be recognized and whereby crude estimates of their stability can be made. The basic "law" of resonance may

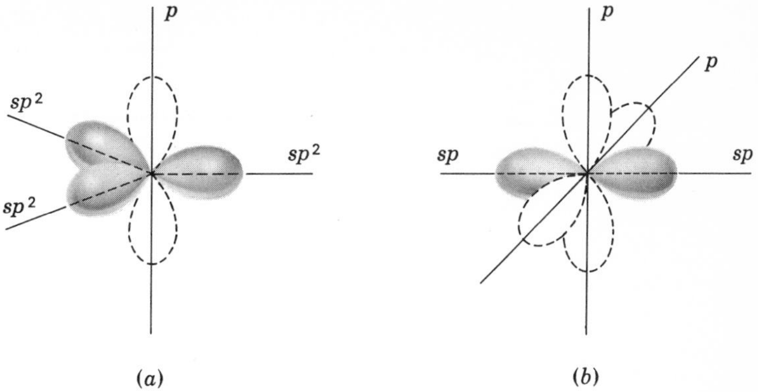
p
sp 2
sp 2
sp 2
p
p
sp
sp
(a)
(b)

figure 1-7
Additional types of hybridization. (*a*) Formation of sp^2 orbitals. (*b*) Formation of sp orbitals.

be phrased: *When a molecule occurs which can be represented by two or more equivalent arrangements of the valence electrons, that molecule will be found to be more stable than would be expected from consideration of only one of the arrangements.* Such a molecule is called a *resonance hybrid* to denote the fact that it is both more stable than any one of these equivalent arrangements, and that it has properties different from any of them.

The so-called "rules of resonance" serve to define what is meant by *equivalent.* For our purposes we can say that structures are equivalent when they differ primarily in electronic arrangement, are of comparable energy, and do not differ appreciably in nuclear positions. These concepts are best illustrated by consideration of an example, such as the nitrate ion (Fig. 1-8). Structures I, II, and III are all equivalent and satisfy the above rules since they differ primarily in electronic energy and not significantly in nuclear position. Structure IV and other permutations of it would require a great change in the O—N—O bond angle and are accordingly of minor importance.[1] Structures V and VI are of higher energy than I, II, and III because V has a large accumulation of charge (electrostatically unfavorable) and one less bond, and VI has one less bond and an incomplete octet on an electronegative atom (oxygen), which is unfavorable. We summarize by saying that the nitrate ion

[1] Any contribution from Structure IV would have to come from a highly strained geometry of IV since the resonance hybrid will have a geometry similar to I, II, and III.

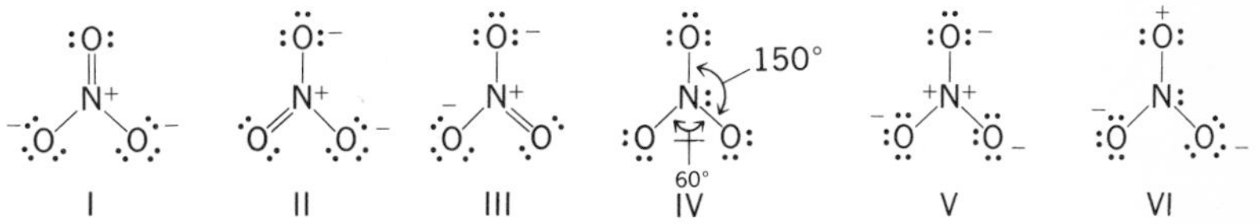
150°
60°
I
II
III
IV
V
VI

figure 1-8
Possible equivalent structures for the nitrate ion, NO_3^-.

is a hybrid of Structures I, II, and III with only minor contributions from IV, V, VI, etc. It is not possible to write a simple structure for the nitrate ion in our customary dot and dash representation that contains all of the above information. It is customary to describe the hybrid by writing the most important contributing structures connected by double-headed arrows. Thus

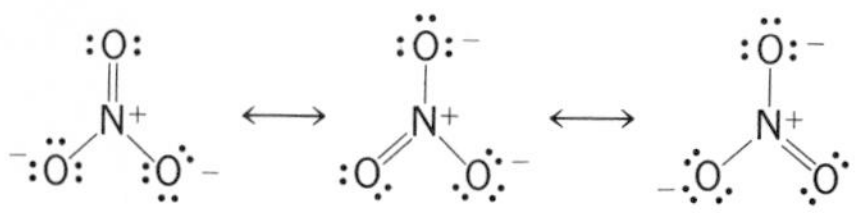

Such a representation tells us several things:

1. The nitrate ion is more stable than a structure such as I would be.
2. Structures I, II, and III *do not exist* since they would, if formed, be immediately converted into the hybrid.
3. The hybrid is symmetrical, with all three N—O bonds and the O—N—O bond angles the same.

1.8 States of Matter At ordinary temperatures a substance like oxygen is a gas because the kinetic energy of the molecules overcomes any tendency of the molecules to adhere to each other. The molecules collide and rebound from each other with an energy which precludes liquefaction and which is directly related to the temperature.

Substances which liquefy at ordinary temperatures do so because the physical attractive forces between the molecules are strong enough to restrain some of the effects of the kinetic motion and hold the molecules nearly in contact. But a liquid takes the shape of its container; that is, it flows. This means that the molecules are still free enough to move about in a limited way. They are found to vibrate, rotate, and undergo translatory motion (to diffuse). At the exposed surface of some liquids, certain of the molecules may have enough velocity to escape into the space above the surface. Then the substances will exhibit a vapor pressure. At lower temperatures most of the surface molecules may be unable to leave the bulk liquid. In this case the liquid will be practically nonvolatile at the temperature in question.

In solids the restraining forces between molecules are great enough to prevent flow. The molecules may be able to vibrate in a limited way, but their average positions relative to neighboring molecules are fixed. If the molecules are arranged in an orderly fashion, the solid is said to be crystalline; if the arrangement is disorderly, the solid is called amorphous. Molecules of certain solids are able to escape from the surface. In such cases, the solid will show a vapor pressure, and is said to sublime. Sublimation from the solid state is analogous to evaporation from the liquid state. (For a solid, "moth crystals" show a relatively high vapor pressure.) The inherent ability of molecules to escape from the surfaces of solids is shown when the solid dissolves. The molecules escape into the surrounding solvent. The ease with which the solid dissolves depends on the nature of the solvent as well as on the strength of the forces holding the solid together. If the molecules of a solid interact with those of a liquid, then there will be a competition between the forces that tend to draw molecules into solution and the forces in the solid that tend to maintain its form. A measure of these latter forces is obtained in the temperature at which the solid begins to flow, or melt. This temperature is called the melting point (though it is often observed as a melting range), and it is quite characteristic of the substance in question. Since the melting point is related to the strength of the intramolecular interactions which maintain the solid condition, and dissolving of the substance is conditional on the moving apart of the molecules, their escape into solution, it follows that there should be a relation between melting point and solubility. It is in fact found that when substances with similar structures are compared, the higher melting ones are usually less soluble than the lower melting in a given solvent.

1.9 Solubility The extent to which one substance dissolves in another is a measure of the interactions between the molecules of the two substances. Substances which are chemically "alike" usually can dissolve to some extent in each other; the molecules of one diffuse freely between those of the other. Water, HOH, and methyl alcohol, CH_3OH, show unlimited solubility in each other; hexane, C_6H_{14}, and octane, C_8H_{18}, are mutually completely soluble; but water and hexane are virtually insoluble in each other. In general, polar substances (substances with polar bonds) dissolve other polar substances; nonpolar substances dissolve nonpolar, but scarcely dissolve polar substances. The strong mutual interactions of polar molecules hold them together so firmly that nonpolar molecules have little chance of introducing themselves between the others; they are "squeezed out," as Professor Joel Hildebrand puts it.

If a nonpolar molecule has in it polar bonds, then its solubility in polar solvents will be aided. Inorganic substances which are polar will usually dissolve in polar solvents such as water or liquid ammonia. One mechanism of solution is *solvation.* (When the solvent is water, the specific term *hydration* is used.) When a substance that ionizes is dissolved in water, the ions separate and become hydrated by the water. The water molecule contains a dipole, as shown below, and many of them cluster around the ion, aligning themselves to form an envelope. When polar nonionic molecules dissolve in water or other polar substances, a similar solvation mechanism seems to operate: the dipoles in solute and solvent interacting to draw the molecules together. Such interactions are not as strong as chemical bonds, and the association between the particles is a reversible one. No real distinction can be drawn between really strong interactions and weak chemical bonds. In general, if the interaction leads to a product with a definite (stoichiometric) composition, it is likely to be considered a chemical reaction.

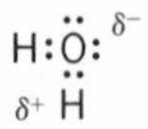

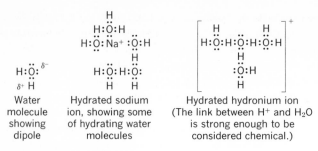

Water molecule showing dipole

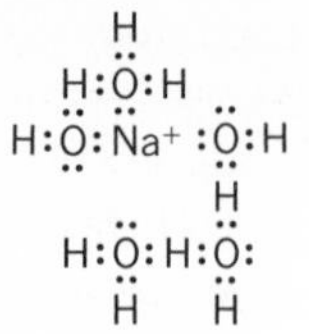

Hydrated sodium ion, showing some of hydrating water molecules

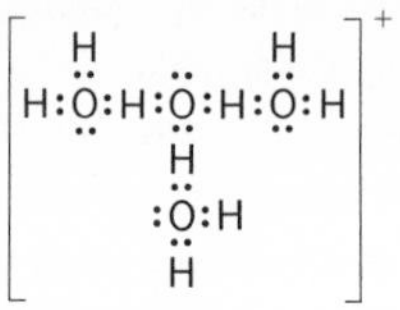

Hydrated hydronium ion (The link between H^+ and H_2O is strong enough to be considered chemical.)

Most organic substances are largely nonpolar, and their solubility in water is usually limited, but they dissolve readily in organic solvents. For example, water will not dissolve grease, but gasoline or kerosene will. Exceptions to these rules occur and lend interest to advanced courses. One important exception is that when covalent molecules are made very large, their solubilities in any solvent are decreased.

1.10 Stability A matter of importance in chemistry which is largely passed over in first-year work concerns substances whose existence can be demonstrated but which cannot be isolated. It can be shown, for example, that when chloroform is treated with base, a reactive species, $:CCl_2$, is generated under certain conditions:

$$HCCl_3 + OH^- \longrightarrow H_2O + Cl^- + [:CCl_2]$$

Brackets, such as those used in the above equation, will be used consistently throughout this text to indicate substances which cannot be isolated. No one has ever isolated $[:CCl_2]$ (dichlorocarbene), yet its existence can be demonstrated in certain reactions. The student should realize at this point that ability to be isolated is not the only criterion for the existence of a molecular structure. Ability to be isolated indicates that the structure has a certain measure of stability. Certain substances are stable up to high temperatures (sodium chloride); others can be isolated only at lower temperatures (ammonium carbonate); some, which can be shown to exist under certain conditions, can nevertheless not be isolated under any presently known conditions ($[:CCl_2]$).

The stability of a molecule depends on the tenacity with which the constituent atoms are bonded together. In a complex molecule certain bonds will be weaker than others, so that the molecule may break apart at the weakest bond. But the stability of a bond may also be influenced by the nature of the environment in which a molecule finds itself. For example, HCl gas, or HCl dissolved in benzene, is not appreciably ionized; but HCl dissolved in water reacts with the water and becomes dissociated:

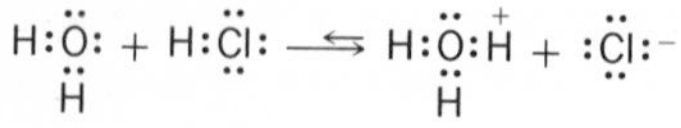

Other reagents exist (such as liquid ammonia or alcohol) which act in a similar way and bring about varying degrees of ionization of HCl. In organic chemistry the courses of a great many reactions are markedly influenced by the environment. The understanding of these influences makes possible a great clarification and simplification of organic-chemical reactions.

1.11 Definition of Group It has been necessary, in the course of the preceding sections, to introduce certain terms which will be used and clarified later. There are certain other terms which need to be introduced at this point in preparation for their future use. A *group* is a part of a molecule (it may be an atom or a group of atoms) which passes through a given reaction unchanged, or which reacts more or less independently of the rest of the molecule, and which may thus be dealt with as a unit. Following are some of the groups which will be encountered frequently:

—Cl	chloro- or chloride
—Br	bromo- or bromide
—I	iodo- or iodide
—OH	hydroxy- or hydroxyl-
$—NO_2$	nitro- 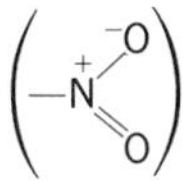
$—SO_3H$	sulfonic acid 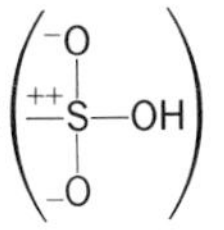

Other groups will be defined from time to time. It is convenient to prepare a table to which each new group is added. After a while the need for such a table will disappear, but it can be very useful at first.

1.12 Definition of Ion The term *ion* will be used in the same sense in which it is used in inorganic chemistry; however, there will be a marked extension in the meaning of the term as it will be used in connection with explanations of certain organic reactions. It will be one of the objects of this text to explain (as far as possible) the chemical mechanisms by means of which organic reactions take place. In simple inorganic reactions, such as the collision of two ions to produce a slightly soluble substance, a very simple equation suffices, for example,

$$Cl^- + Ag^+ \rightleftharpoons \underset{\text{(solid)}}{Ag^+Cl^-}$$

However, most organic reactions are more complicated than this, and a consideration of the steps involved in them will constitute an important part of our study.

In this connection, the term "ion" will be used for a group not only if it has a permanent existence as an ion, as do the ions in the above equation, but also when it exists as an ion during the course of a reaction. Thus, in the course of a reaction ions may be formed whose lifetimes are so short that the kind of observation possible in the inorganic field for demonstrating ions is not available. Such reactions may proceed by an *ionic mechanism,* but neither the reactants nor the products ionize in the outspoken sense of first-year chemistry. For example, such substances do not conduct electricity. When reactions which proceed by an ionic mechanism are first encountered in the text, the meaning of the terms used will be explained in detail (Secs. 3.6 and 3.10).

1.13 Definition of Radical The name *radical* will be used for atoms or groups of atoms which take part in a reaction and which have available one electron which can be used in bond formation. They contain an odd number of electrons and are electrically neutral. In contrast to ions they often carry no charge (although charged forms are known). Most of the radicals encountered in explaining reaction mechanisms have very short lifetimes and cannot be isolated.

1.14 Acids and Bases In the study of elementary chemistry an acid is often defined as a source of H^+ or H_3O^+. Further examination reveals that many of the properties of H^+ and H_3O^+ are shared by compounds in which an atom possesses an incomplete valence shell. Such compounds are called "Lewis acids." Examples of Lewis acids are $AlCl_3$, BF_3, and $FeCl_3$. These substances can form bonds to species having unshared pairs of electrons, for example,

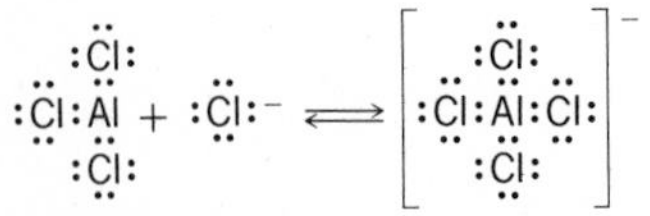

The counterpart to the Lewis acid is the Lewis base, which is a species with an unshared electron pair, such as OH^-, Cl^-, H_2O, or NH_3.

1.15 Electrophilic and Nucleophilic Many chemical reactions can be visualized as reactions of Lewis acids with Lewis bases, as illustrated for aluminum chloride and chloride ion above. However, a much greater range of reactions can be classified as reactions between electron-deficient species (including Lewis acids) and electron-rich species (including Lewis bases). The electron-deficient species are termed "electrophilic" (electron loving) and the electron-rich species are called "nucleophilic" (nucleus loving). As an example, a polarized bond, as in chloromethane, can provide a nucleophilic site (the chloro group) and an electrophilic site (the carbon atom).

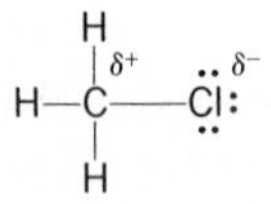

It should be noted that the terms electrophilic and nucleophilic are relative, and a given reagent may be nucleophilic toward a strong electrophile and yet behave as an electrophile toward a very nucleophilic species.

1.16 Curved Arrows Since many chemical reactions involve the transfer or rearrangement of one or more pairs of electrons, it is convenient to have a shorthand representation of this process. The curved arrow ↷ *represents the transfer of a pair of electrons from an atom to a bond, from a bond to an atom, or from one bond to another,* either during the course of a reaction or under the influence of a polarizing reagent. This notation represents a convenient method for keeping track of electrons during a transformation, and in some cases it may correspond to the path actually followed during the reaction. An example of the use of curved arrows is given below in a description of a complex ionization.

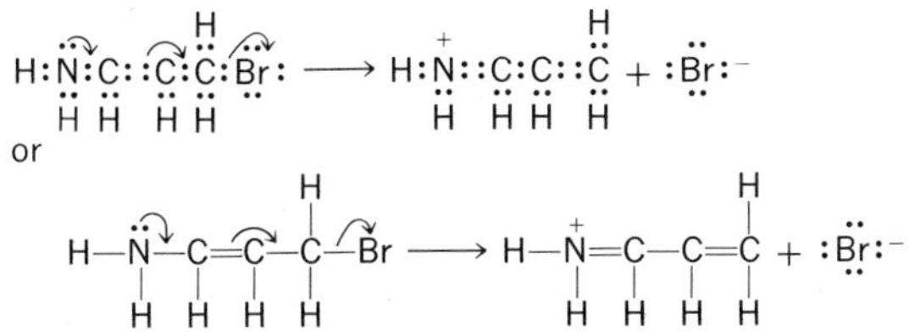

1.17 Definition of Formula Empirical and molecular formulas are already familiar from first-year chemistry. The *empirical formula* shows the simplest correct ratio of the different atoms in the molecule, while the *molecular formula* shows the actual numbers of atoms in the molecule. The molecular formula may be, but is not necessarily, the same as the empirical formula. The empirical formula of mercurous chloride is HgCl; the molecular formula is Hg_2Cl_2. The empirical formula and the molecular formula of formaldehyde is CH_2O; this is also the empirical formula of glucose, whose molecular formula is $C_6H_{12}O_6$. The *structural formula,* or structural symbol, of a compound is the molecular formula written out so that the arrangement

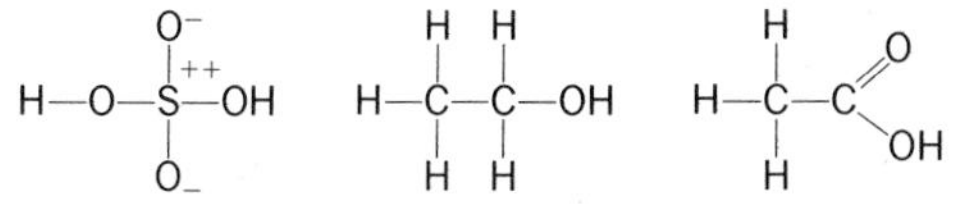

Sulfuric acid Ethyl alcohol Acetic acid

of the constituent atoms is clearly shown. The structural formula also shows the various *functional groups* (reactive entities) present in the molecule and their relative positions. In ethyl alcohol the —OH is a functional group. It is designated specifically as a hydroxyl group. Usually a formula is written out in only enough detail to be readily understood.

Although empirical and molecular formulas are of value in inorganic chemistry, they are for the most part of little use in describing compounds of carbon. This arises from the fact that there are often a number of carbon compounds having the same molecular formula but differing in some way in the structural arrangement of the atoms. This phenomenon is known as "isomerism," and substances of this sort are called "isomers." Although there is only one substance of the formula

H_2SO_4, there are no less than 75 isomers of the composition $C_{10}H_{22}$; these have different structures and are designated by different structural formulas. A major part of the organic chemist's work is concerned with the establishment of structural formulas in cases of this sort.

1.18 Summary of Definitions We have now given the definitions of a number of words which will be used extensively in this book. It is essential that they be memorized. The language of a science is as much a part of the science as the test-tube observations and the concepts of the science. The words are chemistry; the ideas which they embody are chemistry; the observations they describe are chemistry. An *understanding* of all three will develop gradually, but perhaps a brief summary of the more important points will help here.

Group: A part of a molecule which passes through a reaction unchanged, and which it is convenient to consider as a unit. It may consist of one or more atoms.

Ion: A group which carries a charge and which may exist as an ion before or after a reaction, or which may exist for only a very short period during a reaction.

Radical: A group which often carries no charge and which has available one unshared electron.

Electrophilic Reagent: One which is electron-poor and hence enters reactions in which it stands to gain a pair of electrons. Lewis acids belong in this class.

Nucleophilic Reagent: One which is electron-rich and enters reactions with electron acceptors. Lewis bases belong in this class.

Electronegativity: A term applying to the property of an atom, *when bonded covalently to another,* to attract electrons. An atom is electronegative with respect to another bonded to it if it tends to attract the bond electrons more than does the other atom.

The use of these terms in discussing the mechanisms of reactions, and indeed the discussion of mechanisms itself, is very important to any study of organic chemistry. Sciences pass through stages of growth. In their youth they are largely concerned with the collecting of data, often intuitively guided. This phase is followed by the binding together of the data by theory, and this leads into the phase of prediction. In this later stage new data are found under the guidance of predictions made in the course of formulating and testing theory. One great value of discussing mechanisms in an elementary course is that it is possible, by showing similarities in mechanism between large numbers of reactions, to simplify and clarify the data. Another very important function of the discussion of mechanisms is to show how new experiments may be formulated, and how answers may be obtained to questions about mechanism in cases where they are lacking. In other words this kind of treatment illustrates the use of the scientific method as a tool in examining natural phenomena.

1.19 Historical Perspective Organic chemistry, using the term defined as in Sec. 1.1, had its beginning in the nineteenth century when enough knowledge about the compounds of carbon had accumulated so that specialists who worked in this field came to give their speciality a name. The term "organic" originally had vitalist implications, and referred to the chemistry of compounds of carbon produced by living organisms. However, in 1828 when F. Wöhler produced the substance urea from ammonium cyanate, it became evident to thoughtful chemists that this aspect of the theory of vitalism was untenable. This opened up the suggestion of synthesis of all sorts of products of plant and animal origin. An interesting insight into the importance of Wöhler's work is given by Ernest Campaigne [*J. Chem. Education,* **32:** 403 (1955)].

Organic chemistry is not a sharply marked-off discipline, but one contributed to by physical, analytical, and theoretical as well as organic chemists, and also by physicists, biochemists, engineers, and physicians. It seems right to call attention to those scientists who have been responsible for various reactions and concepts as we take them up in this book.

The tetrahedral structure of the carbon atom was postulated less than one hundred years ago. The men chiefly responsible for this were the Dutch chemist Jacobus Hendricus van't Hoff and the French chemist Jules Achille LeBel, who as students had worked in the laboratory of Charles Adolphe Wurtz in Paris and who, in 1874, developed the theory of the spatial arrangement of substituents about the carbon atom quite independently. Their work rested on the earlier work of the French chemist Louis Pasteur.

Electrons were discovered, or perhaps we should say invented, by the English chemist J. J. Thomson in 1897. The implications of the new discoveries about the structure of atoms that developed from this time on bore fruit in a theory of valence proposed by Gilbert N. Lewis in 1916 at the University of California in Berkeley. This theory, involving the idea of the covalent bond, laid the foundation for the modern theories of the chemical bond. There are several of these, resulting from different ways of looking at the data, and we cannot take all of them up in detail in this book. The theory of resonance is one of them, and is associated with the names of Linus Pauling (California Institute of Technology) and a number of his coworkers. The molecular orbital theory is associated with the names F. Hund, E. Hückel, R. S. Mulliken, and J. E. Lennard-Jones.

The terms electrophilic and nucleophilic were introduced by C. K. Ingold and E. D. Hughes, two English pioneers in the field of chemical reaction mechanism. The names Lewis acid and Lewis base honor G. N. Lewis, referred to above.

1.20 Spectroscopic Methods Of the tools utilized by the organic chemist for the study of chemical structure, those described as *spectroscopic methods* have come to occupy a paramount position in recent times. The phenomenon of absorption spectroscopy, which involves the absorption of electromagnetic radiation by molecules, has provided a wealth of information about molecular structure. Some of this information has developed from thorough, theoretical understanding of the interaction between electromagnetic radiation and molecules. Many valuable concepts have resulted from this approach. Organic chemistry is both vast and complex, however, and only in a few simple cases are all the details of this interaction well understood. Consequently, the organic chemist has resorted to the powerful tool *analogy* in approaching this problem. In brief, this involves the observation of a given spectroscopic property for a large number of molecules of structures that have been determined by known methods. Certain patterns are found to recur and correlations of absorption patterns and structure can then be made. (These are often based on sound theoretical principles, but this is not essential to empirical use.) These correlations are applied to compounds of unknown structure and, as their structures are confirmed by other evidence, these in turn are added to the correlation scheme. As the body of known compounds grows, these correlations are continually refined, the inevitable exceptions are rationalized, and new correlations are discovered. By the applications of different spectroscopic methods complete structures can now often be derived without additional information.

In this work we shall confine ourselves to infrared, ultraviolet, and nuclear magnetic resonance spectroscopies. Other methods such as microwave, raman, electronic spin resonance, X-ray, and mass spectroscopy[1] will not be treated here, although recent advances in the latter two have brought them renewed significance in organic chemistry.

1.21 Infrared Spectroscopy That portion of the electromagnetic spectra of wavelengths slightly longer than those of visible light is referred to as the *infrared region*. This encompasses wavelengths from 2.5 to 30 microns (1 micron $= 1\ \mu =$ 10^{-6} meter). This text, however, shall use wavenumbers, cm^{-1}, in the infrared region. These are related to other common units as follows:

$$\text{Wavenumbers (cm}^{-1}) = \frac{10{,}000}{\text{wavelength } (\mu)} = \frac{\text{frequency (sec}^{-1})}{\text{velocity of light (cm/sec)}}$$

The region of most interest in the study of carbon compounds is that ranging from 4000 to 600 cm^{-1} (2.5 to 16.7 microns).

The frequency (and energy) of infrared radiation is of the same magnitude as the vibrational and rotational frequencies of atoms bonded by covalent bonds. Quantum theory tells us that such systems can absorb radiation of essentially the same frequencies as those at which they vibrate, and not at other frequencies. This means that if infrared light is passed through a substance with a variety of different types of bonds, and one looks at the absorption as a function of frequency, a series of areas of strong absorption (absorption bands) corresponding to the different bond types involved will be seen. Since single bonds are the most common in carbon compounds, it is often difficult to tell one from another. They give rise to a complex series of vibrations in the 1500 to 600 cm^{-1} region. This is called the "fingerprint" region since the pattern of absorptions found there is usually characteristic of a particular structure. Double and triple bonds are considerably stiffer than single bonds and are usually present in smaller numbers so that their

[1]Mass spectroscopy is not strictly electromagnetic, but is generally treated with other spectroscopic methods.

absorptions are more easily sorted out and correlated with bond types. These absorptions normally occur in the 2500 to 1500 cm^{-1} region. Finally, hydrogen, because of its relatively low mass, vibrates at a much higher frequency than the heavier atoms common to organic materials. Therefore, the bonds between hydrogen and carbon, oxygen, or nitrogen give rise to maxima in the 4000 to 2800 cm^{-1} region. In subsequent chapters the absorption maxima associated with the different kinds of bonds will be discussed as each type is encountered.

1.22 Nuclear Magnetic Resonance This technique, which was virtually unknown to the organic chemist in the early 1950s, has since assumed a place as one of his most "essential" tools. The method uses the nucleus of certain atoms as a probe to measure the electronic distribution in its vicinity. The nuclei of certain atoms have *spins*, which are analogous to those possessed by electrons and which cause these nuclei to behave like tiny magnets. In the presence of strong magnetic fields some of these magnets will be found lined up with the strong field and some will be lined up opposed to it. From theoretical considerations beyond the scope of this book, it is known that only certain orientations of these magnets are possible and that they will differ in energy from each other. In order to bring about transition from one orientation to another, it is necessary to supply electromagnetic radiation of energy and frequency that corresponds to that of this difference in orientation. The frequency involved depends on the strength of the magnetic field, but at currently available fields it falls in the radiofrequency range of 40 to 300 megacycles. When suitable nuclei are placed in magnetic fields they will absorb radiation in this range, and with proper equipment it is possible to obtain an absorption spectrum called an "nmr spectrum."

To the organic chemist the most important nucleus possessing a spin is hydrogen (although fluorine and phosphorus are also useful). Ordinary carbon 12 and oxygen 16 do not possess nuclear spins, and nitrogen 14 usually does not exhibit the phenomenon for other reasons. The utility of the technique stems from the fact that a hydrogen nucleus, a proton, in an organic compound does not experience the full effect of the applied magnetic field since it is partially *shielded* by electrons in the molecule. These electrons alter the proton's magnetic environment and cause it to absorb at slightly different radiofrequencies, a phenomenon known as a "chemical shift."

The measurement of chemical shift is usually carried out relative to some standard substance, usually tetramethylsilane (TMS). To allow for the fact that measurements are made at different combinations of radiofrequencies and magnetic flux, the chemical shift is expressed in essentially dimensionless units, δ, as follows:

$$\delta\ (\text{ppm}) = \frac{\begin{pmatrix}\text{frequency at which a}\\ \text{given proton absorbs}\end{pmatrix} - \begin{pmatrix}\text{frequency at which the}\\ \text{standard TMS absorbs}\end{pmatrix}}{\text{nominal radiofrequency in megacycles}}$$

The commonly observed range of δ for protons on carbon is from 0 to about 10, although a number of exceptions are known. A low value of δ indicates that the proton is highly shielded by electrons in its environment; a high value indicates that it is *deshielded*.[1]

The absorption spectra for given protons in compounds containing several protons are often found to have more than one maximum per proton due to a secondary interaction called "coupling" or "splitting." This splitting arises from interactions among the spins of nuclei (usually hydrogens) on one carbon atom with those on adjacent carbon atoms.[2] The overall effect of splitting (coupling) can be very complicated and will not be discussed in this course. Our discussion shall be restricted to two extremes.

In one extreme the existence of splitting is shown by the breakup of a single peak into two or more partially separated peaks, indicating the presence of protons on adjacent carbon atoms.

In the special case where the chemical shift between two protons is large relative to the magnitude of the splitting (for practical purposes about one δ unit), and where other protons are not present on adjacent atoms, a particularly simple result is obtained. The hydrogens on one carbon atom will be split into $n + 1$ peaks, where n is the number of identical hydrogen atoms on adjacent atoms. The hydrogens on the other carbon atom will, in turn, be split into $m + 1$ peaks, where m is the number of hydrogens on the first atom. The spacing of these multiplets (i.e., doublets, triplets, quartets, etc.) will be the same, and the separation between any two peaks in such a multiplet is called the "coupling constant" j, which is given in cycles per second (cps). Furthermore, the intensities of the peaks in the multiplets will be given by the coefficients of the binomial expansion, seen most easily from Pascal's triangle:

Multiplet	*Relative intensities*
singlet	1
doublet	1 1
triplet	1 2 1
quartet	1 3 3 1
quintet	1 4 6 4 1

[1] In many references the term tau τ is used, although it is gradually being replaced by δ. The two are related by the expression

$$\delta = 10 - \tau$$

δ shall be used exclusively in this book, but it is useful to remember the above relationship.

[2] In special cases longer-range coupling is observed. It shall not be dealt with in this course, as it is an unusual observation.

For the group $CH_3—CH_2—X$ (where X does not have hydrogens), if the criterion of large difference in δ is observed, a pattern of the following type may be observed:

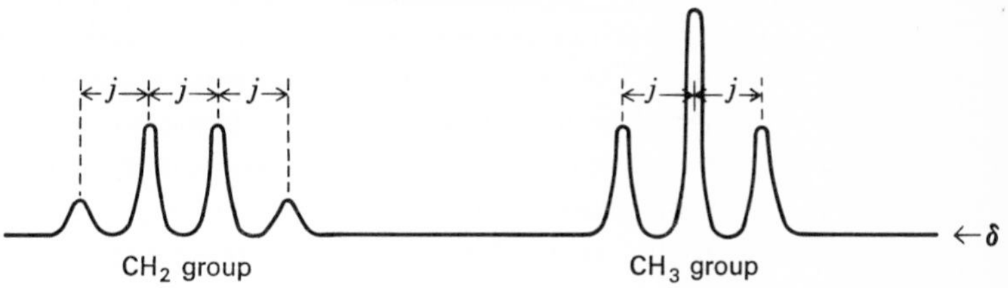
j
j
j
j
j
CH2 group
CH3 group
←δ

figure 1-9
Nuclear magnetic resonance spectrum for CH_3CH_2X.

Note that both groups must show the same coupling constant (if they do not, the above considerations do not apply). Also note that the relative intensities of the peaks for the CH_2 group are in the ratio 1:3:3:1; for the CH_3 group they are 1:2:1.

The second extreme occurs when the chemical-shift difference is very small relative to the coupling. The most important examples of this occur when the protons are identical in chemical shift; they will then merge into a single peak of increased intensity, even though they may be strongly coupled to one another. For example, in ethane the two CH_3 groups show up as one peak, as do all the hydrogens in $Cl—CH_2—CH_2—Cl$, etc. In short, identical (in chemical shift) protons will not give rise to splitting of each others' absorption, although they may still be split by other, nonidentical protons.

An added advantage of nmr is that the (integrated) intensities of the nmr maxima (areas under the curves) are proportional to the relative number of hydrogens in a molecule. This is independent of splitting and other effects as long as the peaks do not overlap. Thus one can gain information about the numbers of hydrogens in a given environment as well as about the environments themselves.

The student will realize that since most carbon compounds contain hydrogens in a variety of different environments, the possibility of "counting" and identifying them as to type and environment by interpretation of an nmr spectrum is of real importance in solving problems of structure in organic chemistry. Indeed, in many situations nmr studies are replacing chemical and other kinds of data in structural work.

1.23 Ultraviolet Spectroscopy The interaction of molecules with visible and ultraviolet light involves the *excitation* of *electrons*. Electrons in atoms and in bonds are excited, or "raised," to orbitals of higher energy. The most commonly utilized region of the electromagnetic spectrum is that ranging from 2000 to 7000 Å (1 Å = 10^{-8} cm). This comprises the visible and near ultraviolet. The energies involved in this range are sufficient to excite unshared or loosely bound electrons, but not those in simple, single covalent bonds. Therefore, only molecules containing atoms with unshared valence electrons or multiple bonds are found to absorb light of these wavelengths. This both limits the generality of ultraviolet spectra as a tool and allows it to provide valuable structural information in those cases where it is observed. The ultraviolet spectra observed from solutions of carbon compounds usually consist of smaller numbers of broad absorption maxima than are seen in the infrared; both the position and intensity of the absorptions in this region provide useful information. The position of the maxima are usually reported as λ_{max}, wavelengths as angstroms Å and/or millimicrons $m\mu$, and the intensities as ϵ_{max} which is the molar absorptivity at standard concentration and cell length.

CHAPTER 1 EXERCISES

1. In classroom demonstrations the carbon atom is often symbolized by a black wooden sphere with four pegs arranged equidistantly around it for bonds. In what ways does this symbol fail to correspond with the actual atom? In what ways is it satisfactory?
2. Why, do you suppose, is HSO_4^- a weaker acid than H_2SO_4? (Resonance stability of the ions is not the only factor involved here.)

★3. Which of the molecules listed below are polar and which nonpolar? Designate any polar bonds.

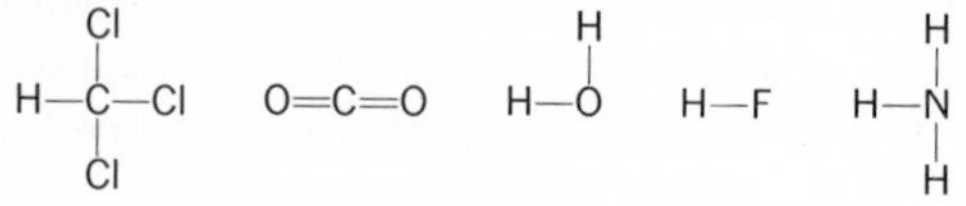

Cl
H—C—Cl
Cl
O=C=O
H
H—O
H—F
H
H—N
H

4. Describe the solubility behavior which you would expect in the following cases: (*a*) hydrogen in water; (*b*) water in ethyl alcohol (C_2H_5OH); (*c*) ethyl alcohol in kerosene (approximately $C_{12}H_{26}$); (*d*) ethyl alcohol in chloroform ($CHCl_3$); (*e*) sodium chloride in liquid ammonia; (*f*) sugar $[C_6H_6(OH)_6]$ in water; (*g*) sugar in kerosene; (*h*) filter paper $(C_6H_{10}O_5)_n$, where n is several hundred, in water, alcohol, chloroform, and kerosene.

★5. Indicate the more electronegative component in each bond: F—F, H—F, H_2SO_4, $AlCl_3$, NaOH, H_3O^+.

6. Write out the electronic structures of the molecules or ions in Exercise 1.5 and indicate all coordinate bonds and formal charges.

★7. Write a structure for nitromethane CH_3NO_2 showing electrons and any formal charges. Is resonance possible in this structure? Explain.

8. Which of the following would have the most electrovalent bond? The most covalent?

   KCl, CH_3OH, H_2, $BeCl_2$, H_2CO_3.

★9. Is CH_2Br_2 a polar molecule? Explain.

10. Are resonance forms isomers? Explain.
11. Write the structure, showing electrons and all formal charges, for

   HSO_4^-, H_3PO_4, NH_4OH, BF_3, B_2O_3 (OBOBO).

12. Assume a reaction between BF_3 and NH_3 to form BF_3NH_3. Show positions of electrons and formal charges, if any. Which of the reagents is electrophilic?

★13. Although NH_3 can be drawn as 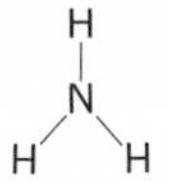
the bond angles are not 120° as the figure suggests. How can this be explained?

14. How many electrons can occupy an orbital? Whereabouts in the orbital are they located?

★15. Draw all the resonance forms for HCO_3^-. Would you expect all the C—O distances to be the same? Explain.

16. Draw an electronic formula for (*a*) a Lewis acid, (*b*) a Lewis base, (*c*) an ionized salt of H_2SO_4, (*d*) a compound having a polarized bond.

★17. There is a known nonpolar substance, C_2H_2, in which carbon and hydrogen exhibit their normal valences. Draw its formula. Compare it to CH_4 in predicted nucleophilic reactivity.

18. Arrange the following compounds in order of increasing electrovalent character, that is, tendency to ionize at one or more bonds: NaF, B_2O_3, CH_3OH, $MgCl_2$.

★19. Can you suggest an explanation for the fact that the O—H bond in the acid

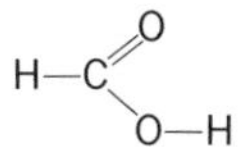

is more ionized in solution (to O^- and H^+) than the O—H bond in

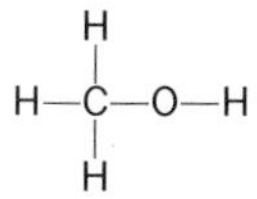

20. Show by equations how H_2O can act both as a nucleophile and as an electrophile.

21. Draw a chart of the electromagnetic spectrum showing the regions involved in visible, infrared, ultraviolet, and radio frequencies. Mark the approximate boundaries of these zones in centimeters.

22. How many peaks would you expect in the nmr spectrum of the following:

★23. Which of the following would be expected to absorb in the ultraviolet region?

CH_4, $H_2C{=}CH_2$, NH_3, H_2SO_4

24. Which spectral method or methods might be useful in distinguishing between the following pairs of compounds?

a. $CH_3CH_2CH_3$ and $CH_2{=}CHCH_3$

b. CH_3CH_2OH and CH_3OCH_3

c. $CH_2{=}CH{-}CH{=}CH_2$ and $CH_3CH{=}CHCH_3$

d. $CH_2{=}O$ and CH_3OH

e. NH_3 and CH_3NH_2

★**25.** Which of the following are Lewis acids or bases?

H_2O, BCl_3, CH_4, KBr, NH_3, CH_3^+, OH^-.

26. Show how water can act as a nucleophilic reagent.

27. Indicate the more electronegative component of each bond in

F_2, HF, $AlCl_3$, NaOH, H_3O^+.

28. Show all the important resonance forms of (*a*) acetate ion ($CH_3CO_2^-$), (*b*) the positive ion $H_2C{=}CH{-}CH{=}CH{-}CH_2^+$, (*c*) carbonate ion ($CO_3^{--}$).

★**29.** Assume that a positive ion can be formed from $CH_2{=}CH{-}CH_2CH_2CH_2Br$ by loss of Br^-. Would it be a resonating ion? Explain.

2

HYDROCARBONS: THE ALKANES

2.1 Methane The simplest known stable compound of carbon and hydrogen is CH_4. This substance is called methane. As would be expected from the facts that the molecule is small and that the bonds are covalent, it is a gas at ordinary temperatures. Methane is found in natural gas and in the products of bacterial decomposition of vegetable matter (marsh gas). The chief interest which methane has for the elementary organic course lies in its being the simplest member of a very long series of compounds of carbon and hydrogen known as "alkanes," sometimes called "paraffins." This series was not discovered in an orderly fashion, starting with methane; instead, many compounds of carbon and hydrogen were discovered before it was observed that they could be classified into a series, and still more work had to be done before the finer chemical and physical implications of the series could be understood (this phase is still developing). Many members of this series have been isolated from petroleum, which consists chiefly of a complex mixture of alkanes.

2.2 Names of Some Alkanes By increasing numbers of carbon atoms in the molecules, the series shown in Table 2.1 may be developed (adhering to a valence number of four for carbon and one for hydrogen).

After the first four members, the method of naming the higher hydrocarbons becomes very simple and regular. The number of carbon atoms present is given by the prefix (from the Greek root), and the suffix *-ane* is appended to indicate that the substance belongs to the family of alkanes (Sec. 2.3). These Greek-root prefixes are easy to remember because they are the same ones used in geometry (for example, *penta*gon, *hexa*gon). The compound $C_{11}H_{24}$ is usually called undecane, though the correct name would be hendecane. Exceptions such as undecane and the names of the first four members of the series, which are not derived from Greek number roots, are found to some extent in the systematic nomenclature. One of the rules of systematic nomenclature (Sec. 2.6) states that common names which have long been in use shall be retained.

TABLE 2.1
SOME ALKANES

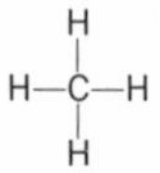	Methane	CH_4[a]
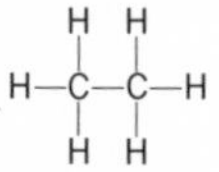	Ethane	C_2H_6
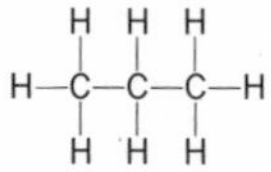	Propane	C_3H_8
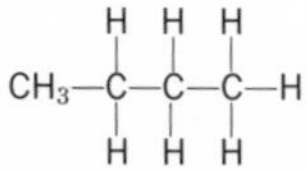	Butane	C_4H_{10}
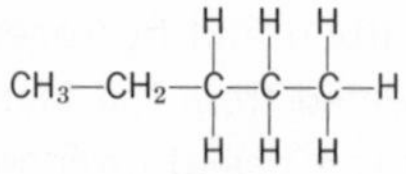	Pentane	C_5H_{12}
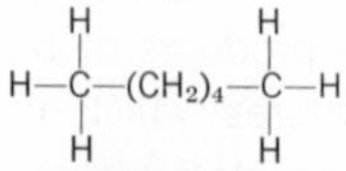	Hexane	C_6H_{14}
$CH_3(CH_2)_5CH_3$	Heptane	C_7H_{16}
$CH_3(CH_2)_6CH_3$	Octane	C_8H_{18}
$CH_3(CH_2)_7CH_3$	Nonane	C_9H_{20}
$CH_3(CH_2)_8CH_3$	Decane	$C_{10}H_{22}$
$CH_3(CH_2)_9CH_3$	Hendecane	$C_{11}H_{24}$

[a] Major constituent of natural gas.

2.3 Homologous Series The names of the first 11 alkanes (Table 2.1) should be memorized because they constitute an important part of the nomenclature of organic chemistry. The formulas do not need to be memorized if it is observed that they are all related by the *type formula* C_nH_{2n+2}, in which n is the number of carbon atoms present (n is a whole number, of course). Since carbon is tetravalent in all these compounds and hydrogen is monovalent, it follows that these molecules (like all compounds containing only carbon, hydrogen, and oxygen) will always have an even number of hydrogen atoms. Consecutive members of this series differ from each other by

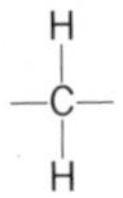

A series which shows this property is known as a *homologous series.*

From each member of the series there can be derived a group by removing one of the hydrogen atoms. The group is designated by the suffix *-yl.* A list of groups which should be memorized is given in Table 2.2.

TABLE 2.2
SOME ALKYL GROUPS

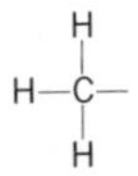	Methyl- (meth′ĭl)
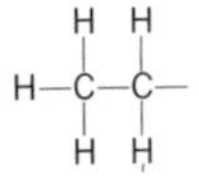	Ethyl-
C_3H_7—	Propyl-
C_4H_9—	Butyl-
C_5H_{11}—	Pentyl-
$C_{10}H_{21}$—	Decyl- (dēs′ıl or dĕk′ıl)

It should be noticed that these groups are not ions. Their chief use is as a convenient device in nomenclature, and when they are written this way, the dash does not imply anything except that a bond may be formed at that point.

Table 2.3 shows some more complex groups that will be encountered later.

2.4 The Writing of Equivalent Structures The reaction of an alkane with any reagent becomes more complicated with the higher members of the homologous series, and this introduces several new concepts which are of great importance and utility throughout organic chemistry. These will have to be discussed before reactions of the alkanes are taken up.

Consider the simple case of the reaction of ethane and bromine (or chlorine). With one molecule of halogen the reaction is written

TABLE 2.3
NAMES OF SOME MORE COMPLEX ALKYL GROUPS

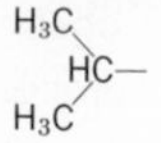	Isopropyl-	*i*-Pr-
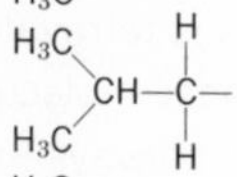	Isobutyl-	*i*-Bu-
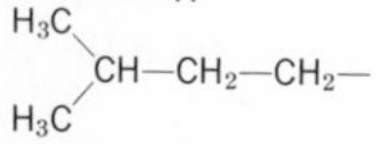	Isoamyl-	*i*-Am-
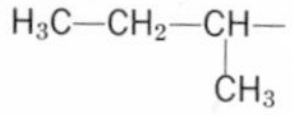	Secondary butyl-	*sec*-Bu-
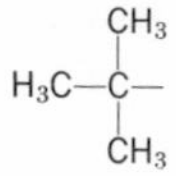	Tertiary butyl-	*t*-Bu-
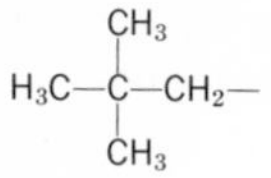	Neopentyl-	

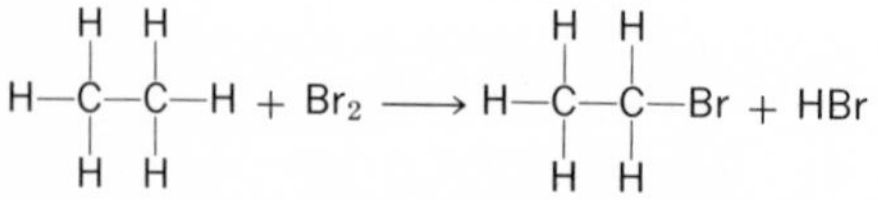

Bromoethane

All attempts to prepare bromoethane, by whatever method, have always produced the same unique substance: there is only one bromoethane.

By adhering to the valence rules, a number of formulas representing apparently different structures can be written for bromoethane:

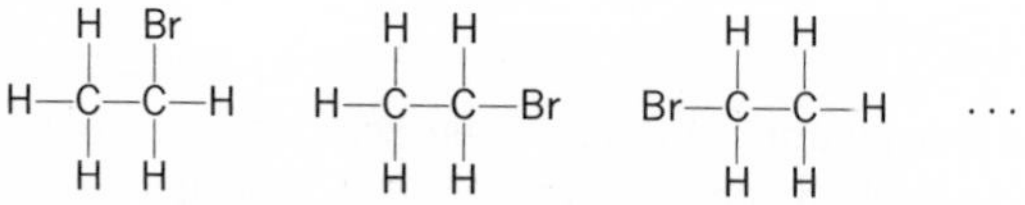

Since it is true experimentally that there is only one bromoethane, then these structures must all represent the same molecule. This can be the case only if each hydrogen in ethane bears the same geometrical and chemical relation to the rest of the molecule as every other hydrogen. These hydrogens are then said to be "equivalent." But this equivalence is possible only if the bonds from each carbon atom are arranged in the tetrahedral manner. In Fig. 2-1 a dashed line represents an atom going behind the plane of the page, a solid triangle represents an atom coming above the plane, and a solid line represents bonds in the plane of the paper.

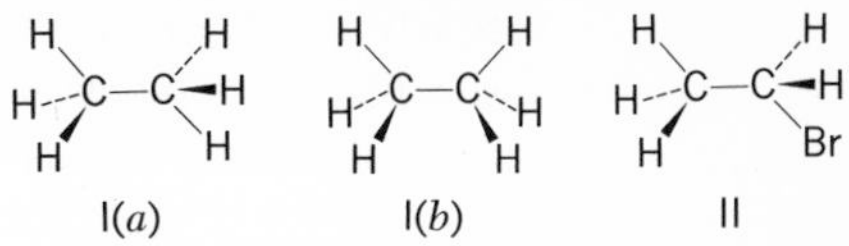
I(a)
I(b)
II

figure 2-1
I(a) and (b) Conformations of ethane. II Bromoethane.

By noting that rotation of the two halves of the ethane molecule about the C—C bond [I(a) ⟶ I(b)] does not alter the structure, one can see that all six hydrogens in I are equivalent.

I(a) and I(b) differ in what is called "conformation." The energy required to produce rotation about the axes of covalent bonds is so low that in most cases the concept of free rotation about single covalent bonds is in agreement with experimental facts. Although there are an infinite number of possible conformations for ethane, only one kind of ethane molecule can be obtained. The nmr spectrum of ethane reveals only a single kind of proton; the protons all appear to be in the same environment (are equivalent) due to rapid rotation about the C—C bond. In some more complex derivatives of methane, nmr spectroscopy has made it possible to study the rate of this rotation by working at exceedingly low temperatures. For all practical purposes, however, single covalent bonds in carbon chains like that of ethane exhibit free rotation (Sec. 7.9).

2.5 Isomerism In bromethane the hydrogens no longer occupy equivalent positions. This can be shown in physical and chemical ways apart from the geometric observation that, according to Formula II, two of the hydrogen atoms are closer to the bromine than the other three. If bromoethane is treated with bromine, two possibilities exist for further monobromination. The product in each case is a dibromoethane. Two, and only two, dibromoethanes are known.

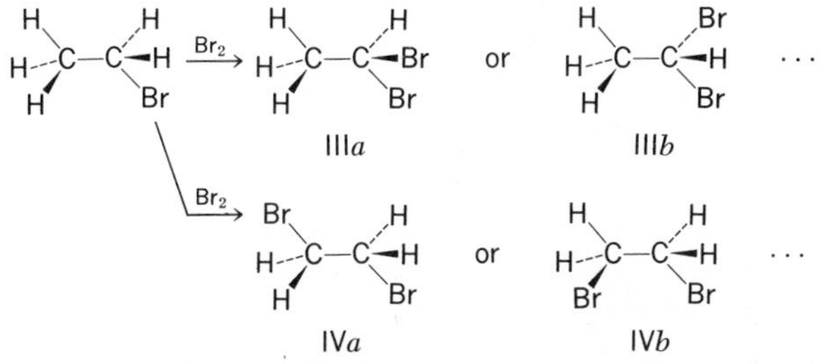
IIIa
IIIb
IVa
IVb

In the formation of dibromoethane both bromines may end up attached to the same carbon atom (III), or one bromine may end up attached to each carbon atom (IV). The resulting two substances are different; one boils at 110°C, the other at 132°C. The chemical behavior of the two are also different. These dibromides are distinguished by name as follows: III is *unsymmetrical* dibromoethane, and IV is *symmetrical* dibromoethane. Compounds related in the sense that they contain the same number and kinds of atoms but in different arrangements are called "isomers." Since drawing three-dimensional formulas is both time- and space-consuming, and since conformations are of limited interest, structures such as IV are usually written

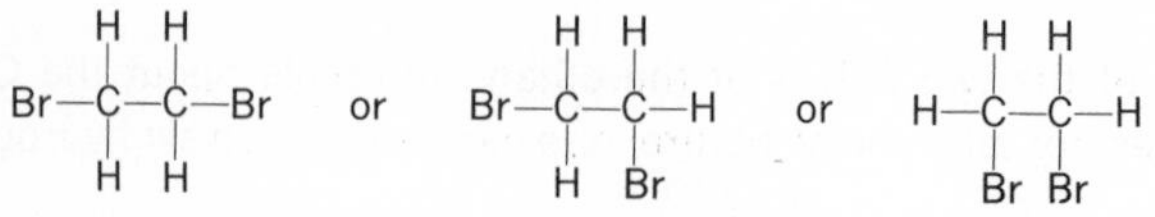

or $CH_2Br—CH_2Br$, but writing $C_2H_4Br_2$ is not sufficiently explicit.[1]

The question that arises in a monobromination reaction like this is which isomer is actually produced? Experiment and analysis of products show that in each case both dibromoethanes are produced. This finding also applies to further halogenation of dibromoethane and to the halogenation of any of the alkanes. In each case all possible products are formed, although in varying proportions.

In the case of the bromination of propane (and all higher homologs) certain other complications arise, for here, even in the hydrocarbons, not all the hydrogens are equivalent.

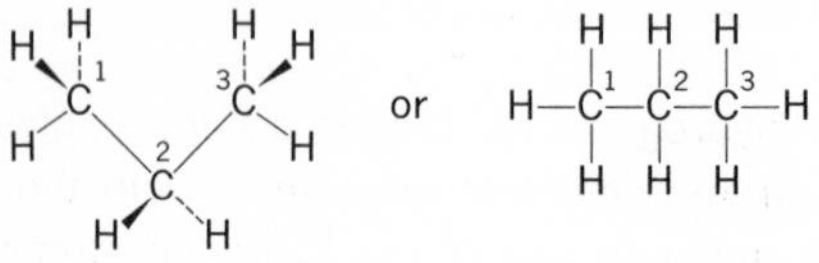

The hydrogens on carbon 2 are differently situated from those on the other two carbons for they are attached to a carbon atom which is attached, in turn, to two other carbon atoms. This environmental difference correlates with a chemical difference in their behavior. It is also found that the nmr spectrum of propane shows two different "kinds" of protons; the two secondary ones are found to have different δ values from the six primary protons. See Sec. 2.17. The hydrogens on carbon 2 are said to be *secondary*. The hydrogens on carbons 1 and 3 are *primary:* they are attached to a carbon atom which is attached to only one other carbon atom. Experiment shows that the proportions of isomers formed by substitution reactions vary with the number of primary, secondary, and tertiary hydrogen atoms present in the molecules. These proportions of isomers can be predicted in simple cases, but with higher homologs the situation becomes exceedingly complex.

[1]The problem of distinguishing between these two isomers is trivial if an nmr machine is available. The nmr (Sec. 1.22) of $BrCH_2CH_2Br$ will consist of a single peak at -4 ppm, while that of CH_3CHBr_2 will exhibit a quartet at about -5.5 ppm and a doublet at 1.7 ppm. All of these peaks are shifted to a higher delta by the presence of the bromines. Unfortunately, nmr was not available when these compounds were first prepared.

2.6 Nomenclature of Carbon Compounds The existence of isomeric substitution products such as the two dibromoethanes introduced a problem in nomenclature that was solved in that instance simply by calling one of them symmetrical and the other unsymmetrical dibromoethane. Where the number of isomers is very large, however, a solution such as this is impractical. Decane, for example, has 75 possible isomers.

In order to cope with situations of this kind, the nomenclature of carbon compounds has been systematized. In 1892 thirty-four chemists from nine countries met at Geneva and created the Geneva nomenclature. The purpose was to establish an official name for each organic compound by means of which it could be known and indexed unequivocally. The system was widely accepted, being made the basis for great compilations of chemical literature. Some errors and complications appeared from time to time with the rapid growth of organic chemistry, and finally, after a working committee had examined the nomenclature system, their report, correcting these errors and suggesting certain simplifications, was adopted in 1930 by the International Union of Chemistry (IUC) meeting at Liège. [See Austin M. Patterson, "Definitive Report of the Commission on the Reform of the Nomenclature of Organic Chemistry," *J. Am. Chem. Soc.*, **55,** 3905 (1933).] With the continuing fantastic increase in chemical data and the need for indexing it and making it accessible, work is now being done on methods of putting it on punch cards and devising a nomenclature adaptable to machine handling.

The first installment of the rules formulated at the above meetings for naming compounds is given below.

Rules for Naming Alkanes

1. Find the longest continuous chain of carbon atoms in the molecule. Use the name of the alkane corresponding to this number of carbon atoms as the basis for the name of the compound.

2. Number the carbon atoms of this continuous chain.

3. The substituents other than hydrogen, which is assumed to fill all undesignated places, are given numbers corresponding to the carbon atoms of the continuous chain to which they are attached. Each substituent receives a name *and* a number. Groups appearing more than once receive two numbers and the prefix di-, three numbers and the prefix tri-, etc.

4. The numbering of the chain must start from that end of the molecule which allows the smallest possible numbers to be used in locating substituents.

Thus 1,1-dibromethane could without sacrifice of *clarity* be called 2,2-dibromoethane, but the former name is preferred. These rules will be illustrated below.

It is found that there are two different butanes, each with the formula C_4H_{10}, and three different pentanes, C_5H_{12}. The source of this difference is immediately evident when the structural formulas are written down.

Isomeric Butanes

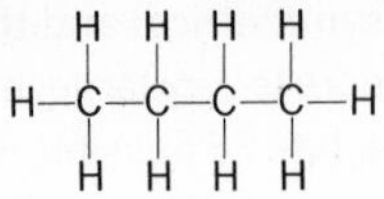

Butane
(*n*-butane, normal butane)

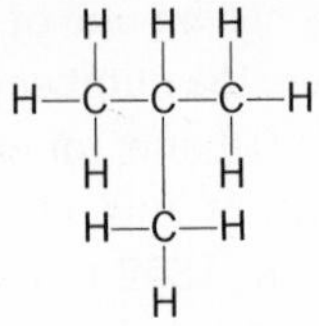

2-methylpropane
(*iso*butane)

Isomeric Pentanes

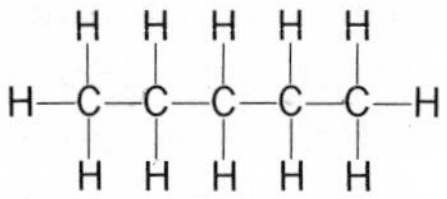

Pentane
(*n*-pentane)

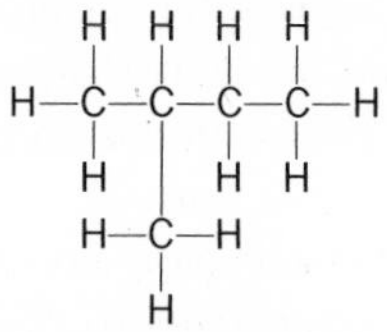

2-Methylbutane
(*iso*pentane)

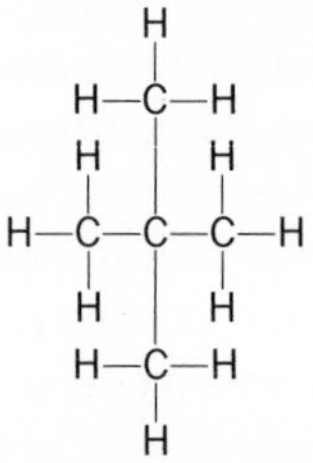

2,2-Dimethylpropane
(*neo*pentane)

The nomenclature of these compounds follows the rules given above, and the student should satisfy himself that this is so. The trivial names given in parentheses are often found in the literature and should be learned. It is considered bad practice to use these names as part of systematic (Geneva) names. For example, you do not say 2-methyl*iso*pentane. The trivial names follow the simple rule that chains without branching are normal, or *n*-, sometimes also spoken of as "straight chains." The chains may or may not be written in a straight line, but if there is no branching they are usually called straight chains.

$$CH_3{-}CH_2{-}CH_2{-}CH_2{-}CH_2{-}CH_2{-}CH_2{-}CH_3$$

n-Octane

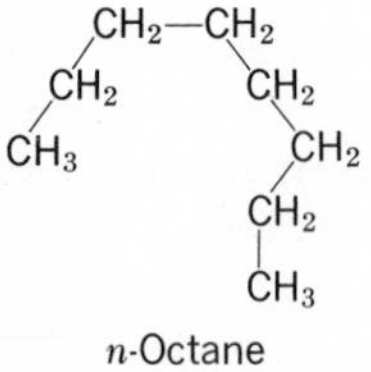

n-Octane

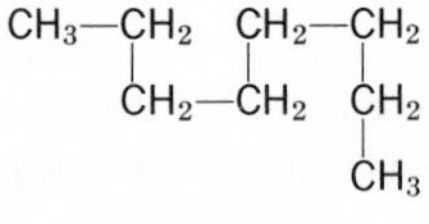

n-Octane

The justification for not writing chains in a straight line lies in the fact that the carbon bonds (sp^3) are at angles to one another (the 109°28′ tetrahedral angle). The molecules of longer-chain compounds, such as *n*-octane, in liquid or gaseous form are seldom stretched out in a straight line. Instead, owing to kinetic motion they undergo continual contortions (changing conformations) under the buffeting of other molecules. The motions of a molecule in a liquid or a gas are exceedingly complex, for not only is this contortion occurring in the chain as a whole, but also there is nearly free rotation about the C—C bonds:

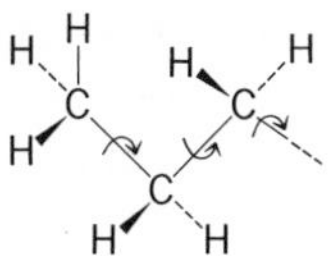

Moreover, there is a certain amount of distortion due to vibrations of the C—C bonds. It is, therefore, quite proper if, for the sake of convenience, one writes a straight-chain compound in a contorted form. The rules of nomenclature will always make it recognizable.

Some further examples will illustrate the use of these rules in deriving names for higher alkanes and their halogen derivatives. It is to be emphasized again that the longest continuous chain in the molecule is not necessarily the one that happens to be written horizontally in the formula.

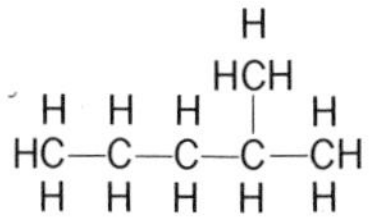

2-Methylpentane

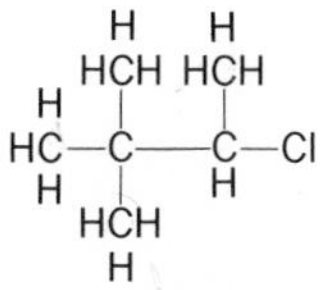

2,2-Dimethyl-3-chlorobutane

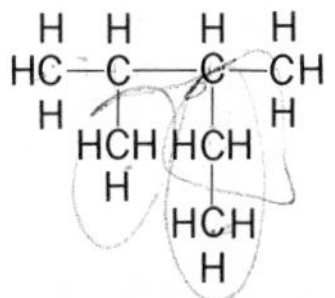

2,3-Dimethylpentane
(note a number and a name for each of the two methyl groups)

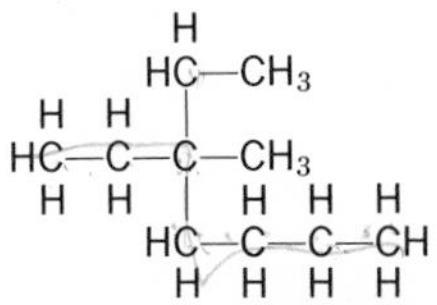

3-Methyl-3-ethylheptane

Even though apparently unequivocal names can be derived by this simple system without using the longest chain as a basis and by leaving out apparently unnecessary numbers, such violations of the rules cannot be tolerated; they often may result in more unwieldy or ambiguous names.[1] Particularly in the case of more complex compounds, the use of a multitude of names for a compound would lead to needless confusion in dealing with that compound.

[1]At this point, as well as throughout the book, the student should make a point of practicing with pencil and paper the application of these rules and the writing of structural formulas (see Exercises).

2.7 Physical Properties The physical properties of the alkanes are related (as are physical properties of all molecules) to molecular structure and molecular weight. Some of these properties are shown schematically in Fig. 2-2. It will be observed that the melting and boiling points change in a regular way with the number of carbon atoms. In general (other factors being equal), melting and boiling points rise with increase in molecular weight. This can be understood from kinetic considerations: more energy (heat) is required to disperse large molecules than smaller ones. There are at least two reasons for this. Not only are the molecules more massive but also they have more *surface area* along which to cling to other molecules. At ordinary temperatures the alkanes with fewer than four carbon atoms are gases; those with more than about twenty are solids. At the short ranges of contact found in solids, liquids, and gases near their boiling points, all molecules experience weak attractive forces toward one another. These *van der Waals'* forces fall off very rapidly with distance.

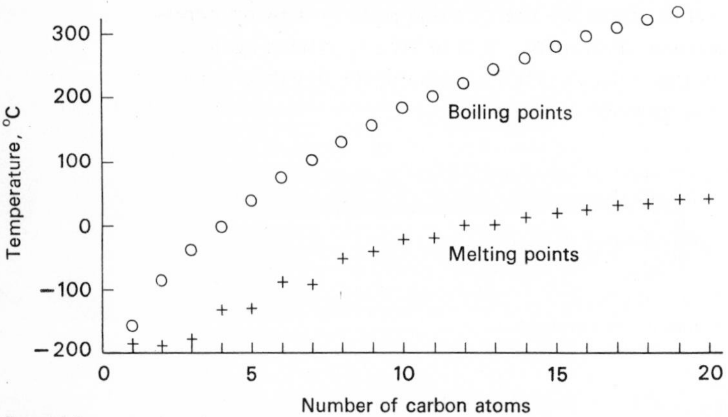
Temperature, °C
300
200
100
0
−100
−200
0
5
10
15
20
Number of carbon atoms
Boiling points
Melting points

figure 2-2
Properties of the normal alkane hydrocarbons.

2.8 Spectroscopic Properties The alkanes, possessing no unshared p electrons and no multiple bonds, do not absorb light in the visible or near ultraviolet regions. Pure hydrocarbons, in fact, are often used as solvents for ultraviolet spectroscopy.

The infrared spectra of alkanes show strong absorption in the regions of 3000 to 2840 cm^{-1} due to C—H stretching vibrations, about 1365 to 1385 and 1440 to 1460 cm^{-1} due to C—H bending, and 800 to 1200 cm^{-1} due to C—C stretching. Since any molecule having alkane groups will show absorption bands in those regions, these characteristics are not unique for pure alkanes. Compounds absorbing only in these regions, however, are most likely alkanes.

The nmr spectra of alkanes show absorption at $\delta = 0.9$ ppm for primary CH_3 protons, at $\delta = 1.25$ ppm for secondary CH_2 protons, and at $\delta = 1.5$ ppm for tertiary CH protons. These are frequently not sharp absorption bands because of complex interaction (splitting) and overlapping of peaks, but they can usually be qualitatively identified because of the low δ value (corresponding to very high shielding) at which they occur.

2.9 Chemical Properties The chemical properties of the alkanes are relatively easily described: these compounds are quite unreactive toward most reagents. They are unaffected by boiling with water, aqueous alkalis and acids, oxidizing reagents such as permanganate and dichromate, and reducing agents. (In this connection recall that alkanes are insoluble in water.) Even hydrofluoric acid can be kept in bottles made of paraffin (a mixture of alkanes). Alkanes are attacked, however, by a number of nonaqueous reagents: halogen gases, nitric oxide and concentrated nitric acid, and oxygen and heat. The reactions which are exhibited by one member of the alkane series are also exhibited by others, for it is a property of a homologous series that the chemical behaviors of its constituents are similar; it is only necessary to learn the behavior of one member to know, or to be able to predict, that of any other. Two points of exception have to be taken to this statement: usually the first member of a homologous series has chemical properties not shared, or shared to lesser degree, by the others; and second, there are actual differences in chemical behavior between the members of a given homologous series, but these are often subtle and usually of no serious import to an elementary course. These differences enrich advanced work in organic chemistry, but they will for the most part be neglected here. The more important chemical properties of alkanes will be illustrated in the equations below.

2.10 Oxidation of Alkanes

$$\begin{array}{c} \text{H} \\ | \\ \text{H—C—H} \\ | \\ \text{H} \end{array} + 2O_2 \longrightarrow CO_2 + 2H_2O + \text{heat of combustion } (210.8 \text{ kcal})^1$$

$$2\begin{array}{c} \text{H}\ \ \text{H} \\ |\quad | \\ \text{H—C—C—H} \\ |\quad | \\ \text{H}\ \ \text{H} \end{array} + 7O_2 \longrightarrow 4CO_2 + 6H_2O + \text{heat of combustion } (2 \times 368.4 \text{ kcal})$$

The alkanes burn. This reaction is the basis for their most important industrial uses: heating, and operating internal-combustion engines. Natural gas, gasolines, and fuel oils consist chiefly of alkanes. If insufficient oxygen is available to the reaction, the products will be water, carbon monoxide, and elementary carbon (soot or carbon black). Observe that the oxidation reactions shown above are not reversible. Most compounds of carbon burn in the presence of air. This constitutes a certain hazard in the organic laboratory.

[1] A kilocalorie is the amount of energy required to raise the temperature of one kilogram of water one degree centigrade; it is equal to 1,000 calories.

2.11 Halogenation Alkanes react readily with the halogens chlorine and bromine (iodine is unreactive), and the products contain halogen attached to carbon by a covalent bond; the halides produced do not ionize in the manner of salts. The equation for the reaction in the case of methane may be written

Overall Reaction[1]

$$\underset{\underset{\displaystyle H}{|}}{\overset{\overset{\displaystyle H}{|}}{H-C-H}} + Cl_2 \longrightarrow \underset{\underset{\displaystyle H}{|}}{\overset{\overset{\displaystyle H}{|}}{H-C-Cl}} + HCl$$

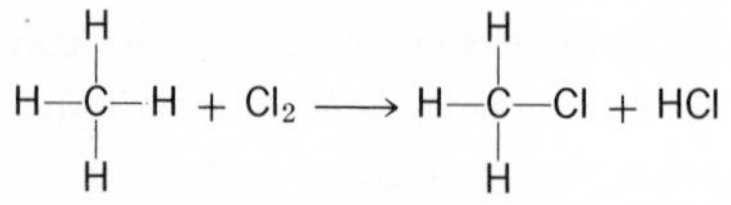

A substitution reaction[2]

The occurrence of this slow reaction is readily detected by the hydrogen chloride evolved (test for acid). *Substitution*, like many covalent bond reactions, not only takes place slowly, but is accompanied by side reactions. In the case of methane, the chloromethane formed has ample opportunity (recall that 1 mole of chlorine contains 6×10^{23} molecules) to react with further amounts of chlorine in the reaction mixture to give di-, tri-, and eventually, tetrachloromethane.

$$\underset{\underset{\displaystyle H}{|}}{\overset{\overset{\displaystyle H}{|}}{H-C-Cl}} + Cl_2 \longrightarrow \underset{\underset{\displaystyle Cl}{|}}{\overset{\overset{\displaystyle H}{|}}{H-C-Cl}} + HCl$$

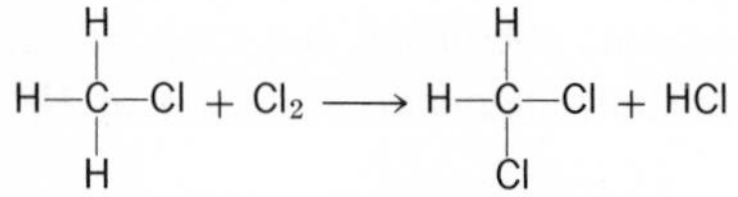

Dichloromethane

$$\underset{\underset{\displaystyle Cl}{|}}{\overset{\overset{\displaystyle H}{|}}{H-C-Cl}} + Cl_2 \longrightarrow \underset{\underset{\displaystyle Cl}{|}}{\overset{\overset{\displaystyle Cl}{|}}{H-C-Cl}} + HCl$$

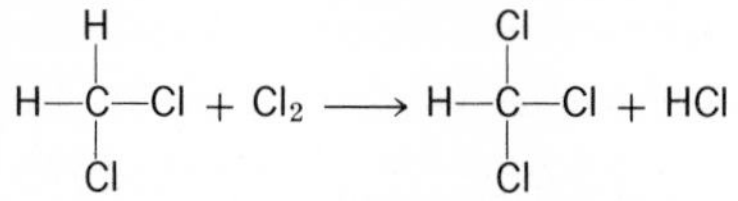

Trichloromethane
(chloroform)

$$\underset{\underset{\displaystyle Cl}{|}}{\overset{\overset{\displaystyle Cl}{|}}{H-C-Cl}} + Cl_2 \longrightarrow \underset{\underset{\displaystyle Cl}{|}}{\overset{\overset{\displaystyle Cl}{|}}{Cl-C-Cl}} + HCl$$

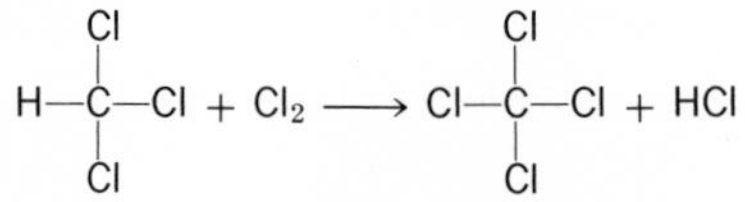

Tetrachloromethane
(carbon tetrachloride)

The probability of four successive substitutions taking place in a single methane molecule is low compared with that of a single reaction, so that on this account alone very little tetrachloromethane would be expected from 1 mole of methane and

---

[1] The term "overall reaction" will be used for reactions which may be subsequently analyzed into their component steps. It will be impractical to analyze all reactions in this manner; however, a large proportion of the reactions in this text will be so analyzed.
[2] Substitution implies replacement of one atom or group by another; it is frequently applied to the halogenation reaction.

1 mole of chlorine. The amount which will be produced, however, can be calculated by the mathematics of probability if we know how often collisions occur between molecules and what percentage of collisions leads to actual reactions in the different cases.

Since the partly substituted products of this reaction are apparently as reactive toward chlorine as methane itself (the amounts of heat given off in each successive substitution are about equal), the reaction cannot be controlled and a mixture of products is always obtained. This is true of the higher homologs also, and substitution is useless as a method of preparation of pure halides from alkanes. A consideration of the number of possible substitution products in the case of propane, for example, serves to emphasize the complexity of the reaction mixtures produced:

$$\underset{\text{Propane}}{CH_3CH_2CH_3 + Cl_2} \longrightarrow \underset{\text{Monochloropropanes}}{\begin{array}{c} CH_3CH_2CH_2Cl \\ + \\ CH_3CHCH_3 \\ | \\ Cl \end{array}} \longrightarrow \underset{\text{Dichloropropanes}}{\begin{array}{c} CH_3CH_2CHCl_2 \\ + \\ ClCH_2CH_2CH_2Cl \\ + \\ CH_3CHCH_2Cl \\ | \\ Cl \\ + \\ CH_3CCl_2CH_3 \end{array}} \cdots \longrightarrow \begin{array}{c}\text{to further} \\ \text{substitution}\end{array}$$

2.12 Rate of Reaction The reaction shown above will proceed slowly in the dark; it goes much faster in the presence of light; it is especially affected by ultraviolet light, and under ultraviolet irradiation may even go so fast as to be explosive. The light is often spoken of as a catalyst, but it is more correct to think of it as a reagent, because light energy is consumed and the reaction is affected by the quality and amount of the light. The chemist is interested in how light speeds up the process or otherwise influences it. This subject cannot be explored very deeply here, but a word must be said about it, because light energy plays a role in so many organic reactions (photosynthesis in plants; irradiation which produces vitamin D; tanning of the skin; and a great many industrial processes).

2.13 Influence of Light: Chain Reactions As we have indicated in Chap. 1, light is a form of electromagnetic radiation. When molecules interact with light they absorb discrete amounts of light energy called "quanta." The amount of energy absorbed in a single event is given by the relationship $E = h\nu$.

Here E is energy, h is Planck's constant, and ν (Greek letter nu) is the frequency of the light (which is related to the wavelength λ by the velocity of light c; $\nu = c/\lambda$). As the organic chemist is usually interested in molar quantities (and not individual molecular quantities) and measures energy in kilocalories and wavelength in millimicrons $m\mu$ this can be rewritten:

$$E = Nh\frac{c}{\lambda}$$

where N is Avogadro's number, or

$$E = \frac{2.86 \times 10^4}{\lambda, m\mu} \quad kcal/mole$$

For visible light λ is about 500 $m\mu$ so that E is approximately 57 kcal/mole, which is close to the energy of the Cl—Cl bond (57.8 kcal/mole).

In the reaction under discussion, it has been found that 1 quantum of light energy, if it is of sufficiently high value (that is, only short wavelengths are effective), can start off the reaction, which *then continues for some time by itself.* In other words, the yield of reacted molecules per quantum of radiation is high. (This is not always the case in photochemical reactions, but frequently is.) What seems to happen may be shown in a set of equations.

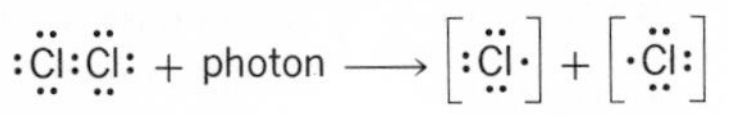

$$:\overset{..}{\underset{..}{Cl}}:\overset{..}{\underset{..}{Cl}}: + photon \longrightarrow \left[:\overset{..}{\underset{..}{Cl}}\cdot\right] + \left[\cdot\overset{..}{\underset{..}{Cl}}:\right] \tag{2.1}$$

$$H:\overset{\overset{\displaystyle H}{..}}{\underset{\displaystyle H}{C}}:H + \left[:\overset{..}{\underset{..}{Cl}}\cdot\right] \longrightarrow \left[H:\overset{\overset{\displaystyle H}{..}}{\underset{\displaystyle H}{C}}\cdot\right] + H:\overset{..}{\underset{..}{Cl}}: \tag{2.2}$$

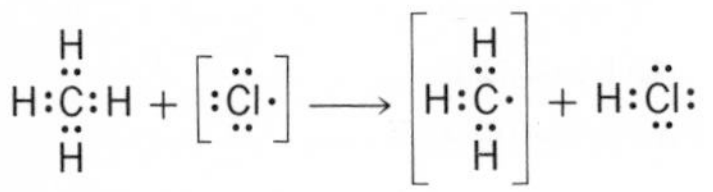

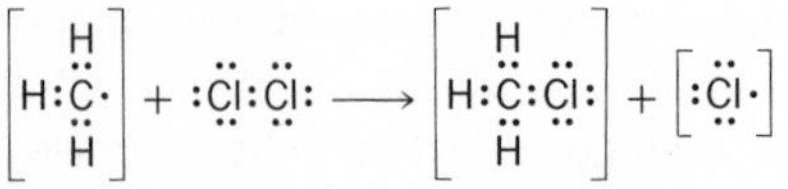

$$\left[H:\overset{\overset{\displaystyle H}{..}}{\underset{\displaystyle H}{C}}\cdot\right] + :\overset{..}{\underset{..}{Cl}}:\overset{..}{\underset{..}{Cl}}: \longrightarrow \left[H:\overset{\overset{\displaystyle H}{..}}{\underset{\displaystyle H}{C}}:\overset{..}{\underset{..}{Cl}}:\right] + \left[:\overset{..}{\underset{..}{Cl}}\cdot\right] \tag{2.3}$$

Equation 2.1 shows how the photon may "initiate" the reaction by forming the two chlorine radicals. Observe that the radical is an *atom.* The two atoms now possess the energy brought by the photon, a good part of which has been used to break the bond that existed between them, the remainder being used to impart kinetic energy to the two atoms and to promote electrons within the chlorine atoms to higher orbitals. In the gas phase these two radicals will seldom recombine since they possess more than enough energy to fly apart again, at least until they have had an opportunity to dissipate some of their excess energy by collision with other molecules. For our purposes the most important reaction that chlorine atoms undergo is shown in Eq. 2.2, which is then followed by Eq. 2.3. Since Eq. 2.3 regenerates a chlorine radical, the whole process of Eqs. 2.2 and 2.3 can repeat over and over again. This process is known as the "propagation stage." Propagation does not continue indefinitely since it is possible for (*a*) two chlorine atoms to recombine (the reverse of the initiation step), (*b*) two methyl radicals or a methyl radical and chlorine atom to combine, or (*c*) for one of the propagating radicals to react with some other species in the reaction mixture and be transformed into a different radical of insufficient energy to continue the process. These steps are known as "termination" steps.

Reactions of this kind are called "chain reactions" because an initial impulse can set a self-propagating set of reactions (Eqs. 2.2 and 2.3) alternating, which continue until stopped by some active termination process, or until all the reagents are exhausted. In practice, one observes that the net number of molecules of chloromethane produced per quantum of light absorbed (called the "quantum yield") is much greater than unity; that is, on the average, each pair of chlorine atoms produced goes through Eqs. 2.2 and 2.3 a number of times.

2.14 Preparation of Alkanes In the discussion of each homologous series we will be concerned, on the one hand, with the preparation of typical members of the series and, on the other, with their chemical and physical properties.

A discussion of the chemical preparation of alkanes from other substances must be somewhat limited by the fact that we have not yet discussed many other substances. Only one reaction for the preparation of alkanes will be dealt with here; the list can be extended later.

2.15 Grignard Reaction The distinguished French chemist V. Grignard discovered in 1900 that bromomethane, CH_3Br, reacts with magnesium suspended in dry ether. The reaction is a vigorous exothermic one during which the magnesium dissolves and an ether-soluble product, now known as a *Grignard reagent,* is formed.

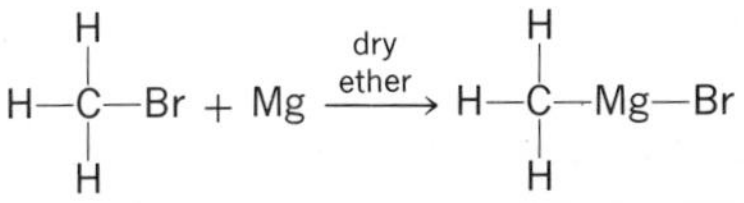
H
H—C—Br + Mg
H
dry
ether
H
H—C—Mg—Br
H

Methylmagnesium bromide
(a Grignard reagent)

Grignard reagents[1] react with acids[2] to form hydrocarbons.

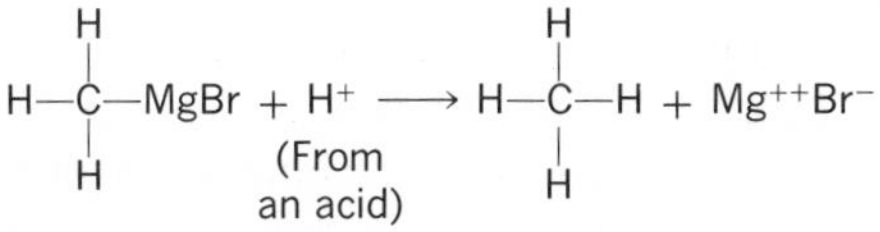
H
H—C—MgBr + H+
H
(From
an acid)
H
H—C—H + Mg++Br−
H

Since the C—H bonds in saturated hydrocarbons such as methane are almost 100 percent covalent in character, the acidic properties

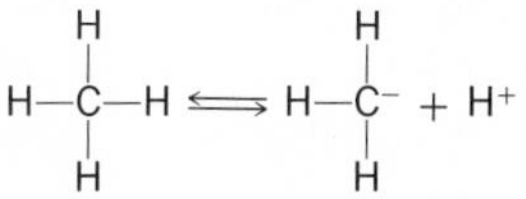
H
H—C—H
H
H
H—C− + H+
H

[1]The structure of the Grignard reagent is not completely known, but it will suffice for our purposes to regard it as R—Mg—X, where R is a hydrocarbon group and X is a halogen. It is important to note, however, that carbon is no longer directly bonded to the halogen, but that a carbon-magnesium bond is involved.

[2]Acids are defined here as *any* substances which can yield protons.

of these substances are not easily demonstrated. Nevertheless, it is useful to regard Grignard reagents as salts of *exceedingly* weak acids; we are then not surprised to find them completely hydrolyzed in water.[1] In laboratory practice, water is often used

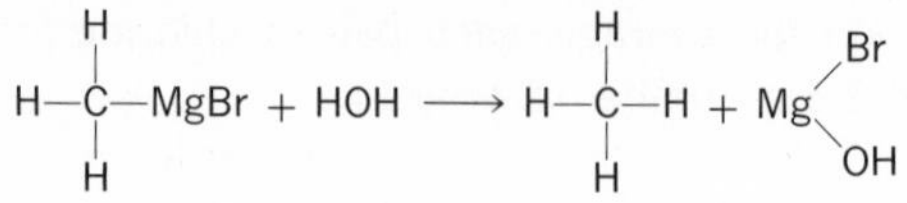

as the source of protons in converting Grignard reagents to hydrocarbons. The student will realize, however, that any acid which is more extensively ionized than the hydrocarbon will serve for this purpose. Even weak acids such as NH_3, alcohols (CH_3OH), and acetic acid, as well as the more familiar mineral acids, may be used.

The halogen atom may be located anywhere on the carbon chain of a saturated molecule with the same result, and chlorine or iodine may replace the bromine used above.

In the many important and useful reactions of Grignard reagents found in later chapters, the polar character of the carbon-to-magnesium bond and its tendency to yield Mg^{++} and $\geqslant C^-$ ions as transitory intermediates, at least, will be observed again. The ready cleavage of the Grignard reagent in this manner is characteristic of these substances as a class and responsible for their great utility in the synthesis of carbon compounds. They are strong, nucleophilic reagents.

2.16 Side Reactions In all the reactions described so far the occurrence of *side reactions* is observed. For example, a Grignard reagent reacts with oxygen from the atmosphere.

$$C_2H_5MgBr + O_2 \longrightarrow C_2H_5OMgBr$$
$$C_2H_5OMgBr + H_2O \longrightarrow C_2H_5OH$$

This particular side reaction is of minor importance, *if reasonable precautions are taken to minimize contact with air,* but phenomena of this sort are very common. Although detailed discussion of side reactions will frequently be omitted in considering synthetic methods, it must be understood by the reader that the majority of reactions involving the formation or cleavage of covalent bonds takes place slowly, incompletely, and with the accompaniment of side reactions. This means that in the actual application of such reactions in the laboratory or in the plant, the final products must be carefully purified. Distillation, crystallization, and other methods of purification make up a major part of the laboratory work of organic chemistry; a discussion of these methods is to be found in laboratory manuals.

2.17 Proof of Structure In the halogenation of an alkane the evidence that a reaction has occurred is obtained by various means. It may be shown that heat is evolved; that the organic product, which can be isolated (for example, by fractional

[1] MgBrOH appears as a mixture of $MgBr_2$ and $Mg(OH)_2$ in the solution.

distillation), is different in chemical and physical properties from the starting materials; that a halogen acid is produced, which may be isolated, identified, and shown to differ from the starting materials. It is, therefore, obvious that it is easy to prove that a reaction has occurred; it is not so easy to show *what* reaction has occurred, for this requires molecular and structural analysis. Some insight into the reasoning involved and the methods used in molecular and structural analysis may be gained from a consideration of the proof of structure of the isomeric butanes.

Granted that there are two compounds in petroleum with the molecular formula C_4H_{10}, one is inclined to ask, reasonably enough: How do we know which of these two is butane and which is isobutane? Often in more complex cases this type of question is a very difficult one to answer, sometimes requiring years of research. In the case of the butanes the problem is essentially this: There have been obtained in pure form two C_4H_{10} isomers, butane *a*, boiling point $-10°C$, and butane *b*, boiling point $-0.6°C$; there are, similarly, differences in other physical properties such as density, refractive index, etc. From a consideration of possible structural formulas based on tetrahedral structures for carbon, it is evident that since one is $CH_3CH_2CH_2CH_3$ and the other is

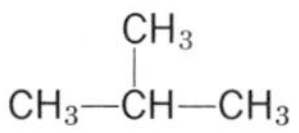
CH3
CH3—CH—CH3

there remains the question of assigning the correct structural formula to each individual (putting the correct formula on the label on each bottle).

There now exist a number of ways by which these two compounds may be assigned their correct structures. The structures were first assigned by methods involving the preparation of these hydrocarbons from more complex compounds of known structure, using reactions which we have not yet encountered. As chemistry has progressed, however, chemists have come to depend more and more on the newer spectroscopic methods, and in all probability, a chemist faced with the problem of distinguishing between these compounds today would resort at once to spectroscopic methods. In this case infrared can provide an answer only with considerable difficulty and uncertainty, but the differentiation is readily made by nuclear magnetic resonance spectroscopy (nmr).

One of the butanes, that boiling at $-0.6°C$, possesses absorptions at $\delta = 0.9$ and $\delta = 1.25$ with relative areas of 3:2 (or 6:4). We can confidently assign the structure $CH_3—CH_2—CH_2—CH_3$ to this isomer. The other butane, bp $= -10°$,[1] possesses absorptions at $\delta = 0.9$ and $\delta = 1.5$ in the ratio of 9:1, so it clearly represents

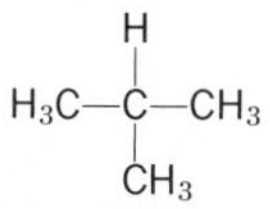
H
H3C—C—CH3
CH3

[1]All temperatures are in degrees Centigrade if not specified.

In addition, the splitting of the nmr absorptions agrees with the proposed structures. The six primary H's of the $-0.6°$-bp isomer appear as a triplet, whereas the nine primary hydrogens of the $-10°$ isomer appear as a doublet. The other hydrogens are correspondingly split; the tertiary hydrogen on the $-10°$ isomer is split into ten maxima which will often be so weak as to show up as a broad hump in the $\delta = 1.5$ region.

Study of a number of hydrocarbons shows that primary protons absorb at δ values of 0.9, secondary protons at 1.25, and tertiary protons at 1.5 (Sec. 2.8). Further, in nmr spectroscopy the areas under the plotted absorption curves are always proportional to the number of protons of each kind.

Proof of structure by spectroscopic methods like this often relies heavily on knowledge of a large number of structures which have been proven by other methods (such as synthesis or degradation) since it is from these compounds that generalization about the δ values for CH_3— groups, etc., are derived.

2.18 Analysis Except in a few instances the common ionic reactions of inorganic chemistry cannot be used directly in the analysis of organic substances, since most of their bonds are covalent. Carbon and hydrogen are determined quantitatively by *combustion analysis*. The organic substance, weighed into a little platinum vessel, is placed in a special tube in a furnace and burned in a stream of purified oxygen.

$$C_nH_{2n+2} + \left(\frac{3n+1}{2}\right) O_2 = nCO_2 + (n+1)H_2O$$

The water that is formed is absorbed in a special drying agent in a detachable tube at the end of the combustion tube and weighed; the carbon dioxide is similarly absorbed in alkali and weighed. Alternative methods of measuring these products of combustion are now being used but the basic analytical process remains the same. Precautions are taken to ensure that if any nitrogen, sulfur, or halogens are present they will be caught in the combustion tube and not turn up in the carbon dioxide or water tubes. For example, sulfur is caught as lead sulfide; and halogens are held up by silver wire. As little as 5 mg of substance can be accurately analyzed. If oxygen is present, it is usually determined by difference. When the percentages of all the other components are added up, if the difference from 100 percent is greater than the experimental error, then the difference is usually oxygen. Other elements can also be determined by special combustion procedures.

For qualitative analysis, a little of the substance is heated with copper oxide in a test tube. If carbon is present, carbon dioxide is formed and can be tested for with lime water at the mouth of the test tube. Hydrogen forms water which condenses on the wall of the tube near the top. For qualitative analysis of other elements a *sodium fusion* is used. A little of the organic substance is dropped cautiously upon a small bead of molten sodium in a test tube. There usually results a puff of smoke and a glow or flash. The mixture is then heated strongly to complete the reaction. Then the tube is broken, the residue taken up in water, filtered, and tested for the various ions by the usual procedure. Nitrogen usually appears as CN^-; sulfur as S^{--}; and halogen as Cl^-, or Br^-, or I^-, which can be detected in the presence of each other. The aqueous extract from the sodium fusion is, of course, quite alkaline. It contains not only sodium carbonate formed from the carbon of the compound, but also sodium hydroxide formed when the excess sodium reacts with water.

Although the combustion analysis has long been the backbone of organic chemistry, techniques such as high resolution mass spectrometry threaten to relegate even this method to a role of lesser importance.

INTRODUCTION TO SYMBOLS AND CONVENTIONS

The symbols R—, R′—, R″—, etc., are taken to mean any aliphatic,[1] or at times, substituted aliphatic, groups. R— and R′— may or may not be different. The symbol R will be made more inclusive at the end of Chap. 8.

The Greek letter Δ under an arrow in an equation indicates that heat is needed for the reaction.

N.R. stands for "no reaction."

(H) stands for reducing agents in equations, such as hydrogen and a catalyst, or tin and hydrochloric acid, etc.; (O) stands for oxidizing agents such as acid or alkaline permanganate, dilute nitric acid, dichromate, etc.; (X) stands for halogen. NOTE: The student should remember that R— is merely a convenient substitute for actual groups, and he should learn, in studying these equations, to use actual groups such as CH_3—, C_2H_5—, etc., rather than the general symbols.

OUTLINE OF ALKANE CHEMISTRY

Alkanes (C_nH_{2n+2}). The methane series, suffix *-ane.*

Preparation

$$R\text{—}MgX + HOH \longrightarrow R\text{—}H + MgXOH$$

Properties

1. When ignited, all alkanes burn in the presence of air or oxygen with the production of CO_2 and H_2O, although at ordinary temperatures they are stable to oxygen (in the absence of a spark).

[1]The term aliphatic is used for all open-chain and cycloalkane groups (Chap. 4); only aromatic groups (Chap. 6) are excluded.

2. R—H + Cl_2 ⟶ R—Cl + HCl
   This reaction is characteristic of all alkanes; it is *useless as a laboratory method of preparing the halogen substitution products of the alkanes.*
3. Most alkanes are stable to the following wide list of reagents: $KMnO_4$, $K_2Cr_2O_7$ + H_2SO_4, HNO_3, conc. H_2SO_4, HCl, NaOH, NH_3, and reducing agents at ordinary temperatures.

Spectroscopic Properties

Group	*ir, cm^{-1}*	*nmr, δ(ppm)*	*uv, λ_{max}(mμ)*
CH_3	2840–3000	0.9	none
C—CH_2—C	2840–3000	1.25	none
C, C—C—H, C (tertiary)	2840–3000	1.5	none

CHAPTER 2 EXERCISES

★**1.** Which of the following are (*a*) identical, (*b*) isomers, (*c*) members of the same homologous series?

a. $CH_3CH_2CH_2Cl$

b. 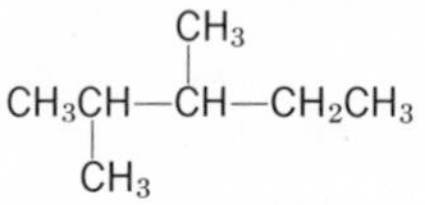

c. CH_3CHOH—CH_2CH_3

d. CH_3CH=CH—CH_2CH_3

e. $CH_3CH_2CH_2CH_2OH$

f. 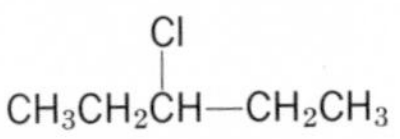

g. 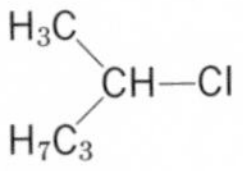

h. CH_3CH_2CH=CH—CH_3

i. 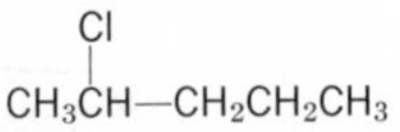

j. 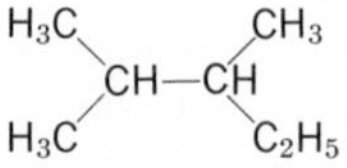

k. 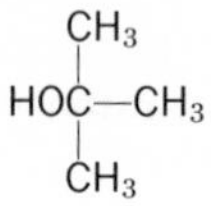

l. 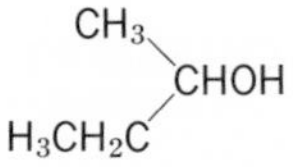

m. 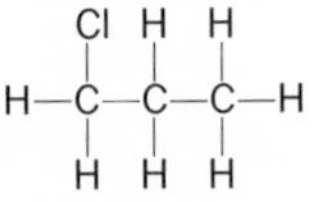

n. $(CH_3)_3COH$

2. Write structures for all isomers corresponding to the following molecular formulas:

a. C_4H_{10} *c.* C_2H_6O *e.* $C_3H_6Cl_2$
b. C_5H_{12} *d.* $C_4H_{10}O$ *f.* C_3H_6BrCl

★3. Draw structures for the following hydrocarbons. Label each hydrogen as primary, secondary, or tertiary. Predict the number of nmr peaks (disregarding splitting) expected for each.

a. 2,3-Dimethylpentane
b. 2,2-Dimethylpropane
c. 3-Ethylpentane
d. Propane

4. How may dichloro derivatives are possible for (*a*) pentane, (*b*) 2-methylpropane, (*c*) 2,2-dimethylpropane?

★5. Would propane, heptane, or $C_{20}H_{22}$ make the most convenient fuel for an automobile engine? Explain.

6. Explain how accurate measurement of the amount of light absorbed in a reaction would help decide whether or not it is a chain reaction. What kind of "reagent" besides light might initiate a chlorination (chain reaction) of methane?

★7. What spectral method would you choose for analysis of a mixture of pentane and 2-methylpentane? Explain.

8. List several reagents that will *not* react with the Grignard reagent. Are these Lewis acids, bases, or neither?

★9. Suppose a hydrocarbon exists consisting of four tetravalent (sp^3) carbon atoms in a ring with eight H atoms attached. How many monobromo substitution products are possible? How many dibromo?

10. Suppose there were two different dichloromethanes. What symmetrical structure or structures would account for this?

11. Write the systematic (Geneva) names for the following:

★*a.* 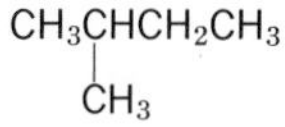

b. 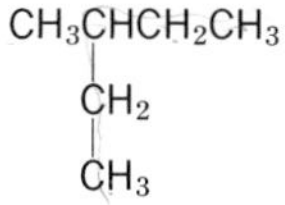

★*c.* 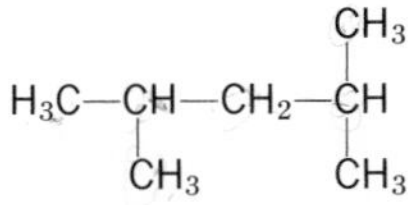

d. 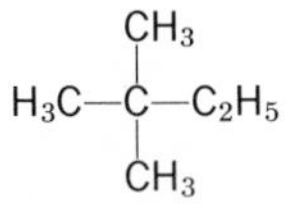

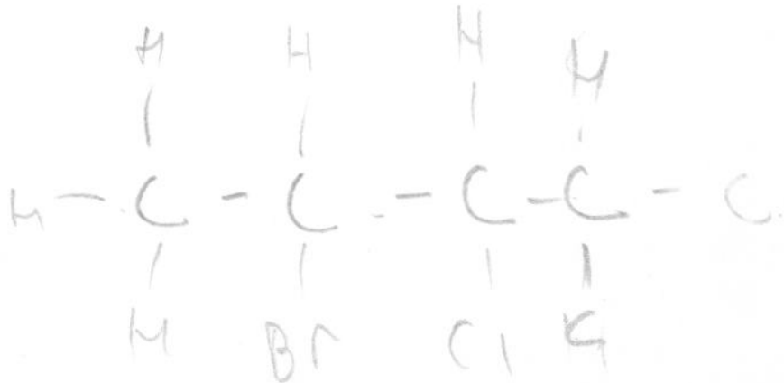

★*e.* $(CH_3)_2CH—CH(C_2H_5)_2$
f. $CH_3(CH_2)_2CH(CH_3)_2$
★*g.* $(C_2H_5)_4C$
h. $CH_3CHBrCHClCH(C_2H_5)_2$

12. Write the systematic names for the compounds (*a*), (*b*), (*f*), (*g*), (*i*), and (*n*) in Exercise 1.

★13. Explain how you would determine whether any reaction is taking place in an experiment in which methane and chlorine (both gases) are mixed. Would the same method work for a mixture of hexane and bromine (both liquids)?

14. Ethane, like the other alkanes, is practically odorless and tasteless. Explain this fact on the basis of the known reactivity of ethane.

★15. Draw a formula for the Grignard reagent prepared from ethyl bromide showing (*a*) any electrovalences, (*b*) any polar bonds and the direction of their polarization, (*c*) any purely covalent nonpolar bonds. If the Grignard reagent is assumed to yield an ethyl ion, will this ion be a nucleophilic or electrophilic reagent?

16. Show clearly how you could distinguish between hydrogen chloride and methyl chloride. Both are gases under ordinary conditions.

★17. The waxy substance found on the leaves and stems of tobacco plants consists largely of a solid paraffin hydrocarbon. Would you expect to be able to remove this material by washing with water? Assuming it to be a normal paraffin chain, estimate the minimum number of carbon atoms in the chain.

18. Write equations showing the bromination of ethane, assuming it to be catalyzed by light.

★19. What explanation may be offered for the fact that the bond length of C—O is shorter than that of C—S? Explain the order of increasing bond length of C—F, C—Cl, C—Br, C—I.

20. What explanation may be offered for the fact that the atomic diameter of O is less than that of S? Why is it less than that of C? Explain why the order of atomic diameters is as it is in the series, C, O, N, F, and F, Cl, Br, I. Be as explicit as possible.

★21. The dipole moment measures disymmetry of charge in a molecule. Thus the polar bond $\overset{\delta+}{C} \longrightarrow \overset{\delta-}{Cl}$ would be the seat of a bond moment of 2.3 Debye units, in the direction shown. The dipole moment is a vector quantity. It is observed that CCl_4 has a molecular dipole moment of 0.0. How would you explain this? Would you expect $CHCl_3$ to show a molecular dipole moment? (The bond moment of C—H is about 0.4 unit.)

★22. The following compounds cannot be readily distinguished from one another by nmr alone. Why? What additional information would enable you to ascertain their structures?

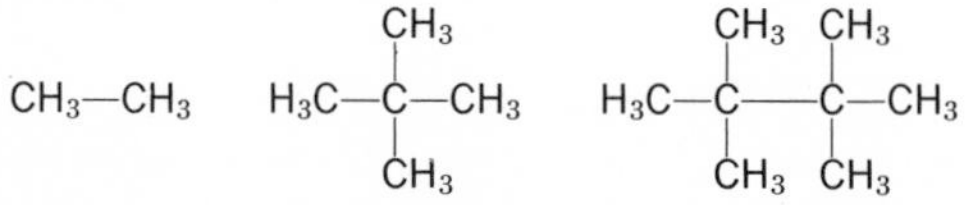

If the limit of quantitative accuracy of nmr is ±5% of the total hydrogen content in a molecule, how long would a straight carbon hydrocarbon chain have to be before one could not be certain that it contained CH_3 groups?

★23. The carbon-to-carbon bond angles of 2-methylpropane have been found to be 111.5°. What explanation can be offered for this variation from the 109°28′ tetrahedral angle?

24. How would the nmr spectra of the two isomeric dibromoethanes be expected to differ? Sketch a *spectrum* for each.

★25. How would you undertake the preparation of CH_3D? Assume that CH_3Cl is available.

26. How could the following be distinguished without reference to spectral methods? By spectral methods?

a. $CH_3CH_2CH_2Br$ and $CH_3CHBrCH_3$

b. *n*-Butane and 2-methylpropane

c. 2-methylbutane and 2,2-dimethylpropane

★27. Suggest a structure for

★*a.* a C_5 alkane that shows only a single peak in its nmr spectrum

★*b.* a C_6 alkane that shows 2 peaks in its nmr spectrum

★*c.* a C_6 hydrocarbon (not necessarily an alkane) that shows only one peak in its nmr spectrum

28. Write equations showing the mechanisms for the reactions

a. $C_2H_6 + Cl_2 + \text{light} \longrightarrow$

b. $CH_3MgBr + NH_3 \longrightarrow$

c. $HCl + H_2O \longrightarrow$

d. $BF_3 + H_2O \longrightarrow$

3

HYDROCARBONS: THE ALKENES

3.1 Introduction The alkenes have the generalized empirical formula C_nH_{2n}. The structural formula of the simplest member of the series[1] is written

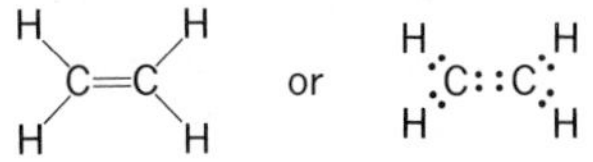

Ethylene
(a gas)

[1]All attempts to prepare CH_2 give rise to other products. CH_2, like CH_3, is unstable. However, there is ample evidence for the *transient* existence of both of these. H_2C: is called *methylene* or *carbene* (Sec. 1.10).

We may learn something about the structure of ethylene from the fact that the gas reacts with chlorine according to the equation[1]

$$C_2H_4 + Cl_2 \longrightarrow C_2H_4Cl_2$$

An *addition* reaction

In other words the chlorine is absorbed by, or added to, the molecule. Only one product is formed, and this turns out to be 1,2-dichloroethane. The nmr spectrum shows only one type of proton, as would be expected from this structure; it follows that the original gas could not have been

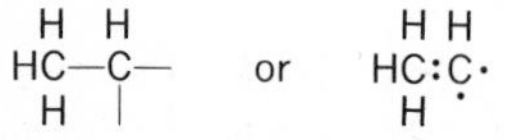

because addition of chlorine to the unshared electrons would have yielded 1,1-dichloroethane. One chlorine is found attached to each carbon. The compound would be unlikely to have a structure with *free valences* or unshared electrons

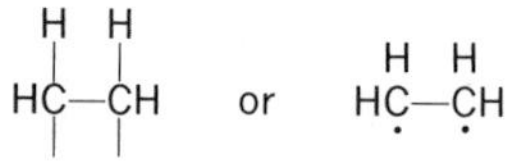

because we have learned that carbon is most stable when it is linked by four covalences. Thus, it seems reasonable to conclude that the structure of the molecule is

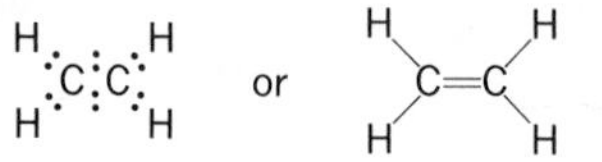

Ethylene is said to contain a *double bond.* Substances containing this type of linkage react by *addition* and are said to be *unsaturated.* The alkanes are saturated compounds; they react with few reagents and always by substitution. The double-bond compounds on the other hand are more reactive and react chiefly by addition.

3.2 Unsaturation Number Because compounds which are unsaturated have fewer groups attached to a given number of carbon atoms, the molecular formula provides information about the question of whether a molecule is saturated or not. This can be done by writing down a molecular formula for a saturated compound with the same number of carbon atoms, and comparing this with the formula under consideration. It is easier to formalize this in the following way: We define N_U as *the unsaturation number which is equal to the number of double bonds or their equivalent*

[1] The term "olefin," often used as a synonym for alkene, originated from this reaction. Ethylene was called olefiant (oil-making) gas because it formed an oil from reaction with chlorine in an instantaneous reaction.

per molecule. This is given by the equation

$$N_U = C + 1 - \frac{1}{2}(H + X)$$

where C = the number of carbon atoms in the molecule, H = the number of hydrogens, and X = the number of halogens, if any. We can see that alkanes have $N_U = 0$, and alkenes have $N_U = 1$.

3.3 Homologous Series of Alkenes Hydrocarbons containing double bonds are distinguished by the suffix *-ene* (compared with the alkane suffix *-ane*) which designates one double bond in the molecule. These compounds make up the class of hydrocarbons known as alkenes (olefins), and the substance described above is ethene, that is, the 2-carbon -ene compound. The common name for ethene is *ethylene.*

3.4 Nomenclature It is apparent that a homologous series of alkenes can be written down, and some examples of these are listed below:

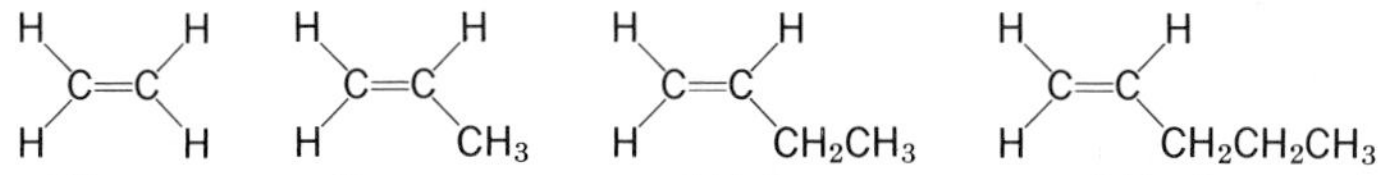

Ethene (ethylene) Propene (propylene) 1-Butene (1-butylene) 1-Pentene

A slight modification of the nomenclature rules given in Chap. 2 is necessary in naming compounds like the alkenes which contain a reactive group.[1] The double bond in the case of the alkenes is very reactive compared with the rest of the molecule, and this fact needs to be emphasized in the revised rules.

RULES FOR NAMING CARBON COMPOUNDS

1. The longest continuous chain of carbon atoms containing the functional group (i.e., double bond in this case) is chosen as the basis for the name. This does not necessarily mean the longest chain in the molecule, and it may involve a ring.
2. The numbering is started at the end nearest the reactive group, and in the case of double or triple bonds proceeds through the carbons of the double or triple bond.
3. The ending *-ene* is used for each double bond present. The position of the double bond is indicated by the use of the smaller of the numbers of the doubly bonded atoms. This number precedes the basic name.
4. Each substituent receives a name *and* a number (Sec. 2.6).

A few examples will illustrate the application of these rules:

[1]Other reactive groups will be encountered later: triple bonds, hydroxyl groups, etc.

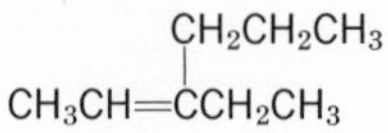

3-Ethyl-2-hexene

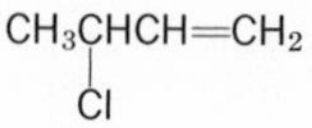

3-Chloro-1-butene

$CH_2{=}CH{-}CH{=}CH{-}CH_3$

1,3-Pentadiene[1]

3.5 Concept of the Double Bond The addition reactions of carbon compounds are of great importance and have been the subject of a good deal of research. A number of questions arise in connection with these reactions. Why is a double bond so reactive compared to a single bond? Why, in other words, doesn't the reaction proceed thus?

$$CH_2{=}CH_2 + Br_2 \longrightarrow CH_2{=}CHBr + HBr$$

This problem has caused considerable speculation among organic chemists. Two more electrons are shared between the doubly bonded atoms than are needed to maintain a stable covalent linkage between them, so that there is a certain amount of unused combining power latent in this arrangement. Although modern electronic theories of the nature of carbon bonds had not then been put forward, Thiele recognized the existence of this unused combining power and spoke of it as a "residual valence" which gives a point of attachment for molecules which might add to the double bond. This hypothesis he symbolized as follows:

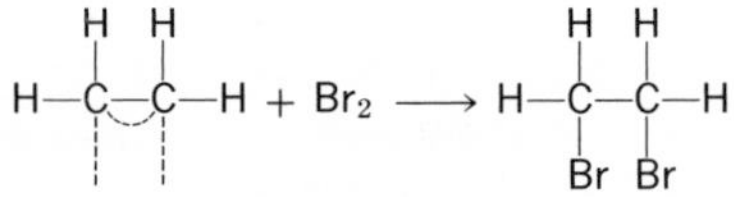

This older theory has been mentioned because parts of it will be recalled later, and because it was one of the earlier attempts to draw pictures in explanation of organic-chemical phenomena.

One modern idea is that the double bond is a resonance hybrid to which forms such as

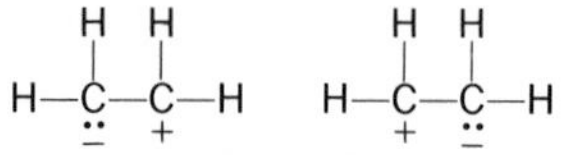

make small contributions. A basis for the idea of residual valence is obvious here, for each of the illustrated forms has bond-forming potentialities at its two carbon atoms. In addition, the fact that one can calculate from the properties of the compound the percentage contribution of each of the above forms to the character of the bond gives a quantitative aspect to the modern hypothesis which was lacking in Thiele's time.

3.6 The Molecular Orbital Model of the Double Bond One may recall from Sec. 1.6 that carbon forms sp^3 hybrids in the formation of covalent bonds with four other groups. In alkenes there are, in effect, not enough groups to go around so that the (energetic) incentive for sp^3 hybridization is lost. An energetically more favorable

[1]Pronounced -dī′ēn. As before, two identical functional groups receive two numbers and the prefix di-, three identical groups receive three numbers and the prefix tri-, etc.

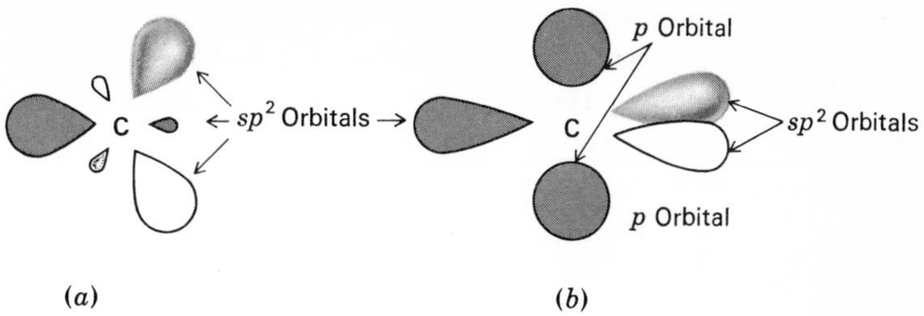

C
← sp^2 Orbitals →
p Orbital
C
sp^2 Orbitals
p Orbital
(a)
(b)

figure 3-1
Molecular-orbital model of the double bond. (*a*) Top view showing three sp^2 orbitals (p orbital omitted). (*b*) Side view showing p orbital.

arrangement for carbon in this situation involves the hybridization of one s and only two p orbitals to give three *equivalent, coplanar* sp^2 orbitals plus one unhybridized p orbital at a right angle to the plane of the other three orbitals. Note that each sp^2 orbital is directed at 120° from the other two, and that one lobe of each orbital is much larger than the other (Sec. 1.6). In forming the double bond, two such carbon atoms overlap along the line of centers of one of their sp^2 orbitals, the remaining sp^2 orbitals being used for bonds with other groups or atoms. This leaves the two p orbitals with one electron each. If these are lined up with their axes parallel as indicated in Fig. 3-2(*a*), it is possible for these to overlap sideways as illustrated in Fig. 3-2(*b*). This type of bond is called a "π bond."

There are, at once, seen to be several consequences of this picture. One is that the two carbon atoms and all four substituents lie in the same plane. Another is that the two carbon atoms are bonded to one another both by a σ bond, formed by overlap of sp^2 orbitals between the two carbon atoms, and by a π bond, part of which is above the plane of the rest of the atoms and the other part of which is below this plane. *This combination of one σ bond and one π bond constitutes a double bond.* It has been seen that in the case of a σ bond (single bond), more or less free rotation may occur about the bond (Sec. 2.4). In the case of the double bond, however, the formation of the π bond requires that the p orbitals lie parallel to one another. In order to rotate about the sigma part of such a bond it would

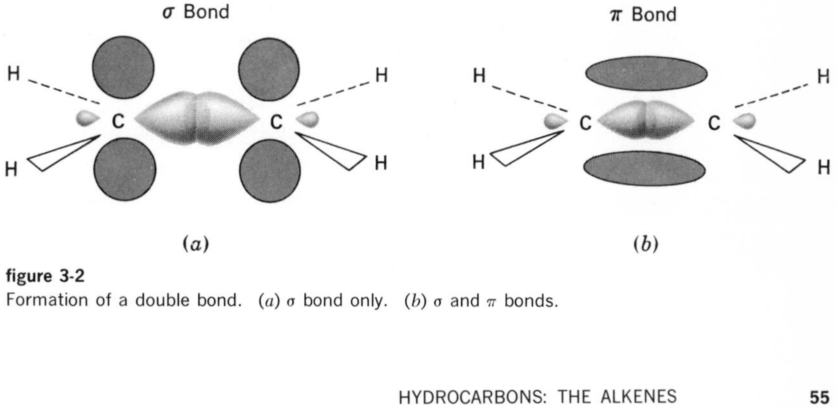

σ Bond
π Bond
H
C
C
H
H
H
H
C
C
H
H
H
(a)
(b)

figure 3-2
Formation of a double bond. (*a*) σ bond only. (*b*) σ and π bonds.

be necessary to break the π bond part. Since the energy required is found to amount to about 40 to 60 kcal/mole, this process does not occur readily at ordinary temperatures. The double bond is thus found to engender a coplanar system of considerable rigidity.

3.7 Geometrical Isomerism In suitable instances the rigidity of the double bond makes for a kind of isomerism known as *geometrical* or *cis-trans* isomerism. For example, two different butenes are known. They have the same groups, undergo the same reactions, but differ in placement of the groups about the rigid double bond. This is not a matter of conformation (see Secs. 2.4 and 7.9), but of isomerism (Sec. 7.2), and the two isomers have separate existences and exhibit different physical (and some chemical) properties.

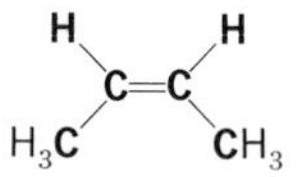
H
H
C=C
H3C
CH3
cis-2-Butene
(bp = 4°C)

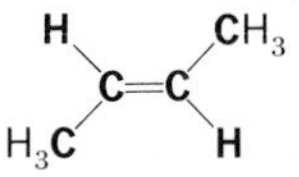
H
CH3
C=C
H3C
H
trans-2-Butene
(bp = 1°C)

The molecule with identical groups on the same side of the double bond is called the "cis" form, and the other, with identical groups across from each other, is called the "trans" form.[1] Note that, as written, the π bonds are above and below the plane of the paper (not shown) and that the atoms in boldface type are all coplanar.

This type of isomerism will be discussed in some detail in Sec. 7.2. It is introduced here as supporting evidence for the structural picture of the double bond developed above. It will be noted that the methyl groups must be farther apart in the trans configuration than in the cis.

A third consequence of the picture is that there is, in the π bond, a region of electron density above and below the plane of the six bonded atoms. This molecular-orbital picture makes it easier to understand the *nucleophilic character of the carbon-to-carbon double bond* which is observed in a study of addition reactions.

3.8 Addition Reactions Reactions which take place by addition usually proceed much more readily than those which require replacement of one atom by another (displacement). The outstanding chemical property of an alkene is its ability to undergo addition reactions, while at the same time possible displacement reactions involving the reagents in question are relegated to a position of minor importance: they become side reactions.

Olefinic double bonds behave as nucleophilic substances in their addition reactions. They combine readily with electrophilic reagents such as the strong acids (H^+), the halogens, and oxidizing agents, and they fail to combine with other nucleophilic reagents such as the Grignard reagents and bases.

[1]These prefixes are from the Latin: *cis,* on this side, as in cisatlantic; *trans,* across, or over, as in transatlantic.

The detailed mechanism of addition to double bonds varies with the nature of the alkene, with the electrophilic reagent, and with such factors as solvent, catalyst, and temperature. However, certain general modes of addition are so common that it is very profitable to use them as working models of the addition reaction.

The most commonly observed mode of addition to double bonds involves the addition of electrophilic reagents, which bear a positive charge or which are strongly polarized so that one portion of the adding reagent is positively charged. These are called "polar" additions. Common polar reagents are H^+, $[:\ddot{\underset{..}{Br}}^+]$, and $[:\ddot{\underset{..}{Cl}}^+]$. A reagent such as HBr has been shown to add to an alkene as follows:

$$HBr \rightleftharpoons H^+ + :\ddot{\underset{..}{Br}}:^-$$

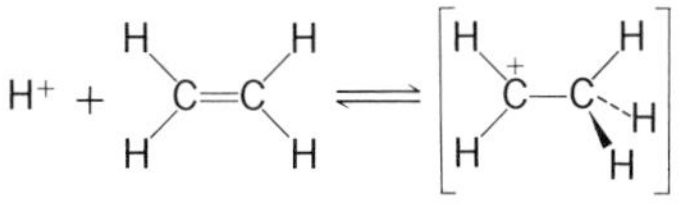

A carbonium ion

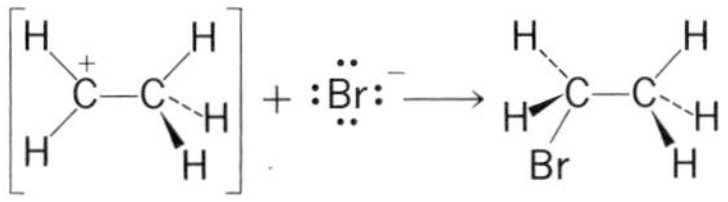

Molecules such as Br_2 and Cl_2 in solvents which stabilize the formation of ions can dissociate as follows:

$$:\ddot{\underset{..}{Br}}:\ddot{\underset{..}{Br}}: \rightleftharpoons :\ddot{\underset{..}{Br}}:^- + [:\ddot{\underset{..}{Br}}^+]$$

$$:\ddot{\underset{..}{Cl}}:\ddot{\underset{..}{Cl}}: \rightleftharpoons :\ddot{\underset{..}{Cl}}:^- + [:\ddot{\underset{..}{Cl}}^+]$$

In each case it is again the positive ion that reacts initially with the nucleophilic alkene to give an intermediate carbonium ion. For example, with Cl_2

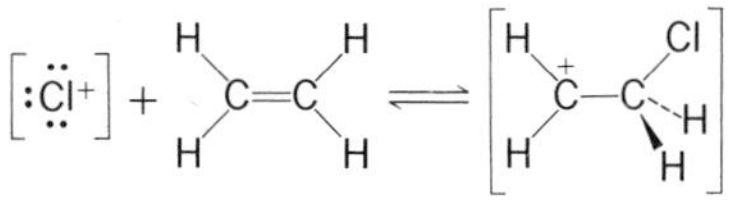

This can further react with the anion to give the isolated product

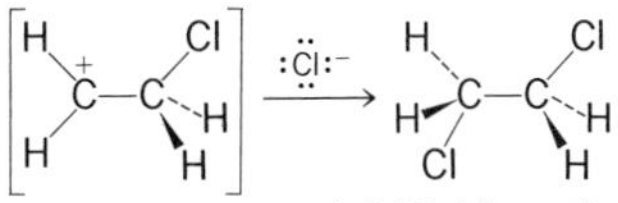

1,2-Dichloroethane

Although it would be expected that free rotation (Sec. 2.4) would allow an attack

by Cl^- on either side of the

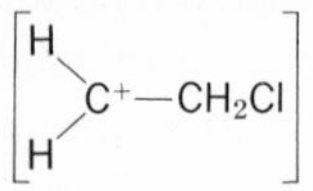

carbonium ion, in fact, it can be shown that the chloride ion actually forms its bond to carbon almost exclusively from the side opposite to that occupied by the $[Cl^+]$ in its initial attack. Apparently the $[Cl^+]$ effectively blocks off approach from the same side, and what is often called "trans addition" results. The occurrence of such trans additions is quite widespread, and in many instances, a rational explanation can be given for it (Sec. 20.6).

If bromine is added to ethylene in the presence of a large excess of sodium chloride, it is found that the major product is $BrCH_2CH_2Cl$, with some $BrCH_2CH_2Br$ being formed, but *no* $ClCH_2CH_2Cl$ at all. This type of evidence is consistent with the mechanism as outlined.

$$Br_2 \rightleftharpoons [Br^+] + Br^-$$

$$[Br^+] + CH_2{=}CH_2 \longrightarrow [BrCH_2\overset{+}{C}H_2]$$

$$[BrCH_2\overset{+}{C}H_2] \xrightarrow{Cl^-} BrCH_2CH_2Cl$$

$$\searrow^{Br^-} BrCH_2CH_2Br$$

A number of reagents, such as HCl, HBr, HI, Cl_2, H_2SO_4 ($H^+OSO_2OH^-$), HOCl (HO^-Cl^+), I—Cl (I^+Cl^-), are found to add to alkenes in similar polar reactions.

The failure of NaCl, for example, to add to alkenes may be attributed to the lack of electrophilic properties in the Na^+ ion. Ions of this type show little if any tendency toward covalent bond formation with nucleophilic agents like alkenes.

In the case of alkenes which are unsymmetrically substituted, such as 1-propene, addition of an unsymmetrical reagent could occur in one of two ways:

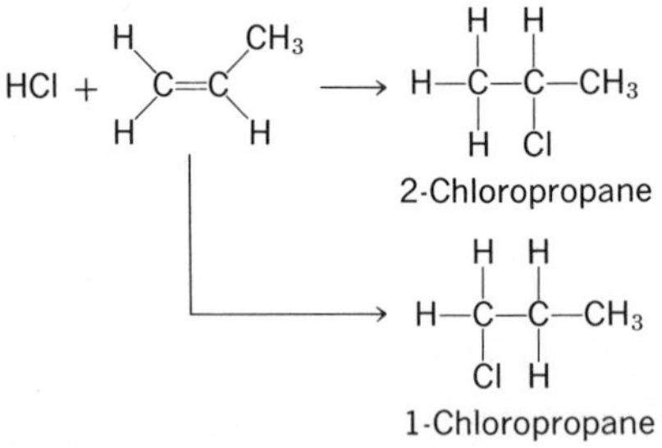
2-Chloropropane
1-Chloropropane

Recourse to experiment shows that under ordinary conditions both products are formed, but much more of the 2-chloropropane than of the 1-chloropropane. In the polar addition reactions of alkenes, in general, it is found that the *more negative portion of the adding reagent goes chiefly to the carbon atom of the double bond which has the smaller number of H atoms.* This rule is called "Markownikoff's rule."

3.9 Inductive Effects As we have seen in Sec. 3.8, the addition of an electrophilic species is believed to proceed by way of a *carbonium ion* in which one of the carbon atoms (in this case one of the carbon atoms of the double bond) acquires a positive charge. These carbonium ions are high energy species because the carbon atom has an incomplete valence shell (octet) (see Sec. 1.2). Any factor that can reduce the positive charge will markedly lower the energy of such a species. It has been found that alkyl groups such as $—CH_3$, $—CH_2CH_3$, etc., are slightly electron releasing (relative to hydrogen) by what is called an "inductive effect." That is, in the presence of a positive charge a certain percentage of the electron density in these groups can be donated (induced) to help reduce the positive charge on an adjacent atom. This effect falls off very rapidly with distance, and for most purposes, we need consider only those groups attached directly to the charged atom.

The consequences of such an effect are that more highly substituted carbonium ions are lower in energy than less substituted ones. Thus, two carbonium ions might be formed from propene by placing a proton on either carbon atom 1 or 2.

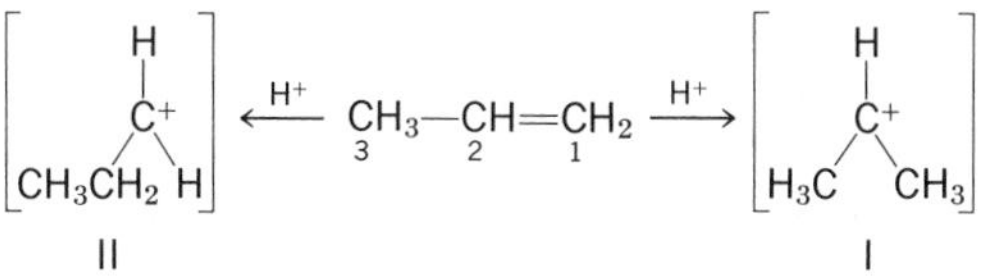
H
C+
CH3CH2 H
II
H+
CH3—CH=CH2
3 2 1
H+
H
C+
H3C CH3
I

Because carbonium ion I has two electron-donating alkyl substituents (CH_3 groups), it is found to be considerably lower in energy than carbonium ion II which is stabilized by only one alkyl substituent (a $CH_3CH_2—$ group). Inductive stabilization is usually indicated pictorially by an arrow along the bond in the direction of electron donation.

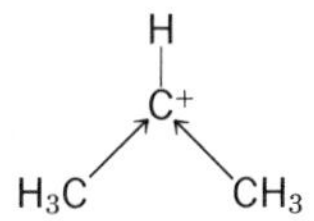
H
C+
H3C
CH3

Markownikoff's rule can be restated in mechanistic terms: *The species adding as the positive ion will add in such a fashion as to give predominantly the most stable carbonium ion.*

3.10 Transition-state Theory Section 3.9 gives a very useful rationale of Markownikoff's rule, but it leaves a fundamental question unanswered. How does the alkene *know* that adding a proton on the CH_2 end will lead to a better ion than adding it to the central CH atom? An even more general phrasing of the above question is: Do reactions always lead to the most stable products? The answer to the latter question is intuitively obvious since essentially all organic compounds in the presence of air are unstable relative to carbon dioxide and water. Finally, the question can be asked: Why do relatively stable compounds give rise to high-energy species like carbonium ions?

We know from elementary thermodynamics that in order for a reaction to proceed appreciably in a given direction, the energy of the products must be lower than the energy of the reactants. This is true regardless of the pathway taken by the reactants on the way to products, and consequently, we can obtain no information about this pathway from these energy considerations. The rate at which this transformation will occur depends very strongly on this pathway, however. The rate may, like the oxidation of most organic compounds at ordinary temperatures, be exceedingly slow, or it may vary from slow to practically instantaneous.

In organic chemistry, reactions almost always involve the breaking of some covalent bonds and the forming of others. The mechanism of such a reaction is a description of the pathway taken by reactants on their way to products. *Transition-state theory* provides our most useful description of this pathway, and it is around the concept of the transition state that our understanding of mechanisms is built. The concepts of transition-state theory may be stated as follows:

1. A chemical reaction may consist of several steps with different rates, but it is the slowest step or steps which effectively limit the rate of the transformation.
2. In the slow step(s) of a reaction the system will have to proceed through a state of high energy in which some bonds are partially or completely broken while new bonds are, at best, only partially formed. This state of high energy is called the "transition state."
3. In any chemical system a certain small percentage of the molecules will momentarily acquire energy greatly in excess of the average energy of the molecules in the system. If these molecules are capable of reaction, or if they collide with other molecules, they may pass over this transition state; if not, they will lose their excess energy to other molecules by collision after a very short time.
4. If a number of transition states are available to a set of molecules, they will prefer the one(s) with the lowest energy, that is, they will take the path of least resistance.

These considerations of transition-state theory are illustrated in Figs. 3-3 and 3-4.

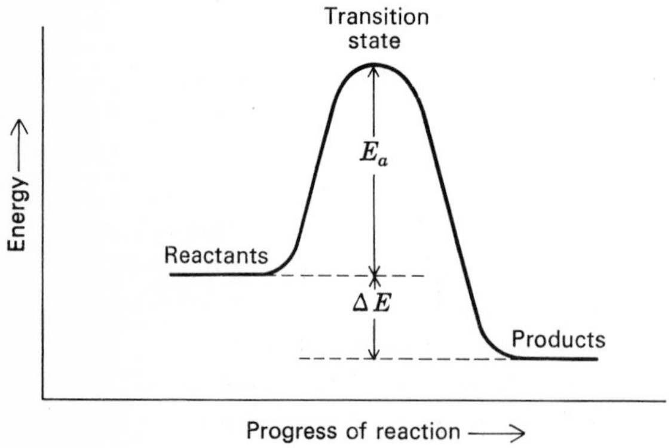
Transition state
E_a
Reactants
ΔE
Products
Energy ⟶
Progress of reaction ⟶

figure 3-3
Energy profile of a transition state.

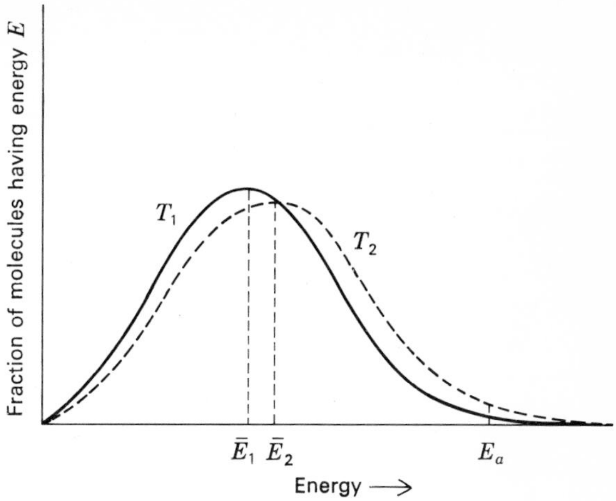
Fraction of molecules having energy E
T1
T2
Ē1 Ē2
Ea
Energy ⟶

figure 3-4
The energy of molecules in a system as a function of the fraction of the molecules having a given energy. (Solid curve = T_1, dotted curve = T_2. T_2 is a higher temperature than T_1.)

In Fig. 3-3, E_a is the *activation energy* of the reaction, that is, the minimum excess energy that the reactants must acquire in order to react. The net energy evolved by the overall transformation of reactants to products is ΔE. A curve such as that in Fig. 3-3 is called the "energy profile" of the reaction. The relationship of E_a to the average energy of molecules in a system is shown schematically for two different temperatures in Fig. 3-4. It can be seen that E_a is higher than the average energies, $\bar{E}_1$ and $\bar{E}_2$, and that only a small fraction of the molecules have such high energies at any one time.

We can now apply the concepts of transition-state theory to the addition of an electrophilic reagent (HBr) to an alkene ($CH_3—CH=CH_2$). The transition state for this reaction is believed to be very similar to the carbonium ion $CH_3—\overset{+}{C}H—CH_3$ in structure.

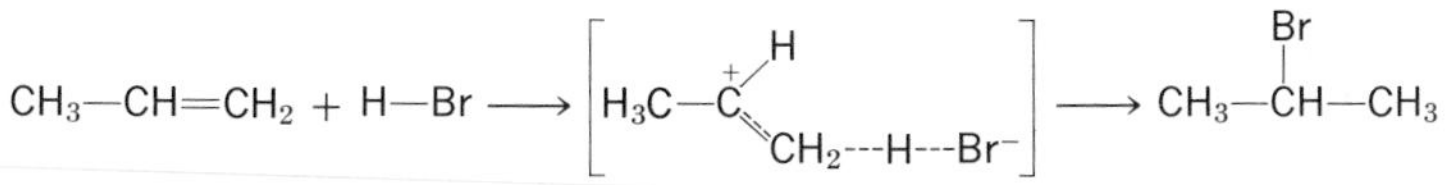

Reactants | Transition state I | Product

As H—Br approaches propene the pi bond in propene starts to weaken and the central carbon atom begins to feel a positive charge. The bond between H and Br also starts to weaken as a new bond is forming between H and the CH_2 end of propene. In the early stages of this process the energy of the system increases as shown by Fig. 3-5 because the partial breaking of bonds costs more in energy

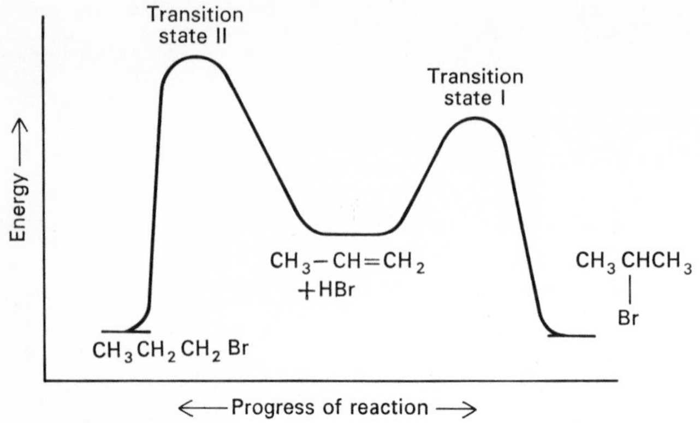
Transition
state II
Transition
state I
Energy ⟶
$CH_3-CH=CH_2$
$+HBr$
CH_3CHCH_3
Br
$CH_3CH_2CH_2Br$
⟵ Progress of reaction ⟶

figure 3-5
Energy profile for addition of HBr to $CH_3{-}HC{=}CH_2$.

than is gained by making a new bond. When the configuration of transition state I is reached the system has a maximum energy. From this point on the reaction proceeds rapidly downhill, releasing the excess energy it acquired plus the ΔE of the overall reaction.

The transition state for antiMarkownikoff addition will be of higher energy since the carbonium ion which it resembles is of higher energy (Sec. 3.7).

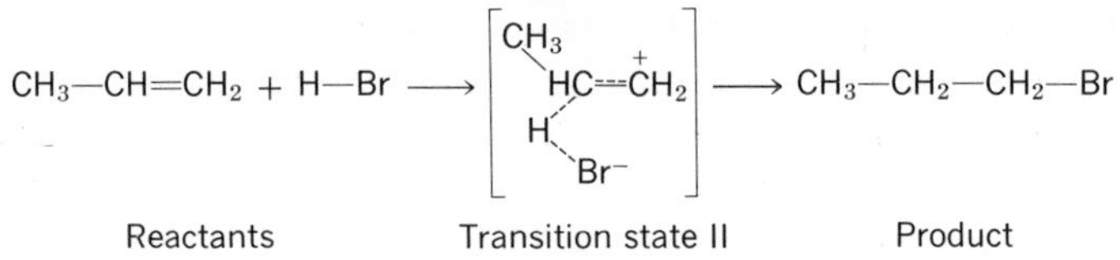
$CH_3-CH=CH_2 + H-Br$
CH_3
$HC{\cdots}\overset{+}{C}H_2$
H
Br^-
$CH_3-CH_2-CH_2-Br$
Reactants
Transition state II
Product

The above can be summarized in the energy-profile diagram shown in Fig. 3-5.

3.11 AntiMarkownikoff Additions Markownikoff's rule is followed in a majority of the addition reactions of alkenes and is a most useful one. It, like all rules, however, has its exceptions; addition reactions are known in which the main products are not the ones expected from the rule. This is due to the fact that different mechanisms are involved in some additions; a classical example is the addition of HBr to alkenes.

With purified alkenes in the absence of peroxides or ultraviolet light HBr adds to unsymmetrical alkenes to give mainly the product expected by the Markownikoff rule. However, in the presence of peroxides, or ultraviolet light, or with old alkene samples (which usually contain peroxides) HBr (but not HCl or HI) gives as the major product the one opposite from that predicted by the rule.

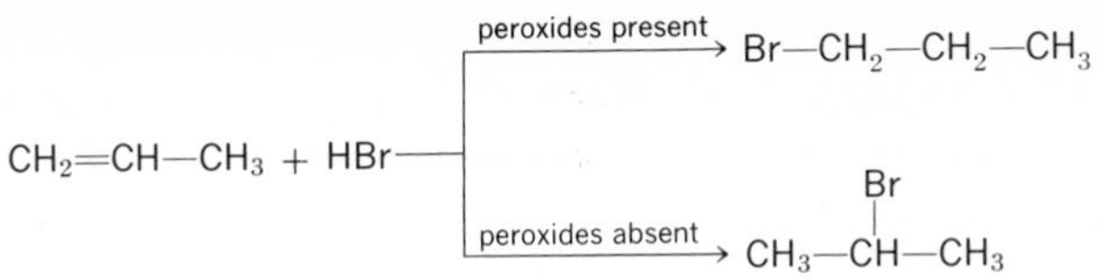
peroxides present
$Br—CH_2—CH_2—CH_3$
$CH_2═CH—CH_3 + HBr$
Br
peroxides absent
$CH_3—CH—CH_3$

The attacking reagent in the presence of light or peroxides is found not to be an ion, but a different-type particle called a "free radical" (Sec. 2.13). A molecule such as HBr may break apart in more than one way, as indicated by Fig. 3.6.

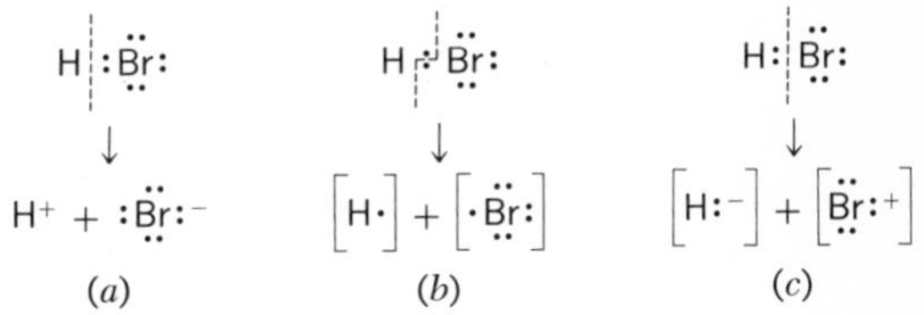
(a)
(b)
(c)

figure 3-6
The breaking apart of HBr by several paths.

The first pathway, shown in Fig. 3-6(*a*), is quite familiar to us, and is what is believed to occur in the normal addition. The third pathway, shown in Fig. 3-6(*c*), is energetically very unfavorable due to the known relative electronegativities of Br and H. The second pathway, Fig. 3-6(*b*), breaking into H· + Br· represents the rupture of the σ bond into the component atoms, more generally called radicals. This can come about by the addition of enough energy (in the form of ultraviolet light for example) to rupture this bond, or by the reaction of HBr with other free radicals as follows:

$$\underset{\text{A peroxide (very weak bond)}}{\mathrm{RO{:}OR}} \longrightarrow \underset{\text{Peroxy free radicals}}{\mathrm{2RO\cdot}}$$

$$\mathrm{RO\cdot + HBr} \longrightarrow \underset{\text{Strong bond}}{\mathrm{R{-}O{-}H}} + [\mathrm{:\ddot{B}r\cdot}]$$

Now, bromine atoms (which are radicals) can react with alkenes as follows:

$$[\mathrm{:\ddot{B}r\cdot}] + \mathrm{CH_3{-}CH{=}CH_2} \longrightarrow \underset{\text{A free radical}}{[\mathrm{CH_3{-}\dot{C}H{-}CH_2{:}\ddot{B}r{:}}]}$$

$$[\mathrm{CH_3{-}\dot{C}H{-}CH_2Br}] + \mathrm{HBr} \longrightarrow \mathrm{CH_3{-}CH_2{-}CH_2Br} + [\mathrm{:\ddot{B}r\cdot}]$$

$$[\mathrm{:\ddot{B}r\cdot}] \longrightarrow \cdots$$

This sequence of reactions can repeat itself over and over again many times. Such a sequence is called a "chain reaction" (Sec. 2.13). The establishment of a reaction chain depends on a critical balance between bond energies and radical reactivities as is found in HBr, but not in HCl or HI. Note that the addition in this "antiMarkownikoff" addition actually follows Markownikoff's rule, but that it is $:\ddot{\underset{..}{Br}}\cdot$ which adds first and not H^+. In many cases free radicals, although they are uncharged, behave as though they were polar addends.

As in the case of the ionic addition reaction, the favored point of attack of the $[:\ddot{\underset{..}{Br}}\cdot]$ is at the terminal carbon of propene. The arguments used in rationalizing this are the same as those for the ionic case (Sec. 3.7): electron repulsion (inductive) by the methyl group and stabilization of the intermediate secondary radical in the same way.

Another example of an apparent antiMarkownikoff addition is found in the addition of BH_3 to alkenes. This reaction, of considerable synthetic value (Sec. 8.5), probably involves the addition of $[H:^- \; B^+H_2]$. As predicted by this assumption, the $[^+BH_2]$ attacks the terminal carbon of propene, yielding the more stable secondary carbonium ion as an intermediate.

$$\underset{}{CH_3CH{=}CH_2} + \underset{\text{From } B_2H_6}{BH_3} \longrightarrow [CH_3CH_2CH_2BH_2] \xrightarrow[\text{with propene}]{\text{further reaction}}$$

$$(CH_3CH_2CH_2)_3B \xrightarrow{H_2O_2} \underset{\text{An alcohol}}{CH_3CH_2CH_2OH} + H_3BO_3$$

3.12 Polymerization of Alkenes Alkenes show another addition reaction which is very important. Under certain conditions alkene molecules can be made to add to each other. These reactions are usually chain reactions started by a small amount of a reagent such as an electrophilic substance (H^+) or a radical. Both ionic and radical chain reactions are known. The sequence may be visualized as shown below.

Radical:

$$R{-}CH{=}CH_2 + \underset{\text{Initiator}}{Y\cdot} \longrightarrow [R{-}\dot{C}H{-}CH_2{-}Y]$$

$$R{-}CH{=}CH_2 + [R{-}\dot{C}H{-}CH_2{-}Y] \longrightarrow [R{-}\dot{C}H{-}CH_2{-}CHR{-}CH_2{-}Y]$$

This sequence repeats until we have $[R{-}\dot{C}H{-}CH_2{-}(CHR{-}CH_2)_n{-}Y]$, where n is the number of alkene units added to the first radical. The sequence can be terminated by reaction of the growing chain with another radical, with oxygen, or by abstraction of a hydrogen atom from some other species. For example,

$$[R{-}\dot{C}H{-}CH_2{-}(CHR{-}CH_2)_n{-}Y] + R'\cdot \longrightarrow R{-}CHR'{-}CH_2{-}(CHR{-}CH_2)_n{-}Y$$

Ionic:

With ionic chains the sequence is analogous,[1]

$$\text{R—CH=CH}_2 + \text{X}^+ \longrightarrow [\text{R—}\overset{+}{\text{C}}\text{H—CH}_2\text{—X}], \ldots$$

except that termination can occur by loss of a proton or by addition of a nucleophile.

$$[\text{R—}\overset{+}{\text{C}}\text{H—CH}_2\text{—(CHR—CH}_2)_n\text{—Y}] + \text{Z}^- \longrightarrow \text{R—CHZ—CH}_2\text{—(CHR—CH}_2)_n\text{—Y}$$

Very large molecules with long chains of repeating units are known as *polymers* (see Chap. 22), and the compounds from which they are formed are known as *monomers*. Polyethylene and polypropylene are two commonly known commercial polymers prepared from ethylene and propylene (although not necessarily by the process indicated). In addition, most of the plastics and fibers known today are synthetic (or naturally occurring) polymers made by analogous methods from a wide variety of monomers.

3.13 Reduction of Alkenes Alkenes can be converted to alkanes by treatment with hydrogen gas in the presence of small amounts of certain catalysts such as finely divided platinum, palladium, rhodium, or nickel. With simple alkenes the reaction usually goes essentially to completion at ordinary temperatures and atmospheric pressure of hydrogen. The reaction may be forced by using both higher temperatures and higher pressures of hydrogen.

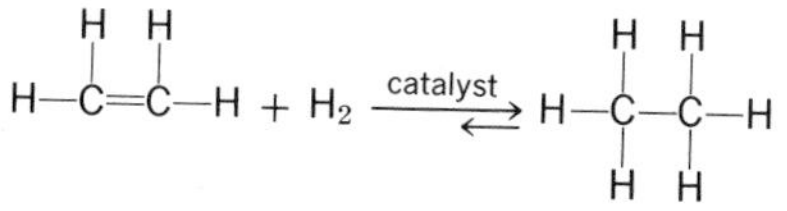

H H
H—C=C—H + H2
catalyst
H—C—C—H
H H
H H

Because the addition of hydrogen to alkenes is essentially quantitative under the usual conditions, measurement of the amount of hydrogen absorbed by weighed samples of alkenes allows determination of the number of double bonds present. Further, this reaction allows a correlation between the homologous series of alkanes and alkenes. Thus, as predicted from their structural formulas, two of the C_4H_8 alkenes on hydrogenation give butane, while a third gives 2-methylpropane.

3.14 Oxidation of Alkenes As might be expected from their nucleophilic character, most alkenes are readily attacked by permanganate, dichromate, ozone, and other oxidizing agents (electrophilic agents). The first two of these reagents are often used in hot aqueous solutions, while ozone, diluted with air, is passed through a cold solution of alkene in an inert solvent such as chloroform.

[1] It is also possible to reverse this sequence and initiate with Z^- in the so-called "anionic polymerizations," but these are less common with alkenes.

Oxidation by Permanganate

$$3CH_3CH{=}CH_2 + 4H_2O + \underset{(\text{dil.})}{2KMnO_4} \xrightarrow[\text{temp.}]{\text{low}} 3CH_3\underset{\text{OH}}{\underset{|}{C}}H{-}\underset{\text{OH}}{\underset{|}{C}}H_2 + 2MnO_2 + 2KOH$$

1,2-Propanediol
(propylene *glycol*)

The reaction with dilute permanganate is used as a test for unsaturation in hydrocarbons. The test is positive if the permanganate color disappears promptly. This is known as *Baeyer's test;* it indicates only that oxidation has occurred. The reaction in the equation above is difficult to stop at the glycol stage as shown, except at low temperatures, when dilute $KMnO_4$ is used. When a carbon atom has been partially oxidized, that is, has been attached to an oxygen atom, further oxidation takes place more easily on this atom than in the case of a saturated (alkane) carbon atom. This weakening of the bonds on a partially oxidized carbon atom is manifested in the case of further oxidation of glycols.

The detailed mechanism of permanganate oxidation is complex and has not been fully elucidated. In cases like this we will endeavor to discuss the basic observations that have been collected in the form of "rules" which serve only to summarize what has been observed in a large number of experiments. The ground rules for permanganate oxidation are as follows:

1. When two adjacent carbon atoms both bear —OH groups, the C—C bond between them will be cleaved, with another —OH group being attached to each carbon atom.
2. Any carbon atom bearing one or more —OH groups and one or more hydrogen atoms will have those hydrogen atoms replaced by —OH groups.
3. Any carbon atom with more than one —OH group attached will, in general, be unstable relative to a carbonyl (C=O) group and a water molecule.

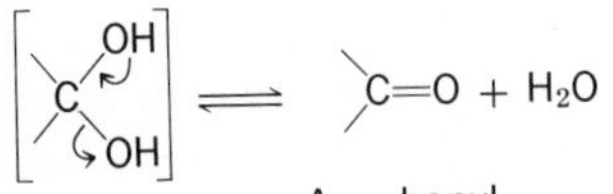

OH
C
OH
C=O + H₂O

A carbonyl group

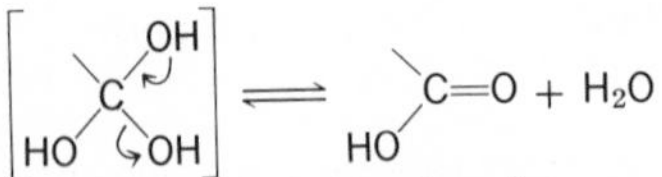

OH
C
HO
OH
C=O + H₂O
HO

A carboxylic acid group

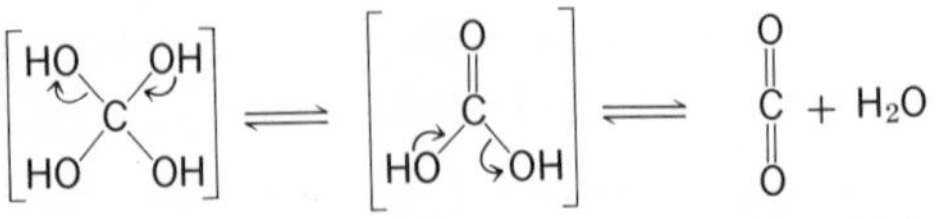

HO
OH
C
HO
OH
O
C
HO
OH
O
C
O
+ H₂O

Carbonic acid   Carbon dioxide

4. For the purpose of deciding whether cleavage of the carbon to carbon bond will occur or whether —H will be replaced by —OH, a carbonyl group ($\diagup$C=O) may be treated under Rules 1 and 2 as a dihydroxyl compound

$$HO-\overset{|}{\underset{|}{C}}-OH \equiv \overset{|}{C}=O$$

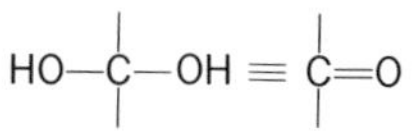

Examples are given below.

$$H_3C-\overset{\overset{H}{|}}{\underset{\underset{OH}{|}}{C}}-\overset{\overset{H}{|}}{\underset{\underset{OH}{|}}{C}}-H \xrightarrow[\text{Rule 1}]{KMnO_4} \left[H_3C-\overset{\overset{H}{|}}{\underset{\underset{OH}{|}}{C}}-OH \right] + \left[HO-\overset{\overset{H}{|}}{\underset{\underset{OH}{|}}{C}}-H \right] \xrightarrow[\text{Rule 2}]{KMnO_4} \left[H_3C-\overset{\overset{OH}{|}}{\underset{\underset{OH}{|}}{C}}-OH \right] + \left[HO-\overset{\overset{OH}{|}}{\underset{\underset{OH}{|}}{C}}-OH \right]$$

$$\text{Rule 3} \downarrow\uparrow \text{ Rule 4 \quad Rule 3} \downarrow\uparrow \text{ Rule 4} \qquad\qquad \downarrow\text{Rule 3} \qquad\qquad \downarrow\text{Rule 3}$$

$$H_3C-\overset{\overset{H}{|}}{C}=O \qquad\qquad O=\overset{\overset{H}{|}}{C}-H \qquad\qquad H_3C-C\overset{\diagup O}{\diagdown OH} \qquad\qquad \underset{2H_2O}{\overset{CO_2}{+}}$$
$$\underset{H_2O}{+} \qquad\qquad\qquad \underset{H_2O}{+}$$

$$CH_2=\overset{\overset{\overset{CH_3}{|}}{}}{C}-CH_3 \xrightarrow{KMnO_4} CO_2 + O=\overset{\overset{\overset{CH_3}{|}}{}}{C}-CH_3 + H_2O$$

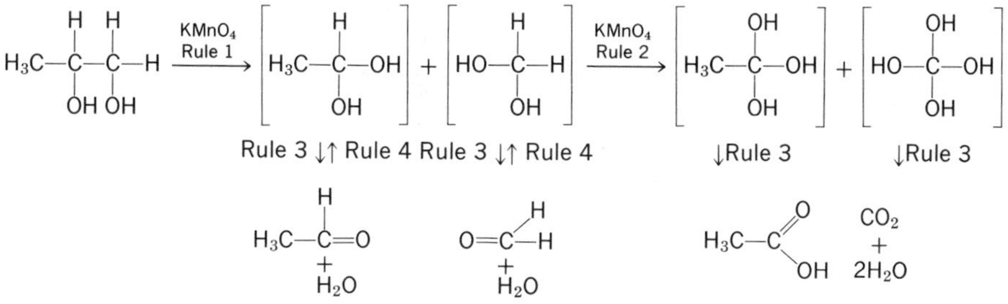

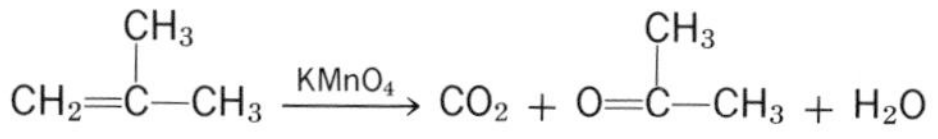

$$CH_2=CH-CH=CH_2 \xrightarrow{KMnO_4} 4CO_2 + 4H_2O$$

It is apparent that if the small fragments from the oxidation can be identified, the structure of the original alkene can then be deduced by mentally putting the pieces together again. For example:

$$C_7H_{14} \xrightarrow{KMnO_4} CH_3CH_2COOH + \underset{O}{\overset{\overset{\overset{CH_3}{|}}{}}{C}}-CH_2CH_3$$

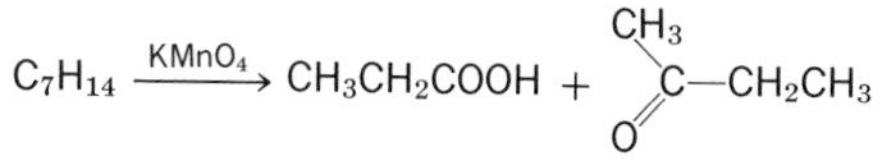

A ketone

The only hydrocarbon structure which will yield the products shown is 3-methyl-3-hexene.

3.15 Balancing Oxidation-Reduction Equations The majority of the oxidation-reduction reactions encountered in the practice of organic chemistry can be most conveniently balanced almost by inspection if one makes the assumption that free (O) is available from the oxidizing agent. For example,

$$2KMnO_4 + H_2O = 2KOH + 2MnO_2 + 3(O)$$

and

$$Na_2Cr_2O_7 + 4H_2SO_4 = Na_2SO_4 + Cr_2(SO_4)_3 + 4H_2O + 3(O)$$

A count of the number of (O)'s required for the oxidation, given a knowledge of the products, makes the balancing very simple. For example,

$$(CH_3)_2C{=}CH_2 \longrightarrow (CH_3)_2C{=}O + CO_2 + H_2O \qquad 4(O) \text{ needed}$$

So, one needs $8KMnO_4$ and $3(CH_3)_2C{=}CH_2$ for a balanced equation.

This method of balancing has pragmatic value. It is very simple, and it is used because we are often not prepared to write mechanisms for these reactions.

3.16 Reaction with Ozone Ozone reacts with alkenes rapidly and quantitatively, and the oxidation usually does not go as far as in the case of permanganate. The first step is a rapid addition of one molecule of ozone to each double bond, followed by rapid rearrangement to cleave the carbon-carbon single bond and produce an ozonide.

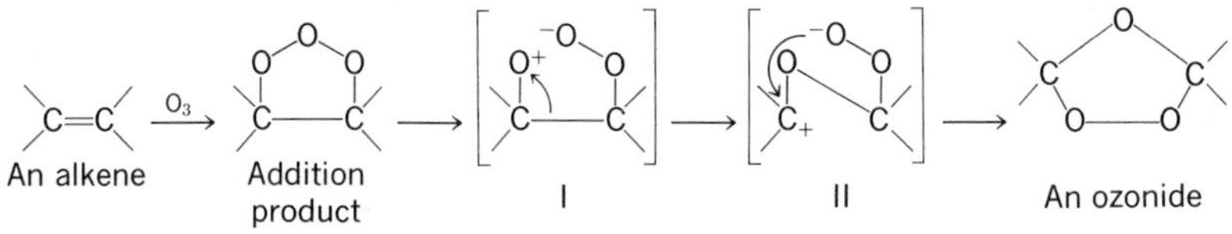
An alkene
Addition product
I
II
An ozonide

In this rather complex reaction mechanism it is noted that the molecule *rearranges,* or reacts internally, forming and breaking bonds in going from the postulated I to the ozonide. Other rearrangements will be encountered later. Fortunately, while rearrangements are known that transform carbon structures into others, they are reasonably rare and well characterized. Butane, for example, does not suddenly become 2-methylpropane under ordinary conditions. Exceptions will be pointed out when they occur.

As a rule, no attempt is made to purify the unstable and sometimes explosive ozonides; instead the reaction mixture is poured into warm water and the ozonide cleaved at once. The water is acidified and some zinc dust is usually added in carrying out this cleavage to prevent excessive oxidation of the cleavage products by the peroxides formed. The reactions are illustrated below:

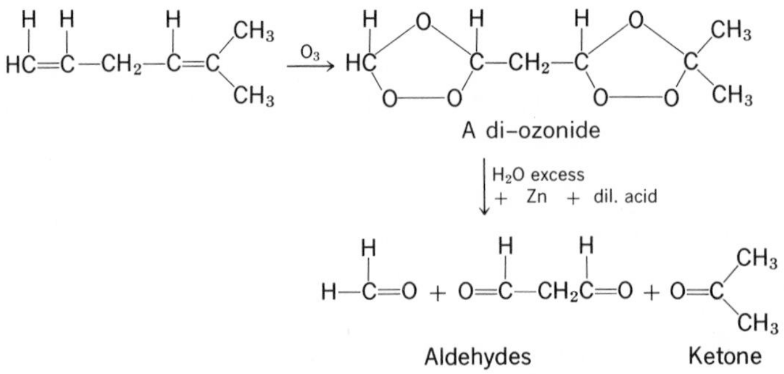
A di-ozonide
H₂O excess
+ Zn + dil. acid
Aldehydes
Ketone

The net effect of the reaction is to replace the

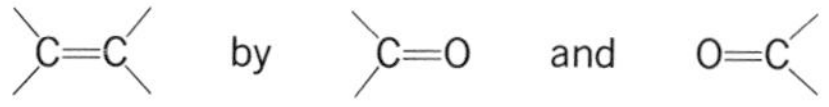

and again, if the fragments can be identified, the formula of the original molecule can be reconstructed. This sequence of reactions is known as *ozone degradation* or *ozonolysis,* and the molecule is said to be *cleaved* or *degraded* into simpler fragments.

3.17 Preparation of Alkenes While alkenes occur as minor constituents of petroleum and in small amounts in a variety of natural products, they are often most conveniently obtained by synthesis. There are two commonly used methods of introducing the double bond.[1]

The first method utilizes alcohols as starting materials. In the aliphatic series, the alcohols are among the most important primary starting materials for most synthetic work (Chaps. 8 and 9). The alcohols are those compounds containing a hydroxyl group attached to a saturated carbon atom.

3.18 Dehydration of Alcohols If 2-propanol, for example, is treated with a dehydrating agent, the following reaction occurs:

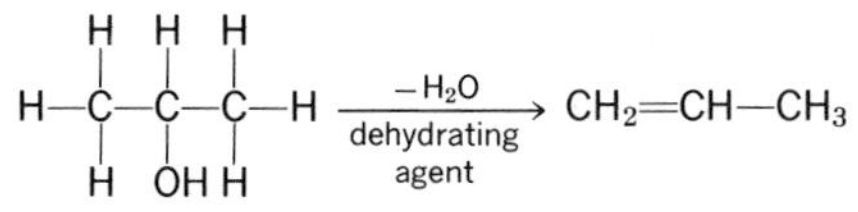
$-H_2O$
dehydrating agent
$CH_2{=}CH{-}CH_3$

2-Propanol

3.19 Mechanism of Dehydration; Whitmore's Hypothesis When alcohols are dehydrated to alkenes in the presence of strong acids or other strongly electrophilic reagents, the reaction is thought to go by certain steps which have been outlined below. These steps are part of a hypothesis based on considerable experimental evidence which was originally postulated by Whitmore.

The First Step:

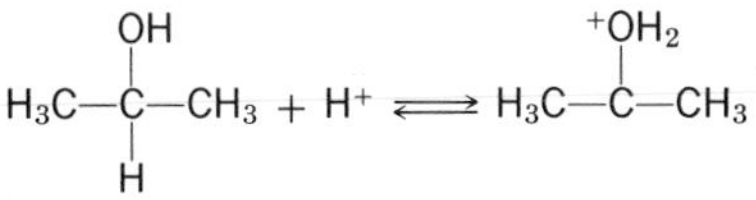

The reaction here is one already familiar from inorganic chemistry: the formation of an *oxonium ion.* It is exactly analogous to the formation of the hydronium ion from a proton in water. The H^+ is written as a free proton purely for convenience.

[1]Note that in the first two one does not "form" a double bond. One bond is already there; one forms a second, π, bond where there already is one bond.

Free protons are not stable in solution, but are always associated with solvent or other molecules. The positive charge on the oxonium ion weakens the carbon-oxygen bond in the alcohol to the point where rupture of this bond at moderate temperature becomes possible.

$$H_3C-\underset{\underset{H}{|}}{\overset{\overset{+}{O}H_2}{C}}-CH_3 \;\rightleftharpoons\; \left[H_3C-\underset{\underset{H}{|}}{\overset{+}{C}}-CH_3\right] + H_2O$$

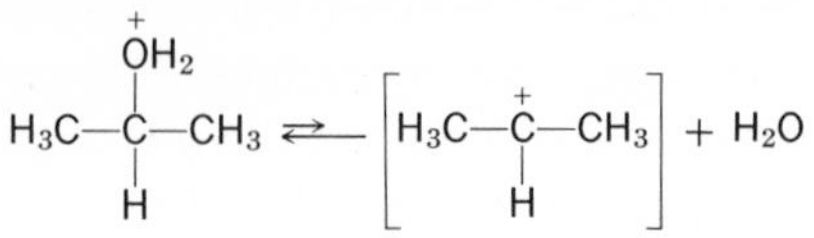

We see that rupture of the carbon-oxygen bond has left us with a carbonium ion. This is the same carbonium ion which would be formed by the addition of a proton to 1-propene (Sec. 3.7).

$$H_3C-\underset{\underset{H}{|}}{\overset{\overset{+}{O}H_2}{C}}-CH_3 \;\rightleftharpoons\; \left[H_3C-\underset{\underset{H}{|}}{\overset{+}{C}}-CH_3\right] \;\rightleftharpoons\; H_3C-\underset{\underset{H}{|}}{C}{=}CH_2 + H^+$$

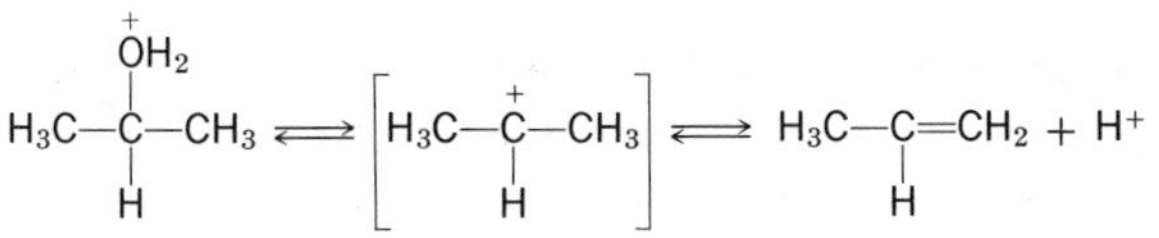

Thus, we can see that dehydration of an alcohol to yield an alkene must be the reverse of a reaction involving addition of water, or hydration. An important corollary of transition-state theory (Sec. 3.10) tells us that a transition state must serve for both the forward and reverse reactions (the law of microscopic reversibility). We can raise or lower the relative energies of reactants and products by application of Le Chatellier's principle, that is, by changing solvent, acid concentration, temperature, and other reaction conditions. Thus operation under conditions of low water content (e.g., conc. H_2SO_4) and (since alkenes invariably boil lower than their parent alcohols) at a temperature where the alkene distils as it is formed favors formation of alkenes from alcohols. As is the case with most carbonium ion reactions, a number of side reactions are also possible as will be seen in Chap. 8.

In the dehydration of more complex alcohols it has been found that the hydroxyl group is always removed, carrying with it a hydrogen atom from an adjacent carbon atom, never from more remote carbon atoms. In some cases, there may be a choice of two or more adjacent carbon atoms from which the hydrogen might come, giving rise to double bonds in different locations.

$$CH_3CH_2-\underset{\underset{OH}{|}}{CH}-CH_3 \;\xrightarrow[\Delta]{H^+}\; \begin{array}{c} CH_3CH{=}CH-CH_3 \\ or \\ CH_3CH_2-CH{=}CH_2 \end{array} + H_2O$$

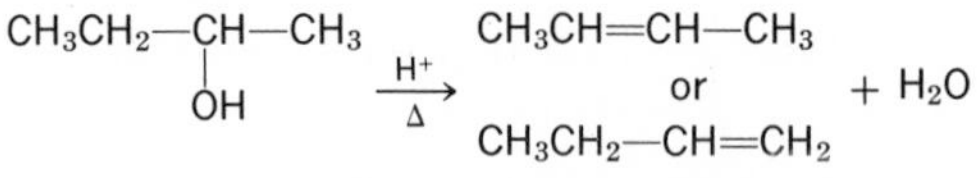

In cases of this sort, the alkene produced will have the greatest possible number of alkyl substituents directly attached to the carbon atoms joined by the double bond. This observation may be explained on the basis that the transition state (Sec. 3.10 and Fig. 3-5) more closely resembles the alkene than it does the alcohol. Factors that tend to stabilize the sp^2-hybridized carbons of the (incipient) double bond would also be expected to stabilize the transition state. It has been established that alkyl groups are electron-donating, so that when they are directly attached to such sp^2-hybridized carbon atoms, the stability of the system is increased.

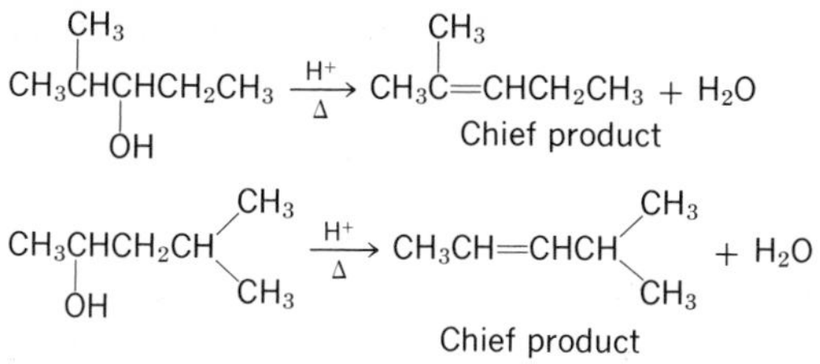
Chief product
Chief product

3.20 Elimination The second method of introducing the double bond into a carbon chain is similar in principle. If simple alkyl halides are treated with suitable reagents, the following type of reaction occurs:

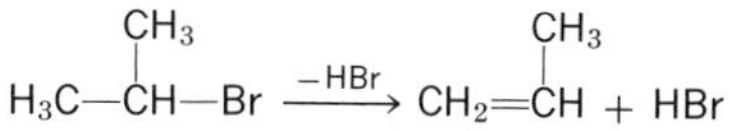

The elimination of HX is most often effected by the use of strong base. In this case the reaction usually involves the *simultaneous removal of a proton and loss of halide ion.* Such a reaction is called "concerted."

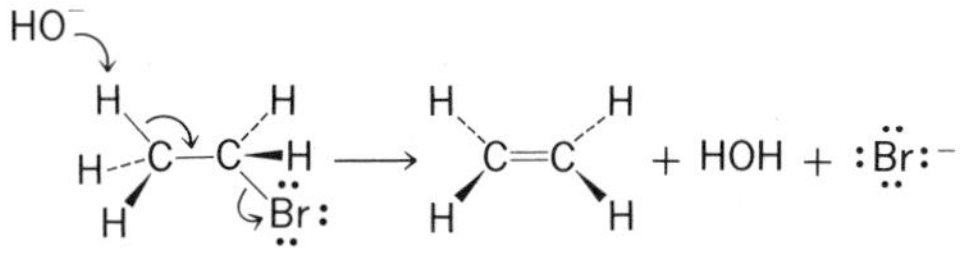

It is known from labeling experiments that the hydrogen and bromide are removed at the same time and not in discrete steps. If, for example, the reaction proceeded by way of two distinct steps

$$HO^- + CH_3CH_2Br \rightleftharpoons HOH + [{}^-CH_2{-}CH_2Br]$$
$$\text{and } [{}^-CH_2CH_2Br] \longrightarrow CH_2{=}CH_2 + Br^-$$

and the reaction were carried out with OD^- in D_2O, one would expect to find some deuterium in some of the bromoethane that had not yet reacted completely. Since this is not found, it is assumed that the $[{}^-CH_2{-}CH_2Br]$ intermediate is not involved.

As in the dehydration of alcohols, we can predict the product in more complex cases by the reasoning given above. The hydrogen again comes from the adjacent, less hydrogenated carbon atom.

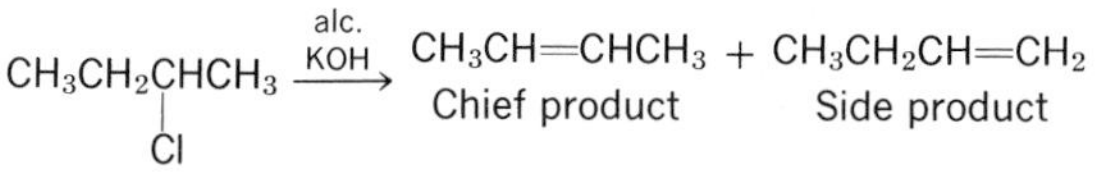
Chief product
Side product

As might be expected, in a compound where all carbons adjacent to the halogenated or hydroxylated carbon are equally hydrogenated, a mixture of nearly equal amounts of all possible products results.

$$CH_3CH_2\underset{\displaystyle Br}{\underset{|}{C}}HCH_2CH_2CH_3 \longrightarrow CH_3CH{=}CHCH_2CH_2CH_3 + CH_3CH_2CH{=}CHCH_2CH_3$$

$$CH_3CH_2\underset{\displaystyle OH}{\underset{|}{C}}HCH_2CH_2CH_3 \longrightarrow CH_3CH{=}CHCH_2CH_2CH_3 + CH_3CH_2CH{=}CHCH_2CH_3$$

3.21 Other Methods of Preparation There are other less generally useful methods available for the synthesis of alkenes. For example, 1,2-dihalogen compounds react with zinc to produce a double bond.

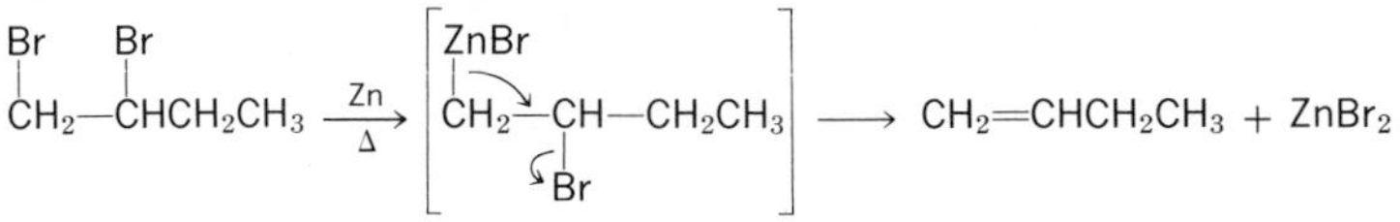

The application of this method is limited by the difficulty in preparing 1,2-dihalides from sources other than the alkenes themselves.

The initial Zn compound [analogous to the Grignard reagent (Sec. 2.16)] provides a negative charge on carbon exactly as does the action of base (Sec. 3.20), and the reaction can be regarded simply as another example of a concerted elimination.

The chief commercial source of the lower alkenes, up through the butenes, is petroleum. Although they do not occur as such in crude petroleum, they are major products of the refining process.

3.22 Homology When we examine the properties of alkenes we find the regular change characteristic of all homologous series (see, for example, Fig. 3-7) and note again the close similarity in chemical behavior of different members. The double bond common to all alkenes gives them all much greater reactivity than the alkanes, and all undergo the same typical addition reactions.

3.23 Occurrence of Alkenes Although simple alkenes such as ethylene and the butenes are synthetic products for the most part and find their main applications in chemical industry, more complex ones occur widely in nature.

The large family of substances called terpenes, occurring in the oils of many plants, contains in addition to more complex cyclic compounds (Sec. 4.8) a number of alkenes; these frequently carry other functional groups also. Examples are

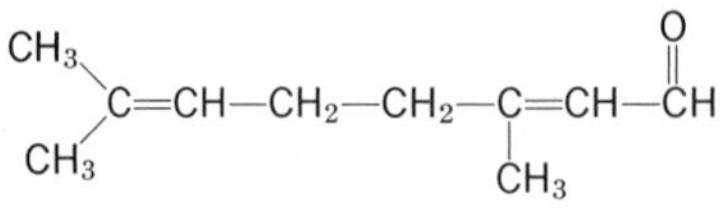

Geranial

and

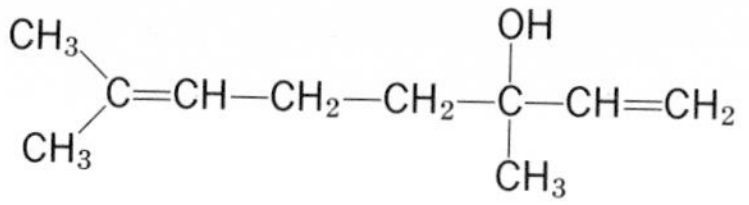

Linalool

Some colored substances containing multiple double bonds are of significance in biochemistry; their structures are analogous to that of β-carotene.

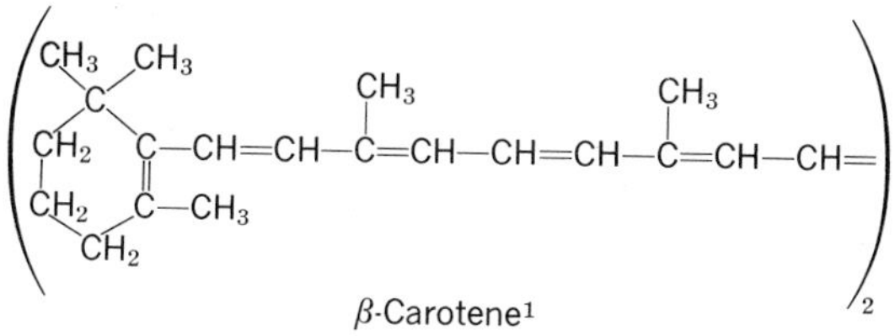

β-Carotene[1]

Natural rubber is a polymer of the diene $CH_2{=}C(CH_3){-}CH{=}CH_2$ (isoprene), and many synthetic rubbers are modeled after this polymer, usually with the cheaper 1,3-butadiene as a basis.

Alkenes containing more than one double bond exhibit the same general reactions toward the reagents discussed in this chapter as do the alkenes.[2] It is generally difficult to attack one double bond while leaving the other untouched.

$$CH_2{=}CHCH_2CH{=}CH_2 + KMnO_4 \longrightarrow 2CO_2 + HOOCCH_2COOH$$

No $CH_2{=}CHCH_2COOH$ can be obtained. This is a general phenomenon; when there

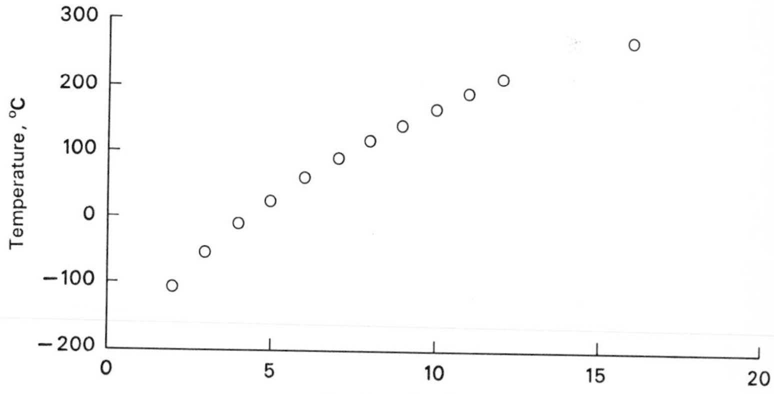
300
200
100
0
−100
−200
Temperature, °C
0
5
10
15
20
Number of carbon atoms

figure 3-7
Boiling points of 1-alkenes.

[1] Carotene was first isolated from carrots. It turned out to be a mixture of isomers. β-Carotene is capable of replacing vitamin A in the diet.

[2] An important exception is alkenes with conjugated double bonds (Secs. 6.4 and 6.26).

are two or more nearly identical reactive groups in a molecule, it is not often possible to attack one without also affecting the other. The student should beware of the temptation to invent reactions involving "selective" reagents in solving problems.

3.24 Spectroscopic Properties of Alkenes Since the electrons in carbon-to-carbon double bonds are more loosely bound than those in single bonds, it might reasonably be expected that raising such electrons to higher energy levels would be less difficult. This is indeed found to be the case; strong absorptions in the ultraviolet are observed in the region 160 to 210 mμ. However, this region has only recently become accessible to commonly available instruments so that the ultraviolet spectra of simple alkenes is of limited utility at this time.

The double bond, consisting as it does of a σ bond plus a π bond, might be expected to be considerably more resistant to stretching than a single σ bond alone. This is very definitely the case, and C=C infrared stretching vibrations occur in the region 1660 to 1640 cm^{-1}. Moreover, the C—H stretching vibration of hydrogens attached directly to double bonds occurs at slightly higher frequencies than those of alkanes, namely 3000 to 3100 cm^{-1}.

In the nuclear magnetic resonance spectra, hydrogens attached to carbon-carbon double bonds characteristically are observed in the region $\delta = 4.5$ to 7.6. Thus alkene protons are easily distinguished from those attached to saturated carbon atoms; the alkene hydrogens are much less shielded than the latter. This deshielding is a consequence of the increased "*s* character" of sp^2 hybrid orbitals (Sec. 3.4). Since *s* orbitals are lower in energy than *p* orbitals, the net effect of an increase in *s* character is to increase the electron density in the vicinity of carbon at the expense of hydrogen so that the latter atom is deshielded.

OUTLINE OF ALKENE CHEMISTRY

Preparation

1. An alkyl halide + alcoholic KOH or NaOH.

$$\mathrm{R{-}CH_2{-}\underset{}{\overset{X}{\overset{|}{C}}}H{-}CH_3 \xrightarrow{\text{alc. KOH}} KX + H_2O + R{-}CH{=}CH{-}CH_3}$$

2. Dehydration of an alcohol.

$$\mathrm{R{-}CH_2{-}\underset{\underset{OH}{|}}{C}H{-}CH_3 \xrightarrow[\Delta]{\text{conc. } H_2SO_4} R{-}CH{=}CH{-}CH_3 + H_2O}$$

Other dehydrating agents besides H_2SO_4 may be used, such as $KHSO_4$ and P_2O_5. This reaction may also be carried out in the vapor state over a heated catalyst such as Al_2O_3 or ThO_2.

3. Treatment of dihalides with zinc

$$R\text{—}CHBr\text{—}CHBr\text{—}R' \xrightarrow{Zn} R\text{—}CH\text{=}CH\text{—}R' + ZnBr_2$$

Properties

1. $RCH\text{=}CH_2 + H_2 \xrightarrow{cat.} RCH_2CH_3$

2. $RCH\text{=}CH_2 + HX \longrightarrow RCHXCH_3$

3. $RCH\text{=}CH_2 + H\text{—}OSO_2OH \longrightarrow RCHCH_3$
           (conc.) OSO_2OH
                     An alkyl hydrogen
                        sulfate

Note that in the bisulfate the sulfur is linked to carbon through oxygen.

4. $RCH\text{=}CH_2 + X_2 \longrightarrow RCHXCH_2X$

5. $RCH\text{=}CH_2 + HOX \longrightarrow RCHOHCH_2X$
                          A halohydrin

6. $RCH\text{=}CH_2 \xrightarrow{dil.\ KMnO_4} RCHOHCH_2OH$

The resultant product belongs to the class of glycols, or 1,2-dihydroxy compounds (1,2-diols).

7. $RCH\text{=}CH_2 \xrightarrow{conc.\ KMnO_4} R\text{—}\overset{O}{\overset{\|}{C}}\text{—}OH + CO_2$
                             A carboxylic
                               acid

$RCH\text{=}CHR' \xrightarrow{conc.\ KMnO_4} RCOOH + R'COOH$

$\overset{R}{\underset{R}{>}}C\text{=}CHR' \xrightarrow{conc.\ KMnO_4} \overset{R}{\underset{R}{>}}C\text{=}O + R'COOH$
                      A ketone

8. $RCH\text{=}CH_2 + B_2H_6 \longrightarrow$ a boron compound
                         $\xrightarrow{H_2O_2} RCH_2CH_2OH$

9. $RCH\text{=}CHR' + O_3 \longrightarrow$ RCH$\underset{\overset{|}{O\text{—}O}}{\overset{O}{\diagup\diagdown}}$CHR'
                       An ozonide
                       $\xrightarrow[Zn]{H_2O} R\overset{O}{\overset{\|}{C}}H + R'\overset{O}{\overset{\|}{C}}H$

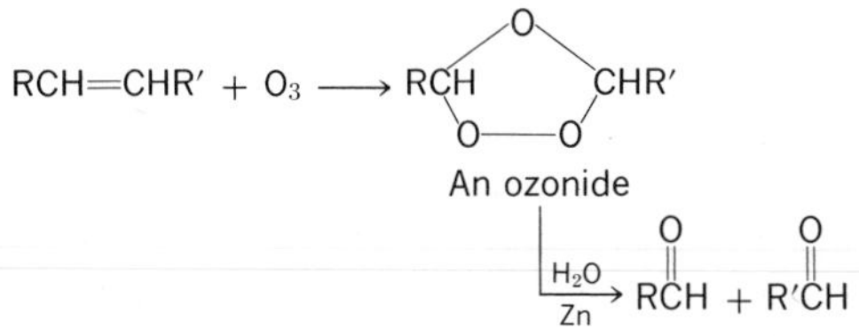

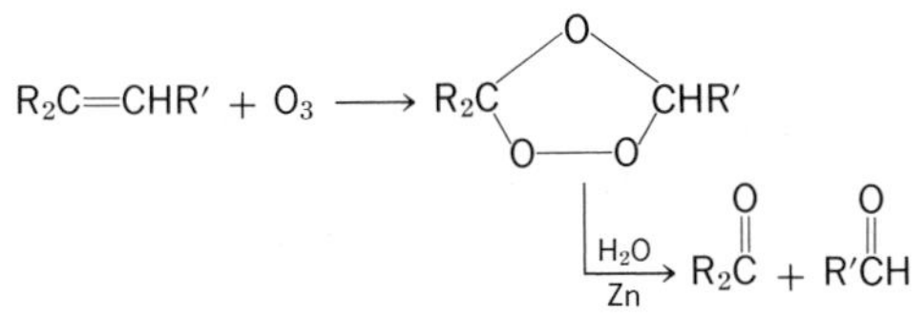

$R_2C\text{=}CHR' + O_3 \longrightarrow$ R$_2$C$\underset{\overset{|}{O\text{—}O}}{\overset{O}{\diagup\diagdown}}$CHR'
                       $\xrightarrow[Zn]{H_2O} R_2\overset{O}{\overset{\|}{C}} + R'\overset{O}{\overset{\|}{C}}H$

Spectroscopic Properties

Group	*ir*, cm^{-1}	*nmr*, δ*(ppm)*	*uv*, $m\mu$
C=C	1640–1660	. . .	below 210
C=C–H	3000–3100	4.5–7.6	

CHAPTER 3 EXERCISES

★**1.** Write the structures of all the possible pentenes and name each according to the Geneva system. Compare the number of pentenes with the number of pentanes.

2. Write equations for the reactions, if any, of the following:

a. 1-Pentene + HCl

b. Ethylene + conc. H_2SO_4

c. 1,3-Butadiene + $KMnO_4$

d. $CH_3CHOHCH_3 + H_2SO_4$ + heat

e. 2-Methyl-2-butene + Br_2

f. 2-Bromo-2-methylpentane + KOH in alcohol

★**3.** Write the structures and names of the substances giving the following products with concentrated potassium permanganate:

★*a.* $(CH_3)_2CO + CH_3CH_2COOH$

b. $CH_3COOH + CH_2(COOH)_2 + CH_3COOH$

★*c.* $2CO_2 + CH_2(COOH)_2$

d. $CH_3COCH_2CH_3 + (CH_3CH_2)_2CO$

★*e.* 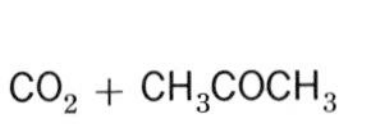

f. CH_3CH_2COOH (only one product)

★*g.* 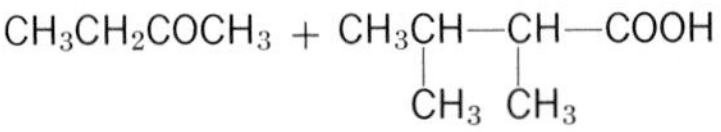

h. $CO_2 + CH_3COCH_3$

★*i.* $CH_3COOH + CH_3CH_2COOH$

4. A substance, C_8H_{14}, (*a*) reacts with H_2 to form C_8H_{16} and (*b*) reacts with $KMnO_4$ to form a single pure acid (no CO_2 is found). Propose a structure for the hydrocarbon and write equations for the reactions.

5. How could you distinguish by chemical and/or spectral (ir, uv, nmr) means between the following pairs?

★*a.* 

b. $CH_3CH{=}CHCl$ and $ClCH_2CH{=}CH_2$

★*c.* $(CH_3)_3C—OH$ and $CH_3CH_2CH_2CH_2OH$
d. $CH_2{=}CH—CH_2CH_3$ and $CH_3CH{=}CH—CH_3$
★*e.* $(CH_3)_2C{=}C(CH_3)_2$ and $(CH_3)_2C{=}CHCH_2CH_3$
f. $(CH_3)_3CCl$ and 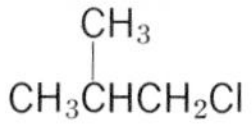

6. Which of the C_6 alkenes will give 2-methylpentane on hydrogenation? Name them.
★7. Show all the cis-trans isomers of geranial (page 72) and linalool (page 73).
8. Outline the step or steps which you would use in performing the following transformations in the laboratory. Give reagents and conditions.

a. 3-Methyl-1-butene ⟶ 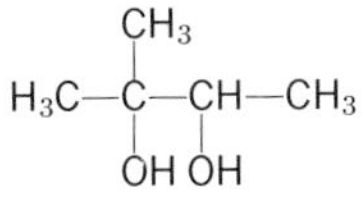

b. 1-Bromo-3-methylbutane ⟶ 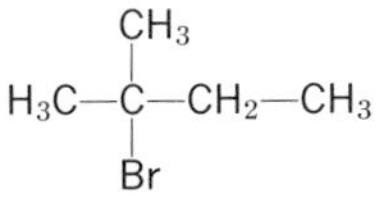

★9. In order to distinguish between Structures I and II for an unknown alcohol, the alcohol was dehydrated with H_2SO_4 and then oxidized with warm potassium permanganate to yield acetone ($CH_3—\underset{\underset{O}{\|}}{C}—CH_3$) and acetic acid ($CH_3—\underset{\underset{O}{\|}}{C}—OH$). These results were interpreted as evidence favoring Structure II. Comment on the structure proof.

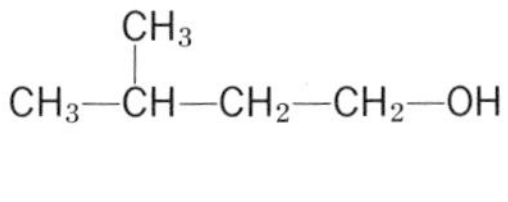
I

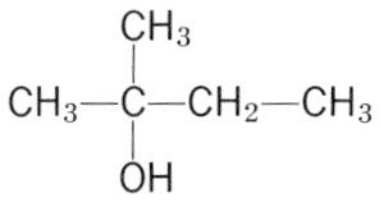
II

10. Outline laboratory tests which would enable you to distinguish by simple chemical or physical tests between the following pairs of compounds:
a. *n*-Pentane and 2-pentene
b. Methyl chloride and 2-chlorooctane
c. Ethyl bromide and 1,2-dibromoethane
d. 2,3-Dimethylbutene and 1-hexene
11. Outline reasonable reaction mechanisms for each of the following reactions:
★*a.* 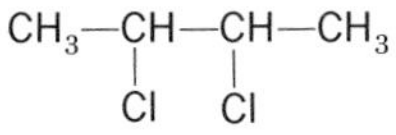 $+ Zn \longrightarrow$ 2-butene

b. Ethylmagnesium bromide + $CH_3CH_2—OH \longrightarrow$ ethane + MgBrOH

c. Isobutane + $Cl_2 \xrightarrow{\text{uv light}}$ 2-chloro-2-methylpropane

d. 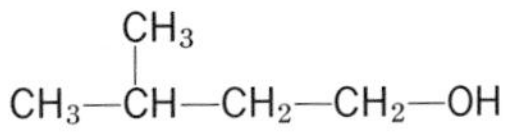 $\xrightarrow{H_2SO_4}$ 2-methyl-2-butene

★*e.* 2-methylpropene $\xrightarrow[\text{peroxides}]{\text{HBr}}$ C_4H_9Br

12. Write equations for the reaction of ICl with 1-butene. Which of the two chloroiodobutanes produced will be present in greater amount? Explain.

★13. Are oxidizing agents nucleophilic? Explain.

14. Write equations for the polymerization of 1-butene by a trace of HCl.

★15. Alkenes are usually removed from gasoline (a mixture of alkanes) in the refining process. Suggest a method of removing butene from a sample of octane. How could you test for the presence of alkenes in gasoline?

16. Predict the number of nmr absorptions expected from the following structures. Consider both cis and trans forms where possible.

a. $CH_3CH{=}CH{-}CH_3$

b. $(CH_3)_2C{=}CH_2$

c. $ClCH{=}CH_2$

d. $ClCH{=}CHCl$

e. 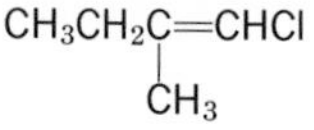

17. Show by equations the mechanism of the following reactions. Not all products are shown; indicate other expected products.

★*a.* 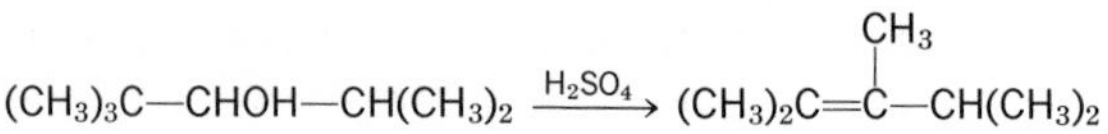

b. $CH_2{=}CH{-}CH_2CH_3 \xrightarrow{\text{DCl}} CH_2DCH{=}CHCH_3$

★*c.* 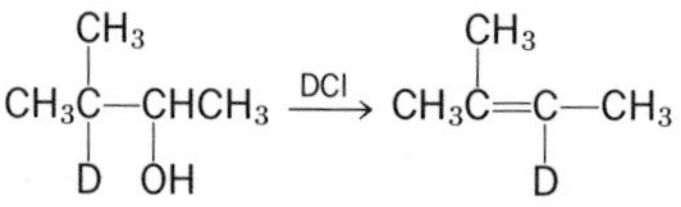

18. Myrcene, $C_{10}H_{16}$, a terpene isolated from oil of bay, absorbs three moles of hydrogen to form $C_{10}H_{22}$. Upon ozonolysis myrcene yields

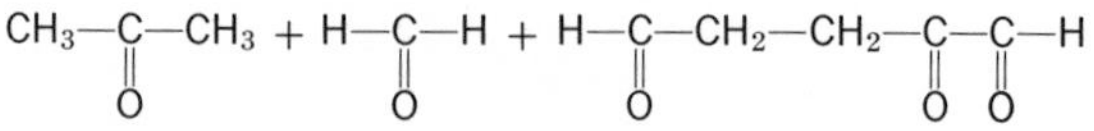

What structures are consistent with these facts?

19. By what laboratory tests would you distinguish between the following pairs of compounds:

★*a.* 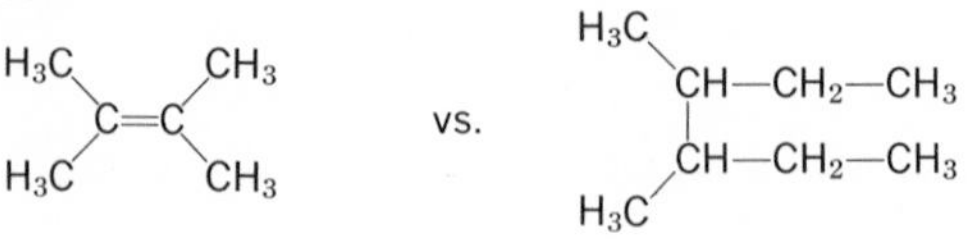

b. $CH_3CH_2CH_2COOH$ vs. $CH_3CH_2CH_2CHO$

★*c.* $CH_3CH_2CH_2MgCl$ vs. $CH_3CH_2CH_2CH_2Cl$

d. 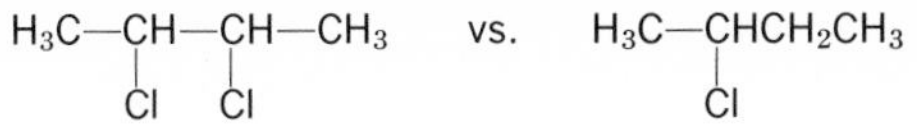

20. A hydrocarbon, C_7H_{12}, behaves as follows. Deduce a structure for it and write equations for the reactions.

$$C_7H_{12} \xrightarrow{HCl} C_7H_{13}Cl \xrightarrow[\text{KOH}]{\text{alc.}} \text{an isomeric } C_7H_{12} \xrightarrow{KMnO_4} \underset{\text{A ketone}}{C_5H_8O}$$

★21. Predict the relative rates of reaction between Cl_2 and the following:

a. $CH_3CH{=}CH_2$ *b.* 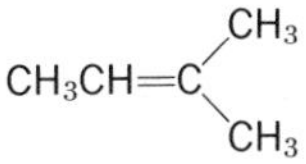

22. Account for the fact that 100% H_2SO_4 is needed to dissolve ethylene, while

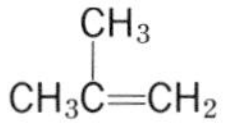

will dissolve in 65% aqueous H_2SO_4. Ethane dissolves in neither.

★23. $CH_3\overset{CH_3}{\overset{|}{C}}{=}CH_2$ with acid gives two isomeric C_8H_{16} dimers. Suggest structures for these, show mechanisms for their formation, and indicate how they might be distinguished by chemical and spectral means.

24. Prepare a mechanism for the conversion of *cis*-2-butene to *trans*-2-butene in the presence of H^+.

25. ★*a.* A substance, $C_{10}H_{18}$, on ozonolysis gives 2 molecules of $CH_3\overset{O}{\overset{\|}{C}}CH_3$ and one molecule of

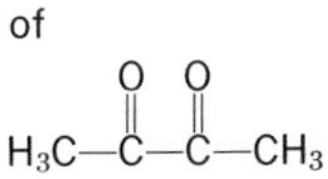

Write a structure for the substance.

★*b.* A substance, $C_6H_{12}Br_2$, reacts with zinc to give a new substance which can be ozonized to $CH_3CH_2CH{=}O$ as the only product. Write structures for the various substances.

26. Ethylene + Br_2 gives different products in the presence or absence of excess Na_2SO_4, although this latter reagent alone fails to react with ethylene. Explain.

27. Show by equations (mechanisms not required) how the following transformations might be accomplished:

★*a.* 1-Butene to 2-butene

b. 2-Hexene to 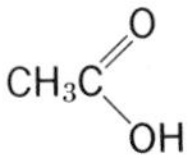 + ?

★*c.* 2,3-Dibromohexane to hexane

d. $CH_3\underset{\underset{OH}{|}}{CH}—CH_2CH_3$ 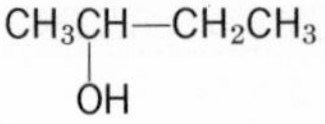 to 2,3-dichlorobutane

★e. 1-Chloro-2-methylpentane to 2-methyl-1-pentene
 f. Propene to CH_3CH_2CHOH
★g. 1-Butene to 1-bromobutane

28. The nmr spectrum of $CH_2{=}CH—CH_2MgBr$ shows only two kinds of H atoms. One would expect(?) three kinds. Explain.

★**29.** Would you expect the Anion I to be more or less stable than Anion II? Explain.

$$CH_3\underset{\text{I}}{\underset{|}{CH}}—\overset{-}{CH}{=}CHCH_3 \qquad CH_3—\underset{\text{II}}{\overset{-}{C}(CH_3)_2}$$

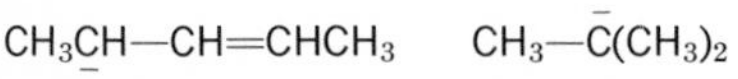

What factors would be involved in assessing the comparative stability of the following cations:

$$CH_3—\underset{\overset{|}{CH_3}}{\overset{+}{C}}—CH{=}CH—CH_3 \qquad CH_3—\overset{+}{C}(CH_3)_2$$

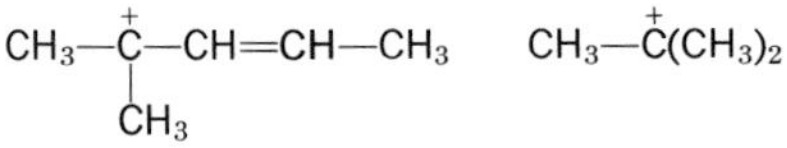

30. Explain by use of a diagram how it is that the reaction involving the more stable intermediate proceeds faster.

★**31.** Which of the following compounds would be expected to show cis-trans isomerism? Draw the *trans* isomer where such isomerism is expected:
 a. 4,5-Dimethyl-4-octene
 b. 1-Chloro-2,3-dimethyl-2-butene
 c. 2,3-Dimethyl-4-ethyl-4-heptene

32. Write structures for all products expected from each of the following reactions. Indicate the *main* product. Mechanisms of reactions are not required.
 a. 3-Bromo-3,4-dimethylhexane treated with alcoholic KOH

 b. $CH_3—\underset{\overset{|}{CH_3}}{CH}—CH_2—CH_2—OH$ 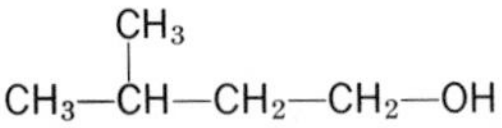
 heated with concentrated H_2SO_4
 c. A mixture of butane and propene treated with a solution of bromine in carbon tetrachloride at room temperature in the dark

4

HYDROCARBONS: THE CYCLOALKANES

4.1 Introduction Petroleum contains, in addition to alkanes, smaller and varying amounts of other classes of compounds. One of these classes is composed of cycloalkanes. These hydrocarbons resemble the alkanes closely in chemical behavior but differ in empirical formula. The generalized empirical formula for these products is C_nH_{2n} as compared to C_nH_{2n+2} for the alkanes; they contain two hydrogen atoms fewer than the corresponding alkanes because there is present one more carbon-to-carbon bond; a *ring* of carbon atoms is present.

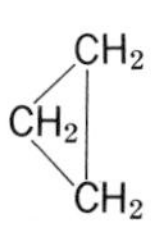

Cyclopropane, C_3H_6
(bp = −32.9°C)

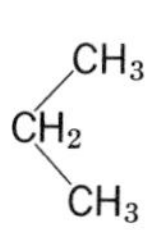

Propane, C_3H_8
(bp = −42.2°C)

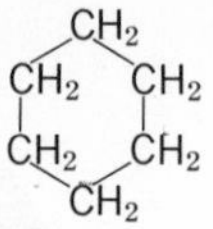

Cyclohexane, C_6H_{12}
(bp = 81.4°C)

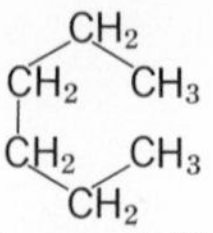

Hexane, C_6H_{14}
(bp = 69°C)

The presence of a ring has the same effect on the calculated unsaturation number as a double bond; consequently, other evidence besides that given by a molecular formula is needed to distinguish these.

4.2 Nomenclature The systematic nomenclature in this series is similar to that used for the alkanes, except that the name chosen is based on the size of the ring rather than on the longest continuous chain present. The prefix *cyclo-* denotes the presence of a ring, and the usual alkane names are used to show the number of carbon atoms in the ring. The carbon atoms of this ring are numbered, the numbering being arranged so that substituents on the ring have the smallest possible numbers. A few examples will illustrate the application of these rules.

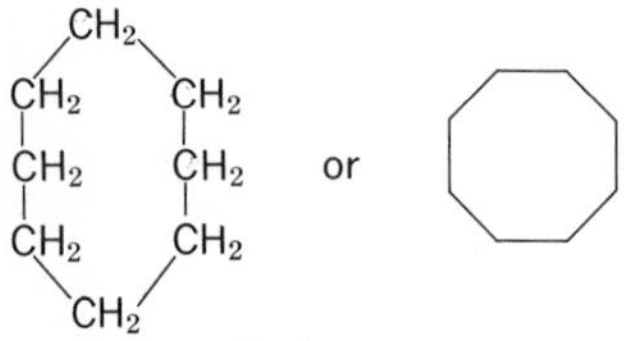

Cyclooctane

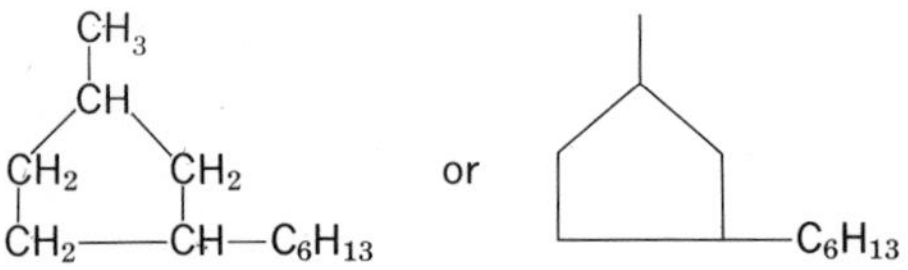

1-Methyl-3-hexylcyclopentane

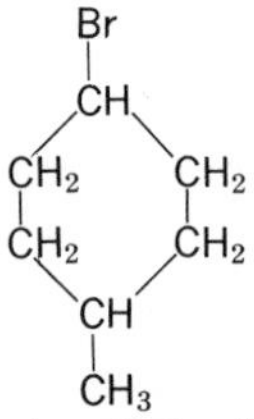

1-Bromo-4-methylcyclohexane

The cycloalkanes most commonly encountered and most easily prepared are those containing rings of five or six carbon atoms. Cycloalkanes with much larger rings have been prepared, and the chemical properties of these substances have been found almost indistinguishable from those of the *open-chain* alkanes of the same number of carbon atoms. Rings smaller than five carbon atoms, as in cyclopropane and cyclobutane, exhibit exceptional behavior.

4.3 Ring Closure Cycloalkanes are prepared in the laboratory by a variety of methods, almost all of which involve reactions which will not be treated until later chapters. A reaction which is of some commercial importance, however, will serve to illustrate some of the important features of cyclization reactions. Certain dienes when treated with acids can undergo reactions analogous to polymerization reactions.

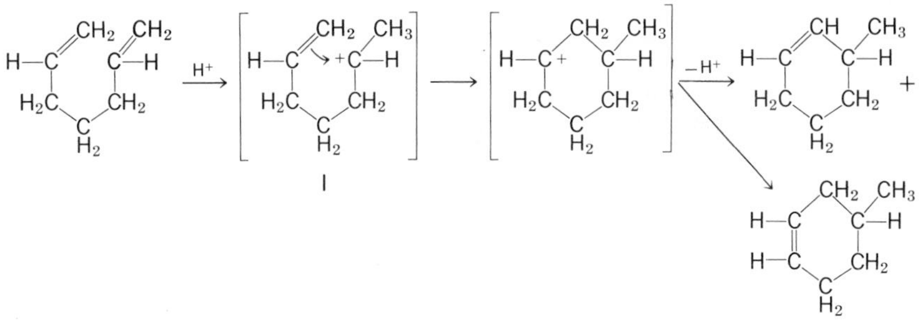

Here, instead of adding to another molecule of alkene, the first-formed carbonium ion I "adds to itself."

Experience has shown that in cyclization reactions, in general, the five- and six-membered rings are most readily formed. When the two atoms to be joined are five or six atoms apart there is a high probability that the normal contortions of the molecule will bring these two points close enough together to enable them to react. As the intervening distance grows greater, this probability constantly diminishes. As a consequence of this, the probability of the reactive site of one molecule colliding with that of some other molecule increases relative to that of the process leading to cyclization; i.e., polymerization becomes the favored process.

Ziegler in 1933 reasoned from this fact that it should be possible to carry out cyclization reactions in such dilute solution that the chance of two ends of the same molecule coming together was better than the chance of the ends of different molecules coming together. Thus more rings than chains should result. Experiment confirmed this prediction. An industrial process now rests on the production of large-ring compounds by ring closure in dilute vapor or solution.

4.4 Carbene Addition to Alkenes A remarkable class of reactive intermediates called "carbenes" or "methylenes" possesses the ability to add to alkenes to generate cyclopropanes.

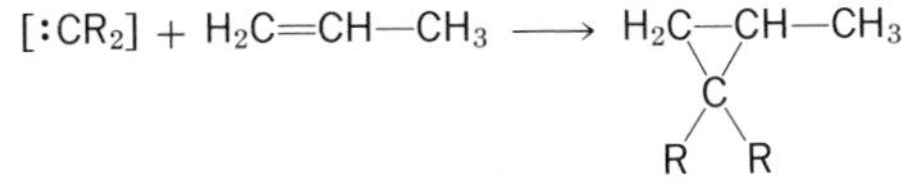

A carbene Propene A cyclopropane

Dichlorocarbene can be prepared from chloroform by treatment with certain strong bases.

$$H{-}CCl_3 + B \rightleftharpoons [:CCl_3]^- + BH$$

Chloroform

$$[:CCl_3]^- \longrightarrow :\ddot{\underset{..}{Cl}}:^- + [:CCl_2]$$

Dichlorocarbene

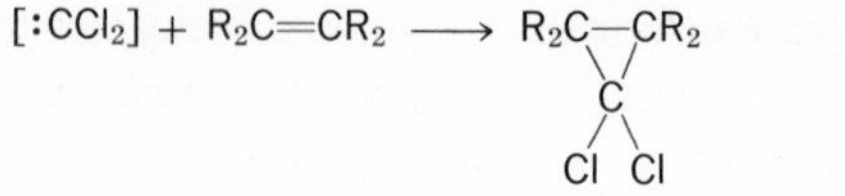

$$[:CCl_2] + R_2C=CR_2 \longrightarrow R_2C-CR_2$$

A dichlorocyclopropane

The corresponding parent compound, $[:CH_2]$, can be prepared by more difficult routes, but for synthetic purposes a reagent that is intermediate between a free carbene and a Grignard-type compound called the "Simmons-Smith reagent" may be used.

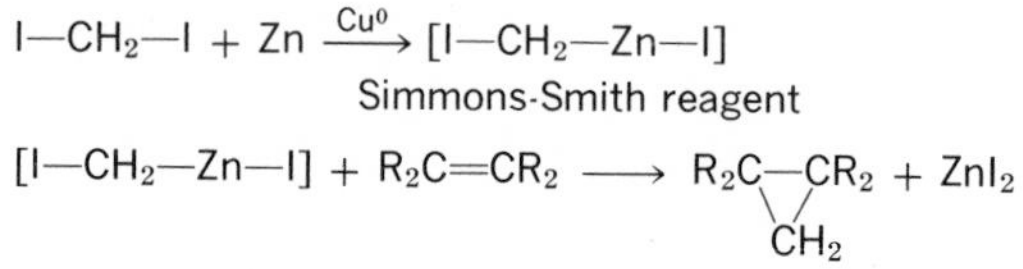

$$I-CH_2-I + Zn \xrightarrow{Cu^0} [I-CH_2-Zn-I]$$
Simmons-Smith reagent

$$[I-CH_2-Zn-I] + R_2C=CR_2 \longrightarrow R_2C-CR_2 + ZnI_2$$
$$CH_2$$

This reaction is fairly general for simple alkenes and represents one of the most useful methods for making substituted cyclopropanes.

4.5 Strain Theory of Baeyer   In the case of rings that have fewer than five members, the probability of the ends being reasonably close is very high. Nonetheless, the extent of formation of three- and four-membered rings by cyclization is generally very low relative to that of the five- and six-membered analogs. A reason for this phenomenon becomes apparent when we consider the three- and four-membered rings themselves. In tetrahedral carbon, the normal C—C—C bond angle is the tetrahedral angle, 109°28′, but in cyclobutane it must be 90° and in cyclopropane it must be still further distorted to 60°!

TABLE 4.1
BAEYER STRAIN IN RING SYSTEMS

Compound	Name	Distortion of valence angle per bond[a]	Strain energy kcal/mole/CH_2 group
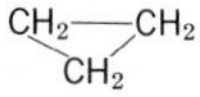 CH_2-CH_2 CH_2	Cyclopropane	24°44′	11
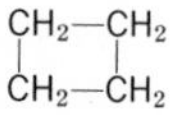 CH_2-CH_2 CH_2-CH_2	Cyclobutane	9°44′	8
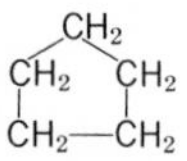 CH_2 CH_2 CH_2 CH_2-CH_2	Cyclopentane	0°44′	~0
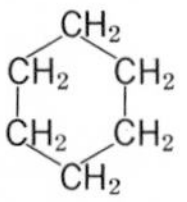 CH_2 CH_2 CH_2 CH_2 CH_2 CH_2	Cyclohexane	−5°16′[b]	~0

[a] Assuming a planar ring.
[b] Distortion in the opposite direction (out) for a planar ring.

The suggestion of the existence of a preferred angle for the valences of carbon and the concept of a strain in certain ring structures was put forward by A. Baeyer and is called the "Baeyer strain theory." Assuming that the ring is planar (flat), Baeyer calculated from simple geometry the distortion involved in forming each bond of different-sized rings. More recently it has been possible to measure the strain of small rings relative to CH_2 groups in an open-chain molecule.[1] These strain energies are given in Table 4.1 in terms of kilocalories per mole per CH_2 group.

The Baeyer strain theory receives experimental support from the fact that both cyclopropane and cyclobutane can be hydrogenated to give saturated alkanes. Cyclohexane and cyclopentane do not react under these conditions.

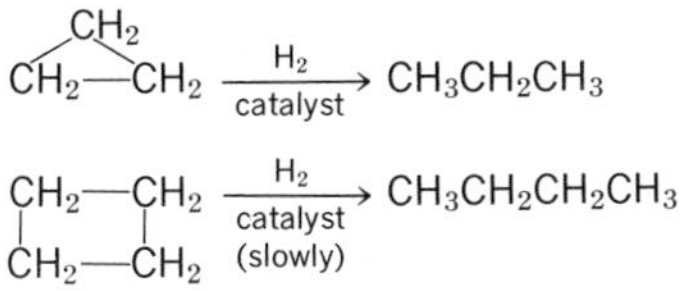
CH2
CH2—CH2
H2
catalyst
CH3CH2CH3
CH2—CH2
CH2—CH2
H2
catalyst
(slowly)
CH3CH2CH2CH3

In addition, cyclopropanes (but not cyclobutanes) react with acids, or halogens in the presence of Lewis acids, to yield ring-opened products. As with alkenes, Markownikoff's rule (Sec. 3.6) is obeyed.

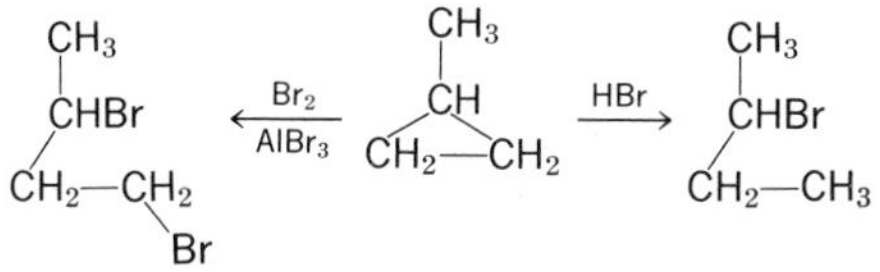
CH3
CHBr
CH2—CH2
Br
Br2
AlBr3
CH3
CH
CH2—CH2
HBr
CH3
CHBr
CH2—CH3

Unlike alkenes, however, cyclopropanes are inert to permanganate under mild conditions, as are other cycloalkanes.

$$\text{Cycloalkane} + KMnO_4 \xrightarrow{25^\circ} \text{N.R.}$$

4.6 Unstrained Rings If one considers rings of six or more atoms to be planar, it is apparent that few of these can have "normal" bond angles since regular heptagons and octagons, etc., cannot have tetrahedral C—C—C angles. In fact, it is found that essentially all saturated cyclic systems of six or more atoms have nonplanar, puckered systems. Cyclohexane, for example, can exist in forms such as the *chair* and *boat* shown below in which all bond angles are 109°28′ and all

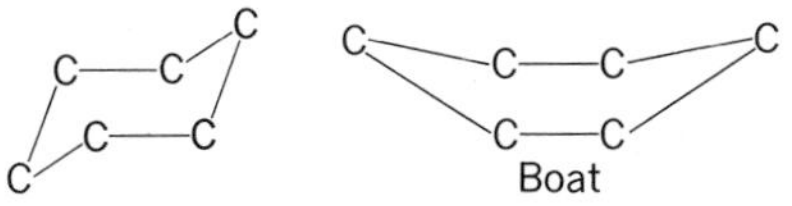
Boat

Chair

[1]This can be done, for instance, by measuring the heat of combustion of cyclopropane and comparing the heat of combustion per CH_2 group with that obtained for unstrained (open-chain) molecules. The greater heat of combustion per CH_2 group in strained systems is attributed to the higher energy content of the strained molecule.

Baeyer strain is absent. Thus cyclohexane and most larger rings are not strained and their chemical properties are those of the alkanes. These different arrangements are referred to as *conformations* (Sec. 2.4).

4.7 Geometric Isomerism One feature of cyclic compounds that differentiates them from their saturated analogs is that rings, like double bonds, have two sides. This makes possible a type of cis-trans isomerism analogous to that in alkenes in which substituents may be on the same side as one another (cis) or on opposite sides (trans). See Secs. 3.5 and 7.2. This relationship is independent of the planarity of the ring and no amount of twisting, flipping, or contortion *short of actually breaking and re-forming chemical bonds* will alter it. This is easy to see in a molecule like cyclobutane which is nearly flat.

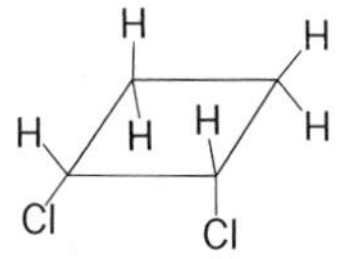

cis-1,2-Dichlorocyclobutane

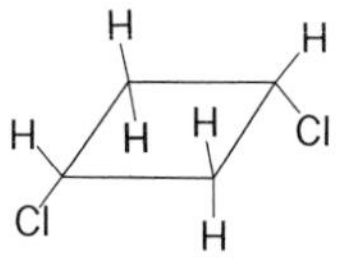

cis-1,3-Dichlorocyclobutane

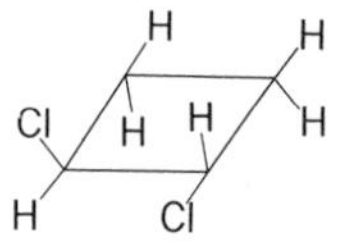

trans-1,2-Dichlorocyclobutane

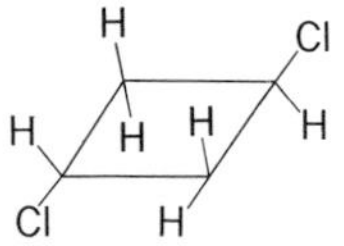

trans-1,3-Dichlorocyclobutane

In the case of larger, puckered rings these relationships are still valid and it is often easier to compare cis-trans isomers in these cases by first drawing the ring as if it were planar. The cis-trans relationships obtained from the planar drawing will always be valid in the puckered conformation, but it is often more difficult to visualize them in the latter.

4.8 Spectroscopic Properties The infrared and ultraviolet spectra of the cycloalkanes are much like those of their open chain analogs. Similarly, the nmr spectra are undistinguished except in the case of cyclopropanes in which the hydrogens are very highly shielded, appearing at $\delta = 0.4$ to 0.6 ppm. Cyclopropanes with at least one CH_2 group in the ring possess a maximum in the infrared at about 3050 cm^{-1}.

4.9 Occurrence Cycloalkanes and their derivatives are found widely in nature (in addition to occurring in petroleum). Large rings of certain sizes and with certain structures have a musk-like odor and are highly prized by perfumers. Cyclopropane is a general anesthetic of some merit. Five-atom and six-atom rings are found as parts of molecules of many kinds in plants and animals. It is found that even when atoms other than carbon occur in rings, the preferred ring sizes are those containing a total of five or six atoms. Because the valence angles of oxygen, nitrogen, and sulfur are reasonably close to that of carbon, Baeyer's strain theory is applicable here also, with, of course, the modification of the theory necessary when it is applied to large rings.

4.10 Historical Notes The English chemist William Henry Perkin has given an account of the early chemistry of cycloalkane synthesis [*J. Chem. Soc.*, 1347 (1929)]. He tells of the skepticism that existed among chemists as late as the 1880s about the feasibility of preparing rings with fewer than six carbons. When, however, synthesis showed that the preparation of the smaller ring cycloalkanes was possible and chemists began to be familiar with them, then instances of their natural occurrence began to turn up. It is interesting that a somewhat similar history with respect to certain seven-membered rings has developed. See, for example, P. L. Pauson, *Chem. Rev.*, **55,** 9 (1955).

The strain theory of Adolf von Baeyer (Sec. 4.5) was proposed by him in 1885 as a conclusion from the observation that five- and six-membered rings predominated in the cyclic compounds known at that time. It was one of Perkin's contributions to show that the smaller rings could be synthesized and show a reactivity attributable to "strain" in the bond angles. Baeyer also conceived of all rings as planar, and it was not until some seven years later that the notion that the ring could pucker with relief of strain in the larger rings was advanced by H. Sachse. The modern notions have developed from these beginnings.

OUTLINE OF CYCLOALKANE CHEMISTRY

Cycloalkanes (C_nH_{2n})

Preparation

Most practical reactions involve chemistry which has not yet been discussed. Cyclopropanes, however, may be prepared by the Simmons-Smith reaction.

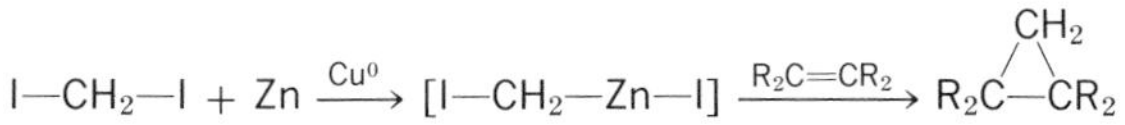

Properties

1. 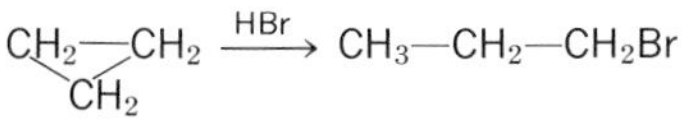

2. 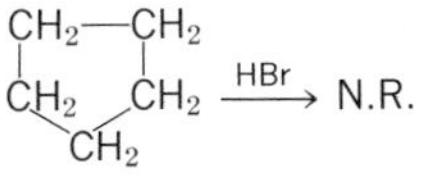

3. Cycloalkanes + $KMnO_4 \xrightarrow{25^\circ}$ N.R.

CHAPTER 4 EXERCISES

1. Name the following:

★*a.*

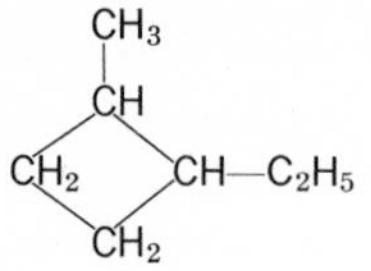

★*b.*

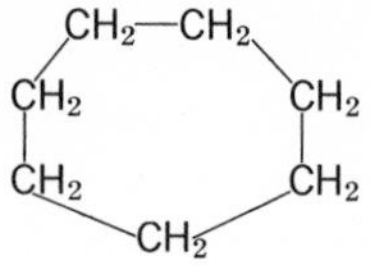

2. Write equations showing all the products of the reaction between the following compounds and bromine. Consider only the products involving 1 mole of Br_2 per mole of substance. (Lewis acids may be necessary in the case of the three-membered rings.)
   a. Methylcyclopropane
   b. 1,2,3,4,5-Pentamethylcyclopentane
   c. Cyclohexane
   d. Hexane
   e. 1,2,3-Trimethylcyclopropane

★3. Explain clearly why a five-membered ring forms more readily than a twenty-membered ring, although both are free of strain.

4. There are found to be two isomeric (cis-trans isomers) 1,2-dimethylcyclohexanes. Would there be two isomeric 1,4-dimethylcyclohexanes? Explain. Draw structures to illustrate your answer.

★5. Which of the compounds in Sec. 4.2 would be expected to occur as cis-trans isomers? Draw formulas of any of these.

6. Discuss in detail why the cis-trans isomerism of 1,2-dichloroethylene is different from that of 1,2-dichlorocyclopentane.

★7. Are there cis and trans isomers of 1,2-dichloro-1-cyclopentene? Your answer should contain reasons and, if possible, simple diagram(s).

8. What experimental conditions (excess of which reagents, concentrations, pressures, etc.) would you use to get a maximum amount of very long chain compounds and a minimum of cycloalkanes in the reaction of 1,6-octadiene with acid?

★9. The following are structural formulas of natural products. Two of them have marked physiological activity: penicillin, an antibiotic; pyrethrin, an insecticide. By inspection of the formulas, predict which two will be very reactive substances.

a.

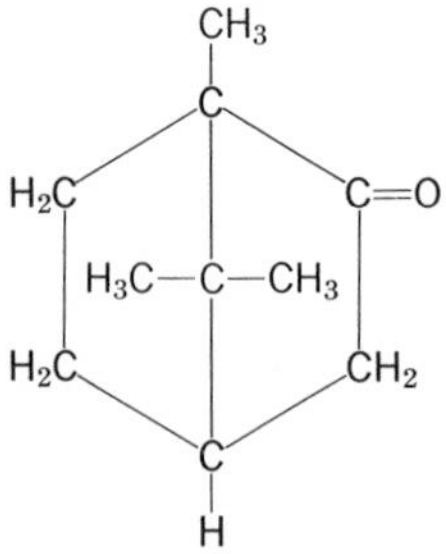

b.

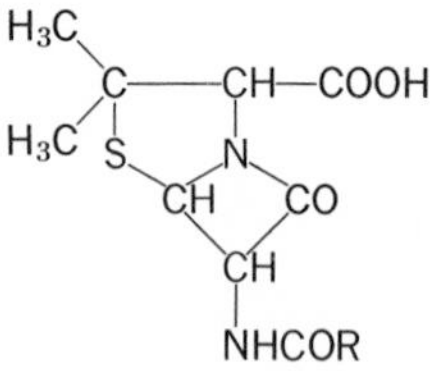

c. 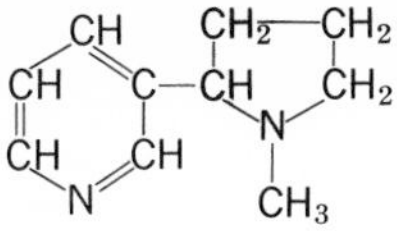

d. 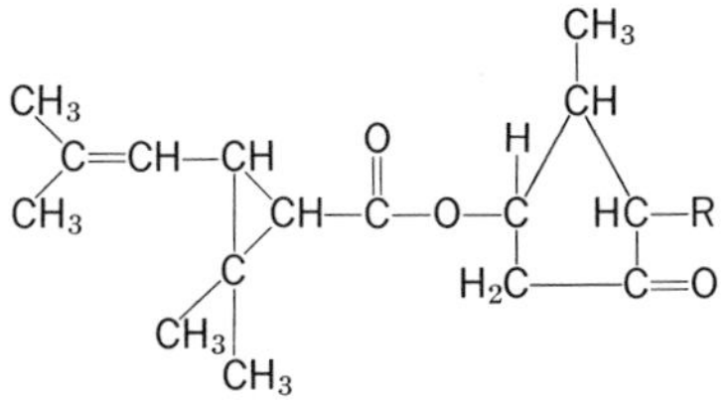

e. 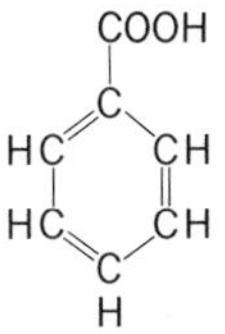

10. A certain hydrocarbon X, $C_{10}H_{16}$, shows only two peaks in the nmr at $\delta = 0.9$ and 1.25 in the ratio of 3:1. Hydrogenation yields a new compound Y, $C_{10}H_{20}$, which has a more complex nmr spectrum with new absorption at $\delta = 1.5$ as well as doublets centered at $\delta = 1.25$ and 0.9 ppm. The ratio of absorptions at 1.5:1.25:0.9 is 1:1:3. When X is oxidized with $KMnO_4$, a compound Z, $C_5H_8O_2$, results as the only organic product (no CO_2 is formed). Compound Z has nmr absorptions similar to those of X, but shifted to somewhat higher δ. Give reasonable structures for X, Y, and Z.

11. By analogy with alkene addition reaction mechanisms predict the main product of the reaction between methylcyclopropane and HCl.

12. The nmr spectrum of cyclohexane shows a difference in the number of H absorptions at room temperature and at −160°C. Explain.

13. A certain hydrocarbon X has the formula $C_{10}H_{16}$. X behaves as follows:

a. $X + H_2 \xrightarrow{Pt} C_{10}H_{18} \xrightarrow{H_2}$ no further reaction

b. $X + HCl \longrightarrow Y \xrightarrow[\text{alcohol}]{KOH}$ Z (an isomer of X)

c. $Z + KMnO_4 \longrightarrow$ 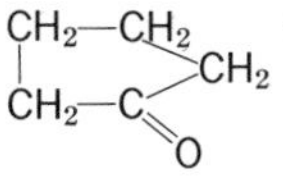 and no other organic products

Give reasonable structures for X, Y, and Z.

14. A certain hydrocarbon, C_4H_6, shows 2H's at 5.8δ and 4H's at 1.9δ. Propose a structure for it.

15. Outline a synthesis of the following. You may use any needed open-chain compounds.

a. 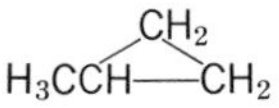

b. 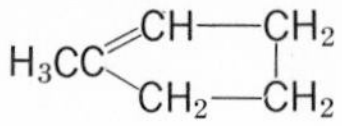

c. 1,2-Dimethylcyclopropane

d. 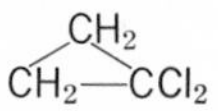

16. Suggest methods of distinguishing the following pairs. Consider both chemical and spectral methods.

a. 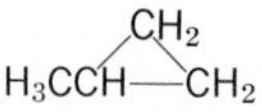 and 1-butene

b. Cyclopentane and pentane

c. 1,5-Hexadiene and 1-methyl-1-cyclopentene

d. 1,2,3,4-Tetramethylcyclobutane and 2,3,3-trimethylpentane

e. 1-Hexene and 2-hexene

★17. Calculate the unsaturation numbers of all the compounds of Exercise 16.

18. Show all the possible cis-trans isomers of

a. 1,2,3-Trimethylcyclopentane

b. 1,3-Dibromo-4-methylcyclohexane

c. 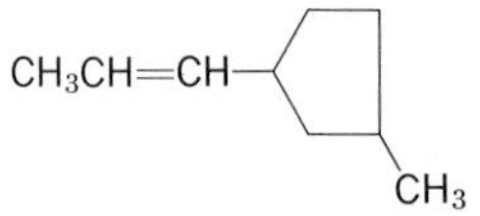

d. 1,2,3-Trichloro-1-cyclohexene

e. Cyclopentadiene

★19. The nmr spectrum of fluorocyclohexane shows marked changes in the character of the absorption due to F with temperature. At lower temperatures the F absorption is seen to break up into a number of distinct bands. How can you account for this? (Fluorine has an nmr absorption similar to hydrogen, but occurring in a different electromagnetic region. The fluorines are split by hydrogens in the same way that other hydrogens are split.)

5

HYDROCARBONS: THE ALKYNES

5.1 Introduction A class of compounds is known which involves two adjacent carbon atoms linked by a triple bond. The simplest member of this series, acetylene, may be written

$$H—C{\equiv}C—H \qquad \text{or} \qquad H{:}C{:}{:}{:}C{:}H$$

As was the case with the double bond, the triple bond is both shorter and mechanically stronger than a single bond, and is even shorter and stronger than a double bond.

The molecular orbital theory gives us a picture of acetylene as follows: Each carbon atom possesses four valence electrons, making a total of ten when added

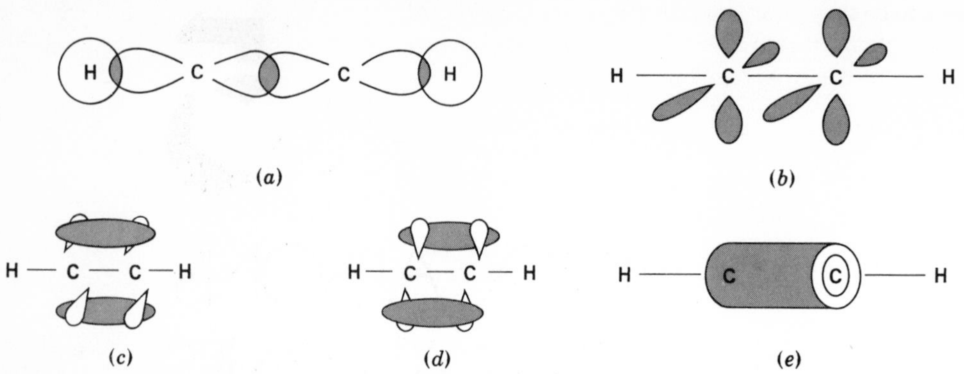
H
C
C
H
H
C
C
H
(a)
(b)
H — C — C — H
H — C — C — H
H
C
C
H
(c)
(d)
(e)

figure 5-1
Molecular-orbital model of acetylene. (*a*) σ framework. (*b*) *p* orbitals. (*c*) A π bond. (*d*) A π bond. (*e*) Overall electron distribution.

to the two from the hydrogens. Each carbon atom can bond to the other carbon atom and to one hydrogen; this requires six electrons; the other four electrons are available for extra bonds. It turns out that the most efficient way to utilize them is to use *sp* hybrid orbitals for the σ bonds so as to leave *p* orbitals available for sideways π overlap. Since *sp* orbitals are oriented 180° from one another, the π framework of acetylene is necessarily linear, as shown in Fig. 5-1*a*. With this framework the *p* orbitals shown in Fig. 5-1*b* can overlap to form two π bonds as shown in Fig. 5-1*c* and *d*. Each π bond contains two electrons so that all ten electrons are now accounted for. Moreover, the overall electron distribution on a time average is cylindrically symmetrical, so that acetylenes are often pictured as two carbon atoms surrounded by a cylinder of electron density, as pictured in Fig. 5-1*e*.

Acetylene is the first member of a homologous series of compounds containing the triple bond. These compounds are known as the *alkynes,* and their type formula is C_nH_{2n-2}. Following the system used in the alkene series, the presence of a triple bond is indicated by the ending *-yne*[1] accompanied by a number. The number given is the lower of the two numbers assigned to the triply bonded carbon atoms. Alkynes, C_nH_{2n-2}, have an unsaturation number N_U of 2 (corresponding to two double bonds, or one bond and one ring, or two rings). See Sec. 3.2.

A few members of the alkyne series and an acetylenic alkene are shown below as examples in nomenclature.

$HC{\equiv}CH$ Acetylene	$H_3C{-}C{\equiv}C{-}CH_3$ 2-Butyne (dimethylacetylene)
$HC{\equiv}C{-}CH_2{-}CH_3$ 1-Butyne (ethylacetylene)	$CH_2{=}CH{-}C{\equiv}C{-}CH_2Cl$ 1-Chloro-4-penten-2-yne

[1]Pronounced "ine" as in line.

It used to be thought that alkynes were not found in nature, but recent investigations have shown that this is not so. For example, cicutoxin, a toxic principle from a variety of hemlock, has the formula

$$HOCH_2CH{=}CH{-}C{\equiv}C{-}C{\equiv}C{-}(CH{=}CH)_2(CH_2)_2\underset{\displaystyle OH}{\underset{|}{C}}H{-}C_3H_7$$

and mycomycin, an antibiotic, has the formula

$$HC{\equiv}C{-}C{\equiv}C{-}CH{=}C{=}CH{-}CH{=}CH{-}CH{=}CHCH_2COOH$$

From an industrial point of view acetylene itself is the most important member of the family, and its importance is very great. It is prepared commercially in enormous quantities from calcium carbide, which is made from coke and lime.

$$3C + CaO \longrightarrow CaC_2 + CO$$
$$CaC_2 + 2HOH \longrightarrow Ca(OH)_2 + HC{\equiv}CH^1$$

5.2 The Acetylenic Hydrogen Atom The hydrogen atoms of acetylene, or in general, hydrogens attached to triple-bond carbon atoms, are sufficiently acidic to be replaced by certain metals:

$$CH_3C{\equiv}CH + Ag(NH_3)_2^+ \xrightarrow{NH_4OH} \underset{\substack{\text{A silver acetylide}\\\text{(white)}}}{CH_3C{\equiv}CAg\downarrow} + NH_4^+ + NH_3$$

$$2CH_3C{\equiv}CH + CuCl \xrightarrow{2NH_4OH} \underset{\substack{\text{A cuprous acetylide}\\\text{(red)}}}{CH_3C{\equiv}CCu\downarrow} + 2NH_4Cl + 2H_2O$$

Acetylene is not acidic enough to react appreciably with bases that would give soluble acetylides in *water*

$$HC{\equiv}CH + NaOH \rightleftharpoons HC{\equiv}C^-Na^+ + H_2O$$

but it will react in less acidic solvents, such as NH_3, or in the absence of solvent.

$$HC{\equiv}CH + \underset{\text{Sodium amide}}{NaNH_2} \xrightarrow[\text{Solvent}]{NH_3} HC{\equiv}C^-Na^+ + NH_3$$

$$2CH_3C{\equiv}CH + 2Na^0 \longrightarrow 2CH_3C{\equiv}C^-Na^+ + H_2\uparrow$$

These metallic compounds are known as *acetylides;* calcium carbide is calcium acetylide. Silver and cuprous acetylides are explosive when dry, and detonate when struck with a feather; sodium and calcium acetylides are stable to heat and shock. The first two reactions given above are useful as convenient tests for the linkage

[1]Not all carbides react the same way with water. For example, Al_4C_3 yields methane.

—C≡CH. They are tests for H on a triply bonded carbon, not for the triple bond. Alkynes without such groups do not react:

$$CH_3—C≡C—CH_3 + Ag^+ \xrightarrow{NH_3} \text{N.R.}$$

Modern theory attributes the acidity of terminal alkynes to the large amount of "*s* character" in the *sp* bonds compared with that in sp^2 and sp^3 bonds; *s* orbitals are closer to the nucleus than *p* orbitals and, accordingly, are more tolerant of negative charges. There is now a body of evidence in favor of this hypothesis, but it is appropriately the subject of more advanced courses.

5.3 Addition Reactions Further evidence of the reactivity of acetylene is given by the variety of its addition reactions. Equations are given below to illustrate some of the more important addition reactions of the triple bond.

$$HC≡CH + 2H_2 \xrightarrow[\text{cat.}]{\text{Pt}} CH_3CH_3$$

The use of a limited amount of hydrogen (or other addendum) leads to the formation of alkenes (or alkene derivatives) in reasonable yields.

$$CH_3CH_2C≡CH \xrightarrow[\text{(limited)}]{H_2} CH_3CH_2CH{=}CH_2 \xrightarrow{H_2} CH_3CH_2CH_2CH_3$$

This reaction makes it possible to correlate the structures of alkynes with the corresponding alkenes.

A reaction cannot always be stopped at an intermediate stage by limiting the amount of a reagent. Only when the rate of formation of the intermediate is markedly greater than the rate of its conversion to subsequent products is such control possible. Even in favorable cases (as with the alkynes), all possible products are produced, though some may be formed in small amounts. Control of this kind is practical in only three or four of the reactions dealt with in this book. This and the following are examples of such controllable reactions.

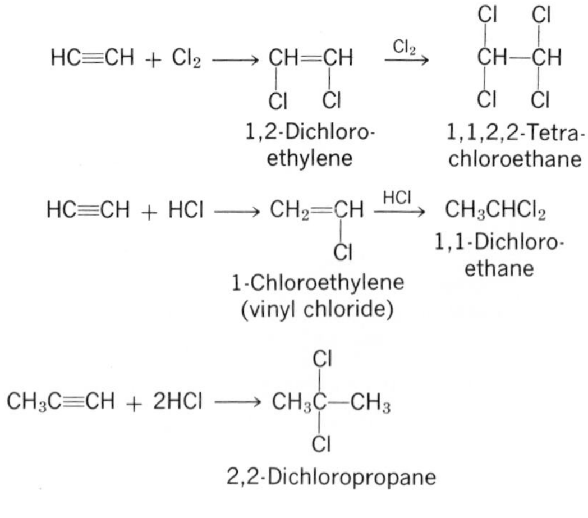
1,2-Dichloro-ethylene
1,1,2,2-Tetra-chloroethane
1-Chloroethylene (vinyl chloride)
1,1-Dichloro-ethane
2,2-Dichloropropane

That successful "half halogenation" of the alkynes can be carried out in good yield rests in part on the fact that the resulting halogenated alkene reacts only slowly by addition. The electron-attracting nature of the halogen decreases (inductively) the nucleophilic character of the double bond. Other effects may also operate.

5.4 Addition of Water Although pure water does not add to alkenes, it can be made to react readily with alkynes in acid solution in the presence of mercury salts.

$$HC{\equiv}CH + H_2O \xrightarrow[\text{dil. } H_2SO_4]{HgSO_4} CH_3C\diagdown{\!\!\!\!}^{O}_{H}$$

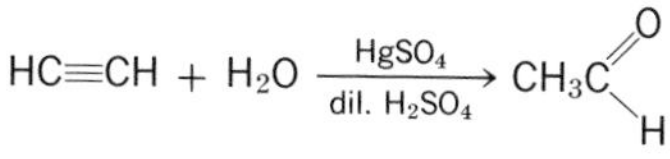

Acetaldehyde

This particular example is an important industrial process. It is carried out by allowing the acetylene to bubble up through towers filled with broken tile (or other inert packing) over which aqueous acidic mercuric sulfate solution flows. The acetaldehyde which is formed dissolves in the aqueous layer and is removed later by distillation, the purified aqueous layer being then returned to the top of the tower. The reaction takes a course, which for simplicity, will be written thus:

$$HC{\equiv}CH + HOH \xrightarrow{Hg^{++}} \left[\begin{array}{c} H\ \ H \\ | \ \ | \\ HC{=}C{-}OH \end{array} \right]$$

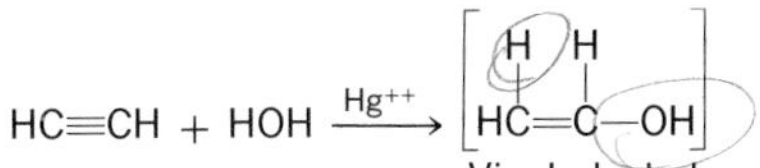

Vinyl alcohol

$$\left[\begin{array}{c} H\ \ H \\ | \ \ | \\ H{-}C{=}C{-}OH \end{array} \right] \longrightarrow CH_3C\diagdown{\!\!\!\!}^{O}_{H}$$

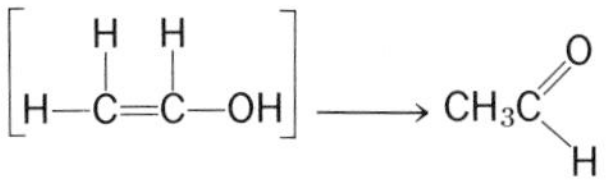

$$CH{\equiv}CH + HOH \longrightarrow CH_3C\diagdown{\!\!\!\!}^{O}_{H}$$

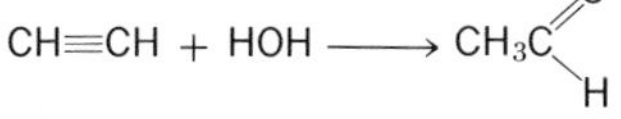

Overall reaction   Acetaldehyde

Vinyl alcohol is unstable and has never been prepared; all attempts to produce it lead to the formation of acetaldehyde. This observation, together with many others to be described later on in the course, indicates that compounds of the form

$$\diagdown_{C}{=}C\diagup^{}_{OH}$$

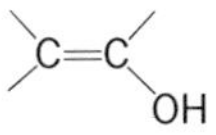

are able to *rearrange* spontaneously into the form

$$\underset{H}{-}\overset{|}{C}{-}C\diagup^{O}$$

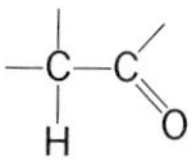

This rearrangement involves a shift of hydrogen as a proton. If the H is replaced in vinyl alcohol by any other group, the resulting compound does not rearrange. Thus

$-\underset{|}{C}=\underset{|}{C}-O-CH_3$ does not arrange.

This rearrangement is readily understandable if we consider the fact that alcohols (Chap. 8) are weakly acidic. Loss of a proton from vinyl alcohol (called an *enol* after *en* alk*ene* and *ol* from alcoh*ol*) gives the resonance-stabilized *enolate* ion (Sec. 11.10), which can then pick up a proton on the carbon atom to yield an aldehyde or ketone.

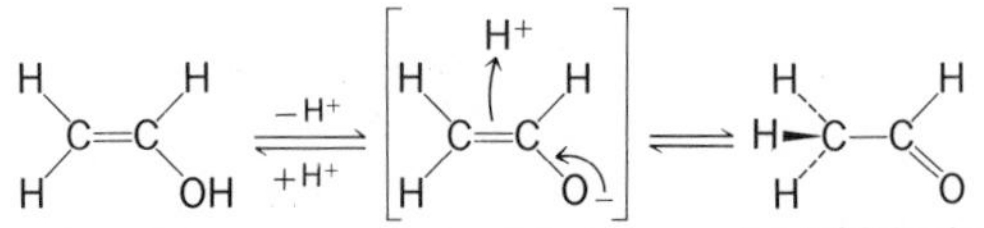
H
H
C=C
H
OH
−H+
+H+
H+
H
H
C=C
H
O−
H
H
H—C—C
H
O

An enol (vinyl alcohol) — An enolate ion — Acetaldehyde

Since aldehydes and ketones are generally more stable than the corresponding vinyl alcohols, the equilibrium is shifted to the right.

Structural isomers which are interconverted by relocating one hydrogen atom are given the special name of *tautomers*. The phenomenon is called "tautomerism." Since the hydrogen is most often transferred as a proton, the interconversion of tautomers is usually catalyzed by acid and/or base.

It is very important to note the distinction between tautomers and contributing structures to a resonance hybrid. Tautomers differ from one another in the position of at least one nucleus (in most cases the hydrogen atom) and they cannot be contributing structures to the same hybrid. Each tautomer may be resonance stabilized, but a different hybrid will be involved for each one.

In the case of higher alkynes, this reaction leads always to the formation of ketones (Markownikoff's rule).

$$CH_3C{\equiv}CH \xrightarrow[\text{dil. } H_2SO_4]{HgSO_4} CH_3COCH_3$$

The addition of water, a nucleophilic reagent, to a multiple carbon-to-carbon linkage would not be expected to occur readily, and the reaction is undoubtedly more complex than the simple addition of electrophilic reagents already discussed (Sec. 3.7). A mercury-containing complex is involved, but since a discussion of this mechanism contributes little to our understanding of the reaction at this point, it will be omitted.

5.5 Oxidation of Alkynes Like all hydrocarbons, alkynes will burn:

$$2HC{\equiv}CH + 5O_2 \longrightarrow 4CO_2 + 2H_2O$$

When pure oxygen is used in this reaction with a special torch, the very hot oxyacetylene flame is produced. This flame is hot enough to "cut" metals by melting them and hot enough for welding purposes. Large quantities of acetylene are used in this way.

Alkynes are oxidized in aqueous solution by permanganate to yield cleavage products. The reaction is similar to that shown by double bonds with permanganate, except that when alkynes are oxidized, the linkage always formed is

$$-\overset{\displaystyle O}{\underset{}{C}}-OH \qquad \text{not} \qquad -\overset{\displaystyle O}{\underset{}{C}}-R$$

The mechanism of this oxidation and rules for balancing equations are the same as already described for alkenes (Sec. 3.15).

$$3CH_3C\!\equiv\!CH + 4H_2O + 8KMnO_4 \longrightarrow 3CH_3C\overset{\displaystyle O}{\underset{OH}{\diagdown}} + 3CO_2 + 8KOH + 8MnO_2$$

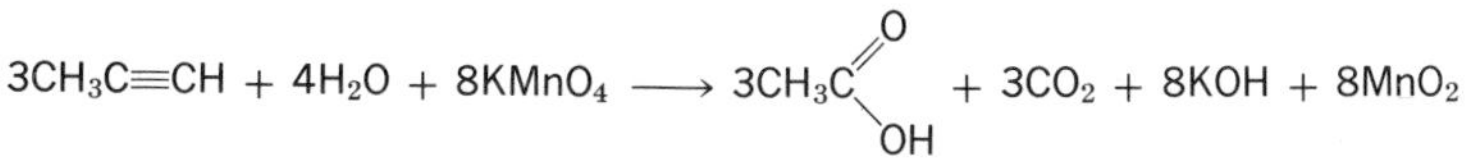

As is the case with the alkenes, this is an important way of proving the location of the unsaturated linkage.

Ozone degradation of alkynes yields acids. The above compound on ozone degradation would yield

$$CH_3C\overset{\displaystyle O}{\underset{OH}{\diagdown}} + \overset{\displaystyle O}{\underset{HO}{\diagdown}}C-H$$

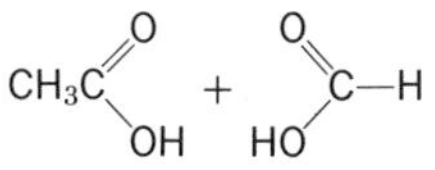

This method offers little or no advantage over permanganate oxidation in this series and finds little practical application.

5.6 Preparation   The preparation of alkynes follows the same pattern as that of the olefins except that now four substituents must be removed from the molecule, two each from adjacent carbon atoms.

$$CH_3\underset{\underset{Br}{|}}{\overset{\overset{H}{|}}{C}}-\underset{\underset{Br}{|}}{\overset{\overset{H}{|}}{C}}H + 2KOH \xrightarrow{\text{alcohol}} CH_3C\!\equiv\!CH + 2KBr + 2H_2O$$

$$CH_3CH_2\underset{\underset{Br}{|}}{\overset{\overset{H}{|}}{C}}Br + 2KOH \xrightarrow{\text{alcohol}} CH_3C\!\equiv\!CH + 2KBr + 2H_2O$$

$$CH_3\underset{\underset{Br}{|}}{\overset{\overset{Br}{|}}{C}}-\underset{\underset{Br}{|}}{\overset{\overset{Br}{|}}{C}}H + 2Zn \longrightarrow CH_3C\!\equiv\!CH + 2ZnBr_2$$

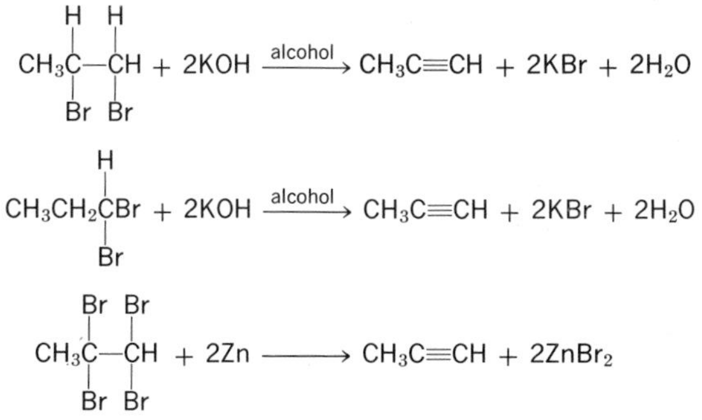

Since 1,2-dihalides are readily prepared from alkenes by addition, the alkynes can be correlated with the corresponding double-bond compound by the first of the

reactions given above. Dihalides with both halogens on the same carbon atom are less readily prepared. Tetrahalides are known only as addition products of alkynes. This last reaction is thus of no value in preparative work; it is included here for comparison in method with the preparation of alkenes (Sec. 3.21).

By analogy with the formation of double bonds from alcohols (Secs. 3.18 to 3.19), it might be expected that the dehydration of glycols with sulfuric acid would lead to the formation of triple bonds. Instead, aldehydes or ketones are found to be produced (also other products), and this is explained by the intermediate formation of structures of the type C=C—OH which, as has been mentioned above, immediately rearrange; thus:

$$\underset{\overset{|}{OH}\quad\overset{|}{OH}}{CH_2-CH_2} \xrightarrow[H_2SO_4]{-H_2O} \left[\underset{\overset{|}{OH}}{CH_2=CH}\right] \longrightarrow CH_3C\overset{\displaystyle H}{\underset{\displaystyle O}{\diagdown}}$$

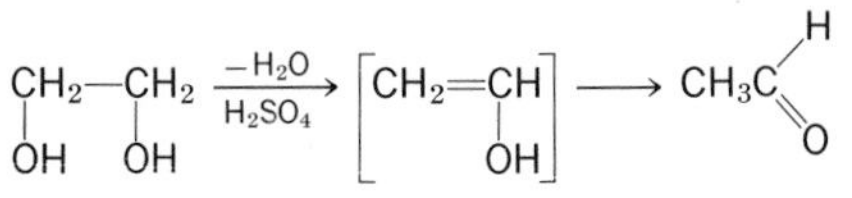

5.7 The Displacement Reaction The higher homologs of acetylene are often prepared in the laboratory from sodium acetylide.

$$CH_3CH_2-I + Na^+{}^-C\equiv CH \longrightarrow CH_3CH_2-C\equiv CH + Na^+I^-$$

The reaction can be carried further by re-forming an anion.

$$CH_3CH_2C\equiv CH + NaNH_2 \longrightarrow CH_3CH_2C\equiv C^-Na^+ + NH_3$$
$$CH_3I + Na^+{}^-C\equiv C-CH_2CH_3 \longrightarrow CH_3C\equiv C-CH_2CH_3 + Na^+I^-$$
(Note that the alkyne anion displaces I⁻.)

This reaction is the first example that we have encountered of a reaction which is a very important one in organic chemistry. This is a reaction in which an anion, or a neutral molecule with an unshared electron pair, forms a covalent bond to carbon while at the same time expelling a second group as an anion or neutral molecule.

In general terms, the reaction may be diagramed thus:

$$X^- + \underset{\overset{\displaystyle |}{R}}{\overset{\displaystyle R}{R}}C-Y \longrightarrow \left[\underset{R\quad R}{X\cdots\overset{\displaystyle R}{\underset{}{C}}\cdots Y}\right] \longrightarrow X-\underset{\overset{\displaystyle |}{R}}{\overset{\displaystyle R}{C}}-R + Y^-$$

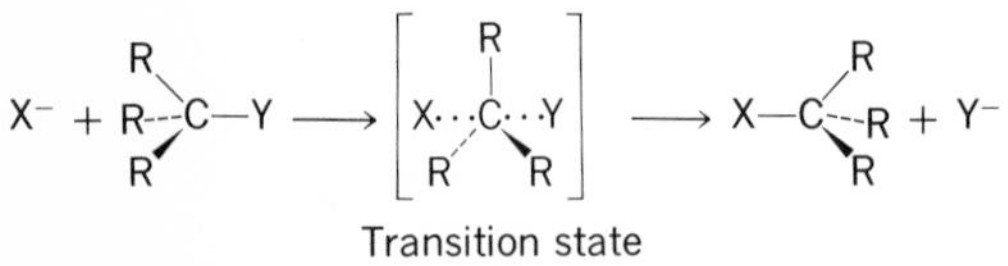

Transition state

figure 5-2
A generalized displacement reaction.

The energy profile (Sec. 3.10) for this reaction is shown in Fig. 5-3.

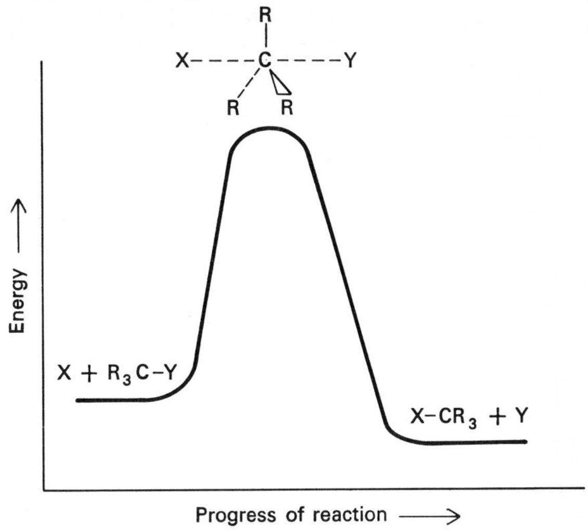
R
X----C----Y
R R
Energy
X + R₃C-Y
X-CR₃ + Y
Progress of reaction

figure 5-3
The energy profile for a general displacement reaction.

The displacement reaction is one of the most thoroughly studied reactions in organic chemistry and certain important generalizations can be made about it:

1. The reaction is *bimolecular,* that is, X commences to bond to the carbon atom *before* the bond between C and Y is completely broken. This means that one does not have to provide all the energy needed to completely break the C—Y bond.
2. The reaction occurs with *inversion,* that is, X attacks the carbon atom from the side of the molecule directly opposite Y. As the reaction proceeds, the originally tetrahedral carbon goes through a planar configuration and then turns inside out as the bond to X is completed.
3. The reaction is sensitive to *steric* effects, that is, it is sensitive to the size of X and the R groups attached to carbon (and to lesser extent that of Y).
4. The reaction depends on the nature of X and on the ability of Y to leave as a stable species (usually an anion). For example, H^- CH_3^-, etc., are rarely displaced because they form such unstable (high-energy) anions.

In the reaction of acetylide ions the nucleophile X is $R'—C{\equiv}C^-$ and the leaving group Y is usually the electronegative Cl^-, Br^-, or I^- since these can easily leave as stable anions. The steric (size) effect is pretty much independent of R′ since it is not at the reaction center, but the steric effect is vitally dependent on the groups attached to the reacting carbon atom. Thus the relative ease with which the reaction

$$R'C{\equiv}C^-Na^+ + RBr \longrightarrow R'C{\equiv}C—R + Na^+Br^-$$

takes place is in the order

$$R = CH_3 > R = CH_3CH_2 > R = \begin{matrix} H_3C \\ {} \\ H_3C \end{matrix}\!\!\!\!\!CH$$

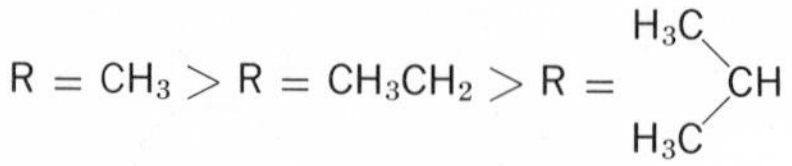

For R=$(CH_3)_3C$— the reaction is not observed at all. (Other side reactions become the major reactions.)

This represents the first reaction we have encountered for selectively generating new carbon-to-carbon bonds, and it is of considerable value in organic syntheses.

5.8 Applications of Acetylene   The synthesis of rubber-like materials from acetylene is only one of a large number of important chemical syntheses that can be carried out using this substance as the raw material. Since acetylene is usually made from calcium carbide, the ultimate starting materials are coal, limestone, water, and energy (for the CaO + C $\xrightarrow{\Delta}$ CaC$_2$ high-temperature step). In the United States, our most abundant and practical raw material for the synthesis of organic compounds has been petroleum, and much of our chemical industry is geared to this substance. Acetylene has become such an important substance that methods for its preparation from petroleum and natural gas have been developed, this in spite of its economical production by way of coal, limestone, water, and energy (Sec. 5.1). By controlled combustion of methane, for example, acetylene is produced in substantial yield. Thus the relatively cheap but inert hydrocarbon methane is converted in one step to the reactive substance acetylene. A large fraction of the acetylene used for chemical purposes, the production of plastics, synthetic rubber, and various intermediates, is produced from methane in the United States.

The choice of raw material for the synthesis of acetylene—and from this many other organic substances—depends on economic and geological factors. In a petroleum-rich economy such as that of the United States, acetylene from natural gas can be economically used. In parts of the world where petroleum is less available than coal, notably in Germany, the calcium carbide process is chosen. In both cases the versatile and reactive substance acetylene plays a key role in the chemical industry.

Examples of some of the reactions of acetylene used in commercial production of polymers (plastics) are given below. In each case it is noted that the reaction is controlled to stop at the double bond stage (Sec. 5.3); polymerization of the alkene derivatives leads to the final products.

$$2HC\equiv CH \longrightarrow CH_2{=}CH{-}C\equiv CH \xrightarrow{HCl} CH_2{=}CH{-}\overset{\overset{\displaystyle Cl}{|}}{C}{=}CH_2$$

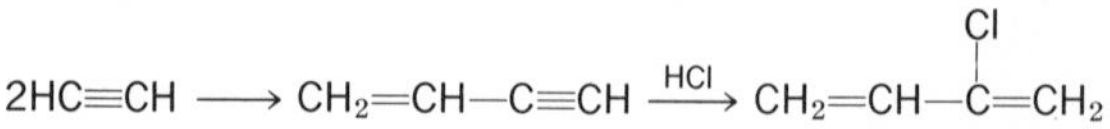

Chloroprene
(used in the manufacture of Neoprene)

$$HC{\equiv}CH \xrightarrow{CH_3COOH} CH_2{=}CHOCCH_3 \text{ (with C=O on the acyl carbon)}$$
Vinyl acetate

$$HC{\equiv}CH \xrightarrow{HCl} CH_2{=}CHCl$$
Vinyl chloride

(Vinyl acetate and vinyl chloride: monomers for vinyl plastics)

$$HC{\equiv}CH \xrightarrow[\text{catalyst}]{HCN} CH_2{=}CH{-}CN \qquad \text{monomer for Orlon}^{®}$$ [1]

5.9 Spectroscopic Properties Alkynes have characteristic stretching frequencies in the infrared due to the stiff triple bond. This occurs in the region 2100 to 2260 cm^{-1} but, unfortunately, is often weak, and in the case of symmetrical alkynes may be completely missing. With monosubstituted alkynes a strong C—H stretching band occurs at 3270 to 3305 cm^{-1}. In the nmr spectrum the acetylenic hydrogen is found at $\delta = 2$ to 3, a region where few other protons are found. This position for the acetylenic hydrogen is much more shielded than intuition would lead us to expect, especially in view of the large amount of "*s* character" of the acetylenic C—H bond (Secs. 3.6 and 5.2). This anomalous shielding of the acetylenic hydrogen is attributed to a magnetic field being induced during the nmr measurement in the cylindrical π cloud of the acetylene. This induced field opposes the applied field so that the hydrogen experiences a weaker field than it otherwise would. The net effect of this is to make the hydrogen absorption appear at a higher field just as if it were more strongly shielded.

In brief, terminal alkynes (with a —C≡C—H group) are readily detected by spectroscopic, as well as by chemical, means. In the case of internal alkynes, especially symmetrical ones, spectral identification is likely to be less positive.

OUTLINE OF ALKYNE CHEMISTRY

Preparation

1. $R{-}CH_2{-}CHX_2 \xrightarrow{\text{alc. KOH}} R{-}CH{=}CHX \xrightarrow{\text{alc. KOH}} R{-}C{\equiv}CH$
2. $R_2{-}CH{-}X + Na^+C^-{\equiv}C{-}R' \longrightarrow R_2CH{-}C{\equiv}C{-}R' + Na^+X^-$

   R, R′ = alkyl or H

Properties

1. $RC{\equiv}CH \xrightarrow[\text{cat.}]{H_2} RCH{=}CH_2 \xrightarrow[\text{cat.}]{H_2} RCH_2CH_3$
2. $RC{\equiv}CH \xrightarrow{X_2} RCX{=}CHX \xrightarrow{X_2} RCX_2CHX_2$
3. $RC{\equiv}CH \xrightarrow{HX} RCX{=}CH_2 \xrightarrow{HX} RCX_2CH_3$

   These three reactions can usually be controlled to yield either alkenes or saturated compounds.

[1] The name "Orlon" is a registered trademark of the DuPont Company.

4. $RC{\equiv}CH \xrightarrow{KMnO_4} RCOOH + CO_2$

$RC{\equiv}CR' \xrightarrow{KMnO_4} RCOOH + R'COOH$

5. $RC{\equiv}CH \xrightarrow[\text{CuCl}]{\text{ammoniacal}} RC{\equiv}CCu\downarrow$
A copper acetylide (red)

$RC{\equiv}CH \xrightarrow[\text{AgNO}_3]{\text{ammoniacal}} RC{\equiv}CAg\downarrow$
A silver acetylide (white)

$RC{\equiv}CR \xrightarrow[\text{reagent}]{\text{either}} \text{N.R.}$

6. $RC{\equiv}CH + Na \longrightarrow RC{\equiv}CNa + \frac{1}{2}H_2$

$RC{\equiv}CNa + R'X \longrightarrow RC{\equiv}CR' + NaX$

7. $RC{\equiv}CH + H_2O \xrightarrow[H_2SO_4]{Hg^{++}} R\underset{\underset{O}{\|}}{C}{-}CH_3$

$HC{\equiv}CH + H_2O \xrightarrow[H_2SO_4]{Hg^{++}} CH_3C\genfrac{}{}{0pt}{}{{\nearrow}O}{{\searrow}H}$

Spectroscopic Properties

Group	*ir, cm^{-1}*	*nmr, δ(ppm)*	*uv, mμ*
—C≡C—	2100–2260 (weak or missing)	. . .	below 200
—C≡C—H	3270–3305	2-3	. . .

CHAPTER 5 EXERCISES

1. Write equations for the following reactions:
 ★*a.* 2-Butyne + $KMnO_4$ + H_2O
 b. 1-Butene-3-yne + $HgSO_4$ + dil. H_2SO_4
 ★*c.* Propyne + HCl excess
 d. 3-Methyl-1-pentyne + Br_2
 ★*e.* Acetylene + $KMnO_4$ + H_2O
 f. 2,3-Dichlorohexane + alcoholic KOH
 ★*g.* Potassium acetylide + iodoethane

2. Write formulas for all the isomers of C_6H_{10} and give systematic names to each.

★3. How could you obtain pure, dry ethylene from a sample contaminated with acetylene? Describe in detail the apparatus you would use.

4. Distinguish between (*a*) $CH_3C{\equiv}CCH_3$ and (*b*) $CH_3CH_2C{\equiv}CH$ without using the acid property of ≡C—H by (*a*) chemical and by (*b*) spectral methods.

5. Distinguish by chemical and/or spectral methods between:

★*a.* Butane, 1-butene, 1-butyne, 2-butyne

b. 2-Pentene, cyclopentene, 2-pentyne

★*c.* Propane, cyclopropane, propyne, cyclopropene

6. A substance, $C_8H_{14}Br_2$, reacts with sodium ethoxide to give a new substance, C_8H_{12}, A, which gives an immediate precipitate with ammoniacal silver nitrate. A reacts with one mole of hydrogen in the presence of palladium to give C_8H_{14}, B. Among the products obtained by oxidizing B with chromic acid is cyclohexane carboxylic acid:

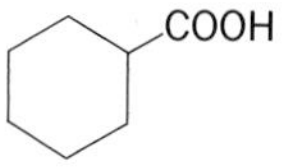
COOH

Write structures for all these compounds and equations for the reactions.

★7. Are all three bonds in a triple bond the same? Explain.

8. Would the molecular orbital picture of dimethylacetylene lead one to expect geometrical isomerism in this compound? Explain.

★9. Write mechanisms for the Lewis acid-catalyzed polymerizations of acrylonitrile and of vinyl acetate.

10. Comment on the anticipated stability of cyclobutyne and of cyclononyne.

11. Reaction of $CH_3\underset{OH}{\underset{|}{C}}H—\underset{OH}{\underset{|}{C}}HCH_3$ with strong acid yields H_2O but no alkene properties can be found in the other product. Explain.

12. Monoalkynes have the same unsaturation as what other classes of compounds? Show typical examples and outline methods, chemical and spectroscopic, of distinguishing them.

6

HYDROCARBONS: THE AROMATIC HYDROCARBONS

6.1 Benzene The simplest member of an important class of compounds, the *aromatic compounds,*[1] is benzene, C_6H_6. Since the great majority of aromatic compounds are best described as derivatives of benzene, it is important to understand the structure of benzene itself. This amazing substance was known for over 100 years before a satisfactory structure could be established. Part of this difficulty may be appreciated if one calculates the unsaturation number for benzene (Sec. 3.2). For C_6H_6, N_U equals 4 and, therefore, benzene has the equivalent of four multiple bonds and/or rings. Such a substance would be expected to be highly

[1]Aromatic compounds are sometimes referred to as arenes (compare alkene, etc.), but the term is less common and will not be used here.

unsaturated, since even if it contained only rings, they would have to be both small and highly strained.

It is then surprising to find that the characteristic reaction of benzene is one of *substitution* and not addition. Thus benzene, on treatment with bromine (or chlorine) in the presence of a Lewis-acid catalyst reacts as follows:

$$C_6H_6 + Br_2 \xrightarrow{FeBr_3} \underset{\text{Bromobenzene}}{C_6H_5Br} + HBr$$

Benzene does not react by addition under ordinary conditions with HBr or HCl; with nitric acid in sulfuric acid, however, it does undergo substitution.

$$C_6H_6 + H{-}O{-}N(=O){-}O^- \xrightarrow[\text{}]{\text{conc. } H_2SO_4} \underset{\text{Nitrobenzene}}{C_6H_5NO_2} + H_2O$$

Strong sulfuric acid also gives sulfonation.

$$C_6H_6 + HO{-}S{-}O_3H \xrightarrow[\Delta]{} \underset{\text{Benzene sulfonic acid}}{C_6H_5SO_3H} + H_2O$$

Benzene is not oxidized by $KMnO_4$ solution, even under conditions more severe than required to completely convert alkenes to acids and ketones. It is, however, attacked by ozone to form a triozonide, which on hydrolysis yields only one product.

$$C_6H_6 + O_3 \longrightarrow \text{triozonide} \xrightarrow[Zn]{H_2O} H{-}\overset{O}{\overset{\|}{C}}{-}\overset{O}{\overset{\|}{C}}{-}H$$

Reduction of benzene occurs only at elevated temperatures and high pressure, whereupon three moles of hydrogen are absorbed and cyclohexane is produced.

$$C_6H_6 + 3H_2 \xrightarrow[\text{high temperature}]{\text{high pressure}} \underset{\text{Cyclohexane}}{C_6H_{12}}$$

In the presence of an ultraviolet light benzene adds three moles of chlorine, presumably by a radical reaction (Sec. 2.13).

$$C_6H_6 + 3Cl_2 \xrightarrow{h\nu} \underset{\text{1,2,3,4,5,6-Hexachlorocyclohexane}}{C_6H_6Cl_6}$$

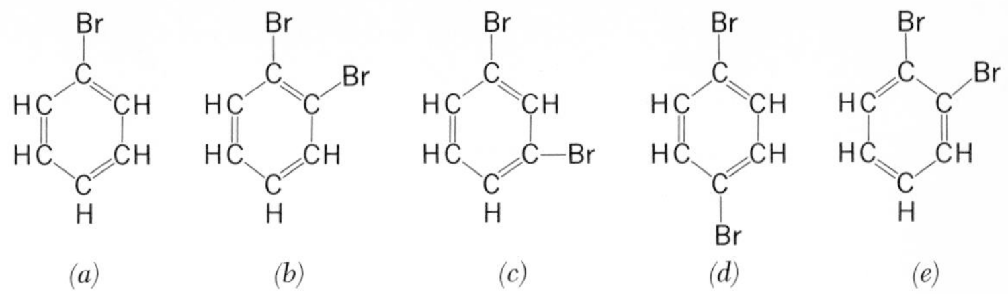

figure 6-1
Mono- and dibromo derivatives of benzene.

6.2 Benzene Structure On the basis of the above evidence one can write the following structure for benzene:[1]

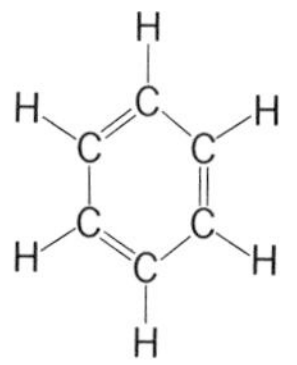

This structure explains the products from the above reactions, but it has several severe shortcomings. For example, it is known that there exists only one monobromo derivative of benzene, C_6H_5Br (Fig. 6-1*a*), and three, and only three, dibromo derivatives (Fig. 6-1*b*, *c*, and *d*).

The dibromo benzenes are known as *ortho* or 1,2-dibromobenzene (Fig. 6-1*b*), *meta* or 1,3-dibromobenzene (Fig. 6-1*c*), and *para* or 1,4-dibromobenzene (Fig. 6-1*d*). All attempts to discover a second 1,2-dibromobenzene (Fig. 6-1*e*), which would be predicted from the structures as shown, have failed. This inconsistency Kekulé explained by assuming a *dynamic equilibrium* between forms such as seen in Fig. 6-1*b* and *e*; the proposed equilibrium involved a rapid shift of double and single bonds in the ring.

[1]Certain other structures such as

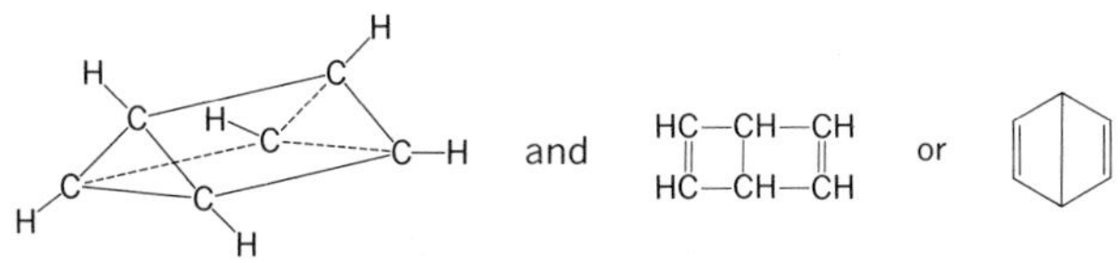

cannot be ruled out by the evidence cited. On the basis of other evidence, however, they can be discarded since they have different geometries and symmetries from that observed for benzene. Recently, representatives of both these classes of compounds have been synthesized and have been found to be about 60 kcal/mole less stable than the corresponding benzene derivative.

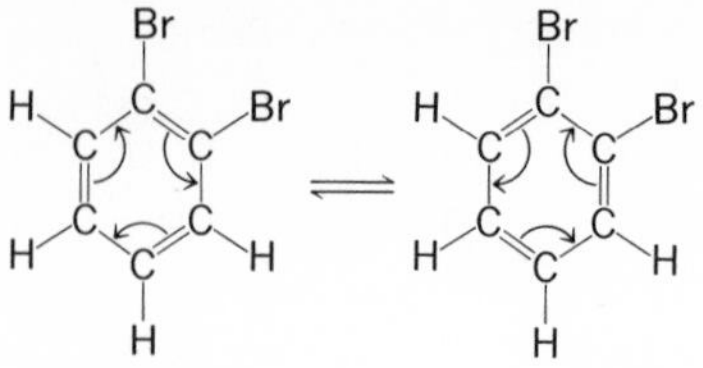

Although his explanation provides the necessary symmetry[1] for benzene, it fails to explain the chemical properties such as resistance to oxidation, reduction, and addition.

Modern Theory of Benzene Structure

In modern terms the properties of benzene are reconciled by the resonance and molecular orbital theories. In the resonance picture, the two Kekulé forms are viewed as primary contributing structures to the resonance hybrid which is benzene. NOTE: In the structures throughout this book, carbon atoms are understood to be located at each corner when not explicitly drawn in.

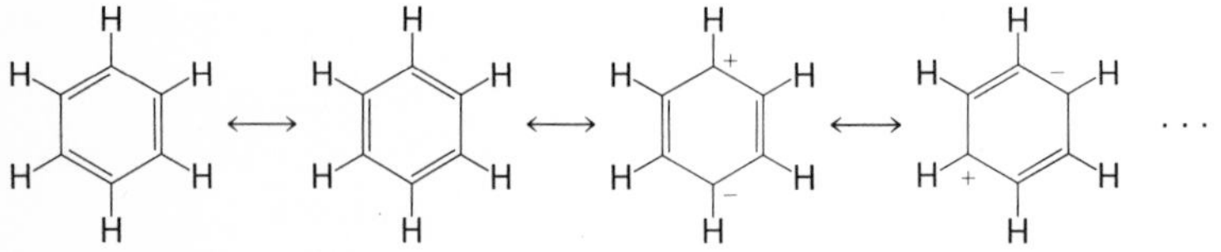

I II III IV

figure 6-2
The resonance structures for benzene.

Structures I to IV represent different electronic arrangements for nearly the same nuclear positions. Since I and II are obviously of equal energy and are of lower energy (i.e., they have one more bond and no separated charges) than III and IV, they contribute most heavily to benzene. The actual benzene molecule is found to have all C—C bond lengths the same (1.39 Å), a value intermediate between that of most double bonds (1.34 Å) and single bonds (1.54 Å). This observation emphasizes the fact that benzene is not sometimes I or sometimes II, but that it is a true hybrid, different (e.g., more stable, more symmetrical) from any or all of the contributing structures.

The molecular orbital theory gives us a picture of benzene that is somewhat easier to visualize. Since the C atoms are equivalent, it would be predicted to have

[1]Benzene has a high degree of symmetry since it has both a plane of symmetry (that is, it would look identical from either side of a plane surface in which it was lying) and a six-fold axis of symmetry (that is, if one ran an axis through the center of the molecule at right angles to the plane in which it lay and rotated the molecule about this axis, it would go through the same arrangement six times per revolution, just as a regular hexagon would). Any substance with this type of symmetry can have three, and only three, disubstitution isomers. It will be found that the use of symmetry properties is a valuable tool in the study of structure.

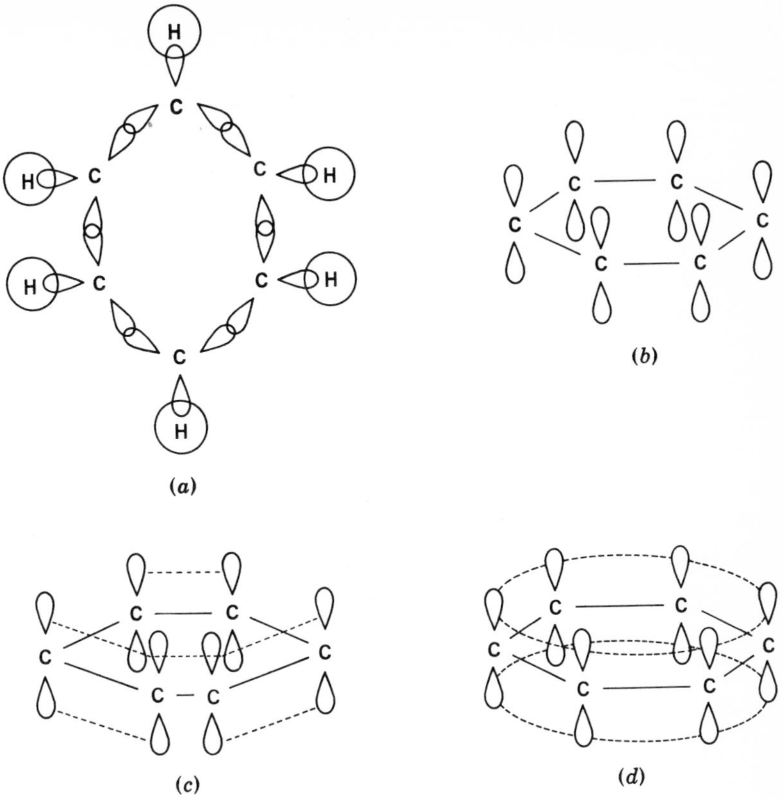
H
C
H
C
C
H
H
C
C
H
C
H
(a)
C
C
C
C
C
C
(b)
C
C
C
C
C
C
(c)
C
C
C
C
C
C
(d)

figure 6-3
Molecular-orbital models of benzene. (*a*) σ framework. (*b*) p orbitals. (*c*) One molecular orbital. (*d*) Net π-cloud distribution of all six π electrons.

a bond framework as shown in Fig. 6-3*a*. This framework consists of sp^2-hybridized carbon atoms. This leaves us with p orbitals above and below the plane of all the atoms, as illustrated in Fig. 6-3*b*. Now if these were to form double bonds as indicated in Fig. 6-3*c*, one of the Kekulé structures shown in Fig. 6-2 would result. The molecular-orbital theory indicates that if, however, electrons can be located in a larger orbital rather than being localized over only two atoms, then their energy will be considerably reduced. This *delocalization* can be achieved for two of the six π electrons in the orbital pictured in Fig. 6-3*d*, which encompasses all six carbon nuclei.[1]

[1]The other four electrons can be placed in two orbitals (which effectively encompass all six nuclei, but which have a region running through the middle of the molecule where the electron density is zero, called a "node") and so are of higher energy than the orbital pictured in Fig. 6-3*d*. One cannot put six electrons in one orbital because of the exclusion principle, but three orbitals can occur in the same regions of space so that the resultant average electron distribution is essentially that of Fig. 6-3*d*.

The symmetry of benzene is apparent in such a structure and many properties of the molecule are readily visualized in terms of it.

6.3 Stabilization of Benzene The properties of benzene may be put on a somewhat more quantitative basis by measurements such as the heats of hydrogenation. Hydrogenation is an exothermic reaction (even for benzene).

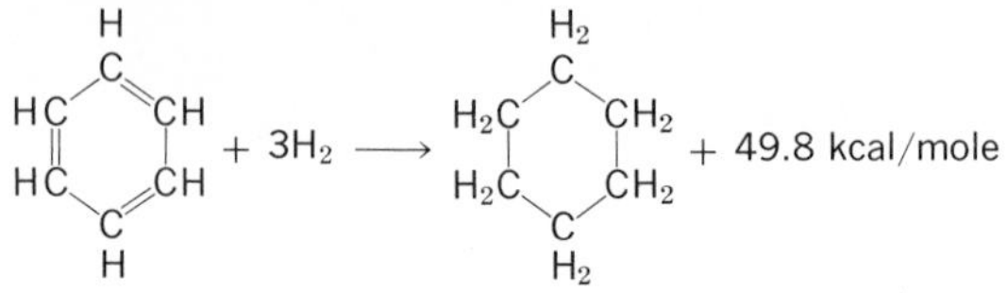

For cyclohexene a similar measurement yields

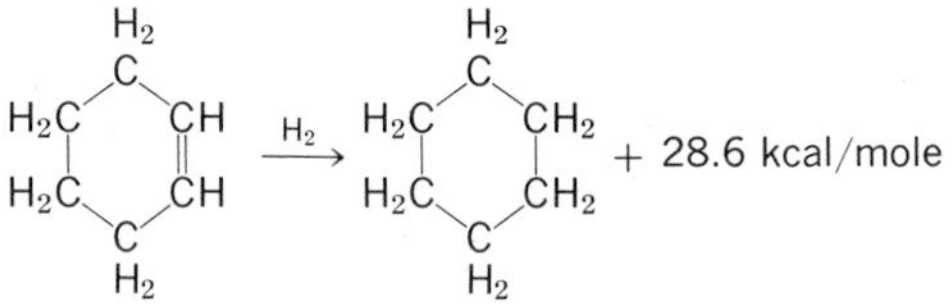

If one assumes that benzene with three "double bonds" should have three times the heat of hydrogenation of cyclohexene, one would expect $3 \times 28.6 = 85.8$ kcal/mole instead of 49.8 kcal/mole. The difference, 36.0 kcal/mole, is called the "resonance" energy of benzene, or in molecular orbital terms, the "delocalization" energy. Since benzene would have to lose a good part of this stabilization energy on undergoing an addition reaction, one can see the reason for the occurrence of substitution reactions in preference to addition reactions. Other measurements, such as heats of combustion ("abnormally" small) of benzene, confirm these observations.

Symbol for Benzene

In writing benzene structures it is awkward to draw out Fig. 6-3*d* or to write all of the resonance structures in Fig. 6-2 every time that benzene must be written. There are several symbols in common use for benzene and related compounds, as shown in Fig. 6-4.

I II III

figure 6-4
Common symbols for benzene structure.

Structures I and II are convenient and are commonly used, especially II. Structure III is clearly incorrect, but is perhaps the most widely used (with the understanding that it represents a hybrid) because it enables us to keep track of the electrons, a distinct advantage in more complex systems. Structure III shall be most commonly used in this book.

6.4 Conjugation Benzene is not the only carbon compound in which we find resonance or delocalization acting as a form of stabilization. Intuition tells us that such interaction might be expected whenever more than two *p* orbitals can line up parallel and adjacent to one another. The utility of such a concept is illustrated by consideration of butadiene. Butadiene, $H_2C{=}CH{-}CH{=}CH_2$, undergoes a reaction which is difficult to interpret.

$$H_2C{=}CH{-}CH{=}CH_2 + Br_2 \longrightarrow \underset{Br}{H_2C}{-}\underset{Br}{CH}{-}CH{=}CH_2 + BrCH_2{-}CH{=}CH{-}CH_2Br$$

1,2-Addition ("normal") 1,4-Addition ("abnormal")

In terms of our resonance or molecular orbital pictures, butadiene can be pictured as follows:

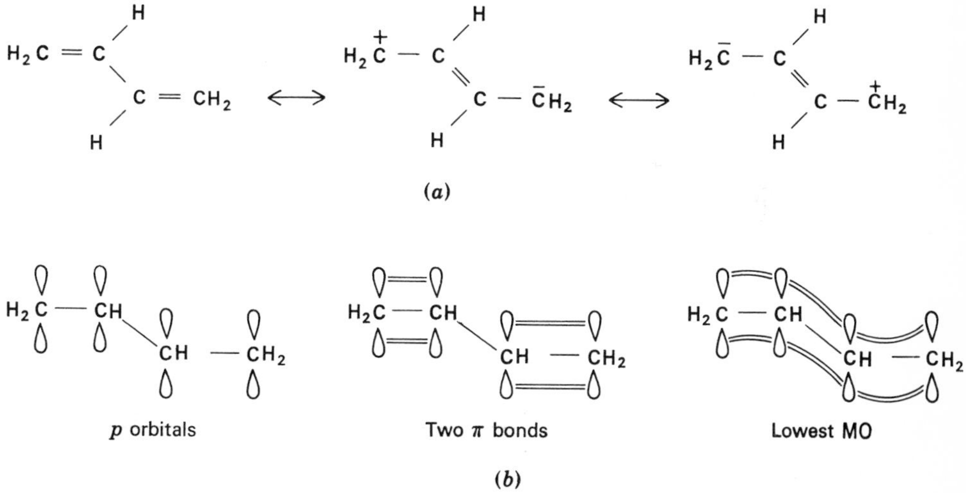

(a)
p orbitals
Two π bonds
Lowest MO
(b)

figure 6-5
(*a*) Resonance picture of butadiene. (*b*) Molecular-orbital picture of butadiene.

The resonance picture indicates that there is a tendency to react at the ends of the molecule, as observed in 1,4-addition, since the hybrid structure for butadiene will have a certain amount of double bond "character" in the central bond. Since the two structures with positive and negative charge separation are, according to

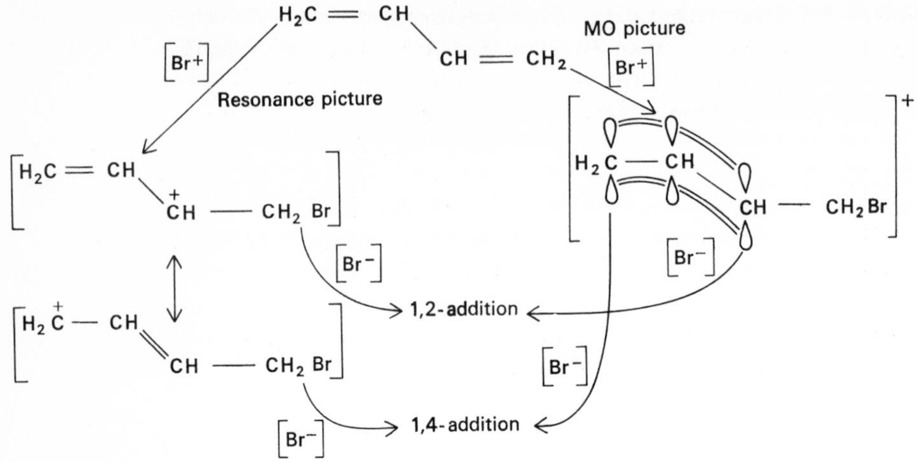
MO picture
Resonance picture
1,2-addition
1,4-addition

figure 6-6
Bromine addition to butadiene.

our "rules of resonance," of higher energy, this stabilization would be expected to be small. All of this is borne out by experience. The measured resonance energy (Sec. 6.3) of butadiene is no more than 3 kcal/mole. The central bond is slightly shorter than ordinary single bonds and there is a small barrier to rotation about this bond.

The molecular orbital picture gives us a roughly similar description. Here, all four p orbitals can combine to give a molecular orbital which encompasses all four carbon atoms, and a second molecular orbital which looks like two π bonds but which contains only two electrons and encompasses all four carbon atoms. The net electron distribution is not symmetrical but indicates more double bond "character" between atoms 1 and 2, and 3 and 4, than between 1 and 3.

When butadiene adds a positive bromine, for example, it can give a resonance-stabilized (or delocalized) ion only if the Br^+ attaches initially at one end. The second bromine, Br^-, can attach to the hybrid (delocalized) ion at either end to give either 1,2- or 1,4-addition as shown in Fig. 6-6.

The interaction of adjacent double bonds and of p orbitals adjacent to double bonds, as pictured in Fig. 6-6, is termed *conjugation*. The curved arrow (Sec. 1.16) is often used to summarize this type of behavior as a more compact notation. For example, additions can be written:

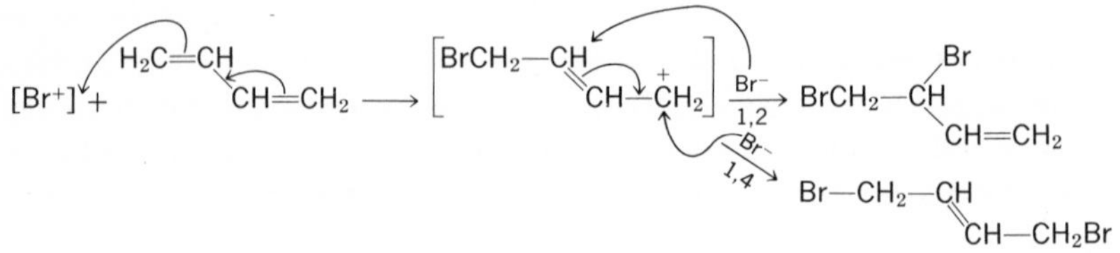

6.5 Definition of Aromatic Benzene is the patriarch of a family of compounds called "aromatic" compounds. The name aromatic was originally applied to benzene and to compounds containing one or more benzene rings in their structure, and referred to the pleasant odors of many naturally occurring substances containing these structures. Today the name is generally applied to cyclic systems possessing some degree of delocalization or resonance stabilization, whether or not benzene rings are involved. Although we will utilize the wider definition, most of our examples will be of the so-called "benzenoid aromatic" types, as opposed to nonbenzenoid aromatic compounds. See Sec. 6.25.

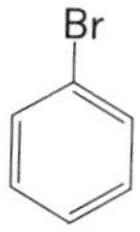
Br

Bromobenzene

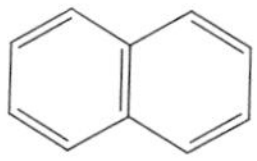

Naphthalene

Benzenoid aromatic compounds

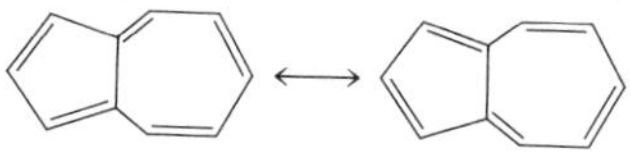

Azulene

A nonbenzenoid aromatic

Note that in the benzenoid compounds above, only one of the possible resonance structures is shown (Sec. 6.2).

6.6 The Nature of Theories; The Approaches Taken in This Textbook Science is based on experimental findings. This means that facts as disclosed by sound experiment remain valid to the extent of their establishment, though they may be refined as time passes and technology is improved. Thus Thiele discovered (through careful analytical work and attention to detail) that in the bromination of butadiene two dibromobutenes, the 1,2-dibromo-3-butene and the 1,4-dibromo-2-butene, are produced. This is a fact that cannot be disputed since it can be confirmed by anyone who is technically capable of repeating the work. However, science progresses as well through the hypotheses and theories that are devised to explain (that is, to rationalize) the experimental findings—to relate them to as many other facts as possible. These theories require that the experimental findings be transformed in such a way as to make discourse possible; then through discourse (spoken or written) the theory can be set up and communicated. It is in the act of this symbolic transformation, and in the resulting symbolism, that scope for imagination, reinterpretation, improvement (and dispute) is present. For example, it is the same experimental fact which has been transformed into the symbols of Kekulé, resonance, and molecular orbital theories.

It must be asked, of course, whether one is better than the others. But the question is misleading unless qualified. Kekulé's theory was best when the other two did not exist. At that time no one had any conception of atomic structure, and the details wrought into the modern theories were not even conceived of, as far as we know. (It is a sign of ignorance, and intolerable arrogance, to derogate old theories which have served their usefulness and which were propounded by pioneers who, in the face of insufficient data, yet had the insights to develop useful theories. No doubt our present theories will seem naive in fifty years.)

Kekulé's theory, then, became superseded by more subtle and more quantitatively approachable theories: the resonance theory and the molecular orbital theory. "Which of these is better?" may be the next question. Again, the qualification is required: better for what purpose? Actually, these two are descriptions of the same experimental facts. The question then is, is one description better than the other for some given purpose. No answer agreeable to every chemist can be given at this time. Instead, both approaches shall be applied to the explanations of mechanisms in this text, since the student will certainly encounter both in the literature. It must be added that one does not, in this elementary text, attempt to do other than employ the symbols used in these two theories for their usefulness in visualizing mechanisms. One cannot attempt to present theory to the depth that would make calculations—say of bond lengths of electron densities—possible.

6.7 Halogenation, Position Isomerism, and Nomenclature The chemical reactions of benzene and its derivatives need some further discussion. If benzene is converted to monobromobenzene, it is found that no matter how this substance is approached, the same product is always formed. This was pointed out above and means that the hydrogens on the ring are all equivalent. It also means that in benzene the Br may be written at any of the corners around the ring which it may be convenient to use.

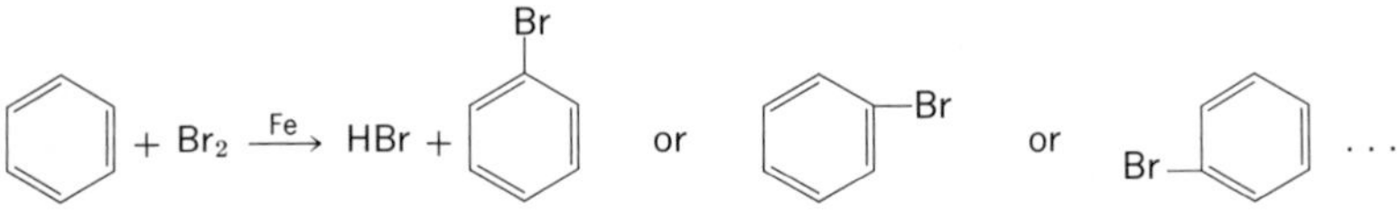

Br
+ Br2
Fe
HBr +
or
Br
or
Br
. . .

In monobromobenzene, however, the hydrogens are not all equivalent. If this is brominated to the dibromo compound, three products would be predicted from the formula

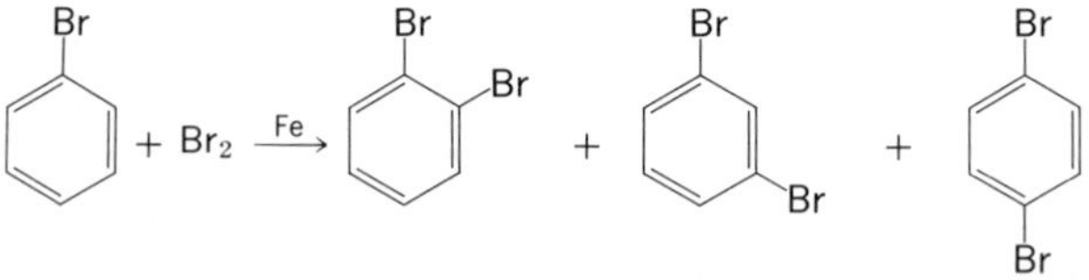

Br
+ Br2
Fe
Br
Br
+
Br
Br
+
Br
Br

1,2-Dibromobenzene (*ortho*dibromobenzene) — 1,3-Dibromobenzene (*meta*dibromobenzene) — 1,4-Dibromobenzene (*para*dibromobenzene)

These products differ in the relative positions of the bromine atoms, and the names for these different positions should be memorized. The relative positions alone are important; *o*-dibromobenzene may be written in many ways:

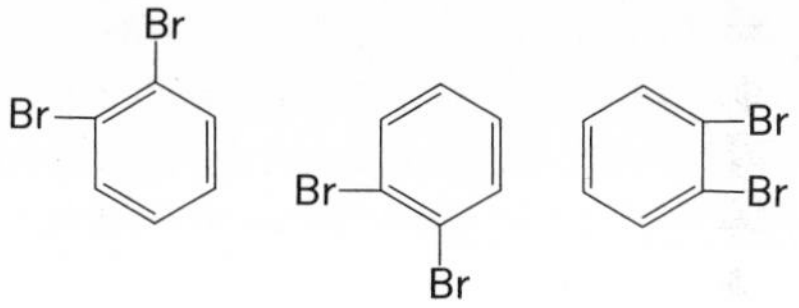

Br
Br
Br
Br
Br
Br

6.8 Körner's Proof of Structure When bromobenzene is treated with 1 mole of bromine, a mixture of products results; this mixture is found on careful analysis by distillation and crystallization to contain all three possible $C_6H_4Br_2$ isomers and no other $C_6H_4Br_2$ compounds. These isomers differ markedly in physical properties, as shown by their melting points. The problem of assigning the correct structural formula to each of the isomers formed in reactions like this is an important one; it was solved by Körner in a very ingenious and simple manner.

Körner assumed only that the benzene ring is symmetrical and that *all possible* isomers will be formed in any substitution reaction. These isomers need not and do not occur in equal amounts; for example, when another bromine atom is substituted into bromobenzene, the ortho, meta, and para isomers are formed. Under certain experimental conditions the ratio of isomers formed is 13.1:1.8:85.1. Since all possible isomers are always formed, as is found experimentally, then when a third bromine is substituted in any one of the dibromobenzenes, it may take up any one of the positions open. This leads to the formation of a different number of products from each of the three isomers.

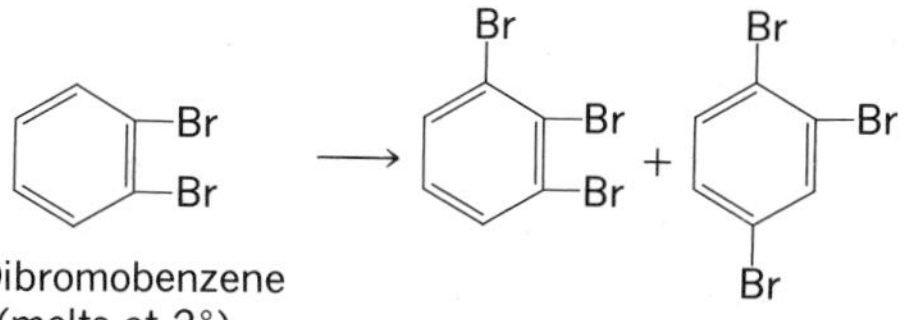

Br
Br
Br
Br
Br
Br
Br
Br
+

o-Dibromobenzene
(melts at 2°)

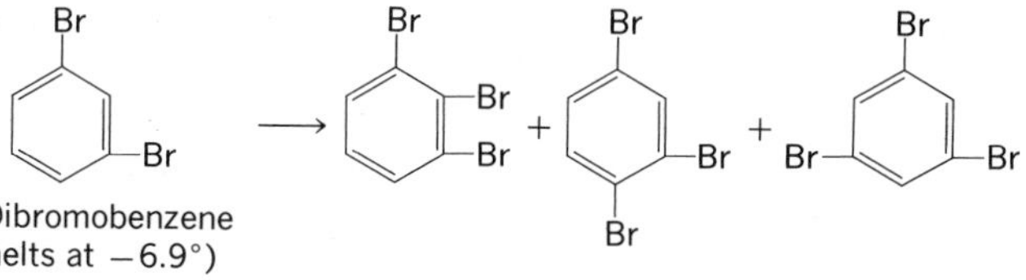

Br
Br
Br
Br
Br
Br
Br
Br
Br
Br
Br
+
+

m-Dibromobenzene
(melts at −6.9°)

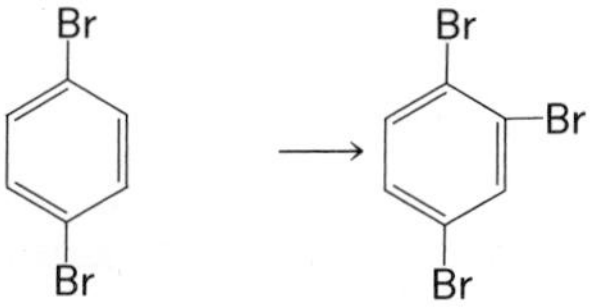

Br
Br
Br
Br
Br

p-Dibromobenzene
(melts at 89°)

It will be seen that the ortho isomer can give rise to two and only two tribromobenzenes, the meta to three, and the para to one. An experiment with the isomer which melts at 2° demonstrated the formation of two and only two tribromobenzenes; therefore, this was *o*-dibromobenzene. In a similar manner the others have been assigned structures.

It is evident also that by comparing the physical properties of the tribromides, structures can be written for them. The *p*-dibromo compound gives only the 1,2,4-tribromo compound:

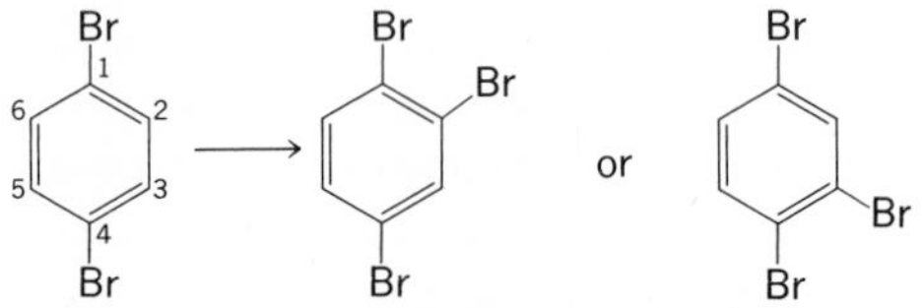

These two are identical

The ortho and meta isomers also give this compound, now recognizable by its physical properties as determined from the product of the bromination of the para compound. Of the two products from the meta compound, one will be like that from the ortho, and this one must be the 1,2,3-tribromo derivative; then the other will be the 1,3,5 derivative. Körner's method gives a powerful tool for determining the positions of substituents on rings. *It must be emphasized that the method depends upon separating and characterizing all the isomeric products formed.* If some tetrasubstitution occurred in the above cases, the tetrabromo compound would have to be taken out and set aside; it can be recognized by a bromine analysis or by a molecular-weight determination. Some of the trisubstitution products are produced only in very small amounts. These have to be detected. A Körner proof is not an easy experiment. It will be realized that the principle of Körner's proof is applicable to any disubstituted benzene; only in those cases, however, in which *all* groups are identical (or the original two groups are identical and the third is different) will the numbers of products in each case be the same as shown in the example above.

It is seldom if ever necessary in modern research to resort to Körner proofs. Since so many reference compounds of established structures are available, it is frequently possible to assign the structure in benzene derivatives by direct comparison of infrared and nmr alone.

6.9 Orienting Effect of Substituents It was stated that in disubstitution reactions, the three possible isomers are not produced in equal amounts. It has been found that two types of behavior are usually exhibited; either the meta is produced in good yield with much less ortho and para, or the ortho and para compounds are produced in good yield with very little meta. Furthermore, which behavior is exhibited depends on the nature of the substituent already present in the ring and not on the nature of the entering substituent. It is as though the substituent on the ring

TABLE 6.1
TABLE OF ORTHO,PARA AND META DIRECTORS

Ortho,para directors	*Meta directors*[a]
—OH	$-\overset{+}{N}(=O)-O^-$
$-NH_2$	$-\overset{++}{S}(-O^-)(-O^-)-OH$
—Cl	$-C(=O)-OH$
—Br	$-C(=O)-H$
—I	$-C(=O)-R$
$-CH_3$	$-C{\equiv}N$
$-C_2H_5$, . . .	
$-CH_2Br$	
$-OCH_3$	
$-CH_2Z$[b]	

[a] Notice that all the meta directors in Table 6-1 have at least one polar double, triple, or coordinate bond on the atom directly attached to the ring.

[b] Where Z may even be a meta director. The CH_2 group effectively insulates the effect of Z.

had influenced the ring so as to make one or another position of substitution more available. This phenomenon is called the "orienting effect" of a group; the substituent already present is said to *direct* the entry of the next substituent. As a result of experiment, substituent groups may be divided into the two classes shown in Table 6.1.

6.10 Mechanism of Orientation The question now arises, "Why should one group orient predominantly ortho and para, and another predominantly meta?" It seems that several factors are involved in this phenomenon. Considering benzene as a reference substance, it is found that when substitution occurs in a benzene derivative carrying *o,p* directors, the reaction is almost always more rapid than the substitution of benzene itself (the halogen-substituted benzenes constitute an exception), whereas when the substituent already present is *m* director, the reaction is slower than with benzene itself. The former substituents are said to *activate* the ring and the latter to *deactivate* it.

Since hydrogen is the group most frequently replaced by substitution in aromatic structures, and since it is more readily removed as H^+ rather than H^-, most of the reagents which are able to substitute on the ring are electrophilic reagents. So any effect which acts to increase the electron density at a given position on the ring should aid substitution there. That is to say, electron-repelling groups should accelerate substitution, and electron-attracting groups should impede it. On this analysis we can say that the *o,p* directors are electron-repelling groups, and that the *m* directors are electron-attracting groups. That this is so can be seen readily in certain cases. We have seen, for example, that alkyl groups are electron-repelling groups. Accordingly, we are not surprised to find that methylbenzene nitrates more readily than benzene. On the other hand, the *m* directors are electron-attracting groups and decrease the availability of the electrons on the ring; this explains the decreased reactivity of nitrobenzene to substitution. The effect of the halogens is somewhat complicated owing to the fact that they deactivate the ring but also cause *o,p* orientation; thus it is evident that while activation and *o,p*-orienting effect are often found in the same group, they are not necessary to each other.

There remains now the matter of the positions taken by the entering groups. In order to discuss this effect it is necessary to digress into the mechanism of organic substitution in general.

6.11 Mechanism of Aromatic Substitution In many respects aromatic substitution resembles the addition of electrophilic reagents to alkenes (Secs. 3.8 to 3.11). Because of the resonance stabilization of aromatic compounds (Sec. 6.2), however, the reaction is much less facile. Bromine, for example, does not react with benzene at a significant rate in the absence of a Lewis-acid catalyst such as $FeBr_3$. Such a catalyst *activates* the bromine by complexing with one end of the bromine molecule to make it a more powerful electrophilic reagent.

$$:\ddot{\underset{\cdot\cdot}{Br}}-\ddot{\underset{\cdot\cdot}{Br}}: + FeBr_3 \rightleftharpoons [:\ddot{\underset{\cdot\cdot}{Br}}]^+ + [:\ddot{\underset{\cdot\cdot}{Br}}:FeBr_3]^-$$

In terms of transition-state theory (Sec. 3.8) $FeBr_3$ lowers the energy of the transition state for aromatic substitution by assisting the rupture of the Br—Br bond. To a first approximation, the substitution of benzene by bromine may be pictured as follows:

$$Br-Br + FeBr_3 \overset{\text{rapid}}{\rightleftharpoons} [:\ddot{\underset{\cdot\cdot}{Br}}]^+ + [FeBr_4]^-$$

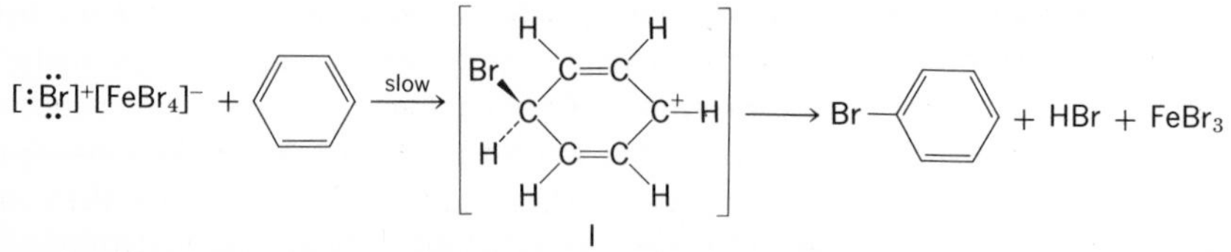
[:Br]+[FeBr4]− +
slow
Br
H
C=C
C
C+—H
C=C
Br
+ HBr + FeBr3

I

Species I, which for the moment we shall call the transition state, is a resonance-stabilized hybrid, the main contributing structures being:

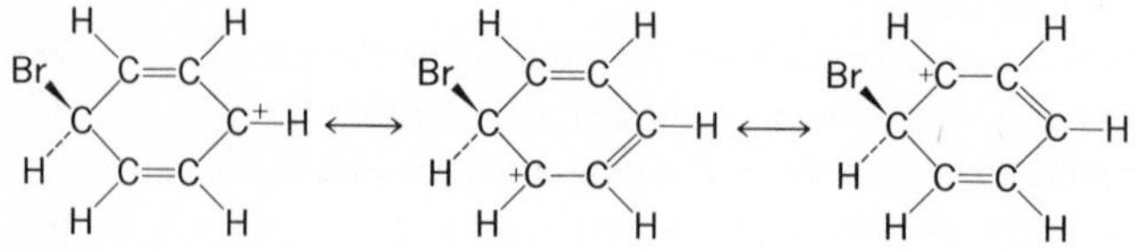
Br
H
C=C
C
C+—H
C—H
+C—C
+C—C

I

The energy profile (Sec. 3.8) for the bromination of benzene may be represented as in Fig. 6-7*a*.

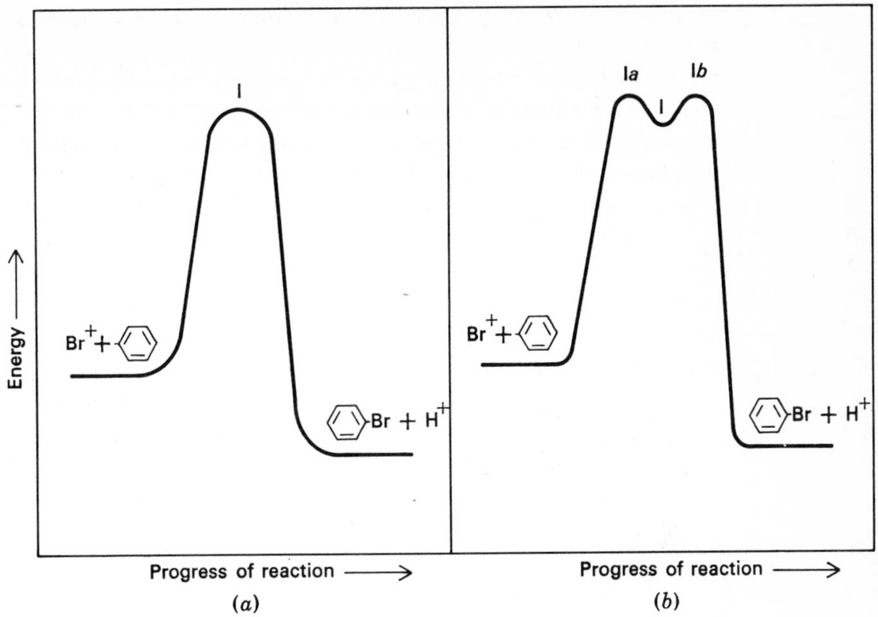

Energy
I
Br⁺ +
Br + H⁺
Progress of reaction
(a)
Ia
Ib
I
Br⁺ +
Br + H⁺
Progress of reaction
(b)

figure 6-7
Alternate energy profiles for the bromination of benzene.

Because I is capable of considerable resonance stabilization, it may not accurately represent the configuration of the transition state in many aromatic substitutions; it can actually represent a small minimum in the potential energy, as illustrated in Fig. 6-7*b*. In this case there will be *two* transition states, I*a* and I*b*, which may be depicted as follows:

Br
H
Br
H

I*a* I*b*

In a situation like this, I is called an "unstable intermediate" in the reaction. In some instances, the dip in the energy profile may be great enough that species similar to I will build up to concentrations where they may be observed spectroscopically (usually ultraviolet absorption), and in extreme cases analogous intermediates may even be isolable. In most cases, however, I will be rapidly converted via transition states I*a* and I*b* to reactants and products, so that the overall course of the reaction is adequately described by Fig. 6-7*a*.[1]

[1]The essential difference between an unstable intermediate and a transition state is that the transition state can only go on to products or fall back to reactants, while an intermediate can undergo a variety of reactions, just like other reactants. This distinction becomes important with the carbonium ions described in Sec. 8.23.

Our analogy with the addition of bromine to alkenes diverges after we reach the configuration described by I. In the alkene case (Sec. 3.8) the next step was the addition of Br^- to give a net addition of Br_2; here we lose H^+ to regenerate the unsaturated system. The reason for this divergence stems from the large amount of resonance energy gained on re-forming the aromatic system. This tends to weaken the H—C and Br—C bonds in I, by the contributions of the following types of structures to the hybrid:

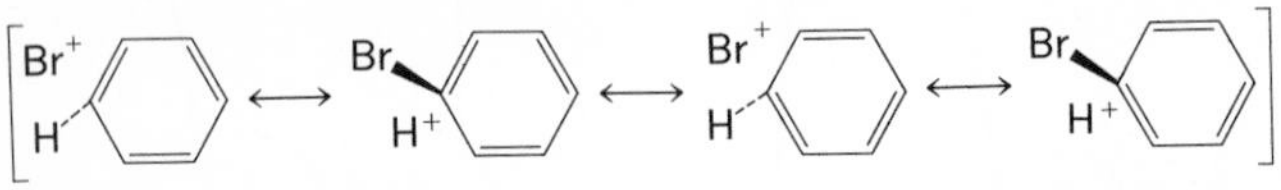

These are relatively minor contributors to I because the geometry (of the H and the Br) is very poor,[1] but their net effect is to weaken the above bonds. Moreover, an approaching Br^- ion does not find a positively charged carbon atom to attach itself to as in the case of alkene addition, but rather a diffuse positive charge spread over the two ortho and one para positions. The net effect of all this is to raise the energy of the transition state for addition relative to that for substitution.

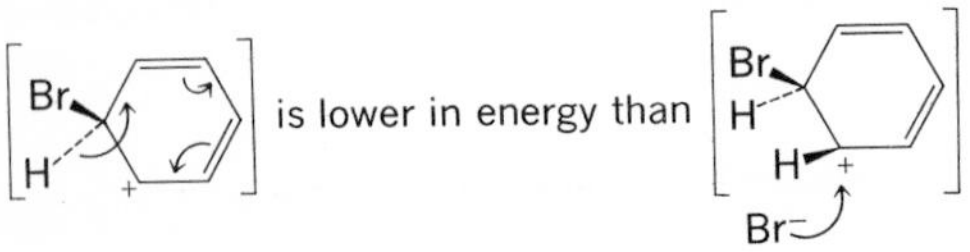

6.12 Directing Effect of Substituents In aromatic substitution reactions the highest energy transition states (rate-determining step) closely resembles in structure the intermediate discussed above. For convenience, then, we can study possible structures for the intermediate and make predictions as to relative reaction rates from these. For benzene substitution we can write resonance forms I*a*, I*b*, and I*c*.

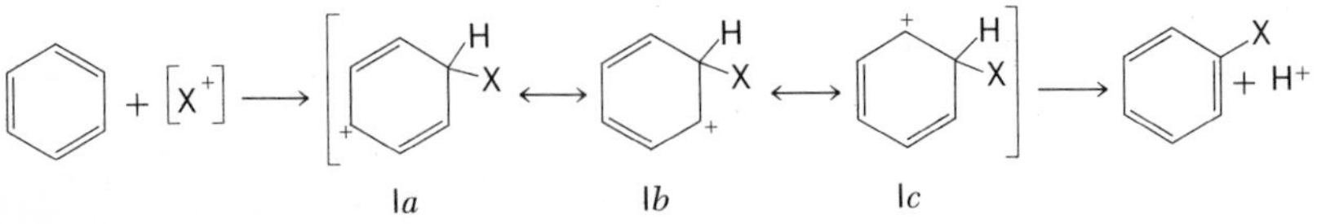

The contributions of these structures to the resonance hybrid lowers the energy of the transition state, thereby accelerating the reaction.

Most *o,p* directing groups (Sec. 6.9) are primarily electron donors and/or have unshared election pairs on the atom directly attached to the ring which can be

[1]In order to form a bond to the sp^2-hybridized ring carbon atom, the attached group must be coplanar with the ring. However, in Intermediate I, either the H or Br must be above the plane of the ring and the other below the plane, since the ring carbon is sp^3 hybridized (tetrahedral). Obviously the geometries cannot be good for both situations.

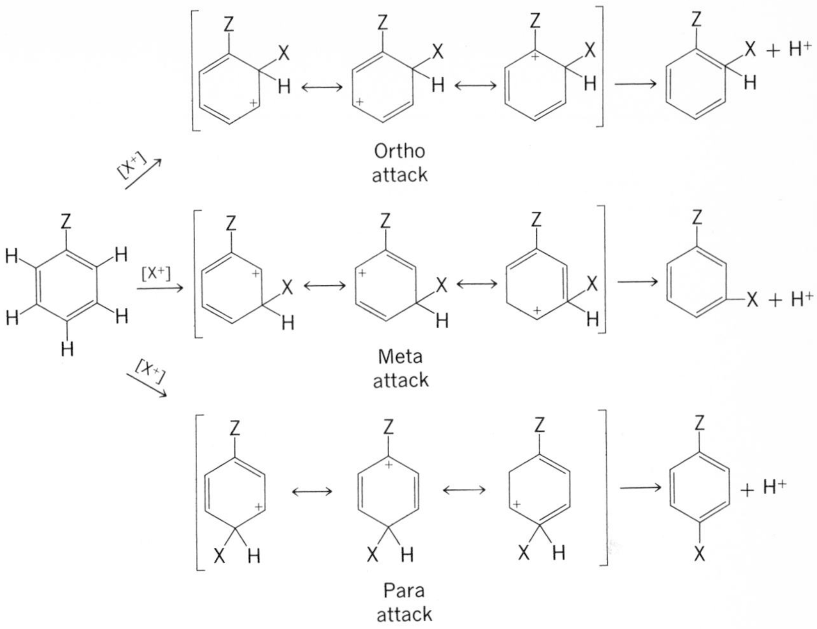
Z
X
H
Ortho
attack
[X+]
H
Meta
attack
Para
attack
X + H+
+ H+

figure 6-8
Forms of the transition state for ortho, meta, and para substitution on a monosubstituted benzene.

donated on demand. If such a group, Z, is attached to benzene, it will lead us to postulate transition states as shown in Fig. 6-8. In *o,p* substitution, one of the contributing resonance structures places the positive charge on the carbon atom bearing the electron-donating group. This electron-donating effect of Z leads to stabilization of these transition states—relieving the electron deficiency on the ring. The energy content is lowered by this delocalization of charge, and reaction by this path accelerated. Attack by electrophilic reagents in the meta position cannot produce resonance structures in which this stabilization is significant; stabilization falls off markedly as the separation of the + charge and the electron-donating group increases. Hence, in the competitive reactions, the ortho and para substitutions win out over the meta.[1]

The effect is even more pronounced when the group Z has an unshared pair of electrons with which it can form a double bond to the ring. Now *o,p* substitution can be stabilized by structures shown below, but no such structure can be drawn for *m* substitution without involving unusual valence for carbon or other violation of the rules (Sec. 1.7).

[1]The difference in stabilization does not have to be too great. A difference of about 1.5 kcal/mole will favor one mode by a factor of 10; a difference of 3 kcal/mole will favor it by 100.

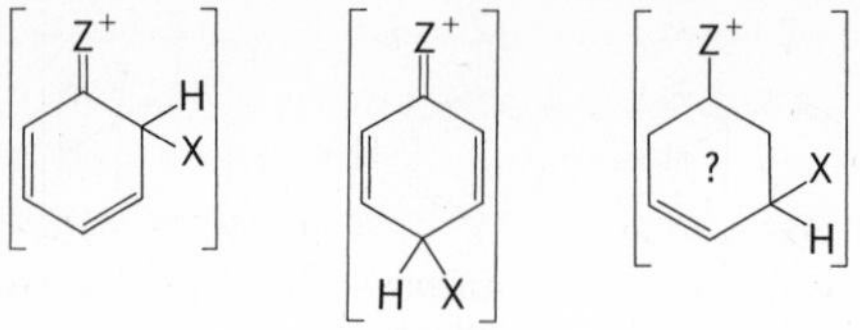

o Attack — *p* Attack — *m* Attack (no possible structure)

The net effect of *o*- and *p*-directing substituents is to activate the ring toward electrophilic substitution at the *o* and *p* positions. The *m* position is not deactivated, in fact, it is usually somewhat activated relative to benzene itself. However, in competition with the strongly activated *o* and *p* positions, the *m* position loses out.

When electron-withdrawing substituents are attached to the ring, the entire ring is deactivated toward electrophilic substitution. However, by the same type of arguments as used above, the *m* position is deactivated less effectively than the *o* and *p* positions, and consequently, these groups are *m*-directing groups. When the group on the ring attracts electrons away from the ring, e.g., $N(CH_3)_3$, only meta substitution fails to put a plus charge on the carbon attached to this group. The juxtaposition of two electron-deficient or positively charged atoms raises the energy content of forms I and III in Fig. 6-9.

In the case of groups containing double bonds attached to electronegative atoms, an important contributing resonance structure puts a + charge on the atom next to the aromatic ring, for example,

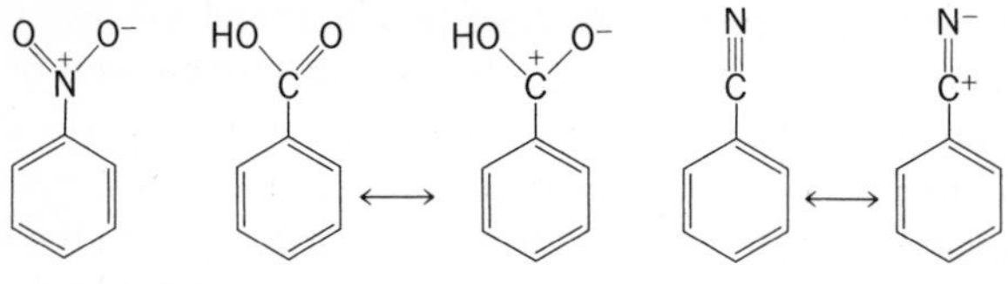

Now, *o,p* substitution would tend to place a positive charge on the carbon atom adjacent to this plus charge, as in Fig. 6-9, and would be strongly disfavored, just as with the electron-attracting group, A.

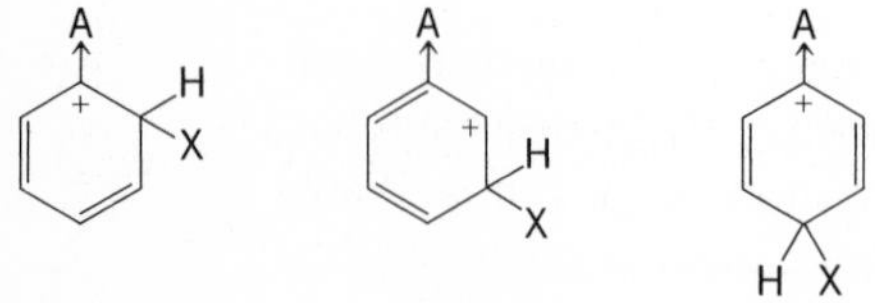

Ortho (higher energy) I — Meta (lower energy) II — Para (higher energy) III

figure 6-9
Resonance structures of a monosubstituted benzene, showing energy difference for meta, ortho, or para substitution.

In certain cases, effects (inductive and resonance) oppose each other. The classic example is that of the halogen-substituted benzene derivatives. By virtue of their electronegativity, halogens are electron-withdrawing groups.

The ortho- and para-substitution transition states, however, offer the possibility of a resonance effect that partially compensates for the inductive electron withdrawal of the halogen atom. The pertinent resonance structures are shown below; a little thought will reveal that the meta-substituted intermediate cannot have a resonance structure in which the + charge is spread to the halogen atom.

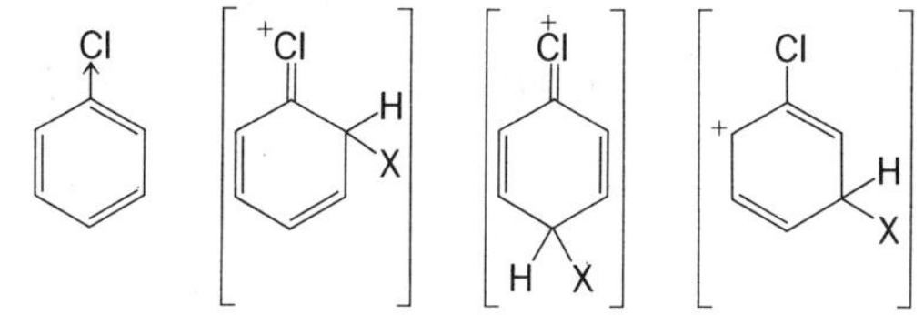

Inductive electron withdrawal — Ortho — Para — Meta (no + on Cl is possible)

Some resonance forms of transition states

The net result is that substitution in all positions is slowed, but that the deactivation is greatest in the meta positions and ortho-para products predominate.

When two substituents are already present in the ring, the difficulty of predicting the site of entry of a third may become great. Such matters are properly left for advanced courses.

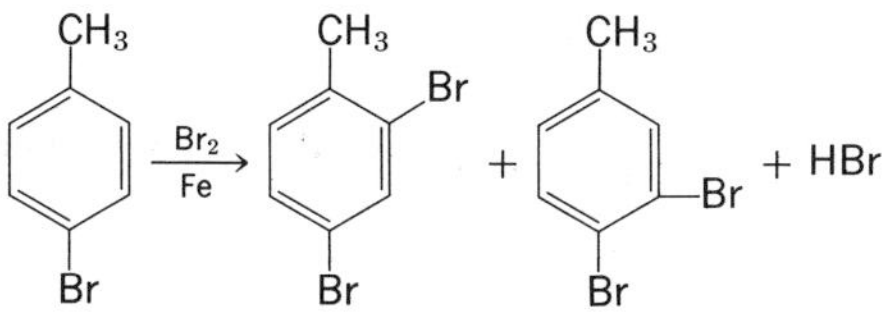

The problem of synthesis in the aromatic series is intimately connected with this matter of directive influence; since the nucleus itself is rarely synthesized, the organic chemist concerns himself chiefly with substitutions in the aromatic hydrocarbons.

6.13 Ratios of Substitution Products Ordinarily, the minor products are neglected in writing the equations of these substitution reactions; however, they must not be forgotten. Körner's proof depends on finding them in the reaction mixture. The problem of identifying and measuring the products of substitution reactions has been much simplified by the development of such modern tools as infrared spectroscopy (Sec. 1.21) and chromatography (Sec. 22.8), and quantitative studies of the composition of mixtures of ortho, meta, and para products are now available and their interpretation is under study.

Some examples of reactions and the use of orienting influence in synthesis are given below.

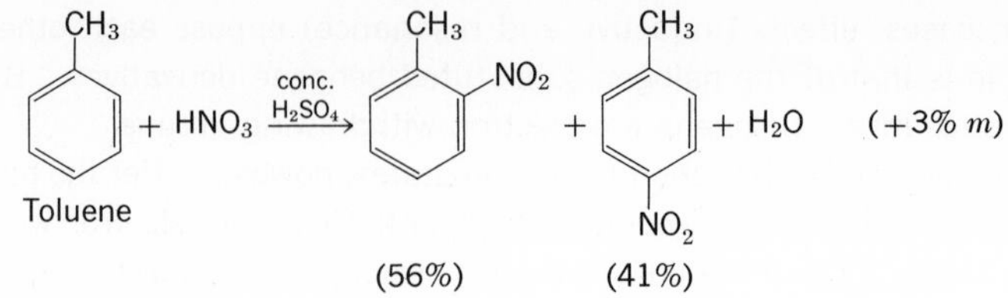
CH3
conc.
H2SO4
NO2
+ HNO3
+ H2O
(+3% m)
Toluene
(56%)
(41%)

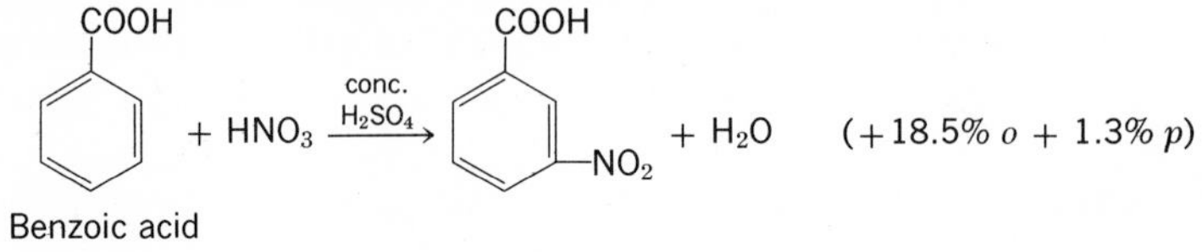
COOH
conc.
H2SO4
+ HNO3
NO2
+ H2O
(+18.5% o + 1.3% p)
Benzoic acid

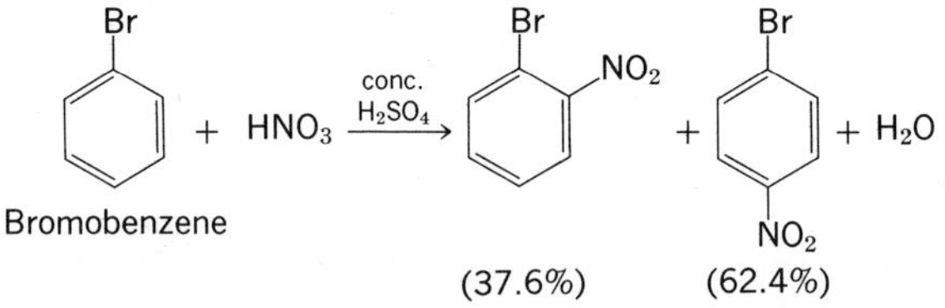
Br
conc.
H2SO4
+ HNO3
NO2
+ H2O
Bromobenzene
(37.6%)
(62.4%)

In the nitration of bromobenzene *very* little meta is produced. If *m*-bromonitrobenzene is desired in good yield, the reaction has to be run as follows:

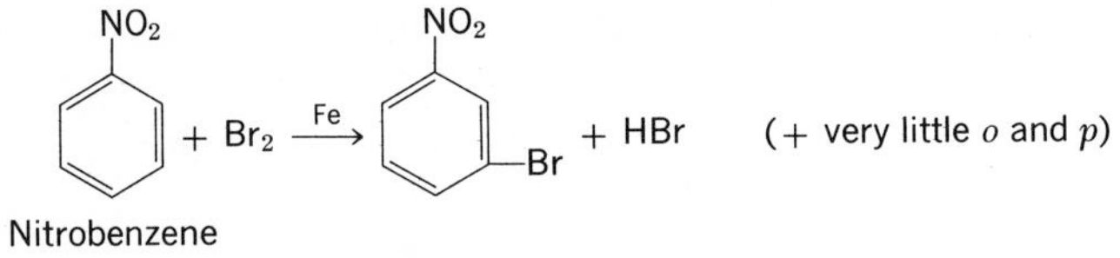
NO2
+ Br2
Fe
Br
+ HBr
(+ very little o and p)
Nitrobenzene

6.14 Nitration A mixture of nitric and sulfuric acids is commonly used for the introduction of a nitro group into an aromatic system. Nitric acid alone is relatively ineffective (unless the ring is highly activated, as by a strongly ortho,para directing group), and sulfuric acid alone leads to sulfonation (Sec. 6.15).

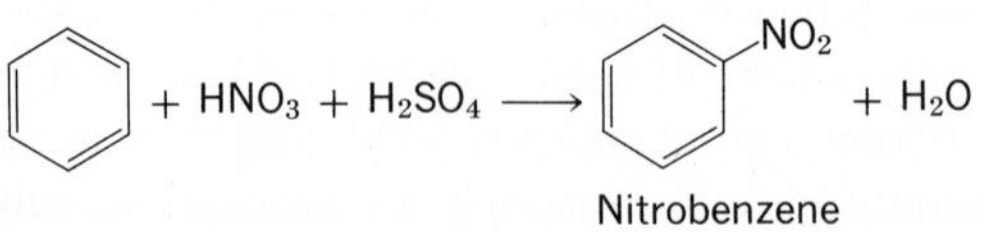
+ HNO3 + H2SO4
NO2
+ H2O
Nitrobenzene

The reaction is another example of an electrophilic attack on the benzene ring and leads to useful and industrially important substances. The reactive intermediate is a nitronium ion ($^+NO_2$) produced from nitric acid (compare carbonium ions from alcohols), and the attack by this ion on the ring follows the pattern already described in general terms (Sec. 6.11).

$$HONO_2 \overset{H^+}{\rightleftharpoons} H\overset{+}{O}NO_2H \rightleftharpoons [^+NO_2]$$

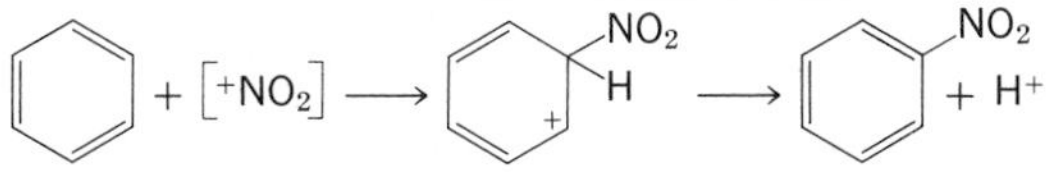

On account of the pronounced deactivating effect (Sec. 6.11) of a nitro group on the ring, further nitration proceeds more slowly, and therefore, the reaction is readily controlled and good yields of mononitro products can usually be obtained.

6.15 Sulfonation The action of sulfuric acid alone, or of fuming sulfuric acid (which contains SO_3 and is named *oleum* because of its oily appearance), leads to the introduction into the benzene ring of a sulfonic acid group in a reversible reaction.

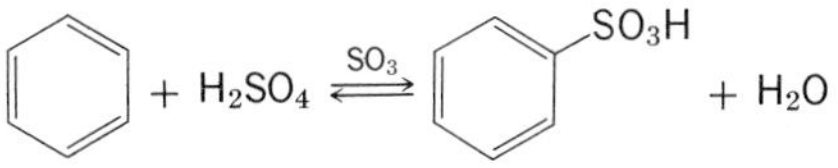

Benzene sulfonic acid

This reaction can be reversed by using an excess of water at a high temperature. The equilibrium can be shifted toward the sulfonic acid by excess (concentrated) acid. As in the case of nitration, the reaction proceeds via an electron-deficient species, most probably the electrophilic SO_3 molecule.

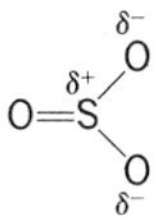

6.16 Friedel-Crafts Alkylation A reaction discovered by Friedel and Crafts, which bears their names, has been found of great value in synthetic work in the benzene series. The reaction between methyl chloride and benzene may be used as an example:

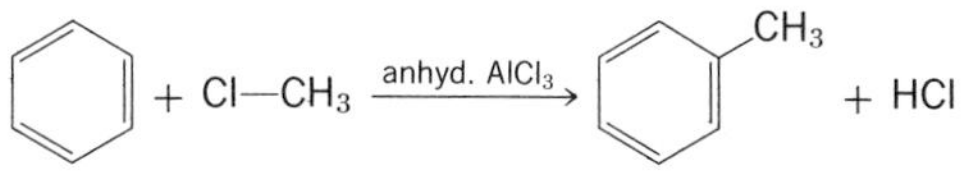

There are certain limitations to the reaction: the reagents must be dry and the ring must not be substituted with a nitro group or other strongly ring-deactivating groups; otherwise it will become inert in the Friedel-Crafts reaction. Groups that react with strong acids (aluminum chloride behaves as a strong acid; see Sec. 6.17) cannot be present in any of the reagents in this reaction. Examples of such groups are —OH, $—NH_2$, —C=C—. The aliphatic reagent must carry the halogen, and chloro derivatives are much more satisfactory (better yields, cheaper) than bromo or iodo compounds. Halogen on the ring is inert. One cannot carry out a reaction, say, with chlorobenzene and methane:

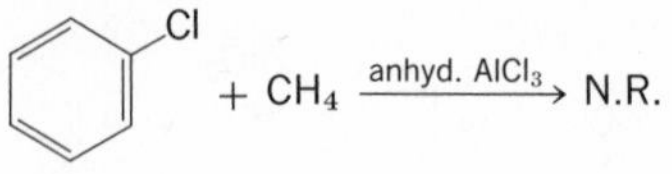
Cl
+ CH4
anhyd. AlCl3
N.R.

The Friedel-Crafts reaction can be carried out with certain other compounds, but these aspects of the reaction will be left for later attention because they involve arrangements of atoms which are more profitably discussed later.

6.17 Mechanism of Friedel-Crafts Reaction The Friedel-Crafts reaction seems to proceed by the following mechanism: Aluminum chloride is a very electrophilic reagent (Lewis acid) owing to the unfilled valence shell of the aluminum.[1]

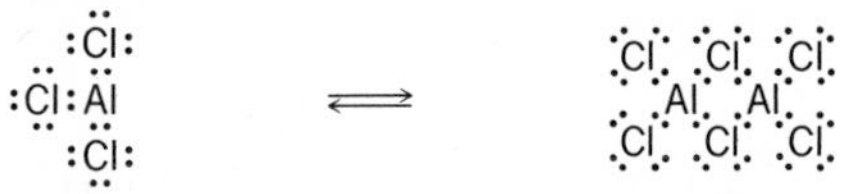

$AlCl_3$
(Al has an incomplete valence shell)

Al_2Cl_6
(more stable form)
(Cl bridges allow some stabilization)

Aluminum chloride reacts with halides, and in the case of alkyl halides weakens the carbon—halogen bond. In the case of secondary and tertiary halides it can actually completely remove the halogen, and even primary halides behave almost as if they were carbonium ions in the presence of $AlCl_3$.

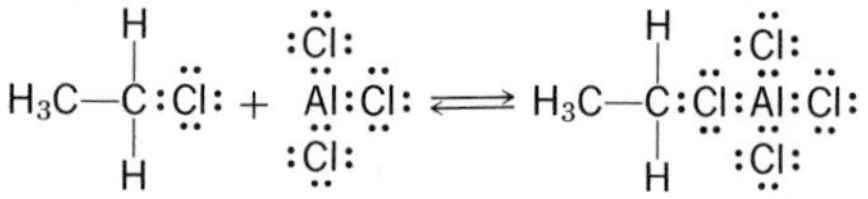

This species, or the actual "free" carbonium ion, is an electrophilic agent and will react with aromatic compounds by substitution, attacking the ring with the electron-deficient carbon of the halide.

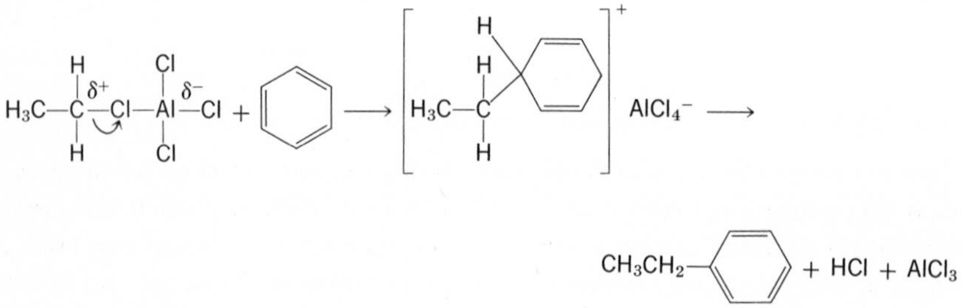
H3C
δ+
δ−
AlCl4−
CH3CH2
+ HCl + AlCl3

[1]Note that $AlCl_3$ has an incomplete octet which it can complete only by placing + charges on the electronegative chlorine atoms, yielding Al_2Cl_6.

The reaction proceeds readily with benzene and with aromatic rings containing activating (ortho,para directing) groups. The reaction does not go well with deactivated aromatic rings (meta directors) because the competing side reactions of carbonium ions, such as elimination (Sec. 9.7), become the major reactions when the aromatic compound is less reactive.

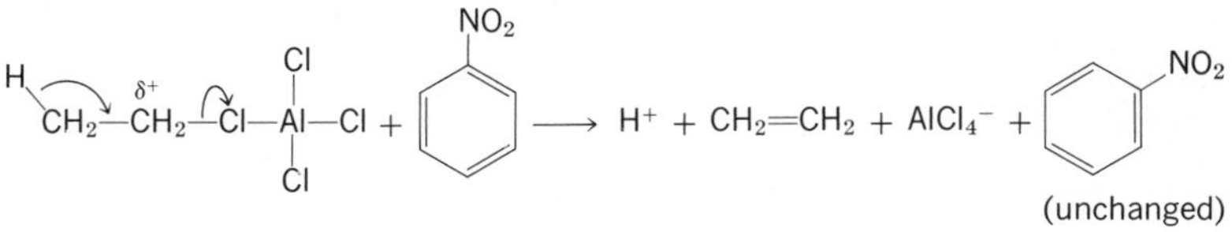

NO2
H
Cl
δ+
CH2—CH2—Cl—Al—Cl +
Cl
⟶ H+ + CH2=CH2 + AlCl4− +
NO2
(unchanged)

The Friedel-Crafts reaction also suffers from the disadvantages that the products are at least as reactive as the starting materials, and that monosubstitution is always accompanied by some di- and trisubstitution, etc.

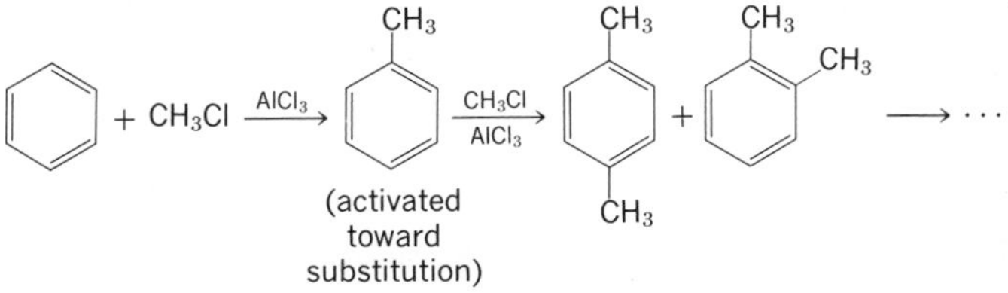

+ CH3Cl
AlCl3
CH3
CH3Cl
AlCl3
(activated toward substitution)
CH3
CH3
+
CH3
CH3
⟶ · · ·

By control of the reaction conditions and by using an excess of the aromatic compound, however, one can usually make monosubstitution predominate.

6.18 Grignard Reaction Halogens directly attached to an aromatic ring are relatively difficult to remove. For example, they are not removed in the Friedel-Crafts reaction above or in displacements (Sec. 5.7). The Grignard reaction, however, is successful and allows a number of syntheses to be carried out with aromatic groups (Sec. 8.7).

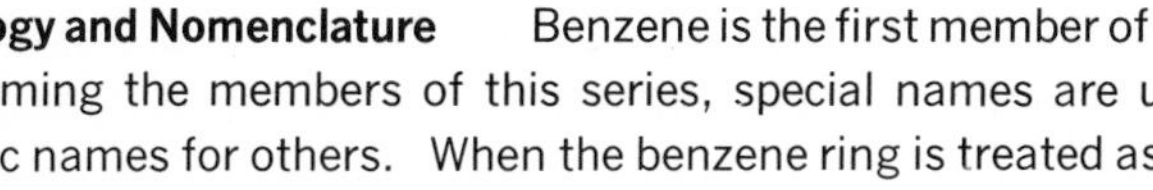

Br
Mg
dry ether
MgBr
H2O
+ MgBrOH

Aromatic Grignard reagents behave much like the corresponding aliphatic Grignard reagent (see Sec. 8.7).

6.19 Homology and Nomenclature Benzene is the first member of a homologous series. In naming the members of this series, special names are used for some and systematic names for others. When the benzene ring is treated as a substituent group, it is called "phenyl."

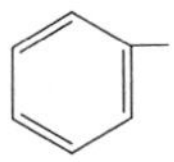

Phenyl group

Some of the hydrocarbon derivatives of benzene are listed below, with their trivial and systematic names:[1]

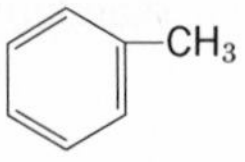

Toluene
(methylbenzene)
(phenylmethane)

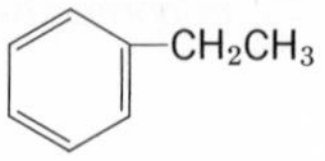

Ethylbenzene
(phenylethane)

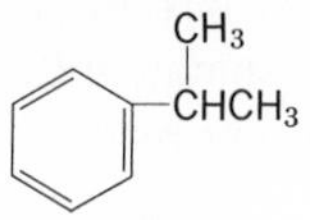

Cumene
(isopropylbenzene)
(2-phenylpropane)

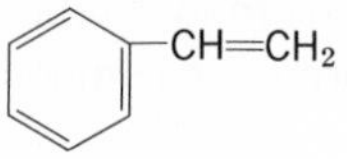

Styrene
(phenylethylene)
(vinylbenzene)

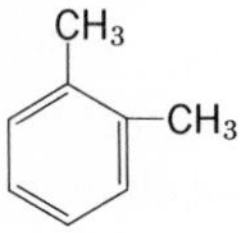

o-Xylene
(1,2-dimethylbenzene)

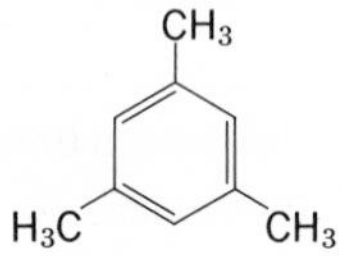

Mesitylene
(1,3,5-trimethylbenzene)

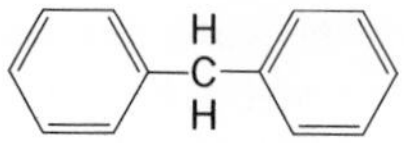

Diphenylmethane

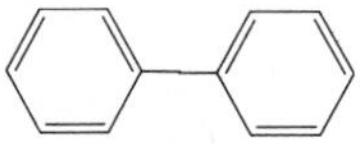

Biphenyl

6.20 Reactions of Aromatic-Aliphatic Substances Interesting problems arise when we consider the chemistry of compounds containing both an aromatic ring and an aliphatic side chain. There are considerable differences in the reactivities of these two structures toward many reagents. In nitration and sulfonation the ring is far more reactive, the side chain as a rule being unaffected.

In the case of halogens, reaction may proceed either in the ring or on the side chain depending on the reaction conditions.

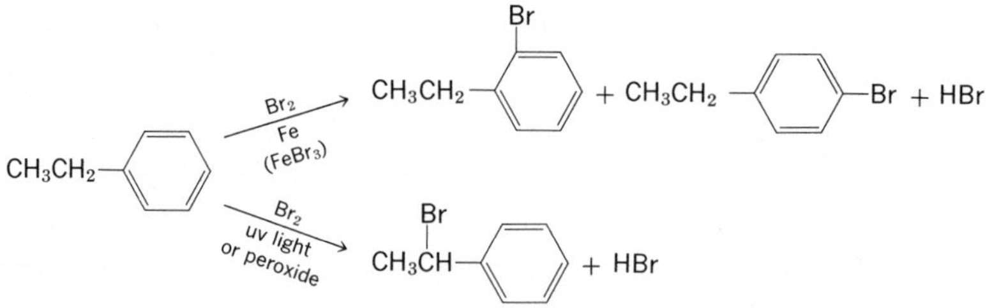

The mechanism of ring substitution has already been discussed (Sec. 6.11). The reaction in the side chain is undoubtedly a free-radical process (see Sec. 2.13). Here the peroxide or ultraviolet light generates bromine atoms (radicals) and a chain reaction ensues:

[1]Besides their use as basic starting materials for chemical syntheses, these substances have many special uses. Styrene polymerizes to polystyrene; the student may have encountered the mixed xylenes ("xylol") as solvents in histological work; biphenyl is used as a heat-transfer agent.

1. $$Br_2 \xrightarrow[\text{uv light}]{h\nu} 2[:\ddot{B}r\cdot]$$

2. 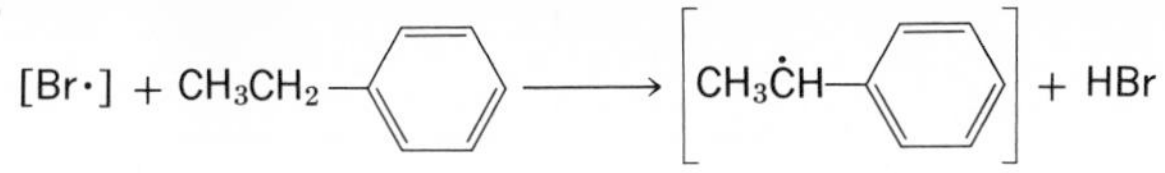

3. 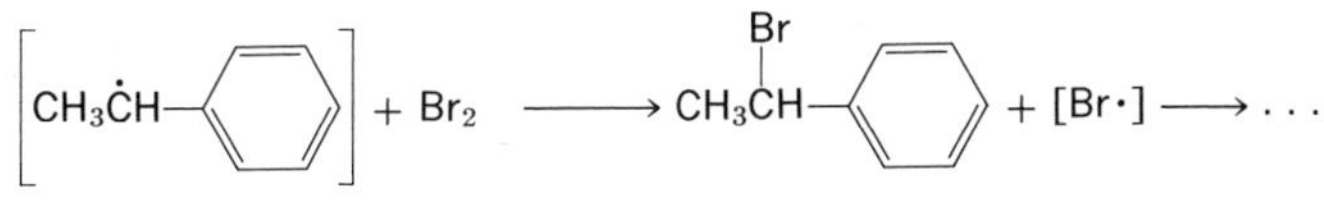

The product of side-chain bromination is predominantly that which results from substitution of the hydrogen on the carbon atom *adjacent to the aromatic ring*. This is understandable if we consider the free radical that is formed in Step 2 above.

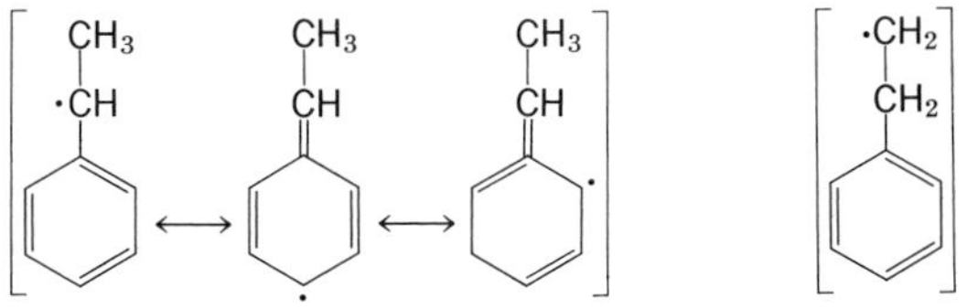

I Resonance stabilized

II No stabilization of $CH_2\cdot$

The free radical I is resonance-stabilized while II acts as an ordinary alkyl free radical. Since free radicals are high-energy intermediates, factors that tend to stabilize these intermediates also tend to lower the transition state leading to them. Therefore, Step 2 is more facile when a stabilized radical such as I is formed than it would be if II were involved.

In the case of simple methyl-substituted benzenes, the reaction may be carried out in stepwise fashion with reasonable yields.

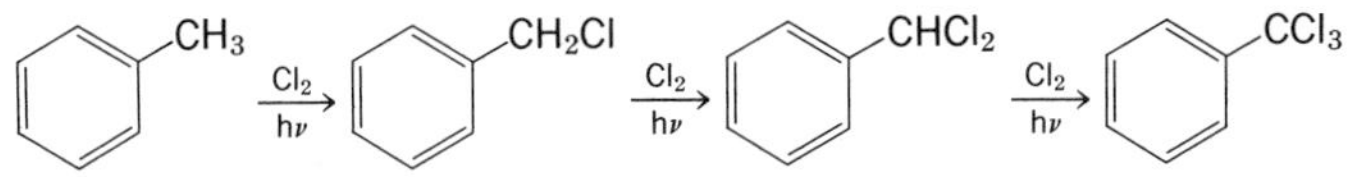

6.21 Oxidation Reaction The great stability of the benzene ring under certain conditions is shown by the fact that upon vigorous oxidation of any benzene derivative in which there is a carbon directly attached to the ring, the side chain may be oxidatively destroyed down to this carbon, which is converted to a carboxyl group:

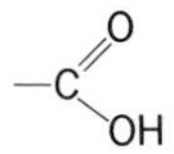

This behavior combined with Körner's proof of orientation permits the detection and location of side chains on the ring. The following examples will illustrate this:

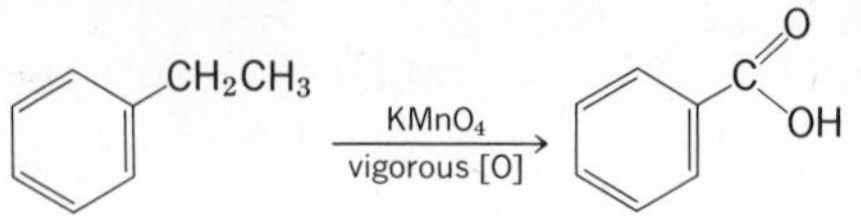
KMnO4
vigorous [O]

one side chain

Benzoic acid

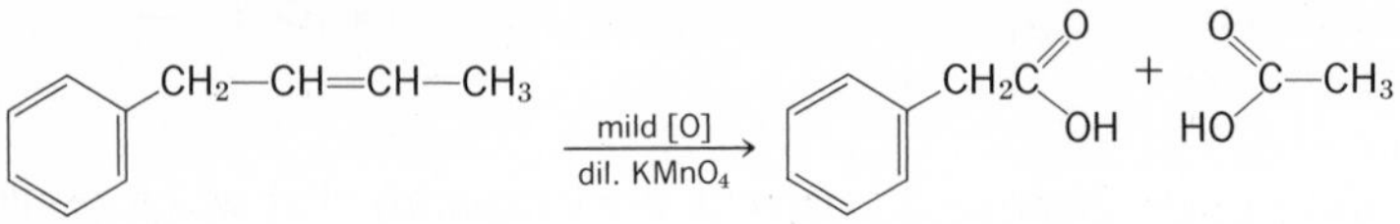
mild [O]
dil. KMnO4

Phenylacetic acid   Acetic acid

Note that this is a normal double-bond reaction (Sec. 3.14).

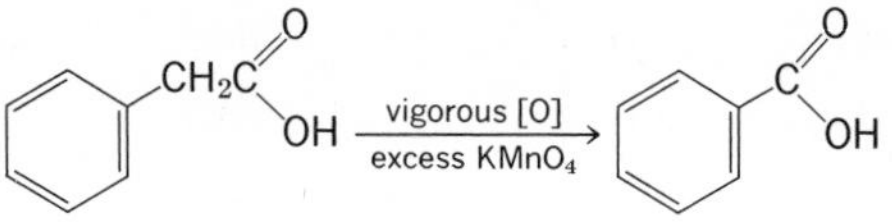
vigorous [O]
excess KMnO4

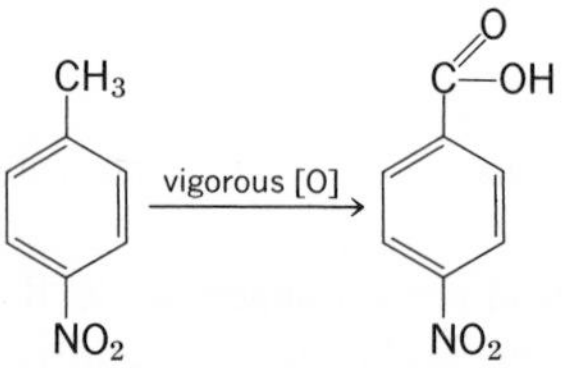
vigorous [O]

The nitro group, being in an oxidized state, is not affected.

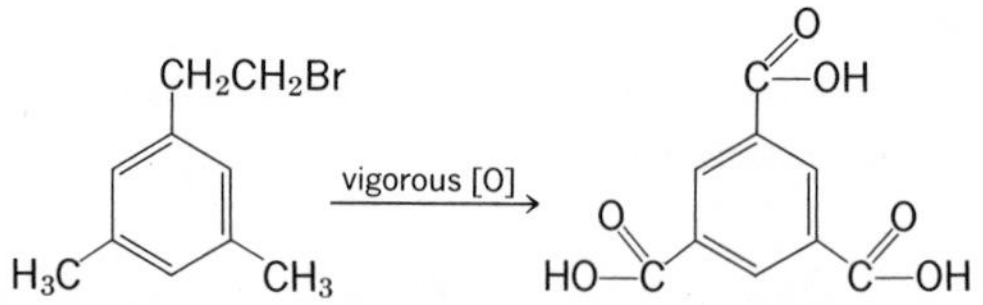
vigorous [O]

three side chains

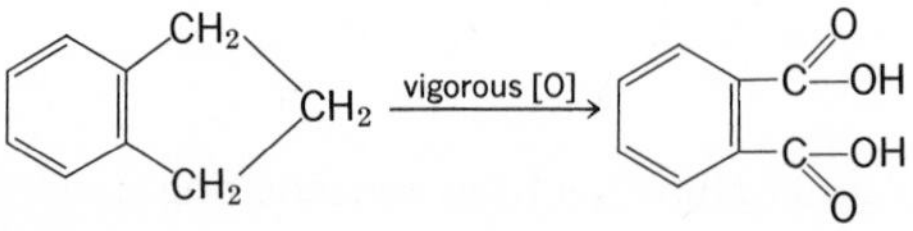
vigorous [O]

The carbon atoms removed from complex side chains such as those in the above examples are converted to CO_2 and small fragments that are ordinarily not isolated. Oxidizable groups, such as double and triple bonds (and alcohol and aldehyde groups, as we shall see later), are more readily attacked by oxidizing agents than are the saturated carbon atoms. When on the side chain of an aromatic nucleus, they can always be made to undergo the normal oxidation (Secs. 3.14 and 5.5) prior to complete destruction of the side chain. Presumably the observed stability of the nucleus is due to ring resonance, and oxidation of the ring would involve its loss;

i.e., additional energy would be required to make up for that not present because of the resonance.

This advantage is lost if a very powerful electron-donating group, such as HO— or H_2—N—, is substituted on the ring. In these instances, it is not possible to oxidize the side chain without attacking the ring.

6.22 Summary of Aromatic Nature The special properties of the aromatic nucleus may be summarized as follows: The nucleus is less unsaturated in behavior than the carbon-to-hydrogen ratio would suggest. The ring shows a remarkable stability toward oxidation, greater than would be expected from the carbon-hydrogen ratio. The ring seems to be an integrated unit justifying the term nucleus. The introduction of a substituent upon the nucleus modifies it in a manner characteristic of the substituent and markedly changes the properties of the whole ring; the effect of substituents is felt throughout the whole nucleus.

To this it may be added that the common reactions of nuclear substitution which are of interest here involve the attack on the ring by electrophilic reagents, so that those substituents which make the ring more nucleophilic increase its reactivity.

6.23 Condensed-ring Compounds In addition to benzene and its homologs, other types of aromatic hydrocarbons are known, mostly as products of coal-tar distillation. The chief component of crude coal-tar distillate is naphthalene, a hydrocarbon which finds many applications in chemical industry, particularly in the field of dyestuffs.

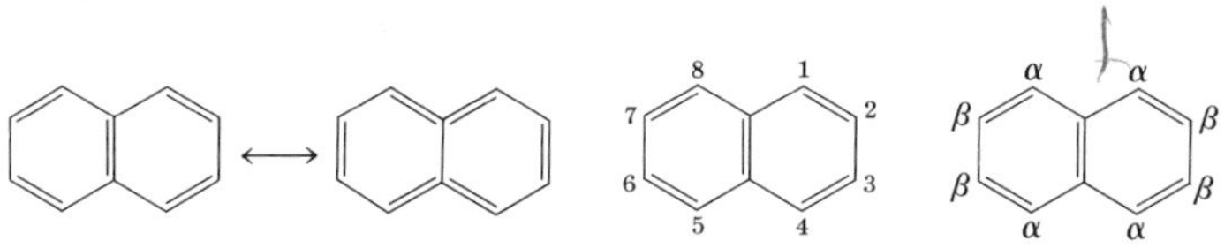

8
1
7
2
6
3
5
4
α
α
β
β
β
β
α
α

Naphthalene[1]

In naphthalene, as in the other more complex fused-ring aromatic hydrocarbons, is found the same problem of assigning positions to the double bonds as in benzene; here again the actual structure is that of a resonance hybrid, with a consequent increase in the stability of the compounds similar to that noted in the case of benzene. Naphthalene, like benzene, undergoes substitution reactions rather than addition. The Friedel-Crafts reaction and other typical aromatic-series reactions are exhibited by naphthalene. In many respects, naphthalene and the other condensed-ring compounds are more reactive than the simple aromatic ring compounds, often giving the above reactions more rapidly and at a lower temperature.

In the case of naphthalene, the problems of isomerism and nomenclature are, of course, more complex than in benzene, there being two monosubstituted naphthalenes and ten disubstitution products theoretically possible. A discussion of

[1]Naphthalene is widely used as the household insecticide moth balls.

directive influences in this series is beyond the scope of this book. It will be sufficient to point out that in most reactions of naphthalene, the main point of attack is the *alpha*, or 1, position. Bromination, for example, leads chiefly to α-bromonaphthalene

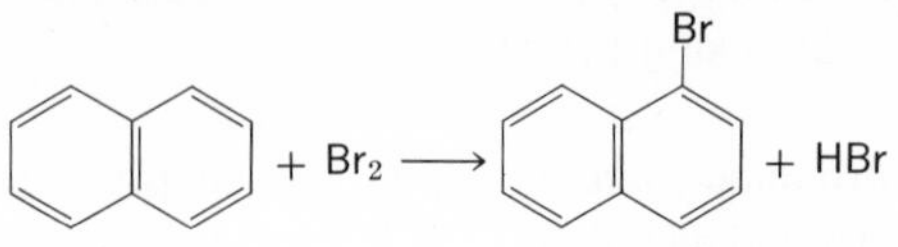
Br
+ Br2 ⟶
+ HBr

α-Bromonaphthalene

but interestingly enough, and in agreement with what has been said above, if one ring of naphthalene is substituted with a nitro group, or other ring-deactivating group, it is the other ring which is more reactive.

Naphthalene and its derivatives behave toward oxidizing agents as though at any given moment one ring is more susceptible to attack than the other.

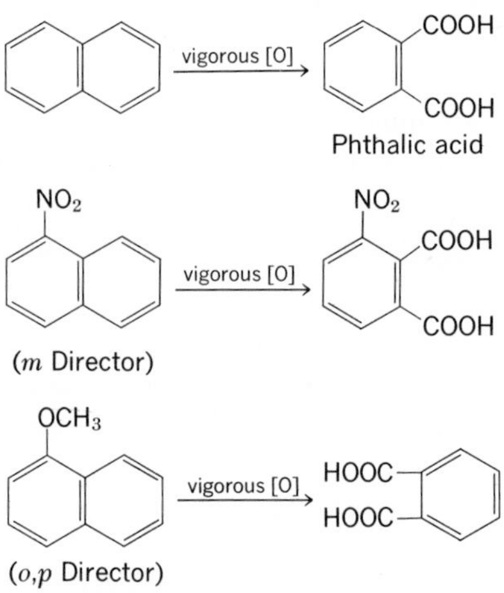
vigorous [O]
COOH
COOH
Phthalic acid
NO2
vigorous [O]
NO2
COOH
COOH
(m Director)
OCH3
vigorous [O]
HOOC
HOOC
(o,p Director)

This oxidation of an aromatic nucleus can be carried out under proper conditions even with benzene.

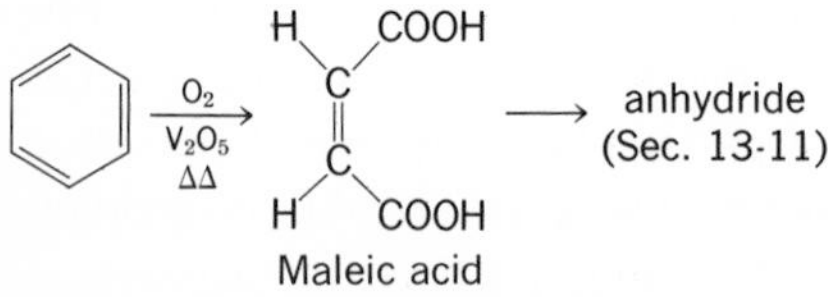
O2
V2O5
ΔΔ
H
COOH
C
C
H
COOH
Maleic acid
anhydride
(Sec. 13-11)

Phthalic and maleic acids prepared commercially in this way are important industrial chemicals.

The more complex condensed-ring aromatic hydrocarbons exhibit for the most part very similar chemical properties and will not be discussed. A few examples of the structures of these compounds are given below.

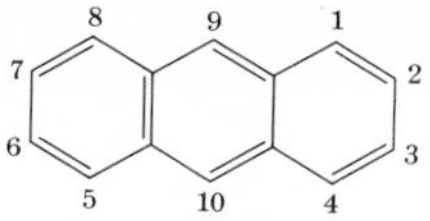

Anthracene

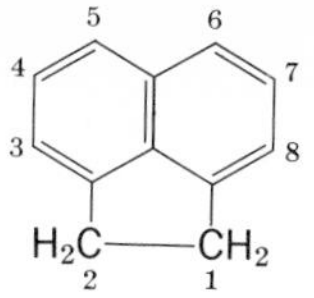

Acenaphthene

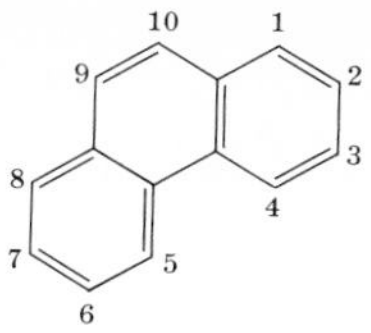

Phenanthrene

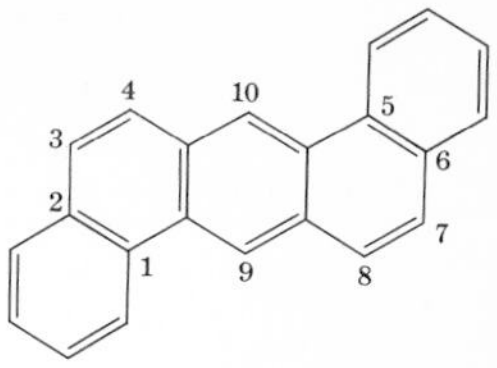

1,2,5,6-Dibenzanthracene

Some of these substances, such as dibenzanthracene, have been found to be the causative agents for a type of skin cancer once rather common among coal-tar workers. Benzopyrene and dibenzopyrene are active carcenogenic (cancer-causing) agents found in cigarette smoke.[1]

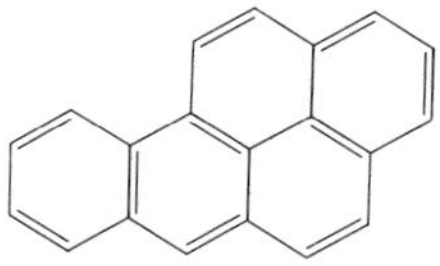

Benzopyrene

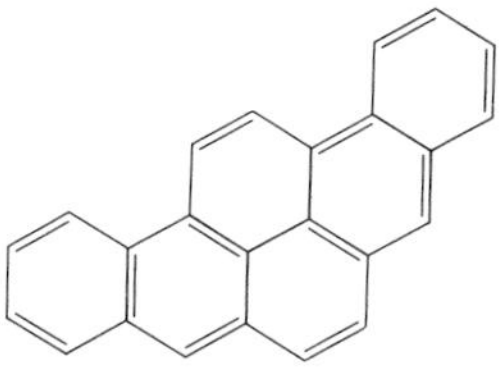

Dibenzopyrene

6.24 Nonbenzenoid "Aromatic" Structures Modern theory has made possible the prediction of the energies (hence the stabilities) of certain π electron systems, analogous to those found in aromatic molecules, but in different structures. These substances have attracted a great deal of interest.

According to a rule proposed by Hückel, based on an early simplified molecular orbital treatment, cyclic structures with $(4n + 2)$ π electrons[2] in a cyclic conjugated (planar) system should show some of the characteristic stability of "aromatic" systems. The experimental test of this prediction presents a real challenge to the ingenuity of the organic chemist.

Cyclooctatetraene, a benzene analog, was synthesized by a complex route in 1910, and more recently, in practical yields by the tetramerization of acetylene.

Cyclooctatetraene

This molecule, which does not fit the $4n + 2$ rule (i.e., it has 8π electrons), is not planar, but puckered like other large rings, and undergoes reactions characteristic of alkenes such as addition and oxidation.

A number of nonbenzenoid species which do fit the rule and which are believed to have aromatic character are known. Many of these are not neutral species but

---

[1] As cited in the Surgeon General's report Smoking and Health, *P.H.S. Publication* (1103) (1964).
[2] Where n = any integer, 0,1,2, etc.

ions, such as

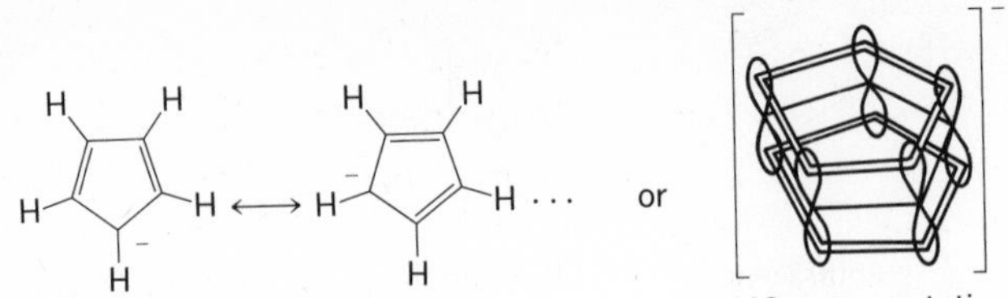

or

Cyclopentadienyl ion

MO representation

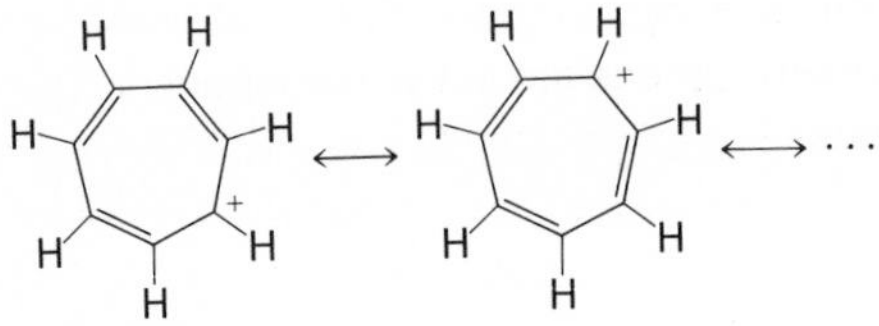

Tropylium ion

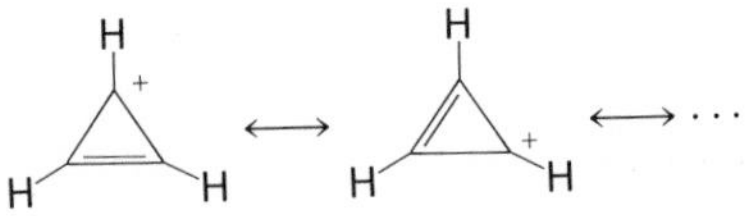

Cyclopropenium ion

There are also the neutral molecules, such as azulene and fulvene.

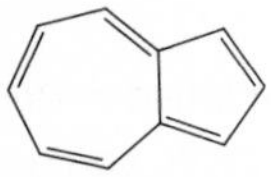

Azulene

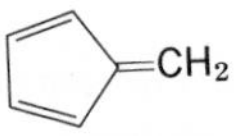

$=CH_2$

Fulvene

Cyclobutadiene, which does not fit the rule, but which cannot be too far from planar, is known only as metal complexes or as a transient intermediate, presumably due to ring strain (see Sec. 4.5).

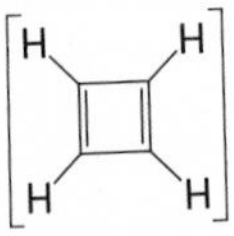

Cyclobutadiene

6.25 Spectroscopic Properties of Aromatic Compounds By virtue of their extensive π electron system, aromatic compounds are found to have considerable absorption in the ultraviolet. Because substitution can markedly enhance both the wavelength and the intensity of this absorption, it is impossible to give a simple set of rules for the location of absorption bands. In general, all aromatics with nonconjugated substituents absorb with moderate intensity in the region 240 to 280 mμ. Conjugating substituents, such as NO_2, —C═CH—, —HC═O, etc., may shift this to considerably greater wavelengths. Indeed, most common dyes and pigments which absorb in the longer wave visible region are substituted aromatic compounds.

The infrared spectra of aromatic compounds are somewhat less complex. Aromatic compounds possess a C—H stretch at 3000 to 3125 cm^{-1}, and ring stretches at 1590 to 1610 cm^{-1} and 1480 to 1500 cm^{-1} regions. Certain other known characteristic absorptions of various substitution patterns are beyond the scope of this book.

In the nmr aromatic hydrogens usually appear in the region $\delta = 7$ to 8. This very deshielded environment of the hydrogens is due to the π electrons, which set up a magnetic field that reinforces the applied field so that the hydrogens experience a field greater than that applied.[1] It is fortunate from the point of view that nmr makes alkene and aromatic hydrogens readily distinguishable, whereas in the infrared this spectral distinction is not readily apparent.

OUTLINE OF AROMATIC HYDROCARBON CHEMISTRY

I. Benzene hydrocarbons.

Preparation

The Friedel-Crafts reaction.

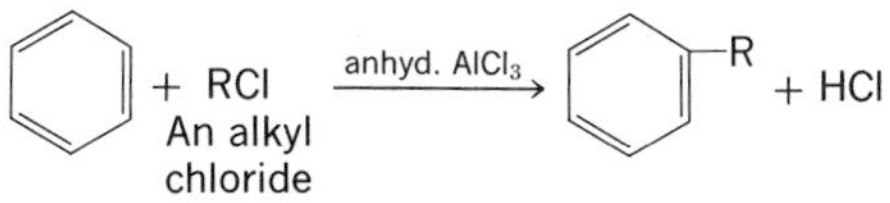
+ RCl
An alkyl chloride
anhyd. AlCl3
R
+ HCl

[1] The detailed mechanism of this induced field is beyond the scope of this book. In a very simplified version, however, it may be said that the applied magnetic field induces a current loop in the π cloud which creates a small electromagnet that opposes the applied field at the center of the benzene ring, but reinforces it around the rim where the hydrogens are.

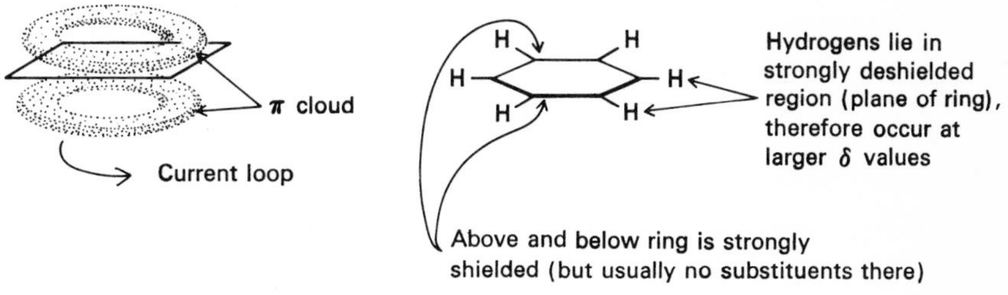
π cloud
Current loop
H
H
H
H
H
H
Hydrogens lie in strongly deshielded region (plane of ring), therefore occur at larger δ values
Above and below ring is strongly shielded (but usually no substituents there)

Properties

A. Halogenation.

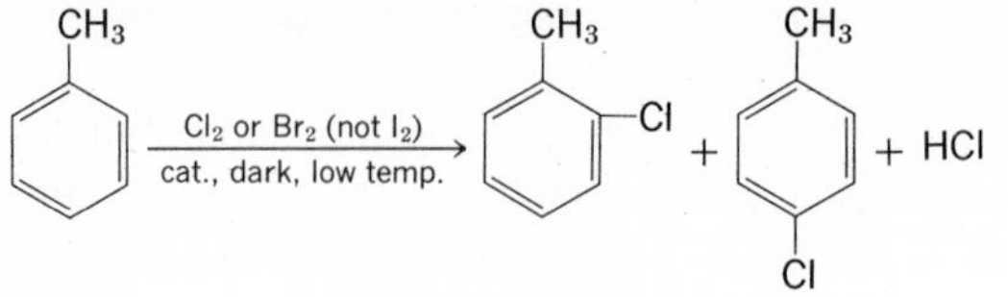

or the bromides and HBr

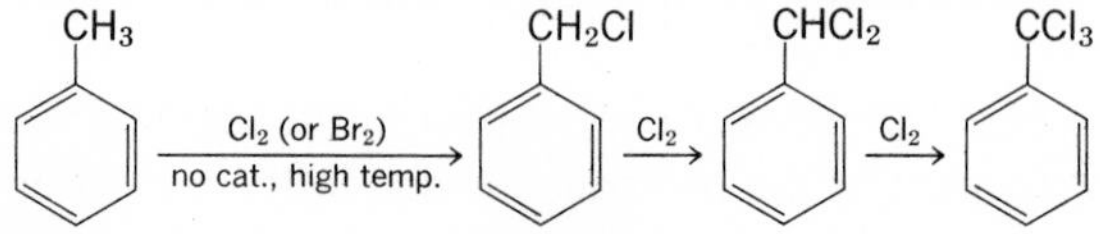

Toluene

Benzyl chloride

B. Nitration.

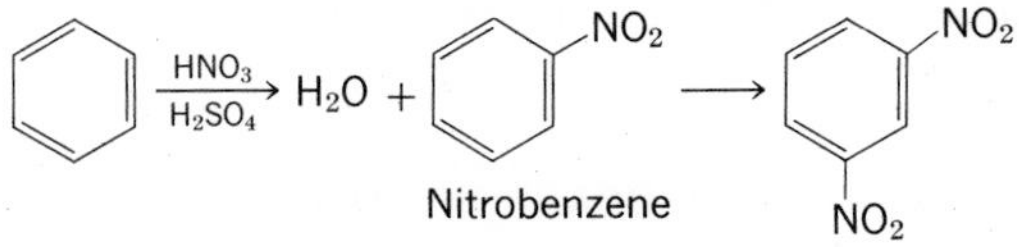

Nitrobenzene

C. Sulfonation.

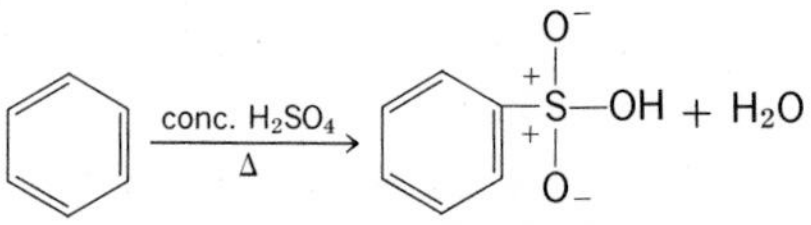

Benzenesulfonic acid

D. Side-chain oxidation.

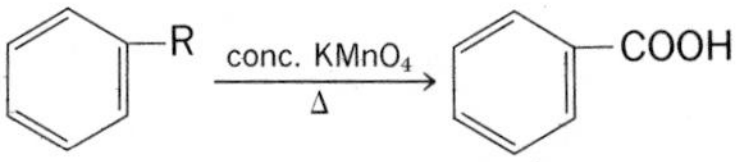

Benzoic acid

II. Condensed-ring hydrocarbons.

Properties

A.

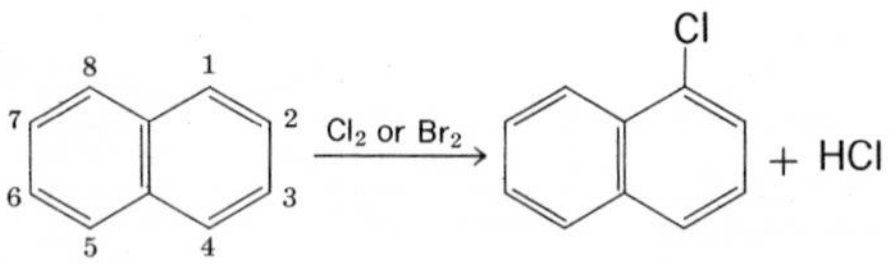

or HBr

B.

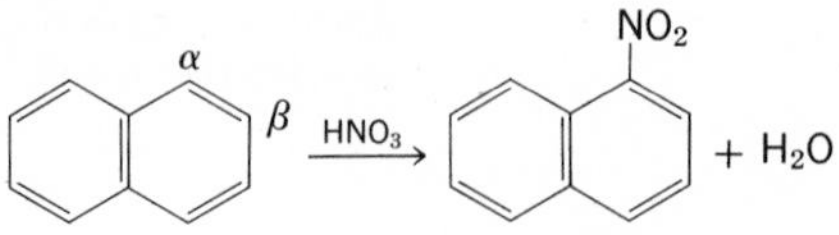

C.

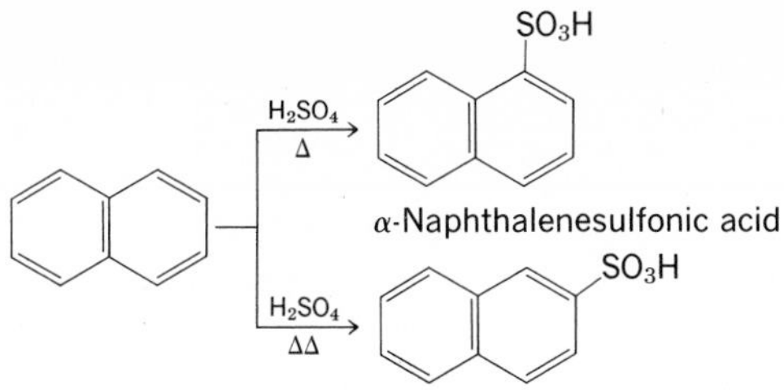

α-Naphthalenesulfonic acid

β-Naphthalenesulfonic acid

In these first three reactions, it is to be noticed that the α position is more susceptible to attack than the β position. The β substituent may actually be the more stable as in the sulfonic acid case.

D.

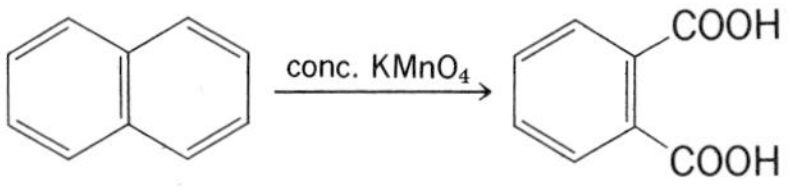

Phthalic acid

This oxidation can also be carried out with a catalyst using air as the oxidizing agent. This is done in commercial practice.

E.

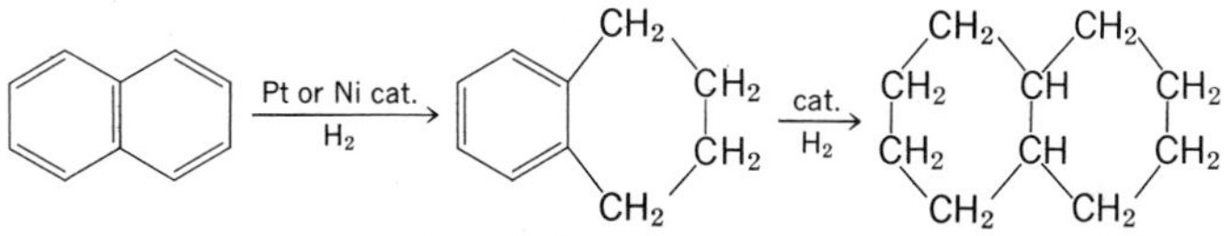

Tetrahydronaphthalene (tetralin)

Decahydronaphthalene (decalin)

Other condensed-ring compounds, such as anthracene, have similar properties.

Spectroscopic Properties

Group	*ir, cm⁻¹*	*nmr, δ(ppm)*	*uv, mμ*
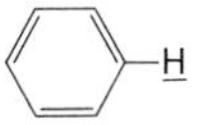 —H	3000–3125	7–8	. . .
	1480–1500 1590–1610	. . .	240–280 and above (enhanced by conjugation)
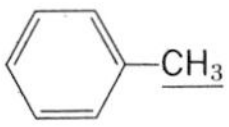 —CH_3	normal C—H	2.4	. . .
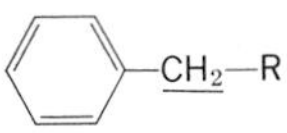 —CH_2—R	normal C—H	3	. . .

CHAPTER 6 EXERCISES

★**1.** Ozonization and hydrolysis of the ozonide of *o*-xylene gives a mixture of glyoxal, CHO—CHO; methylglyoxal, CH_3—CO—CHO; and diacetyl, CH_3—CO—CO—CH_3. Discuss the significance of this observation in connection with the structure of benzene.

2. Which tribromobenzene gives three different tetrabromobenzenes on bromination?

★**3.** The formula

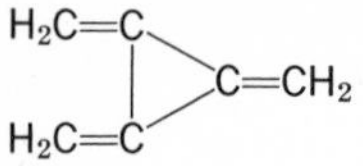

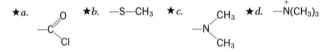

has been suggested as the structure of benzene. Discuss any reactions given by benzene that cannot be accounted for by this formula. Show why the Kekulé formula is preferable and why it more closely describes the properties of benzene.

4. When isoprene (2-methyl-1,3-butadiene) reacts with 1 mole of hydrogen, the following products are obtained:

12 percent 3-methyl-1-butene
13 percent 2-methyl-1-butene
15 percent 2-methyl-2-butene
30 percent 2-methylbutane
30 percent 2-methyl-1,3-butadiene

Explain the origin of each of these products; write equations for their formation.

5. Predict the directive influence of the following groups:

★*a.* 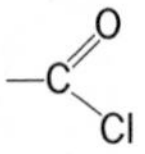 ★*b.* —S—CH_3 ★*c.* 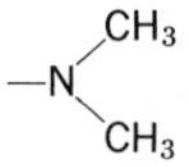 ★*d.* —$\overset{+}{N}(CH_3)_3$

6. By reasoning analogous to that used in Sec. 6.12 for the interpretation of the directive influence of Cl on a bezene ring, explain the fact that 1-chloro-1-propene and HCl gives 1,1-dichloropropane as the main product.

7. Explain the following and decide whether or not they are correct.

★*a.* Bromobenzene is less reactive than bromocyclohexane in reactions involving cleavage of the carbon-bromine bond.

b. 1,3-Cyclohexadiene gives up more energy on hydrogenation to cyclohexane than does benzene on hydrogenation to cyclohexane.

★*c.* Chloroform (trichloromethane) shows weakly acidic properties and will give up a proton to a strong base. Under these conditions methane is unreactive.

d. In the reaction between sodium acetylide and 2-methyl-2-chlorobutane NaCl is formed but no C_5 products.

★*e.* The elements of H_2O can be added to alkynes but no OH groups are formed.

f. Grignard reagents react with some alkynes but not with all alkynes or with any alkenes.

8. Comment on the following "resonance hybrids" of cyclopentadiene. Is this resonance? Would it contribute to the stability of cyclopentadiene?

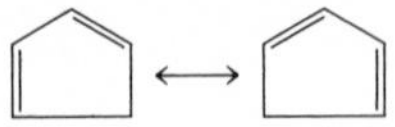

★**9.** Comment briefly on the significance of the following statement. What conclusions regarding the mechanism of the reactions of carbon compounds may be drawn? Ethylene + H_2SO_4 gives the same product with benzene as ethyl chloride + $AlCl_3$.

10. By means of equations, showing the mechanism, account for the following:

a. ϕ-COOH is a meta director and ϕ-CH_2COOH is an ortho, para director.

b. $CH_3CH_2CH_2OH + H_2SO_4 + C_6H_6 \longrightarrow$ 2-phenylpropane.

c. $CH_2{=}CH{-}CH{=}CH_2 + HCl \longrightarrow ClCH_2CH{=}CH{-}CH_3$ Would the trans (lower energy) or cis form predominate?

d. 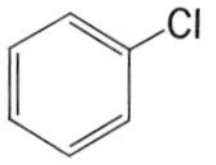 is less reactive than 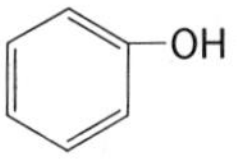 in electrophilic substitutions.

11. Show by equations how you could convert:

★*a.* CH_3CH_2OH to $CH_3COCH_2CH_3$

b. Benzene to 

★*c.* Toluene to 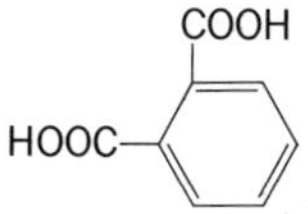

d. Bromobenzene to 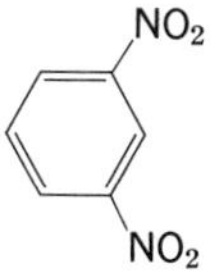

★*e.* Carbon to CH_3CH_2OH

f. $CH_2{=}CH_2$ to $CH{\equiv}CH$

★*g.* Toluene to $C_6H_5{-}CH_2{-}\overset{O}{\overset{\|}{C}}{-}CH_3$

h. *iso*-Butene to 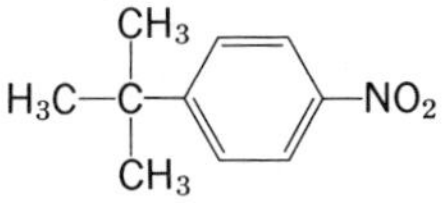

12. How could you distinguish by chemical methods:

a. $HC{\equiv}C{-}CH{=}CH{-}CH_2CH_3$, benzene, and $CH_2{=}CHCH_2CH{=}CHCH_3$

b. 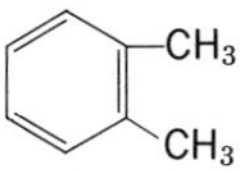 , 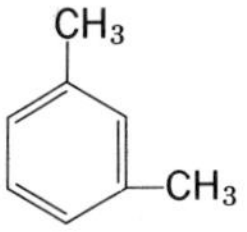 , and 

c.

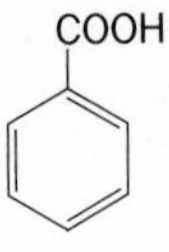

and

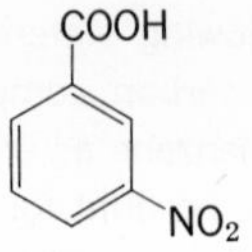

d.

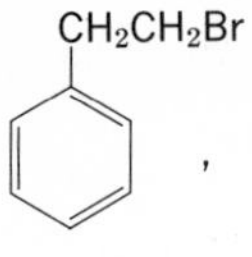

,

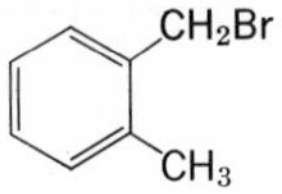

, and

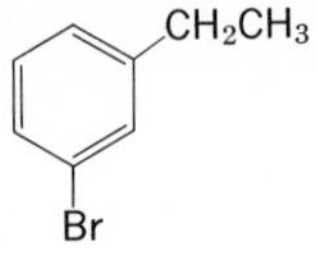

e. $C_6H_5-C(CH_3)=CH_2$ and $C_6H_5-CH=CH-CH_3$

★13. Taking into consideration the properties of $AlCl_3$ discussed in Sec. 6.16, write equations showing the mechanism of the polymerization of ethylene catalyzed by $AlCl_3$.

14. How would you expect the group $-\overset{+}{N}\equiv N$ to direct? How would you expect the group $-CH=CH-CN$ to direct? In each case draw appropriate diagrams to support your analysis.

★15. In the bromination of 1-methyl-5-nitronaphthalene the bromine is found exclusively in one of the aromatic rings. Which one? Explain.

16. Write equations for reactions by means of which benzene could be converted to:

a. Toluene

b. *p*-Bromonitrobenzene

c. 2,4-Dibromotoluene

d.

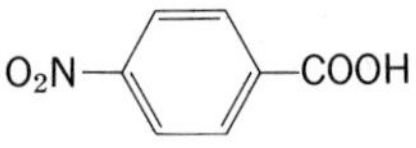

e.

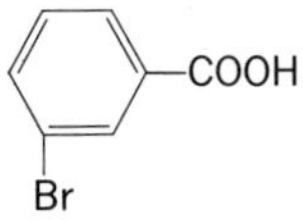

f. TNT (trinitrotoluene)

g.

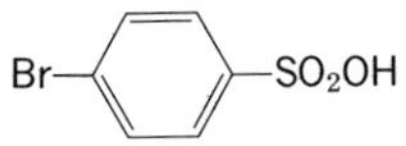

h.

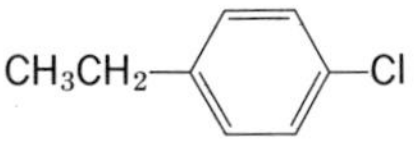

i.

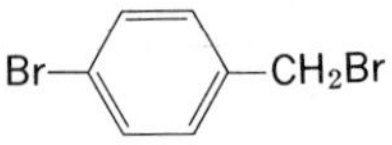

j. *m*-Nitrobenzoic acid

k. Ethylbenzene

l.

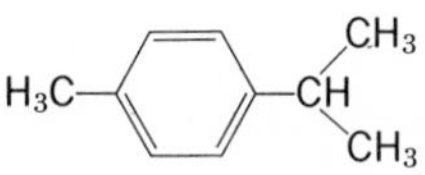

m. Ethylcyclohexane

17. Explain briefly the following:

★*a.*

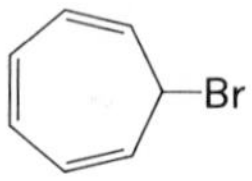

is soluble in water but benzylbromide is not.

★ *b.*

The oxidation of 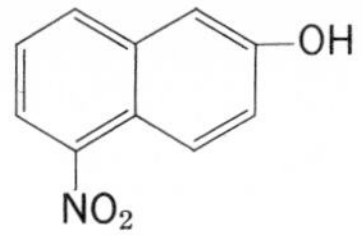
OH
NO2
gives only an acid containing the NO_2 group.

18. A certain compound X, $C_{10}H_{11}Br$, behaves as follows:

a. X gives a positive von Baeyer test and reacts readily with Br_2 to form $C_{10}H_{11}Br_3$.

b. X + $KMnO_4$ (excess) $\longrightarrow$ (Y)C_3H_6O and (Z) $C_7H_5O_2Br$.

c. Y is a ketone.

d. Z on nitration gives two (and only two are possible) mononitro compounds, $C_7H_4O_4BrN$.

State briefly what information about the structures of the lettered compounds is given by each statement and deduce a structure for X.

19. An aromatic compound A, C_9H_8,

★*a.* forms an insoluble salt with $AgNO_3 + NH_3$;

★*b.* reacts when heated with excess $KMNO_4$ to give B, $C_8H_6O_4$.

★*c.* B on nitration gives one and only one product with the formula $C_8H_5O_4NO_2$. Show structures for A and B that fit these data.

20. *a.* $C_6H_5NO_2$ and $C_6H_5NH_2$ react in electrophilic substitution reactions at different rates. Explain, showing which reacts the faster and giving the main products in a typical case.

b. 2-methyl-2-phenyl-1-propanol + H_2SO_4 gives, among other things, two isomeric C_{10} alkenes. Show by equations and mechanisms how you would account for this.

c. Benzyl bromide reacts more readily with sodium acetylide than does phenyl bromide. How can this be explained?

d. Protons are powerful electrophilic agents, yet they do not attack an aromatic system (or at least HCl and benzene do not appear to react). How could one determine whether or not benzene reacts with protons?

★21. A certain hydrocarbon X, $C_{11}H_{12}$, reacts as follows:

a. X + CuCl + NH_3 $\longrightarrow$ an insoluble precipitate.

b. X + 1 mole of H_2 + catalyst $\longrightarrow$ (Y)$C_{11}H_{14}$.

c. Y gives a positive von Baeyer test and will add two more H atoms at a slower rate.

d. Y + HBr $\longrightarrow$ $C_{11}H_{15}Br$ $\xrightarrow{\text{alc. KOH}}$ (Z) $C_{11}H_{14}$, which is different from Y.

e. $Z + KMnO_4 \longrightarrow H_3C{-}C(=O)OH + (A)\ C_9H_{10}O$.

f. $A + \text{excess } KMnO_4 \longrightarrow (B)\ C_6H_4(COOH)_2 \xrightarrow{\text{soda lime}}$ benzene.

g. $B + Br_2 + Fe \longrightarrow$ one and only one substance of the formula

$C_6H_3Br(COOH)_2$

Write structures for the lettered compounds.

22. Compound M, C_8H_{14}, reacts with bromine to give $C_8H_{14}Br_4$. On oxidation M gives $2CO_2$ and $C_6H_{10}O_2$, which is *not* an acid and which resists further oxidation. Write a structure for M and show your reasoning.

★23. Compound A, $C_{12}H_{10}$, shows none of the usual reactions of unsaturated compounds. Suggest a structure for it.

24. What product would you expect from the reaction of benzene with BrCl? Explain. (A carrier catalyst is used, but this does not affect your reasoning here.)

★25. Recently the $—SO_2CF_3$ group has been studied from the point of view of its effect on aromatic substitution. Predict its directive influence and comment on its probable relative effectiveness.

26. $CH_3C{\equiv}CH$ reacts more slowly with HCl than the corresponding alkene, yet the reaction is easily controlled to give the unsaturated adduct $CH_3\overset{\text{Cl}}{\overset{|}{C}}{=}CH_2$. Explain. (Hint: compare effect of Cl on aromatic rings.)

★27. The group CF_3 acts as a meta director. Explain. Predict the product of the reaction of $CF_3CH{=}CH_2$ + HCl.

28. Predict structures for the following (there may be several possible):

a. $C_6H_4I_2$, nmr absorption only at 7.4δ.
b. C_6H_8, nmr absorption at 7.2δ (5 hydrogens) and at 2.3δ.
c. C_9H_{12}, nmr absorption at 2.3δ and at 6.8δ.
d. $C_{12}H_{18}$, shows no absorption in the uv above 220 mμ.

29. How many nmr absorption bands would you expect in the following structures:

★*a.* $CH_3CH_2C_6H_5$
b. Naphthalene
★*c.* $C_6H_5NH_2$
d. 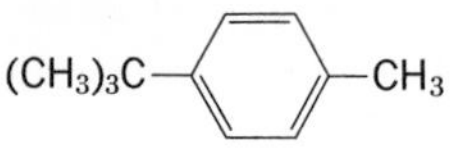
(CH3)3C
CH3
★*e.* Benzoic acid
f. $C_6H_5CH{=}CH_2$

7

STRUCTURE, ISOMERISM, AND SYMMETRY OF MOLECULES

7.1 Introduction The concept of molecular structure has been one of the most fruitful tools of the organic chemist. The strength and directional character of covalent bonds is embodied in the concept that the same atoms may be linked to one another in many different ways. An obvious example of this concept is given by the existence of structural isomers, such as 2-methylpropane and *n*-butane. Other, less obvious types of isomerism also exist, and the study of these more subtle forms of isomerism has proven very rewarding to the organic chemist. In particular, the study of *geometric, optical,* and *conformational* isomerism has provided much insight into the behavior and properties of molecules.

7.2 Geometrical Isomerism Whenever a rigid element exists in a molecule so that rotation about one or more bonds is essentially prohibited, isomers may exist in which a given pair of atoms are on the same side of the rigid element (*cis*), or across (*trans*) from (to) each other. Examples of this type of isomerism are found in ring systems and with double bonds. See Secs. 3.5 and 4.7.

Note that in such isomers all atoms are attached to the same sets of atoms in each case so that these are not structural isomers. Geometrical isomers are *not* ordinarily interconvertible except by processes of sufficient energy to permit rotation about the double bond, or by breaking and re-forming the ring in cyclic systems—processes which are usually difficult relative to other reactions. Thus, distinct and separate behavior may usually be observed for such isomers. Geometrical isomers usually differ in boiling point, melting point, solubility, and other physical and chemical properties.

7.3 Optical Isomerism Many common three-dimensional (nonplanar) objects are found to be identical in all respects except one called "handedness" or chirality.[1] Examples of these are right- and left-hand gloves, right- and left-hand screw threads, etc. Objects which differ from one another only in handedness or chirality are found to be nonsuperimposable on their mirror images. Since sp^3-hybrid carbon atoms are tetrahedral, it is found that many substituted carbon compounds are also chiral. An example of two molecules which differ in this respect is given by 2-chlorobutane.

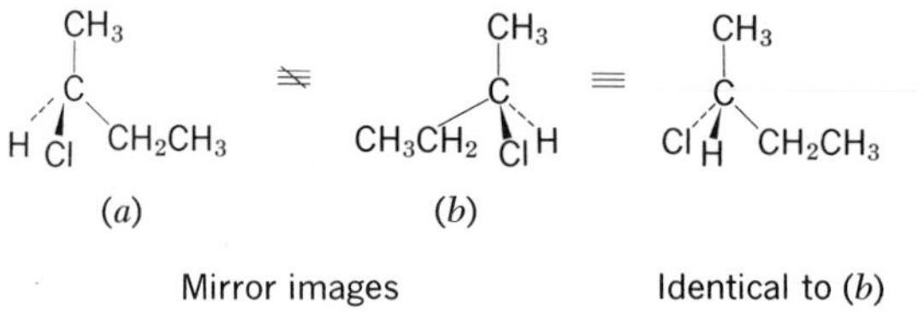
CH3
C
H Cl CH2CH3
(a)
CH3
C
CH3CH2 Cl H
(b)
CH3
C
Cl H CH2CH3
Mirror images
Identical to (b)

Note that the two mirror images are not superimposable, and therefore, not identical; they are, in fact, isomers. One can superimpose the central carbon atom and the methyl and ethyl groups respectively, but then the chlorine and hydrogen come out transposed. Rotations of the molecules will enable one to superimpose the central carbon and any two of the attached groups, but the other two will always come out backward. This type of isomerism is called *optical isomerism* because of the methods by which these isomers can be differentiated experimentally.

One can see that if any two substituents in a molecule like 2-chlorobutane were identical, one could turn the mirror images so that they would be superimposable. This is the case, for example, with 2-chloropropane.

[1] Pronounced "kirality."

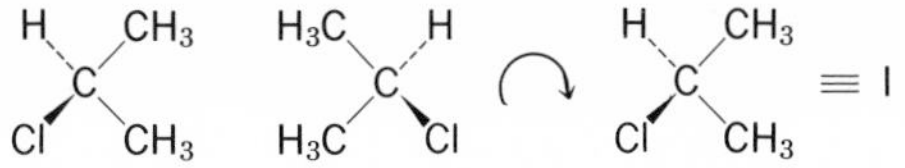

I   II

Mirror images   II rotated by 180° about an axis perpendicular to the plane of the page; I is identical to its mirror image II

Thus, only in the case where carbon atoms possess *four different* substituents will optical isomerism of this type be possible. Such carbon atoms are known as *asymmetric carbon atoms.*[1] Isomers which are mirror images of one another are called "enantiomers" and are said to differ in *configuration* from one another.

Enantiomers differ from one another only in certain special ways. Thus, a bottle of "left-handed" 2-chlorobutane will have the same boiling point, density, freezing point, vapor pressure, solubility, etc., as a similar bottle of "right-handed" 2-chlorobutane. Whenever a chiral (the term used in modern chemical terminology for "handed") substance interacts with a nonchiral substance or agent, the interaction will be the same for either right- or left-handed substances. Only when one chiral substance interacts with another chiral substance does a difference occur. An analogy for this might be taken from a nonchiral object bolted together with a nut and bolt. It makes no difference if a right-hand nut and bolt are used or whether a left-hand pair is chosen. However a right-hand bolt cannot be used with a left-hand nut and vice versa. Here the difference is detectable by lack of fit, even though the pitch and size of the threads are identical—only the chirality is different.

7.4 Polarized Light Polarized light is a form of light which is chiral and as such can interact differently with right- and left-handed molecules. It is from this interaction that the term "optical isomerism" is derived.

Light is transmitted by electromagnetic waves which have an electric component at right angles to the direction of propagation. The intensity of the component varies with the amplitude of the light wave.

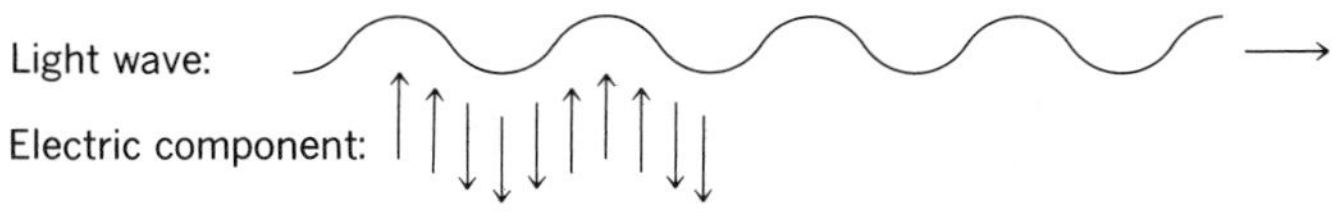
Light wave:
Electric component:

With ordinary light the directions of these components are randomly oriented (in a plane perpendicular to the direction of motion of the wave), seen here end on.

[1]Molecules which are optically active by virtue of asymmetric elements other than asymmetric carbon atoms are known, but we shall not dwell upon this point here.

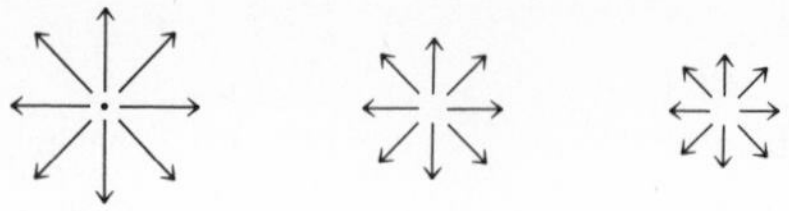

However, by the use of certain optical devices called polarizers[1] the passage of only the components oriented in one direction may be effected.

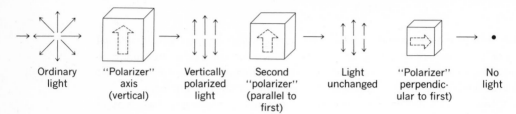

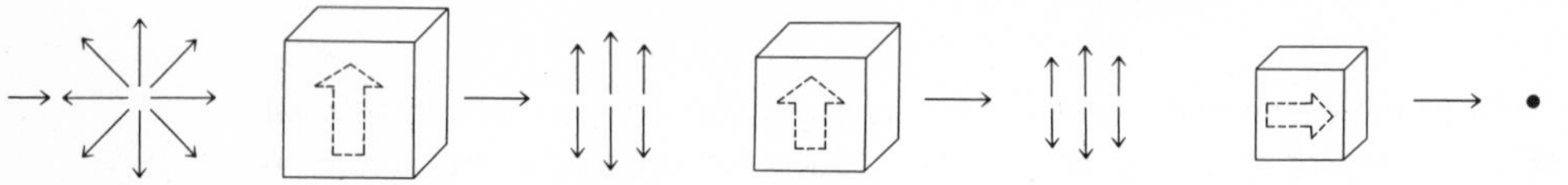
Ordinary light
"Polarizer" axis (vertical)
Vertically polarized light
Second "polarizer" (parallel to first)
Light unchanged
"Polarizer" perpendic- ular to first)
No light

A second polarizer will pass all of this polarized light if it is lined up parallel to the first, and will pass none if lined with its axis at 90° to the first. At intermediate angles it will pass only that component which is parallel to its axis. This second polarizer is called an "analyzer."

When polarized light passes through a medium, the electric component interacts with the electrons in the medium and the effect of such interactions is to induce electric components that are not parallel to the plane of the incident light. With any isotropic (nonchiral) medium there will be an equal number of induced deviations in every direction and there will be no overall effect on the plane of the polarized light emerging. With anisotropic substances, such as certain crystals or solutions of asymmetric molecules, there will be an imbalance of such interactions and the emerging polarized light will still be polarized, but the plane of polarization will be different from that of the incident light. This amounts to a net *rotation* of the plane of the polarized light. Such a rotation can be detected by the angle through which the analyzer must be rotated to permit the maximum passage of light (to line it up parallel to the new plane). This rotation may either be clockwise (positive) or counterclockwise (negative) from the observer's point of view.

It is found that when solutions of one enantiomer of a substance will rotate the plane of polarized light in one direction, comparable solutions of the other enantiomer will rotate the plane an equal amount in the opposite direction. Substances which can rotate the plane of polarized light in this fashion are said to be "optically active." This rotation is a function of the compound, its concentration, path length, temperature, solvent, and wavelengths of the light. The specific rotation, α_{D}^{25}, is standardized so that concentration and path length are taken into account. The temperature is usually set at some specific value (25°C) and the wavelength of the light is commonly that corresponding to the yellow D line of a sodium-vapor lamp. The solvent must be specified.

[1]Polarizers are certain modified crystals (e.g., calcite) or arrays of crystals (polaroid) that possess this ability to pass only this one component of ordinary light.

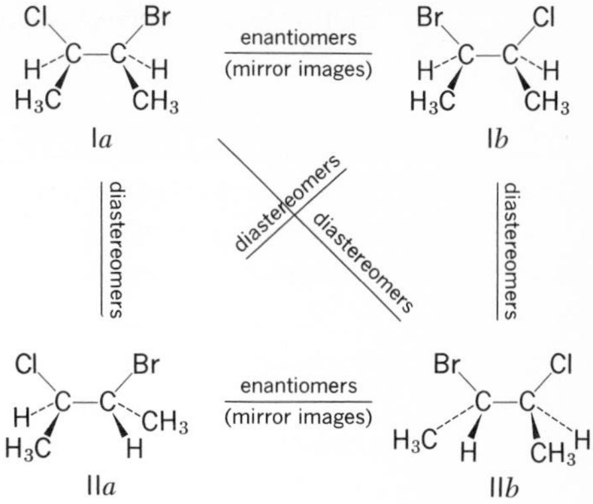
Ia
Ib
enantiomers
(mirror images)
diastereomers
diastereomers
diastereomers
diastereomers
IIa
IIb
enantiomers
(mirror images)

figure 7-1
Optical isomers of 2-bromo-3-chlorobutane.

If one enantiomer is optically pure, its specific rotation will always be equal and opposite to that of its enantiomer. An equimolar mixture of the two will appear optically inactive (that is, it will have zero net rotation) and is called a "racemic" mixture.

7.5 Diastereomers It is possible to have more than one asymmetric carbon atom (or other element) in a molecule. In general, the number of isomers is equal to 2^n, where $n =$ the number of asymmetric carbon atoms. These isomers are all optical isomers of one another, but they cannot all be enantiomers since a given molecule can have, at most, one nonidentical mirror image. These isomers are called "diastereomers" of each other, and there will be n diastereomeric pairs of enantiomers. These relationships can best be seen by examination of an example, 2-bromo-3-chlorobutane (Fig. 7-1).

Diastereomers do not have the same physical properties, and in some cases, may be quite different from one another. The reason for these differences may be visualized as a difference in interaction between two asymmetric carbon atoms, each of which is left-handed or right-handed. Such an interaction is also found when enantiomers are mixed in equal proportions. Such a mixture of enantiomers is called a racemic mixture and will have different physical properties from either pure isomer. The chemical properties of a racemic mixture will often be very similar to those of either pure enantiomer, since in solution the molecules are widely separated. In the case of diastereomers, however, differences in chemical as well as physical behavior are observed. In fact, these chemical differences in diastereomers have enabled organic chemists to develop a much greater understanding of how molecules react. One example of this shall be mentioned later in this chapter.

7.6 Resolution of Enantiomers The separation of a racemic mixture into its two enantiomers is not always a simple task. Since both enantiomers have the same melting point, boiling point, etc., they cannot be separated by the methods used for other types of isomers. Only very rarely does the separation, called resolution, occur spontaneously. In these rare cases all of the left-handed molecules crystallize in one crystalline form and the right-handed ones in the mirror image of this crystal. Then the mechanical separation of the two types of crystal will effect the resolution. Much more commonly, however, the racemic mixture forms only one type of crystal, containing equal numbers of each enantiomer.

The most general method for resolution of enantiomers is to convert them into diastereomers, utilizing a previously resolved or naturally optically active substance. This can be done, for example, by forming diastereomeric salts[1] with an optically active base (commonly the alkaloids brucine or strychnine).

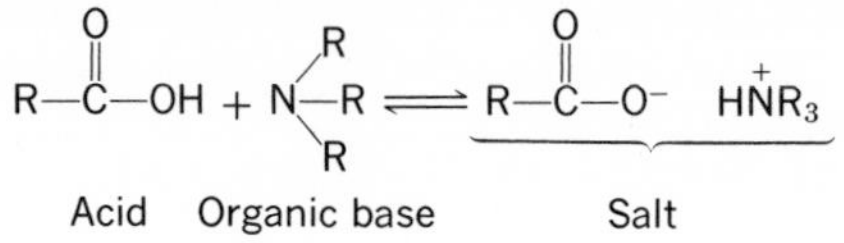
Acid
Organic base
Salt

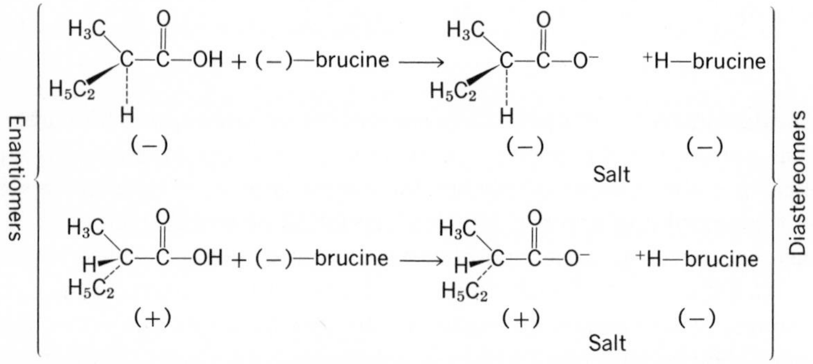
(−)—brucine
+H—brucine
Salt
Enantiomers
Diastereomers

The mixture of enantiomers is thus converted to a mixture of diastereomeric salts. These salts will differ somewhat in physical properties such as solubility, melting point, etc. By fractional crystallization it is often possible to separate the diastereomers (often a very tedious process). Regeneration of the parent acids gives the separate enantiomers.

Another method of resolution utilizes the propensity of living organisms to utilize or convert only one enantiomer of a given pair. In such fashion (e.g., by feeding the material to a mold or bacterial system) it is sometimes possible to selectively destroy or change one enantiomer, leaving the other in a high state of optical purity. This method suffers from lack of generality, but the extreme ease with which it may be effected in those systems where it is applicable have made it the basis of a number of commercial processes, particularly in the antibiotic and antifertility fields.

[1]Any method of chemical combination can, in principle, be used if it is reversible so that the desired substance can be regenerated.

7.7 Meso Forms Not all molecules which possess asymmetric carbon atoms are capable of optical activity. When, for example, a molecule possesses two asymmetric carbon atoms, each with the same substituents, the number of diastereomers possible is less than four. Consider the case of tartaric acid.

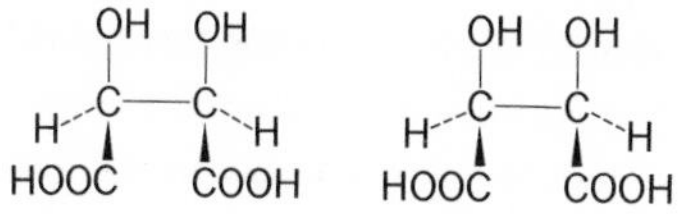

meso-Tartaric acid
(identical with its mirror image)

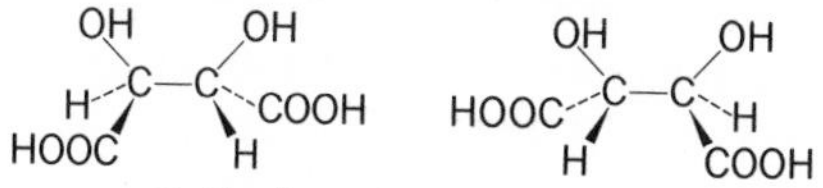

Optically active tartaric acid
(not identical with its mirror image)

Note that *meso*-tartaric acid is identical with its mirror image in spite of the fact that the molecule possesses two asymmetric carbon atoms. This occurs because one half of the molecule is the mirror image of the other; the molecule is said to be *internally compensated.* One can best see this if one draws a plane through the molecule as indicated. Such a plane, which bisects the molecule in such a way that

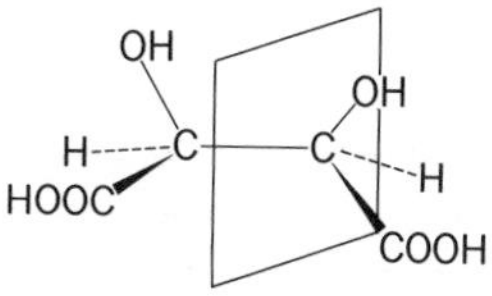

the part of the molecule on one side of the plane is the mirror image of the part on the other side, is called a "plane of symmetry."[1] *Any molecule through which a plane of symmetry can be drawn will not be optically active.* If such a plane cannot be drawn, the molecule *may* be capable of optical activity *unless* certain other types of symmetry are present.

The optically active forms of tartaric acid do not possess planes of symmetry, since the two halves of the molecule are identical but not mirror images. (In the meso form the two halves are mirror images but not identical.) We, therefore, find three forms of tartaric acid: an enantiomorphic pair of right- and left-hand molecules and an inactive meso form. The meso form is a diastereomer of the other two.

7.8 Nomenclature of Optical Isomers With optical isomers the observable property is the rotation of the plane of polarized light (Sec. 7.4) which has both a magnitude and a direction. Rotation in one direction is called plus (+) (in the older

[1]It may also bisect an atom as in the case of 2-chloropropane, where a plane of symmetry bisects the central carbon atom and the hydrogen and chlorine atoms.

literature as d for "dextro") and is measured in degrees. Rotation in the other direction is minus (−) (or l for "laevo").

The property of interest to the study of molecular structure is the *configuration* of molecules. It has long been known that the configuration is related to the observed rotation, but that this relationship is not simple, except of course that changing from one configuration to that of the enantiomer changes only the sign of observed rotation. Because of the difficulty of relating the observed rotation to the actual configuration of molecules, certain molecules with a known rotation were arbitrarily said to have a specific configuration. Other molecules with known signs of rotation were related to these model components by chemical methods and a system of *relative configuration* was built up. Relative configuration was denoted by D or L. In order to specify both the configuration and rotation of a molecule it was customary to write the configuration (D or L) and the rotation (+ or −), for example, D-(+), D-(−), L-(+), etc.

The system of relative configuration as it developed had one major drawback. As more and more molecules were related to the reference molecules, ambiguities developed in relating different groups to one another. It became apparent that the situation was getting worse, so that different investigators would arrive at different configurations from the same molecule.

Since the absolute configurations of a number of molecules had become available from x-ray crystallographic studies, a new system of nomenclature was devised which overcame the ambiguities of the old method. The important features of the new system[1] are summarized below:

1. Configurations are specified as R (for *rectus* or right-handed) and S (for *sinistre* or left-handed) for each asymmetric center, rather than for the molecule as a whole.
2. The substituents on each asymmetric center are assigned a priority, or sequence number, based on the periodic table:
   - *a.* The atom that comes earliest in the periodic table is given the lowest priority and vice versa (e.g., priority increases from $H < C < N < O < F < Si < P < Cl$).
   - *b.* If two or more attached atoms have the same priority, then the priority of the atoms attached to each of these atoms determines their priority (e.g., $—CH_3 < —CH_2CH_3 < —CH_2—NH_2 < —CH_2OH$).
   - *c.* In case 2*b* above, the priority increases with the number of substituents, but position in the periodic table is more important. Thus, $—CH_2—CH_3 < —CH(CH_3)_2 < —C(CH_3)_3 < —CH_2—NH_2 < —CH{<}^{CH_3}_{NH_2}$.
   - *d.* Atoms which are doubly bonded count as two substituents (e.g., $—CH_2CH_3 < —CH{=}CH_2 < —CH_2OH < —CH{=}O$).
   - *e.* If no decision can be reached on the basis of the above, one must go one atom farther out and repeat the process (e.g.,

$$—CH_2CH_2CH_3 < —CH_2CH(CH_3)_2 < —CH_2CH_2OH < —CH_2CH{=}O).$$

[1]The system was proposed by R. S. Cahn, C. Ingold, and V. Prelog in 1956.

f. In choosing between doubly-bonded and multiply-substituted carbon atoms, the substituents on the second atom determine the priority (e.g., $-CH_2{=}O < -CH_2(OH)_2 < -CH_2(OCH_3)_2$).

g. Rules are available for more complex situations, but will not be considered here.

3. Once the priorities for each substituent on the asymmetric center have been assigned, the substituents are numbered 1, 2, 3, and 4, in order of increasing priority. The molecule is then drawn so that the atom of lowest priority, 1, points toward the observer. Thus, for 2-hydroxybutane

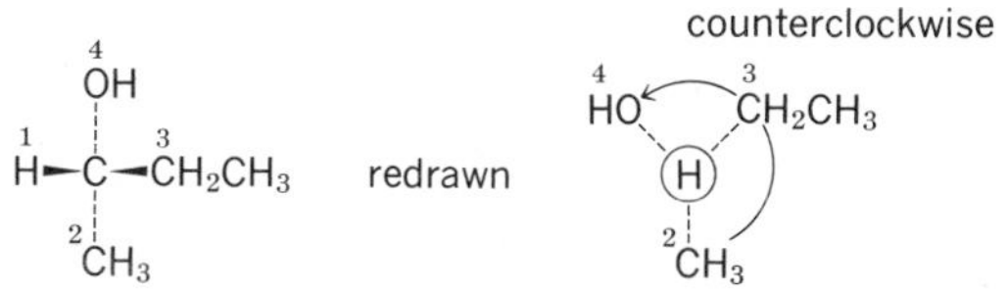

counterclockwise
OH
H
CH_2CH_3
CH_3
redrawn
HO
H
CH_2CH_3
CH_3

The priority sequence is $H < -CH_3 < -CH_2CH_3 < -OH$.

4. An arrow is drawn (see above) in the direction of increasing priority number, starting with number 2. If this arrow proceeds in a *clockwise* direction, the configuration of the active center is *R*. If it proceeds *counterclockwise*, the configuration is *S*. With the molecule above the configuration is clearly *S*.

5. The whole process is repeated for each asymmetric center of the molecule. The configuration is incorporated into the name of the molecule as though it were a substituent. Thus the 2-hydroxybutane molecule above is called 2-*S*-2-hydroxybutane (or *S*-2-hydroxybutane, as there is only one asymmetric carbon atom.)

6. The old terms D and L are retained for *molecular* configurations in certain molecules with a number of asymmetric centers where repeated specification of individual atomic configurations would be tedious. In these cases D and L will be defined in terms of *R* and *S* configurations.[1] See Sec. 21.8.

One can make certain generalizations in terms of absolute configuration that will apply to all molecules with multiple asymmetric centers.

1. Enantiomers will always differ from each other in the configuration of *every* asymmetric carbon atom. If one enantiomer is *R-S-R*, its mirror image must be *S-R-S*.
2. Diastereomers will have at least one center of the same configuration and at least one center of the opposite configuration. Thus, a molecule with three different asymmetric centers which is *R-S-R* will have six diastereomers: *S-S-R*, *R-R-S*, *R-R-R*, *S-R-R*, *S-S-S*, and *R-S-S*.
3. Meso diastereomers will always have one part of the molecule and that is the mirror image of the other, so that it will have configurations such as *R-S* or *S-R*, but will never have configurations *S-S* or *R-R*.

7.9 Conformational Isomerism We have previously treated rotation about single bonds as being "free." As with most "facts" of organic chemical theory, this is

[1] In older textbooks where D and L were used for atomic and molecular configurations, it turns out that D usually corresponds to *R* and L to *S*, but this should always be verified by application of the above rules.

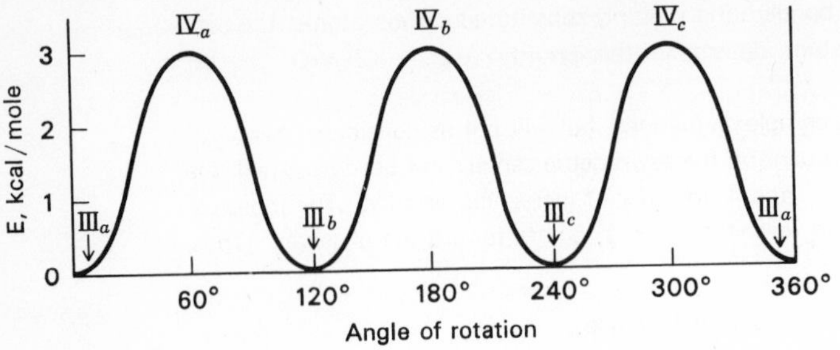

IVa
IVb
IVc
3
2
1
0
E, kcal/mole
IIIa
IIIb
IIIc
IIIa
60°
120°
180°
240°
300°
360°
Angle of rotation

figure 7-2
Rotational barriers in ethane.

not strictly true as a small but finite barrier to rotation does exist. In the case of ethane, for example, a barrier of almost 3 kcal/mole is found. The nature of this barrier is complex, but for practical purposes may be considered to arise from crowding engendered by the finite size of substituents attached to the atoms joined by the bond in question. If one considers the "bow-tie" model of ethane, one can see two different arrangements corresponding to structures III and IV below.

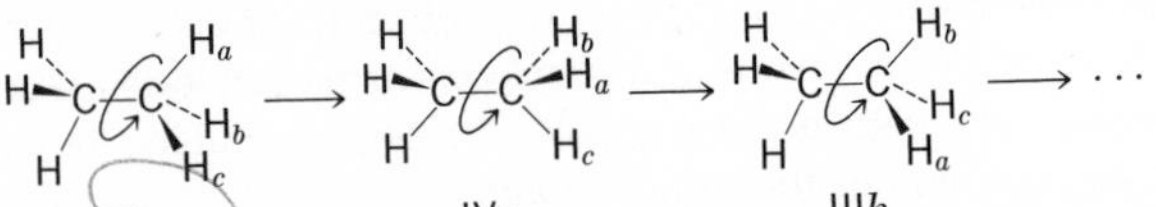

III*a* IV*a* III*b*

Positions III correspond to positions in which the hydrogen atoms are *staggered* with respect to each other, while in position IV the hydrogens are directly in each other's way, or *eclipsed*. For electronic reasons beyond the scope of this book as well as for atomic bulk (*steric* requirements), groups or atoms which are eclipsed are of higher energy than those which are staggered. For ethane, the energy is depicted in Fig. 7-2.

Since the average thermal energy of molecules at 25°C is about 0.6 kcal/mole, many molecules will have sufficient excess energy to cross over the barrier, so that the rotation of ethane is found to be rapid. However, most of the time ethane molecules will be found in configurations III*a*, III*b*, or III*c* and not in IV*a*, IV*b*, or IV*c*. When two or more groups larger than hydrogen are present in the molecule the configurations will not all be of the same energy; the larger groups will prefer positions remote from each other. For example, in *n*-butane V*a* is lower in energy than V*b* or V*c*.

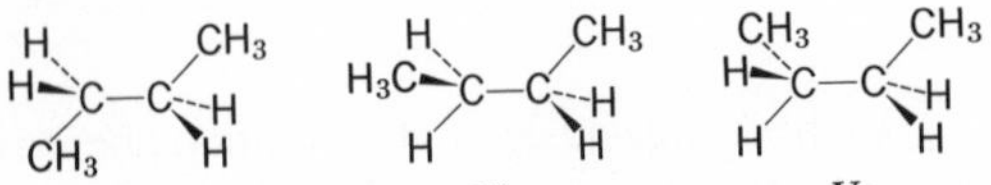

V*a* Favored; V*b*, V*c* Less favorable

These different arrangements are called "conformations" of *n*-butane. The term "conformational isomerism" refers to isomers formed by partially restricted rotation about *single* bonds. Eclipsed conformations represent energy maxima in most systems[1] and do not usually need to be considered.

The importance of conformational isomers in organic chemistry stems from the fact that molecules spend most of their time in the conformations of lowest energy. Various physical and chemical properties such as density, refractive index, infrared and nmr spectra, reaction rates, etc., will reflect the properties of the favored conformations. Other properties, such as many types of asymmetry, will be averaged out by the rapid conformational isomerization. For example, conformational isomers VI*a* and VI*b* are, in principle, capable of optical activity, but rapid conformational equilibration will always result in a racemic mixture of VI*a* and VI*b*. See Sec. 2.4.

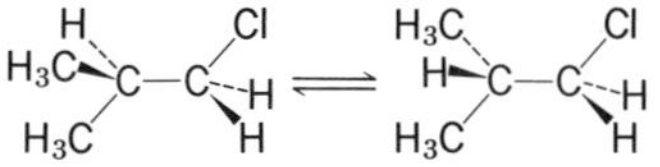
H
Cl
H3C
C—C
H
H3C
H
H3C
Cl
H
C—C
H
H3C
H
VI*a* VI*b*

In cyclic systems the existence of conformational isomerism is most important in six-membered rings where highly staggered arrangements can be formed. Thus, cyclohexane is found in one of two identical "chair" conformations in which all groups are staggered, rather than in a "boat" conformation in which several eclipsed interactions occur.

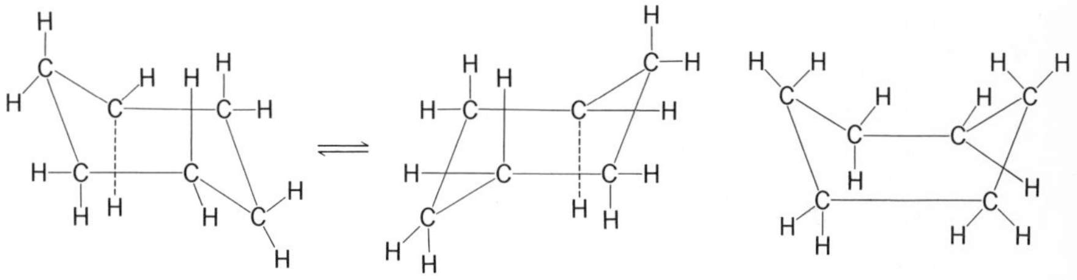
Chair conformations Boat

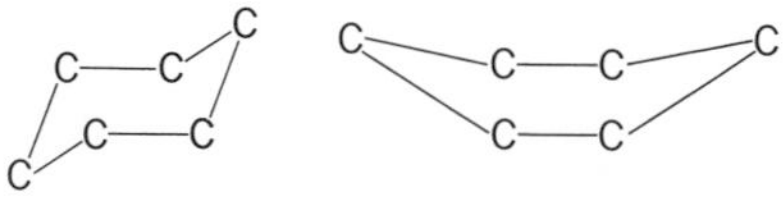
Chair Boat

Skeletons

The hydrogens in a given chair conformation of cyclohexane are not all identical, but fall into two distinct environments. Half of the hydrogens may be found to lie above or below the ring and the other half in a belt around the periphery of the

[1]A number of molecules are known in which interactions in one part of the molecule force groups in other parts of the molecule into eclipsed relationships. This usually raises the overall energy of the molecule.

ring. The former hydrogens are called "axial" (their C—H bonds lie parallel to an imaginary axis through the ring), and the latter "equatorial." The two types of hydrogen are sufficiently different that two types of hydrogen can be seen in the nmr of cyclohexane at very low temperatures. However, at ordinary temperatures the rapid interconversion of the two different chair conformations interconverts axial and equatorial hydrogens rapidly enough to average out the environments of the two types of hydrogens.

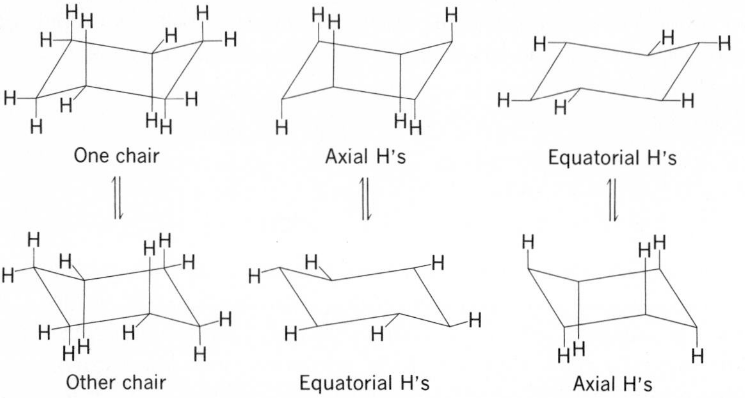
One chair
Axial H's
Equatorial H's
Other chair
Equatorial H's
Axial H's

One important consequence of the difference between axial and equatorial environments is that bulky substituents prefer (on energetic grounds) to be in equatorial positions where they are less crowded. This will tend to favor the equilibrium concentration of one conformer over that of the other.

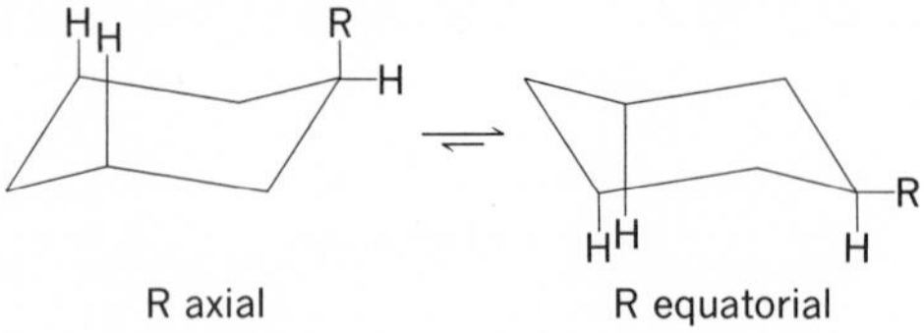
R axial
R equatorial

Because of the rapid equilibration of the two chair forms, conformers cannot ordinarily be isolated. In contrast to this, the cis and trans isomers, and optical isomers are isolable and do not interconvert readily. Although these isomers will exist in different conformations, it is often convenient to draw the rings as though they were flat so as to focus on the geometric or optical isomerism. For example, *trans*-1,2-dibromocyclohexane may be drawn as follows:

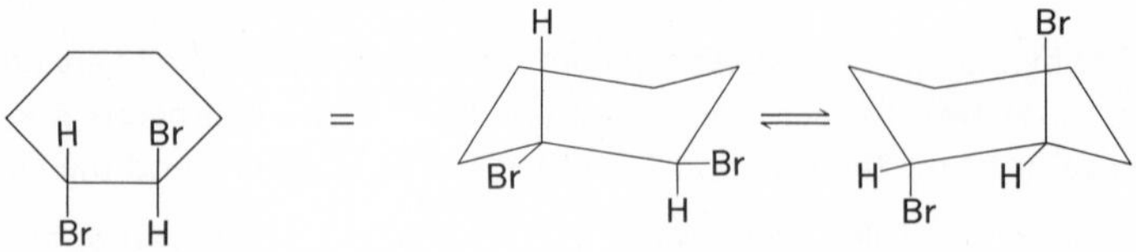

One enantiomer of *trans*-1,2-dibromocyclohexane (*R*,*R*) | Conformational isomers

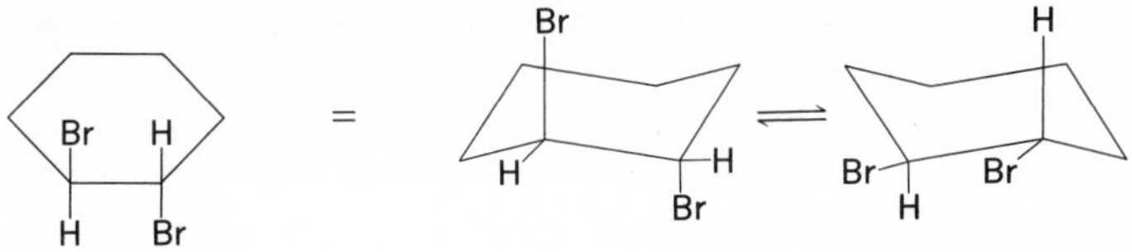

Other enantiomer of
trans-1,2-dibromocyclohexane (*S*,*S*)

Conformational isomers

7.10 The Role of Isomerism in Studies of the Mechanism of Reactions By attempting to understand the way in which reactions proceed as well as what products are obtained, organic chemists have developed an insight into chemistry which simple "cookbook" procedures could not give. The use of optical and geometrical isomerism has served as one tool in the study of the *mechanism* of reactions.

One reaction which has been carefully studied is the addition of bromine to an alkene, which formally is written:

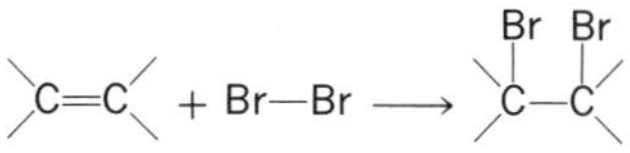

There are several ways in which this reaction might occur, such as addition of both bromine atoms from the same side of the alkene, addition of the two halves of the bromine molecule independently, etc. (see Sec. 3.6). In fact, the following observations have been made:

1. *trans*-2-butene ⟶ *meso*-2,3-dibromobutane (but no racemic 2,3-dibromobutane)

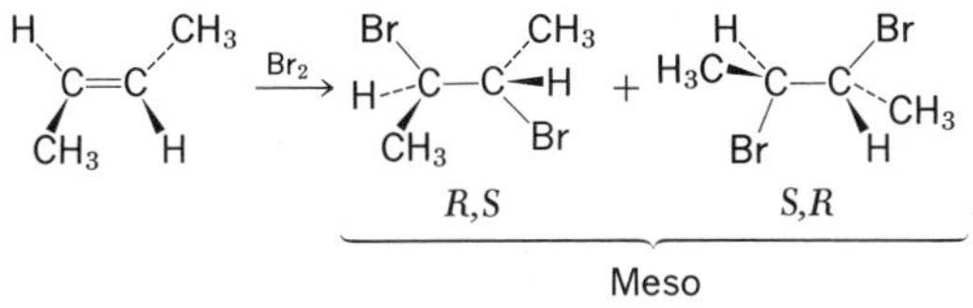

2. *cis*-2-butene ⟶ racemic 2,3-dibromobutane (but no meso)

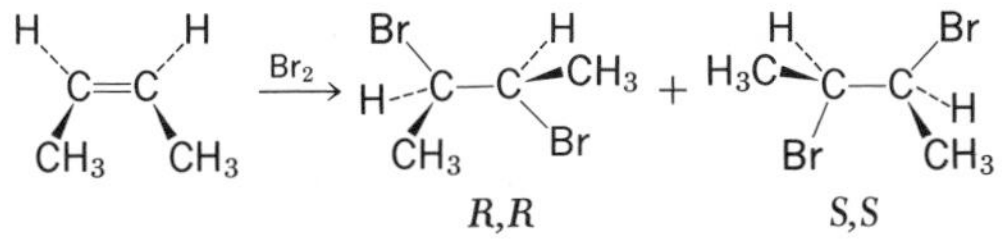

These observations provide us with valuable insight as to the pathway by which the reaction proceeds. It is immediately apparent, for example, that the two bromine atoms cannot come in from the same side of the molecule. If this were the case, the *cis*-butene would give meso, and the *trans* would give racemic. Less obviously but very significantly, the addition cannot involve an intermediate long-lived carbonium ion such as

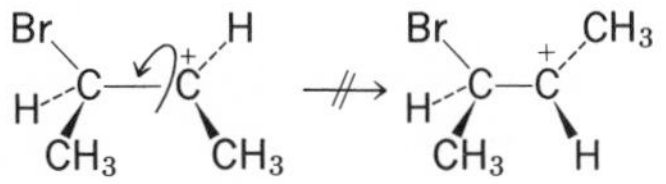

since both cis and trans alkenes would be expected to give the same mixtures of meso and racemic dibromides due to rotation about the central bond.

One mechanism which very succinctly accounts for the observed results involves the formation of an intermediate "bromonium ion," whereby one of the unshared pairs on bromine stabilizes the intermediate.

$$Br_2 \rightleftharpoons Br^+ + Br^-$$

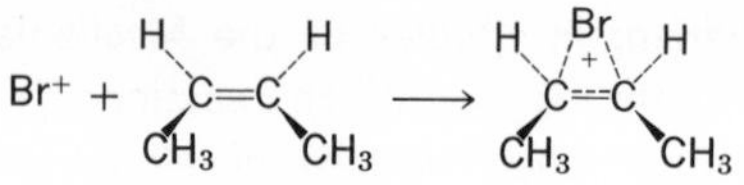

where 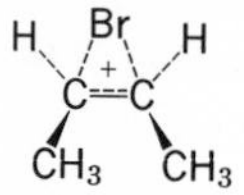

is a resonance hybrid of 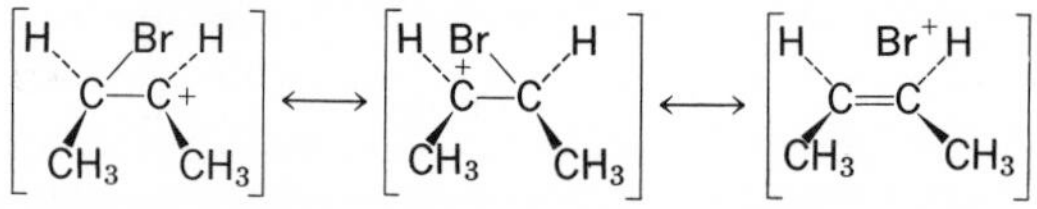

The bridged structure prevents rotation about the central bond and also inhibits attack by Br^- from the same side. When bromide ion does attack, it may do so with equal probability at either end.

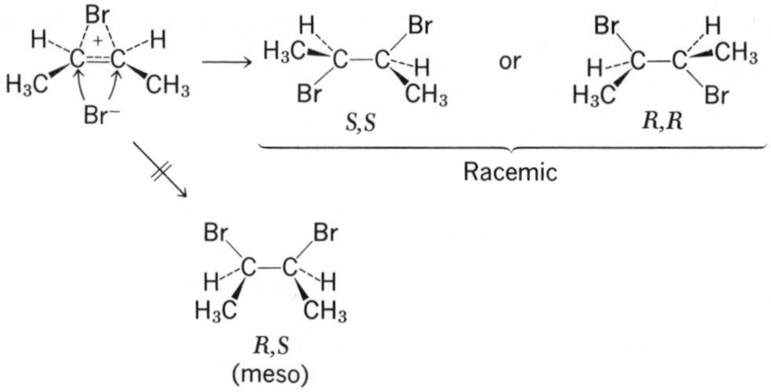
or
S,S
R,R
Racemic
R,S
(meso)

The bromonium-ion concept affords a very reasonable explanation for the observed facts. If we accept the postulate that attack is possible only from the side opposite the bridging bromine, then it is not possible to form the meso (*R*,*S*) dibromide from the above ion. The above bromonium ion is itself optically inactive because it has a plane of symmetry; it is, in fact, a meso ion. Since opening of the bromine bridge must take place with inversion (see Sec. 5.7) at one and only one center, it is clear that the meso ion can only give rise to the racemic pair of dibromides.

Although the bromonium ion affords a simple and reasonable explanation of the majority of currently available facts, it represents a hypothesis and may be subject to alteration or even abandonment if newly uncovered facts render it untenable. There is also a danger in extending such a concept by analogy to related systems such as chloronium ions, since in many cases such analogies are not valid. Nevertheless, for the organic chemist the use of structural concepts has proved an invaluable tool in the understanding of existing phenomena and the investigation of new areas.

7.11 Spectroscopic Properties Because of their greatly different symmetry, geometric isomers usually differ in infrared, uv, and nmr spectra. In many cases geometric isomers can be distinguished by infrared and nmr measurements. For example, trans ethylenes, RCH═CHR′, almost invariably have a band in the infrared at 965 to 990 cm^{-1} which is not shown by the cis isomer. The carbon—carbon double bond absorption of the trans isomer is weaker than that of the cis because the dipole moment does not change as much on vibration in the former and the interaction with the light is weaker.

In the nmr the two hydrogens on the double bond, if not coupled with other hydrogens so as to obscure the pattern, show significantly different coupling constants (Sec. 1.22). For trans isomers the two hydrogens are usually coupled by 12 to 18 cps, while in the corresponding cis isomers the range is 6 to 13 cps.

With cyclopropanes cis and trans hydrogens are coupled by 7 to 13 cps and 3.5 to 8 cps, respectively. Because of very complex interactions with the remaining hydrogens, however, this criterion is readily used only in the case of highly substituted molecules where only two hydrogens are present on the ring. With larger rings a similarly complex situation arises.

Although spectroscopic methods are of considerable utility in the differentiation of enantiomers, diastereomers, and conformers, the treatment of these systems is a matter for more advanced courses.

7.12 Summary Geometric, optical, and conformational isomerism represent three important types of isomerism in chemical systems. The corresponding isomers are called cis-trans isomers,[1] enantiomers (or disastereomers), and conformers, respectively.

CHAPTER 7 EXERCISES

1. Show all the geometrical isomers (disregarding optical isomers) for the following:

a. 2-Pentene

b. 2-Methylcyclopentene

c. 1,2,3-Trichlorocyclobutane

d.

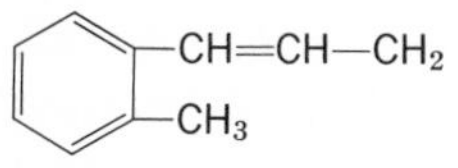

CH═CH—CH2
CH3

[1]Other types of geometric isomerism are known, but in principle they are very similar to cis-trans and will not concern us here.

e. $CH_3C{\equiv}C{-}CH{=}CH{-}CH{=}CH_2$

f. 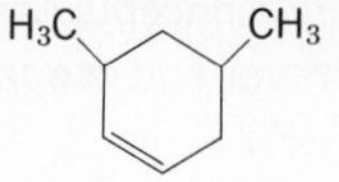
H3C
CH3

g. 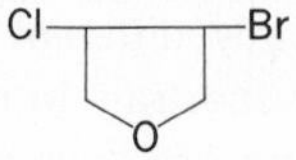
Cl
Br
O

2. Account for the fact that the nmr spectrum of cyclohexane compounds changes at very low temperatures while corresponding cyclobutane compounds show no such spectral change. Describe the kind of spectral change you might expect.

3. Define the following terms:

a. Enantiomer
b. Meso
c. Diastereomer
d. Configuration
e. Conformation
f. Racemic
g. Resolution
h. Asymmetry

4. How could you distinguish between a racemic and a meso form of a substance?

5. Which of the following are capable of existing in optically active forms? In each case draw the active forms and label them *R* or *S*.

a. $CH_3CH_2CH(CH_3){-}CH_2CH_2Br$ (with CH_3 on the third carbon)

★*b.* $H_2C{=}CH{-}CH(OH){-}CH{=}CH_2$

★*c.* $CH_2(Br){-}CH(Br){-}CH(OH){-}CH{=}CH_2$

d. $CHBrClF$

★*e.* 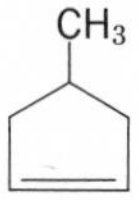
CH3

6. Show how you would resolve a basic substance into its *R* and *S* forms.

★7. Could *meso*-tartaric acid be used in the resolution of a basic substance? Explain.

8. Draw several conformations for the following compounds and predict the most stable one where possible:

a. 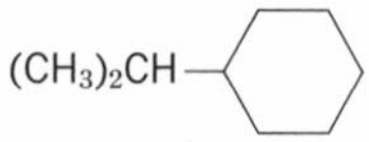
(CH3)2CH

b. $BrCH_2{-}CH_2Br$

c. 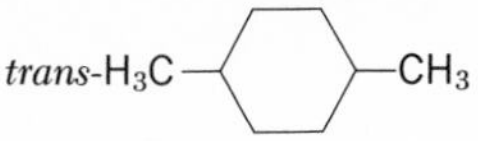
trans-H3C
CH3

d. $CH_3CH(OH)—CH(OH)CH_3$ (OH on each of the two central carbons)

e. 

★**9.** Pure R-$CH_3CH(C_2H_5)—CH_2Br$ reacts (twice) with sodium acetylide to form

$CH_3CH(C_2H_5)—CH_2—C{\equiv}C—CH_2—CH(C_2H_5)—CH_3$ Will the product be optically active? *R*? *S*? *meso*? Explain and show configurations.

10. To the product of Exercise 9 above add one molecule of H_2 to form an alkene. Show all isomers of the alkene produced; all possible isomers are not necessarily produced.

11. Show all possible stereoisomers of the following. Label them as active, meso, cis, trans, etc.

★*a.* $H_3C—CH(Br)—CH(Br)—CH(Br)—CH_3$

★*b.* 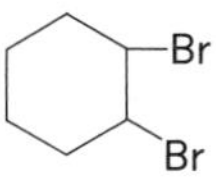
Br
Br

★*c.* 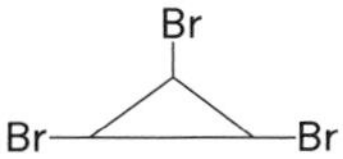
Br
Br
Br

12. Assuming the mechanism of addition of Br_2 as described in this chapter, show the isomers expected in the following reactions. Would the products be optically active? Capable of resolution?

a.  + Br_2

b. $CH_3CH{=}CHC_2H_5 + Br_2$

c. $CH_3CH{=}CH—CH{=}CH—C_2H_5 + Br_2$

★**13.** R-$H_3C—CH(C_2H_5)—Br + HC{\equiv}CNa \longrightarrow$ an *S* product. Show its configuration. What does this imply about the direction of approach of the anion to the asymmetric carbon atom? Explain by a diagram.

14. *trans*-$CH_3CH{=}CHCH_3$ + dil. $KMnO_4$ $\longrightarrow$ $\underset{\displaystyle OH}{CH_3\overset{|}{C}H}-\underset{\displaystyle OH}{\overset{|}{C}H}-CH_3$

Assuming that the OH groups come on to the same side of the alkene, which isomer will result? Will it be optically active?

★**15.** Cyclopentene and dil. $KMnO_4$ give a product that cannot be resolved. What information does this provide about the nature of the reaction? Does it agree with your answer to Exercise 14 above? Explain.

8

ALCOHOLS AND PHENOLS

8.1 Classification of Hydroxylated Compounds Oxygen is divalent in practically all its organic linkages. One or both of its bonds may be attached to carbon. When one bond of oxygen is attached to hydrogen and the other to carbon thus: —C—O—H, there is formed the class of hydroxylated organic compounds. Here the functional group is the hydroxyl (—OH) group.

It is customary to subdivide the class of hydroxylated organic compounds into a number of smaller classes on the basis of chemical behavior: one may distinguish the subclass of aliphatic *alcohols* and the subclass of *phenols*.[1] In the former, the —OH group is attached to an aliphatic carbon; in the latter, directly to a carbon of an aromatic ring (replacing —H). Both these subclasses show the reactions of

[1] There are other hydroxyl-containing groups, such as —COOH; however, the properties of the hydroxyl in this group are so influenced by the —CO— that we classify the whole group, including the —OH, by a special name, carb*oxyl*.

the —OH group, but they differ in certain respects in reactions involving, in one way or another, the rest of the molecule. Again, the subclass of the aliphatic alcohols may be divided, on the basis of reactions which involve the carbon to which the —OH is attached, into *primary*, *secondary*, and *tertiary* alcohols, depending upon the number of groups (aliphatic or aromatic) that are attached to the hydroxylated carbon:[1]

CH_3CH_2OH A primary alcohol

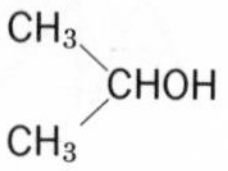

A secondary alcohol

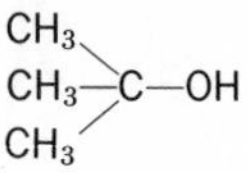

A tertiary alcohol

8.2 Nomenclature The first few members of the homologous series of alcohols are generally named from the groups to which the OH is attached; the higher members are named systematically. In the systematic naming the suffix *-ol* is used to indicate the presence of an OH group, and a number is used to locate the OH on the carbon chain. A few examples will serve to illustrate:

Symbol	*Trivial name*	*Systematic name*
CH_3CH_2OH	Ethyl alcohol	Ethanol
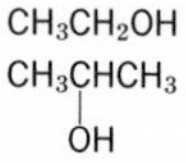	Isopropyl alcohol	2-Propanol
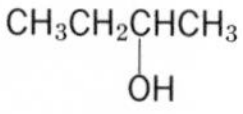	*sec*-Butyl alcohol	2-Butanol
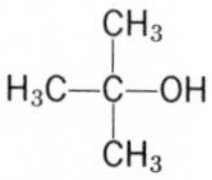	*t*-Butyl alcohol	2-Methyl-2-propanol
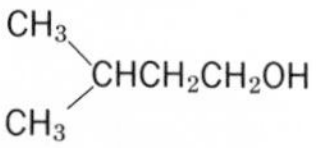	Isoamyl alcohol[a]	3-Methyl-1-butanol
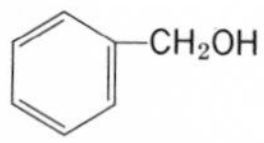	Benzyl alcohol	Phenylmethanol

[a] *Amyl* is in common use for 5-carbon alcohols and alkenes and their derivatives. Amyl alcohols are found in fusel oil, a by-product of the fermentation of starch to ethyl alcohol.

---

[1]The terms primary, secondary (*sec-*), and tertiary (*tert-*) are sometimes used in the same way in the nomenclature of other classes of compounds; thus,

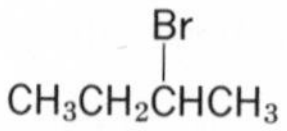

sec-Butyl bromide

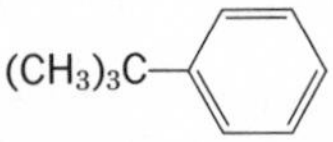

tert-Butylbenzene

It is important to note that either the trivial or the systematic name may be used, but they should not be mixed. Thus, names which use *sec, iso,* and *tert* should not be used with systematic endings, and names such as *t*-butanol, *iso*-propanol, or 2-propyl alcohol should be avoided.

Another system of nomenclature which is sometimes convenient in naming secondary and tertiary alcohols is found in the literature. In this system, all alcohols are considered as derivatives of *methanol,* called "carbinol."

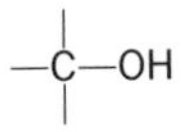
—C—OH

Carbinol

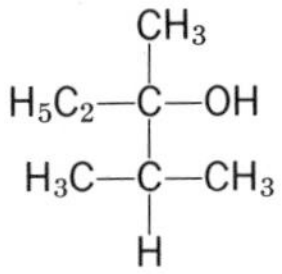
CH3
H5C2—C—OH
H3C—C—CH3
H

Methylethylisopropyl carbinol

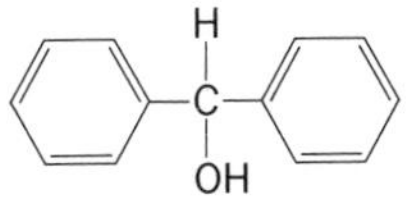
H
C
OH

Diphenyl carbinol

This nomenclature is ordinarily used when the previous nomenclature is awkward. Phenols may be named as derivatives of the parent phenol; thus,

HO
NO2

is *p*-nitrophenol, or 4-nitrophenol; or by the usual numbering methods, calling —OH "hydroxy"; thus,

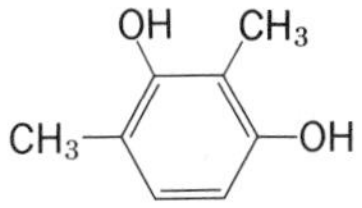
OH
CH3
CH3
OH

is 2,4-dimethyl-1,3-dihydroxybenzene. The student should, however, memorize the following trivial names:

HO
CH3

p-Cresol[1]

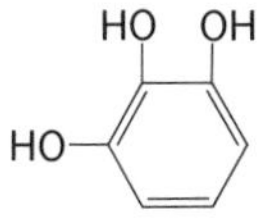
HO
OH
HO

Pyrogallol[2]

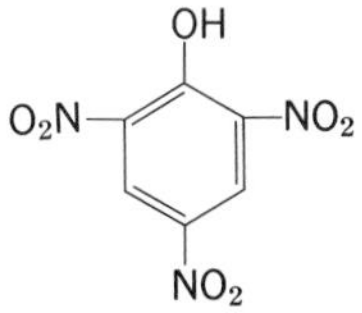
OH
O2N
NO2
NO2

Picric acid[3]

8.3 Preparation In commercial practice the simple alcohols are often prepared by more or less specialized processes such as fermentation.[4] Because of the importance of the reactive OH group, however, it is necessary to study in detail the

[1] The cresols are used as antiseptics. They are less toxic and more bactericidal than phenol.
[2] Pyrogallol is used as a photographic developer and for the absorption of oxygen in gas analysis.
[3] Picric acid finds many uses, some of which are as a tanning reagent, a dye, a fixative in histological preparations, and as an explosive.
[4] 1-Butanol is a major product of the action of bacteria on otherwise waste cornstalks. The amyl alcohols are a by-product of ethanol fermentation. A discussion of these processes is usually regarded as part of biochemistry and will be omitted here.

methods for introducing this group into carbon compounds. Phenols are usually prepared by specialized methods; hence, a consideration of their preparation will be deferred, and only alcohols will be taken up at this point.

8.4 Hydrolysis Since alcohols are alkyl derivatives of water, they can be prepared by the reaction of alkyl halides and alkyl esters with water. This is usually referred to as *hydrolysis* (decomposition, *-lysis,* by means of water, *hydro-*) of halides and esters. Alkyl halides are found to react slowly with water in a reversible reaction.

$$CH_3CH_2Br + H_2O \rightleftharpoons CH_3CH_2OH + HBr$$

The position of equilibrium can be shifted by application of Le Chatellier's principle: an excess of water favors alcohol formation, and an excess of acid favors halide formation. In many cases the reaction is too slow to be of value, and a modification of it is used in the preparation of alcohols.

$$CH_3CH_2Cl + OH^- \xrightarrow{NaOH} CH_3CH_2OH + Cl^-$$

Here we find a *displacement reaction* taking place. See Sec. 5.7. This reaction is not ordinarily a good reaction for the preparation of alcohols. Although it proceeds reasonably well with primary halides, it is always accompanied by some elimination, and with more substituted halides elimination often becomes the dominant reaction.

$$\underset{\text{A primary halide}}{CH_3{-}CH_2{-}Br} + OH^- \longrightarrow \underset{\text{Mostly}}{H_3C{-}CH_2{-}OH} + \underset{\text{Trace}}{H_2C{=}CH_2}$$

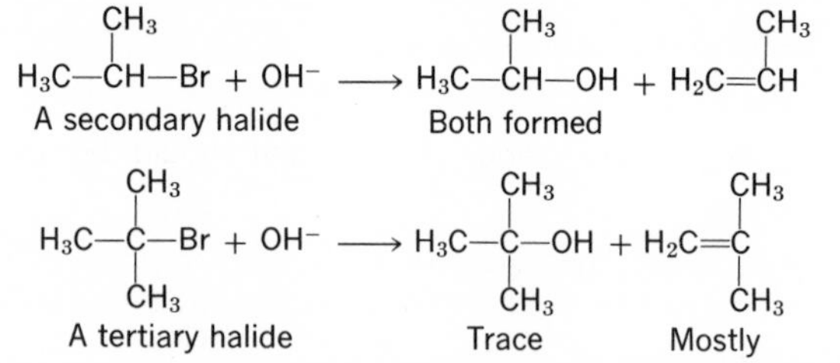
A secondary halide
Both formed
A tertiary halide
Trace
Mostly

The reaction fails completely with most aromatic and vinyl halides.

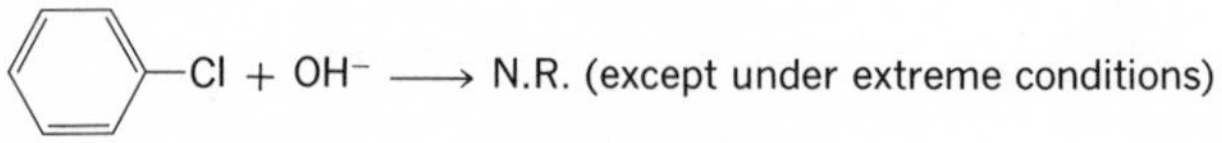
N.R. (except under extreme conditions)

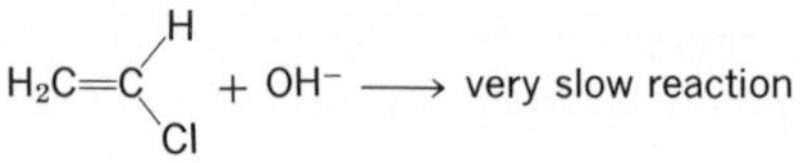
very slow reaction

The mechanisms of the displacement reaction have received considerable study and we shall consider them in detail in Secs. 9.5 and 9.6.

8.5 Hydrolysis of Esters An important source of some alcohols is a class of compounds which *formally* are derived from an alcohol and an acid, minus the elements of one water molecule. These compounds are called esters, some examples of which are given below:[1]

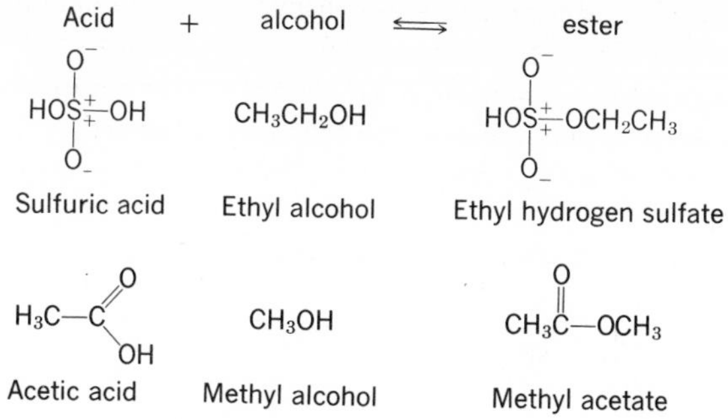

Acid   +   alcohol ⇌ ester

$$HOS_{+}^{\pm}OH \qquad CH_3CH_2OH \qquad HOS_{+}^{\pm}OCH_2CH_3$$

Sulfuric acid   Ethyl alcohol   Ethyl hydrogen sulfate

$$H_3C-C{\overset{O}{\underset{OH}{}}} \qquad CH_3OH \qquad CH_3\overset{O}{\overset{\|}{C}}-OCH_3$$

Acetic acid   Methyl alcohol   Methyl acetate

Although the alkyl halides might be defined as esters,

Acid + alcohol ⇌ "ester"
HCl   CH₃OH   Cl—CH₃

it is more usual to restrict the term ester to compounds in which an oxygen linkage is involved.

The hydrolysis of an ester by the action of water proceeds to equilibrium according to the equation below:

$$H_3C-O-S_{+}^{\pm}-O-H + HOH \rightleftharpoons CH_3OH + HO-S_{+}^{\pm}-OH$$

$$CH_3\overset{O}{\overset{\|}{C}}-OCH_3 + HOH \underset{esterification}{\overset{hydrolysis}{\rightleftharpoons}} CH_3\overset{O}{\overset{\|}{C}}-OH + CH_3OH$$

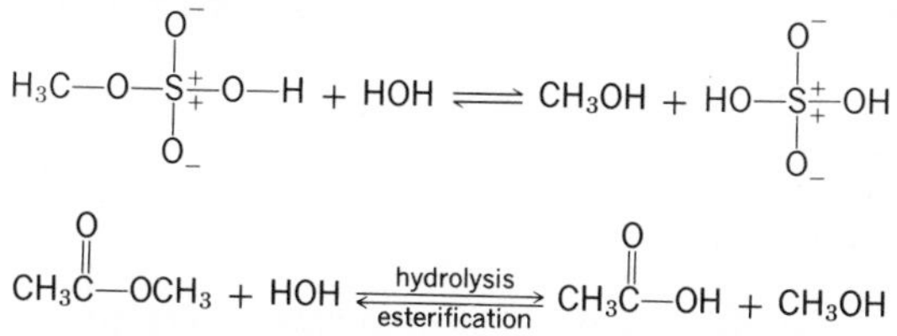

Hydrolysis of a carboxylic ester with water containing an excess of O^{18} (the heavy isotope of oxygen) shows that all the O^{18} is incorporated into the acid and none into the alcohol.[2] The reaction, therefore, takes place with cleavage of the bond between the C=O and —O—, and not between O— and CH₃.

$$CH_3\overset{O}{\overset{\|}{C}}{\vdash}O-CH_3 + HO^{18}H \rightleftharpoons CH_3\overset{O}{\overset{\|}{C}}-O^{18}H + HOCH_3$$

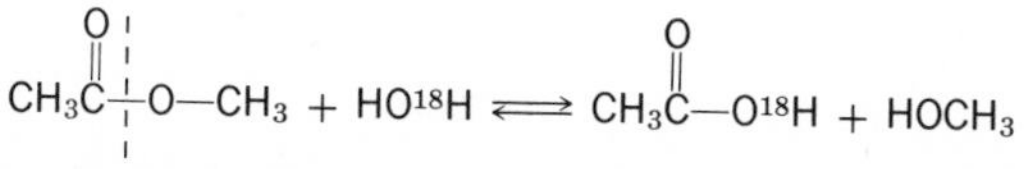

The mechanism of this hydrolysis (as well as the reverse reaction, esterification) is complex, and only the overall reaction will be written until the chemistry of esters is considered in more detail (Sec. 13.3).

---

[1] There exist compounds analogous to these involving oxygen, in which elements in the same column of the periodic table, namely sulfur, selenium, etc., occur in place of one or both of the oxygens.

[2] In organic chemistry it is nearly always possible to find some molecule that does not obey a given generalization. In this case, a few examples are known in which the O^{18} ends up in the alcohol, but these are readily understood in terms of modern mechanistic theory.

Those esters which are available as natural products or from syntheses which do not start with the alcohols themselves are useful materials for the preparation of alcohols. Esters which are natural products are discussed in the chapter on carboxylic acids (Chap. 12). One type of synthetic ester has been encountered before and is of considerable value in the preparation of alcohols. For example, the commercial preparation of ethyl alcohol involves the following sequence.

$$H_2C=CH_2 + HO-\overset{\overset{-}{O}}{\underset{\underset{-}{O}}{\overset{+}{\underset{+}{S}}}}-OH \xrightarrow{55-80°} H_3C-CH_2-O-\overset{\overset{-}{O}}{\underset{\underset{-}{O}}{\overset{+}{\underset{+}{S}}}}-OH$$

Concentrated

$$CH_3CH_2O-\overset{\overset{-}{O}}{\underset{\underset{-}{O}}{\overset{+}{\underset{+}{S}}}}-OH + H_2O + H_2SO_4 \xrightarrow{\Delta} CH_3CH_2OH + 2H_2SO_4$$

Diluted

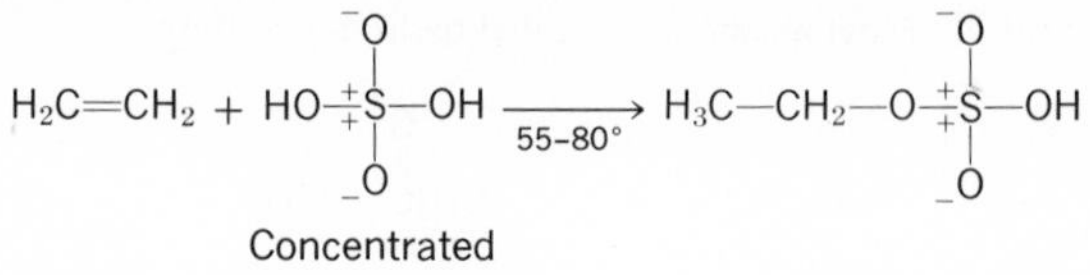

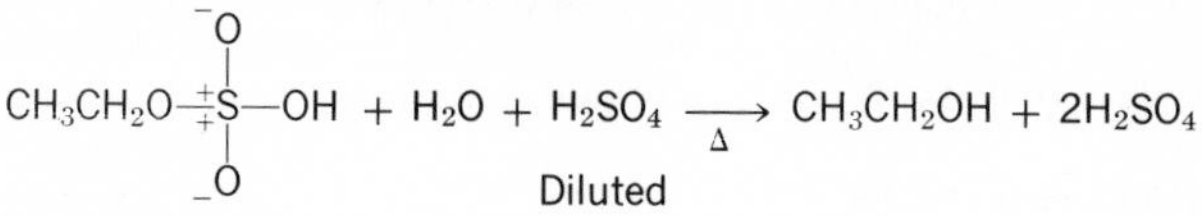

Thus, although ethylene does not hydrate readily with water, the same result can be obtained indirectly via the hydrogen sulfate ester.

In the case of acid sulfates cleavage may occur between the CH_2 and —O—, or between the O— and —SO_3H. Both types of cleavage are known, with the former often predominating under alkaline conditions and the latter under strongly acidic ones. Thus acid sulfates undergo displacement reactions much like the alkyl halides.

$$CH_3CH_2-O-\overset{\overset{-}{O}}{\underset{\underset{-}{O}}{\overset{+}{\underset{+}{S}}}}-OH + OH^- \longrightarrow CH_3CH_2-OH + {}^-O-\overset{\overset{-}{O}}{\underset{\underset{-}{O}}{\overset{+}{\underset{+}{S}}}}-OH$$

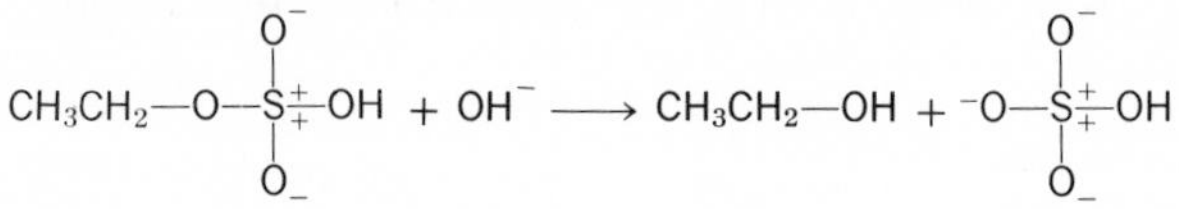

The hydrolysis of alkyl halides and of sulfate esters constitutes an effective method of preparing certain alcohols from alkenes. Thus 2-propanol can be prepared either by

$$H_3C-CH=CH_2 + HBr \longrightarrow H_3C-\overset{\overset{Br}{|}}{C}H-CH_3 \xrightarrow[\Delta]{H_2O} H_3C-\overset{\overset{OH}{|}}{C}H-CH_3 + HBr$$

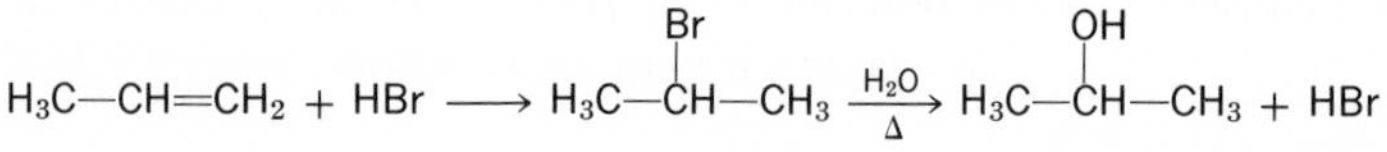

or by

$$H_3C-CH=CH_2 + H_2SO_4 \longrightarrow H_3C-\overset{\overset{OSO_3H}{|}}{C}H-CH_3 \xrightarrow[\Delta]{H_2O} H_3C-\overset{\overset{OH}{|}}{C}H-CH_3 + H_2SO_4$$

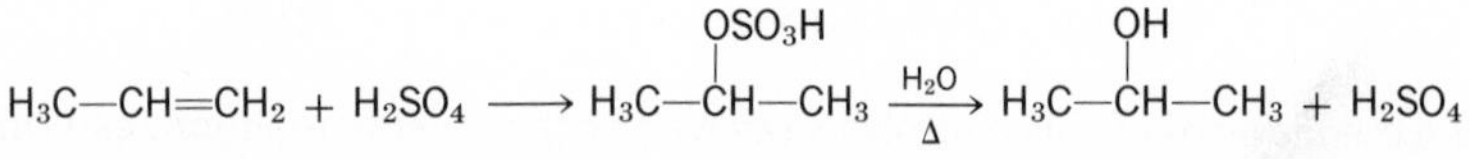

These do not constitute catalytic methods because the concentrations of HBr or H_2SO_4 necessary to effect the first step are much greater than those which can be used for the second step.

It should be noted also that because of Markownikoff's rule, primary alcohols cannot be prepared in this fashion since the proton always adds to the *least-substituted* carbon.

The effective antiMarkownikoff addition of water to alkenes can be brought

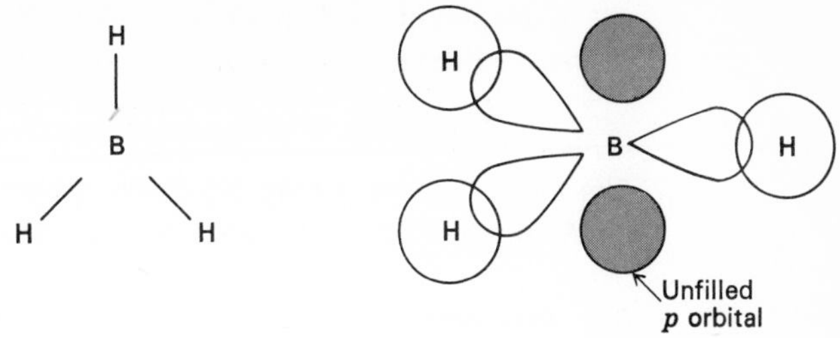
H
B
H
H
H
H
B
H
Unfilled
p orbital

figure 8-1
BH_3, boron hydride.

about by the use of diborane (B_2H_6). This unusual reagent has the structure

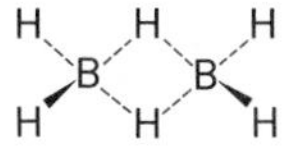

The central hydrogens are involved in a B---H---B three-electron bond, the exact nature of which is beyond the scope of this book. Diborane can be prepared in a variety of ways, one of the most convenient being the reaction of sodium borohydride with boron trifluoride.

$$3NaBH_4 + BF_3 \longrightarrow 2B_2H_6 + 3NaF$$

Diborane is a very reactive species which spontaneously ignites in air and dissociates to give an equilibrium concentration of BH_3.

$$B_2H_6 \rightleftharpoons 2BH_3$$

The molecule BH_3 is planar, sp^2-hybridized (Sec. 3.5), and has an incomplete valence shell (Sec. 1.2) so that it closely resembles a carbonium ion in structure (Sec. 9.6). The unfilled p orbital is readily available for bond formation with electron-rich species like the π electrons of alkenes (Fig. 8-1).

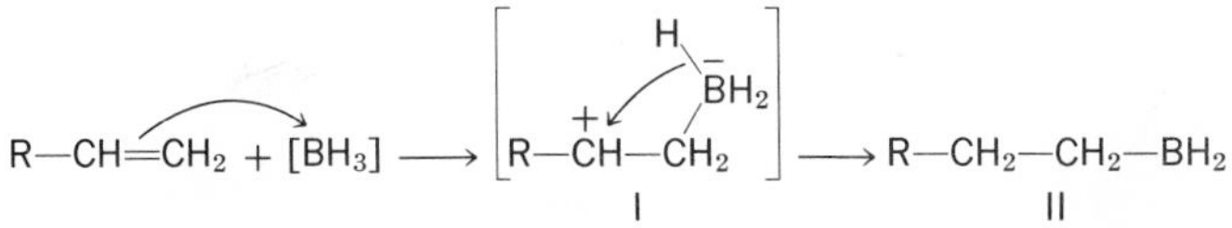

I II

The BH_3 adds according to Markownikoff's rule (Sec. 3.9) to give the most stable carbonium ion-like species (Sec. 3.10), I, which can accept a hydride ion from the BH_3^- moiety to yield II. The sequence can be repeated until all three hydrogens of BH_3 have been utilized.

$$R{-}CH_2{-}CH_2{-}BH_2 + R{-}CH{=}CH_2 \longrightarrow \underset{\text{III}}{(R{-}CH_2{-}CH_2)_2BH}$$

$$(R{-}CH_2{-}CH_2)_2BH + R{-}CH{=}CH_2 \longrightarrow \underset{\text{IV}}{(R{-}CH_2{-}CH_2)_3B}$$

Oxidation of the trialkylborane, IV, with hydrogen peroxide leads to a cleavage of the C—B bond, yielding a C—O bond.

$$(R—CH_2—CH_2)_3B + 3H_2O_2 \longrightarrow 3\ R—CH_2—CH_2—OH + H_3BO_3$$

The entire transformation effectively adds the elements of H_2O to the alkene in antiMarkownikoff fashion, since the OH ends up on the less-substituted carbon atom.

8.6 Preparation by Reduction As discussed in the chapters on alkenes and alkynes, compounds with the groups

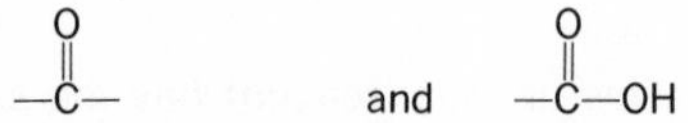

Aldehydes and ketones — Acids

can be prepared by oxidative methods. The aldehydes and ketones will add hydrogen (be reduced) at the double bond in the group

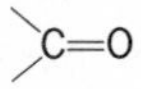

to form, respectively, primary and secondary alcohols.

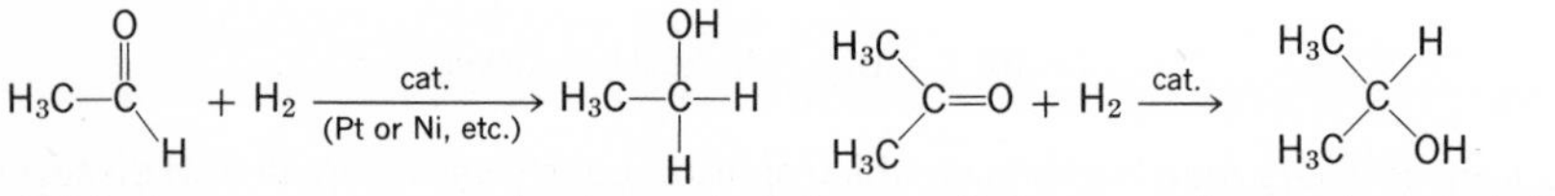

Aldehyde — A primary alcohol — Ketone — A secondary alcohol

The acids, however, though they appear to have a C═O group, do not readily add hydrogen. An explanation of this will be given later (Chap. 12).

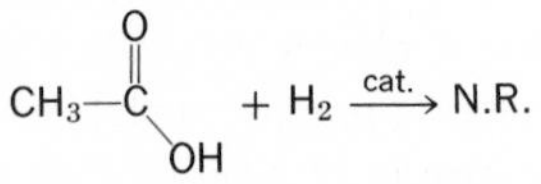

If the acids are converted to esters, they add hydrogen more readily.

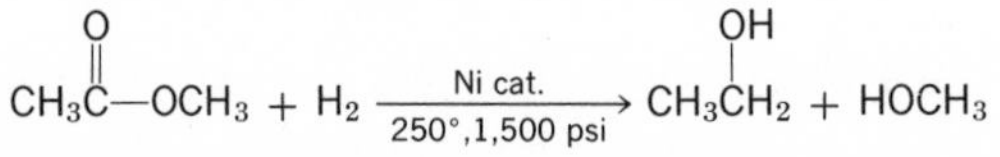

Note that not only does hydrogen add to the C═O group, it also cleaves the C—O—C bond. It is generally not possible to stop at any intermediate stages. This reaction is of considerable commercial importance and will be discussed later.

The carbonyl group of acids can be smoothly reduced by the reagent lithium aluminum hydride, $LiAlH_4$.

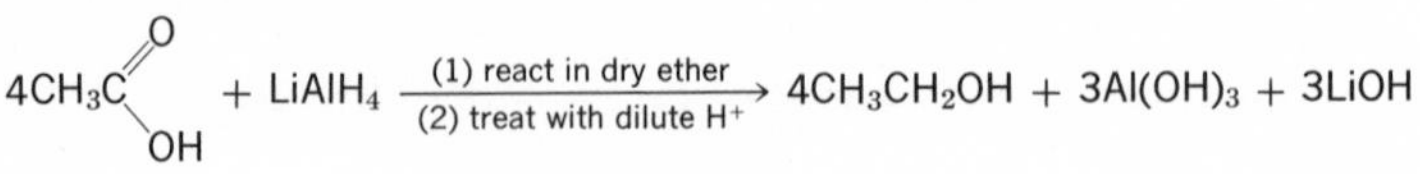

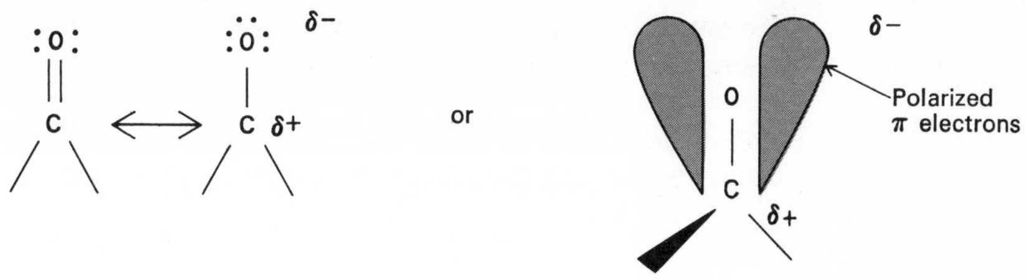
δ−
δ+
or
O
C
δ−
Polarized
π electrons
δ+

figure 8-2
The carbonyl group.

This reagent will reduce all varieties of carbonyl group, but does not ordinarily touch carbon-carbon double bonds.

$$H_2C{=}CH{-}CH_2{-}\overset{O}{\overset{\|}{C}}{-}H \xrightarrow[(2)\ H_2O]{(1)\ LiAlH_4} CH_2{=}CH{-}CH_2{-}CH_2OH$$

Many other mixed hydrides also find use for reduction of carbonyl compounds to alcohols. An important example is sodium borohydride, $NaBH_4$, which reduces aldehydes and ketones, but not acids and esters, and which may be used in the presence of water (with which lithium aluminum hydride reacts violently).

$$H_3C{-}\overset{O}{\overset{\|}{C}}{-}CH_2{-}CH_2{-}\overset{O}{\overset{\|}{C}}{-}O{-}CH_3 + NaBH_4 \xrightarrow[\text{solvent}]{CH_3OH} H_3C{-}\overset{OH}{\overset{|}{C}H}{-}CH_2{-}CH_2{-}\overset{O}{\overset{\|}{C}}{-}O{-}CH_3$$

8.7 Preparation by the Grignard Method The most versatile and general method for the laboratory synthesis of alcohols makes use of Grignard reagents (Sec. 2.15). The carbonyl group, unlike the carbon-carbon double bond, is strongly polarized with the negative end toward the electronegative oxygen atom (Fig. 8-2). The strongly polarized Grignard reagent reacts as a nucleophile toward the electophilic carbon atom of the carbonyl.

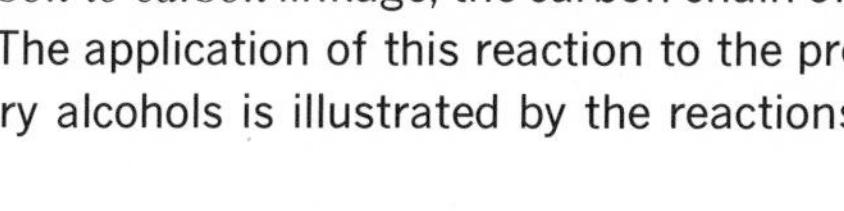
δ+
δ−
Br—Mg—CH3
C
δ+
:O:
δ−
H3C—C—
:O:−
+
MgBr

In this process the alkyl group of the Grignard reagent becomes attached to the carbon atom of the carbonyl group and the magnesium to the oxygen. There is thus formed a new *carbon-to-carbon* linkage; the carbon chain of the Grignard reagent has been extended. The application of this reaction to the preparation of primary, secondary, and tertiary alcohols is illustrated by the reactions given below.

Primary alcohol

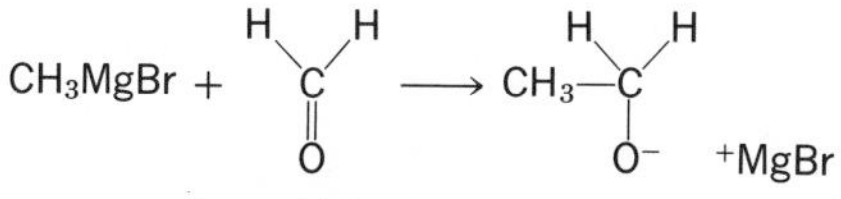
H
H
CH3MgBr +
C
O
CH3—C
O−
+MgBr

Formaldehyde

The salt $C_2H_5O^{-}\ {}^{+}MgBr$ is not ordinarily isolated. It usually forms an ether-insoluble precipitate in the reaction flask and is converted to the parent alcohol (a very weak acid) by the use of stronger acid. Even water is sufficiently acidic for this (Sec. 8.16). In synthetic work ammonium chloride or dilute acid is used for convenience to prevent the precipitation of magnesium hydroxide.

$$CH_3CH_2O^{-}\,{}^{+}MgBr + H^{+} \rightleftharpoons CH_3CH_2OH + {}^{+}MgBr$$

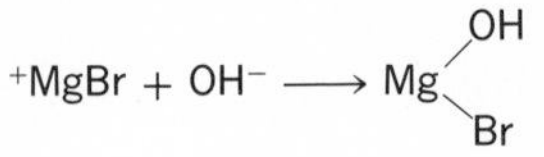

Basic magnesium bromide or a mixture of the bromide and hydroxide

Secondary alcohol

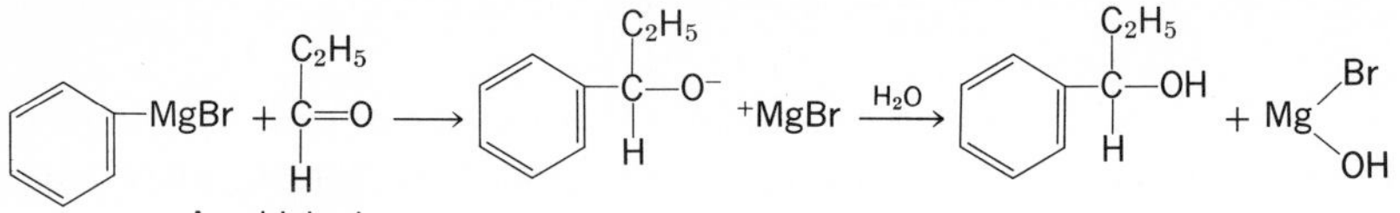
An aldehyde other than formaldehyde

Tertiary alcohol

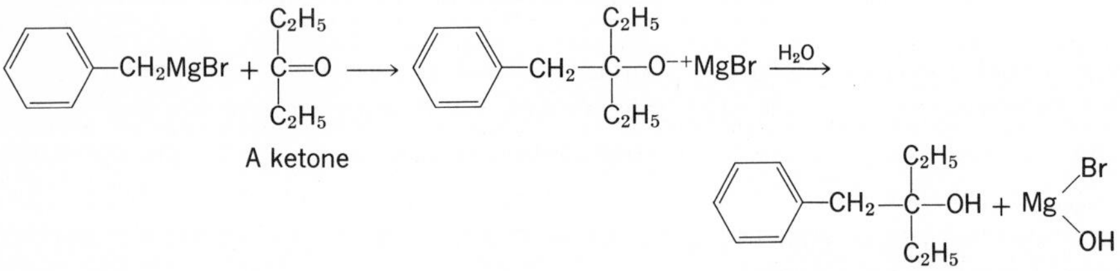
A ketone

Besides the familiar Grignard reagents, other organometallic reagents are known which have found use in synthetic work. Among the most useful of these are the lithium compounds which can be prepared in a manner analogous to the preparation of Grignard reagents; the experimental procedures are often more complex (exclusion of air, use of solvents other than ether, etc.), but the reactions parallel almost exactly those of the magnesium analogs. Being more reactive and less subject to interference by nearby bulky groups, they are sometimes useful in situations where the Grignard reagent fails.

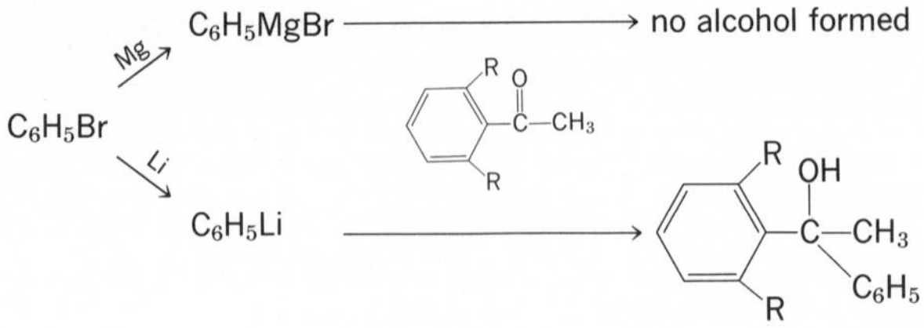
no alcohol formed

Grignard reactions take place readily and are carried out at or near room temperature. The yields are usually good, so that this process has great synthetic value, especially in the synthesis of complicated alcohols or other substances made from alcohols. It should be observed that in this reaction the length of a chain may be increased at a definite point and with maintenance of a functional group.

The importance of having a functional group arises from the low selectivity of reactions of the hydrocarbons. In order to limit reaction in a molecule to a specific site it is necessary to have a reactive site (functional group) in the molecule where selective reactions can occur. In the Grignard synthesis a functional group is maintained so that this reaction may be used as one of a series of steps for building up complex molecules from simple ones.

8.8 Use of Ethylene Oxide The strained three-membered ring in ethylene oxide (Sec. 10.8) and its homologs behave toward organometallic reagents in a manner analogous to the behavior of carbonyl compounds.

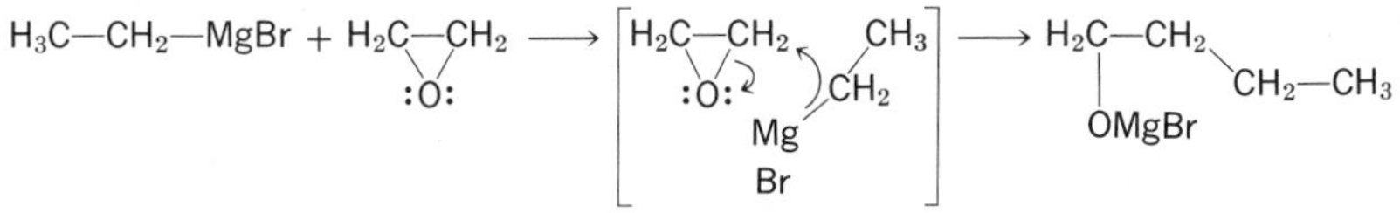
H3C—CH2—MgBr + H2C—CH2
:O:
H2C—CH2 CH3
:O: CH2
Mg
Br
H2C—CH2
CH2—CH3
OMgBr

Complex
(possible structure)

This useful reaction, in which the carbon chain may be extended by two CH_2 groups, presumably involves ring opening assisted by coordination of the magnesium with the oxygen. Since magnesium bromide is always present in such solutions,[1] a side reaction is found to accompany the above.

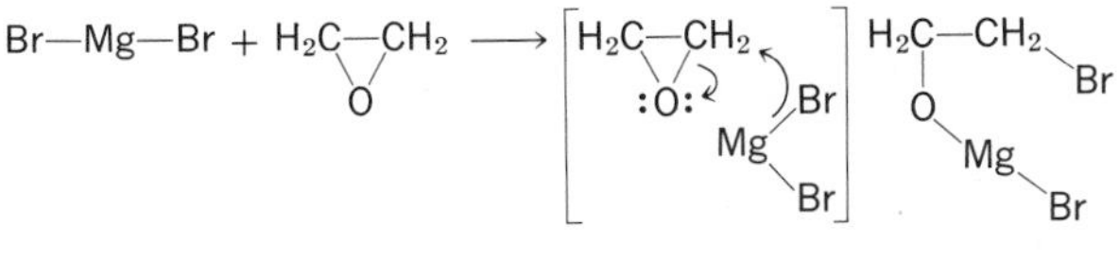
Br—Mg—Br + H2C—CH2
O
H2C—CH2
:O: Br
Mg
Br
H2C—CH2
Br
O
Mg
Br

As would be predicted on the basis of strain theory (Sec. 4.5), four-membered and larger cyclic oxides do not undergo this reaction.

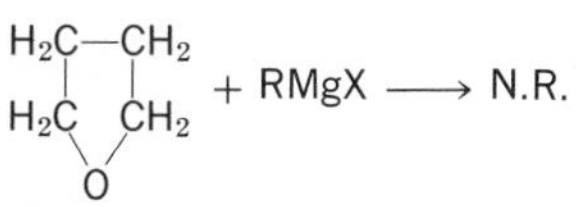
H2C—CH2
H2C CH2
O
+ RMgX ⟶ N.R.

Tetrahydrofuran (THF)
It may be used
as a solvent in
place of ether

[1]The Grignard reagent is a complicated reagent and many apparent variations occur in its structure. For practical purposes it will suffice to say that in the preparation of RMgX from RX and Mg, some MgX_2 will be present unless it is specifically removed.

8.9 On the Solution of Problems   The analysis of a synthetic problem is the most important step in solving it, and this is usually undertaken by working backward. Suppose, for example, the desired substance here is the tertiary alcohol 2-methyl-2-butanol. This can be made from a ketone and a Grignard reagent:

$$\underset{\substack{H\ \ OH \\ | \ \ | \\ H_3C-C-C-CH_3 \\ | \ \ | \\ H \ \ CH_3}}{} \xleftarrow{H_2O} \underset{\substack{H\ \ O^-\ \ ^+MgBr \\ | \ \ | \\ H_3C-C-C-CH_3 \\ | \ \ | \\ H \ \ CH_3}}{} \longleftarrow \underset{\substack{H \\ | \\ H_3C-C-MgBr \\ | \\ H}}{} + \underset{\substack{O \\ \| \\ C-CH_3 \\ | \\ CH_3}}{}$$

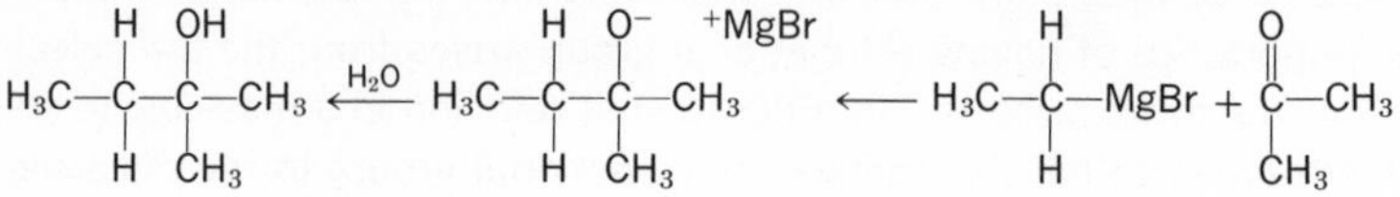

The ketone, in turn, may be made from a secondary alcohol by oxidation (see Sec. 8.21).

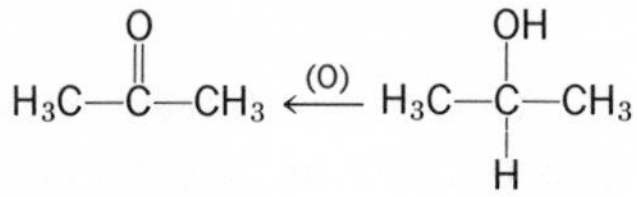

$$\underset{\substack{O \\ \| \\ H_3C-C-CH_3}}{} \xleftarrow{(O)} \underset{\substack{OH \\ | \\ H_3C-C-CH_3 \\ | \\ H}}{}$$
2-Propanol

2-Propanol is a secondary alcohol, and thus can be made from a Grignard and an aldehyde, one a 1-carbon compound and the other a 2-carbon compound.

$$\underset{\substack{OH \\ | \\ H_3C-C-CH_3 \\ | \\ H}}{} \xleftarrow{H_2O} \underset{\substack{O^-\ \ ^+MgBr \\ | \\ H_3C-C-CH_3 \\ | \\ H}}{} \longleftarrow \underset{\substack{O \\ \| \\ H_3C-C \\ | \\ H}}{} + CH_3MgBr$$

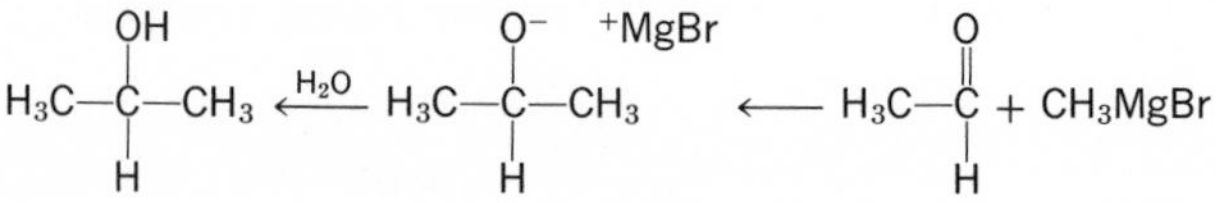

Other syntheses using these principles but following different paths could be devised for making 2-methyl-2-butanol. The important point to be observed in analyzing problems of this type is that the aldehyde or ketone to be used must always contain that carbon which, in the final alcohol, carries the —OH.

8.10 Preparation of Phenols   With concentrated sulfuric acid at elevated temperatures, sulfonation of benzene takes place.

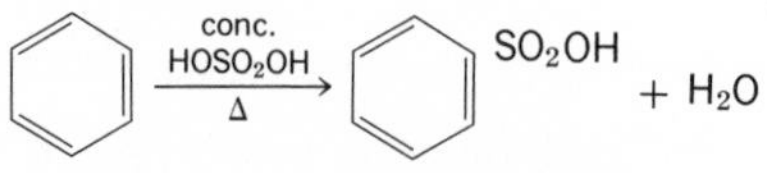

Benzenesulfonic
acid

When this is converted to the sodium salt and fused with sodium hydroxide, there is formed sodium phenolate.

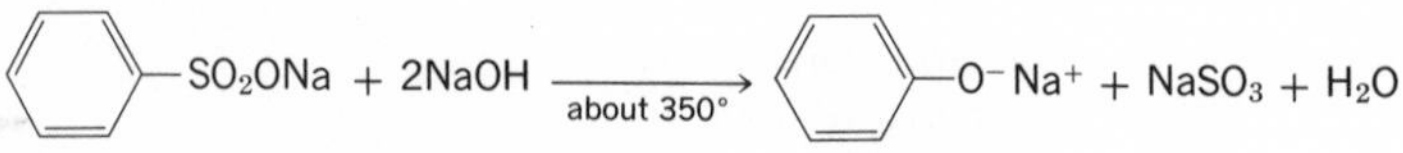

The sodium phenolate is converted to free phenol by strong acids

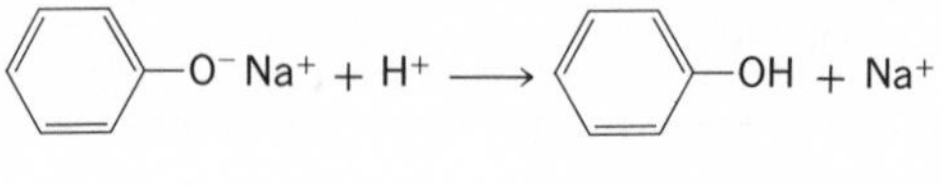

and in commercial practice the benzenesulfonic acid itself serves for this reaction, thus accomplishing in one step the freeing of the phenol (which may be distilled out) and the neutralizing of the benzenesulfonic acid

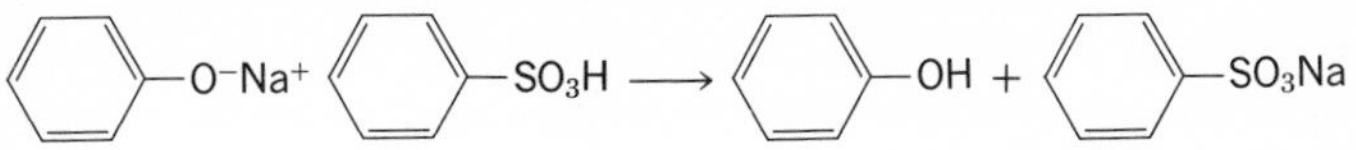

A second method for preparing phenols makes use of *amines,* which are substitution products of ammonia. Aromatic amines are conveniently prepared by reduction of nitro compounds [which can easily be made (see Sec. 6.14)].

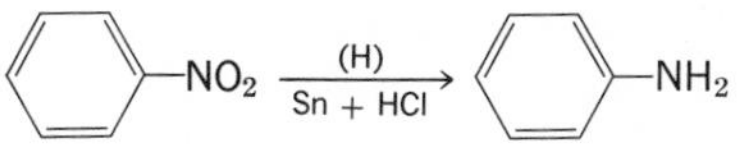

(an amine)
Phenylamine (aniline)

The amine reacts readily with nitrous acid to yield phenol by way of an unstable intermediate compound known as a *diazonium salt.* Only the overall reaction will be given here (see Sec. 14.17).

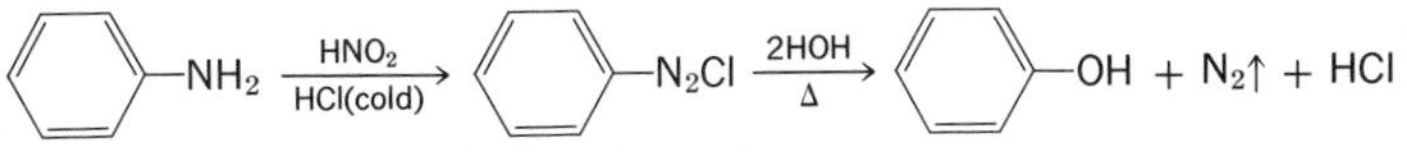

A diazonium salt

The HCl present in the reaction above represents the acid which is used to make nitrous acid from the easily available stable sodium nitrite. It also enters the reaction in a manner explained more fully later.

$$HCl + NaNO_2 \longrightarrow NaCl + HNO_2$$

Aliphatic amines yield nitrogen gas in a reaction similar to that of aromatic amines. The intermediate diazonium compounds are even more unstable than in the aromatic series. With aliphatic amines, side reactions such as alkene formation, nitrite formation, and various rearrangements occur to such an extent that the yield of alcohol is very poor, and hence, the method is seldom used to prepare alcohols.

8.11 Commercial Preparations Two specialized modern synthetic processes are worth mentioning. Methanol is now being made in large quantities by catalytic hydrogenation of carbon monoxide.

$$CO + 2H_2 \xrightarrow[\Delta]{\text{cat.}} CH_3OH$$

The carbon monoxide is largely derived from methane, and this is one of the major synthetic uses of methane. The preparation of acetylene is another (Sec. 5.8). The reaction gives excellent conversion (nearly 100 percent) when carried out at elevated pressure and with a suitable catalyst. Small amounts of higher alcohols are formed in this process as by-products. This process has largely supplanted the old preparation of methanol (wood alcohol) by destructive distillation of wood.

Phenol has been prepared in considerable quantity by a process involving two vapor-phase reactions, which lend themselves particularly well to automatic control. In one reaction benzene, air, and hydrochloric acid are passed as vapor over a catalyst, with the production of chlorobenzene and the release of energy.

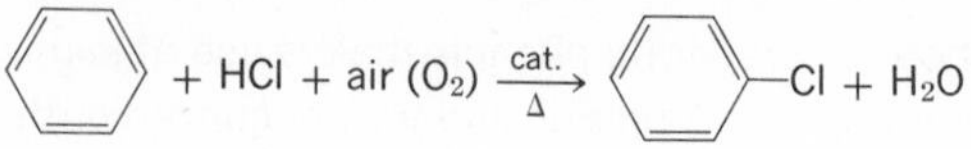
+ HCl + air (O_2)
cat.
Δ
—Cl + H_2O

The yield of this reaction is about 10 percent, but the unchanged starting materials are readily separated and recycled without appreciable loss. The chlorobenzene is then brought into another vapor-phase reaction with steam. This reaction takes up more energy per mole than the first reaction produces.

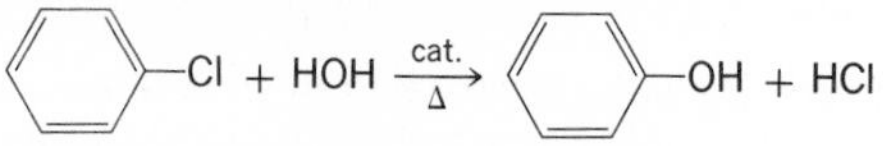
—Cl + HOH
cat.
Δ
—OH + HCl

Phenol[1]

Here the yield is again about 10 percent, but unchanged material can be separated and recycled. The operation is a continuous one: the chlorobenzene from the first reaction is separated and fed continuously into the second; the phenol produced is stripped out, and the HCl is recovered and passed, together with fresh benzene, into the first, the rates in all cases being under careful control and adjustment. It is said, as a measure of the efficiency of the operation, that about 95 percent of the HCl is recovered and reused.

It must be emphasized that this hydrolysis of aromatic halides is limited in its application to simple compounds capable of withstanding the drastic conditions required to bring about the reaction. Actually it is used only to produce phenol itself.

8.12 Properties The alcohols and phenols can be regarded formally and experimentally as derivatives of water in which one H has been replaced by an aliphatic or an aromatic group. It will not be surprising, therefore, to find that many of the chemical reactions of alcohols are already familiar as reactions of water.

8.13 Physical Properties of Alcohols The physical properties of the alcohols show the relation to water particularly clearly in the lower members of the homologous series. Whereas methane, ethane, and propane are practically insoluble in water, methanol, ethanol, and the propanols are completely miscible with water. The solubility of the alcohols in water decreases rapidly with increasing chain length above butanol in the homologous series. Higher alcohols, such as 1-dodecanol, are virtually insoluble in water. This behavior is directly related to the molecular size

[1] Phenol is used in the manufacture of Bakelite-type (phenol-formaldehyde) plastics, in antiseptic formulations, and as an intermediate in the manufacture of medicinals and dyestuffs.

and shape. In methanol, the —OH is $\frac{17}{32}$ of the mass of the entire molecule, and its effectiveness in dragging the small hydrocarbon part into water solution is relatively great. However, the —OH in dodecanol is only $\frac{17}{186}$ of the molecule; the molecule of dodecanol is much more of a paraffin than an alcohol; it is only locally, so to speak, an alcohol. The shape of the molecule plays some role in the solubility, for while 1-butanol is soluble in water only to the extent of 9 g per 100 g water, the more compact 2-methyl-2-propanol is completely miscible. It is evident that solubility is a function of the balance of polar and nonpolar groups in the molecule, as well as a function of shape. Within limits, the more polar the molecule and the more compact, the more soluble it is in water.

8.14 Surface Activity It was pointed out that 1-dodecanol is more of a paraffin than an alcohol. The important effect of this is that the —OH group of dodecanol is unable to drag the rest of the molecule into water solution. If a very small amount of dodecanol is placed on a clean surface of water, the dodecanol spreads out over the water surface until the whole surface is covered with a layer one molecule thick, a *monolayer*. Any surplus alcohol remains in a droplet on the surface. It is found by optical and other measurements of this surface film that it is one molecule thick, with the —OH groups down in the water surface and the hydrocarbon chains lying on the surface. If there are not enough molecules to cover the surface with a complete layer, then many of the molecules lie on their sides, others are gathered together as floating islands one molecule thick, held together laterally by the adhesion of the hydrocarbon chains (Fig. 8-3*a*). If the molecules on the surface are pushed together (they may be pushed together by the use of strips of paper lying on the surface), they become oriented with the —OH ends of the molecules in the

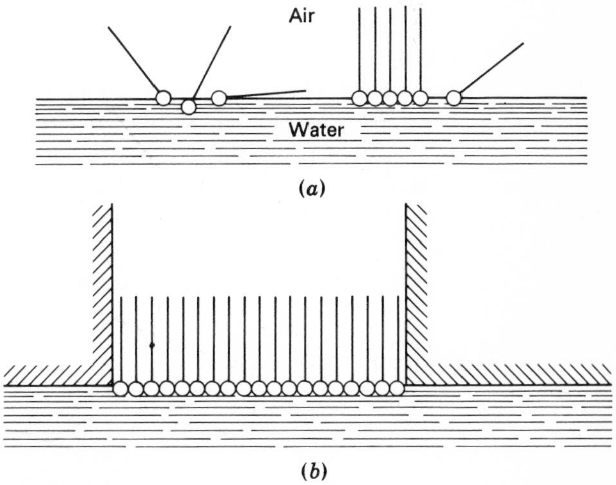
Air
Water
(*a*)
(*b*)

figure 8-3
Surface films of a long-chain alcohol.

water and the hydrocarbon chains vertical, or nearly vertical, and closely packed together (Fig. 8-3*b*). In the figure the molecule is symbolized thus: ————○, where ○ is the —OH and ———— is the $—(CH_2)_{11}CH_3$ chain. It is possible to calculate the cross-sectional area of the molecules in these films from the area of the film and the number of molecules in the film.

The spreading on the surface is a manifestation of the presence of a polar group. This part of the molecule is wetted by the water. But solution does not follow, because the water molecules have more affinity for each other than for the long-chain part, and the latter is squeezed out, so to speak.

8.15 Association Alcohols, along with water and certain other molecules, exhibit boiling points much higher than other compounds of comparable molecular weight. Thus, although boiling points usually parallel molecular weight, 1-butanol boils at 118°C while the isomeric $C_4H_{10}O$ compound, diethyl ether ($CH_3CH_2OCH_2CH_3$), boils at 35°C. Although polar factors such as dipole moment are no doubt involved, they cannot be the only factors at work here since both of these compounds have roughly comparable dipole moments.

TABLE 8.1
BOILING POINTS AND MOLECULAR WEIGHTS OF SOME SIMPLE COMPOUNDS

Substance	*Molecular weight*	*Boiling point, °C*	*Associated?*
H_2O	18	100	Yes
NH_3	17	−33	Not appreciably
CH_4	16	−161	No
CH_3OH	32	65	Yes
C_2H_5OH	46	78	Yes
CH_3OCH_3	46	−24	No

This "abnormal" behavior is found almost exclusively in compounds where hydrogen is attached to one of the three most electronegative elements, F, O, or N. A bond, called a hydrogen bond, between the hydrogen atom attached to one electronegative atom and a second electronegative atom is largely responsible for this phenomenon. These hydrogen bonds are weak (about 5 kcal/mole as opposed to 60 to 90 kcal/mole for ordinary covalent bonds) and are constantly broken and re-formed by the thermal motions of the molecules. Liquids which are held together by hydrogen bonds are said to be *associated.* The hydrogen bonds are generally indicated by a dashed line, as shown for methyl alcohol below:

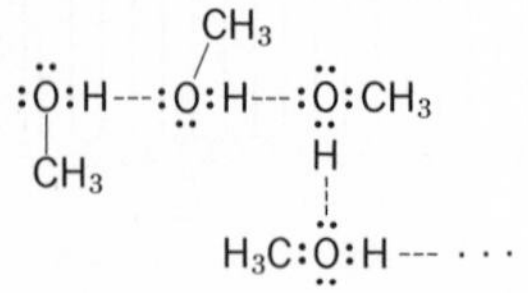

As indicated above, the average lifetime of a given hydrogen bond is short, although the net number of hydrogen bonds is essentially constant. Hydrogen bonds act as a sort of molecular "glue" to hold molecules together so that they exhibit properties such as high boiling points. In essence, the hydrogen bond is an especially strong polar interaction made possible by the small size of the proton and the high electronegativity of certain first-row elements.

Alcohols are able, like water, to form loose compounds with certain salts. Thus, calcium chloride forms a crystalline hydrate, $CaCl_2 \cdot 6H_2O$, and so does copper sulfate, $CaSO_4 \cdot 5H_2O$. Both these salts will form crystalline alcoholates, $CaCl_2 \cdot 3C_2H_5OH$ and $CaSO_4 \cdot CH_3OH$. This phenomenon is known as *solvation;* a special case occurs with water and is called *hydration.* In general, ionic substances in a hydrogen-bonding solvent are solvated (hydrated, in water).

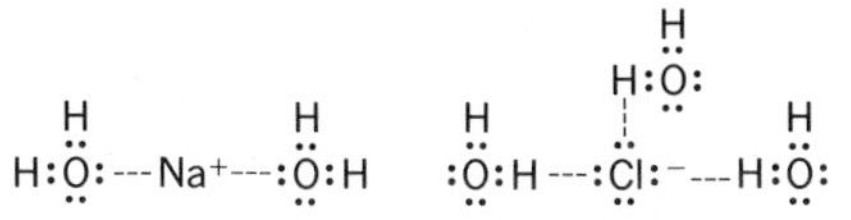

Hydration of NaCl in water

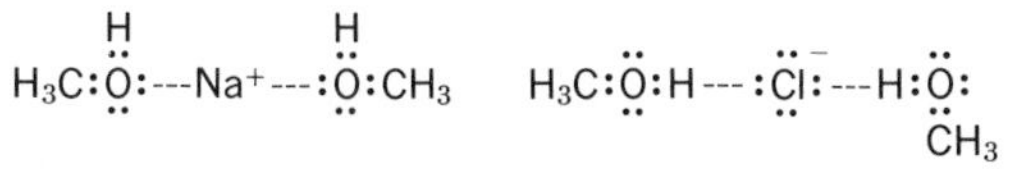

Solvation of NaCl in methyl alcohol

Alcohols are generally less successful than water at solvating ionic substances, and species such as NaCl, etc., are generally much less soluble in alcoholic media than in water, but still much more soluble than in hydrocarbons.

8.16 Ionization Water ionizes, as is well known, in the following manner:

$$2H_2O \rightleftharpoons OH^- + H_3O^+$$

The ion product equals $(OH^-)(H_3O^+) = 10^{-14}$. Water is a very weak base and a very weak acid. Alcohols may ionize in a fashion similar to that of water.

$$2CH_3CH_2OH \rightleftharpoons CH_3CH_2O^- + CH_3CH_2OH_2^+$$

We might guess that the electron-donating power of alkyl groups would tend to destabilize the anion derived from alcohols, that is, ^-OH is more stable than $CH_3CH_2 \rightarrow O^-$ which is more stable than

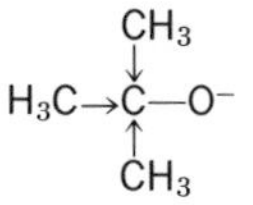

and this is indeed the case; alcohols are found to be weakly acidic, but less so than water. This means that the equilibrium

$$CH_3CH_2OH + {}^-OH \rightleftharpoons CH_3CH_2O^- + H_2O$$

lies well to the left, and that alcohols cannot be converted to their salts (called alkoxides or alcoholates) by hydroxide ion, and, in fact, do not give an acid reaction in water solution. However, in the absence of water, alcohols can be converted to alcoholates by alkali metals or strong anhydrous bases.

$$2CH_3CH_2OH + 2Na^0 \longrightarrow 2CH_3CH_2O^-Na^+ + H_2\uparrow$$

or

$$H_3C{-}\underset{\displaystyle OH}{\underset{|}{CH}}{-}CH_3 + \underset{\text{Sodium hydride}}{NaH} \longrightarrow H_3C{-}\underset{\displaystyle O^-Na^+}{\underset{|}{CH}}{-}CH_3 + H_2\uparrow$$

The reaction of alcohols with alkali (or alkaline earth) metals roughly parallels their acidities, and is less violent than that of water with these metals so that it is more easily controlled.

When the hydroxyl group is attached directly to an aromatic ring, the resulting compound, a phenol, is a stronger acid than water. Hence, in this case, the equilibrium with sodium hydroxide in water solution is displaced on the side of the sodium phenolate (sodium phenoxide). Water solutions of phenols are weakly acidic,

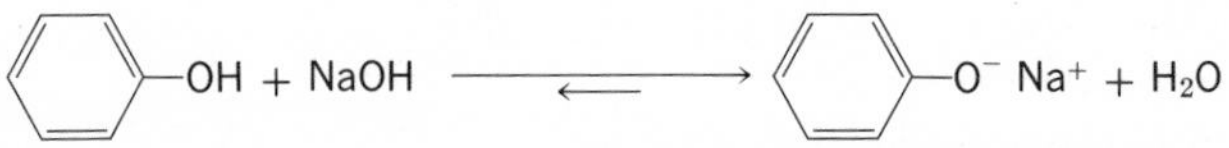

—OH + NaOH
—O⁻ Na⁺ + H₂O

but not as acidic as acetic acid (the familiar weak acid of elementary chemistry) and much less so than strong acids such as hydrochloric acid. Consequently, fairly strong alkaline conditions are needed to drive the equilibrium completely to the right.

8.17 Interpretation of the Acidity of Phenols The acidity of an acid, HX, depends on the stability of the form HX relative to that of the ion X^-. Anything that tends to stabilize X^- relative to HX will increase the acidity of HX since HX will be more willing to form $X^- + H^+$. In the case of phenol, the acid form is partially stabilized by resonance involving contributions from the structures shown below:

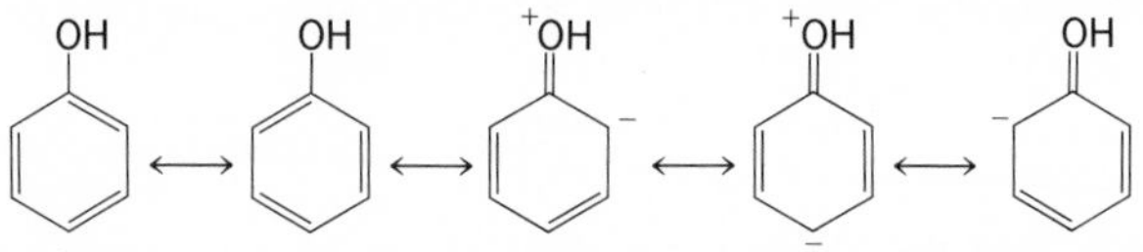

I*a* I*b* I*c* I*d* I*e*

The contributions of Structures I*c*, I*d*, and I*e* are relatively small as they involve the separation of positive and negative charge, a process which raises their energy. In the phenolate ion II, however, such structures play a more important role, since charge separation is not involved.

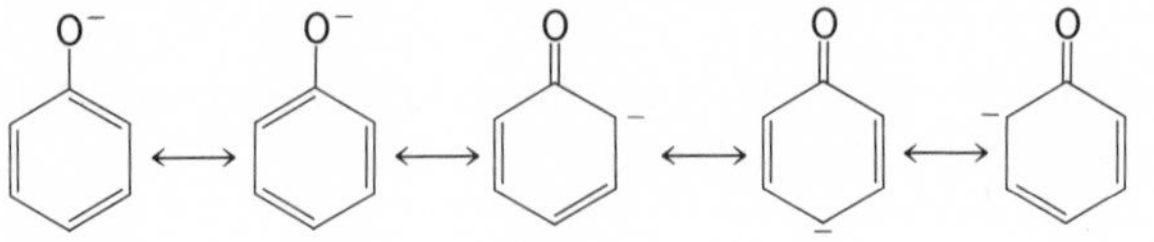

II*a* II*b* II*c* II*d* II*e*

The net result of this is that it is not as difficult to go from I $\longrightarrow$ II as it is from $CH_3OH \longrightarrow CH_3O^-$, since II gets back some extra resonance energy. Thus, phenols are mildly acidic in water. This acidity is enhanced by the presence of electron-withdrawing groups on the ring (meta-directing groups) since they help to stabilize the negative charge even more. This effect is even greater when such structures can farther spread the negative charge by resonance interaction. For instance, picric acid is a very strong acid because the ion can be stabilized as indicated below.

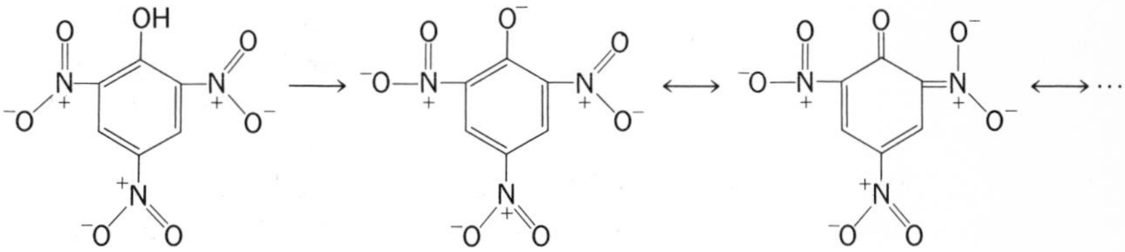
Picric acid   Picrate ion

8.18 Properties of Salts: Separations As we have mentioned previously, most organic reactions involve a competition between several alternate processes. Consequently, much of the organic chemist's time is devoted to finding conditions which conduce to a given reaction and to finding methods by which he can separate the desired product from the various by-products that accompany it. For this reason some of the more general separation methods will be discussed as the background is provided.

Like the alkali salts of most organic acidic compounds, sodium phenolate has the properties of an electrovalent substance, that is, it is soluble in water, insoluble in ether, nonvolatile, etc. Advantage may be taken of these properties in separating phenols from mixtures of compounds including alcohols.

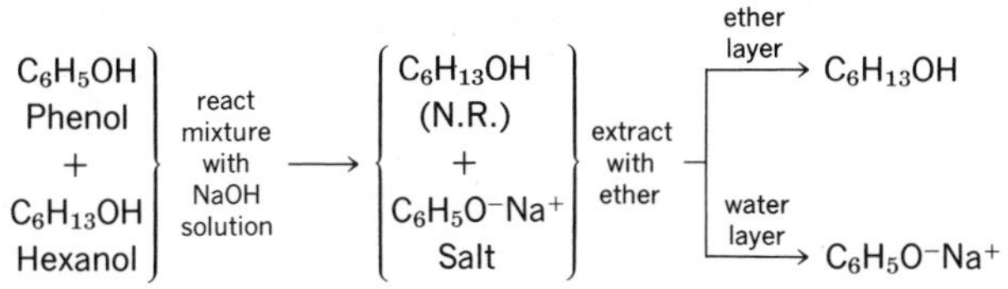
C_6H_5OH
Phenol
+
$C_6H_{13}OH$
Hexanol
react mixture with NaOH solution
$C_6H_{13}OH$
(N.R.)
+
$C_6H_5O^-Na^+$
Salt
extract with ether
ether layer
$C_6H_{13}OH$
water layer
$C_6H_5O^-Na^+$

The phenol is recovered from its salt by the addition of an excess of strong acid.

Separations based on this principle are often possible even though the compounds involved contain hydrolyzable groups, such as halides and esters. The requirement for such separations is that they be carried out rapidly and at low temperatures.

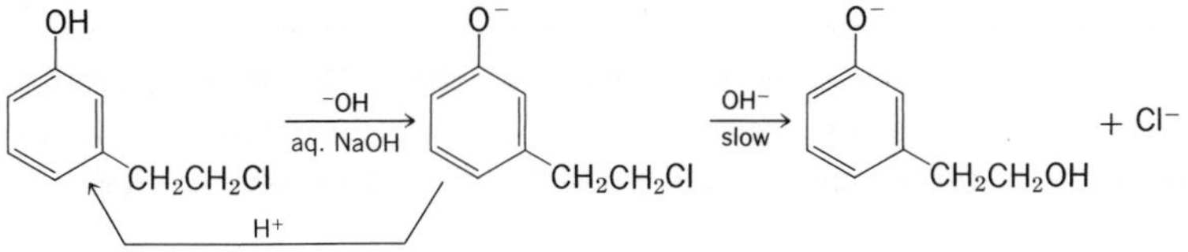

In general, neutralization reactions (ionic) are practically instantaneous, while reactions involving cleavages of covalent bonds are relatively slow.

8.19 Esterification Reaction   As was mentioned in Sec. 8.5, the hydrolysis of esters represents a process by which alcohols may be obtained. This is a reversible reaction. One of the characteristic reactions of alcohols is the reverse, the formation of esters: esterification.

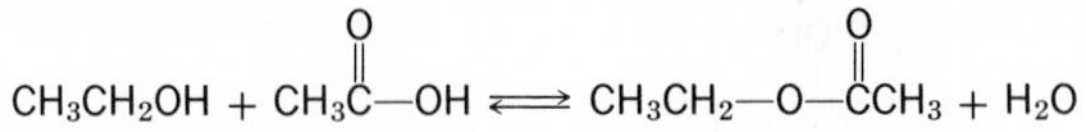

$$CH_3CH_2OH + CH_3\overset{O}{\overset{\|}{C}}{-}OH \rightleftharpoons CH_3CH_2{-}O{-}\overset{O}{\overset{\|}{C}}CH_3 + H_2O$$
$$\text{Ethyl acetate}$$

The uncatalyzed reaction is generally very slow. The reaction is accelerated by strong acids such as HCl, H_2SO_4, etc. This brings the reaction to equilibrium in a matter of minutes or hours rather than days. The position of the equilibrium is not influenced by the catalyst per se, although an acid such as H_2SO_4 is also a dehydrating agent and can shift the equilibrium by tying up one of the products (H_2O). One can also favor the esterification of a given acid by using a large excess of the alcohol or by removing the water in various ways.

Esters are covalently linked compounds, soluble in organic solvents, and (in the absence of other functional groups) neutral. Di- and tribasic acids may form mono-, di-, and (in the case of the latter) triesters. Examples of common esters are given below.

$$CH_3CH_2CH_2\overset{O}{\overset{\|}{C}}{-}O{-}CH_2CH_3 \qquad CH_3CH_2CH_2CH_2{-}O{-}\overset{O}{\overset{\|}{C}}CH_3 \qquad CH_3{-}O{-}\overset{O}{\underset{O^-}{N^+}}$$

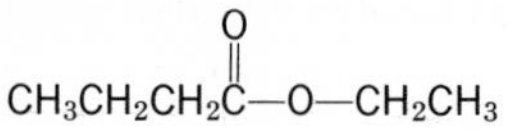

Ethyl butyrate   Butyl acetate   Methyl nitrate (ester of nitric acid)

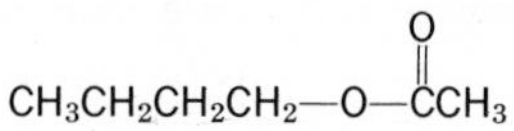

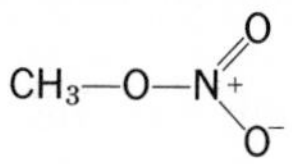

$$CH_3CH_2{-}O{-}\overset{O}{\overset{\|}{C}}CH_2CH_2\overset{O}{\overset{\|}{C}}{-}OH$$

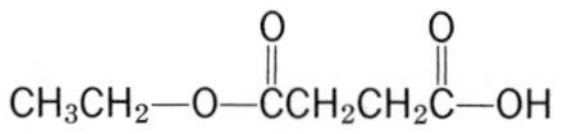

Ethyl hydrogen succinate (acidic)

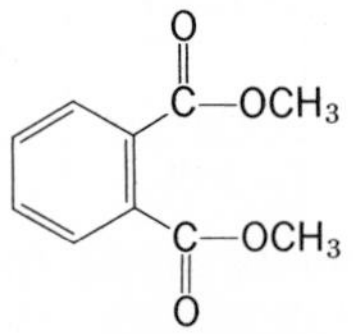

Dimethyl phthalate (neutral)

The mechanism of ester formation, like the hydrolysis, is complex and discussion of it will be reserved for Chap. 13.

In the case of phenols, the direct esterification reaction does not give satisfactory yields since the equilibrium is less favorable. This is most probably due to loss of resonance energy relative to the phenol, since Structure II would be expected to be of higher energy than I due to adjacent positive charges.

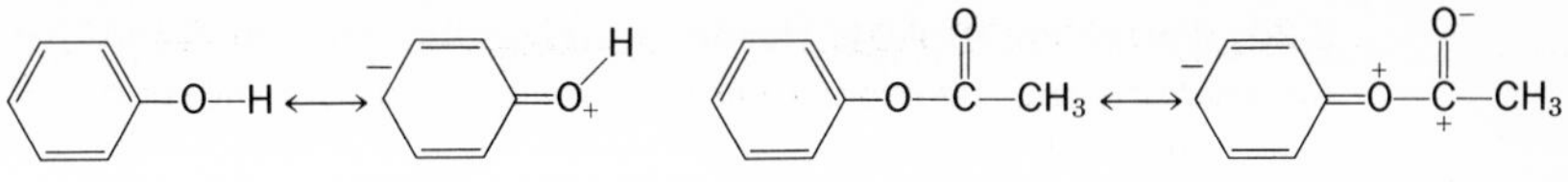

I　　　II

Phenol esters can be obtained by using reagents which are more reactive than the acids themselves. One of the more common reagents used for this purpose is an acid chloride, which can be produced from the acid by sufficiently active chlorinating agents such as PCl_5, PCl_3, and $SOCl_2$.

$$CH_3\overset{\overset{O}{\|}}{C}{-}OH + PCl_5 \longrightarrow CH_3\overset{\overset{O}{\|}}{C}{-}Cl + POCl_3 \text{ or } H_3PO_4$$

An acid chloride

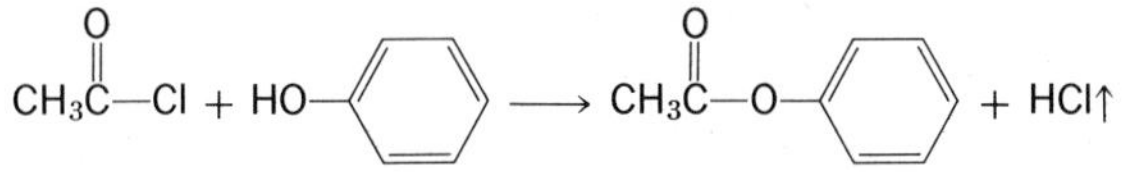

Phenyl acetate

$$CH_3\overset{\overset{O}{\|}}{C}{-}Cl + HO{-}CH_2{-}CH_3 \xrightarrow[\text{exothermic}]{\text{very}} CH_3\overset{\overset{O}{\|}}{C}{-}O{-}CH_2{-}CH_3 + HCl\uparrow$$

This reaction of acid chlorides with simple alcohols gives esters in a very exothermic reaction. The driving force for this reaction arises from the higher energy of acid chlorides plus the fact that one product (HCl) removes itself from the reaction mixture as a gas and thus makes the reaction essentially irreversible. Acid chlorides may be prepared from most acids and are commonly used to increase their reactivity in reactions such as esterification.

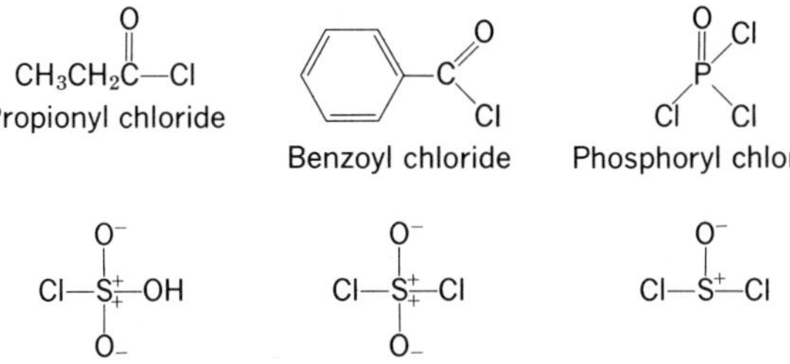

Propionyl chloride　Benzoyl chloride　Phosphoryl chloride

Chlorosulfonic acid　Sulfonyl chloride　Thionyl chloride

Representative acid chlorides are given above. Some of these, such as thionyl chloride, may be used to prepare organic acid chlorides.

$$CH_3CH_2\overset{\overset{O}{\|}}{C}{-}OH + SOCl_2 \longrightarrow CH_3CH_2\overset{\overset{O}{\|}}{C}{-}Cl + SO_2\uparrow + HCl\uparrow$$

Acid bromides and iodides are known, but offer no advantages over the chlorides and are rarely used because of their greater cost.

8.20 Formation of Alkyl Halides Alkyl halides may be prepared from alcohols by reactions that bear considerable similarity to esterifications. Thus, the halogen acids react with alcohols.

$$CH_3-CH_2-OH + HBr \rightleftharpoons CH_3-CH_2-Br + H_2O$$

These reactions are mechanistically different from most esterification reactions (Secs. 8.5 and 12.3) in that they necessarily involve cleavage of the carbon-oxygen bond of the alcohol. The reaction is, thus, a type of displacement reaction, and is effectively the reverse of the formation of alcohols from alkyl halides (Sec. 8.4).

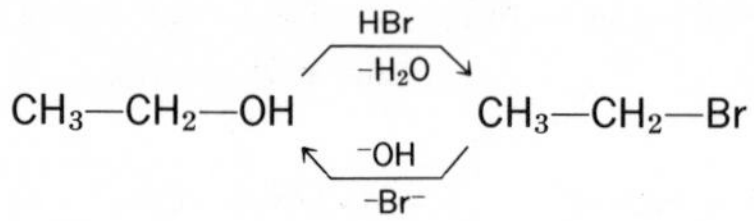

$$CH_3-CH_2-OH \quad \overset{\overset{HBr}{\diagup} \overset{-H_2O}{\searrow}}{\underset{\underset{-Br^-}{-OH}}{\nwarrow \diagup}} \quad CH_3-CH_2-Br$$

Unlike the base-promoted displacement of alkyl halides (Sec. 9.7), this reaction proceeds readily with many secondary and tertiary alcohols.

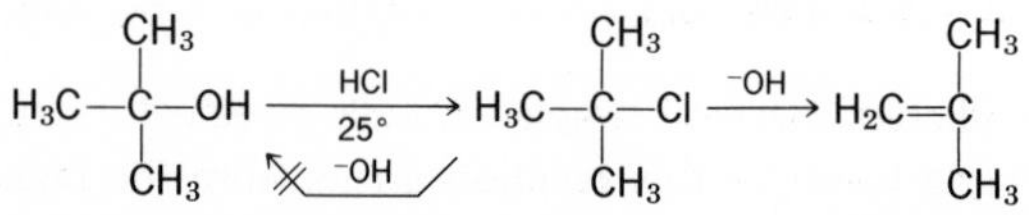

$$H_3C-\underset{\underset{CH_3}{|}}{\overset{\overset{CH_3}{|}}{C}}-OH \xrightarrow[\underset{X \quad -OH}{25°}]{HCl} H_3C-\underset{\underset{CH_3}{|}}{\overset{\overset{CH_3}{|}}{C}}-Cl \xrightarrow{-OH} H_2C=\underset{\underset{CH_3}{|}}{\overset{\overset{CH_3}{|}}{C}}$$

The explanation for this apparent anomaly is found in the intervention of carbonium ions (Secs. 3.6, 3.7, 3.17, and 8.23) in tertiary and many secondary systems. In the presence of strong acid the alcohol can lose a molecule of water as follows:

$$H_3C-\underset{\underset{CH_3}{|}}{\overset{\overset{CH_3}{|}}{C}}-OH + HCl \overset{1}{\rightleftharpoons} \left[H_3C-\underset{\underset{CH_3}{|}}{\overset{\overset{CH_3}{|}}{C}}-\overset{+}{O}H_2 \right] \xrightarrow[slow]{2} \left[H_3C-\underset{\underset{CH_3}{|}}{\overset{\overset{CH_3}{|}}{C}}+ \right]$$

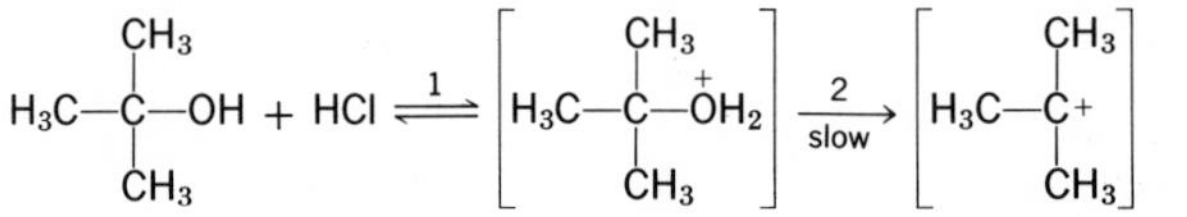

A carbonium ion

$$\left[H_3C-\underset{\underset{CH_3}{|}}{\overset{\overset{CH_3}{|}}{C}}+ \right] \xrightarrow[Cl^-, \, fast]{3} H_3C-\underset{\underset{CH_3}{|}}{\overset{\overset{CH_3}{|}}{C}}-Cl$$

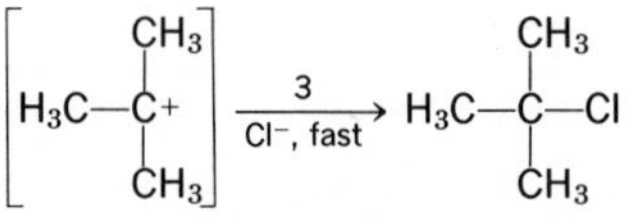

Although the reverse of Steps 2 and 3 is possible, the use of high concentrations of HCl drives the reaction toward completion. As with other carbonium-ion reactions (Secs. 3.17 and 8.23), a variety of side reactions is possible (such as rearrangement, elimination, etc.).

The displacement reaction is one of the most carefully studied reactions in organic chemistry. In Secs. 9.5 and 9.6 the mechanism shall be examined in more detail, but for the present the concern is with the more practical consequences of the reaction.

Phenols, like aromatic halides (Sec. 8.4), undergo the displacement reaction very poorly or not at all.

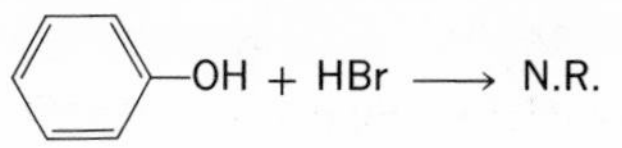

The mechanistic reasons for the failure of these systems will also be discussed in Sec. 9.6.

An alternate method of preparing alkyl halides utilizes halogenated derivatives of certain inorganic acids. In these cases an anion is displaced rather than a water molecule.

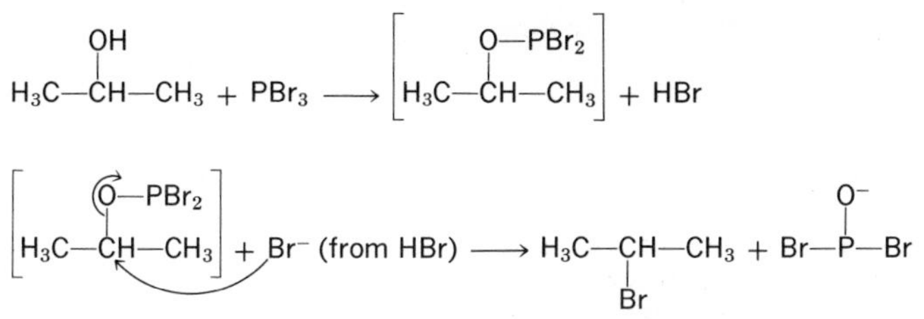

$Br{-}P(O^-){-}Br$ can react further until it is converted to H_3PO_3. The inorganic halides most commonly used for this purpose are PCl_3, PBr_3, $P + I_2$ (PI_3), and $SOCl_2$. It is also possible to use other derivatives of inorganic acids and to carry out the reaction in stepwise fashion:

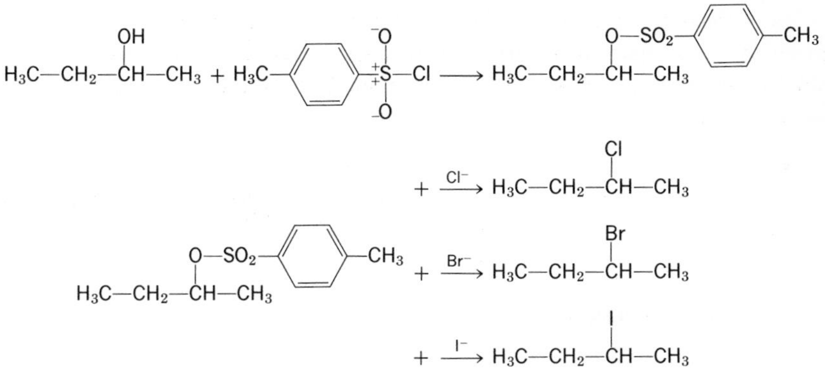

For tertiary alcohols, reaction with concentrated HCl, HBr, or HI usually provides the best yield of halide. For secondary alcohols both halogen acids and inorganic halides may be used, each method having specific advantages in a given case. With primary alcohols HCl will not ordinarily react without a Lewis-acid catalyst such as $ZnCl_2$, and the inorganic halides are used. Phenols give halides with none of these reagents.

8.21 Oxidation Reactions The oxidation of alcohols is analogous to that of the intermediates in alkene oxidation (Sec. 3.14). Primary alcohols upon oxidation undergo the following reactions:

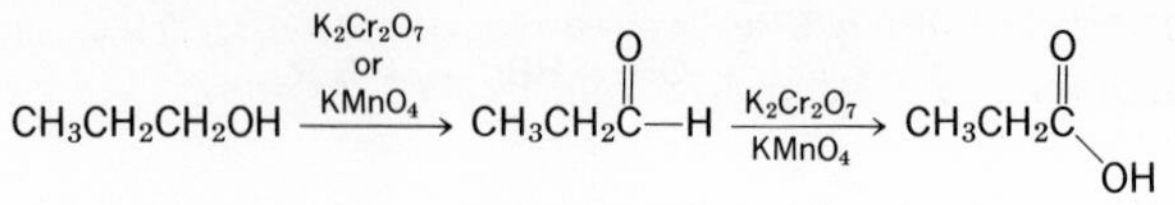

$$CH_3CH_2CH_2OH \xrightarrow[\text{KMnO}_4]{\substack{\text{K}_2\text{Cr}_2\text{O}_7 \\ \text{or}}} CH_3CH_2\overset{\overset{\displaystyle O}{\|}}{C}-H \xrightarrow[\text{KMnO}_4]{\text{K}_2\text{Cr}_2\text{O}_7} CH_3CH_2\overset{\overset{\displaystyle O}{\|}}{C}_{\diagdown OH}$$

| 1-Propanol | Propionaldehyde | Propionic acid |

The carbon bearing the —OH becomes oxidized, first to an aldehyde and then to an acid. Aldehydes are generally more rapidly oxidized than the alcohols so that they are not readily obtained by this process. (However, see Sec. 11.4.)

Secondary alcohols react to give ketones.

$$H_3C-\underset{\underset{\displaystyle H}{|}}{\overset{\overset{\displaystyle OH}{|}}{C}}-CH_3 \xrightarrow[\text{KMnO}_4]{\substack{\text{K}_2\text{Cr}_2\text{O}_7 \\ \text{or}}} H_3C-\overset{\overset{\displaystyle O}{\|}}{C}-CH_3$$

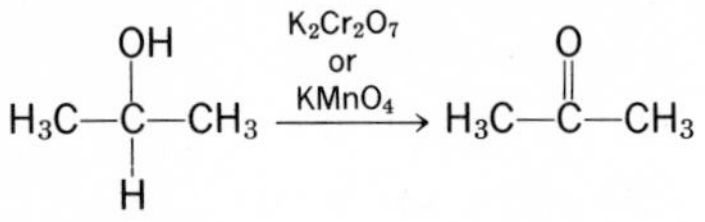

| 2-Propanol | 2-Propanone (acetone) |

Since ketones have no hydrogens on the carbon atom bearing the oxygen, they are resistant to further oxidation and good yields of ketones may be obtained.

Tertiary alcohols do not react under conditions that oxidize primary and secondary alcohols. Under vigorous oxidation conditions tertiary (as well as secondary) alcohols are degraded into small fragments.

All of these oxidation reactions can be carried out when the groups concerned are on the side chain of an aromatic nucleus. The conditions required for the oxidative removal of such a chain are more drastic than those needed in the above reactions.

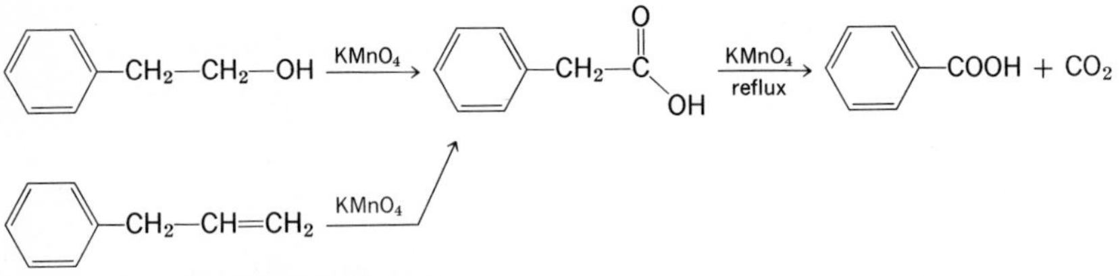

8.22 Oxidation of Phenols; Quinones

Phenol and, particularly, the phenolate ion exhibit marked nucleophilic activity, undergoing substitution reactions readily. As a consequence of this ready availability of electrons in the phenol nucleus, this substance is easily attacked by oxidizing agents (electrophilic agents). The products of the reaction are complex, and extensive degradation of the molecule results. Most phenols, particularly in alkaline solution, are subject to air oxidation, and they develop color on standing. This susceptibility is increased in polyhydroxy aromatic compounds, although in many of these cases well-defined products can be obtained.

For example, colorless 1,4-dihydroxybenzene (trivial name, hydroquinone) is oxidized under mild conditions to a yellow substance called quinone.

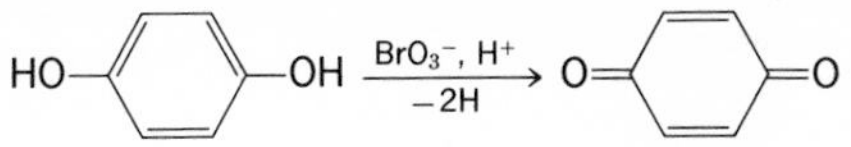
HO
OH
BrO3−, H+
−2H
O
O

Quinone
(p-Benzoquinone)
(yellow)

Quinone may also be obtained as one of the products of the oxidation of phenol and of certain other electron-rich benzene derivatives.

Quinone is the parent of a class of compounds called "quinones." Quinones may be formed from appropriately substituted dihydroxy aromatic compounds under much milder conditions than those required for the oxidation of simple phenols, and the reaction is usually reversible under comparably mild conditions. The two hydroxyl groups need not be located in the same ring, but they must be located in such positions that all sp^2-hybridized carbon atoms can participate in π bonds. Some representative quinones are:

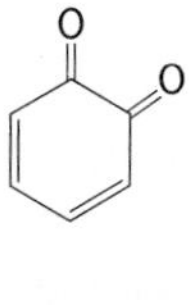

o-Benzoquinone

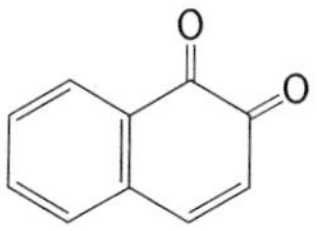

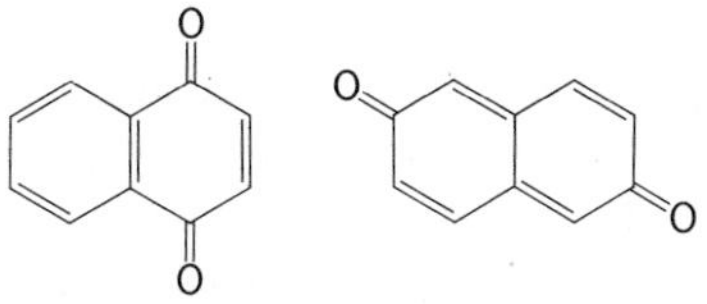

Naphthoquinones

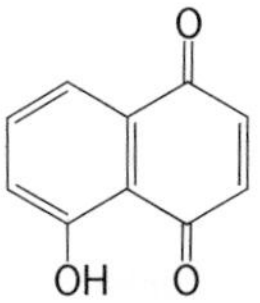
OH

Juglone

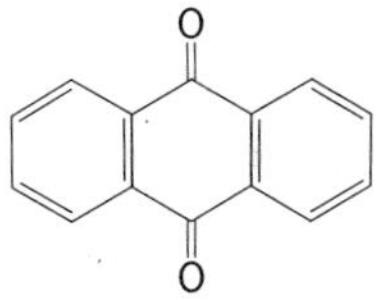

Anthraquinone

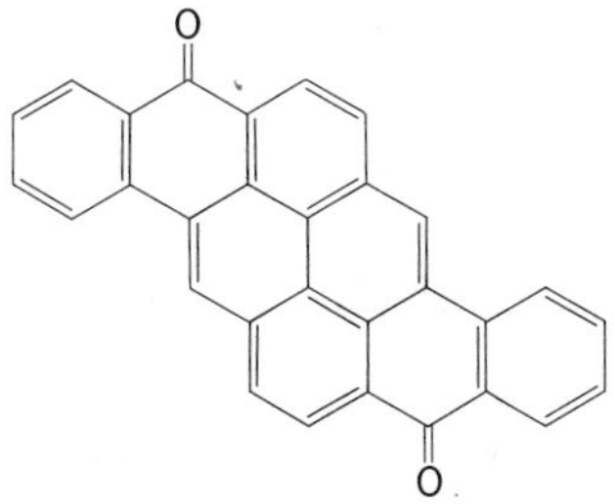

Pyranthrone

Dihydroxy aromatic compounds which cannot give rise to fully bonded structures do not yield quinones, for example, m-dihydroxybenzene.

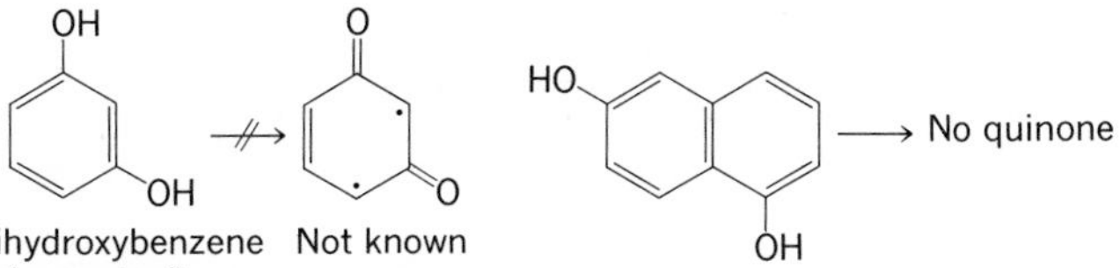
OH
OH
HO
OH
No quinone

m-Dihydroxybenzene (resorcinol)   Not known

Quinones are important as coloring agents (pigments and dyes), and are often found in natural products. They are also important in oxidation-reduction systems since the quinone-hydroquinone reaction is one of the few readily reversible redox reactions in organic chemistry.

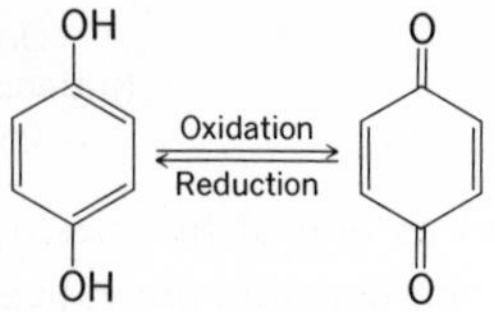
OH
Oxidation
Reduction
OH
O
O

8.23 Dehydration of Alcohols: Mechanisms Alcohols undergo dehydration reactions of two kinds, one within the molecule and one between two molecules. The first is illustrated below:

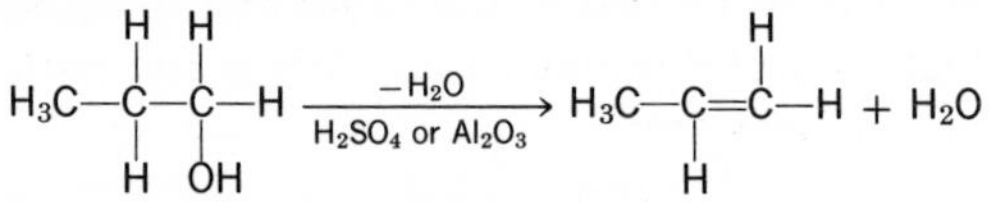

The removal of water may be accomplished by heating the alcohol in the presence of a *strong* dehydrating agent such as sulfuric acid, or by passing the alcohol in vapor form over a dehydrating catalyst, such as alumina (more commonly used as an industrial process). In this reaction, tertiary alcohols react much more readily (that is, at a lower temperature) than secondary alcohols, and these at a lower temperature than primary alcohols. The mechanism of this reaction, in which intermediate positive ion is formed, was discussed in Sec. 3.16.

Alcohols may become dehydrated intermolecularly, a molecule of water being removed between two molecules of alcohol. This may be done with the same reagents as those used for intramolecular dehydration, but with different conditions of temperature, pressure, and proportions of reagents. The first steps of the reaction are familiar (Sec. 8.20).

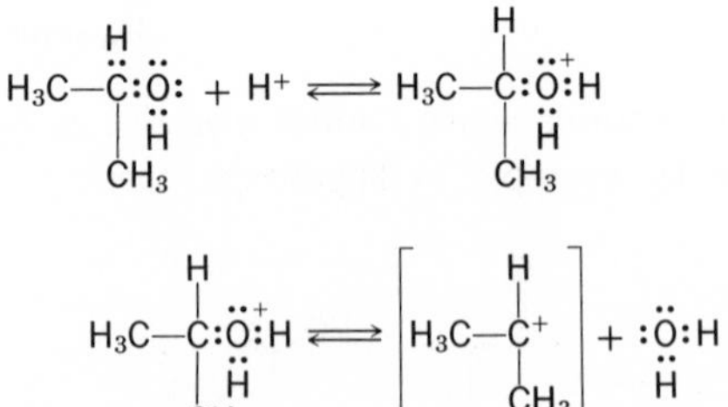

This carbonium ion can react by a variety of pathways, depending on the medium and on the reaction conditions. Some of the more important pathways are as follows:

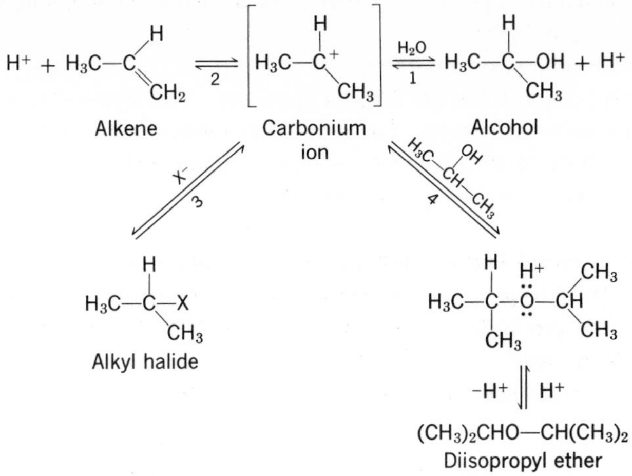

$$H^+ + H_3C-\overset{\overset{\displaystyle H}{|}}{\underset{\displaystyle CH_2}{C}} \underset{2}{\rightleftharpoons} \left[H_3C-\overset{\overset{\displaystyle H}{|}}{\underset{\displaystyle CH_3}{C}}{}^+ \right] \underset{1}{\overset{H_2O}{\rightleftharpoons}} H_3C-\overset{\overset{\displaystyle H}{|}}{\underset{\displaystyle CH_3}{C}}-OH + H^+$$

Alkene   Carbonium   Alcohol
            ion

$$H_3C-\overset{\overset{\displaystyle H}{|}}{\underset{\displaystyle CH_3}{C}}-X$$

Alkyl halide

$$H_3C-\overset{\overset{\displaystyle H}{|}}{\underset{\displaystyle CH_3}{C}}-\overset{\overset{\displaystyle H^+}{\cdot\cdot}}{\underset{\cdot\cdot}{O}}-\overset{\overset{\displaystyle CH_3}{}}{\underset{\displaystyle CH_3}{CH}}$$

$$-H^+ \,\|\, H^+$$

$(CH_3)_2CHO-CH(CH_3)_2$
Diisopropyl ether

All of these reactions will occur to some extent. Reaction 1 is favored by an excess of water, as in dilute aqueous solution. In the absence of excess water, and if the temperature is maintained above the boiling point of the alkene while the alcohol is slowly added (so as to keep its concentration low), Path 2 can be made to predominate. If the anion of the acid HX is a good nucleophilic reagent (e.g., Br⁻ or Cl⁻), or if its concentration is very high (e.g., $HOSO_2O^-$ in concentrated H_2SO_4), then Path 3 is important. If a large excess of alcohol is present and the temperature is high enough to distill off the (unassociated) ether, then Path 4 can be followed. None of these pathways goes to the complete exclusion of the others, but by understanding the mechanism of such a reaction, the chemist can often determine the best set of conditions for a particular reaction.

An important complication of reactions involving free carbonium ions is the possibility of rearrangement. In particular, carbonium ions of low stability can rearrange by way of a 1,2 shift of an alkyl group or hydrogen to a more stable carbonium ion. An example of this is given by neopentyl alcohol.

$$H_3C-\overset{\overset{\displaystyle CH_3}{|}}{\underset{\displaystyle CH_3}{C}}-CH_2-OH \xrightarrow[H_2SO_4]{H^+} \left[H_3C-\overset{\overset{\displaystyle CH_3}{|}}{\underset{\displaystyle CH_3}{C}}\overset{\curvearrowright}{}{}^+CH_2 \right] \longrightarrow \left[H_3C-\overset{\overset{\displaystyle CH_3}{|}}{\underset{\displaystyle CH_3}{\overset{+}{C}}}-CH_2 \right]$$

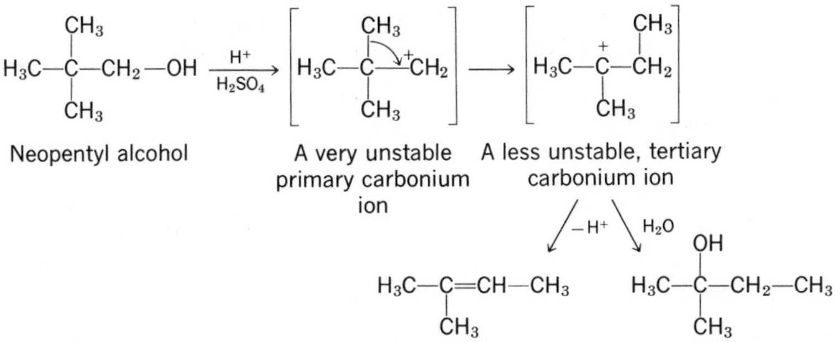

Neopentyl alcohol   A very unstable   A less unstable, tertiary
                   primary carbonium   carbonium ion
                        ion

$$\overset{-H^+}{\swarrow} \qquad \overset{H_2O}{\searrow}$$

$$H_3C-\overset{\overset{\displaystyle }{}}{\underset{\displaystyle CH_3}{C}}=CH-CH_3 \qquad\qquad H_3C-\overset{\overset{\displaystyle OH}{|}}{\underset{\displaystyle CH_3}{C}}-CH_2-CH_3$$

In this migration the migrating group (the CH_3 group in the case above) moves with its electrons to the adjacent position. Other groups such as phenyl and hydrogen can also migrate in this fashion.

In general, carbonium ion rearrangements of primary to secondary, primary to tertiary, and secondary to tertiary may be observed. This represents a downhill transformation in terms of energy. Level transformations of secondary to other secondary, and tertiary to tertiary ions are less common; uphill transformations, such as tertiary to secondary, etc., are extremely rare.

8.24 Differences between Phenols and Alcohols The differences between alcohols and phenols which have so far appeared have been primarily due to the special behavior of an —OH group when it is attached directly to a double bond. Such a system is called an "enol" (from "-ene" + "-ol" for alk*ene* and alcoh*ol*). When H^+ is lost from an enol a resonance-stabilized ion can be formed.

Substance	*Resonance forms*	*Molecular orbital picture*
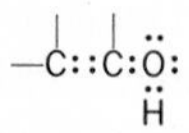 Enol	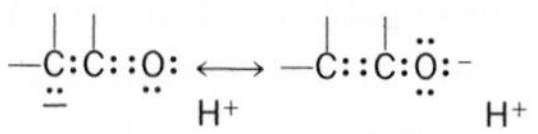 Enolate ion	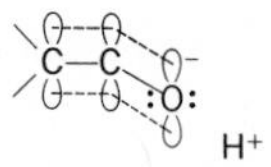 Enolate ion
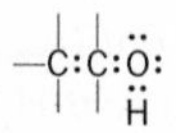 Alcohol	No analogous form 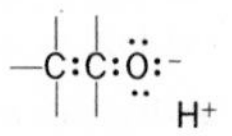 Alcoholate ion	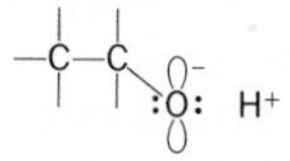 Alcoholate ion

As we have seen in Sec. 5.4, simple enols are usually unstable relative to aldehydes or ketones. In the case of phenols this situation is reversed because the enol form is part of an aromatic ring while the corresponding ketone (keto form) is not.

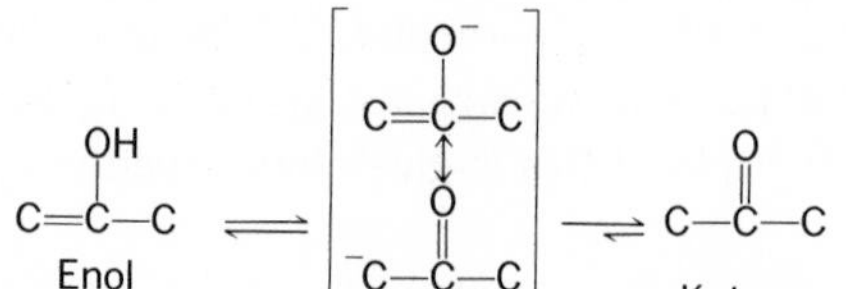

Enol (less stable) Enolate ion Ketone (more stable)

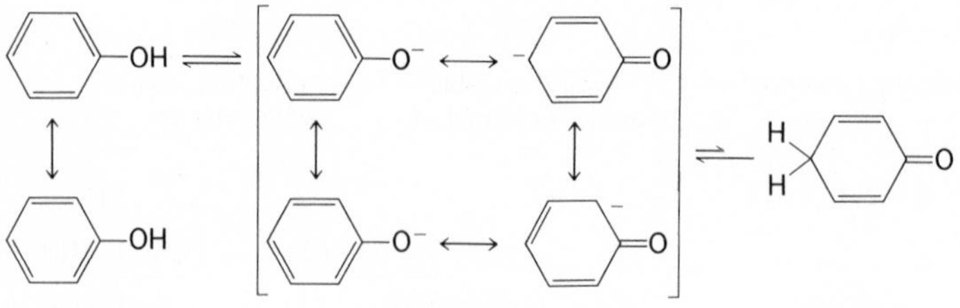

Phenol (more stable) Phenolate ion Ketone (less stable)

In addition to increasing the acidity of the —OH attached to the aromatic ring, the interaction described in Sec. 8.17 has two important effects. One of these is to make the C—O bond in phenols stronger and more difficult to break than in aliphatic alcohols. This accounts, in part, for the failure of phenols to give halides when treated with halogen acids (Sec. 8.20).

The second effect arises from the placement of some of the electrons from the oxygen on the aromatic ring (see Structures I*c*, I*d*, I*e*, II*c*, II*d*, and II*e* in Sec. 8.17). This increases the electron density on the aromatic ring and makes it more susceptible to electrophilic substitution. Phenol undergoes most electrophilic substitution reactions such as sulfonation and halogenation with extreme ease. For example, bromine, even in the presence of water, reacts to give a tribromo derivative.

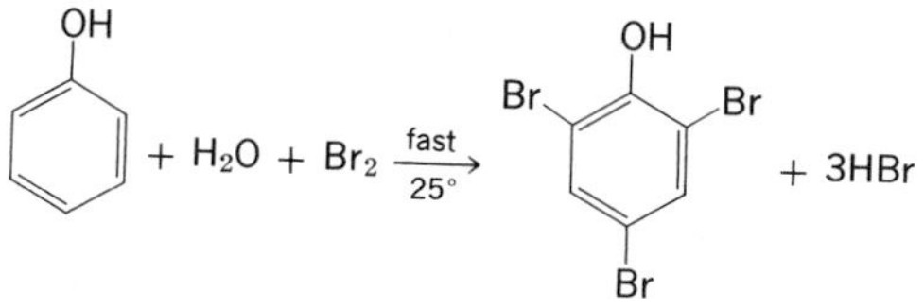
OH
+ H2O + Br2
fast
25°
OH
Br
Br
Br
+ 3HBr

Phenols are also easily oxidized (an electrophilic reaction) (Sec. 8.22) and some substitution reactions, such as nitration, give tarry by-products as a result of this.

Phenols, along with other stable enols, undergo a more-or-less characteristic color reaction with ferric chloride. Although the reaction is complex, it was once used extensively in demonstrating the presence of a phenol. Modern spectroscopic methods have largely supplanted this test (see Sec. 8.27) and it is only mentioned in view of its historical interest. Colors ranging from browns to reds, greens, and purples are observed in some cases, although a number of phenols do not give the test.

8.25 Reduction Reactions The direct reduction of alcohols is not easily accomplished except by the action of an excess of hydriodic acid (but not HBr or HCl).

$$C_{10}H_{21}OH + 2HI \longrightarrow C_{10}H_{21}I + H_2O \xrightarrow[\Delta\Delta]{HI} C_{10}H_{22} + I_2$$

Phenols, of course, do not reduce by this method since the first step fails.

Primary and secondary alcohols may also be reduced by forming sulfonic acid esters followed by reduction with lithium aluminum hydride.

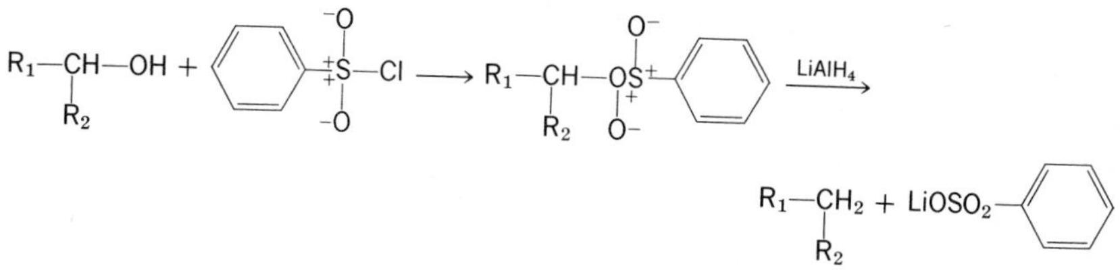
R1—CH—OH +
R2
-O
S—Cl
-O
R1—CH—OS
R2
O-
O-
LiAlH4
R1—CH2 + LiOSO2
R2

This reaction fails with tertiary alcohols and phenols. The phenolic —OH resists reduction by many reagents, but makes the benzene ring more susceptible to reduction.

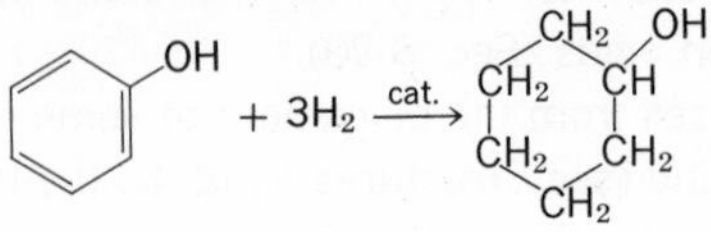
OH
+ 3H2
cat.
CH2
OH
CH2
CH
CH2
CH2
CH2

By distillation with zinc dust it is possible to replace the phenolic —OH by H, but this reagent also attacks many other functional groups and is of limited utility.

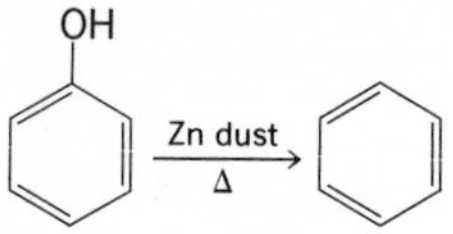
OH
Zn dust
Δ

8.26 Polyhydroxy Compounds Alcohols (and phenols) which contain more than one hydroxyl group are known. Certain of these (such as ethylene glycol) are primarily of industrial importance, and others (glycerol, for example) are also of biochemical significance. Their properties are much the same as those of monohydroxy compounds. A few reactions are given to illustrate this:

$$\begin{array}{l}CH_2OH\\|\\CH_2OH\end{array} + 2HBr \longrightarrow \begin{array}{l}CH_2Br\\|\\CH_2Br\end{array} + 2H_2O$$

Ethylene glycol

$$\begin{array}{l}CH_2OH\\|\\CH_2\\|\\CH_2OH\end{array} \xrightarrow{KMnO_4} \begin{array}{l}COOH\\|\\CH_2\\|\\COOH\end{array}$$

1,3-Propanediol

$$\begin{array}{l}CH_2OH\\|\\CHOH\\|\\CH_2OH\end{array} \overset{HNO_3}{\rightleftharpoons} \begin{array}{l}CH_2ONO_2\\|\\CHONO_2\\|\\CH_2ONO_2\end{array} + H_2O$$

Glycerol (1,2,3-propanetriol) — Glyceryl trinitrate (nitroglycerin)

Two rather specialized oxidizing agents are of importance in the chemistry of aliphatic polyhydroxy compounds. They are periodic acid (pronounced pûr′ī ŏd′ ĭk), HIO_4, and lead tetraacetate, $Pb(O_2CCH_3)_4$. Both of these substances are specific oxidants for 1,2 diol groups; periodic acid finds use in water solutions and lead tetraacetate in organic solvents.

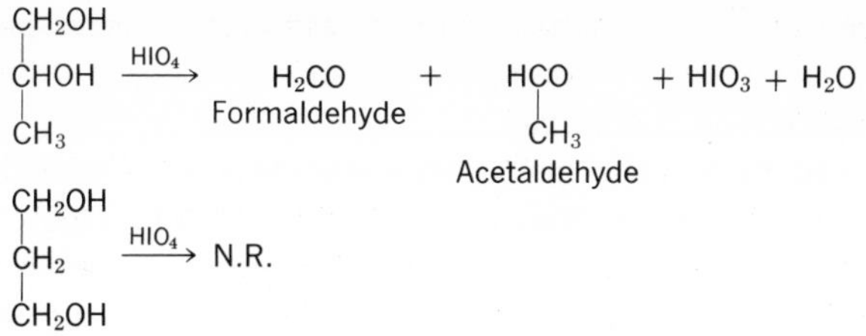
HIO4
Formaldehyde
Acetaldehyde
HIO4
N.R.

The reaction is quite specific for adjacent (vicinal) OH groups. Any species, such as C=O, that can hydrate to an —OH-like function will work, but an $—OCH_3$ group will not. Thus:

$$
\begin{array}{c}
\text{CH}_2\text{OCH}_3 \\
| \\
\text{C}=\text{O} \\
| \\
\text{H}_3\text{C}-\text{C}-\text{OH} \\
| \\
\text{CH}_3
\end{array}
\xrightarrow{\text{Pb(O}_2\text{CCH}_3)_4}
\begin{array}{c}
\text{CH}_2\text{OCH}_3 \\
| \\
\text{C}=\text{O} \\
| \\
\text{OH} \\
+ \\
\begin{array}{c}
\text{H}_3\text{C}-\text{C}=\text{O} \\
| \\
\text{CH}_3
\end{array}
\end{array}
$$

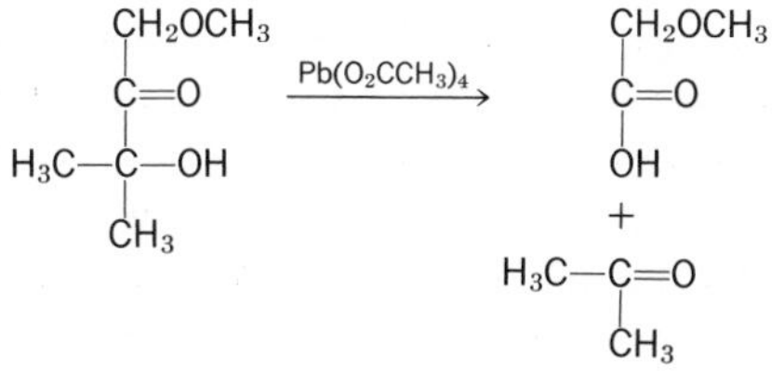
Pb(O2CCH3)4

The requirement for vicinal OH groups is believed to arise from intermediates like

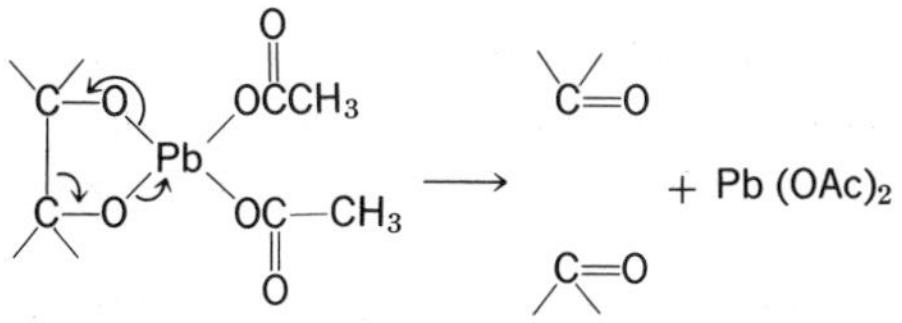
+ Pb (OAc)2

8.27 Spectroscopic Properties of Alcohols and Phenols Aliphatic alcohols have essentially no ultraviolet spectra in the accessible region. For this reason ethyl and methyl alcohols are frequently used as solvents in uv spectroscopy. Phenols, on the other hand, have ultraviolet spectra similar to benzene but usually with greater intensity, longer wavelength, and less fine structure. Thus phenol absorbs strongly at 210 and 270 mμ. Phenols can be readily characterized in the ultraviolet because their spectrum changes markedly on the addition of base due to the formation of the phenolate ion. In the case of phenol in ethanol solution the maxima shift to 235 and 287 mμ. This phenomenon is not only useful in characterizing phenols, but can often be used to obtain their ionization constants.

The most striking feature of both the infrared and nmr of alcohols and phenols is connected with their propensity to hydrogen bond, or associate. Since spectra of both types are usually determined in inert solvents such as CCl_4 or $CHCl_3$, a marked concentration dependence is observed.

In the infrared, both alcohols and phenols have two absorption maxima: a sharp maximum at 3500 to 3700 cm^{-1} and a broad band at about 3350 to 3360 cm^{-1}. The former is strongest in dilute solution, and at higher concentrations diminishes with respect to the broad band. The band at 3500 to 3700 cm^{-1} is thus attributed to the free O—H stretching vibrations, and the broad band to complex hydrogen-bonded stretching vibrations. In addition to the stretching vibrations there is an OH bending vibration at about 1340 to 1420 cm^{-1} and a C—O stretching vibration at about 1000 to 1230 cm^{-1}. Usually primary alcohols are found to absorb at the lower (1000 cm^{-1}) end of the range with secondary and tertiary alcohols, and phenols absorbing at respectively higher frequencies.

In the nmr both the shape and the position of the O—H proton are highly variable, ranging from fairly high delta values (δ = about 6 to 7) in concentrated solution to low values ($\delta = 2$ to 3) in dilute solution. Moreover, the shape (sharp vs. broad, etc.) is dependent both on temperature and the presence or absence of trace impurities. If the presence of an alcohol is suspected (e.g., from an OH in the infrared), it can be verified by running the nmr in a solvent such as CCl_4 and then adding D_2O, shaking the tube and rerunning the spectrum. Since alcohols ionize rapidly, the following equilibrium is rapidly set up:

$$R_3C{-}OH + D_2O \rightleftharpoons HOD + R_3C{-}OD$$

Since deuterium does not show up in the part of the nmr spectrum covered by proton spectrometry, the OH proton will be missing from the spectrum (it is now in the aqueous phase). Thus the disappearance of a peak on shaking with D_2O can be used as a quantitative determination of the number of OH groups in a molecule.

Finally, phenols can readily be differentiated from aliphatic alcohols by the presence or absence of characteristic aromatic absorptions (Sec. 6.25). This, of course, will not differentiate phenols from alcohols possessing aromatic rings.

8.28 Note on Symbols Used in Outlines From this point on, the symbol R will be used to represent any aliphatic or aromatic group, or substituted aliphatic or aromatic group, provided the substituents do not interfere with the reaction being illustrated. In a few cases restrictions will have to be placed on R, as in the preparation of A.1 below, and these will be clearly marked. The symbol ϕ will be used to represent aromatic groups exclusively, and will be used only in those cases in which the reaction is restricted to this class of molecules.

OUTLINE OF ALCOHOL AND PHENOL CHEMISTRY

Preparation

A. General methods for alcohols.
   1. RX + HOH $\longrightarrow$ ROH + HX
      NOTE: ϕX + HOH $\longrightarrow$ N.R.

B. Primary alcohols only.
   1. Reduction of an aldehyde.

$$RC\overset{O}{\underset{H}{\big\langle}} \xrightarrow[\substack{H_2 \\ LiAlH_4, \\ or \\ NaBH_4}]{Ni} RCH_2OH$$

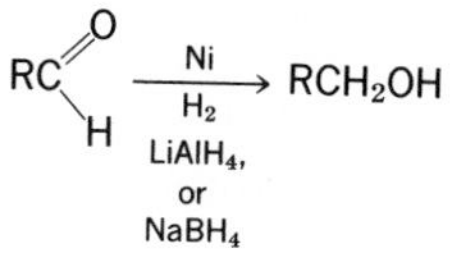

   2. Reduction of an acid.

$$RCOOH \xrightarrow[H_2]{Ni} N.R.$$

$$RCOOH \xrightarrow[(2)\ H_2O]{(1)\ LiAlH_4} RCH_2OH$$

   3. Reduction of esters.

$$RC\overset{O}{\underset{OC_2H_5}{\big\langle}} \xrightarrow[\substack{H_2 \\ or \\ LiAlH_4}]{Ni} RCH_2OH + C_2H_5OH$$

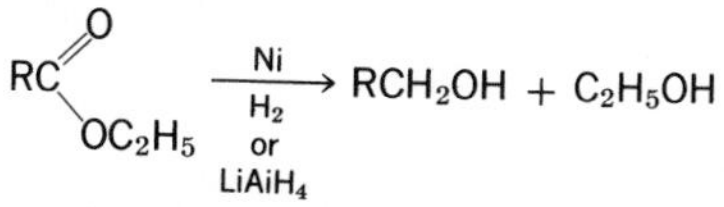

   This hydrogenation requires relatively high temperatures and pressures compared to the conditions required in the addition of hydrogen to alkenes and alkynes. It is used industrially.

   4. Grignard method.

$$RMgX + \underset{H}{\overset{H}{\big\rangle}}C{=}O \longrightarrow RCH_2OMgX \xrightarrow{H_2O} RCH_2OH + MgXOH$$
$$(RLi) \qquad\qquad\qquad\quad (Li) \qquad\qquad\qquad (LiOH)$$

$$RMgX + \overset{O}{CH_2{-}CH_2} \longrightarrow RCH_2CH_2OMgX \xrightarrow{H_2O} RCH_2CH_2OH$$
$$(RLi) \qquad\qquad\qquad\qquad (Li)$$

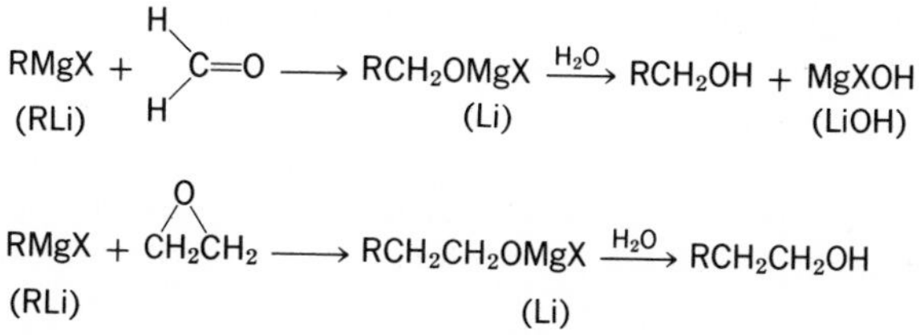

   5. Hydroboration.

$$6RCH{=}CH_2 + B_2H_6 \longrightarrow 2(RCH_2CH_2)_3B \xrightarrow{H_2O_2} 6RCH_2CH_2OH + 2B(OH)_3$$

C. Secondary alcohols only.
   1. Reduction of ketones.

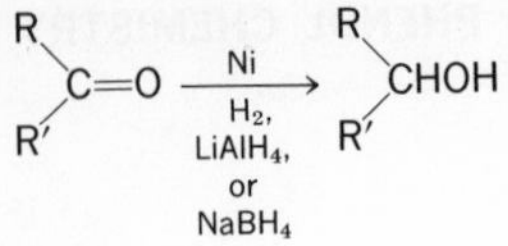

$$\underset{R'}{\overset{R}{\diagdown}}C{=}O \xrightarrow[\substack{H_2, \\ LiAlH_4, \\ \text{or} \\ NaBH_4}]{Ni} \underset{R'}{\overset{R}{\diagdown}}CHOH$$

2. Grignard method.

$$RMgX + R'\overset{O}{\overset{\|}{C}}\diagdown_H \longrightarrow \underset{R'\ (Li)}{\overset{R}{\diagdown}}CHOMgX \xrightarrow{H_2O} \underset{R'}{\overset{R}{\diagdown}}CHOH + \underset{Li\ OH}{MgXOH}$$
$$(R\ Li)$$

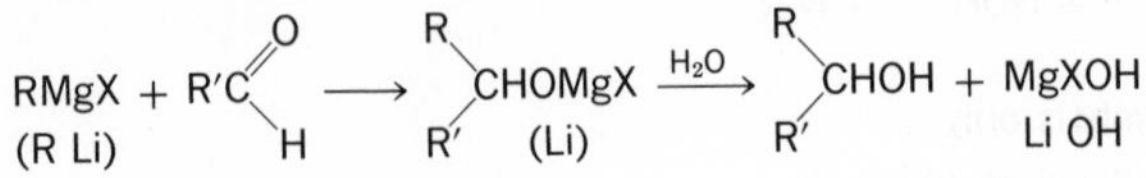

3. Hydration of alkenes.

$$RCH{=}CH_2 \underset{\underset{H_2SO_4}{}}{\overset{\text{conc.}}{\rightleftharpoons}} \underset{\underset{OSO_2OH}{|}}{RCHCH_3} \overset{H_2O}{\rightleftharpoons} RCHOHCH_3 + H_2SO_4$$

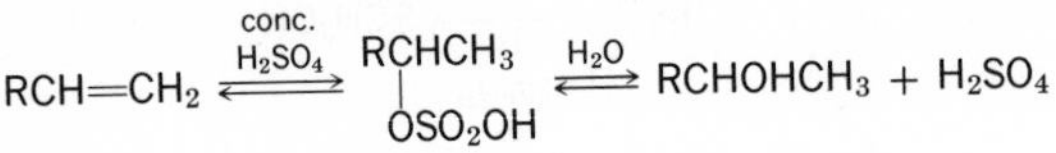

4. Hydroboration.

$$6R_2C{=}CHR + B_2H_6 \longrightarrow 2(R_2CH{-}CHR)_3B \xrightarrow{H_2O_2} 6R_2CH{-}\overset{\overset{\displaystyle OH}{|}}{C}HR + 2B(OH)_3$$

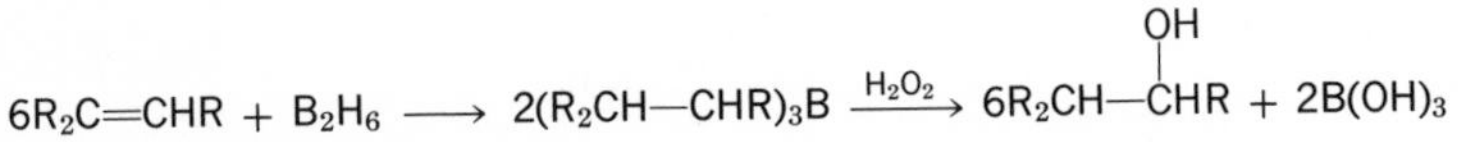

D. Tertiary alcohols only.
   1. Grignard method.

$$RMgX + \underset{R''}{\overset{R'}{\diagdown}}C{=}O \longrightarrow R'{-}\underset{R''\ (Li)}{\overset{R}{|}}COMgX \xrightarrow{H_2O} R'{-}\underset{R''}{\overset{R}{|}}COH + MgXOH$$
$$(R\ Li) \qquad\qquad\qquad\qquad\qquad\qquad\qquad\qquad (Li\ OH)$$

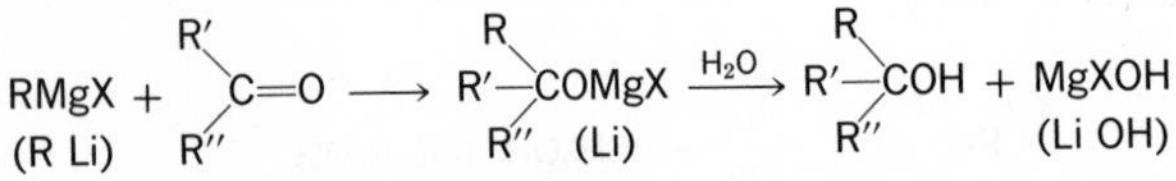

E. Phenols. (None of the above methods is applicable.)

   1. $\phi SO_3Na \xrightarrow[NaOH]{fusion} \phi ONa + Na_2SO_3$

   $\phi ONa \xrightarrow[H_2O]{acid\ (H^+)} \phi OH + Na^+$

   2. $\phi NH_2 \xrightarrow[HCl]{HNO_2} \phi N_2Cl \xrightarrow{H_2O} \phi OH + HCl + N_2\uparrow$

Properties

In all reactions, the type ϕCH_2OH behaves like the type ROH, and, in general, a hydroxyl group on a side chain behaves like an aliphatic hydroxyl.

1. $2R{-}OH + 2Na \longrightarrow 2R{-}ONa + H_2$
   Sodium alcoholates
   or phenolates

2. Salt formation.

$$\phi OH + NaOH \xrightarrow[H_2O]{instantaneous} \phi ONa + H_2O$$

3. Ester formation.

$$ROH + HOSO_2OH \rightleftharpoons ROSO_2OH + H_2O$$

Ester of an inorganic acid

$$ROH + R'COOH \rightleftharpoons R'COOR + H_2O$$

Ester of a carboxylic acid

NOTE: $\phi OH + \begin{cases} HOSO_2OH \\ R'COOH \end{cases} \longrightarrow$ esters in impracticably low yields.

4. Displacement of OH by X^-.

$$ROH + HX \longrightarrow RX + H_2O$$
$$\phi OH + HX \longrightarrow \text{N.R.}$$

5. $R\text{—}OH + R'C(=O)Cl \longrightarrow R\text{—}O\text{—}C(=O)R' + HCl$

Because of the serious side reactions, phenols are not used in reactions 6 and 7. Note that reactions 5, 6, and 7 are not reversible.

6. $ROH + \begin{cases} PX_3 \\ PX_5 \\ P + I_2 \end{cases} \longrightarrow RX + \begin{cases} H_3PO_3 \\ POX_3 + HX \text{ or } H_3PO_4 + HX \\ H_3PO_3 \end{cases}$

7. $ROH + SOCl_2 \xrightarrow{\text{pyridine}} RCl + HCl + SO_2\uparrow$

Thionyl chloride

Thionyl chloride has the advantage that the inorganic products are gases.

8.

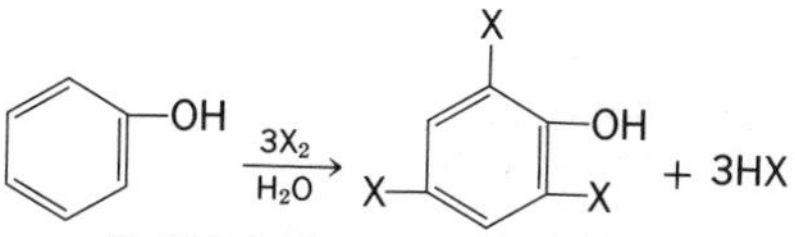

OH
$3X_2$
H_2O
X
OH
X
X
+ 3HX

Rapid quantitative substitution

9. Oxidation with $K_2Cr_2O_7 + H_2SO_4$.

$$RCH_2OH \xrightarrow{(O)} RCHO \xrightarrow{(O)} RCOOH$$

$$RCH(OH)R' \xrightarrow{(O)} RC(=O)R'$$

Further oxidation is difficult and gives acids of lower carbon content, that is,

$$RCH_2COCH_2R' \xrightarrow{(O)} RCH_2COOH + R'COOH \ldots$$

$$R_3COH \xrightarrow{(O)} \text{N.R. or acids of lower C content}$$

$$\phi OH \xrightarrow{(O)} \text{oxidative destruction}$$

The ring is broken up into fragments.

10. 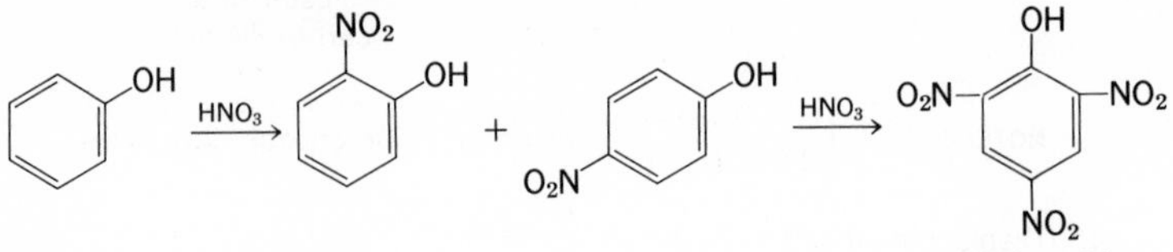

Picric acid

This reaction goes more easily than the nitration of benzene, but is always accompanied by some oxidative destruction of the ring.

11. $ROH \xrightarrow{\text{conc. } H_2SO_4} [R^+]$

$$[R^+] \longrightarrow \text{alkene}$$

(favored by high temperature, absence of excess ROH or H_2O)

$$[R^+] \xrightarrow{ROH} ROR$$

An ether

(moderate temperature, excess ROH)

$$[R^+] \xrightarrow{HSO_4^-} ROSO_2OH$$

An ester

(low temperature, conc. H_2SO_4)

$$[R^+] \xrightarrow{H_2O} ROH$$

Alcohol

(lower temperature, dilute aqueous solution)

This reaction is often accompanied by rearrangement where possible.

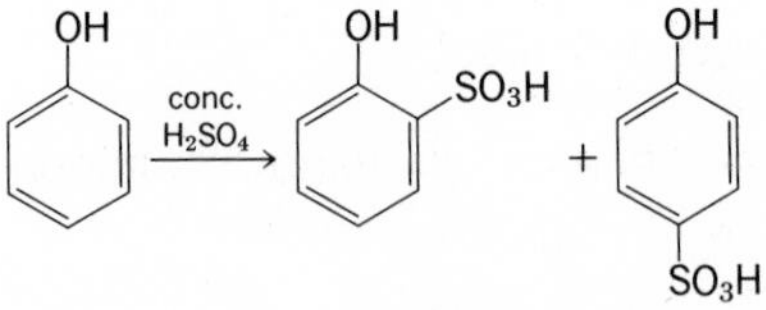

Sulfonic acids

12. 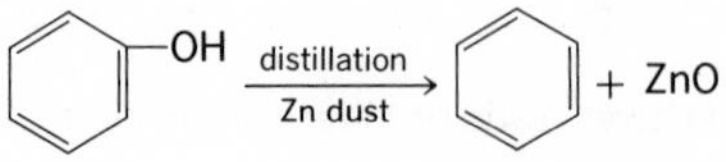

Reaction often fails with substituents on ring other than R.

Spectroscopic Properties

Group	*uv, mμ*	*ir, cm^{-1}*	*nmr, δ(ppm)*
R—OH (except φ)	transparent	3500–3700 (sharp) 3350–3360 (broad) 1340–1420 1000–1230	Variable in position with conc., temp. (2–7); disappears on shaking with D_2O
φOH	λ = 210, 270; shift to λ = 235, 287 on adding base	Similar to ROH, but also has aromatic stretch at 1610–1480	Similar to ROH, but also has aromatic H's at 6–9

CHAPTER 8 EXERCISES

1. Indicate by equations a practicable method for the preparation of each of the following compounds:
   ★*a.* Ethylene glycol
   b. Glycerol
   ★*c.* Phenol
   d. *p*-Chlorobenzyl alcohol
   ★*e.* 2-Phenyl-2-propanol
   f. 2-Butanol
2. 2-Phenyl-1-ethanol is a perfume base (rose oil). Write equations for the synthesis of this substance from benzene and simple intermediates.

★3. In Sec. 8.4, H_2O was spoken of as a Lewis base. What is its conjugate acid? Is H_2O a stronger or a weaker base than HCl? How do you estimate such matters of "strength"? (Your answers to questions like the first two should never be merely "yes" or "no," which are indistinguishable from guesses. You should give reasons, or analogies, which make your answer convincing.)

4. It is theoretically possible to prepare 1-decanol from methanol as the only organic starting material. Write equations to show how this could be done. Assuming the yield at each step is 80 percent of theory, calculate the overall yield of decanol based on the original methanol used.

★5. Write equations illustrating the effect of hydroxyl groups on the behavior of the benzene ring. Does the ring have an effect on the OH group? Explain.

6. A reaction is carried out between $HCOOC_2H_5$ and water containing isotopic oxygen (O^{18}). Write an equation showing the fate of this O^{18} in this experiment.
7. Arrange the following compounds in order of increasing solubility in water. Outline the principles on which your answer is based.
   ★*a.* C_4H_9OH, CH_3OH, C_4H_9Br
   b. *n*-Amyl alcohol, pentane-1,2-diol, 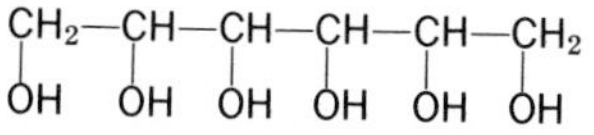
   ★*c.* Phenol, benzene, sodium phenolate, sodium salt of 1-decanol
   d. Sodium sulfate, sodium phenolate
8. The following reactions may be wrong as written. Explain any errors.

a. $CH_3CH(OH)\text{—}CH_2CH_2Br + Mg \longrightarrow CH_3CH(OH)\text{—}CH_2CH_2MgBr$

b. $R\text{-}CH_3CH(C_2H_5)\text{—}CH_2OH \xrightarrow{HCl}$ racemic $CH_3CH(C_2H_5)\text{—}CH_2Cl$

c. 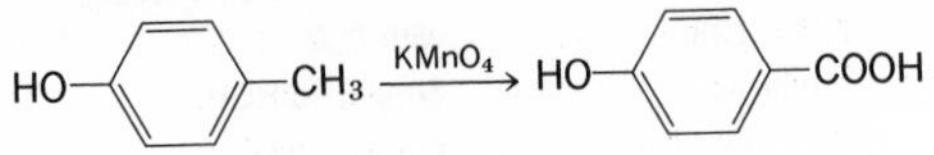

d. $CH_3CH{=}CH_2 \xrightarrow{BH_3} \xrightarrow{H_2O_2} CH_3C(=O)\text{—}CH_3$

e. $(CH_3)_3C\text{—}OH \xrightarrow{H_2SO_4} CH_3CH{=}CH\text{—}CH_3$

f. 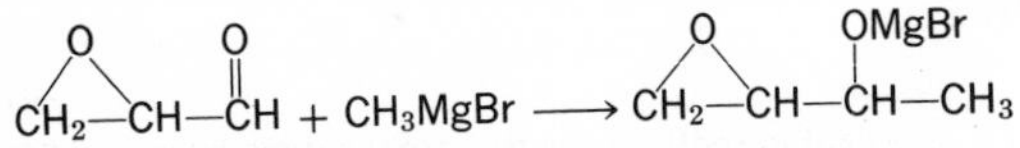

9. Show how you could distinguish by chemical and spectral methods between the following. If spectral distinction is impossible, explain.

★*a.* 1-Pentanol, 2-pentanol, $(CH_3)C\text{—}CH_2OH$

b. Phenol, $C_6H_5CH_2OH$, toluene

★*c.* $CH_3CH_2CH(OH)\text{—}CH_2(OH)$ and $CH_3CH(OH)\text{—}CH_2CH_2(OH)$

d. CH_3CH_2MgBr and CH_3CH_2OMgBr

e. 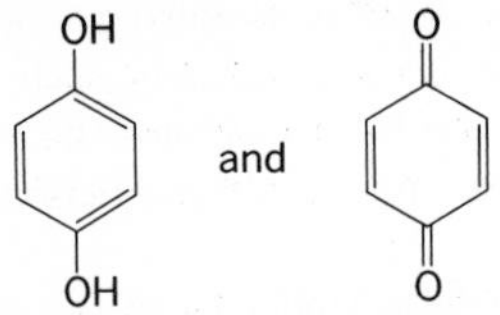

★*f.* 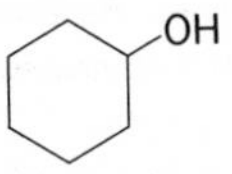 and $CH_3CH_2CH_2CH(OH)\text{—}CH_2CH_2$

g. $CH_2{=}CHCH_2OH$ and $CH_2CH{=}CHOH$

10. Show how the following conversions could be accomplished.

a. $CH_3CH_2CH_2OH \longrightarrow CH_3CH_2CH_2C(CH_3)_2\text{—}Cl$

b. $CH_3CH_2OH \longrightarrow CH_3CH_2CH(O\text{—}C(=O)\text{—}CH_3)\text{—}CH_2CH_3$

c. *t*-Butyl alcohol $\longrightarrow$ $(CH_3)_2CH{-}CH_2OH$

d. 1-Phenyl-1-propene $\longrightarrow$ $C_6H_5CH{=}CH_2$

e. Ethylene $\longrightarrow$ $CH_3\overset{O}{\overset{\|}{C}}{-}OCH_2{-}CH_2{-}O\overset{O}{\overset{\|}{C}}{-}CH_3$

f. Benzene $\longrightarrow$ 
H3C
OH

g. Isopropyl alcohol $\longrightarrow$ 2-methylpropane

h. $CH_3CH_2{-}CH_2{-}O\overset{O}{\overset{\|}{C}}{-}CH_2CH_3 \longrightarrow CH_3CH{=}O$

i. *p*-Bromoethylbenzene $\longrightarrow$ 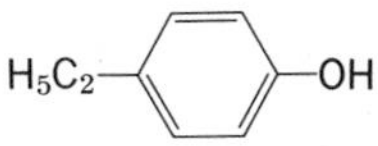
H5C2
OH

j. Acetylene $\longrightarrow$ 1-butanol

k. Toluene $\longrightarrow$ $C_6H_5\overset{OH}{\overset{|}{C}}{-}(CH_2C_6H_5)_2$

11. Show how you could distinguish by simple chemical tests between:

a. Phenol and benzyl alcohol

b. 2-Methyl-3-pentanol and 3-methyl-3-pentanol

c. Methanol and $CH_2{=}O$

d. Ethanol and ethylene

e. Toluene and benzene

f. Cyclohexanol and 1-hexanol

g. $CH_2{=}CHCH_2OH$ and $CH_3CH{=}CHOH$

h. $CH_3CH_2{-}O{-}\overset{O}{\overset{\|}{C}}{-}CH_3$ and $CH_3O{-}\overset{O}{\overset{\|}{C}}{-}CH_2CH_3$

★12. Echinochrome A, the physiologically active pigment of sea urchin eggs, has the formula below. Discuss the possibilities of tautomerism in this substance.

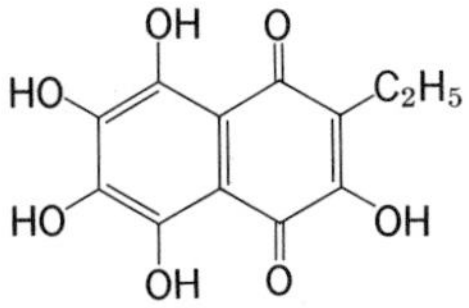
OH
O
HO
C2H5
HO
OH
OH
O

13. Show how you could separate by chemical methods (that is, not relying on the physical properties of the original substances) mixtures of

a. *n*-Hexane and methanol

b. *p*-Nitrophenol and benzene

c. Phenol and cyclohexanol

d. Phenol, cyclohexanol, and cyclohexane
e. *p*-Cresol and benzyl alcohol

Indicate spectral differences, if any, you might find in each case.

14. Write equations for the reactions of 2-butanol, allyl alcohol (2-propen-1-ol), and 1-phenylethanol with each of the following reagents:

a. Dil. $KMnO_4$
b. Conc. $KMnO_4$ + heat
c. H_2 + cat.
d. C_6H_5COCl
e. Excess HBr

15. Indicate by equations how the following conversions could be accomplished:

a. Ethyl alcohol to *n*-propyl alcohol
b. Methylethyl ketone to *tert*-amyl alcohol
c. Ethanol to 2-methyl-2-butene
d. 1-Propanol to 2,2-dimethyl-1-propanol
e. Acetic acid to ethanol
f. 1-Propanol to 3,3-dimethyl-1-pentene
g. 1-Butene to 1-butanol
h. 1,2-Dibromoethane to 2-butanol
i. Ethanol to

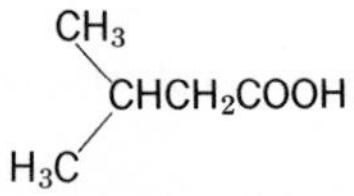
CH3
CHCH2COOH
H3C

j. 1-Butanol to 3-methyl-1-pentanol
k. 1-Butene to 2,3-dichloro-2-butene
l. 3-Hexene to isopropyl alcohol
m. 1-Butanol to 2-methyl-1-butanol
n. Phenol to ethylbenzene
o. Isopropyl alcohol to *n*-propyl alcohol
p. *n*-Butyl alcohol to 3,4-dimethyl-3-hexanol
q. Benzoic acid to benzyl chloride
r. Ethanol to diisopropyl ketone
s. Ethyl acetate to $CH_3—CO—CH(CH_3)_2$
t. Benzene to

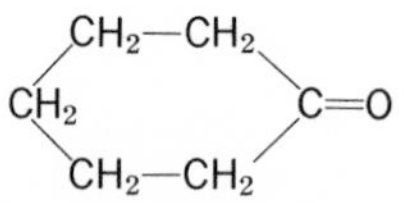
CH2—CH2
CH2
C=O
CH2—CH2

u. 2-Butanol to acetylene

★**16.** An aromatic substance, C_8H_9OCl, gives no reaction with dilute alkali; on gentle oxidation it gives C_8H_7OCl, and on strong oxidation, $C_7H_5O_2Cl$. What may its structure be?

17. A certain aromatic substance, which has the molecular formula $C_9H_{12}O$, reacts with acetic anhydride to give $C_{11}H_{14}O_2$, and with concentrated $KMnO_4$ to give $C_8H_6O_4$. Weak oxidation of $C_9H_{12}O$ gives $C_9H_{10}O$, and this substance cannot be oxidized to an acid containing 9 carbon atoms. Deduce a structure for $C_9H_{12}O$, and give a method for preparing it. What other structures fit the data?

18. Predict relative strength as acids of:

★*a.* 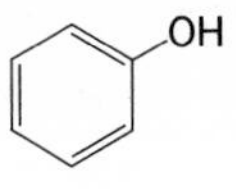 and 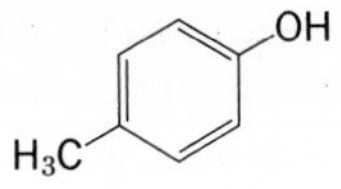

b. 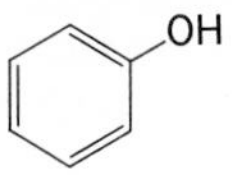 and 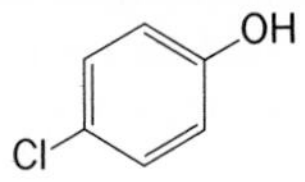

★*c.* CH_3CH_2OH and CCl_3CH_2OH

d.  and 

★*e.* 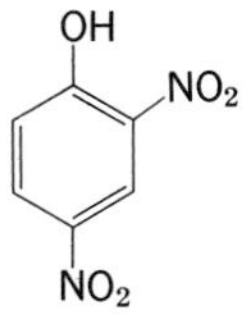 and 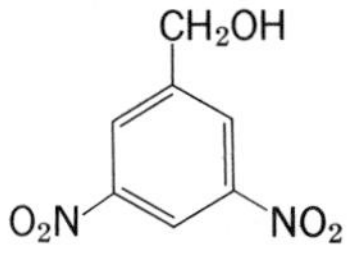

19. Point out errors, if any, in the reactions:

a. 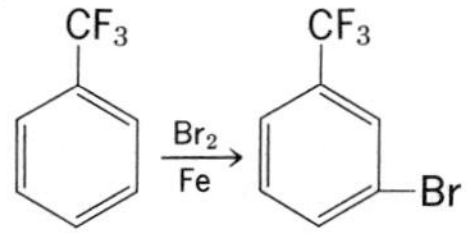
Br2
Fe

b. 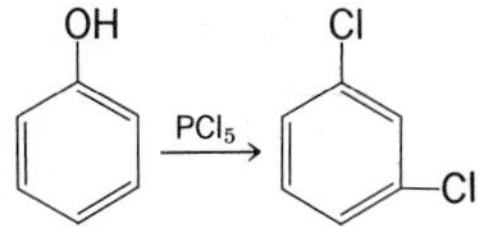
PCl5

c. 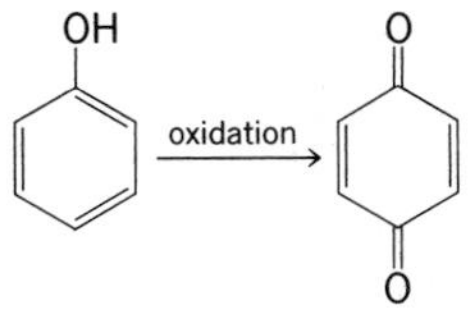
oxidation

d. 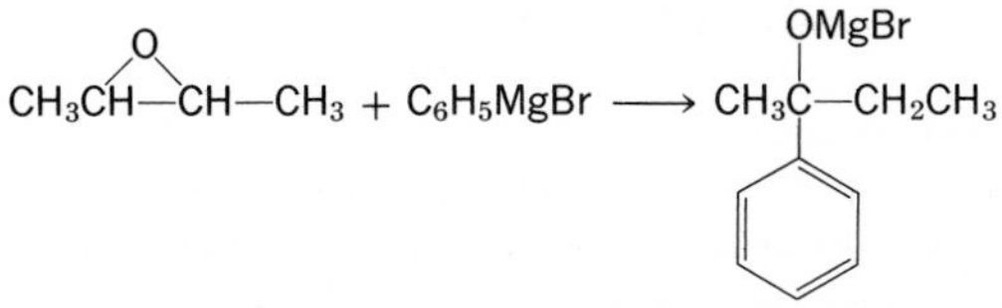

★20. In the reaction between acetic acid and *S*-2-butanol, would the ester be optically active? *R* or *S*? Explain.

21. In the reaction between CH_3MgBr and optically active $CH_3CH\text{—}CH_2$ (epoxide, O bridging the two carbons), would all or any of the alcohols produced be optically active? Explain.

★22. In the reaction between *R*-$CH_3CHOH\text{—}CHOH\text{—}CHOHCH_3$ and excess HIO_4, would the product(s) be optically active? Explain.

23. Does BH_3 addition follow Markownikoff's rule? Explain.

24. Show how you can reasonably account for the products shown in the following equations. They may not be the only products or the major ones. In such a case predict the major ones.

★*a.*

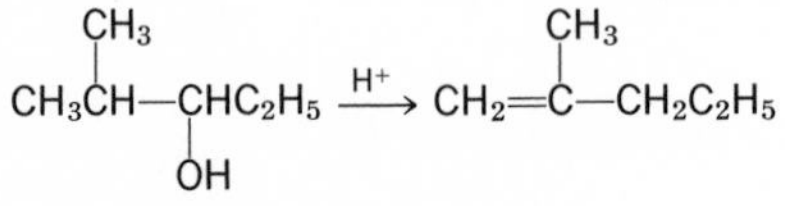

b.

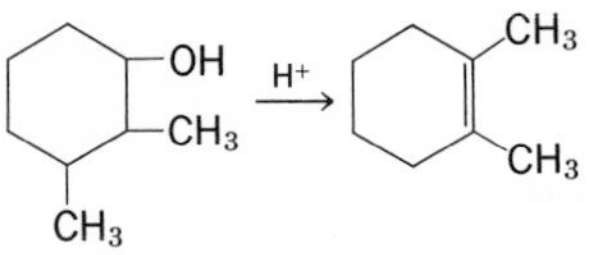

★*c.*

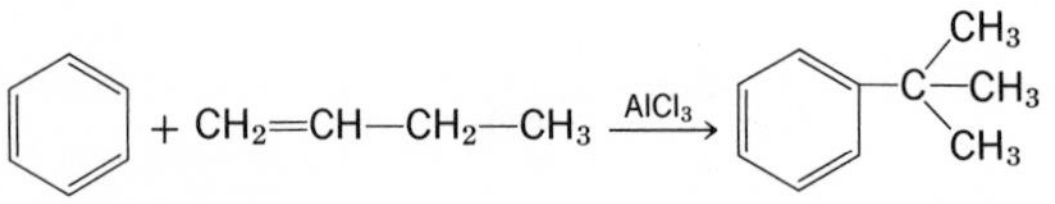

d.

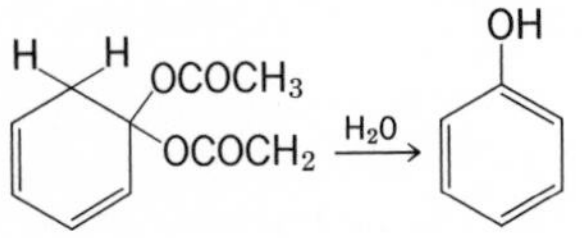

25. A compound Δ, $C_{14}H_{14}O$, gives, on mild oxidation, $C_{14}H_{12}O$. If Δ is treated with dehydrating agents, it loses 1 mole of water, and the product, upon vigorous oxidation, yields 2 moles of benzoic acid. Give the equations for the above reactions, and write the structure of compound Δ.

26. Write formulas for all $C_7H_{14}O$ compounds showing the following reactions:

★*a.* $C_7H_{14}O$ + benzoic acid $\longrightarrow C_{14}H_{18}O_2$

b. $C_7H_{14}O + H_2SO_4 \longrightarrow C_7H_{12}$

★*c.* $C_7H_{14}O + KMnO_4 \longrightarrow$ N.R.

27. In the dehydration of 1-propanol there is observed the formation of some 2-propanol and its derivatives. How could you account for this?

★28. Compound X, $C_7H_{16}O$, reacts with CH_3MgBr to evolve a gas. X is dehydrated to a hydrocarbon, which in turn is ozonized to give CH_3CHO and $C_5H_{10}O$. $C_5H_{10}O$ adds H_2 to give $C_5H_{12}O$. This in turn is dehydrated to a single C_5 alkene. Deduce the structure of X.

29. True or false? Explain.

a. Methanol dissolves in H_2O but hexanol does not.

b. CH_3CH_2OH has a lower boiling point than the isomer CH_3OCH_3.

c. Phenol reacts more slowly than benzene with HNO_3.

d. The von Baeyer test is specific for alkenes and alcohols.

e. Grignard reagents are ionic and water soluble.

f. $LiAlH_4$ is an effective oxidizing agent.

g. Sodium phenolate is volatile with steam.

★30. Show by equations how you might synthesize benzene from cyclohexanol. Would such a process be of industrial interest? Explain.

31. Account for the following by showing reaction mechanisms.

a. Glycerol (1,2,3-propanetriol) $\xrightarrow{H^+}$ $CH_2{=}CH{-}CH{=}O$

b. 3-Methyl-1-pentene + benzene + H_2SO_4 $\longrightarrow$ 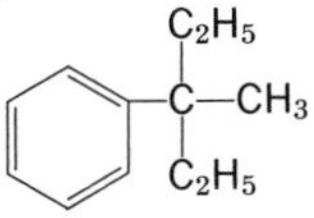

c. 3-Methyl-2-pentanol + $AlCl_3$ $\longrightarrow$ the same product as *b* above

d. *t*-Butyl chloride + KOH + H_2O $\longrightarrow$ a product that gives a positive von Baeyer test

e. $H_2C{=}CH{-}CH_2C({=}O){-}CH_3$ + NaOH + H_2O $\longrightarrow$ a product with no carbon to carbon double bond

f. $CH_2OH{-}CH{=}O$ + CH_3MgBr $\longrightarrow$ no three-carbon products

g. 1-Butanol + H_2SO_4 $\longrightarrow$ a polymer on long heating. Show its probable structures.

★32. Write the formula of a polymer that might be found as a side-product in the reaction of 1-butanol with sulfuric acid.

33. Tertiary butyl alcohol reacts faster with sulfuric acid and slower with strong bases than normal butyl alcohol. Explain.

★34. The dehydration of 1-phenyl-2-butanol leads predominantly to one olefinic product. What is it? Explain.

35. Show by equations the conversion of glycerol by acid to $H_2C{=}CH{-}CH{=}O$.

36. How would you expect the nmr spectra of the following pairs to differ? Indicate approximate δ values and any splitting you can predict.

★*a.* $(CH_3)_3C{-}OH$ and *n*-butyl alcohol

b. 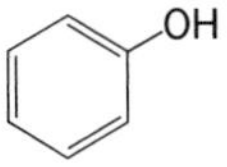 and 

★*c.* CH_3CH_2OH and $CH_3CH_2CH_2OH$

d. 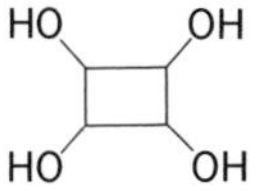 and $CH_2(OH){-}CH(OH){-}CH(OH){-}CH_2(OH)$

★*e.* 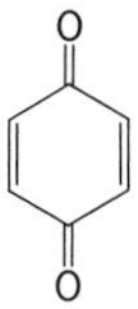 and 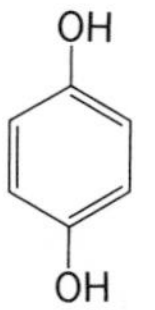

9

HALOGEN COMPOUNDS

9.1 Nomenclature All the organic halogen compounds may be grouped into classes in a manner analogous to the classification of the hydroxy compounds. Of these classes two large groups will be singled out for treatment in this chapter: the aliphatic and the aromatic halides. Only these two broad classes differ *enough* in chemical properties to justify their separate treatment at this point.

Aliphatic and aromatic halides have been discussed already in connection with hydrocarbons and alcohols, and their nomenclature presents few problems. The simpler halides are named by the groups to which the halogen is attached. More complex compounds are given systematic names in which chloro-, bromo-, and iodo- are used as prefixes and the position of the halogen is denoted by number. A few examples will make this clear:

CH_3I

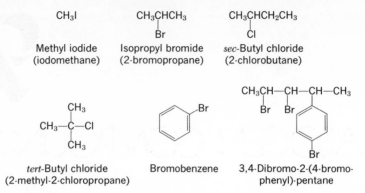

Methyl iodide
(iodomethane)

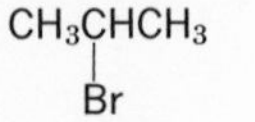
CH3CHCH3
Br

Isopropyl bromide
(2-bromopropane)

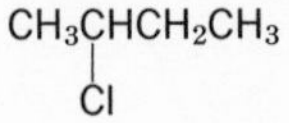
CH3CHCH2CH3
Cl

sec-Butyl chloride
(2-chlorobutane)

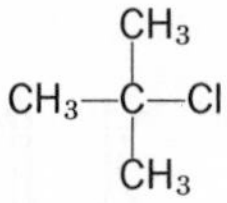
CH3
CH3—C—Cl
CH3

tert-Butyl chloride
(2-methyl-2-chloropropane)

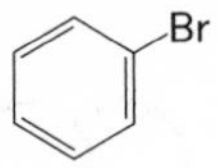
Br

Bromobenzene

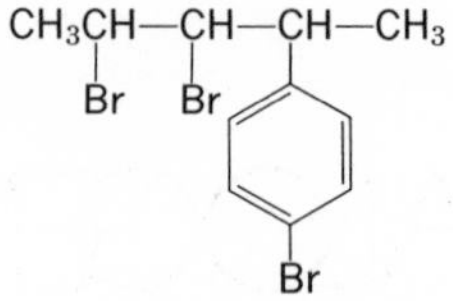
CH3CH—CH—CH—CH3
Br
Br
Br

3,4-Dibromo-2-(4-bromo-
phenyl)-pentane

9.2 Classification The aliphatic, or *alkyl, halides,* in which the halogen is attached to a saturated carbon atom, have to be distinguished from the aromatic, or *aryl, halides,* in which the halogen is attached directly to an aromatic ring carbon, because the two classes of halides show marked differences in the rates of their reactions with many reagents (compare alcohols and phenols). The alkyl halides may be subdivided (as were the alcohols) into primary, secondary, and tertiary halides. These subclasses differ from one another to varying degrees in the rates and sometimes the mechanisms of their reactions, but all three are closer to each other in behavior than to the aromatic halides.

There is another type of difference in reactivity which should be mentioned. This has to do with the relative reactivities of fluorides, chlorides, bromides, and iodides. In most reactions fluorides are much less reactive than chlorides, which are less reactive than bromides, which, in turn, are less reactive than iodides. The reactivity of fluorides in most cases is so low that they are not usually considered with the other halides. The reactivity differences may be attributed to the increasing polarizability of atoms as one goes down in the periodic table so that the nucleus of iodine, for example, exerts less influence on the outer electrons than does that of bromine or chlorine due to screening of the nuclear charge by the inner shells of electrons. Moreover, the carbon-halogen bond strength decreases in the order fluorine $>$ chlorine $>$ bromine $>$ iodine. The choice of a particular halogen for a reaction depends on other features such as cost and tendency toward side reactions. These dictate the common use of bromides and chlorides.

9.3 Preparation Alkyl halides are usually prepared in the laboratory from the corresponding alcohols by methods which have been discussed in the preceding chapter. For example,

$$CH_3CH_2OH + HBr \rightleftharpoons CH_3CH_2Br + H_2O$$

Often the use of inorganic acid derivatives in place of the free acid is preferable because the reaction is irreversible. For example, thionyl chloride yields only volatile by-products.

$$CH_3(CH_2)_{10}CH_2OH + SOCl_2 \longrightarrow CH_3(CH_2)_{10}CH_2Cl + SO_2\uparrow + HCl\uparrow$$

Other useful halides are those of phosphorus, PBr_3, PCl_3, and PCl_5.

Preparation of halides by direct halogenation is useful only in the aromatic series. The reasons for this have already been discussed.

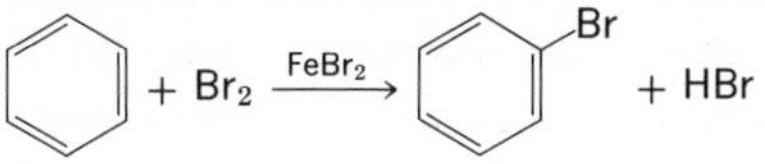
+ Br2
FeBr2
Br
+ HBr

9.4 Reactions A number of reactions of halides have already been considered (Secs. 2.16 and 8.7). Two of them, the formation of Grignard reagents and lithium derivatives, are general and can be applied to almost all halides. With respect to most other reactions there are significant variations in reactivity between "aromatic halides" (halogen directly attached to the benzene ring) and the aliphatic halides (halogen attached to a saturated carbon atom). Vinyl halides (halogen attached *directly* to a double bonded carbon) resemble aromatic halides.

$$CH_3CH_2CH_2Br + OH^- \longrightarrow CH_3CH{=}CH_2 + CH_3CH_2CH_2OH + {}^-Br + H_2O$$

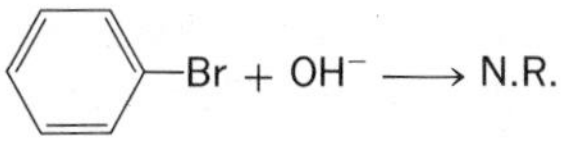
Br + OH⁻ → N.R.

Displacement is an important reaction of aliphatic halides, although except in very special cases it is not observed with aromatic halides. Some of the more important types of displacements involving organic halides are shown below:

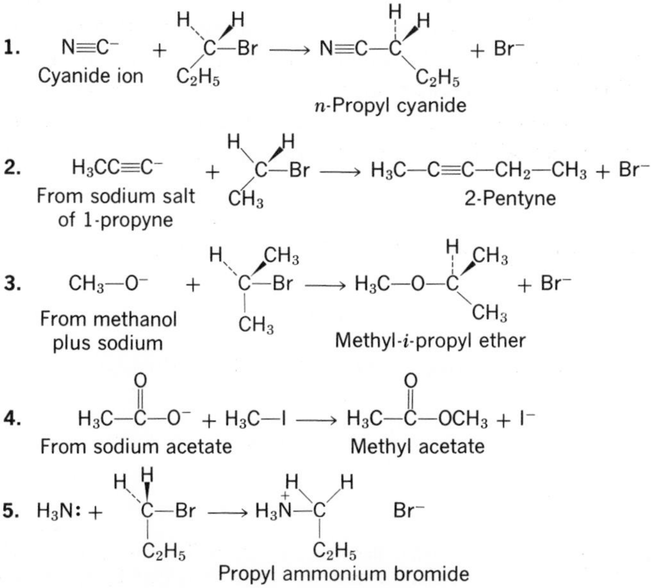
1.
Cyanide ion
n-Propyl cyanide
2.
From sodium salt of 1-propyne
2-Pentyne
3.
From methanol plus sodium
Methyl-i-propyl ether
4.
From sodium acetate
Methyl acetate
5.
Propyl ammonium bromide

Also R_3P: and R_2S: give similar displacement.

6. 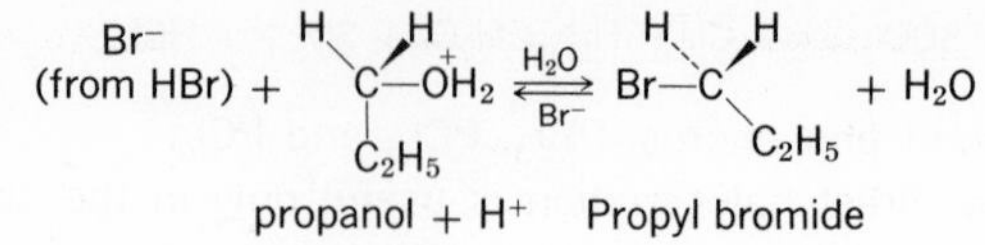
Br⁻
(from HBr) +
H
H
C—OH₂
H₂O
Br⁻
Br—C
+ H₂O
C₂H₅
propanol + H⁺
Propyl bromide

9.5 Displacement Reactions Most displacement reactions, from the point of view of their synthetic applications, are methods for the introduction of anions of weak acids (as groups attached by covalences) to the carbon chain of saturated compounds. They are often successful with primary and secondary aliphatic halides, but fail in the case of aromatic and vinyl halides. Interposition of one —CH_2— group leads to systems which undergo displacement reactions, often with great ease.

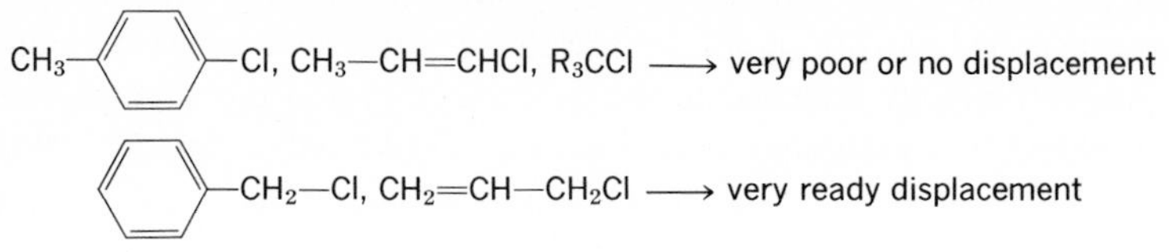
CH₃—
—Cl, CH₃—CH═CHCl, R₃CCl ⟶ very poor or no displacement
—CH₂—Cl, CH₂═CH—CH₂Cl ⟶ very ready displacement

$R_3C—CH_2Cl$ ⟶ slow but satisfactory displacement

Much of the utility of the displacement reaction lies in its ability to form new carbon-carbon bonds [with CN^-, $R—C{\equiv}C^-$, and other anions (Sec. 18.6)] while maintaining a functional group, and in preparing unsymmetrical amines and ethers.

9.6 Mechanisms of Displacement Reactions The displacement reaction has been one of the most extensively studied reactions in organic chemistry. From data such as that shown in Table 9.1 has evolved the concept of two mechanistic extremes of the displacement reaction. It can be seen that primary, and to a lesser extent secondary, halides are susceptible to added base concentration while tertiary halides show no effect. The reaction of primary and many secondary halides is believed to proceed via a *concerted* process. This is illustrated for the reaction of 2-butyl bromide with hydroxide ion.[1] By representing the energy content of the

TABLE 9.1
RATES OF SUBSTITUTION OF ALKYL BROMIDES IN ETHANOL-WATER SOLUTION
ROH (or RO^-) + R′Br ⟶ R—O—R′ + HBr (Br^-)

	R′Br			
	CH_3—Br	CH_3CH_2—Br	$(CH_3)_2CHBr$	$(CH_3)_3CBr$
Relative rate in neutral solution	0	0	0.24	1010
Relative rate with added 0.01 *N* NaOR	21.4	1.7	0.29	1010

[1] This does not mean to imply that neutral molecules such as H_2O cannot give this type of displacement reaction. Ordinarily such displacement is slower than that by the ion via this mechanism, but in many instances it can and does occur.

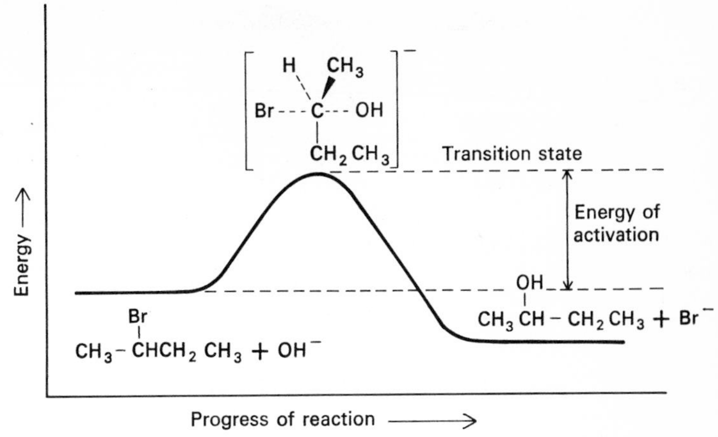
Transition state
Energy of activation
Energy ⟶
Progress of reaction ⟶

figure 9-1
Energy profile for the displacement reaction of 2-butylbromide with hydroxide ion.

reacting species on the vertical axis and the course of the reaction in time on the horizontal axis, the process may be represented by an energy profile similar to that used for other transition states (Sec. 3.10).

The rate of the reaction depends on the concentration of both reactants. This type of reaction is referred to as S_N2 which stands for substitution-nucleophilic-bimolecular.

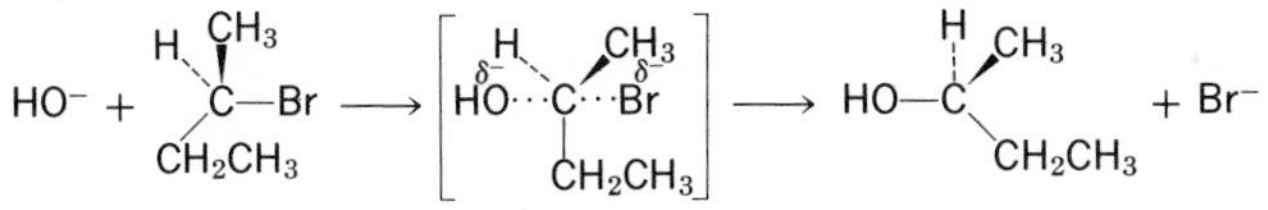

Transition state, S_N2 reaction

As shown in the equation above, the attacking nucleophilic agent approaches the reaction from the opposite side of the carbon atom from that occupied by the bromide ion, and the other groups on the reactive carbon are shifted from one side of the molecule through a planar configuration to the other side; the molecule is said to have been "inverted" with respect to the reacting carbon atom.

The "backside" approach of the reagent is facilitated by the shape of an sp^3 orbital (Fig. 1-6); this orbital has a small lobe on the backside of the atom, and bond formation is conceived of as beginning with the overlap of this portion of the orbital with an orbital of the attacking reagent before the original bond on the other side of the atom has completely broken. Since most of the electron density of the original bond lies between the leaving group and the carbon atom, the electron-rich nucleophile chooses to attack at a site of comparatively low electron density.

If the reacting group is asymmetric, then the configuration of the asymmetric carbon will be inverted during such a displacement. If the entering group and leaving group are of the same priority (Sec. 7.8), then an R molecule will become

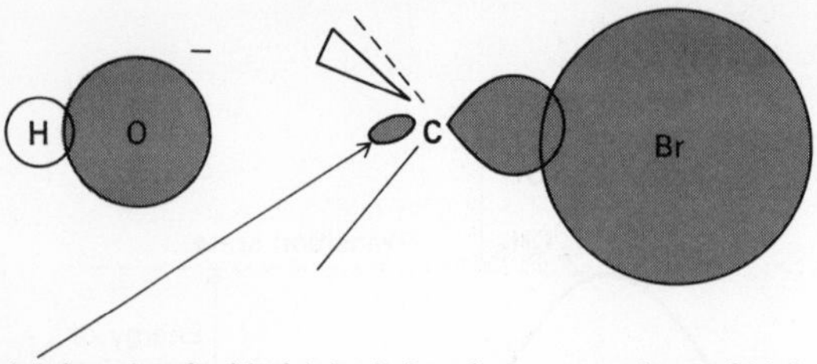
H
O
−
C
Br

Back lobe of sp^3 orbital (shaded regions are regions of high electron density)

figure 9-2
Nucleophilic attack.

an *S* and vice versa. Thus, in the example given above *R*-2-bromobutane is converted into *S*-2-hydroxybutane (since OH and Br both have the highest priority in this system). This phenomenon has been amply demonstrated experimentally and represents one of the criteria for an S_N2 reaction.

This mode of attack is very susceptible to hindrance by the bulk of the substituents attached to the carbon atom which is undergoing the displacement. Thus the rate of displacement decreases on going from methyl (only hydrogens attached) to ethyl (one alkyl group attached plus two hydrogens) to 2-propyl (two alkyl groups and one hydrogen). This can be seen from the first column in Table 9.2. When three alkyl groups are present the rate is prohibitively slow, and it is unlikely that any tertiary halides react by this mechanism.

When the behavior of a typical tertiary halide is examined, it turns out that the rate of the reaction (for a given temperature and solvent, etc.) depends only on the concentration of the halide, and is independent of the concentrations of added nucleophilic reagent. This is termed an S_N1 reaction because it shows first-order kinetics.

TABLE 9.2
APPROXIMATE RELATIVE RATES FOR NUCLEOPHILIC SUBSTITUTION REACTIONS

Group	*Relative rate for* S_N2 (R—Br + $^-$OR′ *in alcohol*)	*Relative rate for* S_N1 (R—X ⟶ R$^+$ ⟶ RY *in formic acid*)
CH_3X	30	(0.6)[a]
CH_3CH_2X	1	(1)[a]
$(CH_3)_2CHX$	0.025	26
$(CH_3)_3CX$	~0	100,000,000
$CH_2{=}CHX$	~0	~0
$CH_2{=}CH{-}CH_2X$	40	260
ϕ—X	~0	~0
ϕ—CH_2—X	120	17,000

[a] These values represent S_N2 processes even in formic acid.

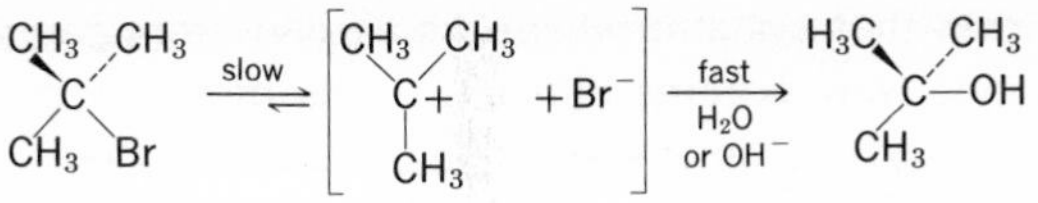
CH3 CH3
C
CH3 Br
slow
CH3 CH3
C+
CH3
+Br−
fast
H2O
or OH−
H3C CH3
C—OH
CH3

S_N1 reaction

The energy diagram for an S_N1 reaction differs from that just described in that a relatively stable ion appears as an intermediate with a finite lifetime; this intermediate may proceed over a new barrier to give a product, or, in principle, it may go back the way it came to starting material. Usually the former pathway (to products) is sufficiently favored that reversal does not significantly affect the rate of the reaction. If a solvent system is chosen in which no nucleophilic species are present, some of these ions may be observed directly by nmr, but, in general, such ions are only transient intermediates.

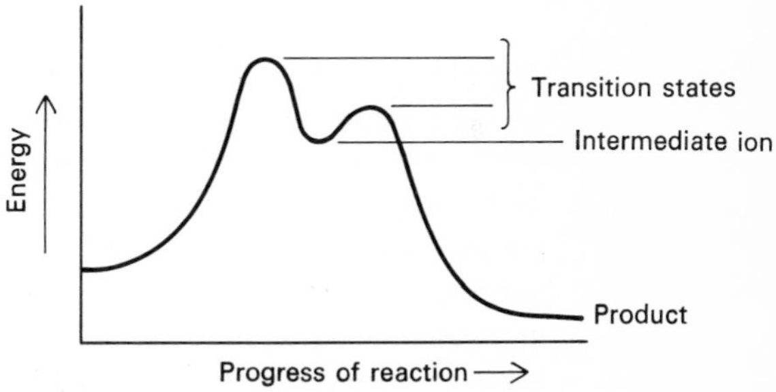
Transition states
Intermediate ion
Energy
Product
Progress of reaction

figure 9-3
Energy diagram for an S_N1 reaction.

In the formation of a carbonium ion it is believed that the bonding orbitals rehybridize to the planar sp^2 configuration (Sec. 3.4), leaving an unoccupied p orbital. The rationale for this is that p orbitals have zero value at the nucleus of the atom, while s orbitals have a finite magnitude there. Placing the electrons in sp^2 orbitals rather than sp^3 increases somewhat the electron density at the nucleus since each orbital has about one-third "s character" in sp^2, rather than one-fourth "s character" as in sp^3.

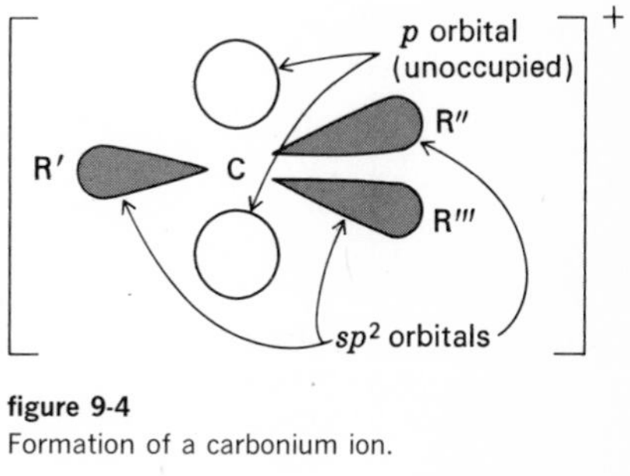
p orbital
(unoccupied)
+
R″
R′
C
R‴
sp2 orbitals

figure 9-4
Formation of a carbonium ion.

It is well known that systems where the carbon atom cannot become planar give carbonium ions with extreme difficulty. The molecule below, which is constrained by the ring system, is one example of such a system.

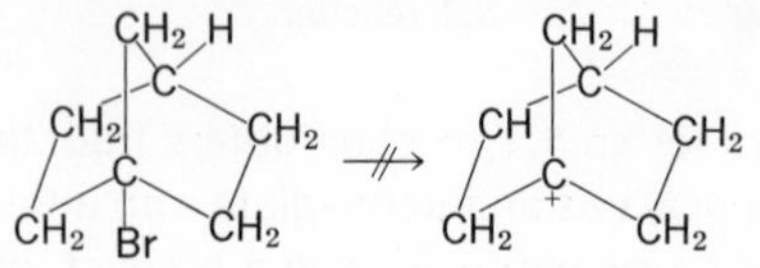

Carbonium-ion reactions are markedly affected by the reaction medium (solvent). Although S_N1 reactions are common in solvents like water, alcohols, and organic acids like acetic and formic, they are extremely rare in nonpolar solvents like hydrocarbons or simple aromatic compounds. The reason for this solvent dependence arises from the inherently high energies of carbonium ions. This is shown in Table 9.3 for some carbonium ions in the gas phase.

Reactions with activation energies (Sec. 3.10) of as much as 60 kcal/mole are very slow at ordinary temperatures. In order to bring energies such as those of free carbonium ions into an accessible range, rather large interactions of the carbonium ions with solvents are necessary. In polar solvents, such as water, alcohols, acids, and the like, such interactions are large enough to bring tertiary and many secondary carbonium ions within reach. This interaction, called "hydration" in water and "solvation" in other solvents, is responsible for the stability of most ions in solution (Sec. 8.15). For example, HCl is a nonionized molecule in benzene solution, but is completely ionized in water.

The behavior of halides according to the $S_N1 - S_N2$ scheme is summarized in Table 9.2.

As can be seen from Table 9.2 and Sec. 8.4, both types of nucleophilic substitution reactions fail or are extremely difficult with aromatic and vinyl halides. Factors which contribute to this lowered reactivity are the increased C—X bond strengths in such systems, geometric hindrance to backside displacement, and lowered carbonium-ion stability because of different hybridization.

TABLE 9.3
ENERGETICS OF GAS PHASE CARBONIUM-ION REACTIONS
$R{-}Y \longrightarrow R^+ + Y^-$ (ΔH, kcal/mole)

R^+	Y^-			
	Cl^-	Br^-	I^-	OH^-
H^+	328	317	308	382
CH_3^+	220	215	204	268
$CH_3CH_2^+$	192	183	176	239
$(CH_3)_2CH^+$	168	150	149	214
$(CH_3)_3C^+$	149	132	132	202
$H_2C{=}CH{-}CH_2^+$	158	150	145	206

The bonds between aryl and vinyl groups and substituents capable of being displaced are usually strengthened by resonance interactions of the type shown below:

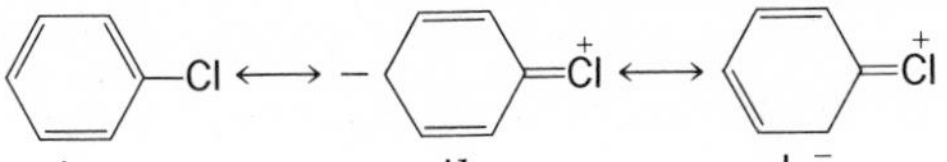

I*a* I*b* I*c*$^-$

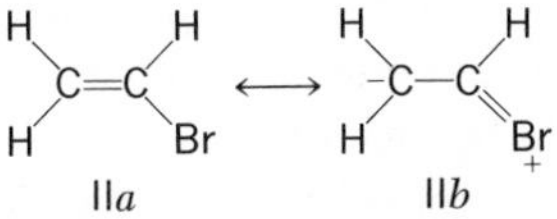

II*a* II*b*

Contributions of structures I*b* and I*c*, and of II*b*, tend to increase the C—Cl or C—Br bond strength and make the displacement of such a group more difficult.

The carbon atom at which the displacement is occurring is sp^2-hybridized in vinyl and aryl (aromatic) compounds. Since the entering nucleophile, in coming in from the backside, must come in between the two substituents, the hindrance to such an approach is greater than in saturated (sp^3) systems. The filled p orbital tends to repel the electron-rich nucleophile from approaching above or below the plane of the three sp^2 orbitals. With aromatic compounds the fact that the two R groups are part of a six-membered ring makes approach from the back essentially impossible.

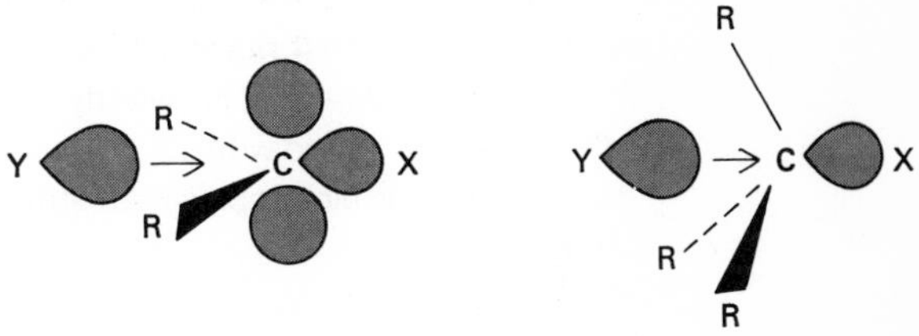
R
Y
C
X
R
sp^2
R
Y
C
X
R
R
sp^3

figure 9-5
Backside nucleophilic attacks for substitution.

Substitution by an S_N1 process is also difficult with sp^2-hybridized carbon atoms. The carbon atom, of necessity, has one less substituent since one of the orbitals is involved in the π bond. The π electrons cannot interact to stabilize the charge because to do so would require contributions to the resonance hybrid of structures like III*b* and III*c*, or IV*b*.

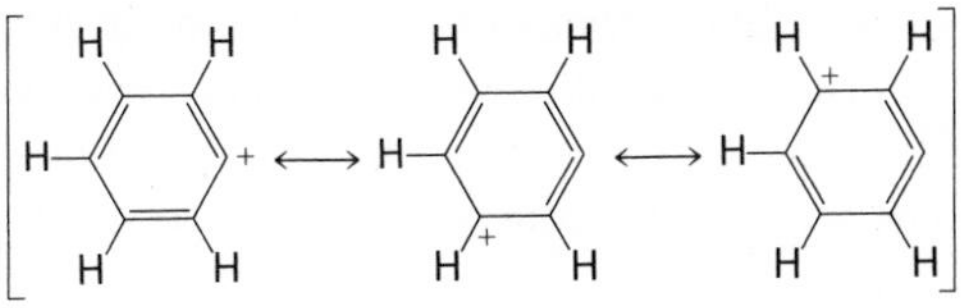

III*a* III*b* III*c*

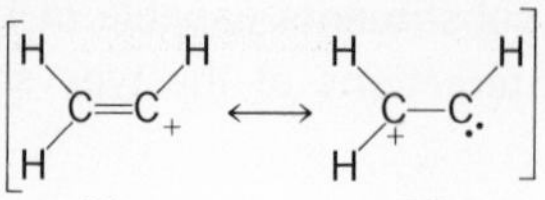

IVa IVb

Structures IIIb and IIIc would require two double bonds to one carbon atom. This would require that the carbon atom be sp-hybridized and linear in order to have two p orbitals available for π bonds. In a ring system of this type this would be energetically prohibitive. With Structure IVb a bond is lost without any compensating factors; accordingly, it would be much higher in energy than IVa.

A third mechanism which might be available to nucleophilic substitution on sp^2-hybridized systems is addition-elimination, analogous to the mechanism of *electrophilic* aromatic substitution.

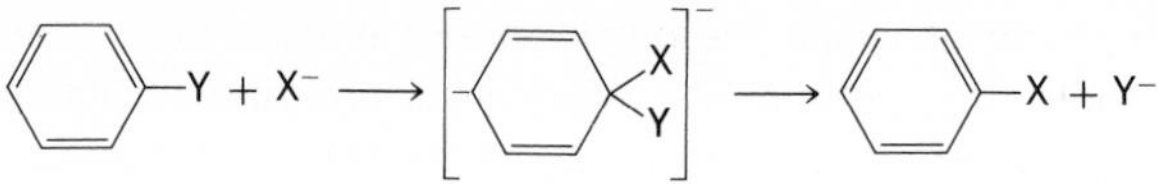

or

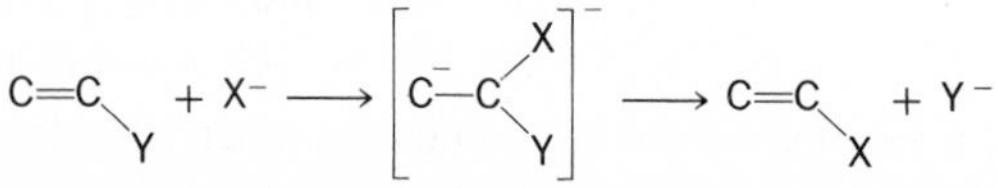

Unlike electrophilic substitution this mechanism is unimportant with unactivated aromatic rings. Most nucleophiles are either not reactive enough to add to the ring (or double bond), or react directly with the substituent rather than the ring. Only with rings substituted with strong meta directors like nitro groups (which are strongly deactivating for electrophilic substitution) do such reactions occur.

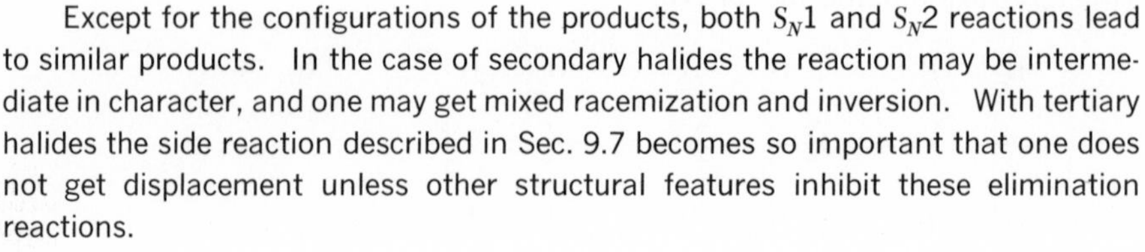

Except for the configurations of the products, both S_N1 and S_N2 reactions lead to similar products. In the case of secondary halides the reaction may be intermediate in character, and one may get mixed racemization and inversion. With tertiary halides the side reaction described in Sec. 9.7 becomes so important that one does not get displacement unless other structural features inhibit these elimination reactions.

9.7 Elimination Reactions The loss of H—X from a saturated molecule, with the formation of a double bond, is often referred to as an elimination reaction (or more properly as a β-elimination reaction). The net result of such a reaction is that a hydrogen atom is removed β to the halide ion and a double bond is formed between this carbon atom and the one formerly bearing the halogen. There is considerable similarity between this reaction and the nucleophilic displacement in that the electrons which formerly bonded the β-hydrogen atom displace the halide ion internally.

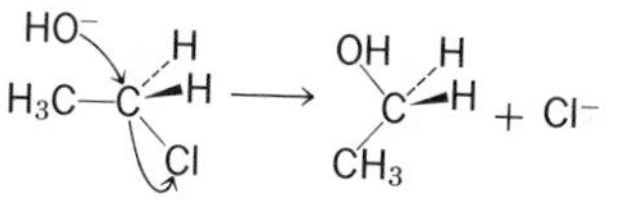

S_N2 displacement

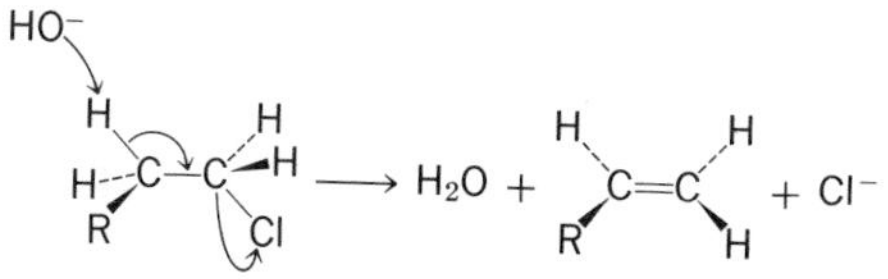

E_2 elimination

It is not surprising to find that displacement and elimination almost always accompany one another. The relative proportions of each reaction are determined by such factors as structure of the halide molecule, solvent, concentration, temperature, structure and basicity of the nucleophile, etc. With anions derived from weak acids (bases) tertiary halides tend to undergo elimination reactions in preference to substitution, and it is not possible to make cyanide or acetylide derivatives with tertiary halides in most cases. On the other hand, primary halides do not undergo elimination readily. It is possible to effect elimination with primary halides by using high concentrations of base (KOH in ethanol), hindered bases such as *t*-butoxide, and higher temperatures.

Like displacement reactions, eliminations of two sorts are known: E_1, first-order eliminations, and E_2, second-order eliminations.

E_1 *elimination:*

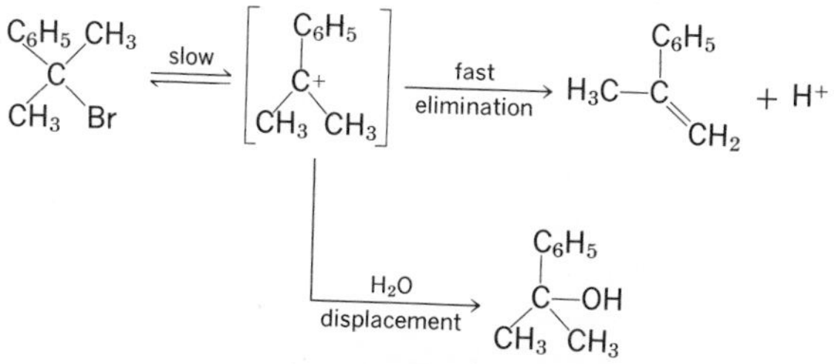

E_1 eliminations are most likely to occur with tertiary halides, and to a lesser extent with secondary halides. The tendency toward E_1 can be increased by the use of Lewis acids such as silver or mercury ions which help to pull the halides off. As in the case of S_N1, the presence of base is not necessary in the slow or rate-determining step but only to neutralize the acid which would be generated.

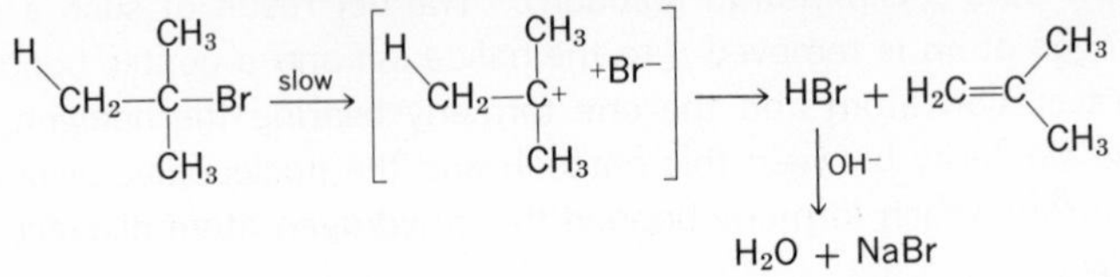

H
CH₃
CH₂—C—Br
CH₃
slow
H
CH₃
CH₂—C+
CH₃
+Br⁻
HBr + H₂C=C
CH₃
CH₃
OH⁻
H₂O + NaBr

E_2 elimination is similar to S_N2 in that it is second-order and requires the presence of base in the slow step. Unlike S_N2 it is the preferred reaction with tertiary halides and is much less successful with primary.

$$HO^- + CH_3{-}CH_2{-}Br \xrightarrow{S_N2} CH_3CH_2OH$$

$$\xrightarrow{E_2} H_2C{=}CH_2 + H_2O$$

$$HO^- + (CH_3)_3CBr \xrightarrow{S_N2} \text{not observed}$$

$$\xrightarrow{E_2} H_2C{=}C(CH_3)_2 \qquad \text{only product}$$

The E_2 elimination is favored over S_N2 by crowding of groups so that the nucleophile cannot get close enough to the backside of the halide to start bond formation. Thus the base $(CH_3)_3CO^-$ (*t*-butoxide) is more effective than $CH_3CH_2O^-$ or OH^-. The reverse, of course, is true for S_N2.

The E_2 elimination is subject to certain steric requirements of its own. Because the removal of hydrogen ion from the β carbon is occurring at the same time the carbon-halogen bond is breaking, the electrons from the carbon-hydrogen bond must come in from the backside, as in

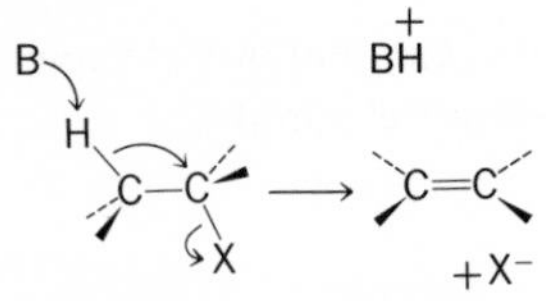

B
H
C—C
X
C=C
BH
+X⁻

Thus, four atoms must be coplanar and trans. In open-chain systems there is usually little difficulty in obtaining such an arrangement, but the structure of the products will often be controlled by this geometry. For example, the two diastereomers of 1-bromo-1,2-diphenylpropane give different alkenes on elimination.

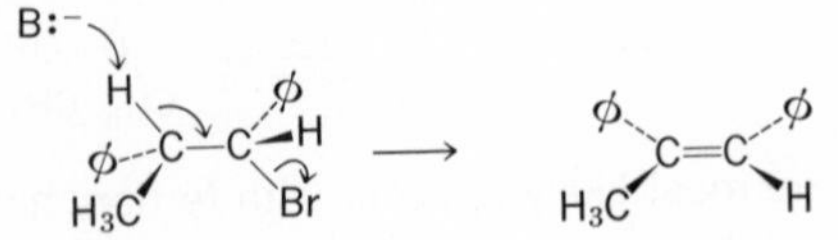

B:⁻
H
ϕ
C—C
H
H₃C
Br
ϕ
C=C
ϕ
H₃C
H

1-*R*,2-*S*,1-Bromo-1,2-diphenylpropane — *cis*-1,2-Diphenylpropene

also 1-*S*,2-*R*,1-Bromo-1,2-diphenylpropane ⟶ *cis*-1,2-diphenylpropene

1-*S*,2-*S*,-Bromo-1,2-diphenylpropane — *trans*-1,2-Diphenylpropene

also 1-*R*,2-*R*,1-Bromo-1,2-diphenylpropane ⟶ *trans*-1,2-diphenylpropene

The use of stereochemical arguments like that above has added much to our understanding of mechanism, and knowledge of mechanism enables us to predict the stereochemistry in new systems.

In cyclic systems, hydrogens which are cis to the leaving group cannot undergo the E_2 reaction even though they would otherwise be favored by the rule given in

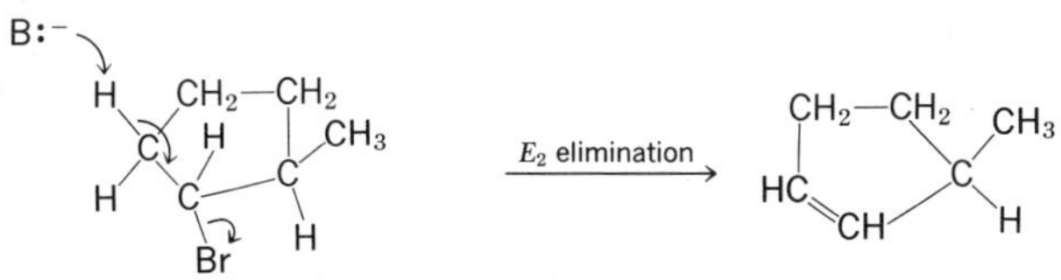

trans-1-Bromo-2-methylcyclopentane — 3-Methylcyclopentene

Sec. 3.16. Thus, with *trans*-1-bromo-2-methylcyclopentane, in which the tertiary hydrogen (the one predicted by the rule to be lost) is cis to the bromine, the product is predominantly the isomer in which the secondary hydrogen is lost, 3-methylcyclopentene.

In the E_1 reaction (favored by Ag^+, which helps pull the bromine off, and by low base concentration) the product is the 1-methylcyclopentene predicted by the rule (Sec. 3.16).

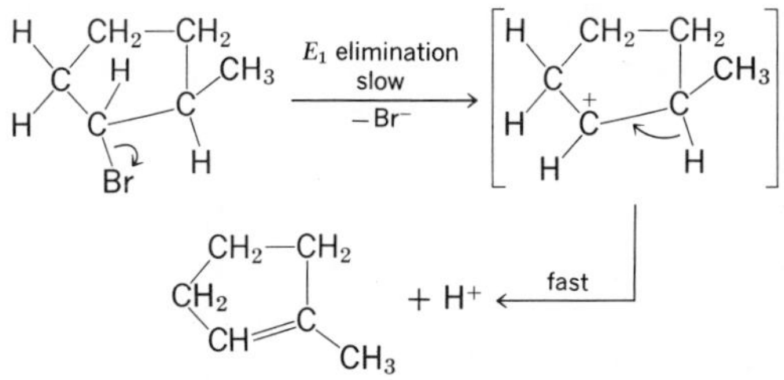

1-Methylcyclopentene

9.8 Fluorides Because of the great reactivity of fluorine and HF, carbon compounds containing fluorine are not prepared by the general methods used for the other halides. Fluorine compounds are usually made by displacement, or by the "universal reagent," SF_4, which replaces carbon-oxygen linkages by C—F bonds.

Displacement.

$$2C_2H_5Br + HgF_2 \text{ (or AgF)} \longrightarrow 2C_2H_5F + HgBr_2$$

$$3CCl_4 + 2SbF_3 \xrightarrow{SbCl_5} 3CCl_2F_2 + 2SbCl_3$$

$$SF_4 + 2ROH \longrightarrow 2RF + SOF_2$$
$$SF_4 + R_2C{=}O \longrightarrow R_2CF_2 + SOF_2$$
$$SF_4 + RCOOH \longrightarrow RCF_3 + \cdots$$

Although monofluorides are hydrolyzed rather easily, the polyfluorides are often very stable substances. (In RCH_2F the C—F bond length is 1.42 Å; in $RCHF_2$ the bond length is 1.36 Å.) The polyfluorides are finding application in industry where stability is desired. For example, a substance for use in mechanical refrigeration as a carrier of heat should be stable and noncorrosive. Dichlorodifluoromethane (Freon-12) is widely used for this purpose, being nontoxic, odorless, and nonflammable. Also, it has suitable physical properties. It finds application also as a propellant in aerosol sprays.

Initiated by the need for chemically resistant materials in the uranium purification process, much research has been carried out on the "fluorocarbons," alkanes and alkenes in which all the H's are replaced with F. Polytetrafluoroethylene, one of the products of this research, is finding applications in chemically resistant gaskets, etc., under the name Teflon®.[1] The great stability of the carbon-to-fluorine bond is here being utilized for special purposes.

9.9 Polyhalides Other organic halides with several halogen atoms per molecule are also found useful for special purposes. Their chemical properties resemble those of the simpler monohalides and do not merit close study at this point. The applications of these substances, a few of which are discussed below, usually depend on their physical properties.

Carbon tetrachloride, CCl_4, being nonflammable, was formerly used as a fire extinguisher, since its heavy vapors will smother a fire by screening it from oxygen of the air. At high temperatures the tetrachloride may be hydrolyzed to HCl and the toxic substance $COCl_2$ (phosgene), so that this type of extinguisher now finds little use. Chloroform, $CHCl_3$, formerly much used as an anesthetic, is now largely displaced by the less toxic ether. Iodoform, CHI_3, is an effective antiseptic.

A practical application of the property of toxicity of polyhalides is made in the effective insecticide DDT, dichlorodiphenyltrichloroethane [1,1,1-trichloro-2,2-di-(4-chlorophenyl)-ethane]. Unfortunately this substance is not readily biodegradable and its use is now curtailed.

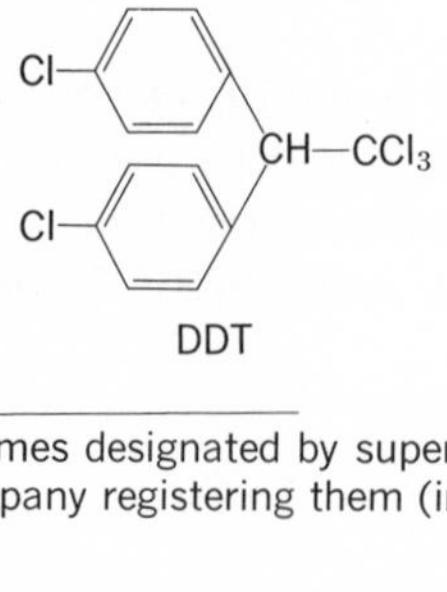
Cl
CH—CCl3
Cl

DDT

[1]Names designated by superscript ® are registered trademarks and are the property of the company registering them (in this case duPont).

9.10 Spectroscopic Properties The halocarbons do not have remarkable spectroscopic properties. They are for the most part transparent in the ultraviolet and infrared except for difficult to distinguish bands at the low frequency end of the spectrum. In the nmr Cl, Br, and I affect the electronegativity of the carbons to which they are attached so that the proton absorptions are shifted downfield. Fluorine, on the other hand, has a magnetic moment and interacts with protons to cause splitting. The nmr of fluorocarbons is invaluable in studies of these compounds but will not concern us here.

The table below lists the expected chemical shift value δ for halocarbons. It should be noted that other electronegative groups can cause similar chemical shifts.

TABLE 9.4
δ VALUES

X	CH_3X	RCH_2X	R_2CHX	CH_2X_2
Cl	3.0	3.6	4.0	4.7
Br	2.7	3.3	3.6	5.1
I	2.2	3.1	4.2	6.1

OUTLINE OF HALIDE CHEMISTRY

Preparation

1. Direct halogenation. Not applicable in aliphatic series. For halogenation of aromatic hydrocarbons see Sec. 6.7.
2. Alcohol and halogen acid.

$$R{-}OH + HBr \rightleftharpoons RBr + H_2O$$

NOTE: $\phi OH + HBr \longrightarrow$ N.R.

3. Alcohol + inorganic halide.

$$3ROH + PX_3 \longrightarrow 3RX + H_3PO_3$$
$$4ROH + PCl_5 \longrightarrow 4RCl + HCl + H_3PO_4$$
$$6ROH + 3I_2 + 2P \longrightarrow 6RI + 2H_3PO_3$$
$$ROH + SOCl_2 \longrightarrow RCl + SO_2 + HCl$$

NOTE: Not useful for replacement of aromatic OH.

4. Addition of halogen acid to an alkene (or alkyne).

$$R{-}CH{=}CH_2 + HX \longrightarrow RCHX{-}CH_3$$

Properties

NOTE: It is to be observed in the reactions below, that in aromatic compounds, where the halogen is not on the ring, the halide reacts as does the aliphatic analog, but that a halogen on the ring is in nearly all cases very unreactive at ordinary temperatures. R = Alkyl, ϕ, or H.

1. $R_2CHX + NaOH \longrightarrow ROH + NaX + \text{alkene}$

   $R_3CX + NaHCO_3 + H_2O \longrightarrow R_3COH + \text{alkene}$

2. $R_2CH{-}\underset{\displaystyle R}{\underset{|}{C}}H{-}X + KOC(CH_3)_3 \longrightarrow R_2C{=}CH{-}R + HX + HOC(CH_3)_3$

   $R_2CH{-}\underset{\displaystyle R}{\underset{|}{C}}H{-}X + \text{conc. NaOH} \xrightarrow[\text{alcohol}]{\text{hot}} R_2C{=}CH{-}R + \text{displacement}$

3. $R_2CHX \xrightarrow[\text{or NaCN}]{\text{KCN}} R_2CH{-}CN + KX \text{ or } NaX$

4. $R_2CHX + NaC{\equiv}CR' \longrightarrow R_2CHC{\equiv}CR' + NaX$

5. $R_2CHX + NaOR' \longrightarrow R_2CHOR' + NaX$

6. $R_2CHX + NaO\overset{\displaystyle O}{\overset{\|}{C}}R' \longrightarrow R_2CHO\overset{\displaystyle O}{\overset{\|}{C}}R' + NaX$

   NOTE: Reactions 1 through 6 do not take place with ϕX; reactions 3 through 6 do not take place with R_3—CX where R-alkyl.

7. $RX + \text{aq. or alc. } AgNO_3 \longrightarrow ROH + \text{alkene}$

   R = *tert* > *sec* > *pri;* ϕX does not react.

8. $RX \xrightarrow[\text{dry ether or THF}]{\text{Mg}} RMgX$ formation of Grignard reagent

   Works with all types of R.

   $RX \xrightarrow{\text{Li}} RLi$

9. In any of the cases above in which ϕX gives no reaction, a reaction similar to that for RX will take place if there is a nitro group ortho or para to the halogen.

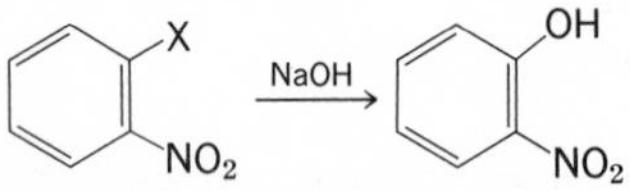
X
NO2
NaOH
OH
NO2

+ NaX

CHAPTER 9 EXERCISES

1. Write equations for the reaction, if any, between the following:

★*a.* 2-Bromopropane + NaCN

b. 1-Phenyl-2-chlorobutane + NaOH in alcohol

★*c.* 1-Bromo-2-methylbenzene + Mg in ether

d. *t*-Butyl chloride + NaOH

★*e.* Iodoethane + H_2O + NaOH

f. Benzyl bromide + NaSH

★*g.* 2-Butanol + thionyl chloride

h. Toluene + bromine and uv light

★*i.* 2-Methyl-1,3-butadiene + HCl

j. Ethyl bromide + sodium acetate

2. Show by equations how you could convert

a. 1-Bromobutane to 2-methyl-2-bromobutane

b. 1-Butanol to 1-chloro-1-phenylbutane

c. *n*-Propyl chloride to $(CH_3)_2CHOCH_3$

3. Show by equations a practical method for the preparation of each of the following substances from simple starting materials.

★*a.* Diphenylmethane

b. 1,2-Diphenylethane

★*c.* *p*-Bromobenzyl chloride

d. *p*-Dibromobenzene

★*e.* *n*-Butyl bromide

f. $CH_3CH_2C{\equiv}N$

★*g.* $CH_3C{\equiv}C{-}C_6H_5$

4. $CH_3CH{=}CH{-}CH_2{-}Cl$ reacts under certain conditions with KOH in a reaction, the rate of which is dependent on both chloride and OH^- concentrations to yield a single product. In water alone a mixture of two products (OH^- displacements) is observed. Explain.

5. Show how you could distinguish by chemical and/or spectral means between

★*a.*

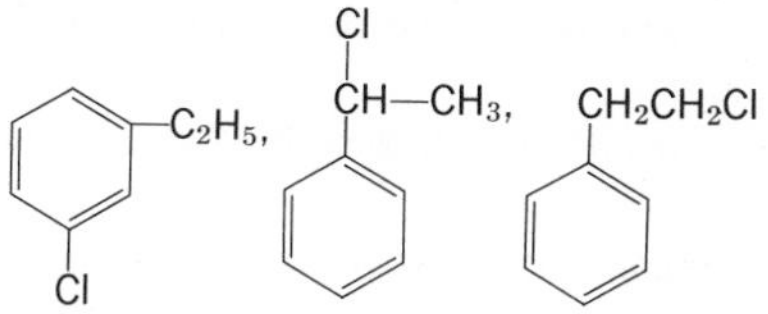
C2H5,
Cl
Cl
CH—CH3,
CH2CH2Cl

b. $(CH_3)_3C{-}Cl$, $CH_3CH_2CH_2CH_2Cl$

★*c.* *cis* and *trans*-

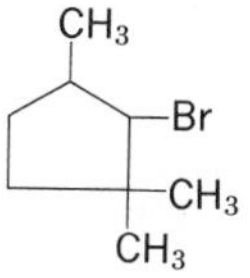
CH3
Br
CH3
CH3

d.

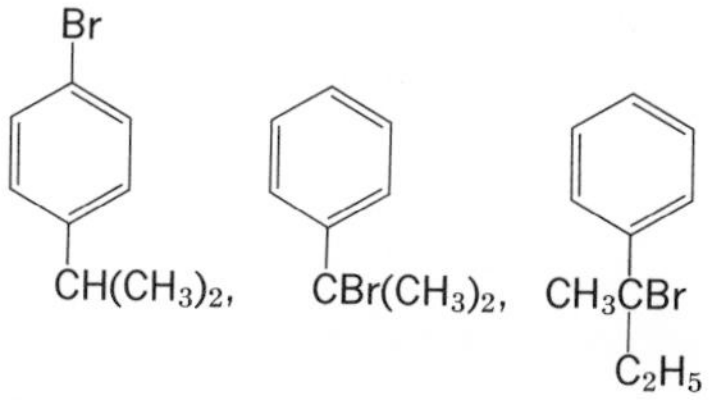
Br
CH(CH3)2,
CBr(CH3)2,
CH3CBr
C2H5

★**6.** In the reaction of 1-bromobutane with KOH to form alcohols a certain amount of dibutyl ether is produced. Show by equations the probable mechanism of its formation.

7. Comment on the following. Are they true or false? Explain using equations, mechanisms, and diagrams.

a. p-$NO_2C_6H_4Br$ reacts with KCN, but bromobenzene does not.

b. Dextrorotatory 2-bromobutane and H_2O yields dextrorotatory 2-butanol.

c. 2,2-Dimethyl-1-bromopropane reacts with KCN in 90% acetic acid and not in dry benzene.

d. One of the isomers of $(CH_3)_3C$— 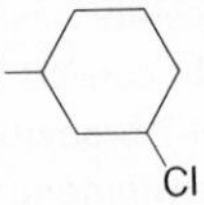 undergoes E_2 elimination more readily than the other.

e. *tert*-Butyl alcohol reacts with HBr and HCl at the same rate while methanol reacts with these same reagents at different rates.

★8. $CH_3C{\equiv}CH$ reacts more slowly with HCl than the corresponding alkene, yet the reaction is easily controlled to give the unsaturated adduct

$CH_3\underset{\underset{Cl}{|}}{C}{=}CH_2$. Explain.

9. The "all trans" isomer of 1,2,3,4,5,6-hexachlorocyclohexane reacts 20,000 times more slowly with alcoholic KOH than any other isomer. Explain.

★10. Compound X, $C_9H_{10}Br_2$, reacts with hot sodium hydroxide solution to form $C_9H_{11}OBr$ which resists any further action of the alkali. The product with concentrated H_2SO_4 eliminates H_2O to form an alkene that, on ozonolysis, yields CH_3CHO. Deduce a formula for X.

11. Show the resonance forms of chlorobenzene. What effect would this resonance have on the electrophilic or nucleophilic character of the carbon carrying the Cl? Compare it with chlorocyclohexane.

★12. Explain why S_N2 displacements are more difficult with 2-methyl-2-chlorobutane than with 2-methyl-1-chlorobutane.

13. What would be the stereochemical outcome of the reactions of optically active 1-chloro-2-methylbutane with the following:

a. OH^-, H_2O

b. Mg, then H_2O

c. $AlCl_3$, Δ

d. CN^-

e. Alcoholic KOH, followed by H_2

★14. Suggest structures for the halogen compounds, given the nmr spectra indicated.

a. $C_6H_4I_2$ — 4 protons at 2.6 δ

b. $C_2H_3F_3O$ — 1 proton at 5.0 δ, 2 at 4.0 δ (a quartet)

c. C_4H_9Br — 2 protons at 3.4 δ, 1 at 1.4 δ, 6 at 0.8 δ

15. Compare $C_6H_5CH_2CH_2Br$ with

$C_6H_5\underset{\underset{Br}{|}}{C}HCH_3$

in their reactions with nucleophilic agents. Predict relative tendencies for S_N1 vs. S_N2, etc.

★16. Account for the fact that certain optically active alcohols can be racemized in dilute acid solution and not in dilute base.

10

ETHERS

10.1 Definition; nomenclature It was pointed out in the chapter on alcohols that oxygen is usually divalent in organic linkages. When both bonds are attached to carbon atoms which are not otherwise attached to oxygen we have the class of organic oxides known as *ethers*. These have the formula type R—O—R. Ordinary anesthetic ether is $C_2H_5OC_2H_5$. As a rule, ethers are known by trivial names rather than Geneva names; the Geneva (systematic) names use -oxy- prefixed by the appropriate group name:

	$CH_3OCH_2CH_3$	$CH_3CH_2OCH_2CH_3$
Group name:	Methyl ethyl ether	Diethyl ether (ether)
Systematic name:	(methoxyethane)	(ethoxy ethane)

A number of compounds containing methoxy groups are found in substantial quantities in nature. Two examples are:

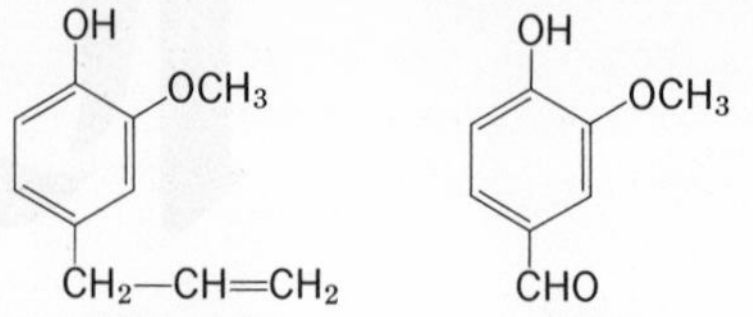

Eugenol (oil of cloves) Vanillin (vanilla bean)

Most ethers, however, are synthetic products; all but the low molecular weight symmetrical ethers are prepared by an application of the method devised by Williamson in 1851.

10.2 Williamson Synthesis In the *Williamson synthesis*, an alkyl halide is allowed to react with the sodium or potassium salt of an alcohol or phenol.

$$CH_3{-}CH_2{-}Br + {}^{-}OCH_3 \longrightarrow CH_3CH_2OCH_3 + Br^-$$

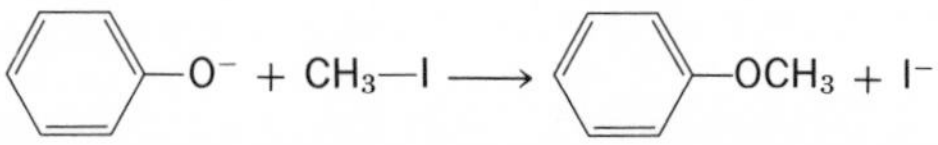

Anisole

The student will recognize this as an example of the displacement reaction. The usual limitations with aromatic and tertiary halides exist (Sec. 9.5).

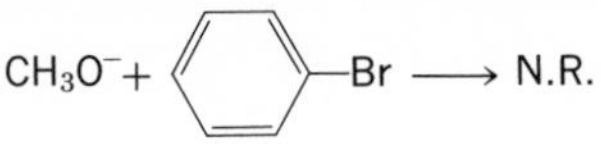

$$CH_3O^- + (CH_3)_3C{-}Br \longrightarrow H_2C{=}C(CH_3)_2 + CH_3OH + Br^-$$

However, tertiary ethers can be prepared in modest yields by changing partners.

$$(CH_3)_3C{-}O^- + CH_3Br \longrightarrow (CH_3)_3COCH_3 + Br^-$$

10.3 Industrial Preparation of Ethyl Ether The Williamson synthesis is too expensive for the industrial preparation of ether. Instead, another reaction is used utilizing ethyl alcohol and sulfuric acid.

$$\underset{(bp=80°C)}{CH_3CH_2OH} + H^+ \xrightleftharpoons{H_2SO_4} CH_3CH_2\overset{+}{O}H_2 \quad \approx H_3\overset{+}{O}$$

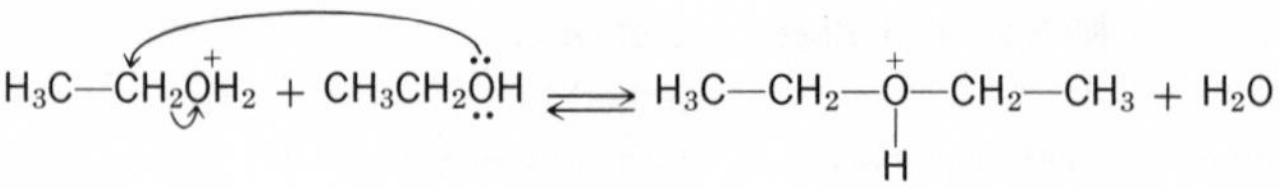

$$H_3C{-}CH_2{-}\overset{+}{\underset{\displaystyle H}{O}}{-}CH_2{-}CH_3 + CH_3CH_2OH \rightleftharpoons \underset{(bp = 34.5°C)}{CH_3{-}CH_2{-}O{-}CH_2CH_3} + CH_3CH_2\overset{+}{O}H_2$$

By distilling the ether off as it is formed, the process can be run continuously until the by-product H_2O dilutes the sulfuric acid too much. Unfortunately, a number of side reactions characteristic of carbonium ions are possible, such as

$$[CH_3{-}CH_2\overset{+}{O}H_2] \xrightarrow{HSO_4^-} \underset{\text{Ethylene}}{H_2C{=}CH_2\uparrow} + H_2SO_4 + H_2O$$

$$[CH_3{-}CH_2\overset{+}{O}H_2] \overset{HSO_4^-}{\rightleftharpoons} \underset{\text{Ethyl hydrogen sulfate}}{CH_3CH_2OSO_3H} + H_2O$$

In the case of ethyl ether the first side reaction becomes important only at high temperatures (>160°C), and the second is reversible. Consequently the reaction can be controlled to give good yields of diethyl ether.

With alcohols having a great tendency to eliminate or rearrange, this method becomes less satisfactory. The method is also unsuited to the preparation of unsymmetrical ethers.

10.4 Cleavage of Ethers As a group the ethers, both aromatic and aliphatic, are characterized by their inertness to many common reagents under ordinary conditions; they do not react with alkalies, even on boiling; they are not readily affected by oxidizing or reducing agents under ordinary conditions; they are stable to most acids. However, several strong mineral acids will bring about a cleavage of the ether linkage.

$$C_2H_5OC_2H_5 + \underset{(48\%)}{HBr} \xrightarrow[\text{hot}]{\text{reflux}} C_2H_5OH + C_2H_5Br$$

$$C_6H_5OCH_3 + HI \longrightarrow \underset{\text{Phenol}}{C_6H_5OH} + CH_3I$$

$$2CH_3OC_2H_5 + HI \longrightarrow \underline{CH_3I + CH_3OH + CH_3CH_2OH + CH_3CH_2I}$$

$$\xrightarrow[\Delta]{\text{excess HI}} 2CH_3I + 2C_2H_5I$$

An excess of the mineral acid reacts as would be expected with the aliphatic alcohols formed by cleavage of the ethers, forming alkyl halides. With an aromatic aliphatic ether a phenol is always produced since the aromatic C—O bond is not cleaved. Diaryl ethers are not readily cleaved.

$$\phi{-}O{-}\phi + HI \longrightarrow \text{N.R.}$$

In all probability the cleavage of ethers by acids (hydrogen ion donors) depends on the intermediate formation of an oxonium ion followed by an S_N1 or S_N2 reaction.

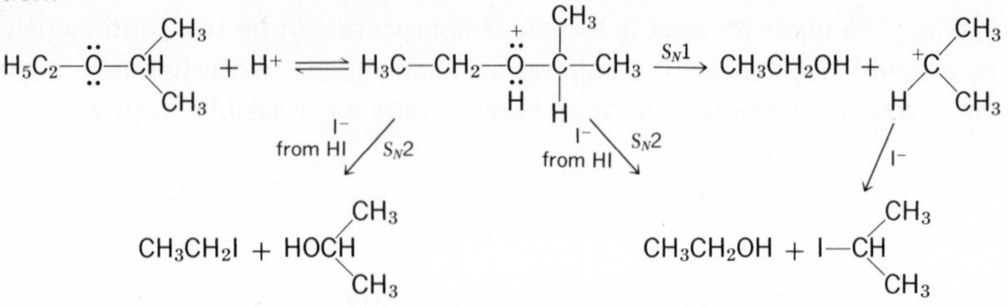

The protonated ether may either form carbonium ions (S_N1), which then react with nucleophiles, or may be attacked by nucleophiles (S_N2). From these mechanistic considerations it would appear that the best acid for cleaving ethers would be a very strong acid with a very nucleophilic counter-ion. The acid most closely fulfilling this requirement is HI. Other acids are as strong or stronger (H_2SO_4, $HClO_4$) but lack a nucleophile as good as I^-. In general, ethers are cleaved better by HI than HBr, which is better than HCl and H_2SO_4. Cleavage with other acids is usually not practical for most ethers unless assisted by strong Lewis acids.

10.5 Reactions of Aromatic Ethers The ether linkage attached directly to the aromatic ring has a pronounced effect on the reactions of the aromatic ring, although much less so than a free —OH attached to the ring. The ether group is strongly *o,p* directing and activating although the problem of controlling the reaction (Sec. 8.24) is less serious.

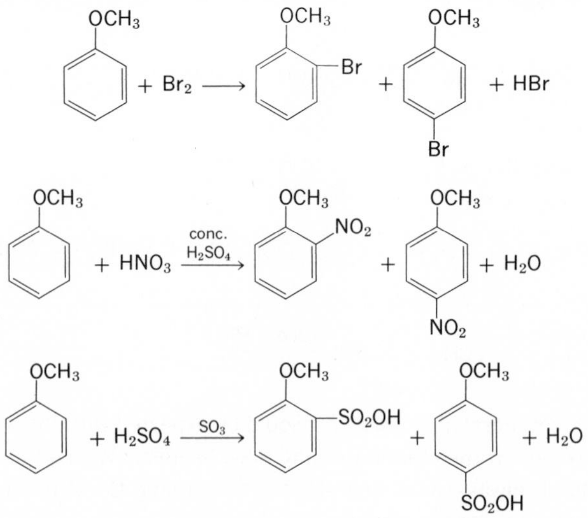

The stability of ethers makes it possible to "protect" a phenolic or alcoholic hydroxyl group from some undesired reaction. For example, phenols cannot be oxidized without destruction of the aromatic ring. The following scheme circumvents this difficulty by protecting the OH group during oxidation.

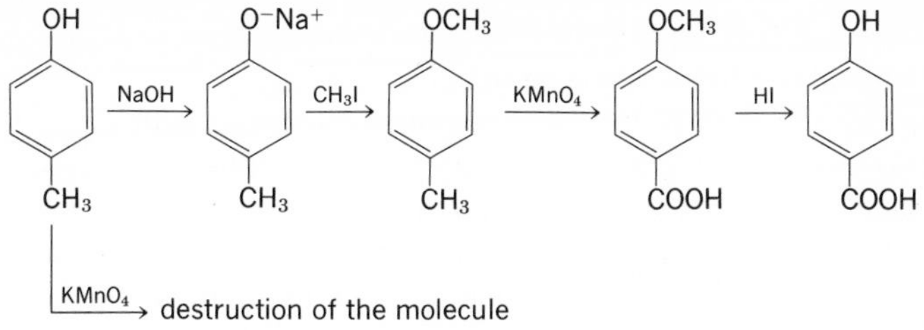
OH
NaOH
O⁻Na⁺
CH₃I
OCH₃
KMnO₄
OCH₃
HI
OH
CH₃
CH₃
CH₃
COOH
COOH
KMnO₄
destruction of the molecule

10.6 Ethers as Solvents The unreactivity of ethers is utilized rather extensively. Higher ethers, which are both unreactive and high-boiling, are used as solvents and as softeners for plastics, etc. Ethyl ether is itself almost insoluble in water and is, at the same time, an effective solvent for a wide variety of organic compounds. A good working rule for predicting the solubility behavior of carbon compounds is as follows: most carbon compounds are soluble in ether unless they have marked ionic character (salts), very many hydroxyl groups (sugars), or large molecular weights (macromolecules). Most ether-soluble carbon compounds containing four or more carbon atoms are not appreciably water soluble. Substances with marked ionic character (salts) are usually water soluble as are those with many hydroxyl groups (sugars, glycols). Very large molecules are insoluble in almost all solvents. On the whole, solubility in ether goes oppositely to that in water. This relationship is used routinely in the separation of different classes of compounds from one another. The mixture is shaken with water and ether in a separatory funnel. The covalently linked substances are concentrated in the ether, while the others largely pass into the water phase. The two layers are allowed to separate and the lower layer is drawn off through the stopcock. The separation is usually not complete because compounds are rarely encountered that are absolutely insoluble in either phase—rather they tend to be distributed between the two layers with nonionized molecules largely in the ether layer and ionized ones largely in the water. It is common practice, therefore, to wash the ether phase by shaking it in a separatory funnel with several small portions of water and to wash the water phase similarly with additional ether, combining the ether fractions and the water fractions. Other organic solvents which are not soluble in water can be used for these separations, but ether is particularly suitable because it is inert and also volatile, so that it is easily separated from most substances dissolved in it. For an example of a separation utilizing ether see Sec. 8.18.

10.7 Properties of Ethers; Hazards There are several hazards in the use of ether in the laboratory. Ether, like most organic liquids, is very flammable. On account of its high vapor pressure and low ignition temperature it is especially dangerous. It may catch fire from a piece of hot metal (such as wire gauze) even in the absence of a flame (the ignition temperature of ether is about 200°), and since it does not dissolve in water but floats on it, an ether fire is difficult to extinguish by flooding with water. (A CO_2 extinguisher is used on an ether fire, unless sodium or magnesium or other active metal is present.) Ether vapor is heavier than air and may roll long distances along a desk top. For this reason flames must be kept at a considerable distance whenever ether is being handled.

Ethers tend to form peroxides on standing in contact with air for a long time. These peroxides boil at higher temperatures than the ethers themselves and, thus, accumulate in the residue during distillation. They are violently explosive and have been responsible for many serious accidents. Precautions must, therefore, be taken to remove them prior to any distillation of an ether, particularly if the ether has stood in storage for some time. This can be accomplished quite simply by shaking the ether with a reducing agent such as aqueous, slightly alkaline, ferrous sulfate.

The ethers are characterized by a higher narcotic activity[1] than the parent alcohols, and one of them is widely used as a general anesthetic. Dimethyl ether is an anesthetic, but its effect is very temporary. Diethyl ether (anesthetic ether) gives a more lasting anesthesia, but it must be administered continuously when anesthesia of long duration is required. Both the narcotic activity and the toxicity of the ethers increase as the homologous series is ascended.

10.8 Cyclic Ethers Three cyclic ethers are of interest as industrial chemicals: ethylene oxide, dioxane, and tetrahydrofuran (THF).

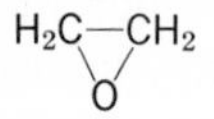
Ethylene oxide

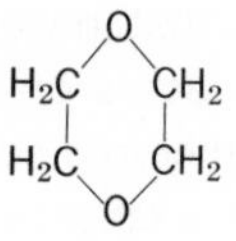
Dioxane

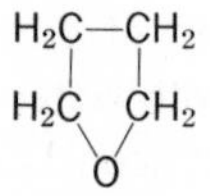
Tetrahydrofuran (THF)

Dioxane, an ether derived from ethylene glycol, is a useful solvent, having the property of complete solubility in water. THF is a particularly useful solvent for Grignard reactions, being superior to ethyl ether in a number of instances. Ethylene oxide is unusually reactive for an ether because of the strain of the molecule. The normal valence angle of oxygen is not very different from that of carbon, and the ideas of strain in Sec. 4.5 apply to rings containing elements other than carbon. The effect of strain is to make the ring open with bases, with Grignard reagents, and particularly, with acids.

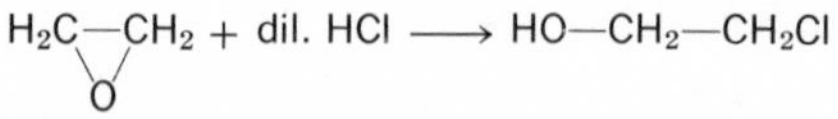

One application of this cleavage is of particular synthetic importance.

[1]A *narcotic* is a drug which produces stupor, insensibility, or sleep. An *anesthetic* produces insensibility; a *general* anesthetic affects the entire body, a *local* anesthetic only a limited region of the body.

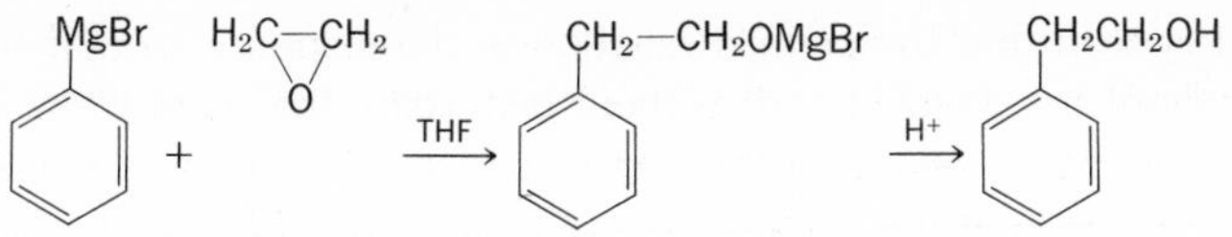

This represents a general method for adding two CH_2 groups at a time and is widely used.

10.9 Epoxides Ethylene oxide itself is produced on a large scale either by the internal displacement reaction of ethylene chlorhydrin with base

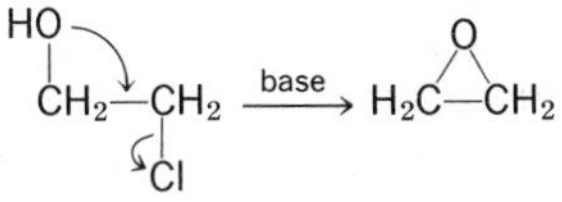

or by a catalytic reaction with oxygen

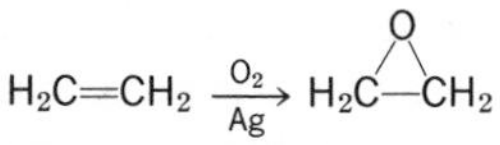

Homologs of ethylene oxide, called "epoxides," can also be produced by the first of these methods. It is interesting to note that in spite of the strain involved in the three-membered ring, closure occurs with good yield. Apparently, the necessary closeness of the nucleophilic OH to the carbon atom undergoing displacement is an important factor here.

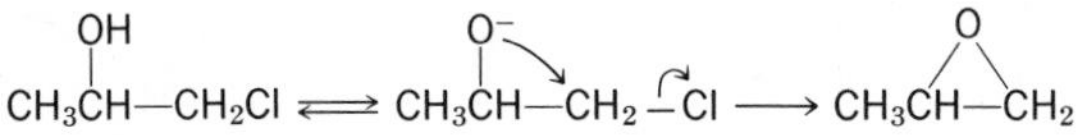

For reactions occurring by an S_N2 mechanism (often the case) the geometry of the structure is important (Sec. 9.6). Unless the two groups can adopt a trans relationship, ring formation is unsuccessful because the backside approach of the O^- is prevented.

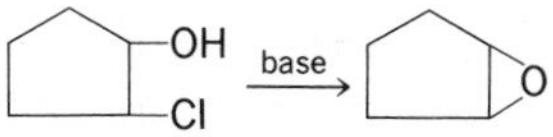

Trans

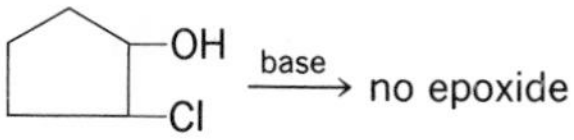

Cis

An alternate, useful laboratory synthesis of epoxides involves the use of peracids, for example, perbenzoic acid.

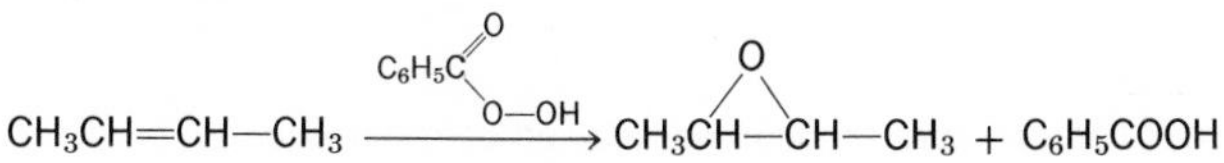

Cis Cis

Here it is seen that the peracid reaction is stereospecific in that the O comes onto one side of the double bond[1] without disturbing the cis or trans relationship of attached groups. Opening of the ring by either acidic or basic reagents usually occurs with inversion at the carbon where the C—O bond is broken, and retention at the other carbon (the one still bearing the oxygen of the epoxide).

Base-catalyzed openings of epoxides are normally S_N2 reactions in which attack occurs preferentially at the least-substituted carbon atom.

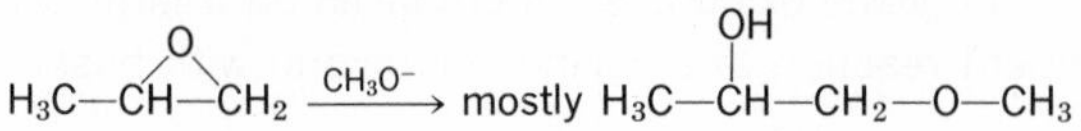
CH3O−
mostly

With a cis epoxide the configurations of the two carbon atoms in the epoxide ring are (in many cases) opposite. If the two carbon atoms are identically substituted, the cis epoxide will be a meso form. Thus, inversion of the configuration of one of the two carbon atoms will yield a product with both carbon atoms of the same (*R* or *S*) configuration. Attack at the other carbon atom will give the respective (*S* or *R*) combination. Thus a pair of diastereomers will be formed. (It is left for the reader to deduce the situation for the corresponding trans epoxides.)

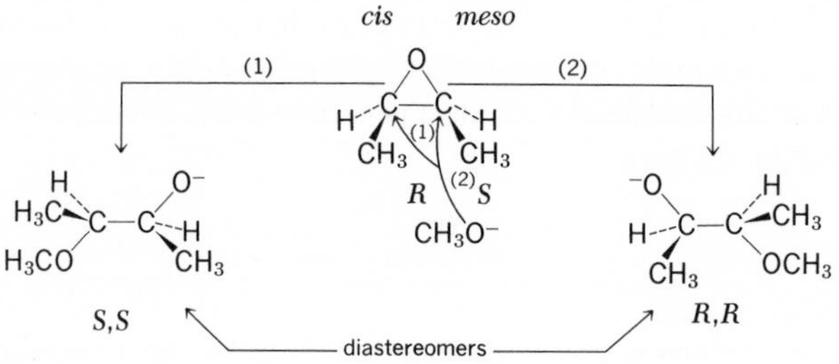
cis
meso
(1)
(2)
S,S
R,R
diastereomers

Acid-catalyzed reactions of epoxides apparently do not involve a free carbonium ion; the products show inversion about the center at which displacement occurs. It is possible, however, to predict the direction of acid-catalyzed reactions by assuming that the transition state has a good deal of carbonium-ion character and following the usual rules for relative energies of carbonium ions (Sec. 3.9).

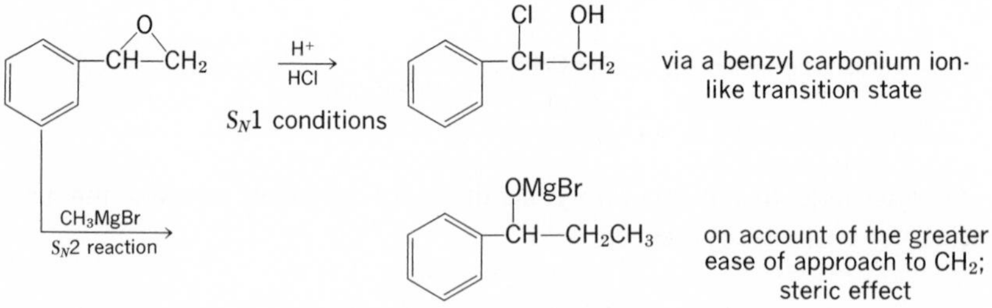
H+
HCl
SN1 conditions
via a benzyl carbonium ion-like transition state
CH3MgBr
SN2 reaction
on account of the greater ease of approach to CH2; steric effect

[1]The mechanism of the peracid reaction will not be considered here.

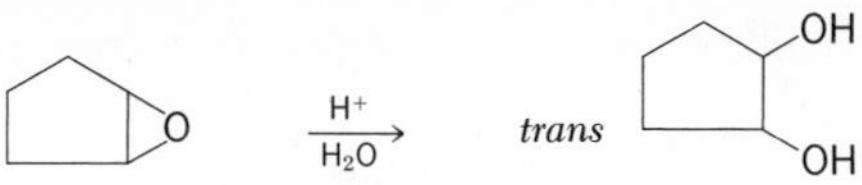
O
H+
H2O
trans
OH
OH

S_N1 conditions

no racemization or free C^+ ion involved

These reactions illustrate the way a knowledge of reaction mechanisms and a choice of experimental conditions based on this allow control of structure and stereochemistry of products even in relatively complex reactions.

10.10 Spectroscopic Properties Ethers are transparent in the ultraviolet and can be used as solvents. Aliphatic ethers show a strong band (usually rather broad) at about 1150 cm^{-1} and aromatic aliphatic ethers at 1250 cm. In the nmr the hydrogens next to an oxygen are displaced toward lower field. Thus an aliphatic CH_2 is found at $\delta = 1.2$ to 1.3 while a CH_2O proton is found at $\delta = 3.4$ ppm.

OUTLINE OF ETHER CHEMISTRY

Preparation

1. "Continuous" method (for aliphatic symmetrical ethers only).

$$2ROH \xrightarrow[\text{regulated temp.}]{\text{conc. } H_2SO_4} ROR$$

2. Williamson's reaction.

$$RO^-Na^+ + X{-}R' \longrightarrow ROR' + NaX$$
$$\phi O^-Na^+ + X{-}R' \longrightarrow \phi OR' + NaX$$

NOTE: In this reaction ϕX cannot be used and R′ does not represent tertiary alkyls.

Properties

1. $ROR' \xrightarrow{HI} ROH + R'I \xrightarrow{HI} RI + R'I$

   Excess HI yields two iodides except where $R = \phi$.

2. $$ROR' + \begin{cases} \text{Na} \\ \text{bases} \\ \text{reducing agents} \end{cases} \longrightarrow \text{N.R.}$$

   In general ethers are stable to alkaline reagents.

3. Ethers are generally stable to oxidizing agents such as CrO_3, but when exposed to air slowly form peroxides which are very explosive when heated.

4. 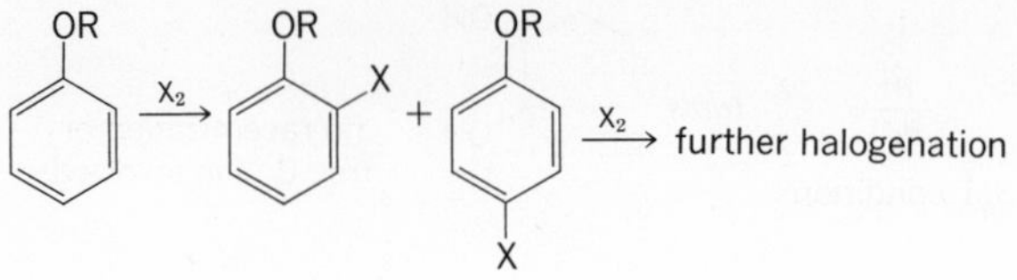
OR
X₂
OR
X
+
OR
X₂
further halogenation
X

5. 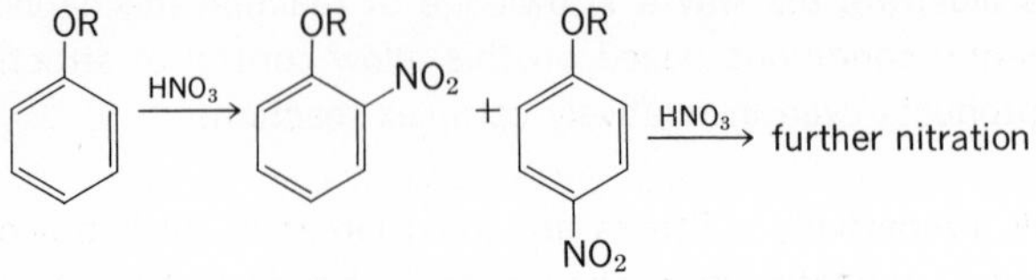
OR
HNO₃
OR
NO₂
+
OR
HNO₃
further nitration
NO₂

6. $H_2C\overset{O}{—}CH_2 + RMgX \longrightarrow R—CH_2—\underset{OMgX}{CH_2} \xrightarrow{H_2O} RCH_2CH_2OH$

This reaction is given only by three- and four-membered cyclic ethers.

CHAPTER 10 EXERCISES

★1. Why would nitric acid not be used to cleave phenyl methyl ether?

2. Write equations for the reactions of

$$H_2C{=}CH—\underset{OCH_3}{CH}—CH_2OH \quad \text{with:}$$

a. Conc. $KMnO_4$
b. Sodium
c. CH_3COCl
d. Aqueous sodium hydroxide
e. Excess HI

★3. In the cleavage of ethers by HI, would you expect any alkene formation? Explain.

4. What product would you expect from the reaction of

cis- 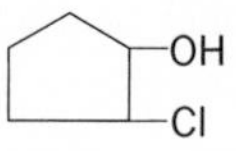
OH
Cl

with hydroxide ion (S_N2 reaction)?

★5. Predict the products of an S_N1 reaction of HCl + $CH_3\overset{O}{CH—CH}—CH_3$ for both the cis and trans forms of the epoxide. (Assume a free carbonium ion intermediate.)

6. Assume the epoxide in question in Exercise 5 is optically active. Predict the products of the reaction and their optical activity.

7. How could you separate a mixture of

★*a.* Phenyl ethyl ether and *p*-ethylphenol
★*b.* Diethyl ether and ethanol
c. Pentane, 1-pentyne, 1-methoxy-3-pentanol

8. How could you distinguish between the following compounds by chemical and/or spectral methods? Indicate the number and approximate location of nmr bands for each.

a. Ethanol and dimethyl ether

b. *p*-Methoxybenzyl alcohol and 

c. Phenyl ethyl ether and methyl benzyl ether

d. $CH_3OCH_2CH_2CH_2OH$ and $CH_3OCH_2CHCH_3$ (with OH on the CH)

9. Show how you could accomplish the following conversions:

★*a.* 1-Butanol ⟶ 2-methoxybutane

b. 1-Propanol ⟶ diethyl ether

★*c.* Phenyl ethyl ether ⟶ benzene

d. Phenol ⟶ *p*-bromophenol

★*e.* Diethyl ether ⟶ 2-butanol

10. Predict the structure of compounds showing nmr absorption as indicated.

a. $C_3H_6O_3$ nmr absorption at 5.0δ only; no ir absorption from 3300 to 3700 cm^{-1}.

b. C_6H_8O nmr absorption at 5.7δ (2H) and at 7.8δ (6H). Absorption is observed in the uv region.

★11. Point out the reasons for the difficulty in obtaining ditertiary ethers.

12. Complete the following equations showing mechanisms and main products. Include stereochemistry when appropriate.

a. $C_6H_5CH_2OCH_2CH_3 + HI$

b. 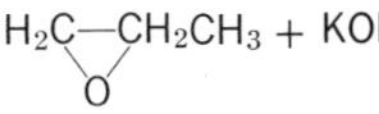 + KOH (S_N2)

c. 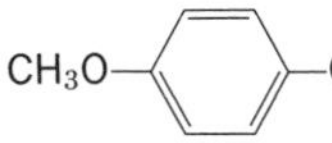 + HI

d. 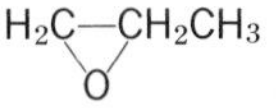 + C_2H_5MgBr

e. 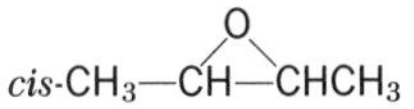 + HCl (S_N1)

f. 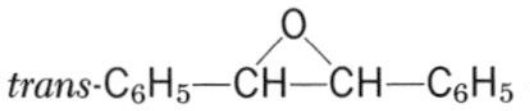 + $NaOCH_3$ (S_N2)

g. *trans-* 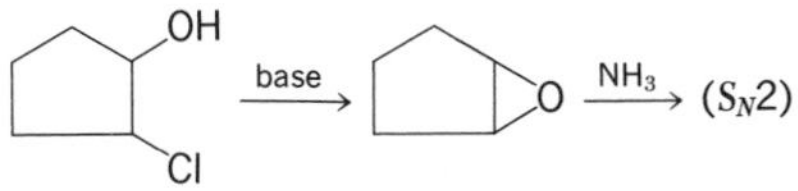

★13. A substance having the molecular formula $C_4H_{10}O_3$ reacts with acetyl chloride to give a substance $C_8H_{14}O_5$. One of the products formed by treating it with HI is CH_3I. Oxidation of it yields a dibasic acid containing four carbon atoms. Give the name and structure of a substance having these properties. Show your reasoning clearly.

14. An aromatic compound, $C_{13}H_{12}O$, on treatment with HI splits into two fragments, one of which gives a phenol color test and the other of which contains iodine. The iodine-containing fraction on oxidation gives an acid, $C_7H_6O_2$. Write the structure of the original compound. Show your reasoning.

★**15.** A certain compound Y, $C_5H_{12}O_2$, reacts with acetyl chloride to form $C_7H_{14}O_3$. With HI in *excess*, Y gives Z, $C_4H_8I_2$, and another iodine-containing compound. Upon hydrolysis with aqueous NaOH and subsequent oxidation, Z gives an acid with a keto group in position 3 relative to the carboxyl group. Write a structure for Y. Show your reasoning clearly.

16. Why is diethyl ether more soluble in concentrated hydrochloric acid than in pure water?

★**17.** There are two possible routes to methyl *t*-butyl ether: (*a*) *t*-Bu-O-Na and CH_3Br; (*b*) *t*-Bu-Br and CH_3ONa. Which would you choose, assuming that all materials are equally available? Explain.

18. A certain compound shows the following properties:

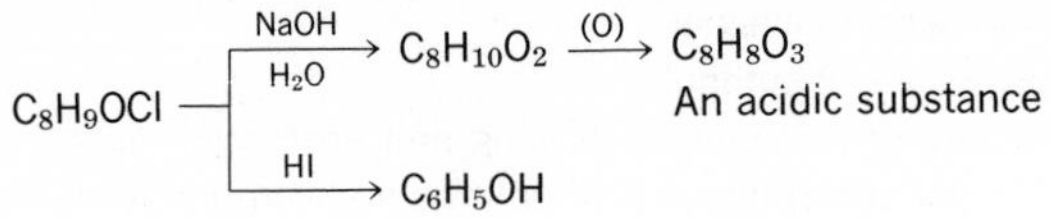
NaOH
H2O
C8H10O2
(O)
C8H8O3
An acidic substance
C8H9OCl
HI
C6H5OH

Give a possible structure of C_8H_9OCl. Include all reasoning.

11

CARBONYL COMPOUNDS: ALDEHYDES AND KETONES

11.1 Definition Compounds containing the carbonyl (C=O) group attached to carbon or hydrogen atoms are called either "ketones" or "aldehydes." The name used depends on the groups attached to the two remaining bonds of the carbon atom. Those molecules which contain the group

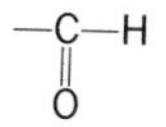 (written —CHO, never C—OH)

attached to carbon or hydrogen only

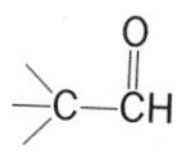 and 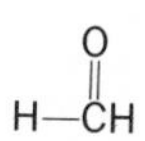

are classed as *aldehydes;* those compounds containing two carbon substituents are *ketones*

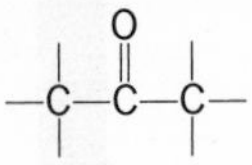

Any structures involving groups other than carbon or hydrogen attached directly to the carbonyl, such as

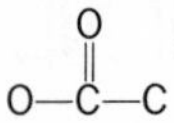 or 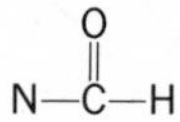

are not included in this category although many basic reaction mechanisms (Sec. 11.7) are the same.

The different names might be expected to denote different chemical properties. The observed differences, however, with one notable exception, namely oxidation (see Sec. 11.8), are differences in degree rather than in kind. Most reagents that react with carbonyl groups react with aldehydes as well as ketones. There are relatively minor differences between the reactions of aliphatic and aromatic carbonyl compounds.

11.2 Nomenclature The simple aldehydes are usually named after the acids which may be formed from them on oxidation.

Aldehyde		*Oxidation product*	
HCHO	Formaldehyde	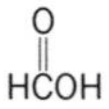	Formic acid
CH_3CHO	Acetaldehyde	CH_3COOH	Acetic acid
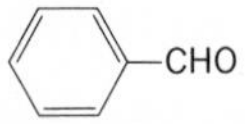	Benzaldehyde	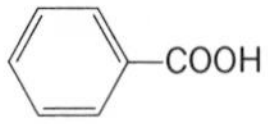	Benzoic acid

The ketones are referred to by trivial names in the case of simple types, or are named as a substituted ketone —CO—.

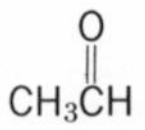
Acetaldehyde

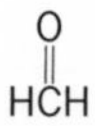
Formaldehyde

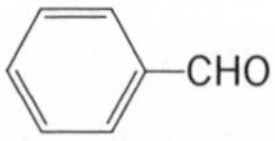
Benzaldehyde

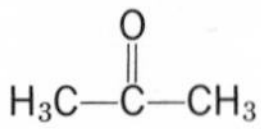
Acetone (dimethylketone)

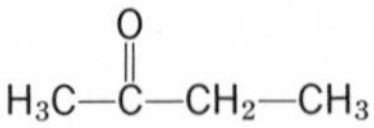
Methyl ethyl ketone (MEK)

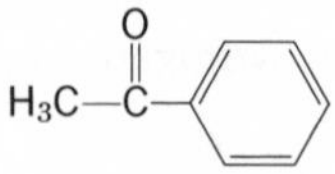
Acetophenone (methyl phenyl ketone)

The Geneva nomenclature uses the ending *al* to designate the aldehyde group and *one* for the ketonic carbonyl; these are accompanied by numbers as usual, although in the case of the aldehyde group, which perforce is at the end of the carbon chain, the number 1 is omitted.

CH_3CH_2CHO

Propanal

$$CH_3\overset{\overset{\displaystyle O}{\|}}{C}CH_2-\!\!\!\!\bigcirc$$

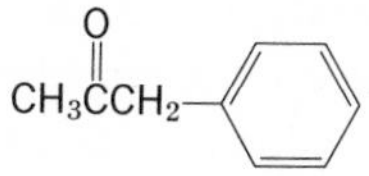

1-Phenyl-2-propanone

$H_2C{=}CH{-}\underset{\underset{\displaystyle CH_3}{|}}{\underset{\underset{\displaystyle CH_2}{|}}{CH}}{-}CHO$

$CH_3-\overset{\overset{\displaystyle O}{\|}}{C}-CH_2-CH_2-\overset{\overset{\displaystyle O}{\|}}{C}-CH_2CHO$

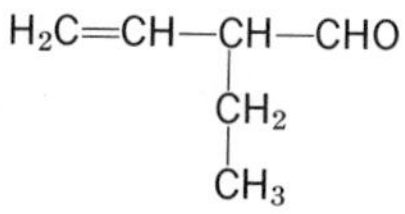

2-Ethyl-3-butenal

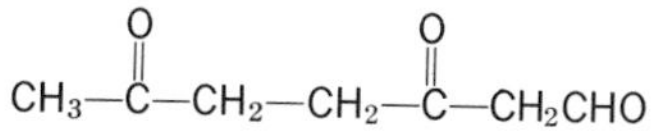

3,6-Heptanedioneal

11.3 Properties Aldehydes and ketones are neutral substances. The lower members of both series are very soluble in water and also in organic solvents. The higher members (more than 4 C atoms) are not soluble in water. The lower aliphatic aldehydes have sharp, rather unpleasant odors (formaldehyde, acetaldehyde) but those with 8 to 12 carbon atoms have, in dilute solutions, flowery odors, a property which makes them valuable in perfumery.

11.4 Preparation of Aldehydes Oxidation of primary alcohols leads to aldehydes as the first-formed products. Unfortunately, the aldehydes are even more suscep-tible to oxidation than the alcohols from which they are derived.

With the lower aldehydes it is found that the boiling points of the aldehydes are sufficiently lower than those of the alcohol so that the oxidation may be run under conditions where the aldehyde distills off as it is formed. (This low boiling point arises from the inability of aldehydes to associate via hydrogen bonds. See Sec. 8.15.)

$$CH_3CH_2{-}CH_2OH \xrightarrow[\text{or KMnO}_4]{K_2Cr_2O_7} H_3CCH_2{-}\overset{\overset{\displaystyle O}{\|}}{CH}\uparrow$$

$$\xrightarrow[\text{or KMnO}_4]{K_2Cr_2O_7} CH_3CH_2\overset{\overset{\displaystyle O}{\|}}{C}OH$$

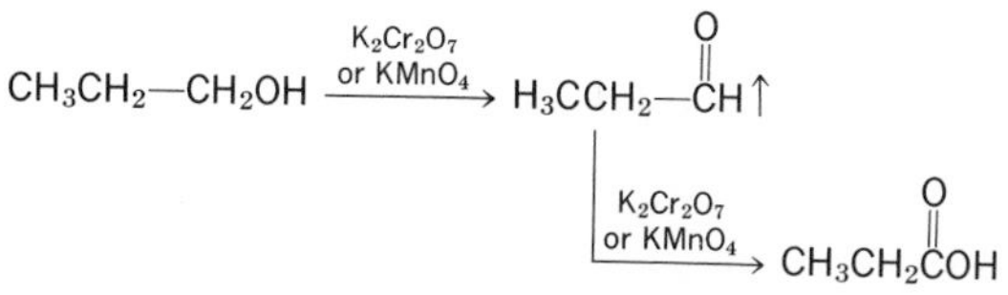

In the case of less volatile aldehydes this method is not applicable. Because of the importance of aldehydes a number of methods have been devised for their synthesis. Three of the more general methods are mentioned below.

Rosenmund reduction:

$$R{-}\overset{\overset{\displaystyle O}{\|}}{C}Cl + \text{poisoned catalyst} \xrightarrow{H_2} R\overset{\overset{\displaystyle O}{\|}}{C}H$$

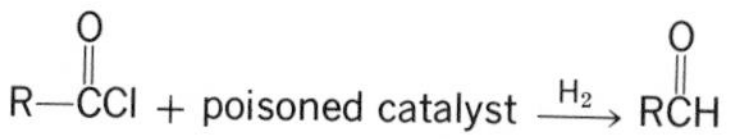

$$R-\overset{\displaystyle O}{\overset{\|}{C}}-SR + \text{Raney nickel catalyst} \longrightarrow R-\overset{\displaystyle O}{\overset{\|}{C}}H + NiS\ldots$$

The first of these involves reduction of the readily available acid chloride with a catalyst that has been decreased in effectiveness by "poisoning" with sulfur. The second utilizes the strong affinity of metallic nickel for sulfur and a thiol ester, which can be obtained from a carboxylic acid and a mercaptan (Sec. 15.2). For aromatic aldehydes a modified Friedel-Crafts reaction may be used. Aldehydes may also be prepared by ozonolysis of alkenes (Sec. 3.14).

$$CH_3CH_2CH{=}CH-CH_2CH_3 \xrightarrow{O_3} \xrightarrow[H_2O]{Zn} 2CH_3CH_2CHO$$

11.5 Preparation of Ketones Ketones are formed in better yields by oxidation than are aldehydes as there is much less tendency for further oxidation, and the product need not be distilled out. Thus

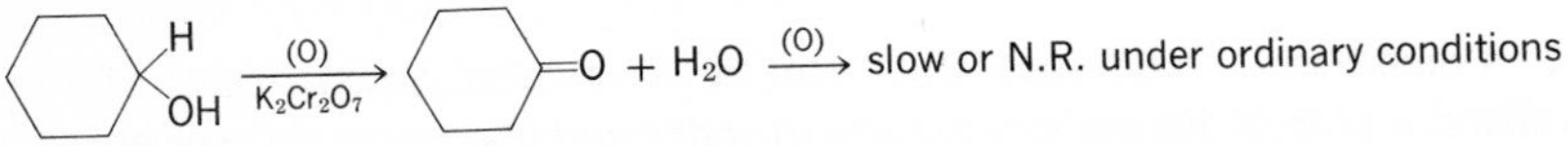
H
OH
(O)
K2Cr2O7
=O + H2O
(O)
slow or N.R. under ordinary conditions

Several methods for the preparation of ketones are of limited utility as they give mixtures of products except in symmetrically substituted cases. These include ozonolysis of alkenes, hydration of alkynes, and pyrolysis of acids.

Ozonolysis:

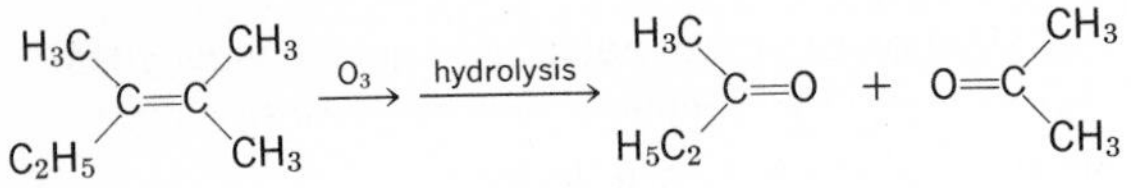
H3C
CH3
C=C
C2H5
CH3
O3
hydrolysis
H3C
C=O + O=C
H5C2
CH3
CH3

Hydration:

$$H_3C-C{\equiv}C-CH_2-CH_3 \xrightarrow[H_2O]{Hg^{++},\ H^+} H_3C-\overset{\displaystyle O}{\overset{\|}{C}}-CH_2-CH_2CH_3 + H_3C-CH_2-\overset{\displaystyle O}{\overset{\|}{C}}-CH_2-CH_3$$

Pyrolysis:

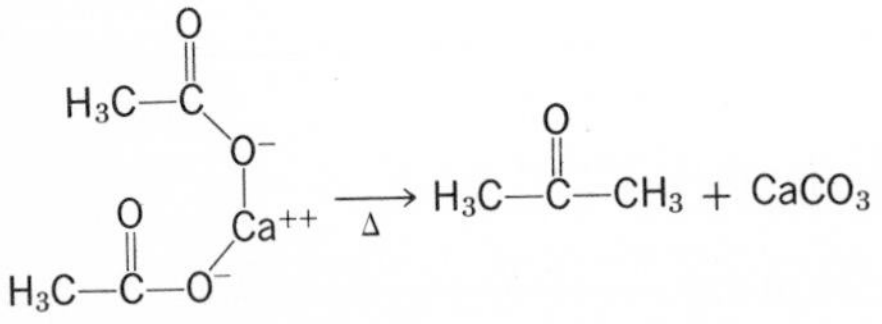
O
H3C—C
O⁻
O
Ca++
Δ
H3C—C—O⁻
O
H3C—C—CH3 + CaCO3

$$\text{or } 2H_3C-\overset{\displaystyle O}{\overset{\|}{C}}-OH \xrightarrow[\Delta]{ThO_2 \text{ or } MnO} H_3C-\overset{\displaystyle O}{\overset{\|}{C}}-CH_3 + CO_2 + H_2O$$

This method requires high temperatures and will give mixtures when unsymmetrical ketones are desired. Thus a mixture of acetic and propionic acids gives a mixture of three ketones.

$$H_3C-\overset{\overset{\large O}{\|}}{C}-OH + H_3C-CH_2-COOH \xrightarrow[\Delta]{ThO_2} H_3C-\overset{\overset{\large O}{\|}}{C}-CH_3 + H_3C-\overset{\overset{\large O}{\|}}{C}-CH_2CH_3 + H_3C-CH_2-\overset{\overset{\large O}{\|}}{C}-CH_2CH_3 + H_2O + CO_2$$

The pyrolytic method is of considerable utility for making cyclic ketones.

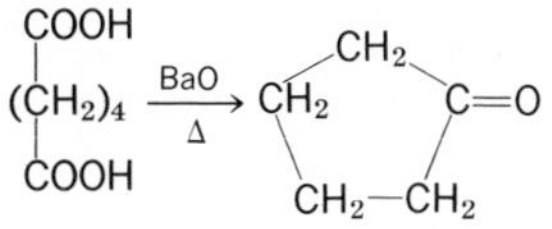

COOH
(CH2)4
COOH
BaO
Δ
CH2
CH2
C=O
CH2—CH2

Adipic acid Cyclopentanone

By this method large rings of up to 30 carbon atoms have been prepared.

11.6 Other Methods of Preparation Any reaction expected to lead to 1,1-diols

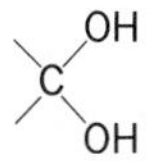

OH
C
OH

yields aldehydes or ketones instead, by spontaneous loss of water (Sec. 3.11). The hydrolysis of some 1,1-dihalides gives good yields of carbonyl compounds. This method is essentially limited to halides available by halogenation of the side chain of aromatic compounds.

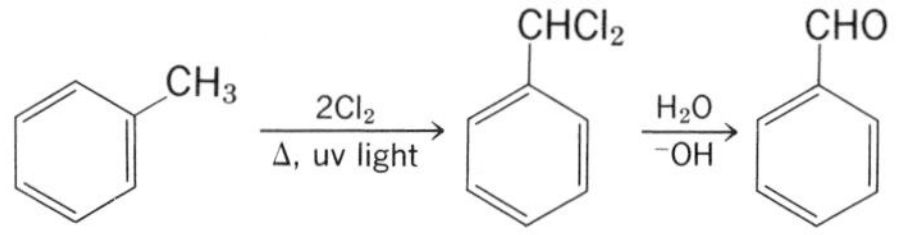

CH3
2Cl2
Δ, uv light
CHCl2
H2O
-OH
CHO

and with

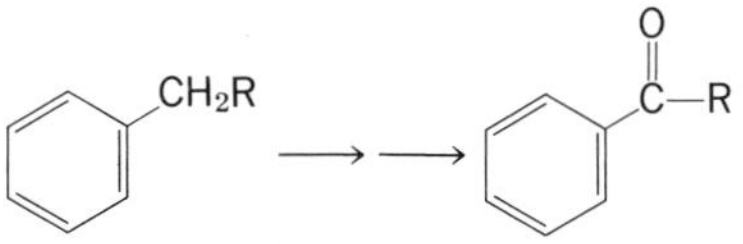

CH2R
O
C—R

The Friedel-Crafts reaction provides an excellent route to aromatic ketones.

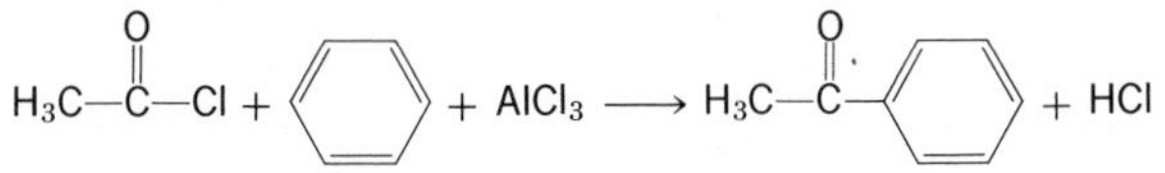

O
H3C—C—Cl +
+ AlCl3
H3C—C
+ HCl

An acid chloride

The aluminum chloride coordinates with both the chlorine and the oxygen of the carbonyl group.

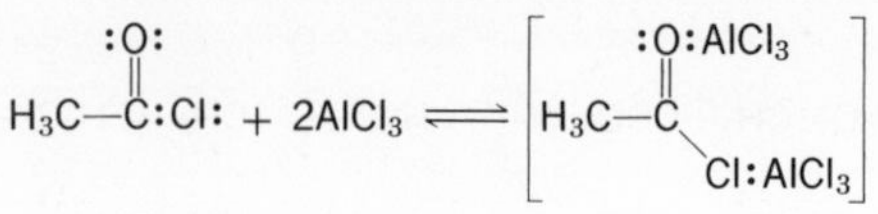

$$\text{H}_3\text{C}-\overset{\displaystyle :\!\text{O}:}{\overset{\|}{\text{C}}}:\!\text{Cl}: \;+\; 2\text{AlCl}_3 \;\rightleftharpoons\; \left[\text{H}_3\text{C}-\overset{\displaystyle :\text{O}:\text{AlCl}_3}{\underset{\text{Cl}:\text{AlCl}_3}{\text{C}}}\right]$$

However, coordination with the oxygen is so strong that one equivalent of aluminum chloride is essentially unavailable as a catalyst.

$$\left[\;\text{H}_3\text{C}-\overset{\displaystyle :\overset{+}{\text{O}}:\bar{\text{A}}\text{lCl}_3}{\overset{\|}{\text{C}}}-\text{Cl} \;\longleftrightarrow\; \text{H}_3\text{C}-\overset{\displaystyle :\text{O}:\bar{\text{A}}\text{lCl}_3}{\underset{+}{\text{C}}}-\text{Cl}\;\right]$$

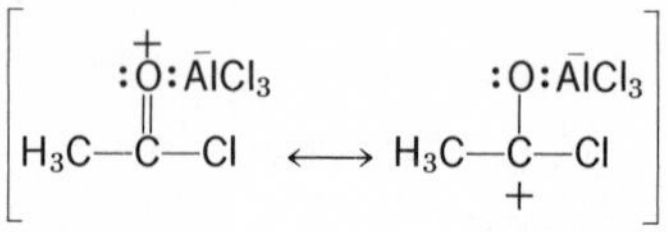

When more than one equivalent of aluminum chloride is used, the removal of the chloride is assisted by coordination with the excess aluminum chloride, as was observed with alkyl halides (Sec. 6.17).

This reaction works with a large number of acid chlorides, RCOCl, where R may be alkyl, aromatic, or even more complex groups. Like the reaction with alkyl halides (Sec. 6.17) the reaction does not work with aromatic rings substituted with meta-directing (deactivating) groups, and nitrobenzene may even be used as a solvent for the reaction. Unlike the alkyl halides, the reaction yields products (R—CO—Ar) which are deactivating to further substitution, so that disubstitution is not a problem. The carbonyl group of the product, like that of the starting acid chloride is coordinated with the aluminum chloride. This serves to further deactivate the ring, but it keeps the aluminum chloride tied up, and, therefore, unavailable to catalyze the removal of Cl. For this reason it is necessary to utilize more than one equivalent of aluminum chloride in this reaction, rather than the trace required for the reaction of alkyl halides. The overall reaction can be summarized

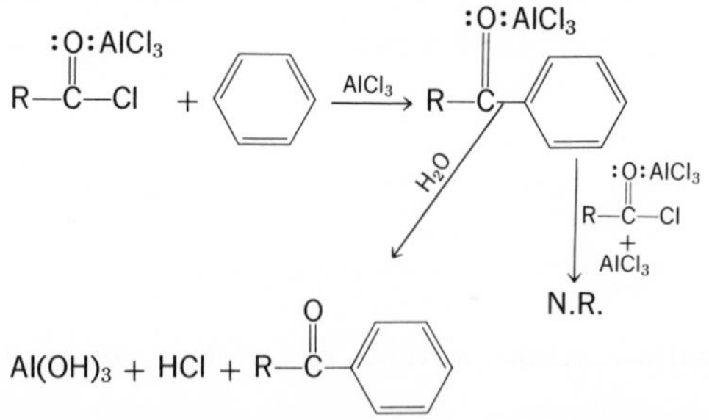

Aldehydes can be prepared by a variation of the above procedure. Since the appropriate acid chloride, formyl chloride, is not a stable substance, it can be generated *in situ* by the use of carbon monoxide and hydrogen chloride.

$$\text{CO} + \text{HCl} \;\rightleftharpoons\; \left[\text{H}-\overset{\displaystyle \text{O}}{\overset{\|}{\text{C}}}-\text{Cl}\right]$$

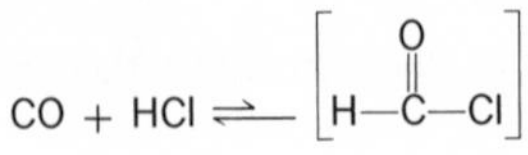

Although the actual course of the reaction may be somewhat more complex than this (for example, CuCl is often necessary as a cocatalyst for the reaction), the overall result is similar to the above ketone synthesis with the same restrictions.

$$CO + HCl + AlCl_3 + \bigcirc \longrightarrow \bigcirc\!-\!\overset{\displaystyle O}{\underset{}{C}}\!-\!H + HCl$$

(coordinated with
AlCl₃)

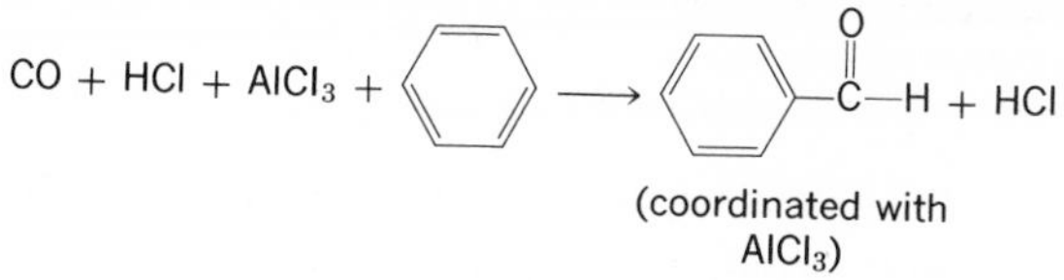

11.7 Addition Reactions The synthetic importance of the carbonyl group stems in large part from the facility with which it undergoes addition reactions. The carbonyl group in ketones and aldehydes is a resonance hybrid, the most important contributing structures being l*a* and l*b*.

$$\underset{\text{l}a}{\bigg\rangle}C\!=\!\ddot{O} \longleftrightarrow \underset{\text{l}b}{\overset{+}{\bigg\rangle}}C\!-\!\ddot{O}:^- \longleftrightarrow \underset{\text{l}c}{\overset{-}{\bigg\rangle}}\ddot{C}\!-\!\ddot{O}^+$$

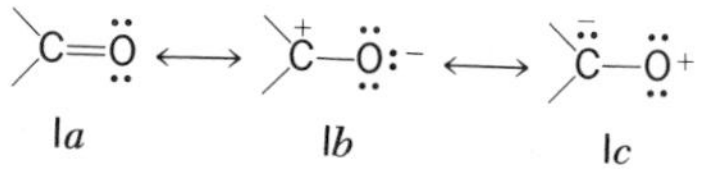

Structure l*c* is unimportant because it places a positive charge on an electronegative oxygen atom, and it lacks the redeeming second bond of l*a*. The carbonyl group behaves as though it were about halfway between l*a* and l*b*, possessing a strong dipole with the negative end on oxygen. This polarization would lead one to expect that carbonyl groups would react with electrophilic reagents on oxygen and nucleophiles on carbon. In fact, a large percentage of the reactions of the carbonyl group involve both electrophilic attack on oxygen and nucleophilic attack on carbon, although the timing of the two modes of attack varies considerably. Because the electrophilic portion bonded to oxygen is usually hydrolytically unstable, most of the products derived from such additions are classified in terms of the nucleophilic agent, the electrophilic reagent being relegated to the role of "catalyst." Thus the addition of many species to the carbonyl group is acid-catalyzed.

$$\bigg\rangle C\!=\!O + HCN \rightleftharpoons \left[\overset{+}{\bigg\rangle}C\!-\!\ddot{O}\!-\!H \longleftrightarrow \bigg\rangle C\!=\!\overset{+}{\ddot{O}}\!-\!H \right]$$

$$N\!\equiv\!C\!-\!\overset{}{\underset{}{C}}\!-\!OH \xleftarrow[\;CN^-\;]{}$$

A cyanohydrin

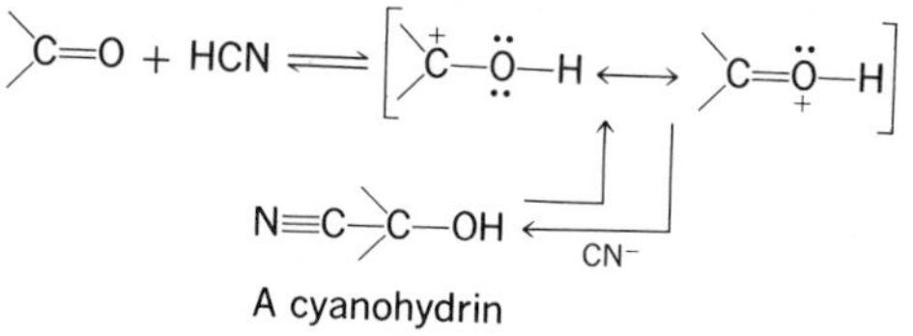

The addition of HCN to ketones and aldehydes yields products known as *cyanohydrins* in which a new carbon-carbon bond has been formed. The cyanide group may be decomposed by hydrolysis to form a carboxylic acid.

$$\underset{H_3C}{\overset{H_3C}{\bigg\rangle}}C\!=\!O \underset{}{\overset{HCN}{\rightleftharpoons}} \underset{H_3C}{\overset{H_3C}{\bigg\rangle}}\underset{CN}{\overset{OH}{\underset{}{C}}} \xrightarrow[\Delta]{\overset{H^+,}{H_2O}} \underset{H_3C}{\overset{H_3C}{\bigg\rangle}}\underset{\underset{OH}{C\!\diagdown}}{\overset{OH}{\underset{}{C}}}\!\!\overset{O}{}$$

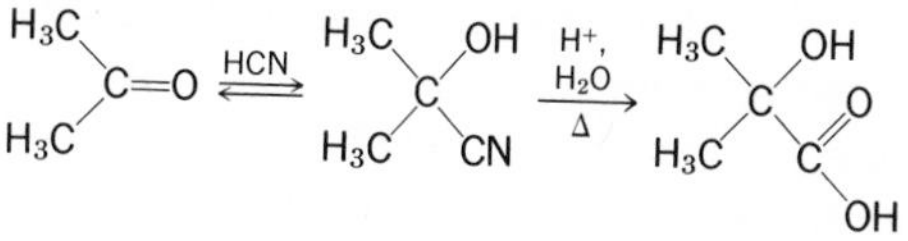

In the addition of HCN, as in many carbonyl addition reactions, the interplay of electrophile and nucleophile is complicated because they also interact with each other. Thus, in strong acid the rate of formation of cyanohydrins is slow because the CN⁻ concentration is suppressed.

$$H^+ + CN^- \rightleftharpoons HCN$$

In alkaline solution the CN⁻ is high but the H⁺ is low and the rate again drops off. The maximum rate thus occurs at some intermediate value.

Bisulfite ion adds to reactive carbonyl groups in a similar manner. The reagent as usually used is a saturated solution of sodium bisulfite in water, since under these conditions the addition product often separates as a crystalline precipitate.

$$CH_3\overset{\text{O}}{\underset{\|}{C}}H + Na^+ + HSO_3^- \rightleftharpoons \left(H_3C-\overset{O^-}{\underset{|}{C}H}-SO_3H \right) \rightleftharpoons H_3C-\overset{OH}{\underset{|}{C}H}-SO_3^-Na^+$$

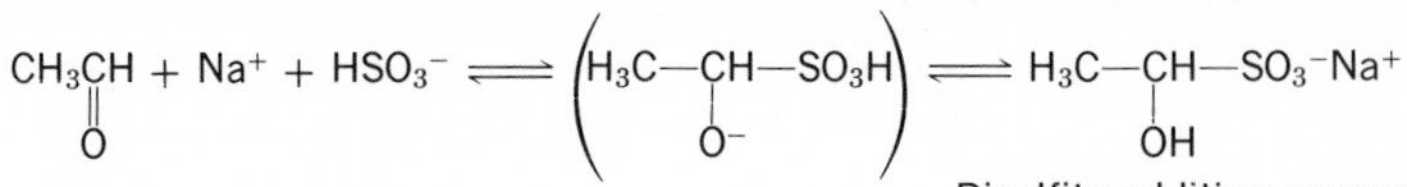

Bisulfite addition compound

The position of the equilibrium in this reversible addition is far on the side of the free carbonyl compound in the case of most ketones, and is a useful reaction only in the case of aldehydes, methyl ketones, and cyclic ketones. The reaction is driven nearly to completion with most aldehydes by the use of a large excess of bisulfite.

The sodium-salt character makes bisulfite addition compounds of great value in separating aldehydes from other groups of compounds. Like most salts, they are nonvolatile, insoluble in ether, and soluble in water (although they may often be salted out of water by saturation with $NaHSO_3$). They can be decomposed by heating in acid (or alkaline) solution to regenerate the original ether-soluble aldehyde (plus SO_2 or SO_3^{--}, respectively). Since liquid aldehydes are often hard to keep because they polymerize easily (Sec. 11.15), it is sometimes useful to preserve them in the form of the bisulfite addition compound.

Ammonia and a number of its derivatives act as nucleophilic reagents, adding in the expected manner to carbonyl groups of both aldehydes and ketones.

$$H_3C-\overset{H}{\underset{\|}{\underset{O}{C}}} + NH_3 \xrightarrow[\text{catalysis}]{H^+} \left[H_3C-\overset{H}{\underset{OH}{\underset{|}{C}}}-\overset{+}{N}H_3 \right] \rightleftharpoons \left[H_3C-\overset{H}{\underset{OH_2}{\underset{|}{C}}}-NH_2 \right]$$

$$\text{polymers} \longleftarrow H_3C-CH=NH \rightleftharpoons H_3C-\overset{H}{\underset{|}{C}}=\overset{+}{N}H_2 \xrightarrow{-H_2O]}$$

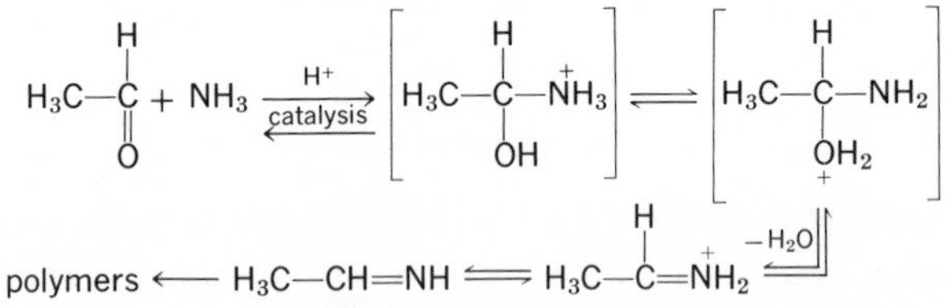

An aldimine

In the case of ammonia and alkyl-substituted amines the addition product is unstable (any carbon atom bearing more than one OH or —NH is usually unstable). It decomposes by an elimination process much like the reverse of that by which it is formed, to yield an unsaturated amine derivative, an aldimine (or ketimine with ketones). With ammonia and alkyl derivatives of ammonia these products are unstable and react further to give more complex products. With ammonia substituted by electronegative groups, however, stable and useful derivatives of aldehydes and ketones are obtained. The more useful reagents and the products derived from them are shown below:

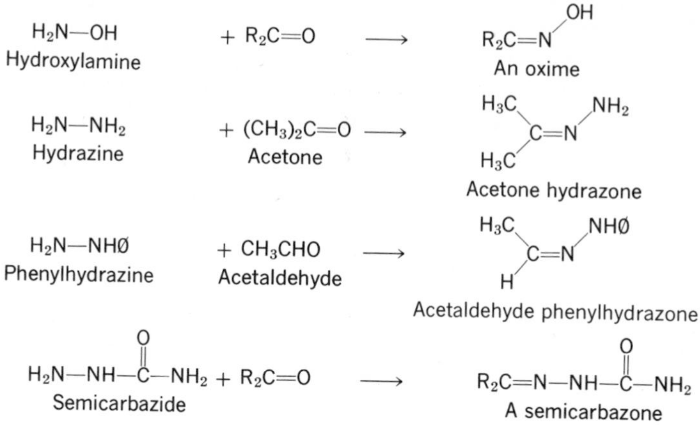

$$H_2N-OH \quad + R_2C=O \quad \longrightarrow \quad R_2C=N{\nearrow}^{OH}$$
Hydroxylamine   An oxime

$$H_2N-NH_2 \quad + (CH_3)_2C=O \longrightarrow \quad \underset{H_3C}{\overset{H_3C}{>}}C=N{\nearrow}^{NH_2}$$
Hydrazine   Acetone   Acetone hydrazone

$$H_2N-NH\emptyset \quad + CH_3CHO \quad \longrightarrow \quad \underset{H}{\overset{H_3C}{>}}C=N{\nearrow}^{NH\emptyset}$$
Phenylhydrazine   Acetaldehyde   Acetaldehyde phenylhydrazone

$$H_2N-NH-\overset{O}{\overset{\|}{C}}-NH_2 + R_2C=O \quad \longrightarrow \quad R_2C=N-NH-\overset{O}{\overset{\|}{C}}-NH_2$$
Semicarbazide   A semicarbazone

The above reactions are catalyzed by electrophilic reagents, most commonly by acids. Complexing of the carbonyl group with the electrophilic reagent is believed to be the first step in the reaction. This is followed by nucleophilic attack by the amino group of the reagent $R'NH_2$.

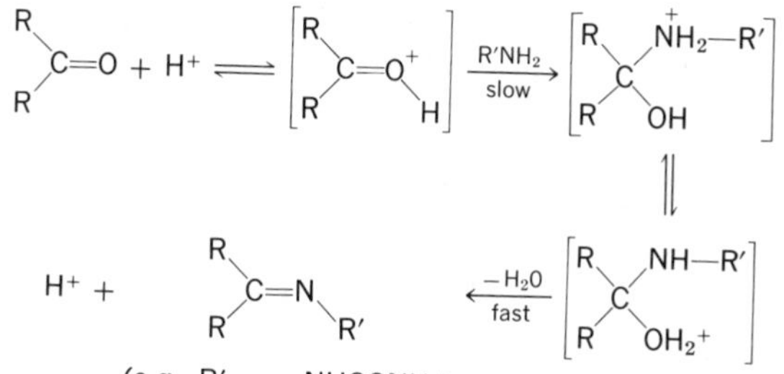

$$\underset{R}{\overset{R}{>}}C=O + H^+ \rightleftharpoons \left[\underset{R}{\overset{R}{>}}C=\overset{+}{O}{\nwarrow}_H\right] \xrightarrow[\text{slow}]{R'NH_2} \left[\underset{R}{\overset{R}{>}}\underset{OH}{\overset{\overset{+}{N}H_2-R'}{C}}\right]$$

$$H^+ + \underset{R}{\overset{R}{>}}C=N{\nwarrow}_{R'} \xleftarrow[\text{fast}]{-H_2O} \left[\underset{R}{\overset{R}{>}}\underset{\overset{+}{O}H_2}{\overset{NH-R'}{C}}\right]$$

(e.g., $R' = -NHCONH_2$)

It is interesting to note that the rate of this reaction is accelerated by acid only up to a certain concentration; addition of more acid actually retards the reaction. This is due to the equilibrium

$$R'NH_2 + H^+ \rightleftharpoons R'\overset{+}{N}H_3$$

where the protonated species, $R'NH_3^+$, cannot add to the carbonyl, as it would provide five bonds to nitrogen.[1] At some optimum $[H^+]$ the most effective balance

---

[1] First-row elements such as nitrogen have only four orbitals available in their valence shell so that only four bonds can be formed. Formation of a fifth bond would involve the next higher shell which is much higher in energy.

between protonation of the carbonyl group and concentration of the free amine form of the reagent will be reached and the maximum rate for the reaction will be observed. The exact [H^+] will depend on the basicity of the amine function: a weakly basic reagent such as 2,4-dinitrophenylhydrazine will work best at a low pH (strongly acid solution), and a stronger base such as hydroxylamine will be favored by less acidic media.

Oximes, hydrazones, phenylhydrazones, and semicarbazones are often sharply melting, easily purified derivatives which are useful for the characterization of aldehydes and ketones. They can often be hydrolyzed back to the parent aldehyde or ketone, taking advantage of the above equilibria to favor the hydrolysis reaction.

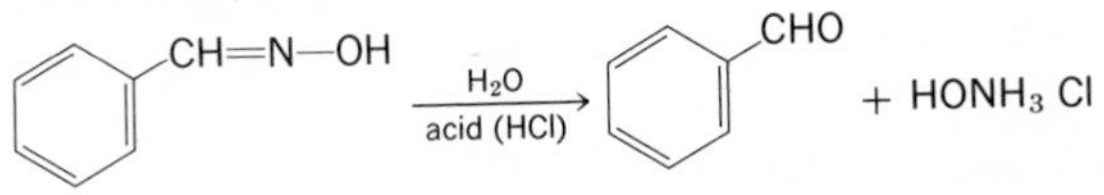
CH=N—OH
H2O
acid (HCl)
CHO
+ HONH3 Cl

When treated with water or alcohols, in the presence of acidic or basic catalysts, most aldehydes and ketones come rapidly to equilibrium in another typical addition reaction. This is the formation of hydrates (with water) and hemiacetals (with alcohols).

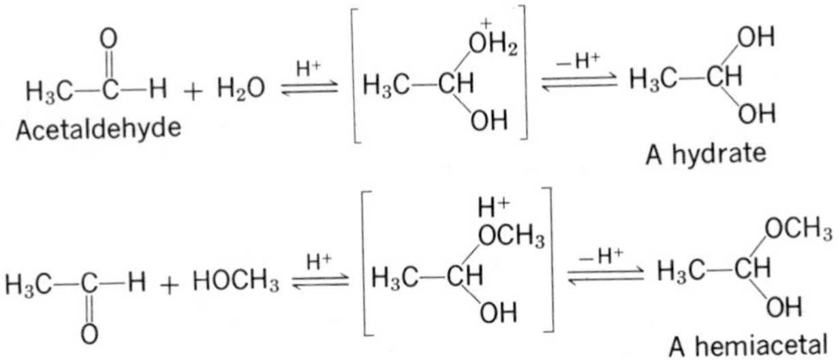
Acetaldehyde
A hydrate
A hemiacetal

In the presence of base similar products are formed by a different route.

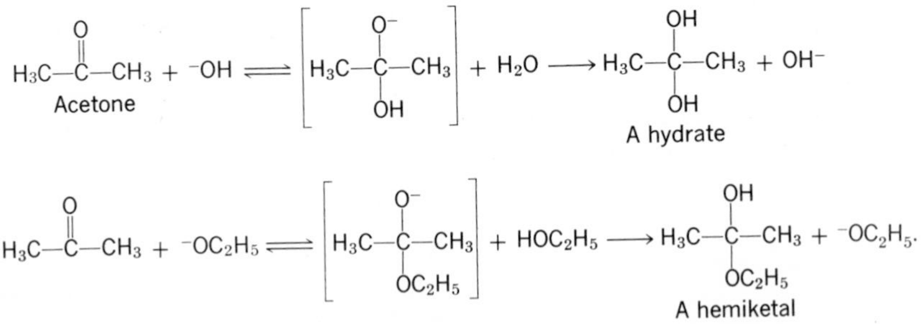
Acetone
A hydrate
A hemiketal

The position of the equilibrium is in many cases far on the side of the free carbonyl compound, and the reaction is so rapid that these addition products are too unstable to be of practical value, except in the case of cyclic hemiacetals. In the case of ketone hydrates the equilibrium is usually so unfavorable that it can only be detected by oxygen-18 exchange.

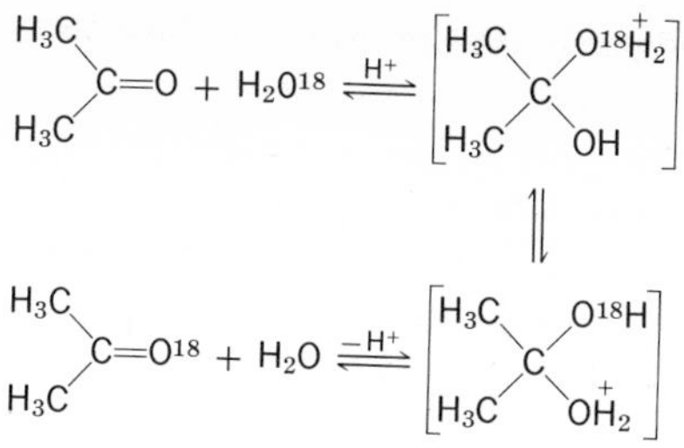

$$\begin{array}{c} H_3C \\ C{=}O + H_2O^{18} \underset{H^+}{\rightleftharpoons} \\ H_3C \end{array} \left[\begin{array}{c} H_3C \quad O^{18}\overset{+}{H_2} \\ C \\ H_3C \quad OH \end{array} \right]$$

$$\big\updownarrow$$

$$\begin{array}{c} H_3C \\ C{=}O^{18} + H_2O \underset{-H^+}{\rightleftharpoons} \\ H_3C \end{array} \left[\begin{array}{c} H_3C \quad O^{18}H \\ C \\ H_3C \quad \overset{+}{O}H_2 \end{array} \right]$$

In the presence of excess alcohol and an acid catalyst (but *not* with a basic catalyst), a second reaction may occur which is analogous to ether formation, resulting in the production of stable acetals.

$$\begin{array}{c} OCH_3 \\ H_3C{-}CH \\ OH \end{array} + H^+ \rightleftharpoons \left[\begin{array}{c} OCH_3 \\ H_3C{-}CH \\ \overset{+}{O}H_2 \end{array} \right] \underset{-H_2O}{\rightleftharpoons} \left[\begin{array}{c} H_3C{-}\overset{+}{CH}{-}\overset{..}{O}CH_3 \\ \updownarrow \\ H_3C{-}CH{=}\overset{+}{\underset{..}{O}}CH_3 \end{array} \right.$$

$$\begin{array}{c} OCH_3 \\ H_3C{-}CH \\ OCH_3 \end{array} \underset{-H^+}{\rightleftharpoons} \left[\begin{array}{c} H \\ {}^{+}OCH_3 \\ H_3C{-}CH \\ OCH_3 \end{array} \right. \Big] \; \overset{CH_3OH}{}$$

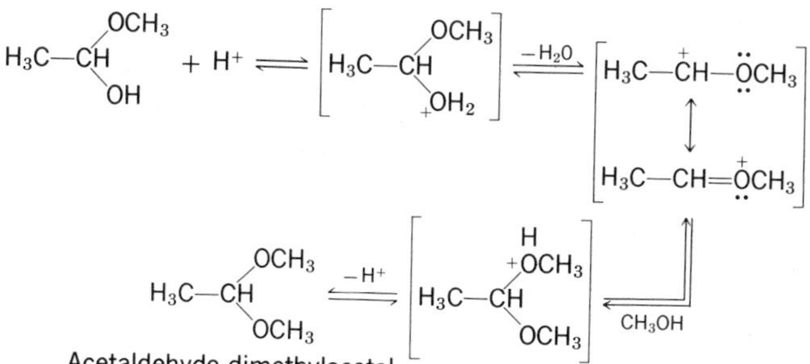

Acetaldehyde dimethylacetal

The carbonium-ion intermediate formed above is stabilized by resonance involving the oxygen. This lowers the activation energy to the point where acetals may be formed and hydrolyzed by dilute acids under far milder conditions than those used for ether formation and cleavage. Like ethers, however, acetals (but not hemiacetals) are stable to alkali.

The less-reactive carbonyl groups of ketones do not yield the corresponding ketals under these conditions because of unfavorable equilibria. Ketals can be prepared by special methods, and are found to be chemically quite similar to acetals.

If the carbonyl group is substituted with strongly electron-withdrawing groups, the hydrates can be isolated in some cases. Thus chloral and hexafluoro acetone form stable hydrates.

$$Cl_3C{-}CHO + H_2O \rightleftharpoons CCl_3\overset{\displaystyle OH}{\underset{\displaystyle OH}{CH}}$$

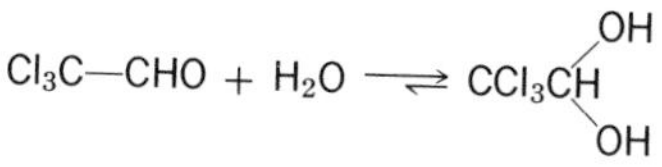

Chloral hydrate

$$F_3C{-}\overset{\displaystyle O}{\overset{\|}{C}}{-}CF_3 \rightleftharpoons (F_3C)_2C\overset{\displaystyle OH}{\underset{\displaystyle OH}{}}$$

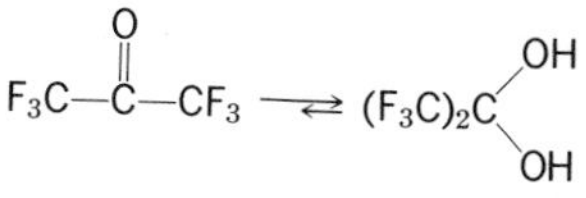

HFA hydrate

The stability of these hydrates derives from the relative instability of the positive carbonyl carbon in the vicinity of another electropositive carbon atom.

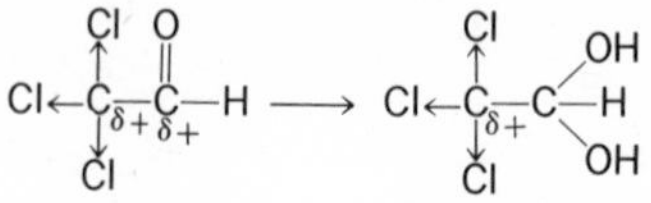

Unfavorable More favorable

11.8 Oxidation Reactions The greatest difference between aldehydes and ketones lies in their behavior toward oxidizing agents. This has been made the basis of several tests and methods of analysis for aldehydes. Very mild oxidizing agents such as Cu^{++} and Ag^{+} in alkaline medium will oxidize most aldehydes to the corresponding acids. Ag^{+} ions are obtained in an alkaline medium by forming the silver ammonia ion. The majority of the silver is tied up in this complex ion but is readily available through the equilibrium:

$$[Ag(NH_3)_2]^+ \rightleftharpoons Ag^+ + 2NH_3$$

A solution of ammoniacal silver nitrate reacts with aldehydes with the deposition of metallic silver.

$$CH_3CHO + 2[Ag(NH_3)_2]^+ + 2OH^- \longrightarrow CH_3COO^- + NH_4^+ + 2Ag + H_2O + 3NH_3$$

This reaction is known as the *Tollens' test.* If a glass surface on which the silver deposits is very clean, the silver comes down as a mirror. Advantage is taken of this reaction in silvering glass to make mirrors; common aldehydes such as formaldehyde or glucose are used. In *Fehling solution* cupric copper is held in solution as the cupritartrate complex ion or, in *Benedict solution,* as a complex citrate ion. Both these ions are soluble in basic solutions, and they yield Cu^{++} in such low concentration that there is not enough to exceed the solubility product of cupric hydroxide.

When many aldehydes are warmed with Fehling or Benedict solution, a brick-red precipitate containing Cu_2O deposits.

$$CH_3CHO + \text{Fehling solution} \longrightarrow CH_3COOH + Cu_2O\downarrow$$

The Cu_2O may be measured, and in some cases the reaction is used as a quantitative analytical process for the determination of "reducing" sugars (which are aldehydic) in blood and urine, etc. It has been a regular practice to test for "sugar in the urine" with one of these reagents. Sometimes the cuprous precipitate comes down as yellow cuprous hydroxide, but on warming this is converted to the brick-red cuprous oxide. The reaction is not generally quantitative nor stoichiometric but has been extensively used.

Stronger oxidizing agents may, of course, be used to convert aldehydes to acids;

the above reagents are useful in that they are specific.[1] For example, the aldehyde group can be oxidized without affecting a double bond present in the same molecule.

$$CH_3CH{=}CH{-}CHO \xrightarrow[\text{or Tollens' reagent}]{\text{Fehling solution}} CH_3CH{=}CH{-}COO^-$$

11.9 Reduction Reactions   The carbonyl group of aldehydes and ketones is readily reduced. Hydrogen in the presence of nonpoisoned nickel will reduce aldehydes and ketones to primary and secondary alcohols respectively. By the use of $CaCrO_2$ catalyst the carbonyl group can be reduced in the presence of a $C{=}C$ group. The latter reaction can also be accomplished by lithium aluminum hydride (Sec. 8.6) in ether, or by the more mild $NaBH_4$ in alcohol.

$$H_2C{=}CH{-}CH_2{-}CH_2{-}\overset{\overset{\text{O}}{\|}}{C}{-}CH_3 \xrightarrow[\text{or NaBH}_4]{\text{LiAlH}_4} H_2C{=}CH{-}CH_2{-}CH_2{-}\overset{\overset{\text{OH}}{|}}{C}H{-}CH_3$$

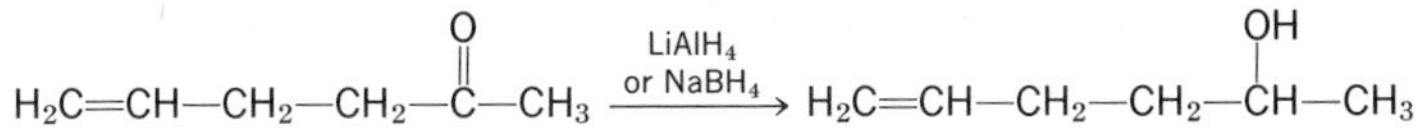

11.10 Substitution Reactions   The carbonyl group is observed, in all compounds containing it, to exert an activating effect on neighboring groups; this effect is of the sort that would be predicted on the assumption that a carbonyl group tends to draw electrons from nearby atoms. This is a consequence of the dipolar nature of the carbonyl group.

$$-\overset{\overset{\text{O}}{\|}}{C}- \longleftrightarrow -\overset{\overset{\text{O}^-}{|}}{\underset{+}{C}}-$$

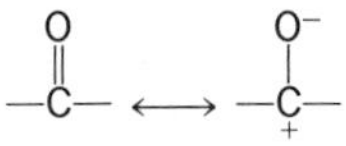

The effect is felt by groups attached to the carbonyl group, but it naturally is strongest at the positions closest to the positive carbon atom. The result of this is to make hydrogens attached to the α carbon atom[2] more acidic than hydrogens in other parts of the molecule.

It is obvious that other factors contribute to the acidity of hydrogens on the α carbon atom since these hydrogens are *found to be many orders of magnitude more acidic than the proton attached directly to the carbonyl group in aldehydes.*

$$\overset{\text{H}}{\underset{\text{H}}{\diagdown}}\hspace{-0.2em}C{-}\overset{\overset{\text{O}}{\|}}{C}{-}R \longrightarrow \left[-\overset{}{\underset{\underset{\text{H}}{|}}{C}}{-}\overset{\overset{\text{O}}{\|}}{C}{-}R \right] + H^+$$

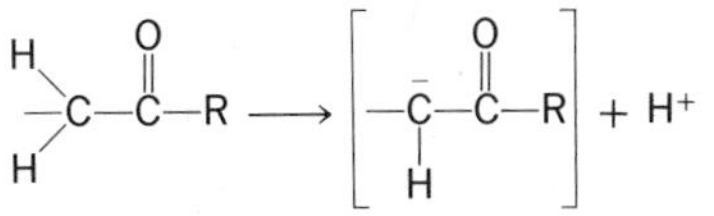

[1]Surprisingly, benzaldehyde does not react with Fehling solution although Tollens' test is positive. It appears that Tollen's test is the more reliable.
[2]The notation α-, β-, etc., is used to designate the positions of carbon atoms with respect to a *functional* group. Thus the ethyl alcohol derivative 2-chloro-1-ethanol (the functional group is —OH) might be designated β-chloroethyl alcohol, CH_2ClCH_2OH. Since the carbonyl group is the functional group of aldehydes and ketones, the α carbon is that next to the carbonyl group. Ketones have two α carbons, which may be distinguished by a prime placed on one of them:

$$CH_3\overset{\overset{}{\underset{\underset{\text{OH}}{|}}{C}}}{H}{-}CHO \qquad\qquad CH_3CHCl{-}CO{-}CHClCH_3$$

α-Hydroxypropionaldehyde α,α'-Dichlorodiethyl ketone

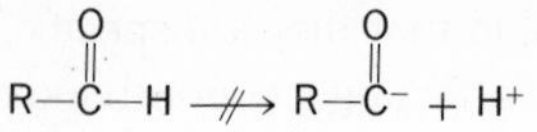

This effect is attributed to resonance stabilization of the ion obtained on the loss of an α proton.

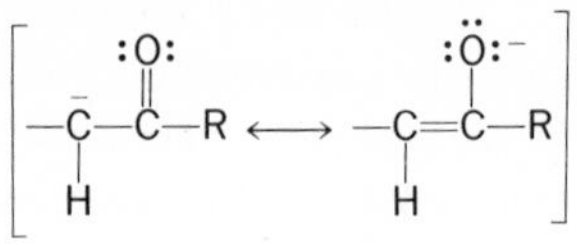

It can be seen that the above structure satisfies all the rules for resonance. The resonance-stabilized ion is called the "enolate" ion. This ion can be derived by removal of an α proton from a ketone or aldehyde, or by removal of the O—H proton from an enol (Sec. 5.4).

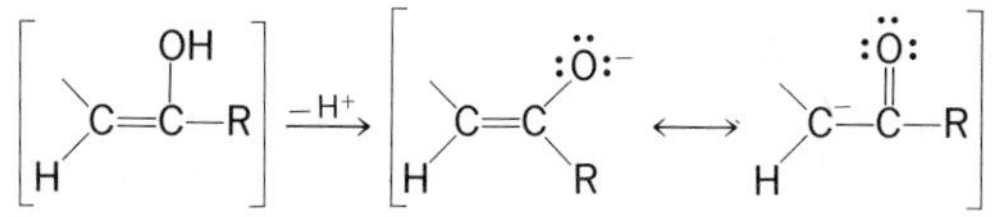

An enol Enolate ion

The actual concentrations of enolate (and enol) present at equilibrium are very small with most ketones except with the very strongest bases (e.g., $NaNH_2$). However, the intervention of the enolate ion can be demonstrated by deuterium exchange in the case of ketones (aldehydes are usually too reactive; see Sec. 11.12).

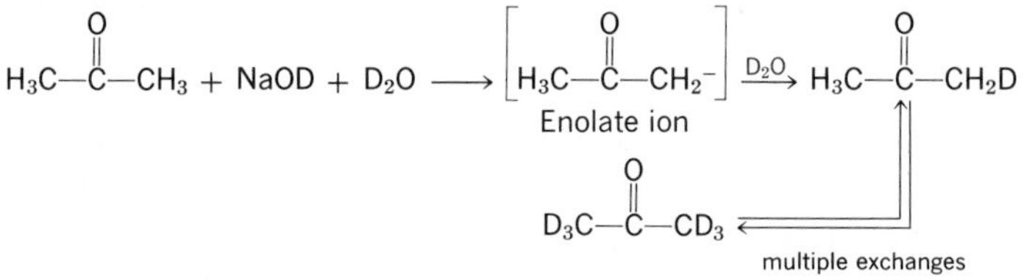

Enolate ion
multiple exchanges

11.11 Haloform Reaction In alkaline solution (conditions that favor ionization of H), aldehydes and ketones react with halogens, including iodine, in a rapid substitution reaction called the "haloform reaction"; this is of a rather specific nature.

$$I_2 + {}^{-}OH \longrightarrow IOH + I^-$$

$$I{-}OH \rightleftharpoons I^+ + OH^-$$

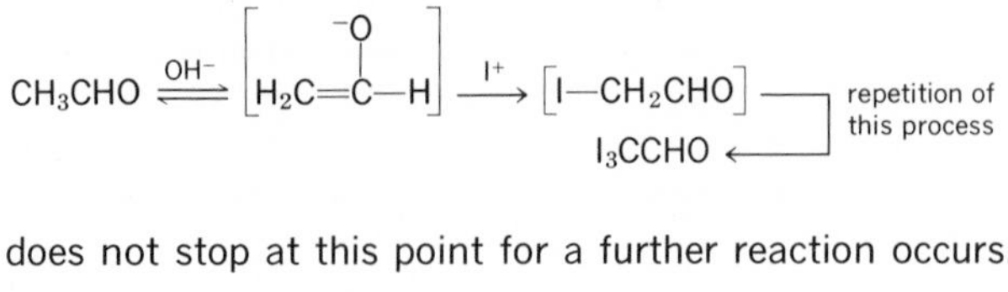

repetition of this process

The reaction does not stop at this point for a further reaction occurs.

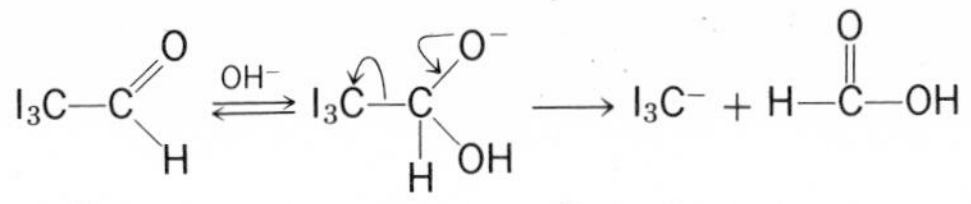

$$I_3C^- + HC(=O)OH \rightleftharpoons \underset{\text{Iodoform}}{I_3CH} + \underset{\text{Formate ion}}{HC(=O)O^-}$$

Carbonyl compounds containing the group

$$H_3C{-}C(=O){-}$$

behave in this way, yielding iodoform and an acid with one less carbon atom; other ketones will be halogenated, as might be expected, but the products do not break down in this manner.

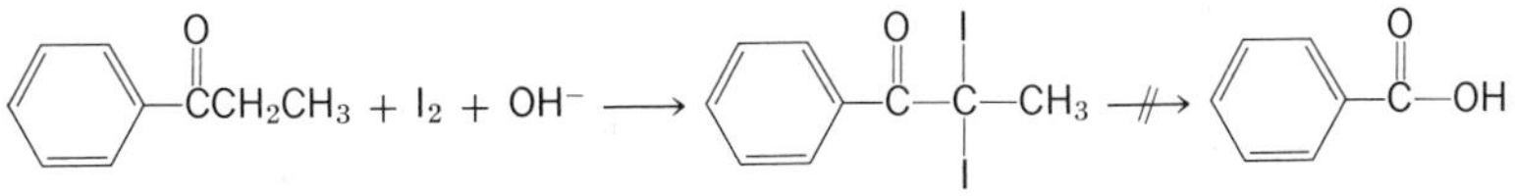

Because of the ease with which the nicely crystalline, insoluble iodoform is identified (it has a characteristic medicinal odor), the iodoform test is a very useful one. As an example, it could be used to distinguish between 2-pentanone and 3-pentanone.

$$CH_3C(=O)CH_2CH_2CH_3 \xrightarrow[\text{NaOH}]{I_2} CH_3I\downarrow + NaOC(=O)CH_2CH_2CH_3$$

$$CH_3CH_2C(=O)CH_2CH_2 \xrightarrow[\text{NaOH}]{I_2} \text{no iodoform} \qquad \text{(a substitution reaction occurs)}$$

Any compound capable of forming the

$$CH_3C(=O){-}$$

group under the conditions of the iodoform test will yield iodoform. The only exceptions are acetic acid and its derivatives. The most important group of this sort is the CH_3CHOH— group, which is oxidized to the corresponding

$$CH_3C(=O){-}$$

group by the action of halogens in alkaline medium. (Recall the oxidizing behavior of hypochlorite, as in "bleaching powder.") Hence secondary alcohols with a terminal CH_3CHOH— group (and ethyl alcohol) also yield iodoform.

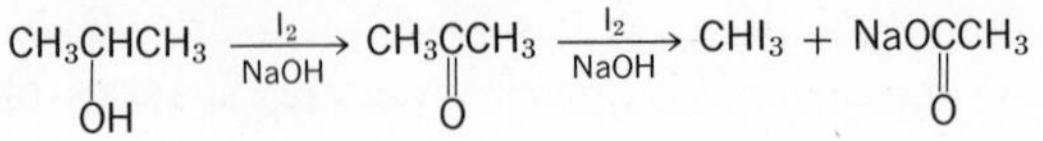

Chloroform and bromoform may be prepared in exactly analogous reactions.

$$CH_3CH_2OH \xrightarrow[NaOH]{Cl_2} CH_3CHO \xrightarrow[NaOH]{Cl_2} \underset{\text{Chloral}}{CCl_3CHO} \xrightarrow[NaOH]{H_2O} CHCl_3 + NaOCHO$$

The iodoform reaction often proves to be of synthetic utility. For instance, the acid $(CH_3)_3CCOOH$ can be prepared from pinacolone.

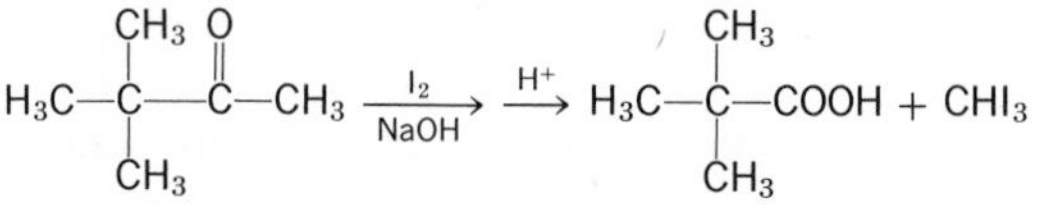

This acid would be difficult to prepare by other means.

11.12 Aldol Reaction; Mechanism The ion produced by the action of alkali on carbonyl compounds having an α hydrogen atom is nucleophilic in its behavior, as are the ions derived from most weak acids, and accordingly adds to the carbonyl group. This reaction results in the formation of a new carbon-to-carbon bond between two molecules.

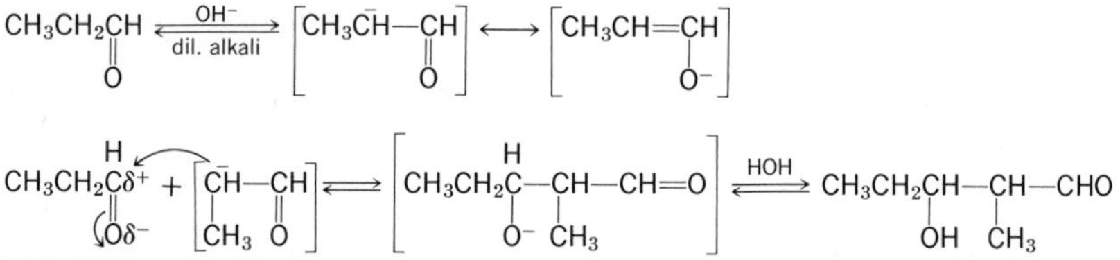

Aldehyde Enolate ion An aldol (*ald*ehyde-alcoh*ol*)

Aldols are very easily dehydrated, and gentle warming is often enough to remove the elements of water, forming unsaturated aldehydes.

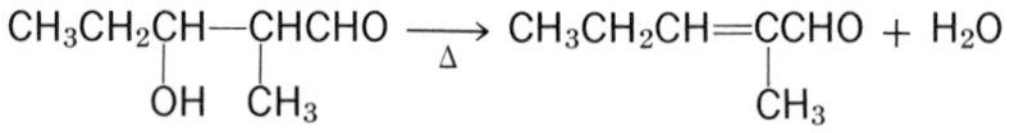

This dehydration is facilitated by the acidic tendencies of the α hydrogen atom, since the new enolate ion can lose OH^-.

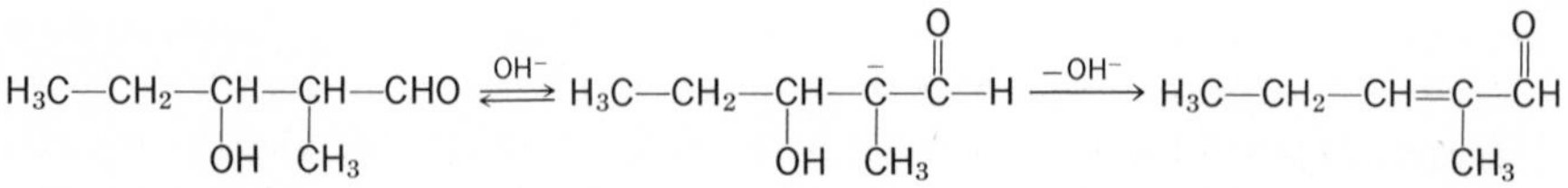

It will be noted that the resulting unsaturated aldehyde contains a conjugated system. The stabilizing effect of conjugation tends to favor the unsaturated aldehyde.

Reactions of this sort which lead to the formation of new carbon-to-carbon bonds in a controllable manner, usually with the elimination of some small molecule such as H_2O, are called *condensation reactions*. The reaction under discussion, which was first studied in the case of aldehydes, was called the "aldol condensation." Ketones, especially the methyl ketones and cyclic ketones, undergo this same reaction; the products are more complicated since many ketones have two different α carbon atoms.

In the case of acetaldehyde itself, a straight-chain compound results from an aldol condensation.

$$\underset{}{CH_3CHO} \longrightarrow \underset{\text{Aldol}}{\underset{OH}{CH_3\overset{}{CH}}\text{—}CH_2CHO} \longrightarrow \underset{\text{Crotonaldehyde}}{CH_3CH\text{=}CHCHO}$$

In all other cases (except CH_2O) *branched-chain* unsaturated and hydroxy aldehydes are the products of the reaction. Since these may be readily converted to a variety of useful substances by reactions already studied, the aldol condensation has real synthetic value in the laboratory and in industry. It is one of the reactions used in nature to form new C—C bonds, probably because the mild conditions required to bring it about are consistent with those found in living cells.

In reactions of this sort, where two identical molecules react, there is the possibility of using two different molecules. This naturally leads to a mixture of products. In the case of a mixed aldol, the working up of the reaction mixture is greatly simplified if one of the aldehydes has no α hydrogen. An example of such a reaction is given below.

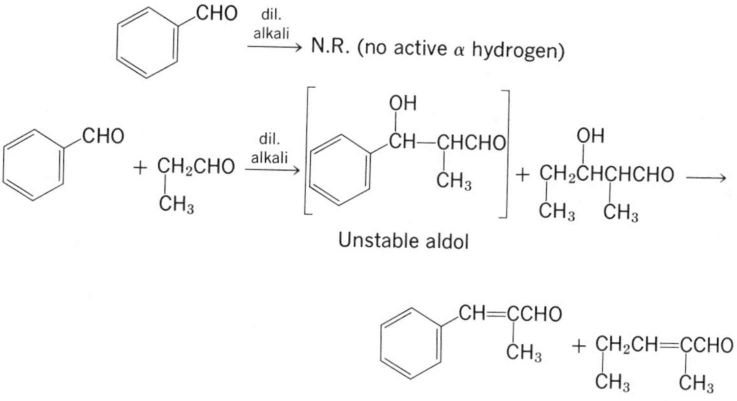
CHO
dil.
alkali
N.R. (no active α hydrogen)
CHO
+ CH2CHO
CH3
dil.
alkali
OH
CH—CHCHO
CH3
+ CH2CHCHCHO
OH
CH3 CH3
Unstable aldol
CH=CCHO
CH3
+ CH2CH=CCHO
CH3 CH3

In this case, instead of four possible products there are only two, and the separation of the resulting mixture is correspondingly easier.

As the final product of an aldol condensation is still an aldehyde, the student may wonder why this does not, in turn, add another molecule to its carbonyl group. Such further addition does, in fact, occur as a side reaction in all aldol condensations. If a more concentrated solution of alkali is used, it becomes the main reaction, and, as a result, long-chain polymers (resins) are formed.

The simple addition product is formed in the case of acetaldehyde by a reaction of the aldol with unchanged aldehyde to form an acetol-hemiacetal. This species is sufficiently unreactive to inhibit further condensation, but regenerates aldol and aldehyde (in equal proportions) on workup.

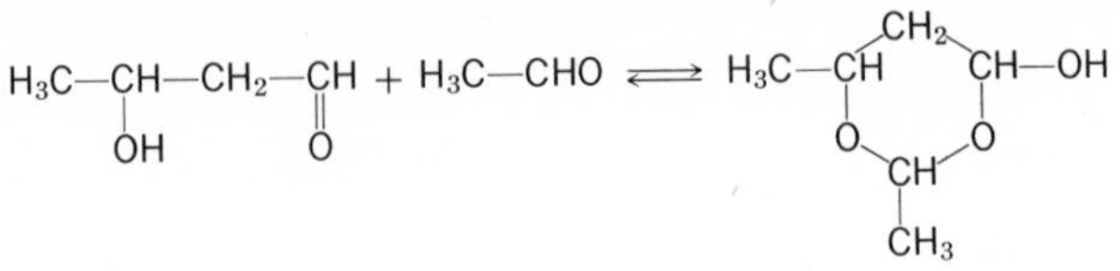

11.13 Perkin Reaction A number of modifications of the addition of an α carbon atom to a reactive carbonyl group have been discovered. For example, Perkin found that aromatic aldehydes could be condensed at elevated temperatures with carboxylic acid anhydrides at the α carbon atom. The reaction gives best yields in the presence of sodium salts of weak acids, that is, basic salts. At the temperatures required, the intermediate addition product cannot be isolated, and the water which is formed hydrolyzes the anhydride immediately, producing an unsaturated acid as the final product.

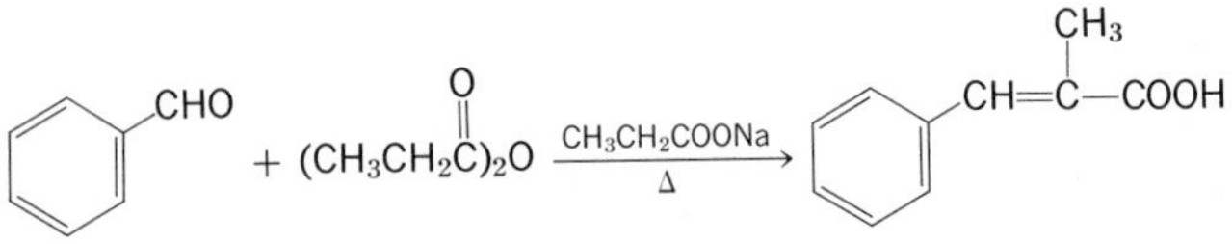

This reaction, the *Perkin synthesis,* is unsuccessful with aldehydes carrying an α hydrogen since they condense with themselves too readily under these conditions, yielding resins.

11.14 Cannizzaro Reaction Although aromatic aldehydes and others having no α hydrogen do not react with themselves under the mild conditions of the aldol condensation, an increase in the concentration of the alkaline catalyst brings about a reaction known as the *Cannizzaro reaction.* The mechanism of the reaction, which is rather complex and may involve the formation of esters as intermediates, will not be discussed.

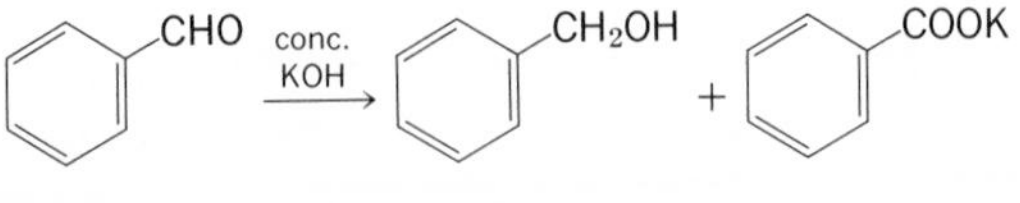

This reaction, which takes place at room temperature, is an internal oxidation-reduction reaction; one molecule of benzaldehyde acts to reduce, and is itself oxidized by, a second molecule. The chief advantage of the reaction lies in its specific nature. Carbonyl groups alone are reduced. For example, a nitroaldehyde will undergo the reaction without involving the nitro group.

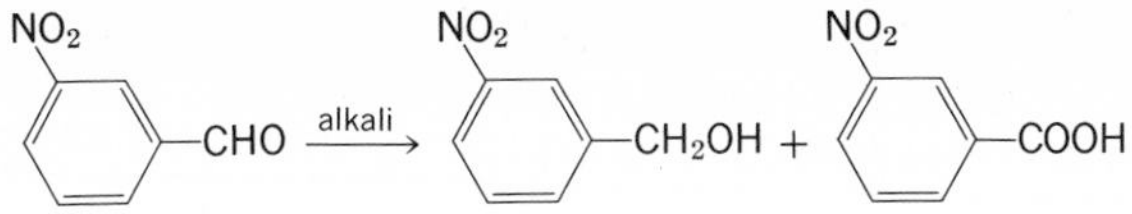
NO2
CHO
alkali
NO2
CH2OH +
NO2
COOH

11.15 Polymerization In addition to polymerization of the aldol type, some of the lower, more reactive members of the aliphatic aldehyde series form trimers by a reversible reaction.

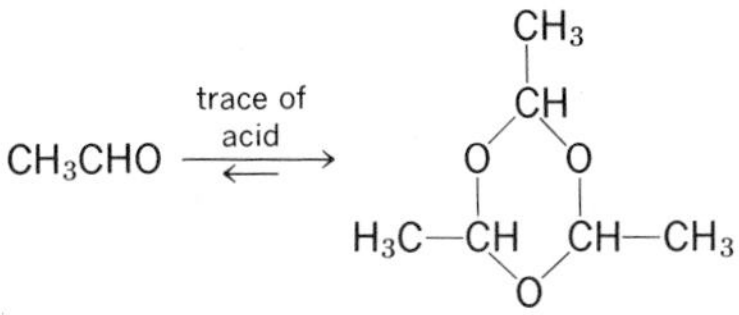
CH3CHO
trace of
acid
CH3
CH
O
O
H3C—CH
CH—CH3
O

Paraldehyde

Because paraldehyde has no free aldehyde group, it is a stable substance and is a useful form in which to preserve acetaldehyde. If a mixture of paraldehyde and sulfuric acid is heated, acetaldehyde distills out, driving the equilibrium in the direction of aldehyde production. Formaldehyde forms a useful polymer in a similar manner. The reaction is less important in the case of the higher aldehydes.

Most aliphatic aldehydes form resins of the aldol type or other polymers even on standing at room temperature. Formaldehyde is an exception in that its 40 percent solution in water (sold as formalin) is stable and does not polymerize readily. This is probably due to the addition of water to the exceedingly active carbonyl to form a hydrate, stable in this case.

$$CH_2O \underset{\longleftarrow}{\overset{H_2O}{\longrightarrow}} CH_2\begin{matrix} OH \\ OH \end{matrix}$$

The commercial polymer, Delrin,[1] is a polymer of formaldehyde, having the repeating unit

$$\text{—}(O\text{—}CH_2\text{—})_n\text{—}$$

In all the preceding discussion, emphasis has been laid on aldehydes rather than ketones. Although ketones will undergo most of the carbonyl-group reactions, even the most reactive ones do so much more slowly than most aldehydes. For example, acetone undergoes an aldol condensation more slowly than acetaldehyde. This may be looked upon in part as due to the electron-repelling effect of the two alkyl groups of ketones, which diminishes the polar character of the carbonyl group compared to corresponding aldehydes containing only one alkyl group.

---

[1] Delrin is a registered trademark of E. I. duPont de Nemours & Co., Inc.

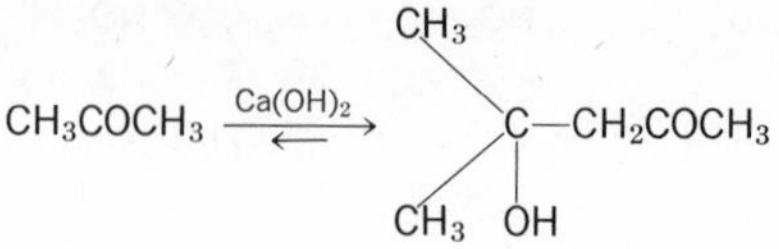

Diacetone alcohol

Here the equilibrium is very unfavorable, and the reaction is run with special equipment which removes the diacetone alcohol from contact with the catalyst as fast as it forms. In contrast to aliphatic aldehydes, ketones show almost no tendency to polymerize on standing.

11.16 Spectroscopic Properties Carbonyl compounds exhibit a wealth of spectroscopic phenomena, so much so that spectroscopic methods have largely supplanted chemical tests (Tollens', Fehling's) for identifying and characterizing them.

In the infrared, the carbonyl group of ketones and aldehydes gives rise to strong absorption in the vicinity of 1740 to 1650 cm^{-1}. So many compounds have been investigated that it is possible to restrict absorptions to very narrow ranges depending on the structure of the aldehyde or ketone. The following table gives these values:

	C=O *stretch*, cm^{-1}	
Type	*Ketone*	*Aldehyde*
Aliphatic and cyclic	. . .	. . .
Six-membered or larger rings	1725–1705	1740–1720
Conjugated, C=C—C(=O)	1685–1666	1705–1685
Aryl, ϕC(=O)—	1700–1680	1715–1695

For aldehydes a very characteristic C—H stretch appears at 2880 to 2650 cm^{-1}. Electron-withdrawing substituents and ring strain move the carbonyl absorption to higher frequencies.

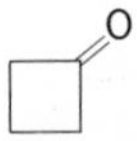

1775 cm^{-1}

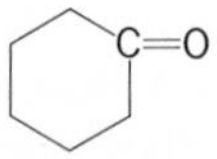

1750 to 1740 cm^{-1}

The nmr of carbonyl compounds is less revealing but quite useful. The aldehyde hydrogen falls at $\delta = 9$ to 10 ppm, away from almost all other hydrogens. This extremely low field value reflects the positive nature of the carbonyl carbon atom and the fact that the aldehyde hydrogen is located in a plane far removed from the π electron density. The α hydrogens in aldehydes and ketones are also shifted slightly downfield (higher δ) from ordinary aliphatic hydrogens by about 0.2 ppm.

The ultraviolet spectra of carbonyl compounds provide useful information. Simple aldehydes and ketones absorb only very weakly in the accessible region of the ultraviolet, but conjugated ones have very strong maxima. The positions of these maxima provide a clue as to the substitution of the conjugated double bond. Thus

H_2C=CH—$\overset{\displaystyle O}{\overset{\|}{C}}$—$CH_3$ has a maximum (in alcohol) at 215 mμ. Addition of methyl or methylene substituents to the double bond adds about 10 mμ per substituent.

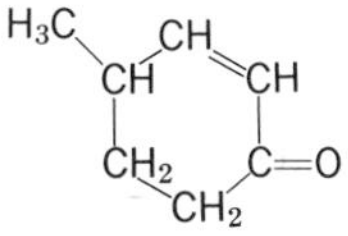

Calc. λ_{max}^{EtOH} = 225 mμ Calc. λ_{max}^{EtOH} = 235 mμ

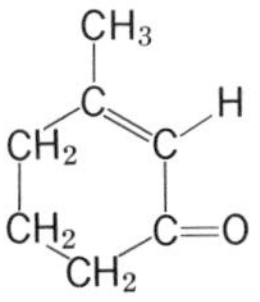

Thus, two ketones as above possess λ_{max} indicative of their substitution pattern.

OUTLINE OF ALDEHYDE AND KETONE CHEMISTRY

Preparation

1. Oxidation of an alcohol. Applicable to side-chain aromatic alcohols also.

$$RCH_2OH \xrightarrow[H_2SO_4]{K_2Cr_2O_7} RCH{=}O + Cr_2(SO_4)_3$$

$$RCHOHR' \xrightarrow[H_2SO_4]{K_2Cr_2O_7} R{-}\underset{\underset{O}{\|}}{C}{-}R' + Cr_2(SO_4)_3$$

2. For ketones only; passing the vapor of an acid over a heated catalyst, ThO_2.

$$\begin{array}{c} R{-}\overset{O}{\overset{\|}{C}}{-}OH \\ + \\ R{-}\underset{O}{\underset{\|}{C}}{-}OH \end{array} \xrightarrow[\Delta]{ThO_2} \begin{array}{c} R \\ \diagdown \\ C{=}O + CO_2 + H_2O \\ \diagup \\ R \end{array}$$

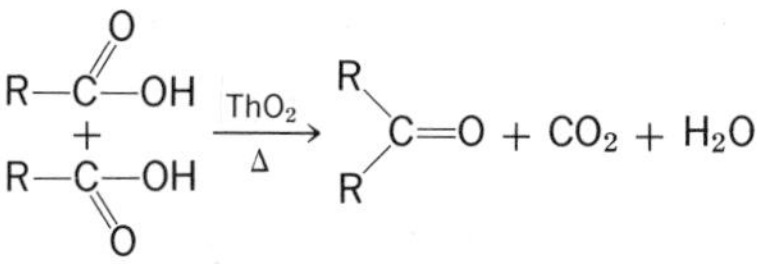

Pyrolysis of the calcium salts yields ketones.

$$(RCOO)_2Ca \xrightarrow{\Delta\Delta} R_2CO + CaO + CO_2$$

3. Hydrolysis of certain dihalides.

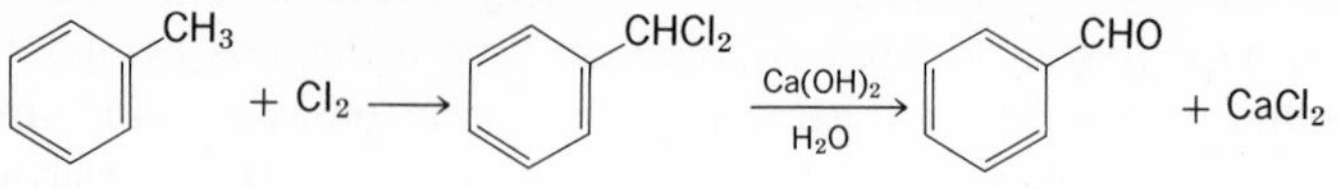

4. The Friedel-Crafts reaction.

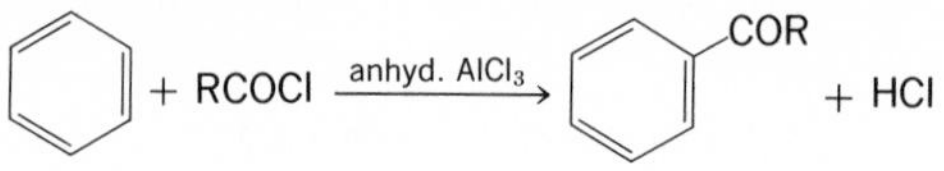

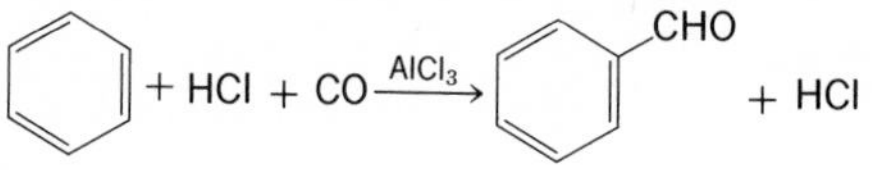

5. Reductive methods (aldehydes only).

a. Rosenmund

$$R{-}\overset{O}{\overset{\|}{C}}Cl \xrightarrow[\text{Pd poisoned}]{H_2} R\overset{O}{\overset{\|}{C}}H + HCl\uparrow$$

b. Raney nickel

$$R{-}\overset{O}{\overset{\|}{C}}{-}SR' \xrightarrow[H_2]{\text{Raney-Ni}} R\overset{O}{\overset{\|}{C}}H + R'H + NiS_n$$

Properties

1. $RCHO + HCN \rightleftharpoons RCH\begin{smallmatrix}OH\\CN\end{smallmatrix} \xrightarrow{H_2O} RCH\begin{smallmatrix}OH\\COOH\end{smallmatrix}$

Cyanohydrin

$$RCOR' + HCN \rightleftharpoons RC\begin{smallmatrix}R'\\-OH\\CN\end{smallmatrix}$$

2. $RCHO + NaHSO_3 \rightleftharpoons RCH\begin{smallmatrix}OH\\SO_3Na\end{smallmatrix}$

3. Grignard reagents add to aldehydes and ketones of all types (Sec. 8.7).

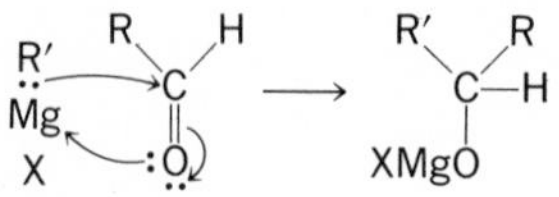

4. Reduction.

$$RCHO \xrightarrow[H_2]{\text{Ni cat.}} RCH_2OH$$

$$RCOR' \xrightarrow[H_2]{\text{Ni cat.}} RCHOHR'$$

5. Oxidation.

   a. With $K_2Cr_2O_7 + H_2SO_4$.

$$RCHO \xrightarrow{(O)} RCOOH$$

   Ketones resist oxidation.

   $RCH_2COCH_2R' \xrightarrow{(O)}$ N.R. or acids of lower C content, i.e., $RCOOH$, $R'COOH$, RCH_2COOH, $R'CH_2COOH$

   b. With Fehling solution.

$$RCHO + 2Cu(OH)_2 \longrightarrow RCOOH + Cu_2O\downarrow + 2H_2O$$

   NOTE: $\left.\begin{matrix}\phi\ CHO \\ RCOR' \\ \phi\ COR\end{matrix}\right\} \xrightarrow{Cu(OH)_2}$ N.R.

   c. With Tollens' reagent.

$$RCHO + 2AgOH \longrightarrow RCOOH + \underset{\text{(mirror)}}{2Ag\downarrow} + H_2O$$

$$RCOR \xrightarrow{AgOH} \text{N.R.}$$

6. $\underset{}{RCH{=}O} + \underset{\text{Hydroxylamine}}{H_2NOH} \longrightarrow \underset{\text{An oxime}}{RCH{=}NOH} + H_2O$

   $RCOR' + H_2NOH \longrightarrow \underset{\substack{\| \\ NOH}}{RCR'} + H_2O$

7. $RCH{=}O + \underset{\text{Phenylhydrazine}}{H_2NNHC_6H_5} \longrightarrow \underset{\text{A phenylhydrazone}}{RCH{=}NNHC_6H_5} + H_2O$

   $RCOR' + H_2NNHC_6H_5 \longrightarrow \underset{\substack{\| \\ NNHC_6H_5}}{RCR'} + H_2O$

8. 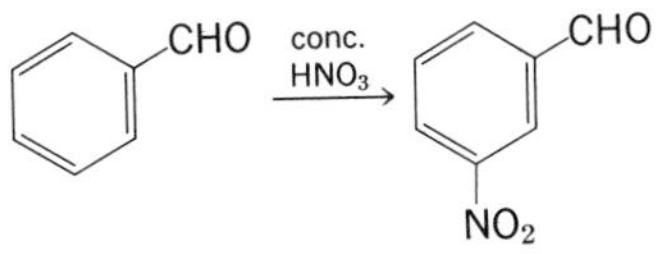
CHO
conc.
HNO3
CHO
NO2

   Same for aromatic ketones. Note that the carbonyl group *on the ring* directs meta.

9. The Cannizzaro reaction.

$$2\phi CHO \xrightarrow{\text{conc. aq. NaOH}} \phi CH_2OH + \phi COOH$$

$$(\phi COOH \xrightarrow{NaOH} \phi COONa)$$

10. The aldol condensation.

$$RCH_2C(=O)H + RCH_2CHO \xrightarrow{\text{dil. alkali}} RCH_2CH(OH){-}CH(R)CHO \xrightarrow[\Delta]{} RCH_2CH{=}C(R)CHO$$

11. The iodoform reaction.

$$RC(=O)CH_3 \xrightarrow{NaOI} RC(=O)CI_3 \xrightarrow[H_2O]{NaOH} RCOONa + CHI_3$$

NOTE: Reactions 2, 6, and 7 are reversed by hydrolysis in the presence of acids. This permits recovery of aldehydes and ketones from these derivatives.

CHAPTER 11 EXERCISES

★1. Write equations for three different preparations of acetophenone. Which seems the simplest and most direct approach?

2. Indicate by equations a practicable method for the preparation of each of the following compounds:

a. Ethyl isopropyl ketone
b. *p*-Bromoacetophenone
c. *p*-Bromobenzaldehyde
d. 3-Methylbutanal
e. Acetaldehyde

★3. Indicate how allyl alcohol, propionaldehyde, and acetone could be distinguished from each other by qualitative tests. by spectroscopic methods.

4. Indicate by equations how you could accomplish each of the following conversions:

★*a.* Propionic acid to 3-pentanol
b. Benzene to *o*-nitrobenzaldehyde
★*c.* Benzene to phenyl methyl ether
d. Acetophenone to 2-phenyl-3-methyl-3-butanol
★*e.* Benzaldehyde to 1-phenyl-1-ethanol
f. Benzaldehyde to 2-phenyl-1-ethanol
★*g.* Propionic acid to propanal
h. Ethyl alcohol to propionic acid
★*i.* Toluene to diphenyl ketone
j. Acetaldehyde to lactic acid
k. Acetic acid to 3-methyl-2-butanone
l. Ethyl propionate to ethyl methyl ketone
m. Calcium carbide to acetone
n. *n*-Propyl iodide to acetone
o. Benzene to *p*-tolualdehyde
p. Toluene to benzyl benzoate
q. Benzene to cinnamaldehyde
r. Butanol to ethylacetylene
s. Ethanol to 2-methyl-3-pentanolal
t. Ethanol to crotonic acid ($CH_3CH{=}CHCOOH$)

5. Indicate a good method for accomplishing each of the following conversions:

★*a.* Acetone to acetaldehyde (two ways)
b. Acetone to isopropyl alcohol
★*c.* Acetone to 2,2-dichloropropane
d. Acetone to propane
★*e.* Acetone to propyne
f. Acetone to 1,2-dichloropropane
★*g.* Acetone to isopropyl chloride (two ways)
h. Acetone to 2-methyl-2-chloropropane
★*i.* Acetone to 2-methylpropane

6. Acetone can undergo deuterium exchange under acidic as well as basic conditions:

$$H_3C-\overset{\displaystyle O}{\overset{\|}{C}}-CH_3 + D^+ + D_2O \longrightarrow H_3C-\overset{\displaystyle O}{\overset{\|}{C}}-CD_3$$

Because of the absence of base, this reaction cannot proceed by way of the enolate ion. Suggest a pathway by which exchange might occur.

★7. A certain compound X, $C_{14}H_{24}O$, shows infrared maxima at 3600 and 3355 cm^{-1}, and at 1650 cm^{-1}. Ozonolysis of X yields Y, $C_{12}H_{20}O_2$, plus another substance that gives a positive iodoform test. Y possesses maxima in the infrared at 3600 and 3355 cm^{-1}, and at 1715 cm^{-1}, but not in the region 2880 to 2650 cm^{-1}. On heating Y loses H_2O to yield a substance Z, $C_{12}H_{18}O$, with infrared maxima in the region 1675 cm^{-1}, but not in the region 3650 to 3300 cm^{-1}. Z shows a maximum in the ultraviolet at 245 mμ and on ozonolysis forms cyclohexanone and 1,2-cyclohexanedione. Deduce structures from the above data and write equations for the reactions.

8. A compound, $C_{14}H_{12}O$, yields, upon oxidation, benzoic acid as the sole organic product. With phenylhydrazine it gives $C_{14}H_{12}NNHC_6H_5$. It does not react with Tollens' reagent. Write the structure of the compound and the reactions involved. Show your reasoning clearly.

★9. A certain compound Q, $C_{18}H_{20}O$, reacts with phenylhydrazine. It is formed when *m*-methylbenzylmagnesium bromide is reacted with compound S, and the product of the reaction is hydrolyzed with water and dilute HCl. Compound S has the formula $C_{10}H_{11}N$ and is produced by treating T, $C_9H_{11}Br$, with KCN. Compound T reacts with dilute NaOH or moist silver oxide to give $C_9H_{12}O$, which may be oxidized to *m*-ethylbenzoic acid. What are the structures of Q, S, and T? Give equations to explain the above reactions.

10. A certain substance is thought to have the formula

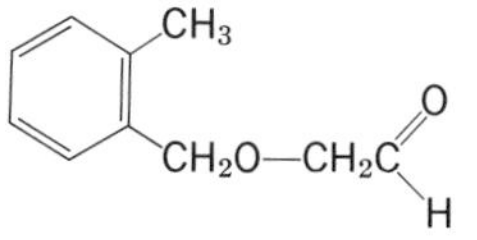

CH3
O
CH2O—CH2C
H

Devise a proof of this structure.

★11. A certain compound has the formula $C_5H_8O_2$. On complete reduction (heating with HI), it gives *n*-pentane. With phenylhydrazine it gives a derivative containing 17 carbons, and with NaOH + I_2 it gives iodoform. On oxidation a substance, $C_5H_8O_3$, is formed which reacts with phenylhydrazine to give a derivative containing 11 carbons. The original substance is unstable to Fehling solution. Deduce its structure. Show your reasoning clearly.

12. A substance, $C_5H_{10}O_2$, reacts with acetyl chloride to give an acetyl derivative containing 58.3 percent carbon and 8.3 percent hydrogen. 1.50 g of the original substance gives 11.6 g of CHI_3 on treatment with NaOH + I_2, had no ir absorption at 2880 to 2650 cm^{-1}, and has no nmr peaks at $\delta = 9$ to 10.

13. How would you distinguish among the following by (*a*) chemical methods; (*b*) by spectral methods?

★*a.*

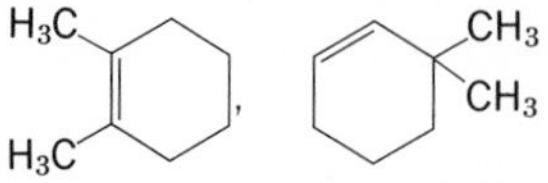

b. C_6H_5CHO, CH_3CHO, $C_6H_5COCH_3$

★*c.* $C_6H_5CH_2—COCH_3$, $(C_6H_5)_2CO$, $C_6H_5CH{=}CH—CO—CH_3$

14. Give the expected products from the following reactions:

a. $(CH_3)_3CCHO$ + conc. $OH^- \longrightarrow$

b. $C_6H_5CH_2CHO$ + EtOH, conc. HCl $\longrightarrow$

c. $C_6H_5CH_2CHO$ + (1) dil. OH^-, (2) $NaBH_4 \longrightarrow$

d. $C_6H_5CHO + CH_2(COOEt)_2 \longrightarrow$

e.

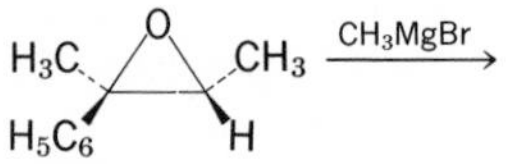

Show stereochemistry.

15. *cis*-1,2-Dihydroxycyclohexane is converted, with dimethyl sulfate and NaOH, to the monomethyl ether, then the other free —OH is oxidized to the ketone (2-methoxy cyclohexanone). The product is now reduced to the 1-hydroxy-2-methoxycyclohexane, which is then hydrolyzed with conc. HI to 1,2-dihydroxycyclohexane.

★*a.* Write the configurations of all the substances formed in each step.

★*b.* Letter them clearly, and state the configurational and stereomeric relationships. That is, which, if any, are diastereomers, or which enantiomorphs, meso forms, cis and trans isomers?

c. Which, if any, of these reactions produce racemic mixtures?

16. The following reaction sequence was carried out starting with an optically pure compound, *R*-$CH_3CHClCH_2CH_3$. Comment briefly on the stereochemical result of each step in the reaction sequence.

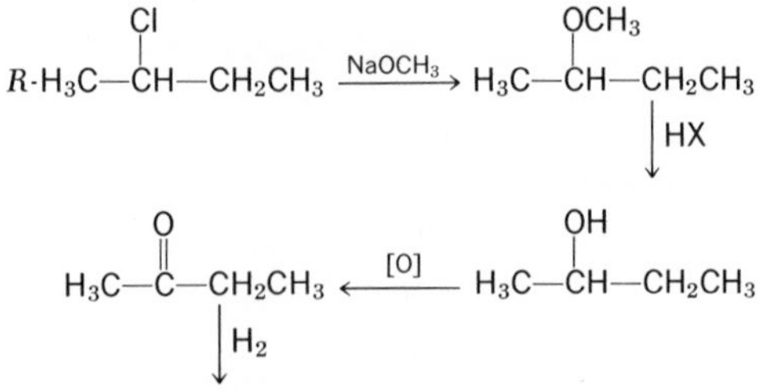

$$H_3C{-}\underset{}{CH}(OH){-}CH_2CH_3 \xrightarrow{HCl} H_3C{-}CH(Cl){-}CH_2CH_3$$

17. How could you distinguish by simple chemical and/or spectral tests between

★*a.* Propanol, propanal, and propanone

b. 2-Pentanone and 3-pentanone

★*c.* Ethylene oxide and acetaldehyde

d. Benzaldehyde and 2-phenylethanal

18. Would the following reactions occur as outlined? If so, indicate a reasonable mechanism. If not, show, with mechanism, the expected products.

a. $C_6H_5CH(OCH_3)_2 + NaOC_2H_5 \xrightarrow{C_6H_5} C_6H_5CH(OC_2H_5)_2$

b. $C_6H_5CH(Cl){-}CH(OH){-}CH_3 + Ag^+ \longrightarrow H_5C_6{-}CH_2C(=O){-}CH_3$

c. $C_6H_5CH(OCH_3)_2 + C_2H_5OH \xrightarrow{H^+} C_6H_5CH(OC_2H_5)_2$

★**19.** Explain the fact that either low or high H^+ concentration inhibits the reaction of acetone with hydroxylamine.

20. Why is NH_3 a poor catalyst for the aldol condensation?

★**21.** Suggest a synthesis from phenol of

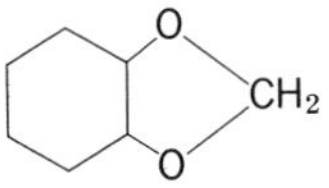

22. Show by equations how you might account for the following observed transformations.

a. $H_2C{=}C(CH_3){-}CH_2CHO \xrightarrow{H^+} H_3C{-}C(CH_3){=}CH{-}CHO$

b. $CH_3C(=O)CH_3 + H_2O^{18} \longrightarrow CH_3C(=O^{18}){-}CH_3$

c.

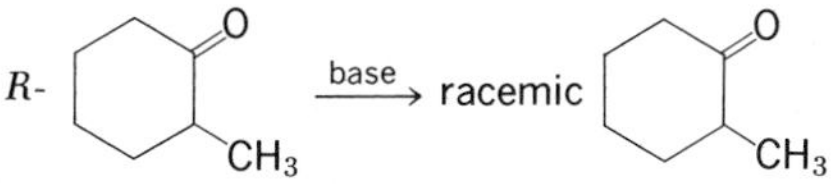
R-
O
CH3
base
racemic
O
CH3

★**23.** Acetone reacts with Cl_2 and with I_2 in basic solution at the same rate. Usually halogens react at different rates in substitution reactions. How can you account for this?

24. The product of the reaction of $(CH_3)_2CHCHO$ with dil. base does not dehydrate easily—in contrast to most aldols. Explain.

★**25.** DDT (Sec. 9.9) is made from $CCl_3CHO + C_6H_5{-}Cl$ in the presence of H_2SO_4. Outline a mechanism for this reaction based on your knowledge of carbonyl-group properties.

12

CARBOXYLIC ACIDS

12.1 Definition Substances containing the carboxyl group

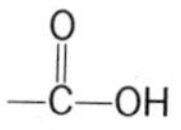

(*carb*onyl and hydr*oxyl,* usually written out as —COOH) make up a large and important class of compounds known as the *carboxylic acids*. These compounds are acidic, owing to the hydrogen of the carboxyl group which is found to be capable of ionization and which may be transferred to a base such as water. The relative extent of ionization is dependent on the strength of the base (for example, water) relative to the base strength of the carboxylate ion.

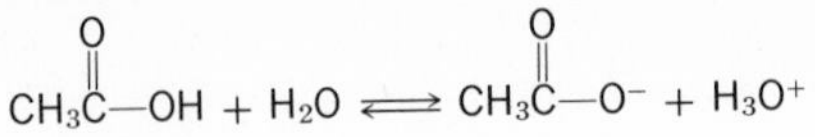

A discussion of the ionizing ability of variously substituted carboxylic acids makes a good point of departure from which to begin a discussion of the reactions of these acids. This is because the degree of ionization of a carboxylic acid is very much influenced by the nature of the rest of the molecule and shows, in a very clear way, the effects of substitution of various groups in organic molecules.

12.2 Ionization and Structure of Acids The degree of ionization of an acid, its "strength," is measured in terms of an ionization constant:

Generalized equation

$$HA + H_2O \rightleftharpoons H_3O^+ + A^-$$

Simplified equation

$$HA \rightleftharpoons H^+ + A^- \qquad \text{(water is left out)}$$

Simple definition of ionization constant

$$K_a = \frac{(\text{concentration of } H^+) \times (\text{concentration of } A^-)}{(\text{concentration of HA})}$$

or, in the usual symbols:

$$K_a = \frac{[H^+][A^-]}{[HA]}$$

From this equation it may be seen that the more ionized the acid, the larger will be the numerator, and hence, the larger the constant.

The constants defined above apply to aqueous solutions. They measure the relative distribution of the proton (H^+) between the two bases (A^- and H_2O). For the sake of ease of comparison of the figures, the effect of the water is taken as constant.

In first-year chemistry, acetic acid was written $HC_2H_3O_2$. This formulation was designed to call specific attention to the fact that acetic acid has only one ionizable hydrogen: it is a *monobasic* acid; its equivalent weight is equal to its molecular weight. Organic chemists write the formula for acetic acid

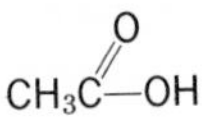

This symbol clearly implies that one of the hydrogens, the one attached to oxygen,

TABLE 12.1
IONIZATION BEHAVIOR OF SOME ACIDS

Substance	K_a	*Substance*	K_a
$(C_6H_5)_3CH$	10^{-33}	$CH_3CHClCOOH$	1.47×10^{-3}
C_2H_5OH	7.3×10^{-20}	$CH_2ClCOOH$	1.55×10^{-3}
HOH	2×10^{-16}[a]	$(COOH)_2$	3.8×10^{-2}[b]
C_6H_5OH	1.3×10^{-10}	$CHCl_2COOH$	5×10^{-2}
$HOCOOH(H_2CO_3)$	3×10^{-7}	CCl_3COOH	2×10^{-1}
CH_3CH_2COOH	1.4×10^{-5}	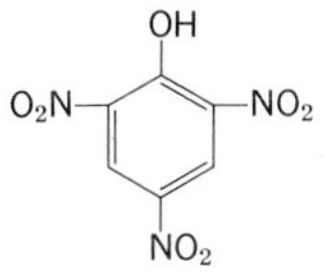OH O₂N NO₂ NO₂	1.6×10^{-1}
CH_3COOH	1.86×10^{-5}		
$CH_2{=}CHCOOH$	5.6×10^{-5}		
C_6H_5COOH	6.6×10^{-5}		
$ClCH_2CH_2COOH$	8.59×10^{-5}		
CH_2ICOOH	7.5×10^{-4}		
$CH_2BrCOOH$	1.38×10^{-3}	H_2SO_4	Large

[a] Equilibrium constant = $[H_3O^+][OH^-]/[H_2O]$.
[b] First ionization constant.

is different from the other three, which are attached to carbon. But it does not show why acetic acid ionizes more completely than, say, an alcohol or phenol ionizes. It must be that the —OH group of the carboxylic acid has properties quantitatively different from those of the hydroxyl group of an alcohol or phenol. Some idea of the reason for this may be gained from an examination of Table 12.1, in which a number of compounds containing the —OH group are listed together with their ionization constants.

The strength of an acid as given by K_a is a measure of the position of the acid-base equilibrium, as indicated above. An important factor which controls the position of this equilibrium is the relative stability of the species on both sides of the equation. Resonance theory provides an explanation for the acidity of the carboxyl group. The group is stabilized by resonance (as will become evident further on) as follows:

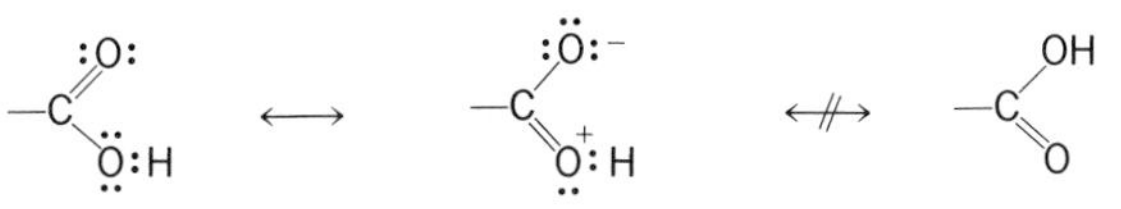

I*A* (main structure) — I*B* (important structure, but charge-separated) — Not involved (would involve nuclear motion)

The two structures I*A* and I*B* are not equivalent and although I*B* is important, it involves considerable charge separation and differs in energy from I*A* so that optimum reduction in energy is not achieved.

In the case of the anion, derived by loss of a proton, this difference is removed.

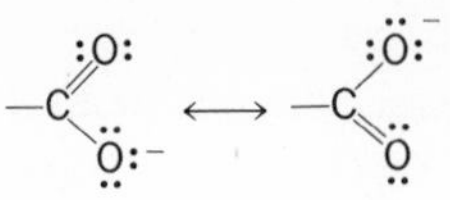

II*A*   II*B*

Here II*A* and II*B* are identical and *the stabilization of II relative to that of I is greater* than in many other systems. Both I and II are stabilized by resonance, but because the stabilization is more efficient in II, *the difference in energy between I and II is diminished.* Since it is this difference in energy that determines the position of the equilibrium, the carboxyl group has a larger tendency to lose a proton than, say, the alcoholic —O—H group, and hence, is a stronger acid.

The molecular orbital picture illustrates the nature of the carboxylate anion. Here the three *p* orbitals contain four electrons in two molecular orbitals: both

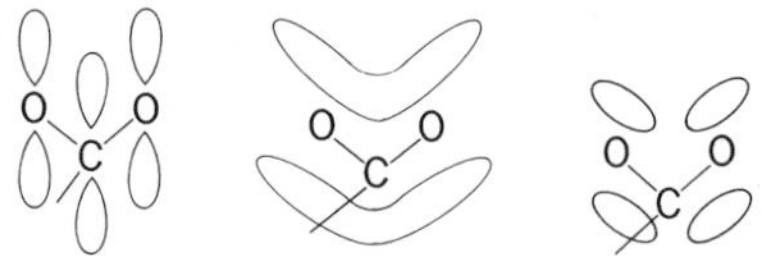

p Orbitals   One MO   Second MO

orbitals encompass the three atoms, but the second, higher energy one is zero at the carbon atom. The net result is that the ion is symmetrical, with the highest electron density at the two oxygen atoms and with lowest at the carbon atom. X-ray diffraction confirms the equivalence of the two C—O bonds, each being intermediate between single and double bonds in length.

As can be seen from Table 12.1 the stabilization of the carboxylate anion relative to the carboxyl group is about 10^{15} times greater than that of the RO^- vs. ROH group. This corresponds to about 21 kcal/mole of extra stabilization.

The carboxylate anion is further stabilized by electron-withdrawing substituents such as aryl groups, C=C, and halogen, and is destabilized by electron-releasing substituents (alkyl groups). As can be seen from examination of Table 12.1, substitution of the CH_3 group of acetic acid by the strongly electron-withdrawing Cl_3—C group enhances the acidity by almost 10^4.

12.3 Carbonyl Properties As was the case with benzene, where resonance stabilization suppresses many of the olefinic properties, many of the characteristic carbonyl group reactions are not observed with the carboxyl group. Moreover, since many of the carbonyl reactions are brought about by more or less basic reagents, the acidic properties of the carbonyl group often supervene.

For example, the usual carbonyl group reagents, such as hydrazine, hydroxylamine, etc., form salts.

$$R-\overset{\overset{\displaystyle O}{\|}}{C}-OH \xrightarrow{NH_2OH} R\overset{\overset{\displaystyle O}{\|}}{C}-\bar{O}\quad \overset{+}{N}H_3OH$$

Similarly, in reactions which would be analogous to the aldol condensation, the base simply removes the acidic OH proton. The removal of a second proton from the negatively charged carboxylate ion is inhibited by the existing negative charge. The next step of addition to the carbonyl is inhibited for similar reasons, that is, reactions between ions of like charge are unfavorable.

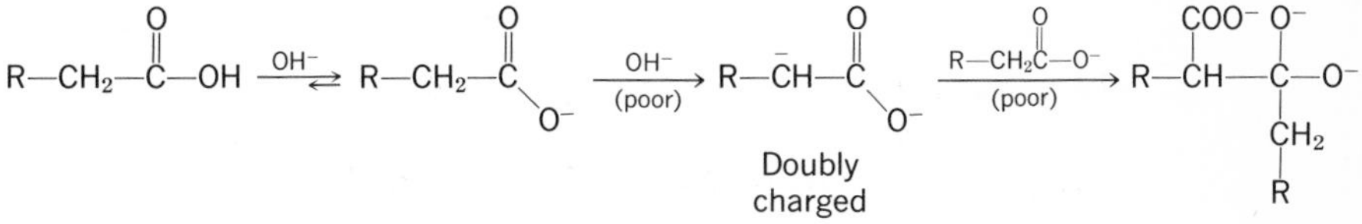

$$R-CH_2-\overset{\displaystyle O}{\overset{\|}{C}}-OH \underset{\text{OH}^-}{\rightleftharpoons} R-CH_2-\overset{\displaystyle O}{\overset{\|}{C}}\diagdown_{O^-} \xrightarrow[\text{(poor)}]{\text{OH}^-} R-\overset{-}{C}H-\overset{\displaystyle O}{\overset{\|}{C}}\diagdown_{O^-} \xrightarrow[\text{(poor)}]{R-CH_2\overset{\displaystyle O}{\overset{\|}{C}}-O^-} R-\overset{\overset{\displaystyle COO^-}{|}}{\underset{\underset{\displaystyle R}{|}}{\underset{\displaystyle CH_2}{|}}}\overset{O^-}{\underset{}{C}}-O^-$$

Doubly
charged

Reduction of the carboxyl group can be accomplished readily by only one reagent, lithium aluminum hydride.

$$4R-\overset{\displaystyle O}{\overset{\|}{C}}-OH + 3LiAlH_4 \longrightarrow 4R-CH_2O^- + \text{lithium aluminum oxides}$$

Catalytic hydrogenation is ineffective except under the most vigorous conditions; acetic acid, in fact, is commonly used as a hydrogenation solvent.

12.4 Association A comparison of the physical properties of the carboxylic acids with those of other carbon compounds of the same molecular weight brings out another point (Table 12.2).

The acids, in general, have higher boiling points than other classes of compounds of similar molecular weights. This is due to a tendency to associate and form double molecules (dimers) to a greater extent even than the other hydroxyl compounds already discussed. Molecular-weight determinations of acetic acid in nonionizing solvents or in the vapor state support this hypothesis: values approaching twice the formula values are obtained. This is interpreted as supporting the following structure:

$$H_3C-\overset{\overset{\displaystyle \ddot{O}---H-O}{\diagup}}{\underset{\underset{\displaystyle O-H---\ddot{O}}{\diagdown}}{C}}\overset{\diagdown}{\underset{\diagup}{C}}-CH_3$$

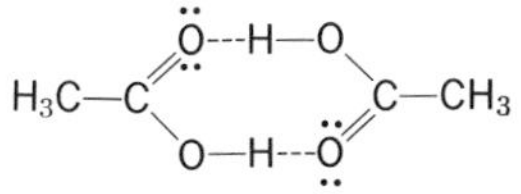

TABLE 12.2
A COMPARISON OF BOILING POINTS OF MOLECULES OF SIMILAR WEIGHT

Name	Structure	Boiling point °C	Associated?
n-Heptane	$CH_3CH_2CH_2CH_2CH_2CH_2CH_3$	98	No
n-Propyl acetate	$CH_3COOCH_2CH_2CH_3$	102	No
n-Amyl chloride	$CH_3CH_2CH_2CH_2CH_2Cl$	106	No
Valeric acid	$CH_3CH_2CH_2CH_2COOH$	187	Yes
n-Hexyl alcohol	$CH_3CH_2CH_2CH_2CH_2CH_2OH$	156	Yes (Chap. 8)

Here again we find hydrogen bridges between two oxygen atoms. The hydrogens are not symmetrically disposed and, although they may be rapidly transferred back and forth, do not satisfy the conditions for resonance. The order of magnitude of the energies of these hydrogen bonds is comparable to that of the average energies of molecules so that these bonds are continually being broken and re-formed.

12.5 Nomenclature Because of the wide occurrence of carboxylic acids and their easy recognition in natural products, members of this class have been known for many years, and many have received special nonsystematic names. Even today there is little uniformity in the naming of the substances. Table 12.3 contains the names and formulas of a number of the more commonly encountered acids; the straight-chain aliphatic acids are often called *fatty acids* from their occurrence in natural fats in the form of esters. These names should be memorized.

TABLE 12.3
SOME CARBOXYLIC ACIDS

Structure	*Common name*	*Systematic name*
$HCOOH$	Formic acid	Methanoic acid
CH_3COOH	Acetic acid	Ethanoic acid; methanecarboxylic acid
CH_3CH_2COOH	Propionic acid	Propanoic acid; ethanecarboxylic acid
$CH_3CH_2CH_2COOH$	Butyric acid	Butanoic acid; propanecarboxylic acid
$CH_3CH_2CH_2CH_2COOH$	Valeric acid	Pentanoic acid; butanecarboxylic acid
$CH_3(CH_2)_{14}COOH$	Palmitic acid	Hexadecanoic acid
$CH_3(CH_2)_{16}COOH$	Stearic acid	Octadecanoic acid
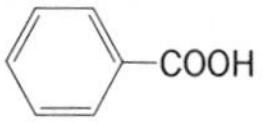	Benzoic acid	. . . ; benzenecarboxylic acid

Both types of systematic names are in common use. It should be emphasized that the name using the *-oic acid* ending takes as its basis the longest chain *counting the carboxyl carbon as one of the chain: The carboxylic acid ending regards the carboxyl carbon as a substituent.* This latter system is often more convenient when several COOH groups are present in a molecule or when the group is a substituent on a ring.

In addition to the nomenclature systems already given, substituted acids are often named after the fatty acid from which they are derived, using the Greek letters α, β, γ, etc., to locate the substituent. These are used with the common, or trivial, names.

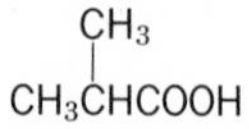

α-Methylpropionic acid
*Iso*butyric acid
2-Methylpropanoic acid

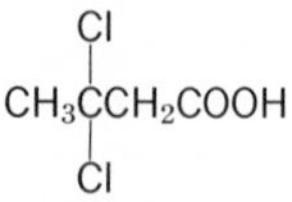

β,β-Dichlorobutyric acid

12.6 Preparation by Oxidation Of the methods available for the preparation of carboxylic acids, two have already been discussed and need only be mentioned here. The oxidation of primary alcohols, aldehydes, alkenes, alkynes, and the side chains of aromatic systems gives rise in all cases to acids. Primary alcohols (or aldehydes) are the most convenient to use because they are no more complex than the acids desired. An alkene, for instance, must contain at least one more carbon atom than any acid that can be obtained from it. Since acids, once formed, are very resistant to the further action of oxidizing agents (resonance again), there are no special requirements as to reagents or conditions for this oxidation.

The oxidation of ketones, although often requiring rather drastic conditions, can be carried out with the production of a mixture of acids.

$$CH_3CH_2COCH_2CH_2CH_2CH_3 \xrightarrow{(O)} \left.\begin{array}{l} CH_3COOH \\ CH_3CH_2COOH \\ CH_3CH_2CH_2COOH \\ CH_3CH_2CH_2CH_2COOH \end{array}\right\} \text{a mixture}$$

Only in the case of cyclic ketones is the reaction useful; here only one product is produced if the ring is symmetrical. The conditions required to produce reaction are essentially the same as in the case of the open-chain ketones; the difference lies in the greater simplicity of the product. The adipic acid used in nylon manufacture may be made in this way.

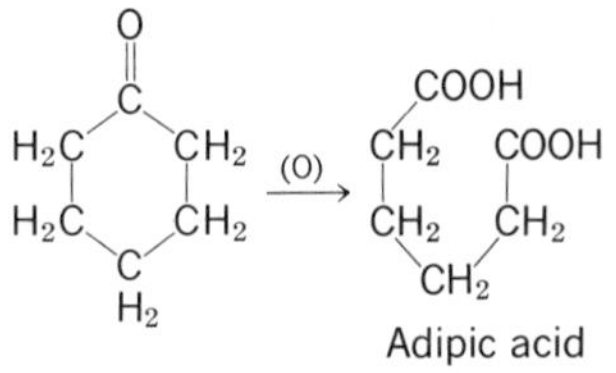

Adipic acid

12.7 Hydrolysis of Cyanides The use of the cyanide group for the production of acids from halides is often an advantageous method since one more carbon is gained. It is not applicable in some cases, such as those in which the halide is tertiary, where elimination reactions may predominate (Sec. 9.7), or those in which the halide is too inert, as on an aromatic ring (Sec. 9.5). In the latter case, a high temperature will produce reaction, and this finds commercial application in special cases.

$$CH_3CH_2Br + K^+CN^- \xrightarrow[\Delta]{} CH_3CH_2CN \xrightarrow[\text{acid, }\Delta]{H_2O} CH_3CH_2COOH + NH_4^+$$

The hydrolysis of most cyanides takes place very slowly unless catalyzed by strong acids such as H_2SO_4; even then prolonged boiling is often required.

12.8 Grignard Method An improvement over the cyanide method, especially on a laboratory scale where expense is less of a factor, uses the Grignard reagent. Grignard reagents add readily to the carbonyl group of CO_2 even at a low temperature. Solid CO_2 (dry ice) is a convenient source of dry CO_2.

$$CH_3CH_2Br \xrightarrow[\text{ether}]{Mg} CH_3CH_2MgBr \xrightarrow{[\overset{\delta-}{O}{=}\overset{\delta+}{C}{=}O]} CH_3CH_2C\begin{matrix}{}^{\nearrow\!\!\!\!\!\!\!=}O \\ \searrow O^- \;{}^+MgBr\end{matrix}$$

The product of this reaction is a magnesium salt of a carboxylic acid. As such it is insoluble in dry ether and precipitates, thus preventing the addition of a second mole of Grignard reagent to the remaining carbonyl. (Because of resonance this carbonyl is rather unreactive in any case.) In order to set the acid free from its magnesium salt, a stronger acid must be used; water alone will not complete this conversion (compare Grignard synthesis of alcohols).

$$CH_3CH_2C\begin{matrix}{}^{\nearrow\!\!\!\!\!\!\!=}O \\ \searrow O^- \;{}^+MgBr\end{matrix} \xrightarrow{HCl} CH_3CH_2COOH + MgBrCl$$

12.9 Hydrolysis of Esters Natural fats and oils are particularly valuable sources of the fatty acids. The acids exist in fats in the form of esters of glycerol; crude fats are mixtures of such esters. In order to hydrolyze fats, an aqueous alkali is used, and the mixture of fat and alkali is boiled. The reaction is driven close to completion because of the formation of the alkali metal salt of the fatty acids present. The reaction is then called "saponification" ("soap making").

$$\begin{array}{l} CH_3(CH_2)_{14}COO{-}CH_2 \\ \qquad\qquad\qquad\quad\;\; | \\ CH_3(CH_2)_{14}COO{-}CH \\ \qquad\qquad\qquad\quad\;\; | \\ CH_3(CH_2)_{14}COO{-}CH_2 \end{array} \xrightarrow[H_2O \atop \Delta]{NaOH} 3CH_3(CH_2)_{14}COONa + \begin{array}{c} CH_2OH \\ | \\ CHOH \\ | \\ CH_2OH \end{array}$$

Sodium palmitate (a soap) — Glycerol (glycerine)

Only those acids which are to be found in animal or plant fats and oils can be obtained practically in this way. Because of the abundance of waste fat in the United States in normal times, this is a useful commercial process. Interestingly enough, except for a few of the lower members of the fatty-acid series, only those acids having an even number of carbon atoms are obtained in quantity from most fats.[1] If odd-numbered fatty acids are desired, they must usually be synthesized.

Glycerol esters of fatty acids which are solid at room temperature are called *fats* (butterfat), and those which are liquid are *oils* (cottonseed oil). Fatty acids are

[1] Exceptions to this classical statement exist, however. By the use of more recently developed and refined methods, branched-chain acids with uneven numbers of carbon atoms are found in tubercle, leprosy, and other bacteria, and in wool fat; 13-, 15-, and 17-carbon acids have been found in butter.

found also in nature linked with alcohols other than glycerol. Esters of fatty acids with monohydroxy long-chain alcohols are *waxes*. Beeswax consists largely of $CH_3(CH_2)_{14}COOC_{30}H_{61}$.

12.10 Reactions: Salt Formation Carboxylic acids, like other acids, are able to combine with bases to form salts.

$$CH_3COOH + OH^- \longrightarrow CH_3COO^- + H_2O$$

Because most organic acids are strong relative to H_2O (Table 12.1), their neutralization by strong bases can be carried out quantitatively and the amount of base utilized by a given acid sample readily determined. A useful quantity called the "neutralization equivalent," which is the same as the equivalent weight, is obtained in this way. A neutralization equivalent, defined as the number of grams of an acid required to neutralize one gram-equivalent of alkali (one liter of one normal solution), is frequently used[1] to characterize carboxylic acids. The equivalent weight is always a simple fraction of, or equal to, the molecular weight of a compound (see Table 12.4).

An examination of the table of ionization constants, Table 12.1, reveals that all carboxylic acids are effectively stronger than carbonic acid. Hence, the salts of carbonic acid will be largely decomposed into CO_2 and H_2O on addition of an organic acid.

$$CH_3COOH + NaHCO_3 \longrightarrow CH_3COONa + CO_2\uparrow + H_2O$$

The evolution of CO_2 from a solution of sodium bicarbonate is a useful test for the presence of a COOH group. Phenols, although stronger acids than water or alcohols, are generally weaker than carbonic acid and do not give the *bicarbonate test*.[2]

TABLE 12.4

Acid	*Neutralization equivalent*	*Molecular weight*
CH_3CH_2COOH	74	74
$CH_3CH(CH_2COOH)COOH$	66	132

[1] It should be remembered that in determining a neutralization equivalent, the proper indicator must be used. With a strong base (OH^-) and a weak (compared to HCl or H_2SO_4) acid equivalence will be reached at a pH greater than 7 and phenolphthalein, which changes color in the region pH 8 to 9, is commonly used for this reason.

[2] By a common-ion effect the H^+ from acetic acid reverses the ionization of carbonic acid.

$$CH_3COOH \rightleftharpoons CH_3COO^- + H^+$$
$$HCO_3^- + H^+ \rightleftharpoons H_2CO_3$$
$$H_2CO_3 \rightleftharpoons CO_2\uparrow + H_2O$$

Because of its instability, carbonic acid cannot exist readily at concentrations given by the third equation, which is determined by the solubility of CO_2. Any acid which generates higher concentrations of H_2CO_3 will release CO_2 from solution.

12.11 Esterification

The formation of esters from alcohols and acids is formally analogous to the formation of salts, although experimentally it is fundamentally different.

$$CH_3COOH + HONa \rightleftharpoons CH_3COONa + HOH$$
$$CH_3COOH + HOCH_3 \rightleftharpoons CH_3COOCH_3 + HOH$$

The difference between these two reactions is the difference in the kinds of bonds being formed and broken. The detailed mechanism of esterification is discussed in Sec. 13.3. The neutralization of acetic acid with sodium hydroxide is a very rapid reaction depending only on the combination of the freely available H^+ and OH^- ions to form water. The reaction with methanol does not appear to involve ionization in the usual sense at all, but is the typically slow incomplete cleavage and formation of covalent bonds found in most organic reactions. As was found to be true for the reverse reaction of hydrolysis (Sec. 8.5), the reaction occurs with rupture of the C—O bond of the acid rather than that of the alcohol.

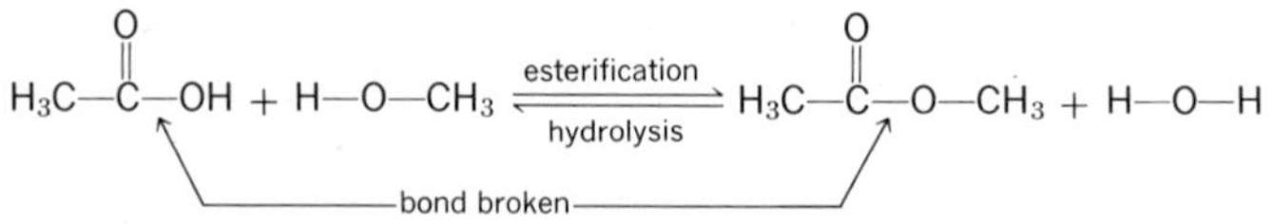

esterification
hydrolysis
bond broken

As will be seen (Sec. 13.3), the reaction is catalyzed by strong acids such as hydrochloric or sulfuric acid. The reaction is, of course, catalyzed by the organic acids themselves, but very much less effectively.

12.12 Formation of Acyl Halides

A reaction which, like the esterification above, involves the OH group of carboxylic acids rather than the easily liberated H^+ is the reaction with phosphorus halides or thionyl chloride.

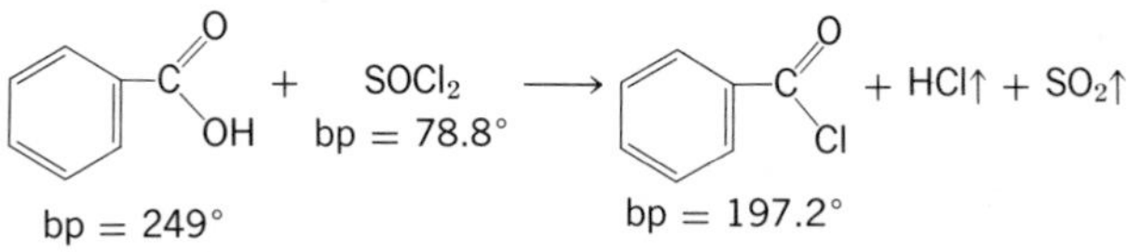

+ SOCl2
bp = 78.8°
+ HCl↑ + SO2↑
bp = 249°
bp = 197.2°

$$\underset{bp = 141°}{CH_3CH_2COOH} + PCl_3 \longrightarrow \underset{bp = 80°}{CH_3CH_2COCl} + \underset{bp = high}{P(OH)_3}$$

$$\underset{bp = very\ high}{CH_3(CH_2)_{16}COOH} + PCl_5 \longrightarrow \underset{bp = high}{CH_3(CH_2)_{16}COCl} + \underset{bp = 107°}{POCl_3} + HCl$$

The first method is usually preferable since the products SO_2 and HCl both are gases and are easily removed from the reaction. The acid chlorides, having no possibility for hydrogen-bridge formation, are not associated and have boiling points much lower than the acids (CH_3COOH, bp = 118°; CH_3COCl, bp = 52°). They are used extensively in the preparation of esters by an irreversible reaction.

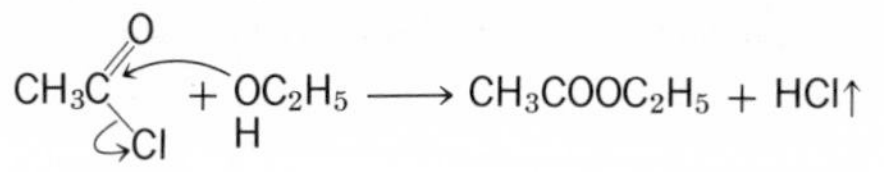

12.13 Formation of Anhydrides Acyl halides may be regarded formally as mixed anhydrides, in which one acid is organic (RCOOH) and the other inorganic (HX) since on hydrolysis these two acids are formed. If the two are carboxylic acids, the corresponding structure is called an acid anhydride.

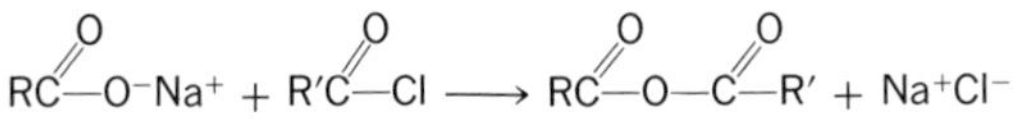

Acid anhydride

These substances react with alcohols, for example, in a manner analogous to the acid chlorides.

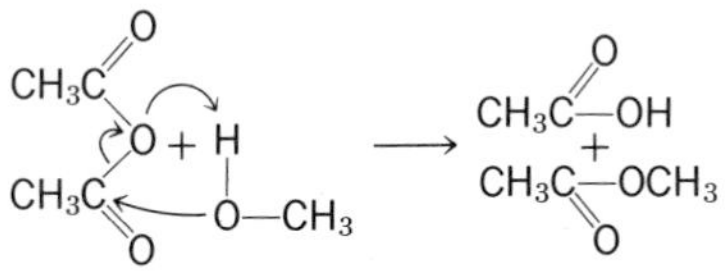

See Sec. 13.3 for a more detailed discussion of the mechanism of these processes.

12.14 Substitution Reactions The carboxyl group itself has much less effect on the α carbon atom than does the carbonyl group of aldehydes and ketones. Thus, the aldol and related condensations fail. It is possible to substitute halogen at the alpha carbon through the agency of the acid halide via an ionic or free radical reaction.

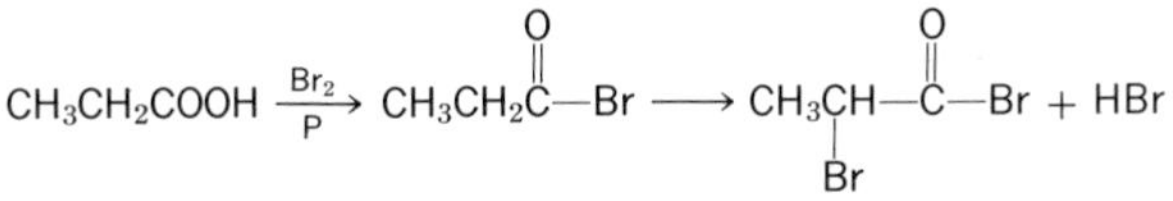

A second halogen can be introduced with much more difficulty.

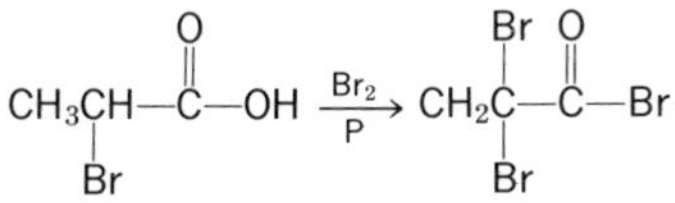

If water is added, the acid halides formed are destroyed and the bromoacid is isolated directly.

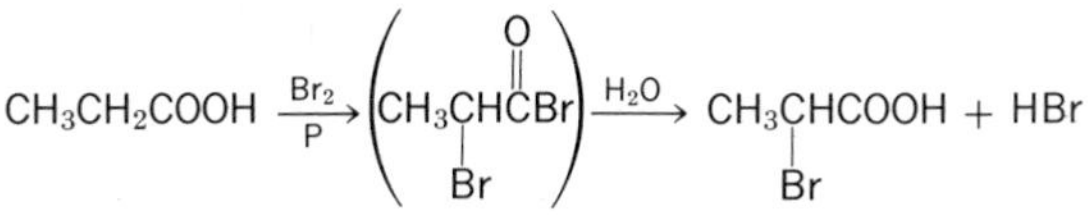

Acids having no α hydrogen do not substitute readily, except benzoic acid and its derivatives which exhibit the normal benzene nucleus reactions.

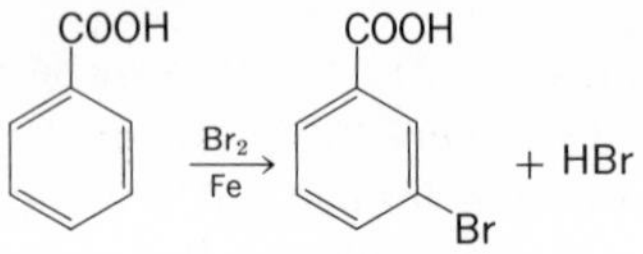

m-Bromobenzoic acid

12.15 Dicarboxylic Acids For the most part, those acids which contain more than one carboxyl group exhibit the same reactions as the monobasic acids. This is particularly true when the two groups are well separated from one another in the molecule. As two carboxyl groups (or for that matter, any two groups) are placed closer together in a carbon chain, they begin to affect one another. The most significant effect is the response of such acids to the application of heat.

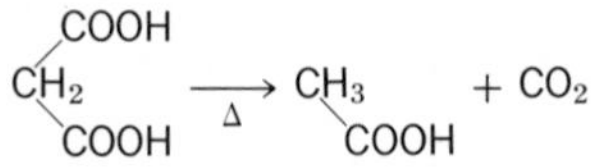

This decomposition represents a case in which the enol of a carboxylic acid is formed, assisted by the resonance energy gained on formation of the enolate and by the irreversible loss of CO_2 (see Sec. 18.5).

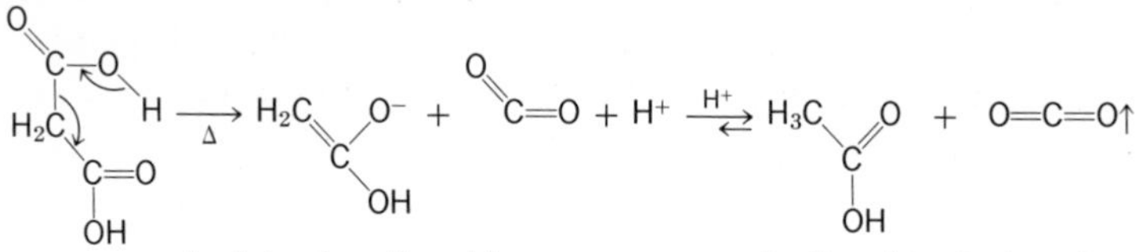

Enolate of acetic acid (resonance stabilized) Acetic acid Carbon dioxide

An example of instability generated by placing two carboxyl groups adjacent to one another is given by oxalic acid which is both thermally and oxidatively unstable.

$$2CO_2 + H_2O \xleftarrow{(O)} \begin{matrix} COOH \\ | \\ COOH \end{matrix} \xrightarrow[\Delta]{} HCOOH + CO_2$$

When two or three carbon atoms are interposed between the two acid groups, a cyclic anhydride forms readily.

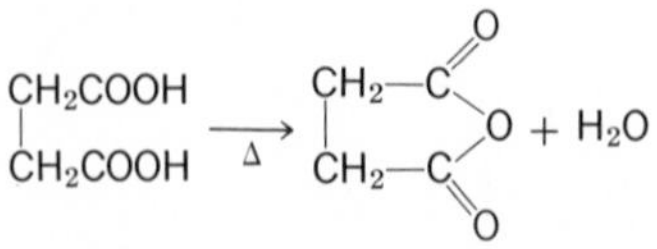

Succinic acid Succinic anhydride

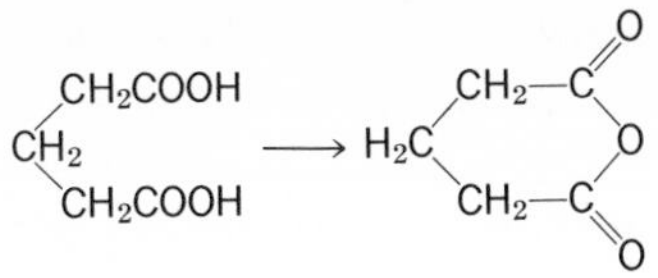

Higher members of this series form polymeric anhydrides or undergo other reactions in preference to anhydride formation.

The reasons for this behavior are brought out by a consideration of the theory of ring formation in carbon compounds (Sec. 4.3). It will be recalled that the preferred valence angle of carbon (and also of oxygen) is such as to make the five- and six-membered rings the most strain-free and most easily formed. We have here another example of the same phenomenon. Malonic acid does not form a stable anhydride since this would have to be a four-membered ring and might be expected to decompose in any event. Succinic and glutaric acids readily cyclize to the five- and six-membered ring anhydrides, while adipic and the higher homologs have their carboxyl groups so far apart that the probability of interaction becomes low.

We observe the same situation in the aromatic series.

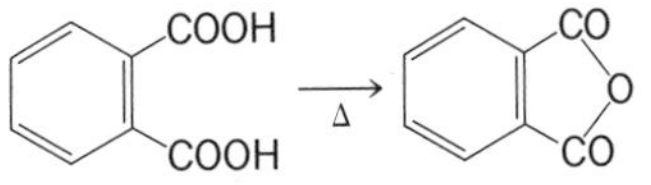

Phthalic acid Phthalic anhydride

This reaction takes place so readily that phthalic anhydride, rather than the acid, is the product obtained in the commercial oxidation of naphthalene in the vapor state. If the carboxyl groups are meta or para to one another, no cyclic anhydride formation occurs.

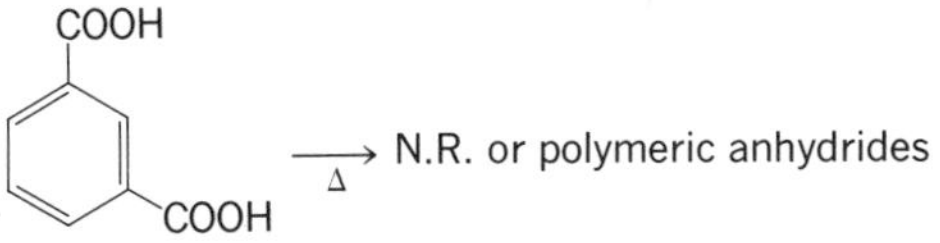

Although in the meta case the anhydride would have only a six-membered ring, the rigid benzene nucleus prevents the two carboxyls from coming close enough to form the cyclic anhydride. This fact is often useful in locating the relative position of substituents on a ring system; the side chains are oxidized to produce acids, and heat is applied to test for presence of any 1,2-dicarboxylic acids.

12.16 Spectroscopic Properties The infrared spectra of the carboxyl group might be expected to show a strong carbonyl band at somewhat higher frequency than ketones plus an OH stretching band in the usual OH regions. Because of the extensive association between these two functions, however, one usually observes broad OH stretching in the region 3300 to 2500 cm^{-1}. The C=O stretch absorbs in the region 1725 to 1700 cm^{-1}.

The OH proton in the nmr often shows up at very low field ($\delta = 10.4$ to 12) and its absorption is dependent on concentration and other exchangeable groups in the molecule. Because of its lability this proton may not show up at all due to rapid exchange with other protons. Often, shaking of a CCl_4 or $CHCl_3$ solution of an acid with D_2O will replace the OH proton by D, removing it from the spectrum.

OUTLINE OF CARBOXYLIC ACID CHEMISTRY

Preparation

1. Oxidation of primary alcohols.

$$RCH_2OH \xrightarrow{(O)} RCOOH$$

Note that an aldehyde is formed as an intermediate product.

2. Hydrolysis of cyanides.

$$RC{\equiv}N \xrightarrow[H_2O,\ \Delta]{NaOH} RCOONa + NH_3$$

$$RC{\equiv}N \xrightarrow[\Delta]{conc.\ HCl} RCOOH + NH_4Cl$$

3. Saponification of fats. Obviously this method can be used to prepare only those acids which occur naturally in fats.

$$\begin{array}{l} RCOOCH_2 \\ \vert \\ RCOOCH \\ \vert \\ RCOOCH_2 \end{array} \xrightarrow[H_2O,\ \Delta]{NaOH} 3RCOONa + CH_2OH{-}CHOH{-}CH_2OH$$

4. For aromatic nuclear acids only.

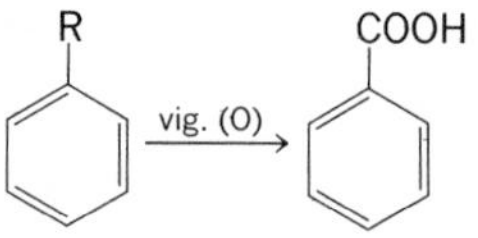
R
vig. (O)
COOH

5. Grignard method.

$$RMgX \xrightarrow{CO_2} RCOOMgX \xrightarrow{HX} RCOOH + MgX_2$$

Properties

Aliphatic and aromatic acids show no differences in properties except in regard to halogenation, nitration, and sulfonation (that is, reactions due to the benzene ring).

1. $$RCOOH + \begin{cases} MOH \\ M_2CO_3 \end{cases} \longrightarrow RCOOM + \begin{cases} H_2O \\ H_2O + CO_2 \end{cases}$$

where M is 1 equivalent of a metal.   Ammonium or substituted ammonium hydroxides also form salts with acids.

2.   Esterification.

$$RCOOH + R'OH \xrightarrow{H^+} RCOOR' + H_2O$$

NOTE: $\phi OH + RCOOH \longrightarrow$ N.R.

3.   Formation of acid (or acyl) chlorides.

$$RCOOH + \begin{cases} PCl_5 \\ PCl_3 \end{cases} \longrightarrow RCOCl + \begin{cases} HCl + POCl_3 \\ H_3PO_3 \end{cases}$$

$$RCOOH + SOCl_2 \longrightarrow RCOCl + HCl\uparrow + SO_2\uparrow$$

4.   Acids are stable toward oxidizing and most reducing agents, unless there is some other oxidizable or reducible group in the molecule.   Formic acid and oxalic acid are the only exceptions.

5.   Halogenation.

   a.   Saturated aliphatic or aromatic acids having one or more hydrogen atoms on the α carbon atom.

$$RCH_2COOH \xrightarrow{Cl_2} \underset{\underset{Cl}{|}}{R}CHCOOH + HCl$$

$$RCH_2COOH \xrightarrow[\underset{(I_2)}{Br_2}]{P} \underset{\underset{Br}{|}}{R}\overset{\overset{O}{\|}}{C}HCBr \xrightarrow{H_2O} \underset{\underset{Br}{|}}{R}CHCOOH$$
$$\qquad\qquad\qquad\qquad (I)\quad\;\; (I)$$
$$\qquad\qquad\qquad (I)$$

   b.

$$+ HCl$$

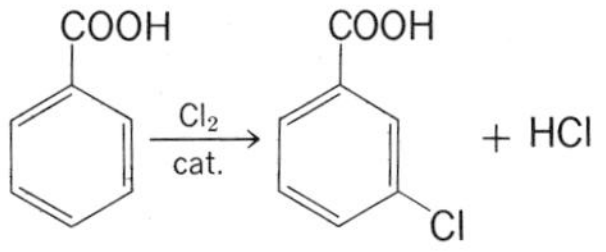

6.   Nitration and sulfonation of nuclear aromatic acids is normal; the carboxyl group is a meta director.

7.   Dibasic acids behave in general as do the monobasic, except as shown below.

$$R_2C\Big\langle\begin{matrix} COOH \\ COOH \end{matrix} \xrightarrow{\Delta} R_2CHCOOH + CO_2$$

$$(CH_2)_{2 \text{ or } 3}\Big\langle\begin{matrix} COOH \\ COOH \end{matrix} \xrightarrow{\Delta} \begin{matrix} CH_2\!-\!\!-\!\!-CO \\ | \qquad\qquad \\ (CH_2)_{1 \text{ or } 2}\!-\!CO \end{matrix}\Big\rangle O + H_2O$$

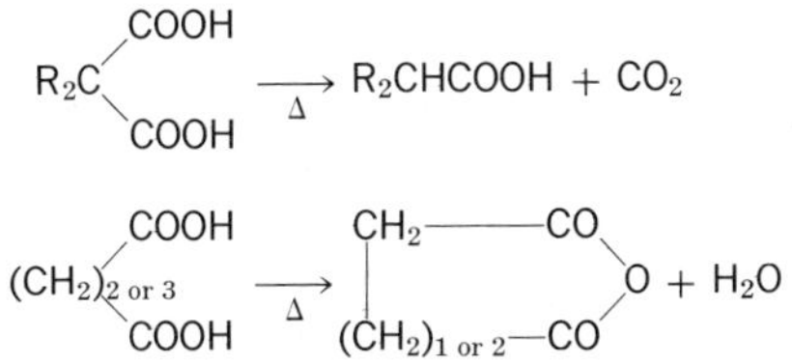

CHAPTER 12 EXERCISES

1. The ionization constant K_a for acids is a measure of the tendency of an acid HA to ionize as follows:

$$\mathrm{HA} \rightleftharpoons \mathrm{H^+ + A^-} \qquad K_a = \frac{[\mathrm{H^+}][\mathrm{A^-}]}{[\mathrm{HA}]}$$

In this expression H^+ is the solvated proton (H_3O^+ in water, $CH_3OH_2^+$ in methanol, etc.) and A^- is the conjugate base of the acid HA (e.g., OH^- for H_2O, $^-C{\equiv}CH$ for $HC{\equiv}CH$, etc.). Below are given approximate ionization constants for a series of acids.

Acid (HA)	K_a	*Acid* (HA)	K_a
Picric acid	4.2×10^{-1}	CH_3CH_2OH	8×10^{-20}
Benzoic acid	6.3×10^{-5}	$HC{\equiv}CH$	10^{-21}
Acetic acid	1.8×10^{-5}	NH_3	10^{-22}
Phenol	1.1×10^{-10}	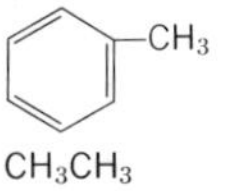 CH3	10^{-37}
H_2O (HOH)	1×10^{-14}	CH_3CH_3	small
	1×10^{-16}		

a. Which is a stronger base, $HC{\equiv}C^-Na^+$ or $CH_3CH_2O^-Na^+$?

b. What would be the approximate equilibrium constant for the reaction

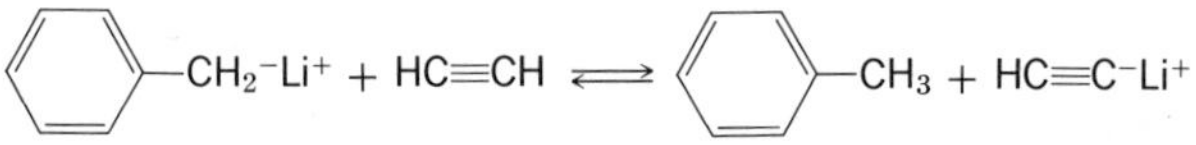
CH2-Li+ + HC≡CH
CH3 + HC≡C-Li+

c. Why cannot one make the following Grignard reagents:
$H_2N{-}CH_2{-}CH_2{-}CH_2{-}MgBr$ and $HC{\equiv}C{-}CH_2{-}CH_2MgCl$?

2. Which of the following statements are *necessary* consequences of the table in Exercise 1?

a. Benzoic acid is a stronger acid than acetylene.

b. Acetylide ion ($HC{\equiv}C^-$) is a stronger base than acetate ion (CH_3COO^-).

c. Acetylene loses a proton less rapidly than benzoic acid (for example, when treated with amide, NH_2^-).

d. Sodium benzoate hydrolyzes slightly in water while sodium acetylide is extensively hydrolyzed in water.

3. Can you apply any of the reasoning given in the discussion of the acid strength of acetic acid to the question of the acid strength of phenol? Of nitric acid? Explain.

4. Are the following statements true or false? Explain.

a. All acids are stable to oxidizing agents.

b. The molecular weight of an organic acid is obtained by titration.

5. Predict the relative acid strengths of the following:

★*a.* C_6H_5—OH and 

b. C_6H_5—C(OH)=$CHCH_2NO_2$ and C_6H_5—C(OH)=CH—NO_2

★*c.* 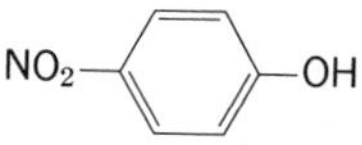 and 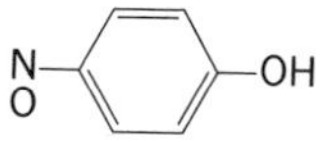

d.

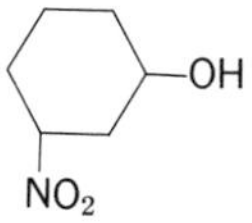 and 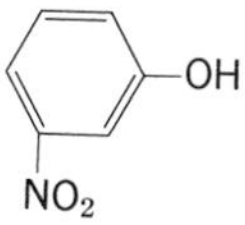

★*e.*

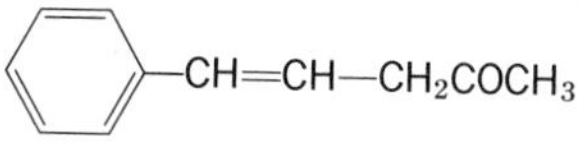 and 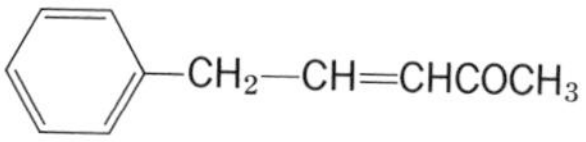

f. CH_3CH_2C(=O)—OH and CH_3C(=O)—CH_2OH

★*g.* 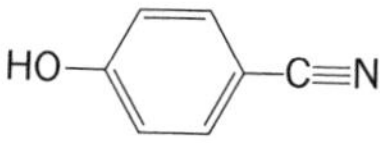 and 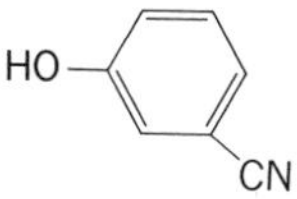

6. Predict relative base strengths of the following:

a. CH_3C(=O)—$CH_2CH_2O^-$ and $CH_3CH_2CH_2$C—O^-
b. CH_3^- and $C_6H_5^-$
c. HO^- and CN^-
d. $CH_3C{\equiv}C^-$ and $CH{\equiv}CCH_2CH_2^-$

e. 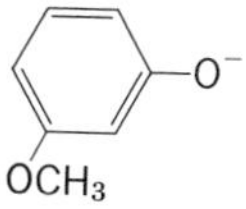 and 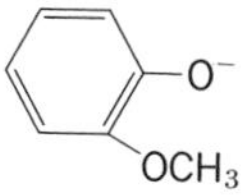

7. Show by equations how you could prepare the following acids:

★*a.* Phenylacetic acid

b. CH_3CH(OH)—CH_2COOH

★*c.* 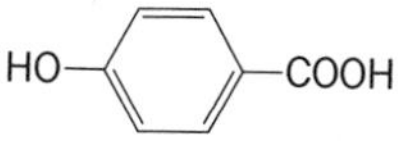

d. $CH_3C{\equiv}C$—COOH

★*e.* 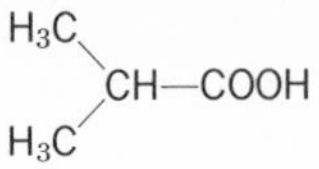

f. $CH_3(CH_2)_{14}COOH$

★*g.* 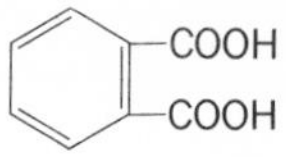

h. Cyclopentanone

★*i.* 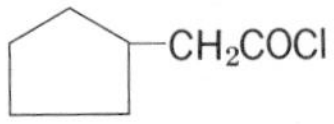

8. What explanation may be offered for the failure of the second carbonyl group in CO_2 to react with a Grignard reagent?

★9. Give the structures of all the substances having the following properties: empirical formula, $C_4H_6O_5$; molecular weight, 134; 0.200 g-equiv to 29.8 ml of 0.1*N* alkali; reaction with acetyl chloride gives $C_6H_8O_6$.

10. A compound A, $C_{14}H_{14}O$, gives, on oxidation, $C_{14}H_{12}O$. If A is treated with dehydrating agents, it loses 1 mole of water, and the product, upon vigorous oxidation, yields 2 moles of benzoic acid. Write the structure of A and give equations to show your reasoning.

★11. Compound A, $C_5H_{10}O$, has infrared maxima at 1650, 3080, 3355, and 3650 cm^{-1} (plus the usual C—H + fingerprint maxima). When A is oxidized it forms a C_4 β-keto acid and CO_2 is given off. Write the structure of A and show your reasoning.

12. A certain acid has a neutralization equivalent equal to 90; it gives no oxime, forms an acetyl derivative, and forms CHI_3 with NaOH and I_2. Deduce a possible structure. Are other structures possible? If so, write them.

★13. A certain hydrocarbon X, C_8H_{14}, is oxidized by $KMnO_4$ to yield an acid Y, $C_8H_{14}O_4$, with a neutralization equivalent of 87. When Y is heated alone there is no reaction; if heated with $Ca(OH)_2$ there is formed a neutral substance Z, $C_7H_{12}O$. Z shows an absorption in the infrared at 1745 cm^{-1} but no maxima above 3000 cm^{-1} and none in the region 2650 to 2880 cm^{-1}, and it reacts with hydroxylamine. Both X and Z, by appropriate methods, could be separated into a meso and a racemic pair of isomers. What is a structure for X? Show your reasoning.

14. A certain acidic substance A, $C_9H_{14}O_5$, has a neutralization equivalent of 202. When heated with NaOH solution and acidified, it forms B, $C_8H_{12}O_5$, which has a neutralization equivalent of only 94. B does not react with acetic anhydride, but with NaOH and I_2 yields iodoform and C, $C_7H_7O_6Na_3$. When C is treated with dilute acid, isolated, and heated dry, it forms D, $C_6H_{10}O_4$. D, when vaporized over ThO_2, forms a ketone, C_5H_8O. Deduce structures for the lettered compounds. Give your reasoning.

15. Show by equations how the following conversions could be accomplished:

★*a.* Benzoic acid ⟶ toluene

b. Propionic acid ⟶ pentane

★*c.* Propionic acid ⟶ $CH_3CH_2CH_2COOCH_2CH_2CH_3$

d. Benzoic acid ⟶ 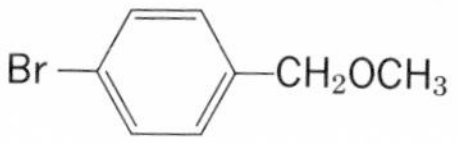

★*e.* $HOOC(CH_2)_4COOH$ ⟶ 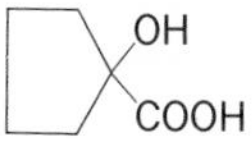

f. $(CH_3CO)_2O$ ⟶ 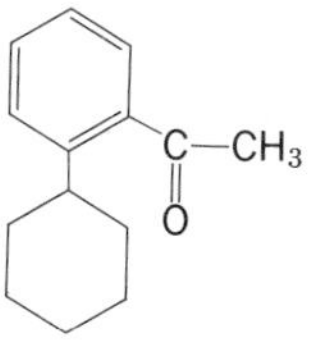

★*g.* Benzoic acid ⟶ 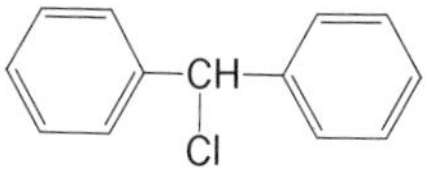

h. Acetic acid ⟶ $CH_3CH(OH)—CH_2CH(=O)$

16. Show how the spectra (nmr and/or ir) of the following pairs would be expected to differ:

a. $CH_3CH_2CH_2COOH$ and $CH_3CH{=}CHCOOH$

b. CH_3CH_2COOH and $CH_3C(=O)—O—CH_3$

c. C_6H_5COOH and $C_6H_5CH_2COOH$

d. $HCOOH$ and CH_3COOH

★17. How many stereoisomers, geometric and optical, would be expected in 1,2-cyclobutane dicarboxylic acid? Some of these form cyclic anhydrides. Which?

★18. *S*-$C_6H_5CH(CH_3)—COOH$ is racemized in base but *R*-$C_6H_5CH(CH_3)CH_2CH_2COOH$ is not. Explain.

★19. Suggest structures for compounds with the following properties: $C_4H_8O_3$; nmr absorption at 1.3δ (3H), at 3.7δ (2H), at 4.1δ (2H not split), and at 10.9δ (1H).

20. Taking into consideration the information so far available about the effects of alkyl and carbonyl groups in influencing adjacent atoms, explain whether you would expect *tert*-butyl chloride and acetyl chloride to react by the same mechanism with the same reagents.

13

CARBOXYLIC ACID DERIVATIVES: ESTERS, HALIDES, ANHYDRIDES, AMIDES, AND NITRILES

13.1 Definition A derivative of a substance is generally defined as a product prepared from, or convertible into, the substance itself by a simple operation, such as hydrolysis, esterification, substitution, etc. The derivatives of the carboxylic acids are a group of compounds all of which, on hydrolysis, can be converted to carboxylic acids. They are named after the acids from which they are derived in the manner shown in Table 13.1 for the case of acetic acid.

13.2 Relationships Although at first glance this may seem a rather heterogeneous collection of compounds formally related in a simple way to the acids, a more careful examination reveals a close relationship of a fundamental nature. All contain a carbonyl group except the last, and this contains the somewhat

TABLE 13.1
DERIVATIVES OF ACETIC ACID

Derivative	*Formula*	*Name*
Ester	$CH_3\overset{\overset{O}{\|}}{C}-OC_2H_5$	Ethyl acetate
Acyl chloride	$CH_3\overset{\overset{O}{\|}}{C}-Cl$	Acetyl chloride (ethanoyl chloride)
Acyl anhydride	$CH_3\overset{\overset{O}{\|}}{C}-O\overset{\overset{O}{\|}}{C}CH_3$	Acetic anhydride
Amide	$CH_3\overset{\overset{O}{\|}}{C}-NH_2$	Acetic amide (acetamide)
Nitrile	$CH_3C{\equiv}N$	Acetonitrile (methyl cyanide)

analogous $-C{\equiv}N$ group. These compounds contain carbonyl or carbonyl-like groups of the same oxidation state,[1] and it is this feature which most distinguishes them from such molecules as aldehydes and ketones.

These acid derivatives possess an electronegative substituent which can exist as a more or less stable anion, or which can be protonated to form a good "leaving group."[2] Accordingly, these molecules can undergo a displacement reaction of the addition-elimination type.

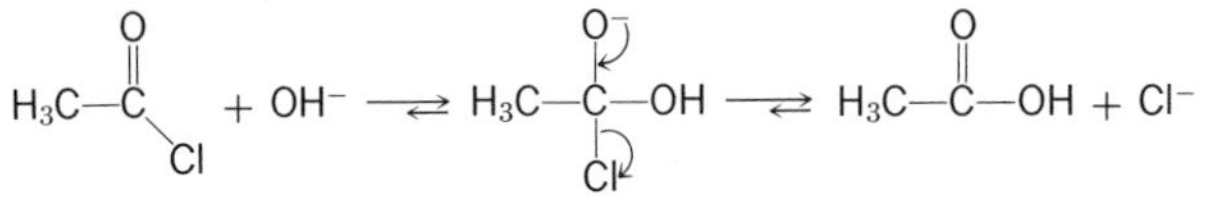

13.3 Mechanisms of the Displacement Reactions of Acid Derivatives In general, the displacement reaction is a little more complicated than shown above in that acid and/or base catalysis involving several proton transfers may be involved.

---

[1]In organic chemistry the usual concepts of oxidation state developed in elementary chemistry are difficult to apply and it is customary to use a modified definition as follows: *Count one unit or degree of oxidation for each multiple bond or for each bond to an element more electronegative than carbon.* In the case of carbon-carbon multiple bonds the oxidation is shared between bonded atoms (i.e., one-half unit each). In the case of multiple bonds to an electronegative atom, each bond counts one unit. Thus, the following groups are at the same oxidation levels:

(1) $H_2C{=}CH_2$ (one-half unit/C atom), H_3C-CH_2Br; (2) $H_3C-\overset{\overset{O}{\|}}{C}-H$, $H_2C{=}CH-OH$, $HC{\equiv}CH$;

(3) $H_3C-\overset{\overset{O}{/\!/}}{C}-OCH_3$, $H_2C{=}C(OCH_3)_2$; (4) $H_3C-\overset{\overset{O}{\diagdown}}{C}-OCH_3$, $(CH_3O)_4C$, CCl_4.

[2]Nitriles may be said to possess a "potential electronegative substituent" on addition of a nucleophile such as OH^-. See Sec. 13.14.

Several of these reactions shall be considered in some detail so as to indicate how they may proceed, because one can better understand the role of catalysis in these reactions in this way. In many cases the formulation given will not be the only one or necessarily the best one.

1. "Uncatalyzed" addition:

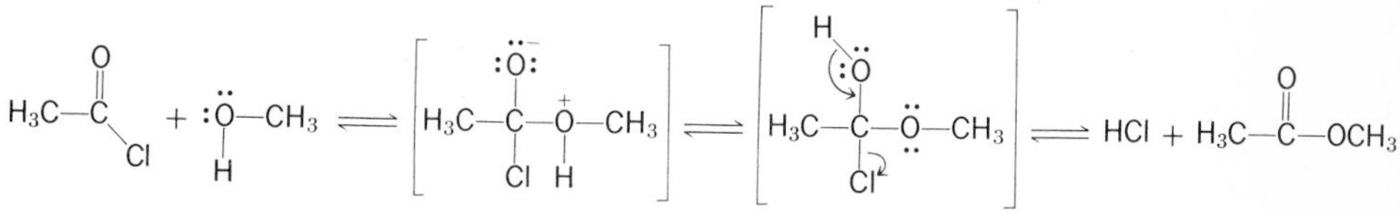

Overall reaction:

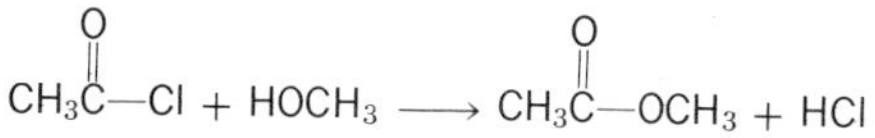

2. Acid catalyzed:

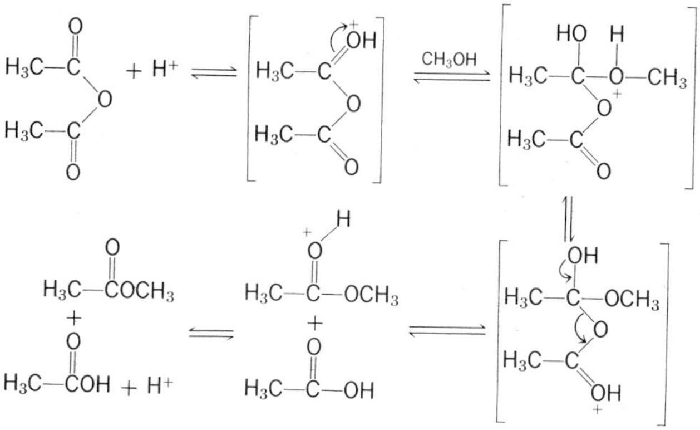

Overall reaction:

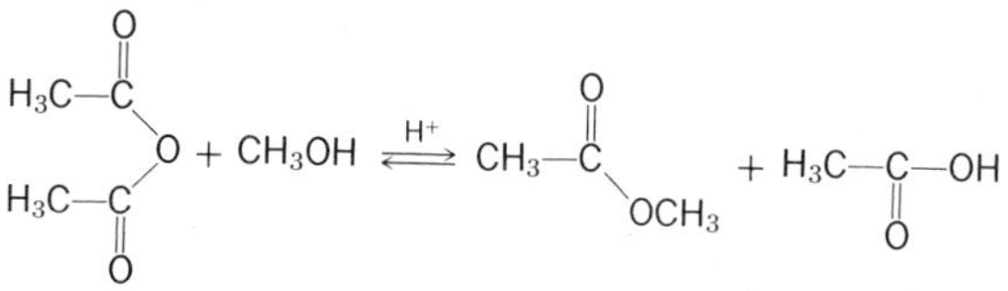

3. Base catalyzed (see Sec. 8.5):

$$B\colon + CH_3OH \rightleftharpoons CH_3O^- + \overset{+}{B}H$$

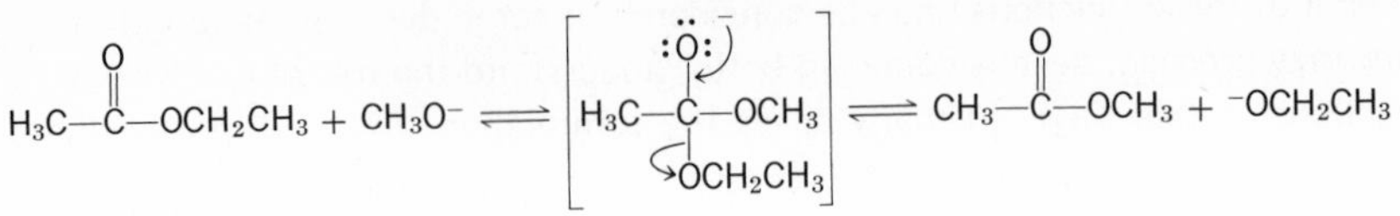

$$^{-}OCH_2CH_3 + BH^+ \rightleftharpoons HOCH_2CH_3 + B:$$

Overall reaction:

$$CH_3OH + H_3C{-}C({=}O){-}OCH_2CH_3 \xrightleftharpoons{B:} CH_3CH_2OH + H_3C{-}C({=}O){-}OCH_3$$

The overall mechanism may be generalized

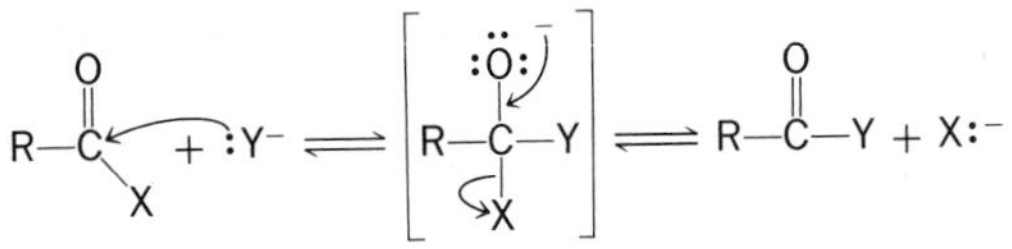

where it is realized that catalysis and proton transfers and one or more intermediates are involved. In principle, all of these reactions are reversible, but in practice, the equilibrium in some cases lies so far in one direction as to make the reaction essentially irreversible.

To a first approximation the equilibrium will favor the acid derivative which has the poorest leaving group. That is, R—$CONH_2$ will be favored over R—COOR, and so on with R—COCl being the least favored. On this basis carboxylic acids fall between esters and amides. All of these considerations can be affected by concentration (Le Chatellier principle) and by secondary ionization. Thus the hydrolysis of an amide can be driven to completion as follows:

$$H_2O + R{-}C({=}O){-}NH_2 \rightleftharpoons R{-}C({=}O){-}OH + NH_3 \qquad \text{Equilibrium lies to left}$$

$$R{-}C({=}O){-}OH + OH^- \rightleftharpoons R{-}C({=}O){-}O^- + H_2O \qquad \text{Equilibrium lies far to right}$$

$$R{-}C({=}O){-}NH_2 + OH^- \rightleftharpoons R{-}C({=}O){-}O^- + NH_3 \qquad \text{Overall equation}$$

The following classes of reaction are important in the chemistry of acid derivatives.

Hydrolysis:

$$R{-}C({=}O){-}X + H_2O \xrightarrow{H^+ \text{ or } OH^-} R{-}C({=}O){-}OH({-}O^-) + HX(X^-)$$

where $X = Cl, O\!-\!\overset{O}{\overset{\|}{C}}\!-\!R$, or NH_2 (for $R\!-\!C\!\equiv\!N$ see Sec. 13.14).

$$R\!-\!C\!\equiv\!N + H_2O \xrightarrow{H^+ \text{ or } OH^-} R\!-\!\overset{O}{\overset{\|}{C}}\!-\!NH_2 \longrightarrow R\!-\!\overset{O}{\overset{\|}{C}}\!-\!OH(\!-\!O^-) + HX(X^-)$$

Ammonolysis:

$$R\!-\!\overset{O}{\overset{\|}{C}}\!-\!X + NH_3 \longrightarrow R\!-\!\overset{O}{\overset{\|}{C}}\!-\!NH_2 + HX$$

where $X = Cl, -O\!-\!\overset{O}{\overset{\|}{C}}\!-\!R, -OR$, or NHR.

Alcoholysis:

$$R\!-\!\overset{O}{\overset{\|}{C}}\!-\!X + HOR' \xrightarrow{\text{cat.}} R\!-\!\overset{O}{\overset{\|}{C}}\!-\!OR + HX$$

where $X = -Cl, -O\!-\!\overset{O}{\overset{\|}{C}}\!-\!R, -OR''$, or $-NH_2$.

13.4 Carbonyl Group Reactivity The "displacement" reaction of the above carbonyl groups represents a type of reaction not possible for aldehydes and ketones. A number of carbonyl reactions of ketones and aldehydes take a different course with carboxylic acid derivatives. Thus, hydrazones, oximes, etc., are not formed with these derivatives, but a simple displacement occurs.

Ketone:

$$R_2C\!=\!O + \underset{\text{Hydroxylamine}}{H_2NOH} \longrightarrow R_2C\!=\!N\!\diagup^{OH} + H_2O$$

Ester:

$$R\overset{O}{\overset{\|}{C}}\!-\!OR' + H_2NOH \longrightarrow \underset{\text{Hydroxamic acid}}{R\!-\!\overset{O}{\overset{\|}{C}}NHOH} + HOR'$$

Certain reagents, such as the Grignard reagent, do add to the carbonyl group of carboxylic acid derivatives much as they add to ketones and aldehydes.

$$H_3C\!-\!\overset{O}{\overset{\|}{C}}\!\diagdown_{OCH_2CH_3} + CH_3MgBr \longrightarrow \left[H_3C\!-\!\overset{O^-MgBr^+}{\overset{|}{\underset{\underset{OCH_2CH_3}{|}}{C}}}\!-\!CH_3 \right]$$

This adduct is not stable, but undergoes an elimination, followed by addition of a second molecule of Grignard reagent.

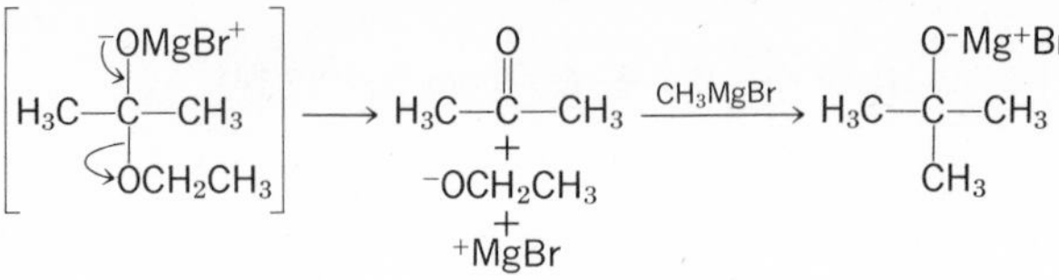

$$\left[\begin{array}{c} \text{OMgBr}^+ \\ | \\ \text{H}_3\text{C}-\overset{|}{\underset{|}{\text{C}}}-\text{CH}_3 \\ | \\ \text{OCH}_2\text{CH}_3 \end{array} \right] \longrightarrow \underset{\overset{+}{-}\text{OCH}_2\text{CH}_3}{\overset{\text{O}}{\underset{\underset{+\text{MgBr}}{+}}{\text{H}_3\text{C}-\overset{\|}{\text{C}}-\text{CH}_3}}} \xrightarrow{\text{CH}_3\text{MgBr}} \underset{\text{CH}_3}{\overset{\text{O}^-\text{Mg}^+\text{Br}}{\text{H}_3\text{C}-\overset{|}{\underset{|}{\text{C}}}-\text{CH}_3}}$$

It is not usually possible to stop this reaction halfway. If one uses only one equivalent of CH_3MgBr, half of the ester is unchanged and the rest is converted to the final product. By working at low temperatures and adding the Grignard reagent to the ester, one can obtain the ketone in some cases, but the method is not generally practical. One can use organocadmium compounds to reduce the reactivity of the organometallic reagent

$$2\text{CH}_3\text{MgCl} + \text{CdCl}_2 \longrightarrow (\text{CH}_3)_2\text{Cd} + \text{MgCl}_2$$

$$(\text{CH}_3)_2\text{Cd} + \text{CH}_3\text{CH}_2\overset{\overset{\text{O}}{\|}}{\text{C}}\text{Cl} \longrightarrow \text{CH}_3\text{CH}_2\underset{\text{CH}_3}{\overset{\overset{\text{O}}{\|}}{\text{C}}} \xrightarrow{(\text{CH}_3)_2\text{Cd}} \text{N.R.}$$

This reaction is not feasible with esters, anhydrides, or amides.

On the above basis it is possible to obtain a qualitative ranking of the various carbonyl derivatives according to their tendency to undergo addition reactions (not considering subsequent steps).

$$\overset{\overset{\text{O}}{\|}}{\text{RCH}} > \overset{\overset{\text{O}}{\|}}{\text{RCCl}} > \text{R}_2\text{C}{=}\text{O} > \overset{\overset{\text{O}}{\|}}{\text{RC}}-\text{O}-\overset{\overset{\text{O}}{\|}}{\text{CR}} > \text{R}-\overset{\overset{\text{O}}{\|}}{\text{C}}-\text{OR}' > \overset{\overset{\text{O}}{\|}}{\text{RCNH}_2} > \text{RC}{\equiv}\text{N}$$

One might expect electronegative substituents to *enhance* the carbonyl group reactivity rather than the *reduction* in reactivity observed above. This effect is attributable to resonance of the type shown below, which tends to *reduce* the positive charge on carbon:

$$\text{R}-\overset{\overset{:\text{O}:}{\|}}{\text{C}}-\overset{..}{\underset{..}{\text{Cl}}}: \longleftrightarrow \text{RC}{=}\overset{:\overset{..}{\text{O}}:^-}{\underset{..}{\text{Cl}}}^+ \qquad \text{R}-\overset{\overset{:\text{O}:}{\|}}{\text{C}}-\overset{..}{\underset{..}{\text{O}}}-\overset{\overset{\text{O}}{\|}}{\text{C}}-\text{R} \longleftrightarrow \text{R}-\text{C}{=}\overset{:\overset{..}{\text{O}}:^-}{\underset{+}{\text{O}}}-\overset{\overset{\text{O}}{\|}}{\text{C}}-\text{R}$$

$$\text{R}-\overset{\overset{:\text{O}:}{\|}}{\text{C}}-\overset{..}{\underset{..}{\text{O}}}-\text{R}' \longleftrightarrow \text{R}-\text{C}{=}\overset{:\overset{..}{\text{O}}:^-}{\underset{+}{\text{O}}}-\text{R}' \qquad \text{R}-\overset{\overset{:\text{O}:}{\|}}{\text{C}}-\overset{..}{\text{N}}\text{H}_2 \longleftrightarrow \text{R}-\text{C}{=}\overset{:\overset{..}{\text{O}}:^-}{\underset{+}{\text{N}}}\text{H}_2$$

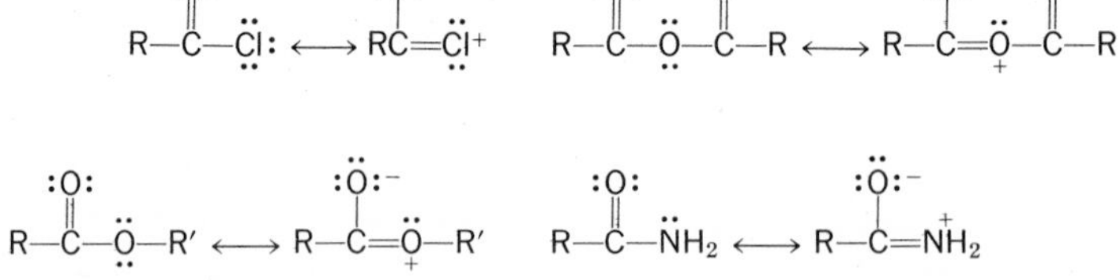

This effect is most pronounced in amides and least in acid chlorides.

Although acetic acid derivatives have been used in the illustrative equations, the student should realize that the same reactions take place with the derivatives of other acids, monobasic and dibasic, aromatic and aliphatic. Even inorganic acids often exhibit analogous behavior, although the mechanisms are frequently different.

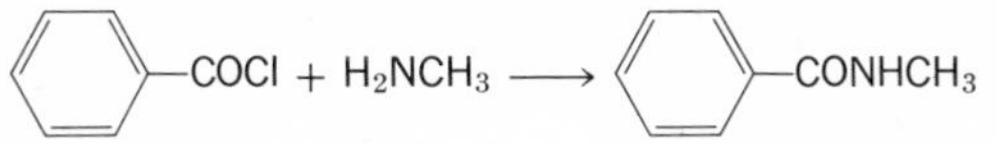

Benzoyl chloride — *N*-Methyl benzamide (a substituted amide)

$$HOSO_2Cl + HOC_2H_5 \longrightarrow HOSO_2OC_2H_5 + HCl$$

Chlorosulfonic acid (an acid chloride of sulfuric acid)

$$CH_2(COOCH_3)_2 + 2NaOH \longrightarrow CH_2(COO^-Na^+)_2 + 2CH_3OH$$

Methyl malonate

ESTERS

Esters are named for both the acid and the alcohol parts that are present in the molecule. The alcohol is given its group name (Sec. 2.3) and the acid takes the suffix *-ate,* which usually implies a link through oxygen. See Table 13.1 for an example.

13.5 Preparation The *preparation* of esters by alcoholysis has already been discussed in connection with the reactions in Sec. 13.3. Esters can also be prepared by a direct reaction between acids and alcohols.

Overall reaction

$$CH_3\overset{O}{\overset{\|}{C}}\text{—}OH + HOC_2H_5 \overset{H^+}{\rightleftharpoons} CH_3\overset{O}{\overset{\|}{C}}OC_2H_5 + H_2O$$

Esterification reaction

The reaction is catalyzed by hydrogen ion. The usual procedure is to use an excess of one of the reactants and catalytic amounts of strong acids (such as sulfuric). The reaction goes to an equilibrium which is not completely to the right unless provisions are made to remove the water as it is formed. The mechanism of the reaction is similar to other acid-catalyzed carbonyl addition reactions.

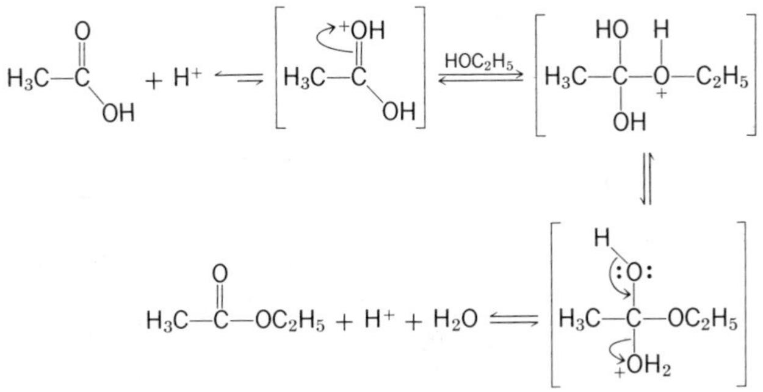

HOC₂H₅

Here it should be noted that the first step *involves the protonation of the already acidic carboxyl group.* This step is very slow in the absence of a strong mineral acid, although the reaction will proceed very slowly catalyzed by the protons derived from other molecules of the carboxylic acid.

Base catalysis is ineffective because of the conversion of the carboxyl group to the carboxylate ion which shifts the equilibrium very far to the left; addition to the resonance-stabilized anion is very slow. Advantage is taken of this fact in the hydrolysis of esters. With basic catalysis the equilibrium is driven completely to one side by the use of more than one equivalent of base.

$$R{-}\overset{\overset{\displaystyle O}{\|}}{C}{-}OR' + \underset{\text{excess}}{NaOH} \xrightarrow[\Delta]{H_2O} R{-}\overset{\overset{\displaystyle O}{\|}}{C}{-}O^-Na^+ + HOR'$$

13.6 Properties While the fatty acids often have disagreeable odors, the esters, both natural and synthetic, are commonly used as flavoring agents. Butyric acid, for example, has an odor like that of rancid fat or sweat, while the odor of valeric acid is reminiscent of manure. But methyl butyrate has a pineapple-like odor, and isoamyl valerate has the odor of apples. A number of esters of this sort actually occur in flowers and fruits and impart the flavor and aroma to them. Perhaps the largest scale industrial application of the esters, however, is due to their properties as solvents. In the paint, lacquer, and plastics industries they are very widely used. Amyl acetate (banana oil) is a common ingredient of lacquers and finishes.

The saponification of esters is a reaction of considerable significance in organic chemistry.

$$CH_3CH_2\overset{O}{\overset{/\!/}{C}}OCH_3 + NaOH \longrightarrow CH_3CH_2COONa + HOCH_3$$

$$\begin{array}{l} CH_2\overset{O}{\overset{/\!/}{C}}OC_2H_5 \\ | \\ CH_2\overset{O}{\overset{/\!/}{C}}OC_2H_5 \end{array} + 2NaOH \longrightarrow \begin{array}{l} CH_2COONa \\ | \\ CH_2COONa \end{array} + 2HOC_2H_5$$

This reaction is irreversible and can be carried out quantitatively. It will be observed that one equivalent of base is consumed by each ester group; the equation is reminiscent of the equation for the neutralization of an acid by a base. There is an important fundamental difference, however.

Neutralization of an acid by a base is an ionic reaction and takes place almost instantly, making it possible to titrate acids conveniently and accurately (see definition of neutralization equivalent, Sec. 12.10). The saponification equivalent, defined as the number of grams of ester reacting with one equivalent of a base, may be determined with accuracy only by an indirect method. A weighed sample of ester is boiled under reflux with an accurately measured excess of alkali to bring the saponification to completion in a reasonable time. The excess of base remaining after completion of the reaction is then measured by titration with a standardized solution of an acid, and the amount consumed determined by difference.

A saponification equivalent thus determined is a simple fraction of, or is equal to, the molecular weight of the ester and is a useful guide in its identification.[1]

A commercial application of the reaction is, as the name implies, in the manufacture of soap. The natural fats, esters of glycerin, called *glycerides* of fatty acids, are saponified on a large scale with sodium hydroxide to yield solutions of a mixture of fatty-acid sodium salts which constitute the ordinary variety of "soap."

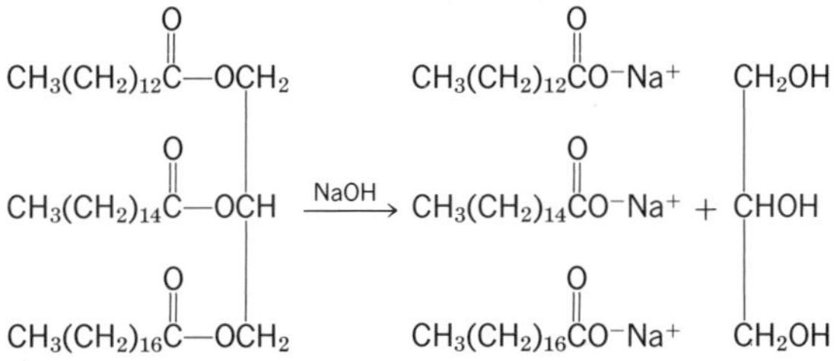

CH3(CH2)12C—OCH2
CH3(CH2)14C—OCH
CH3(CH2)16C—OCH2
NaOH
CH3(CH2)12CO−Na+
CH3(CH2)14CO−Na+
CH3(CH2)16CO−Na+
CH2OH
CHOH
CH2OH

A "mixed" glyceride — Sodium soaps — Glycerol

13.7 Surface Activity It was pointed out in Sec. 10.6 that the solubility of a substance is influenced by the balance of polar and nonpolar groups in the molecule. This is particularly evident with the fatty acids and their derivatives, for the polarity of the carboxylic acid group can be decreased by ester formation (and esters are less soluble in water and more soluble in organic solvents than the parent acids) or increased by salt formation (and the salts of the alkali metals are more soluble in water and less soluble in organic solvents than the parent acids). When the hydrocarbon chain of the acid or derivative is of the order of 11 to 17 carbon atoms in length, solubility behaviors appear which are of the greatest importance and which deserve discussion at this point.

A fatty acid such as lauric acid, $CH_3(CH_2)_{10}COOH$, is virtually insoluble in water. Despite the strong attraction of the —COOH group for water, the insolubility of the hydrocarbon chain is sufficient to prevent solution of the acid. If, however, a small amount of the acid is placed on a clean water surface, either as a tiny crystal or as a drop of dilute solution in some volatile solvent, it spreads out over the surface, with the carboxylic acid groups in the water and the hydrocarbon parts of the molecules floating on the surface or projecting up into the air. This behavior was described in Sec. 8.14. A cross section of such a surface is pictured in Fig. 8-1.

When the fatty acid is converted into a sodium or potassium salt, the polarity and hydrophilic ("water-loving") nature of the end group is greatly increased. The

[1] In the analysis of plant and animal fats and oils which are largely esters, an arbitrary *saponification number,* or *saponification value,* is often used instead of the more logical saponification equivalent. The saponification number is defined as the number of milligrams of potassium hydroxide required to saponify one gram of fat. Other similar arbitrary constants are also used in the fat industry. For example, the *iodine number,* or the number of grams of iodine combining with one hundred grams of fat, is used as a measure of the amount of unsaturation in a given sample.

sodium laurate, $CH_3(CH_2)_{10}COONa$, now becomes "soluble." At least it passes into solution. It is found on examination of such a solution that the sodium laurate molecules are not present to any great extent as single molecules (that is, they are not in true solution) but that actually they are dissolved in the form of aggregates, tiny particles which may give the soap solution an opalescent appearance because they are of the size to scatter light. These particles, or *micelles,* are illustrated in cross section in Fig. 13-1*a*. The *hydrophobic* ("water-hating") hydrocarbon ends of the molecule are inward, and presented to the water are the carboxylate ends. Some of these molecules are ionized, which gives the micelle a charge, making it a giant ion, and this charge prevents the micelles, which repel each other, from coalescing and precipitating. These solutions feel "soapy"; they can be made to foam and form bubbles; they have a lubricating action and a *detergent* (cleaning) action. This behavior is characteristic of compounds whose molecules have a polar group attached to a fairly long (but not too long) hydrocarbon chain. The salts of the fatty acids from about C_{12} to C_{18} are "soaps." The term *syndet* (*syn*thetic *det*ergent) is usually used for substances other than the soaps having similar properties: such substances, for example, as sulfates of fatty alcohols,

$$CH_3(CH_2)_{10}CH_2OSO_2ONa$$

When air is forced into soapy solutions, the air bubbles are immediately coated with an oriented layer of molecules which stabilize them, so that a *foam* is produced. If a bubble is blown with soap solution, layers of molecules stabilize the outside and the inside of the film (Fig. 13-1*b*). When dirty hands are washed in soapy water, the soap molecules go to the surface of the oily, water-insoluble dirt particles and coat them, the hydrocarbon ends of the molecules dissolving in the oil. Then when the hands are rubbed (an essential part of washing), the oil is rubbed off in small droplets. These immediately become coated and stabilized by the soap molecules, and then because they are small and charged, they remain suspended in the water long enough to be washed down the drain (Fig. 13-1*c*). A little grit in the soap helps the breaking up and emulsification of the dirt.

Soaps and detergents are classed as *surface-active* substances because their molecules are able to cover surfaces rapidly with more or less oriented layers of molecules (monolayers). Many of these substances are *wetting agents,* because with their polar-nonpolar nature, they are able to form monolayers which give a transition between polar and nonpolar surfaces: hydrocarbon oils are not wetted by water, but if a little soap is in the water, wetting becomes possible because the soap molecules line up at the surface (the *interface*) between the oil and water (polar group in the water, nonpolar in the oil) and so bind the two phases together. Wetting is evidence of interaction between the wetting liquid and the surface wetted; wetting agents promote this interaction.

Surface-active agents are sometimes added to lubricants; the polar group fastens to the metal surface to be lubricated—it may even form a chemical bond with it, in some cases—and the hydrocarbon end serves to anchor down a film of

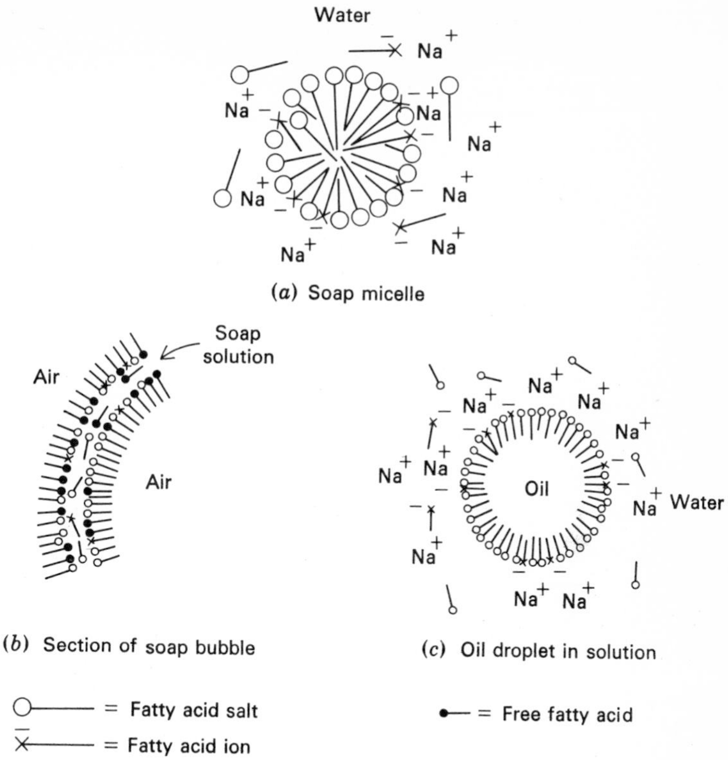
Water
Na+
Na+
Na+
Na+
Na+
Na+
Na+
Na+
(a) Soap micelle
Air
Soap solution
Air
(b) Section of soap bubble
Na+
Oil
Water
(c) Oil droplet in solution
= Fatty acid salt
= Fatty acid ion
= Free fatty acid

figure 13-1
Surface activity of soaps and oils in water, shown in cross section.

the lubricating oil, holding it between the metal pieces to be lubricated. The agent makes the oil wet the metal better.

This discussion is given at some length because it illustrates with examples from everyday experience how many common phenomena may be given a reasonable explanation on the basis of molecular structure. The whole matter has been greatly simplified here. Studies of these phenomena belong to the field of colloid chemistry.

13.8 Hydrogenolysis of Esters The commercial preparation of a type of detergent illustrates another property of esters. Although carboxylic acids resist the action of reducing agents, the esters do not.[1]

$$CH_3\overset{\overset{O}{\|}}{C}OCH_3 \xrightarrow[\text{Ni cat.}]{H_2} CH_3CH_2OH + CH_3OH$$

[1]Esters can be reduced using sodium and alcohol as the source of hydrogen. This reaction is known as the method of Bouveault and Blanc. Lithium aluminum hydride (Sec. 8.6) also reduces esters and fats, smoothly and at low temperature, to alcohols.

When this reagent is applied to a typical fat, the following reaction occurs:

$$\begin{array}{l} CH_3(CH_2)_{14}COOCH_2 \\ \quad\quad\quad\quad\quad\quad | \\ CH_3(CH_2)_{12}COOCH \\ \quad\quad\quad\quad\quad\quad | \\ CH_3(CH_2)_{16}COOCH_2 \end{array} \xrightarrow[\text{Ni cat.}]{H_2} \begin{array}{c} CH_3(CH_2)_{14}CH_2OH \\ + \\ CH_3(CH_2)_{12}CH_2OH \\ + \\ CH_3(CH_2)_{16}CH_2OH \end{array} + \begin{array}{l} CH_2OH \\ | \\ CHOH \\ | \\ CH_2OH \end{array}$$

The mixture of long-chain alcohols so obtained can then be separated by fractional distillation and the various members converted by reaction with sulfuric acid into detergents.

$$CH_3(CH_2)_xCH_2OH + H_2SO_4 \rightleftharpoons CH_3(CH_2)_xCH_2OSO_2OH + H_2O$$
$$CH_3(CH_2)_xCH_2OSO_2OH + NaOH \longrightarrow CH_3(CH_2)_xCH_2OSO_2O^-Na^+ + H_2O$$

Note that sodium hydroxide is used to neutralize the sulfuric acid ester in a rapid ionic reaction without causing hydrolysis (a much slower reaction involving covalent bonds).

Calcium and magnesium salts of fatty acids are very insoluble in water, so that when soap is used in water containing these ions ("hard" water), there is formed a precipitate of the calcium and magnesium soaps. After all the hardness is removed from the water, the soap can carry out its function, but a good deal of soap may be used up in "softening" the water, and the usually curdy precipitate can be a nuisance. Some other detergents such as the alcohol sulfates (for example, sodium lauryl sulfate) do not form insoluble salts with calcium and magnesium, and so these find much use as "soapless soaps" for washing in hard water. Also the parent acids are stronger, and so the sodium salts are less hydrolyzed. Their solutions in water are, therefore, more nearly neutral than soap solutions, and for this reason they find favor in delicate washings.

This hydrogenolysis of fats is not to be confused with the simpler hydrogenation of vegetable oils used in the manufacture of shortening (Crisco, Spry, etc.) and oleomargarine. Vegetable oils (cottonseed oil, for example) contain, besides the usual saturated acids, rather large amounts of unsaturated fatty acids esterified with glycerin. A common constituent is oleic acid, $CH_3(CH_2)_7CH{=}CH(CH_2)_7COOH$. These glycerides of unsaturated fatty acids have lower melting points than those derived from the corresponding saturated acids. The addition of hydrogen to the double bonds of such glycerides converts them to the familiar lardlike solid fats often preferred in cooking. As would be expected, the saturated fats are far more stable than those containing unsaturation: they become rancid through oxidation less easily.

ACYL HALIDES

13.9 Preparation Of the acyl halides, only the chlorides need be considered here since they are the most easily prepared and are exceedingly reactive. Their methods of preparation are based on the replacement of the hydroxyl of a carboxylic acid by halogen. This cannot be accomplished with hydrochloric acid; apparently the equilibrium in this "esterification" is too unfavorable to permit such a preparation, probably because the chloride ion is not sufficiently nucleophilic (basic) to displace hydroxide ion. Only the more vigorous PCl_3, PCl_5, and $SOCl_2$ will react in the desired sense.

$$3CH_3\overset{\overset{\displaystyle O}{\|}}{C}{-}OH + PCl_3 \longrightarrow 3CH_3\overset{\overset{\displaystyle O}{\|}}{C}{-}Cl + H_3PO_3$$

$$CH_3\overset{\overset{\displaystyle O}{\|}}{C}{-}OH + PCl_5 \longrightarrow CH_3\overset{\overset{\displaystyle O}{\|}}{C}{-}Cl + HCl + POCl_3$$

$$CH_3CH_2C(=O)OH + SOCl_2 \longrightarrow CH_3CH_2C(=O)Cl + SO_2\uparrow + HCl\uparrow$$

The mechanism here involves formation of mixed anhydrides such as in many

$$H_3C{-}\overset{\overset{\displaystyle O}{\|}}{C}{-}OH + SOCl_2 \longrightarrow H_3C{-}\overset{\overset{\displaystyle O}{\|}}{C}{-}\underbrace{O{-}SOCl}_{\text{Leaving group}} + HCl$$

cases by displacement on the inorganic P or S atom. The resulting inorganic derivative is more readily displaced by chloride ion.

$$H_3C{-}\overset{\overset{\displaystyle O}{\|}}{C}{-}OSOCl + Cl^- \longrightarrow H_3C{-}\overset{\overset{\displaystyle O}{\|}}{C}{-}Cl + SO_2Cl^-$$

13.10 Properties The hydrolysis, alcoholysis, and ammonolysis of acid chlorides take place very rapidly, in many cases with considerable violence.

$$CH_3\overset{\overset{\displaystyle O}{\|}}{C}Cl + HOH \longrightarrow CH_3COOH + HCl + \text{heat and noise (a hiss)}$$

$$CH_3\overset{\overset{\displaystyle O}{\|}}{C}Cl + C_2H_5OH \longrightarrow CH_3COOC_2H_5 + HCl + \text{heat and noise}$$

$$CH_3\overset{\overset{\displaystyle O}{\|}}{C}Cl + 2NH_3 \longrightarrow CH_3CONH_2 + NH_4Cl + \text{heat, noise, and white clouds of } NH_4Cl$$

The reaction between acid chlorides and alkylated (not acylated) ammonia also takes place readily as long as at least one hydrogen is found on nitrogen. Acylation reduces the basicity of nitrogen so that diacylation does not occur.

$$CH_3\overset{\overset{\displaystyle O}{\|}}{C}Cl + 2CH_3NH_2 \longrightarrow CH_3\overset{\overset{\displaystyle O}{\|}}{C}\underset{\underset{\displaystyle H}{|}}{N}CH_3 + CH_3NH_3^+Cl^-$$

$$CH_3\overset{O}{\overset{\|}{C}}NHCH_3 + CH_3\overset{O}{\overset{\|}{C}}Cl \longrightarrow N.R.$$

Acyl chlorides react, as would be predicted, with the salts of weak acids (nucleophilic reagents). The reactions are again carbonyl addition reactions.

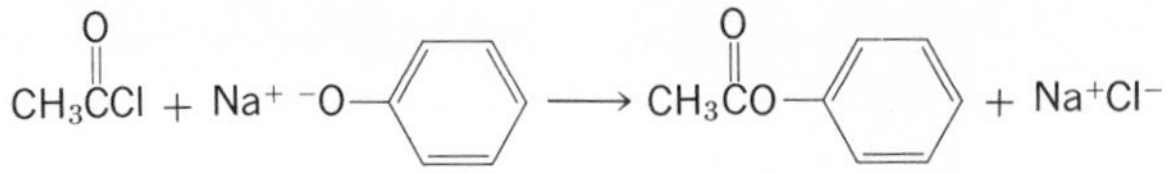

Phenyl acetate

$$CH_3\overset{O}{\overset{\|}{C}}Cl + Na^+ \; ^-O\overset{O}{\overset{\|}{C}}CH_3 \longrightarrow CH_3\overset{O}{\overset{\|}{C}}-O-\overset{O}{\overset{\|}{C}}CH_3$$

Acetic anhydride

The reactivity of acyl halides is sometimes said to be due to a reactive halogen atom. A little consideration, however, brings us to the conclusion that the halogen in acetyl chloride, say, should be *less* reactive than that of methyl chloride in reactions depending on chloride-ion formation. The electron-attracting carbonyl group would not be expected to favor release of the chlorine with its electron pair. This is indeed the case, and all acid chloride reactions seem to depend on initial carbonyl additions as already described.

The presence of groups capable of releasing electrons to the carbonyl carbon, for example, the phenyl group, decreases the reactivity of the acid chloride. Insolubility in water coupled with reduced reactivity makes possible the reaction of many acid chlorides ($\phi COCl$, $CH_3C_6H_4SO_2Cl$) with amines in the presence of aqueous sodium hydroxide. This is called the "Schotten-Baumann reaction."

ACID ANHYDRIDES

13.11 Preparation Acid anhydrides are commonly made by the reaction of acyl halides with the sodium salts of acids (Sec. 12.13); only symmetrical anhydrides are of interest here, although mixed types can be prepared by this method.

The acid anhydrides bear a close resemblance to the acid chlorides in almost all respects. Although they cannot be prepared in this simple fashion, acyl chlorides are really mixed anhydrides of carboxylic acids and hydrogen chloride.

$$CH_3\overset{O}{\overset{\|}{C}}-OH + HCl \not\longrightarrow CH_3\overset{O}{\overset{\|}{C}}-Cl \qquad \text{(not a feasible reaction)}$$

Compare

$$CH_3\overset{O}{\overset{\|}{C}}-OH + HO\overset{O}{\overset{\|}{C}}CH_3 \xrightarrow[\Delta]{\text{cat.}} CH_3\overset{O}{\overset{\|}{C}}-O-\overset{O}{\overset{\|}{C}}CH_3$$

Only in the case of certain dibasic acids (Sec. 12.15) do acid anhydrides form readily when the acids alone are heated. Usually a dehydrating agent is required.

Anhydrides react with the same reagents as the corresponding chlorides, less vigorously in most cases.

$$CH_3C\underset{OCOCH_3}{\overset{O}{\diagup}} + HOH \xrightarrow{H^+} CH_3\overset{O}{\overset{\|}{C}}OH + CH_3COOH$$

$$CH_3C\underset{OCOCH_3}{\overset{O}{\diagup}} + HOCH_3 \longrightarrow CH_3\overset{O}{\overset{\|}{C}}OCH_3 + CH_3COOH$$

$$CH_3C\underset{OCOCH_3}{\overset{O}{\diagup}} + HNH_2 \longrightarrow CH_3\overset{O}{\overset{\|}{C}}NH_2 + CH_3COOH$$

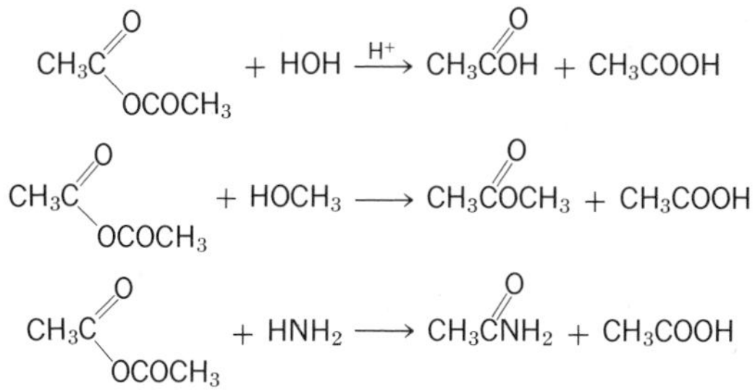

Acyl chlorides and anhydrides are widely used as acylating reagents, that is, for the introduction of acyl groups into various molecules. It should be emphasized that in all their reactions it is the group

$$R-C\overset{O}{\diagdown}$$

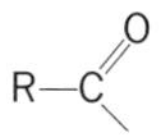

that is "transferred." The transfer takes place to oxygen in ester formation, to nitrogen in amide formation, and to carbon in the Friedel-Crafts reaction.

THE NITROGEN-CONTAINING ACID DERIVATIVES: AMIDES AND NITRILES

13.12 Preparation of Nitriles   The two types of acid derivatives that contain nitrogen will be considered together here. The aliphatic nitriles, or cyanides, are usually prepared by the reaction of halides with alkali cyanides. The reaction is less successful with secondary halides and unsuccessful with tertiary halides. The latter tend to form alkenes by side reactions under the influence of basic medium.

$$Na^+N\equiv C^- \qquad \overset{H}{\underset{CH_3}{\overset{|}{\underset{|}{C}}}}\overset{H}{\diagup}-Br \longrightarrow N\equiv C-\overset{H}{\underset{CH_3}{\overset{|}{\underset{|}{C}}}}\overset{H}{\diagup} + Na^+Br^-$$

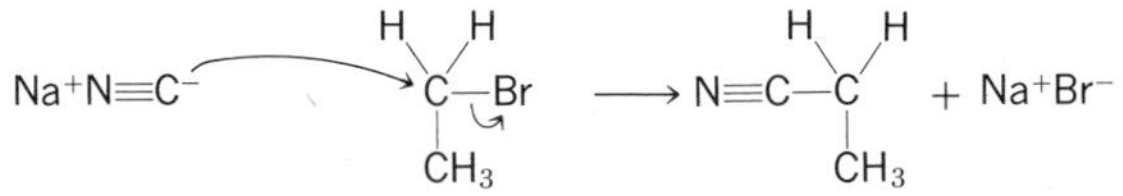

Sodium cyanide   Ethyl bromide   Ethyl cyanide

The use of special solvents, such as dimethylformamide and dimethysulfoxide, makes this a very practical reaction. Since aromatic halides are too unreactive to undergo displacement, the nitriles are usually prepared from the acids (Sec. 13.13) or the amines (Sec. 14.17).

13.13 Properties   The reactions of the cyanide group are in many respects like those of the carbonyl group; it acts as an electrophilic reagent, and additions take place on the carbon atom. With Grignard reagents a typical addition to the electrophilic carbon atom takes place, leading eventually to the formation of ketones.

$$\overset{\delta^+}{H_5C_2}-\overset{\delta^-}{C}\equiv N$$

$$H_3C-MgBr \longrightarrow \left[C_2H_5\overset{..}{C}::\overset{..}{N}^- \right] MgBr^+$$
$$\overset{\delta^-}{} \quad \overset{\delta^+}{} \qquad\qquad\qquad CH_3$$

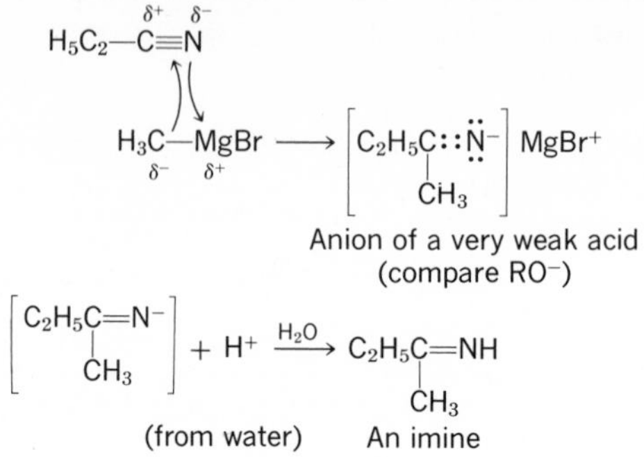

Anion of a very weak acid
(compare RO^-)

$$\left[\overset{C_2H_5C=N^-}{\underset{CH_3}{|}} \right] + H^+ \xrightarrow{H_2O} \overset{C_2H_5C=NH}{\underset{CH_3}{|}}$$

(from water) An imine

The imine is readily hydrolyzed by the action of water or dilute acids to form a ketone.

$$\overset{C_2H_5C=NH}{\underset{CH_3}{|}} \xrightarrow{H_2O} \overset{C_2H_5C=O}{\underset{CH_3}{|}} + NH_3$$

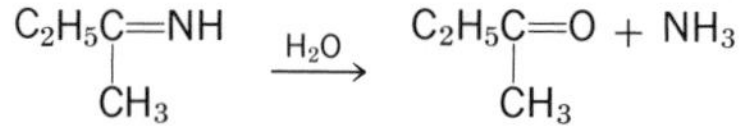

Water adds to nitriles in a similar fashion; here the subsequent reactions are somewhat more complex.

The reaction occurs readily only in the presence of strong acid or strong base. In either case the corresponding amide is an intermediate which can be isolated if desired.

$$H_3C-C\equiv N + H_2O \xrightarrow{H^+ \text{ or } OH^-} H_3C-\overset{O}{\overset{||}{C}}-NH_2 \xrightarrow[H^+ \text{ or } OH^-]{H_2O,} H_3C-\overset{O}{\overset{||}{C}}-OH + NH_3 \text{ or } H_3C-\overset{O}{\overset{||}{C}}-O^- + NH_4^+$$

$$\underset{P_2O_5}{\overline{\qquad\qquad\qquad\qquad\qquad}}$$

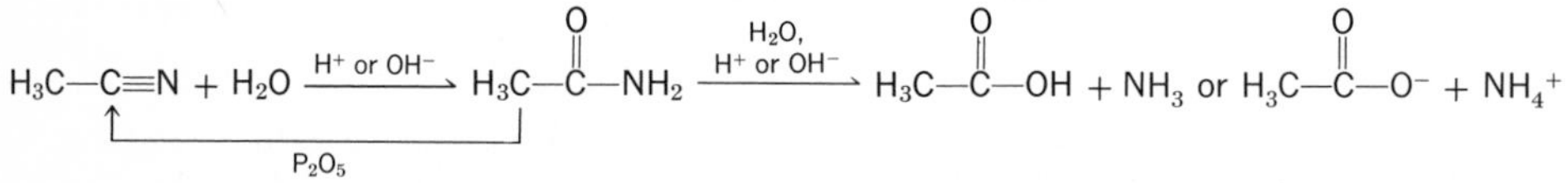

The reaction can be readily reversed by the use of a strong dehydrating agent such as P_2O_5. The mechanism of the hydrolysis (shown for base catalysis) is:

$$H_3C-C\equiv N \longleftrightarrow H_3C-\overset{+}{C}=\overset{..}{N}^- \xrightarrow{OH^-} \left[H_3C-\underset{OH}{\overset{..}{\underset{|}{C}}}=\overset{..}{N}^- \right] \xrightarrow{H_2O} \left[H_3C-\underset{OH}{\overset{|}{C}}=NH \right] \rightleftharpoons H_3C-\overset{O}{\overset{||}{C}}-NH_2$$

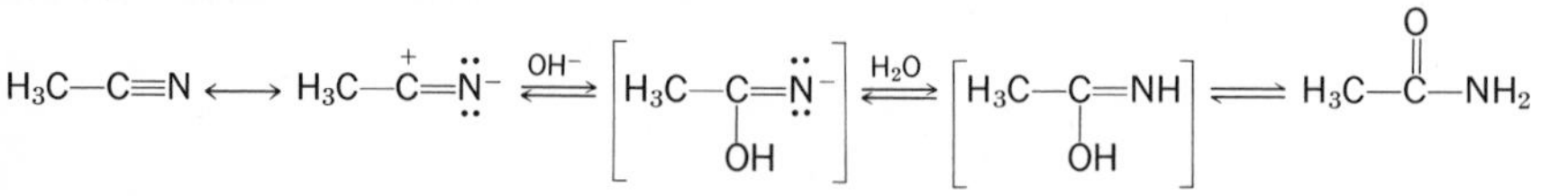

The intermediate enol

$$CH_3C\overset{OH}{\underset{NH}{\big\langle}}$$

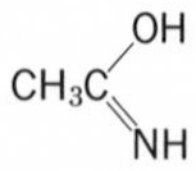

rearranges, as do many compounds having a hydroxyl group on a doubly bonded carbon atom, resulting in the formation of an amide.

As a rule, the reaction can be controlled to give reasonable yields of the intermediate amide or carried through to completion with the formation of the acid.

The process is catalyzed by acids in a manner that cannot be discussed profitably here. As shown in the equation, the separate steps in the hydrolysis can be reversed by heating with dehydrating agents.

Hydrogen may also be added at the triple bond, the reaction being formally analogous to the hydrogenation of aldehydes.

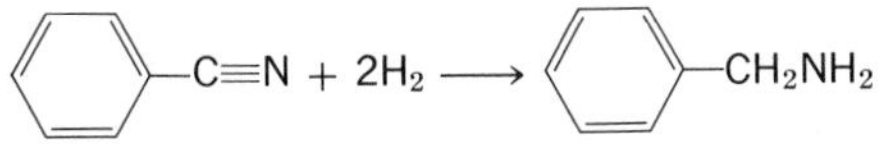

—C≡N + 2H2 ⟶
—CH2NH2

This conversion is effected very smoothly by $LiAlH_4$.

It should be pointed out that in this, as in the other addition reactions (including hydrolysis above), the nitriles react much more slowly than the other acid derivatives discussed previously. Prolonged boiling with acids, for example, is required to effect complete hydrolysis.

13.14 Preparation of Amides Perhaps the most generally used amide synthesis is the reaction of esters, acyl chlorides, or anhydrides with ammonia.

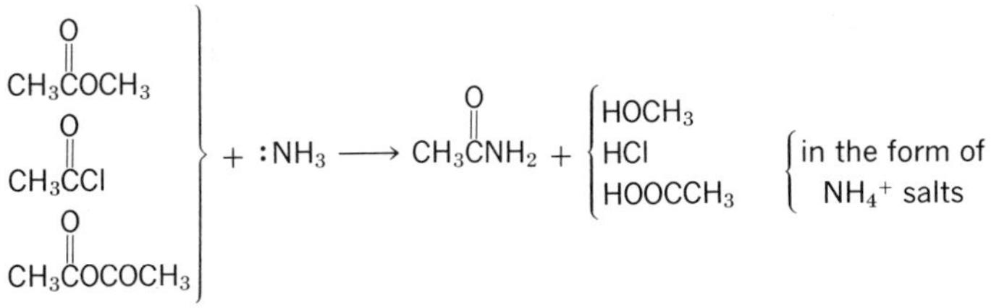

CH3COCH3
CH3CCl
CH3COCOCH3
+ :NH3 ⟶ CH3CNH2 +
HOCH3
HCl
HOOCCH3
in the form of
NH4+ salts

Because of the effect of the carbonyl group on the nitrogen atom of amides, these substances are not nucleophilic and have little tendency to combine with a second molecule of an acid derivative. The carbonyl group tends to draw the odd pair of electrons of the nitrogen toward itself, thus diminishing their availability for further reaction, and at the same time, reducing the reactivity of the carbonyl group.

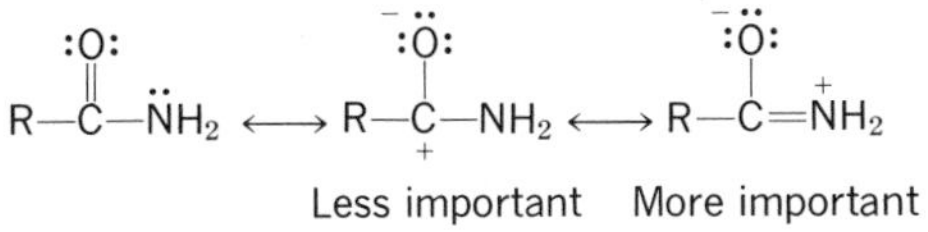

Less important
More important

Thus, amides do not exhibit the basicity or strong nucleophilicity characteristic of amines ($R{-}NH_2$, $R{-}NRH$, etc.).

$$R{-}\overset{O}{\overset{\|}{C}}{-}NH_2 + R{-}\overset{O}{\overset{\|}{C}}{-}Cl \longrightarrow \text{N.R.}$$

When amides of appropriate dibasic acids are heated, imides, or inner amides, are formed in a manner analogous to the formation of inner anhydrides. Compounds in which two acyl groups are attached to nitrogen are called imides.

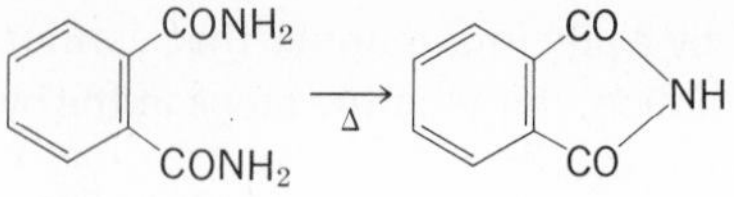

Phthalamide Phthalimide

The Baeyer strain theory applies here as an aid in predicting the ease of formation of imides. Imides react much as the amides do.

The hydrolysis of amides and imides referred to above is catalyzed by either acids or bases but is very slow in pure water.

$$CH_3C(=O){-}NH_2 \xrightarrow[HCl]{H_2O} CH_3C(=O){-}OH + NH_4Cl$$

$$CH_3CH_2C(=O){-}NH_2 \xrightarrow[NaOH]{H_2O} CH_3CH_2COONa + NH_3\uparrow$$

In the second case ammonia is evolved; the detection of ammonia when an unknown substance is heated with alkali is often used as a laboratory test for the amide group. The study of this group is particularly important in the chemistry of proteins.

13.15 Hydrogen Bonding Amides share the relatively high melting and boiling points of carboxylic acids. This is, in large part, due to hydrogen bonding interactions between the carbonyl and the NH group.

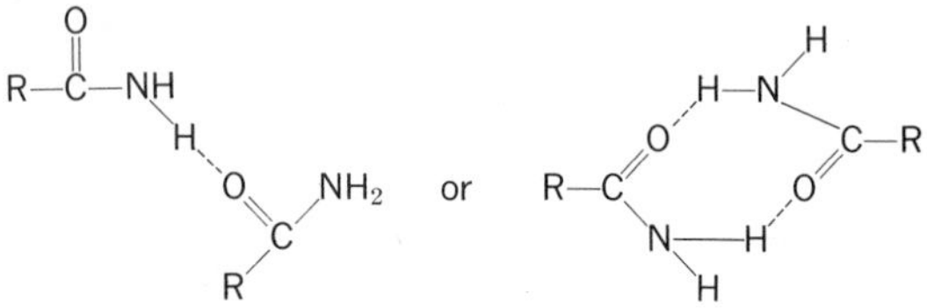

This property is of vital importance in living organisms and is responsible for many of the properties of proteins, which are special types of polyamides.

13.16 Hofmann Rearrangement Occasionally a reaction is found in which the products obtained are different from those expected of a simple replacement or cleavage. One such reaction of acid amides, the Hofmann rearrangement, has been extensively studied and deserves consideration here. When an amide is treated with sodium hypobromite, the following series of reactions has been found to take place:

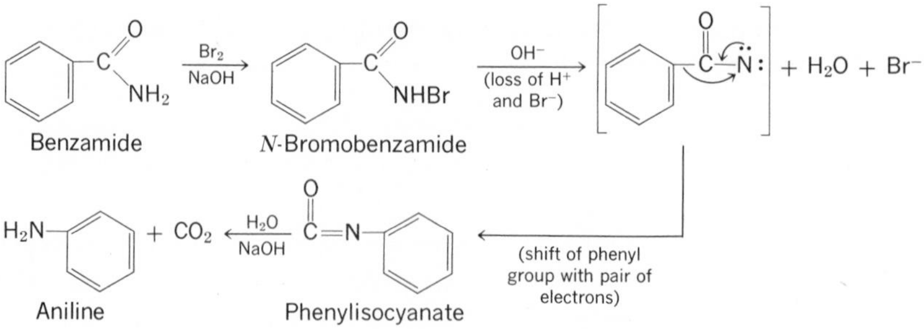
Br2
NaOH
OH−
(loss of H+ and Br−)
+ H2O + Br−
Benzamide
N-Bromobenzamide
+ CO2
H2O
NaOH
(shift of phenyl group with pair of electrons)
Aniline
Phenylisocyanate

The first phase is a substitution of a hydrogen atom on nitrogen by bromide to yield an *N*-bromoamide; substances having this structure have been isolated from the reaction mixture. Under the action of the alkali present, H^+ and Br^- are eliminated to form an unstable intermediate, represented here by the formulation in brackets, in which the nitrogen atom is lacking one pair of electrons. The nitrogen atom gains the needed pair of electrons from the adjacent unsaturated carbon atom *together with the group* attached to the carbon atom (in this case a phenyl group). The rearranged product, an isocyanate, is usually not isolated but allowed to hydrolyze in alkaline solution to form the amine as shown. In the isocyanate we find another example of the cleavage by hydrolysis of carbon-to-nitrogen multiple bonds. This is a rather general behavior.

The importance of the Hofmann rearrangement lies not only in the curious nature of the reaction, but in the fact that in this way one carbon atom is removed from the molecule with the formation of a pure amine.

$$\underset{\text{Propionamide}}{CH_3CH_2\overset{\overset{\displaystyle O}{\|}}{C}NH_2} \xrightarrow[NaOH]{Br_2} \underset{\text{Ethylamine}}{CH_3CH_2NH_2} + Na_2CO_3$$

One of the most compelling demonstrations of the intervention of an intermediate in a given reaction sequence is the generation of the same intermediate from different starting materials, and demonstration of its similar behavior. The Curtius rearrangement complements the Hofmann rearrangement in this respect. Acyl azides can be formed from acid chlorides, anhydrides, or esters by an acylation reaction.

$$R{-}\overset{\overset{\displaystyle O}{\|}}{C}{-}Cl + \underset{\text{Sodium azide}}{NaN_3} \longrightarrow \underset{\text{An acyl azide}}{R{-}\overset{\overset{\displaystyle O}{\|}}{C}{-}N_3} + NaCl$$

Acyl azides are derivatives of the very unstable hydrazoic acid, HN_3; they are best represented as resonance hybrids.

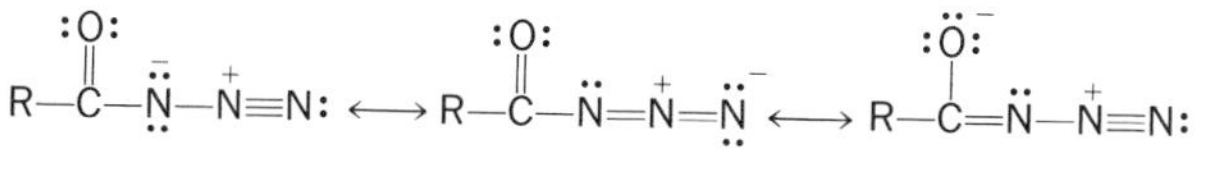

On heating to about 100°C these azides lose a molecule of the exceptionally stable diatomic molecule, N_2 (nitrogen gas), leaving behind the electron-deficient intermediate of the Hofmann rearrangement.

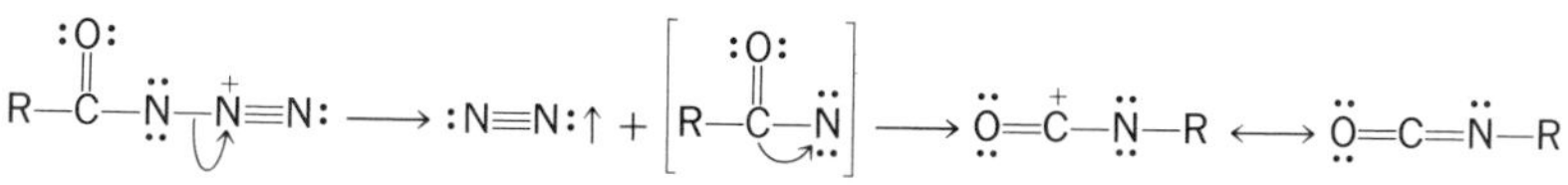

Intermediate An isocyanate

Since the Curtius rearrangement need not be run in aqueous media, the isocyanate can be isolated if so desired (Sec. 16.4).

13.17 Analogous Rearrangements There are known a number of reactions in which rearrangements analogous to the Hofmann rearrangement occur. Although a detailed study of these is beyond the scope of an elementary text, it is instructive to compare a few of them to the Hofmann rearrangement.

The dehydration of ditertiary 1,2-diols (*pinacols*), often called the pinacol rearrangement, occurs as follows:

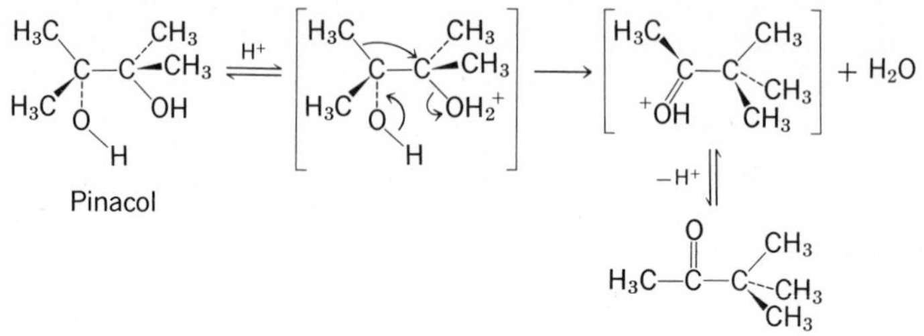
H3C
CH3
C—C—CH3
H3C
O
OH
H
Pinacol
H+
+OH2
+ H2O
−H+
Pinacolone

Here we note that a methyl group migrates to the carbon atom, which is becoming electron deficient (carbonium-ion like). It migrates with the pair of electrons which bonded it to its previous location (Sec. 8.23).

Other reactions which produce carbonium ions are known in which rearrangement occurs. The reaction of primary amines with nitrous acid leads to carbonium ion intermediates (Sec. 14.10).

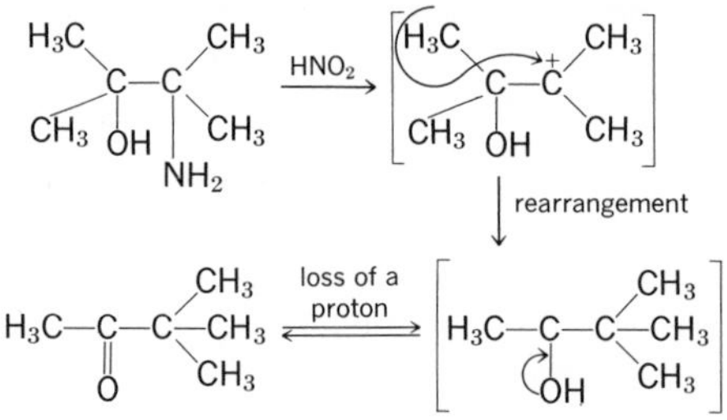
HNO2
NH2
rearrangement
loss of a
proton

The Beckmann rearrangement of oximes (Sec. 11.9) provides another interesting example. When oximes are treated with strong acid or with reagents such as PCl_5, the following type of rearrangement occurs:

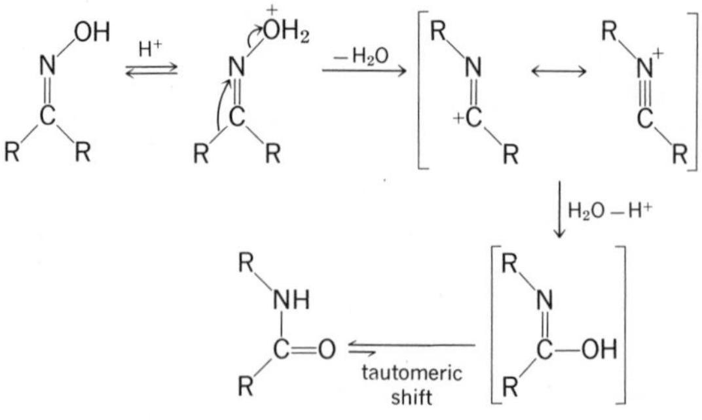
H+
−H2O
H2O −H+
tautomeric
shift

The overall result is that the oxime rearranges to an amide.

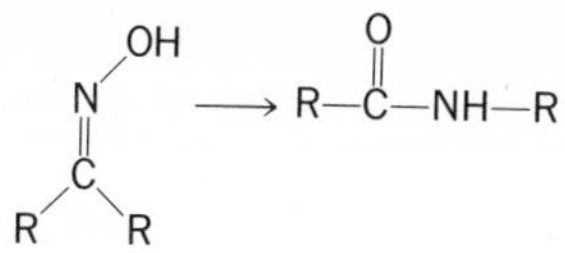

When the two R groups are different, two isomeric oximes may be obtained in many cases. These can have a given R group on the same side of the double bond as the OH group or on the opposite side, that is, they are geometric isomers (Sec. 7.2). Since oxime double bonds have only three substituents, the term *syn* is used in place of *cis* for the former situation, and *anti* in place of *trans* for the latter.

With the Beckmann rearrangements it is always the group that is *anti* which moves to the nitrogen. Thus,

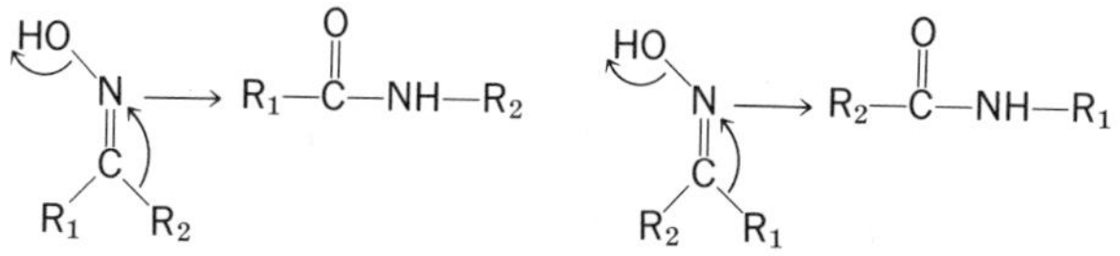

In all these reactions the rearrangement is seen to occur as a result of the production of an electron-deficient site—either a carbonium ion or electron-deficient nitrogen species in a situation in which normal displacement (Secs. 5.7, 9.5, and 9.6) or elimination (Sec. 9.7) is interfered with for some reason. In such cases the electron deficiency is satisfied by the shift of an electron pair, *group and all,* to produce a new intermediate which can stabilize itself by a more usual reaction, such as loss of a proton or formation of a multiple bond involving a previously unshared pair.

13.18 Spectroscopic Properties The infrared is particularly useful in ascertaining the presence of acid derivatives. Very characteristic maxima occur in the carbonyl region, which, coupled with strong maxima in other regions, make identification fairly easy.

Acid derivative	*C═O Region maxima*		*Other (conjugated or unconjugated), cm⁻¹*
	Unconjugated, cm⁻¹	*Conjugated, cm⁻¹*	
R—C(═O)—Cl	1815–1785	1800–1770	. . .
R—C(═O)—O—C(═O)—R	two peaks {1840–1800 1780–1740	1820–1780 1760–1720	1170–1050
R—C(═O)—OR′	1750–1735	1730–1715	1190–1245
R—C(═O)NHR′	two peaks {1640–1690 1510–1650	← similar ← similar	one or two bands in 3100–3550 region
R—C≡N	2245	2220–2229	. . .

The nmr of these derivatives is less definitive as the downfield shift of hydrogen next to the C=O group is not uniquely determined. With amides the NH protons appear in the $\delta = 5.0$ to 8.0 region and are slowly (not instantaneously) removed by exchange with D_2O.

The uv spectrum is of little utility *per se* in these systems. As usual, extended conjugation is necessary for uv absorption in the usable range.

OUTLINE OF ACID-DERIVATIVE CHEMISTRY

ESTERS

Preparation

1. $$R'OH + R{-}\overset{\overset{\displaystyle O}{\|}}{C}{-}OH \underset{}{\overset{H^+}{\rightleftharpoons}} R\overset{\overset{\displaystyle O}{\|}}{C}{-}OR' + H_2O$$

$$\underset{}{ROH + HOSO_2OH} \overset{\text{cold, conc.}}{\rightleftharpoons} \underset{\text{An acid sulfate}}{ROSO_2OH} + H_2O$$

NOTE: $\phi O\overset{\overset{\displaystyle O}{\|}}{C}R$ cannot be made by the method above.

2. $$R{-}\overset{\overset{\displaystyle O}{\|}}{C}{-}Cl + R'OH \longrightarrow R{-}\overset{\overset{\displaystyle O}{\|}}{C}{-}OR' + HCl$$

$$R{-}\overset{\overset{\displaystyle O}{\|}}{C}{-}Cl + \phi OH \longrightarrow R{-}\overset{\overset{\displaystyle O}{\|}}{C}{-}O\phi + HCl$$

3. $$\begin{matrix} R{-}\overset{\overset{\displaystyle O}{\|}}{C}\diagdown \\ \quad O \\ R{-}C\diagup \\ \quad \diagdown\!\!\diagdown O \end{matrix} + R'OH \longrightarrow R{-}\overset{\overset{\displaystyle O}{\|}}{C}{-}OR' + R{-}COOH$$

4. $$R{-}\overset{\overset{\displaystyle O}{\|}}{C}{-}O{-}Na^+ + \underset{\text{Primary or secondary}}{R'Br} \longrightarrow R{-}\overset{\overset{\displaystyle O}{\|}}{C}{-}O{-}R' + NaBr$$

Properties

1. Hydrolysis.

$$R{-}\overset{\overset{\displaystyle O}{\|}}{C}{-}OR' + H_2O \overset{H^+}{\rightleftharpoons} R{-}COOH + R'OH$$

NOTE: ϕOH esters undergo this reaction irreversibly.

2. Saponification.

$$R-\overset{\overset{\displaystyle O}{\|}}{C}-OR' + H_2O \xrightarrow[\text{or alcohol}]{\text{NaOH in } H_2O} R-COONa + R'OH$$

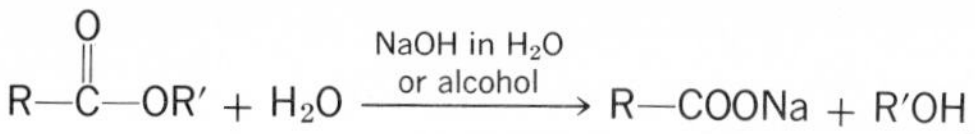

3. Ammonolysis.

$$R-\overset{\overset{\displaystyle O}{\|}}{C}-OR' + NH_3 \rightleftharpoons R-\overset{\overset{\displaystyle O}{\|}}{C}-NH_2 + R'OH$$

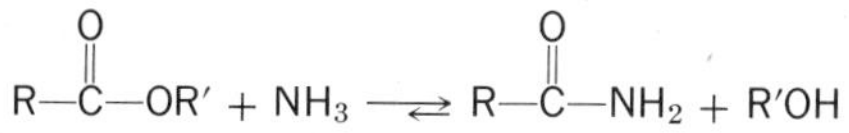

4. Reduction.

$$R-\overset{\overset{\displaystyle O}{\|}}{C}-O-C_2H_5 \xrightarrow[Ni]{H_2} R-CH_2OH + C_2H_5OH$$

$$R-\overset{\overset{\displaystyle O}{\|}}{C}-O-C_2H_5 + LiAlH_4 \longrightarrow R-CH_2OH + C_2H_5OH$$

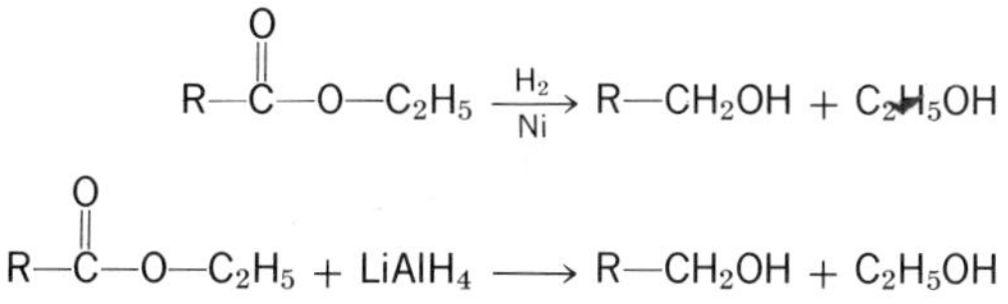

It is possible to reduce alkenes and ketones in the presence of esters by hydrogenation under milder conditions. Ketones and aldehydes are reduced by $NaBH_4$ while esters are not.

5. Grignard reagents.

$$R-\overset{\overset{\displaystyle O}{\|}}{C}-OR' + R''MgX \longrightarrow \rightarrow R-\overset{\overset{\displaystyle R''}{|}}{\underset{\underset{\displaystyle R''}{|}}{C}}-OH + R'OH$$

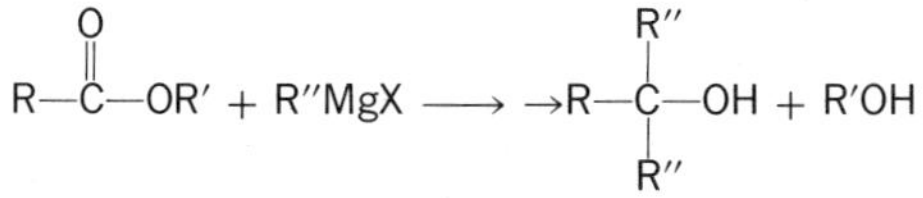

ACID HALIDES AND ANHYDRIDES (ACYLHALIDES AND ANHYDRIDES)

Preparation

1. $RCOOH + PCl_5 \longrightarrow RCOCl + HCl + POCl_3$
2. $RCOOH + PCl_3 \longrightarrow RCOCl + HCl + H_3PO_3$
3. $RCOOH + SOCl_2 \longrightarrow RCOCl + HCl + SO_2$
   HCl *will not* replace the —OH of the carboxyl group with —Cl as it will the alcoholic —OH. The above reagents will replace the —OH groups of alcohols with —Cl.

4. $R-COCl + Na^+ + \bar{O}COR \longrightarrow R-\overset{\overset{\displaystyle O}{\|}}{C}\overset{\displaystyle \diagdown}{\underset{\underset{\displaystyle R-C}{|}}{O}} \underset{\underset{\displaystyle O}{\|}}{} + NaCl$ 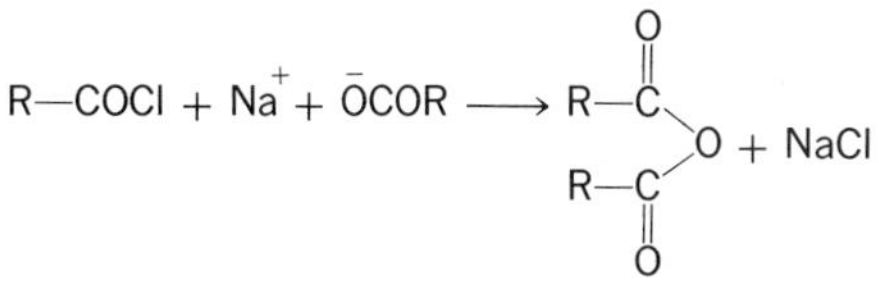

5. 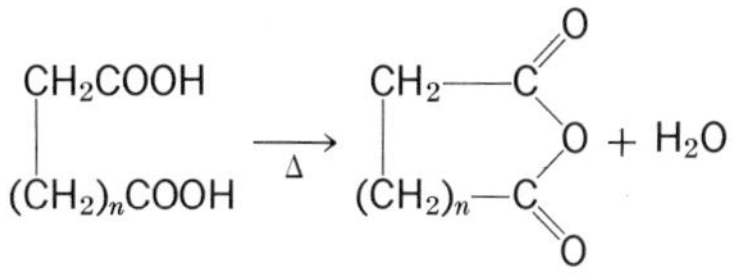
$$\begin{array}{c} CH_2COOH \\ | \\ (CH_2)_nCOOH \end{array} \xrightarrow{\Delta} \begin{array}{c} CH_2-C \\ | \qquad \diagdown O \\ (CH_2)_n-C \diagdown \\ \qquad O \end{array} + H_2O$$

where $n = 1,2$.

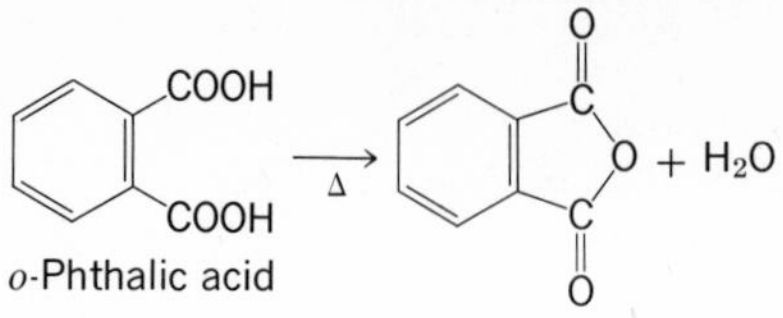

o-Phthalic acid

6.

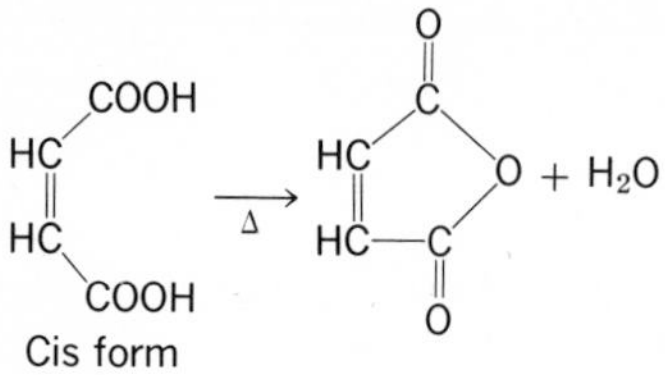

Cis form

Properties

1. Hydrolysis.

$$RCOCl + H_2O \longrightarrow RCOOH + HCl$$

$$(RCO)_2O + H_2O \xrightarrow[\text{or } \Delta]{H^+} 2RCOOH$$

2. Formation of esters.

$$RCOCl + R'OH \longrightarrow RC\overset{O}{\overset{\|}{-}}OR' + HCl$$

This reaction usually liberates heat and can be used as a test for OH groups.

$$(RCO)_2O + R'OH \longrightarrow R\overset{O}{\overset{\|}{-}}C-OR' + RCOOH$$

3. $RCOCl \xrightarrow{2NH_3} RCONH_2 + NH_4Cl$

$(RCO)_2O \xrightarrow{2R'NH_2} RCONHR' + RCOONH_3R'$ (salt)

$RCOCl \xrightarrow{2\overset{R'}{\underset{R'}{}NH}} RCON\overset{R'}{\underset{R'}{}} + \overset{R'}{\underset{R'}{}NH_2Cl}$ (salt)

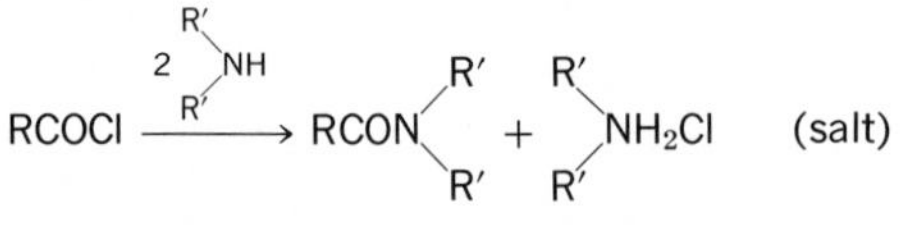

$RCOCl \xrightarrow{R''-N\overset{R'}{\underset{R'''}{}}} N.R.$ (no stable product)

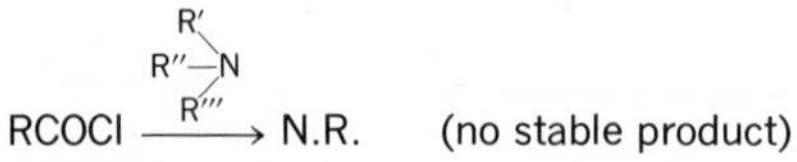

4. $RCOCl \xrightarrow{Cd(R')_2} RCOR' + CdR'Cl$

5. Acid halides and anhydrides may be halogenated. The reactions are the same in all cases as with the corresponding acids.

AMIDES

Preparation

1. $RCOONH_4 \xrightarrow[\Delta\Delta]{} RCONH_2 + H_2O$
   Amide

2. $RCOOR' + NH_3 \rightleftharpoons RCONH_2 + R'OH$

3. $RCOCl + NH_3 \longrightarrow RCONH_2 + HCl$ (or NH_4Cl)
   A rapid reaction; generates heat.

4. $(RCO)_2O + NH_3 \longrightarrow RCONH_2 + RCOOH$ (or $RCOONH_4$)
   Reactions 2, 3, and 4 give substituted amides with primary and secondary amines.

Properties

1. $RCONH_2 + NaOH \xrightarrow[\Delta]{} RCOONa + NH_3$ (test for $CONH_2$ group)

   $RCONH_2 + H_2O + HCl \xrightarrow[\Delta]{} RCOOH + NH_4Cl$

   Substituted amides are hydrolyzed in the same way to yield amines.

2. $RCONH_2 \xrightarrow{P_2O_5} RC{\equiv}N + H_3PO_4$
   This reaction requires a strong dehydrating agent.

3. $RCONH_2 \xrightarrow[NaOH]{Br_2} RNH_2 + Na_2CO_3 + NaBr$

   This is the Hofmann acid-amide rearrangement (see text).

4. 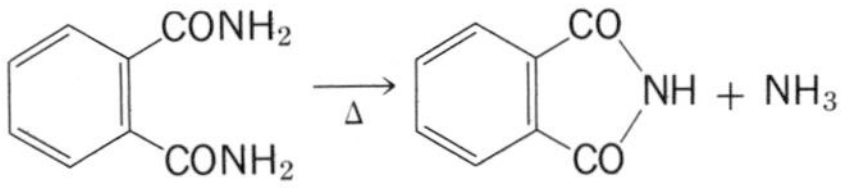
CONH2
CONH2
Δ
CO
NH + NH3
CO
   Phthalimide

CYANIDES (NITRILES)

Preparation

1. $RX + NaCN\ (KCN) \longrightarrow RC{\equiv}N + NaX\ (KX)$
   NOTE: ϕX and tertiary R groups do not give nitriles.

2. $RCONH_2 \xrightarrow[\Delta]{P_2O_5} RCN + H_3PO_4$

Properties

1. $RCN \xrightarrow[H_2O]{NaOH} RCOONa + NH_3$

   $RCN \xrightarrow[HCl]{conc.} RCOOH + NH_4Cl$

2. $RCN \xrightarrow[\text{or } LiAlH_4]{(H)} RCH_2NH_2$

CHAPTER 13 EXERCISES

1. Complete equations for the following reactions. Show the mechanism of each reaction.

★*a.* Methyl acetate + $C_2H_5NH_2 \longrightarrow$

b. Benzoyl chloride + $H_2O \longrightarrow$

★*c.* Acetic anhydride + $C_6H_5{-}NHCH_3 \longrightarrow$

d. 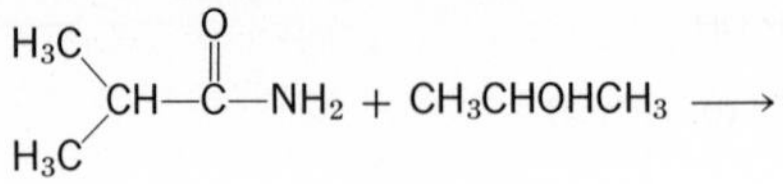

★*e.* $C_6H_5C(=O){-}CH_3 + C_2H_5MgBr \longrightarrow$

f. $C_6H_5C(=O){-}OH + C_2H_5MgBr \longrightarrow$

★*g.* Benzoic anhydride + $NH_3 \longrightarrow$

2. Show how you could distinguish by chemical and/or spectral methods between:

a. 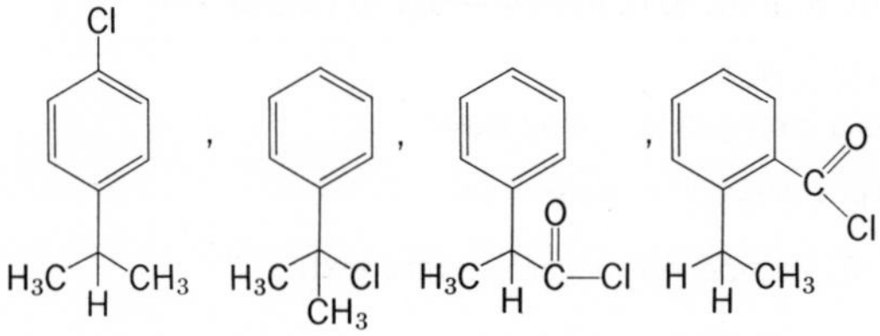

b. $H_3C{-}C(=O){-}O{-}C(=O){-}CH_3$, $H_3C{-}C(=O){-}CH_2CH_3$, $H_3C{-}C(=O){-}O{-}CH_3$

c. $C_2H_5C(=O){-}NH_2$, $C_2H_5CH_2{-}NH_2$, $C_2H_5C{\equiv}N$

d. 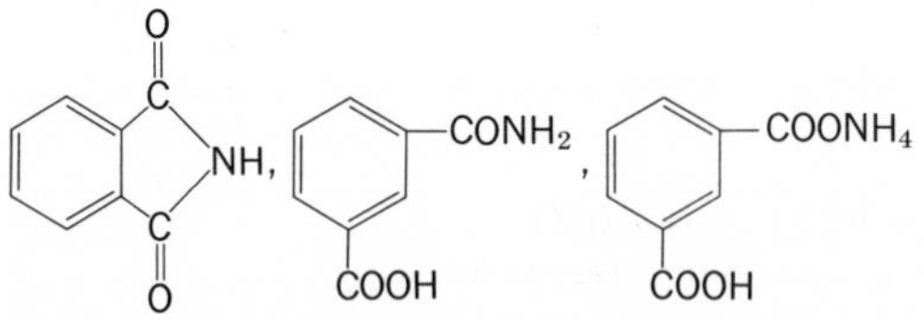

e. $CH_3CH_2COOCH_3$, $CH_3COOCH_2CH_3$, $CH_3OCH_2CH_2COOH$

f. *cis* and *trans* $HOOC{-}CH{=}CH{-}COOH$

g. $CH_3CH(CH_3){-}CH_2COOH$ and $CH_3CH(C_2H_5){-}CH_2COOH$

(HINT: Consider optical isomerism.)

h. 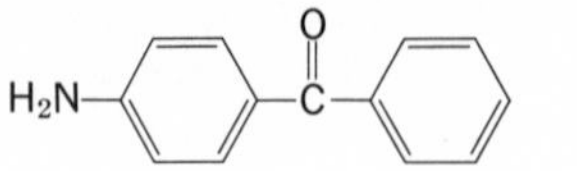 and 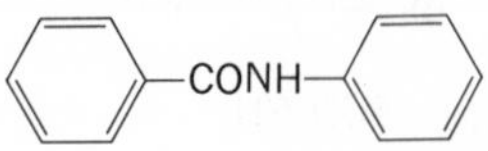

★**3.** Give the names and structures of four substances having the molecular formula $C_4H_8O_2$, and indicate how one could distinguish between them by qualitative tests. By spectroscopic methods.

4. Predict the relative reactivity toward NH_3 of

★*a.* $CH_3COCl + CH_3COI$

b. $C_6H_5COOCH_3 + HCOOCH_3$

c. $CH_3COONH_4 + CH_3CONH_2$

★**5.** Two compounds are known with the structure A below. They have essentially the same chemical properties, reactions, etc., but give different ir spectra. What portion of the spectra would you expect to show differences and why? Draw structures for the two compounds.

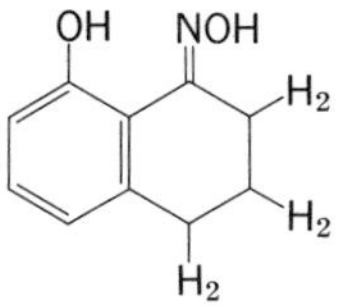

OH
NOH
H2
H2
H2

A

6. Arrange in order of increasing frequency of C═O ir absorption:

a. CH_3CHO, CH_3COOCH_3, $CH_3CON(CH_3)_2$

b. CH_3COOH, CH_3COONa

c. CH_3CHO, C_6H_5CHO

d. $CH_3COCH_2CH{=}CH_2$, $CH_3COCH{=}CHCH_3$

7. What differences, if any, would you expect in the nmr spectra of the following pairs?

★*a.* $CH_3CH_2CH_2OH$ and $CH_3CH(OH){-}CH_3$

b. Right- and left-handed (*R* and *S*) $CH_3CHOHCH_2CH_3$

★*c.* $CH_3CH_2CHD_2$ and $CH_2D{-}CH_2{-}CH_2D$ (D on each terminal CH_2)

d. $H_3C{-}C(CH_3)_2{-}CH_3$ and $CH_3CH_2CH(CH_3)_2$

★*e.* *Cis*- and *trans*- HOOC—CH═CHCOOH

8. An aromatic substance has the empirical formula $C_9H_{10}O_2$ and the saponification equivalent 150. On saponification and acidification it gives an acid with neutralization equivalent 136, which, in turn, is oxidized by permanganate to an acid with neutralization equivalent 83. Give a possible structure for the original substance.

★**9.** A yield of 100 percent of methyl acetate was obtained by the addition of $(CH_3)_2C(OCH_3)_2$ to a mixture of methanol and acetic acid. Explain.

HINT: See Secs. 13.6 and 11.7.

10. Compound Q, $C_{10}H_{12}O_6$, possesses bands in the infrared at 2540 to 2680 cm^{-1} (broad), at 1740 to 1715 cm^{-1}, and at 1200 cm^{-1}. Compound Q has a neutralization equivalent of 114, but when heated with excess NaOH and acidified, a new compound R, $C_8H_8O_6$, which has a neutralization equivalent of 66, is produced. If R is strongly heated, it produces S, $C_8H_6O_5$. In the infrared S had bands at 2680 to 2540 cm^{-1} (broad), 1820 cm^{-1}, and 1760 cm^{-1}. Heating S in ethanol solution, without catalyst, produces T, which is isomeric with Q and has a similar infrared spectrum. Both Q and T are optically inactive, but only T can be resolved into a pair of enantiomers. Deduce structures for the lettered compounds.

11. Predict the products of the following reactions involving rearrangements. Consider stereochemistry where pertinent, assuming a trans, S_N2 approach of the migrating group, as in Sec. 13.17.

★*a.* 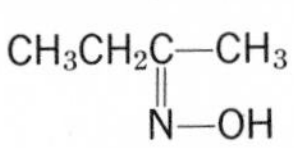
CH3CH2C—CH3
N—OH
(OH and CH_3 are *syn*) $\xrightarrow{H^+}$

b. *cis*- 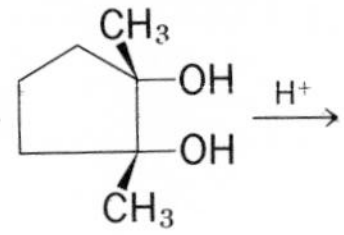
CH3
OH
H+
OH
CH3

★*c.* 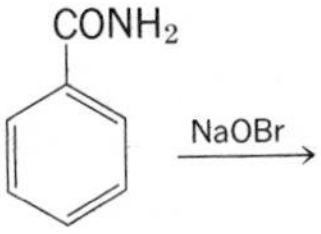
CONH2
NaOBr

d. 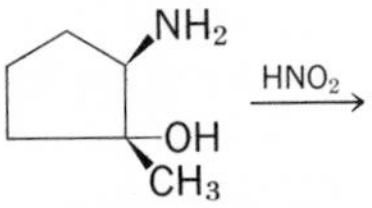
NH2
HNO2
OH
CH3

HINT: The ring carbon may migrate.

★*e.* *trans*-$C_6H_5CH{=}CHCON_3$ $\xrightarrow[\Delta]{100°}$

f. 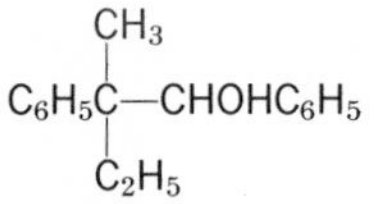
CH3
C6H5C—CHOHC6H5
C2H5
(*R*, *R* configuration) $\xrightarrow{H^+}$

Assume phenyl group migrates.

12. Write equations, showing mechanisms, for the following reactions:

a. $(CH_3CH_2CO)_2O + C_6H_5NH_2$ (excess)

b. $CH_3COOC_6H_5 + C_2H_5OH$ + trace of base

c. 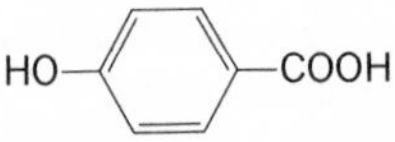
HO—
—COOH
+ benzyl alcohol + trace of H_2SO_4

d. $CH_3COOC_6H_5 + NH_2C_6H_5$

e. 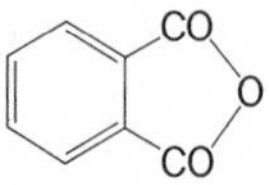
CO
O
CO
+ benzyl alcohol

f. $CH_3CON(CH_3)_2 + NH_3$

g. A fat (glyceride) + $LiAlH_4$ (excess)

13. Predict reactions, if any, based on analogy with known $C=O$ reactions. Show mechanisms.

★a. Acetone + H_2S

b. Benzoyl chloride + Grignard reagents

★c. Ethylacetate + phenylhydrazine

d. Benzamide + H_2O + base

★e. Phenyl acetate + ethanol

f. $CH_3CH(OCOCH_3)_2$ + dil. base

★g. Acetic anhydride + C_6H_6 + $AlCl_3$

14. A rearrangement analogous to those discussed in Secs. 13.16 and 13.17 occurs in the following reactions. Write the equations for the reactions, showing the mechanism.

a.
$$\begin{array}{c} CH_3 \\ | \\ CH_2-C-CH_2NH_2 + HNO_2 \\ | \\ CH_2-CH_2 \end{array}$$

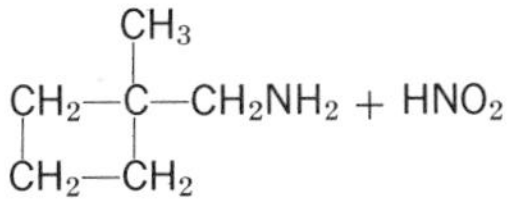

b.
$$\begin{array}{c} C_2H_5 \\ | \\ H_5C_2-C-CH_2OH + H_2SO_4 \\ | \\ C_2H_5 \end{array}$$

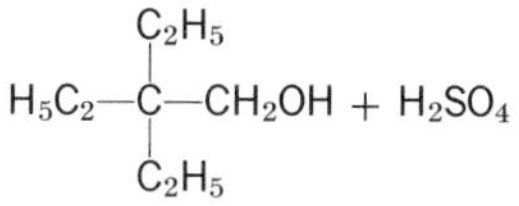

c.
$$\begin{array}{c} CH_3 \\ | \\ H_3C-C-CH=CH_2 + HCl \\ | \\ CH_3 \end{array}$$

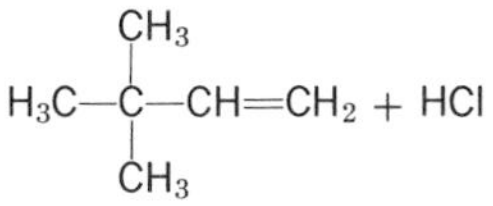

15. The following reactions have not been discussed in the text. Predict the course of the reactions by the use of the general principles advanced in this chapter.

a. $C_6H_5CHO + H_2S$

★b. $C_6H_5CHO + NaCH(COOR)_2$

c. $CH_3COCl + CH_3MgCl$

★d. $CH_3COCH_3 + NaC\equiv CH$

e. $C_6H_5COOR + NH_2NHC_6H_5$

★f. $CH_3CN + CH_3OH$

g. $CH_3CH=CHCOOR + HBr$

16. A certain compound X has the formula $C_8H_5O_2Cl$ and shows the following behavior:

a. X has maxima in the ir at 1740 and 1200 cm^{-1} but none above 3000 cm^{-1}.

b. X reacts slowly with NaOH to form Y, $C_6H_{14}O_2$, plus another substance.

c. 0.12 g of Y with excess Na^0 evolves 22 ml of gas (calculated at standard temperature and pressure).

d. Y reacts on heating with sulfuric acid to form Z, C_6H_{10}.

e. Z shows absorption in the infrared at 1650 cm^{-1} and a maximum in the uv at 220 mμ.

f. The nmr of Z shows peaks in the region $\delta = 5$ to 6, $\delta = 2.0$, and $\delta = 1.0$ in the ratio of 5:2:3.

Deduce structures for each of the lettered compounds.

★17. A certain compound A, $C_{13}H_{16}O_4$, has infrared maxima at 2600 (broad), 1720, 1710, and 1200 cm^{-1}, and gives a neutralization equivalent of 236. On hydrolysis A forms (B), $C_4H_{10}O$, with maxima at 3600 and 3350 cm^{-1} but none in the 1600 to 1800 cm^{-1} region, and (C), $C_9H_8O_4$, with maxima at 2600 (broad) and 1710 cm^{-1}. B is easily oxidized to a substance (D), C_4H_8O, which has ir maxima at 1730 and 2720 cm^{-1} (sharp), plus an nmr peak at $\delta = 9.5$. D reacts with dilute alkali to form $C_8H_{16}O_2$ which is stable to heat (does not lose H_2O). C has a neutralization equivalent of 90 and loses water to form $C_9H_6O_3$ (ir maxima at 1810, 1750, and 1100 cm^{-1}). Vigorous oxidation of C yields phthalic acid. Deduce structures for the lettered compounds.

18. When methylbenzoate, containing "labelled" oxygen in the carbonyl group

$$\underset{\displaystyle C_6H_5\overset{\textstyle \|}{C}-OCH_3}{O^{18}}$$

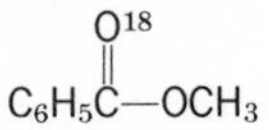

reacts with H_2O and NaOH, there is found in the unreacted ester at equilibrium

$$\underset{\displaystyle C_6H_5\overset{\textstyle \|}{C}-OCH_3}{O}$$

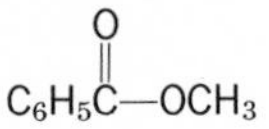

ordinary oxygen in the carbonyl group. Is this expected if the mechanism is a simple displacement as shown?

$$HO\overset{\displaystyle O}{\overset{\|}{C}}-OCH_3 \longrightarrow HO\overset{\displaystyle O}{\overset{\|}{C}} + OCH_3^- \\ \underset{\textstyle C_6H_5}{|} \qquad\qquad \underset{\textstyle C_6H_5}{|}$$

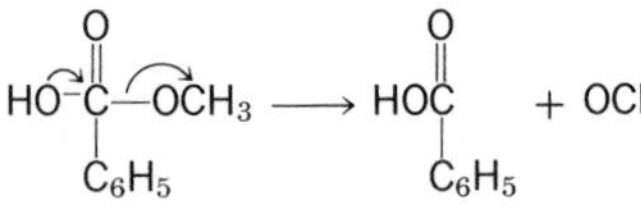

Assuming $CO_2{}^{18}$ is available, how would you synthesize $C_6H_5-\overset{O^{18}}{\overset{\|}{C}}-OCH_3$? 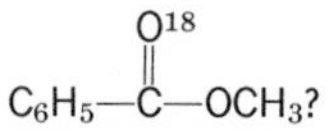

★19. Calculate the saponification equivalent of a glyceride containing 3 C_{10} fatty acid groups. What would be the effect on the saponification equivalent of a 50% impurity of (a) hydrocarbon or (b) a C_{10} carboxylic acid?

20. Outline a mechanism for the hydrogenolysis of esters and for the reduction of esters by $LiAlH_4$ (H:$^-$ reaction).

★21. Explain why the preparation of nitriles from tertiary halides is unsatisfactory.

22. How can

$$\underset{\displaystyle CH_3\overset{\textstyle \|}{C}-C_2H_5}{NOH}$$

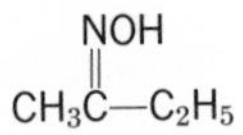

exist in two geometrical isomers? There is only one group on the N.

★23. Ultraviolet spectra of $CH_3CH{=}CH-CH_2COOCH_3$ and $CH_3CH_2CH{=}CH-COOCH_3$ are markedly different. Explain. Would the ir spectra differ? How? Would nmr be useful in distinguishing these? Explain.

14

AMINES AND DIAZONIUM COMPOUNDS

14.1 Definition; Classification; Nomenclature Many years ago, chemists, in examining certain plant tissues, found that they contained basic substances, often of a poisonous nature. These substances, because of a superficial chemical resemblance to the then known alkalis, were called *alkaloids*. Many of them are of great importance today by reason of their physiological activity; one need only mention quinine, morphine, and nicotine to bear this out. The plant alkaloids belong chemically to the much larger group of organic bases known as the *amines;* a study of the chemistry of simple amines forms the basis for the study of the chemistry of the alkaloids and of a substantial proportion of the synthetic drugs in current use.

Amines are essentially "substituted ammonias" in which one or more of the

TABLE 14.1
TYPES OF AMINES

Formula	*Name*	*Class*
$CH_3CH_2NH_2$	Ethylamine	Primary amine
CH_3CH_2 $\;\;NH$ CH_3CH_2	Diethylamine	Secondary amine
CH_3CH_2 $CH_3CH_2—N$ CH_3CH_2	Triethylamine	Tertiary amine
$(CH_3)_3C—NH_2$	*tert*-Butylamine (2-methyl-2-aminopropane)	Primary amine
Related compound: $(CH_3CH_2)_4NCl$	Tetraethylammonium chloride	Quaternary salt

hydrogen atoms of ammonia has been replaced by alkyl or aryl (not acyl) groups. Acyl-substituted ammonias, or amides, have already been studied in the preceding chapter. The amines are divided into classes on the basis of the number of groups attached to nitrogen without regard to the structure of the groups (Table 14.1). The Geneva names make use of the compounding element *amino-* (methylamino, etc.) as in the following example:

$$\begin{array}{l} CH_3CHCH_3 \\ \quad\;\; | \\ \quad\;\; N(CH_3)_2 \end{array}$$

2-Dimethylaminopropane
(dimethyl*iso*propylamine)

Trivial names are often used and are frequently encountered with the aromatic amines:

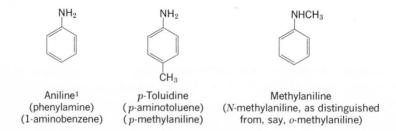

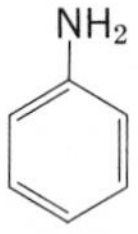

Aniline[1]
(phenylamine)
(1-aminobenzene)

p-Toluidine
(*p*-aminotoluene)
(*p*-methylaniline)

Methylaniline
(*N*-methylaniline, as distinguished from, say, *o*-methylaniline)

14.2 Basicity of Amines All amines have basic properties. In the primary, secondary, and tertiary amines, the basic property, the tendency to bind a proton, is due to the pair of unshared electrons on the nitrogen.

[1]Aniline, which was originally obtained from coal tar along with a great many other bases, is an important intermediate in the dyestuff industry ("aniline dyes" and "coal-tar dyes"; see Sec. 14.17). Aniline, like many other aromatic bases, is *extremely* toxic.

$$H_3C:\underset{\textstyle H}{\overset{\textstyle H}{\ddot{\underset{\cdot\cdot}{N}}}}: + H^+Cl^- \rightleftarrows H_3C:\underset{\textstyle H}{\overset{\textstyle H}{\underset{\cdot\cdot}{\overset{\cdot\cdot\,+}{N}}}}:HCl^-$$

When a proton is bound, there is formed a positive ion, the originally electrically neutral amine taking on the charge of the proton. When ions are formed in this way, whether the atom bound is hydrogen or not, they are designated by the name ending *onium*. The ion formed here is a substituted amm*onium* ion. Another onium ion which has been encountered is the hydronium ion, H_3O^+, which belongs to the class of oxonium ions.

14.3 Effect of Structure The basicity of the amines is dependent upon the nature of the substituent groups. A base is defined as a substance which binds protons, and since, as was pointed out in Chap. 12, substituents on or close to the carboxyl group have a marked influence on the *release* of protons by carboxylic acids, it is to be expected that the *binding* of protons would be subject to the influence of substituents on or near to the nitrogen atom. It follows, then, that those groups which are electron-attracting and facilitate the release of a proton from the carboxyl group will oppose the binding of a proton when they are substituted on nitrogen, and vice versa.

It can be seen from Table 14.2 that the nature of the substituent has a marked effect on the basicity. The introduction of a methyl group in proceeding from NH_3 to CH_3NH_2 increases the availability of the electron pair on N and results in a stronger base. This is in accord with our expectation since it agrees with the role of methyl groups in addition reactions (Sec. 3.8) and in aromatic substitution (Sec. 6.12). The introduction of three methyl groups, however, to form $(CH_3)_3N$ produces a base weaker than methylamine. This is attributed to the effective bulk (steric hindrance) of the methyl groups preventing solvent molecules from getting close enough to

TABLE 14.2
SOME DERIVATIVES OF AMMONIA ARRANGED IN ORDER OF DECREASING BASICITY

$(CH_3)_4N^+OH^-$ (OH^- is the base)
$(CH_3)_2NH$
CH_3NH_2
$(CH_3)_3N$
$\left\{\begin{matrix} NH_3 \\ C_6H_5CH_2NH_2 \end{matrix}\right\}$
$C_6H_5NHCH_3$
$C_6H_5NH_2$
$(C_6H_5)_2NH$
CH_3CONH_2
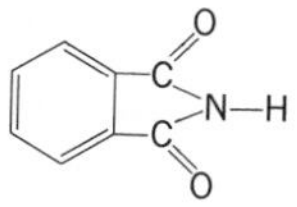

the site of positive charge in $(CH_3)_3NH^+$ to help stabilize the charge.[1] It is important to note that generalizations based on simple considerations such as electron release must be applied with caution as a large number of factors contribute to the behavior of any real molecule and it is sometimes difficult to assess their relative importance.

Aromatic amines in which nitrogen is directly on the aromatic ring appear less basic than ammonia. This we find consistent with other properties of aromatic amines, and we attribute it to resonance interaction between the ring system and the electrons on nitrogen in a manner analogous to that observed with phenol.

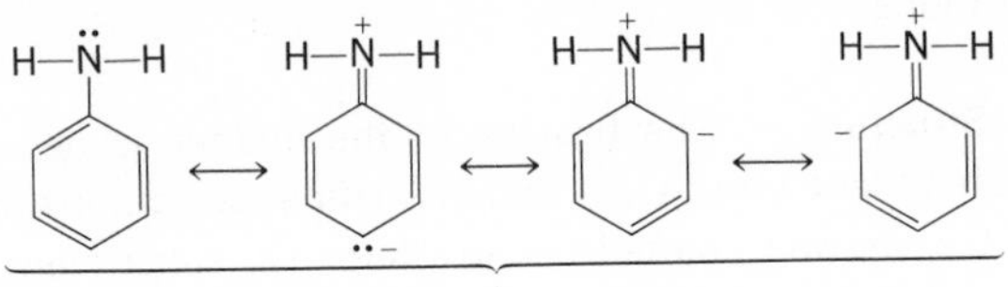

I

The net effect of this resonance interaction is to increase the electron density on the ring and to decrease the availability of electrons on nitrogen. In the case of the ammonium salt II, the contribution of structures such as II*a*, is energetically prohibited by the unavailability of any more orbitals in the second shell of nitrogen. Use of orbitals from the third shell would so raise the energy of the system that structures with five bonds to nitrogen need not be considered.

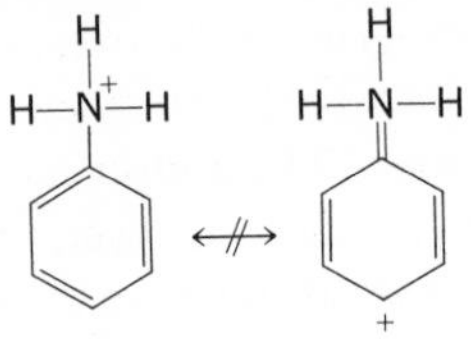

II IIa

The lowered basicity of aromatic amines arises from the resonance stabilization of the free amine relative to the ammonium salt which is not stabilized in this way.

Other groups on nitrogen which can act either by inductive (electron attraction due to electronegativities) or resonance effects to withdraw electrons also decrease the basicity of the N atom. The R—C=O group for example gives rise to resonance as shown, while the N-protonated form cannot be stabilized.

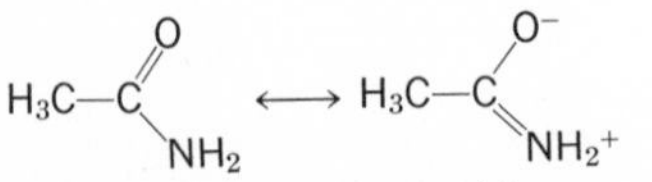

Resonance in acetamide

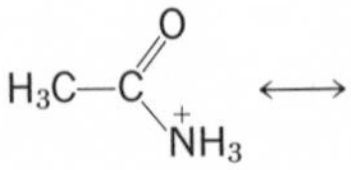

no reasonable structures

An even greater effect in the same direction is found when two R—C=O groups are attached to the same N atom.

[1] Most ions, even Na^+, etc., derive a large part of their stability from solvation by solvents such as H_2O:, H_3N:, RÖH, etc. The insolubility of many salts in hydrocarbons also arises from the unavailability of such solvation.

TABLE 14.3
THE DISSOCIATION CONSTANTS OF SOME AMIDES AND RELATED COMPOUNDS

Substance	*Dissociation constant*[a] K_b *(basic dissociation)*	K_a *(acid dissociation)*
CH_3CONH_2	3×10^{-15}	
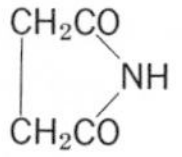	. . .	3×10^{-11}
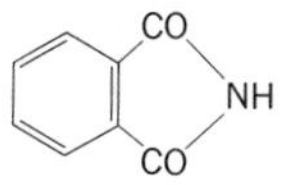	. . .	5×10^{-9}
$C_6H_5SO_2NH_2$	. . .	Acidic

[a] The basic dissociation constant for an amine K_b is for the equilibrium:

$$R—NH_2 + H_2O \rightleftharpoons R\overset{+}{N}H_3 + OH^-$$

$$K_b = \frac{(RNH_3^+)(OH^-)}{RNH_2}$$

H_2O is incorporated into K_b.

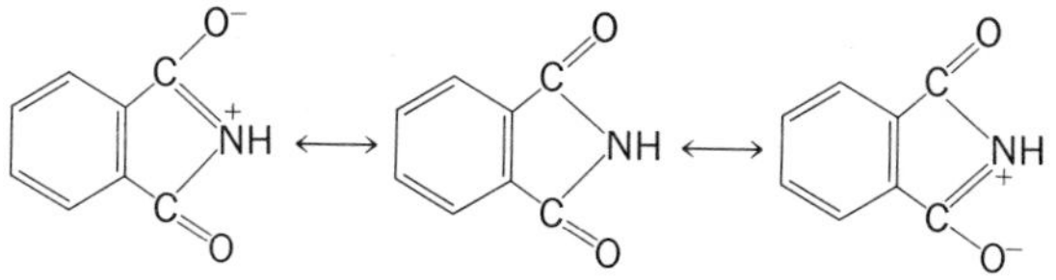

Resonance in phthalimide

Although acetamide and its analogs are essentially neutral compounds, phthalimide is actually acidic and forms salts with strong bases. Here the delocalization of electrons is sufficient to stabilize the anion to a significant extent. Other strongly electron-attracting groups, notably RSO_2—, have a similar effect, and simple sulfonamides, for instance, are also weak acids.

The effects of groups on the basic properties of amines is analogous, naturally, to the effects of the same groups on the acidity of OH compounds. The following comparisons are illustrative:

CH_3CH_2OH a weaker acid (and a stronger base) than C_6H_5OH
CH_3NH_2 a weaker acid (and a stronger base) than $C_6H_5NH_2$
CH_3CH_2OH a weaker acid (and a stronger base) than CH_3COOH
CH_3NH_2 a weaker acid (and a stronger base) than CH_3CONH_2
CH_3CONH_2 a weaker acid (and a stronger base) than $C_6H_5SO_2NH_2$

The "proton-binding" nature of salt formation in amines is often emphasized by the manner in which the formulas of amine salts are sometimes written.

$$(CH_3)_2NH + HCl \longrightarrow (CH_3)_2NH \cdot HCl$$

Dimethylamine hydrochloride
(dimethylammonium chloride)

Since the salts of amines with strong acids such as HCl are rather completely ionized, the properties of these salts resemble those of other electrovalent compounds; they are high-melting, nonvolatile, water-soluble, ether-insoluble solids. By the addition of a strong base such as NaOH, the salt formation can be reversed, regenerating the ether-soluble, often volatile amine. The base used must be stronger (that is, have greater proton-binding power) than the amine; OH^- is a stronger base than any simple amine.

$$(CH_3)_2NH_2Cl + NaOH \rightleftharpoons (CH_3)_2NH + NaCl + H_2O$$

The phenomenon here parallels exactly the behavior of the salts of carboxylic acids, and we may be guided by a similar rule. The stronger base "steals" the proton from the weaker one.

The formation of salts and their regeneration forms the basis for the separation of amines from mixtures of other classes of compounds. The natural alkaloids—quinine, nicotine, etc.—are isolated by extraction of the crude source material with dilute acid, thus converting the water-insoluble amines into water-soluble salts and separating them from most of the water-insoluble organic material present. On addition of alkali, the alkaloid is set free and may be extracted with ether or distilled, leaving behind water-soluble products.

14.4 Alkylation of Ammonia Ammonia is a nucleophilic reagent, as was pointed out in Sec. 13.3, and reacts with alkyl halides,

$$\overset{\delta+}{R}-\overset{\delta-}{Br}$$

in a displacement reaction, the primary attack on the ammonia being at the pair of unshared electrons on the nitrogen.

$$H_3N\!: + \overset{\delta+}{H_3C}-\overset{\delta-}{Br} \longrightarrow [H_3N\text{---}CH_3\text{---}Br] \longrightarrow H_3\overset{+}{N}CH_3 + Br^-$$

The overall reaction is

$$CH_3Br + NH_3 \longrightarrow CH_3NH_3^+Br^- \text{ (or } CH_3NH_2 \cdot HBr)$$

Since ammonia is a base comparable in strength to the primary amine, reaction takes place between amine salt and ammonia. This liberates free primary amine which in turn reacts with the alkyl halide. The result is the production of a mixture of amines in all alkylation reactions of this type.

$$CH_3NH_2\cdot HBr + NH_3 \rightleftharpoons NH_3\cdot HBr + CH_3NH_2$$

$$CH_3NH_2 + \overset{\delta+}{CH_3}-\overset{\delta-}{Br} \longrightarrow (CH_3)_2NH_2^+ + Br^-$$

The overall reaction then becomes

$$CH_3Br + NH_3 \longrightarrow \begin{cases} CH_3NH_2\cdot HBr \\ (CH_3)_2NH\cdot HBr \\ (CH_3)_3N\cdot HBr \\ (CH_3)_4N^+Br^- \end{cases}$$

In laboratory practice, mixtures of this sort are often difficult to separate by physical methods such as distillation, and in such cases chemical procedures are used. (In commercial synthesis this particular separation is best accomplished by fractional distillation.) The chemical separation illustrated below (Sec. 14.6) with the methylamines is of general applicability and is described in detail because it illustrates some of the important properties and reactions of amines.

14.5 Quaternary Ammonium Compounds In the quaternary ammonium compounds four covalent carbon-to-nitrogen links are present.

$$(CH_3)_3N\!: + CH_3Cl \longrightarrow \left[\begin{array}{c} CH_3 \\ H_3C\!:\!\ddot{\underset{..}{N}}\!:\!CH_3 \\ CH_3 \end{array}\right]^+ + Cl^-$$

Tetramethylammonium chloride

A quaternary salt, such as tetraethylammonium chloride, is the salt of a strong acid, HCl, and a strong base, tetraethylammonium hydroxide. This latter substance is comparable with the hydroxides of the alkali metals in basicity; its physical and chemical properties are in many respects like those of sodium and potassium hydroxides, and its salts are highly ionized. As would be predicted, both the base and its salts are soluble in water and insoluble in ether. This property is discussed below. The quaternary salts are really not amines, though they are classed with the amines for convenience. For a substance to be a true amine, it should be able to bind a proton because of the presence of a pair of unshared electrons on the nitrogen. An important difference between an amine salt and the quaternary ammonium salts is that an amine salt, such as methylammonium chloride (Sec. 14.2) is easily converted by a stronger base to the free amine by the removal of a proton. The four covalent carbon-to-nitrogen bonds in the quaternary compound are all stable, and resist cleavage (see Sec. 14.15).

The R_4N^+ ion, like the inorganic Na^+ and K^+ ions, does not have an unshared pair of electrons and is not basic. The hydroxide $R_4N^+\ {}^-OH$, like NaOH, is a strong base, but other salts (like $R_4N^+Cl^-$, etc.) are not.

14.6 Separation of Primary, Secondary, and Tertiary Amines The reaction mixture (Sec. 14.4) is first treated with an excess of strong nonvolatile base such as sodium hydroxide.

$$\left.\begin{array}{l} CH_3NH_2 \cdot HBr \\ (CH_3)_2NH \cdot HBr \\ (CH_3)_3N \cdot HBr \\ (CH_3)_4N^+Br^- \end{array}\right\} \xrightarrow{NaOH} \begin{array}{l} \left.\begin{array}{l} CH_3NH_2 \\ (CH_3)_2NH \\ (CH_3)_3N \end{array}\right\} \text{volatile, ether-soluble} \\ (CH_3)_4N^+OH^- \quad \text{nonvolatile, ether-insoluble} \end{array}$$

The quaternary salt, being the salt of a base of almost the same strength as NaOH, merely remains in solution as a positive ion, the total number of R_4N^+ and Na^+ ions equaling the total number of Br^- and OH^- ions.

A simple distillation or extraction with ether will now leave behind in water solution the quaternary ions and remove the three classes of amines in the free state. These are treated with an acid chloride; *p*-toluenesulfonyl chloride is often used, since the amides so formed are stable solids and usually easy to purify.

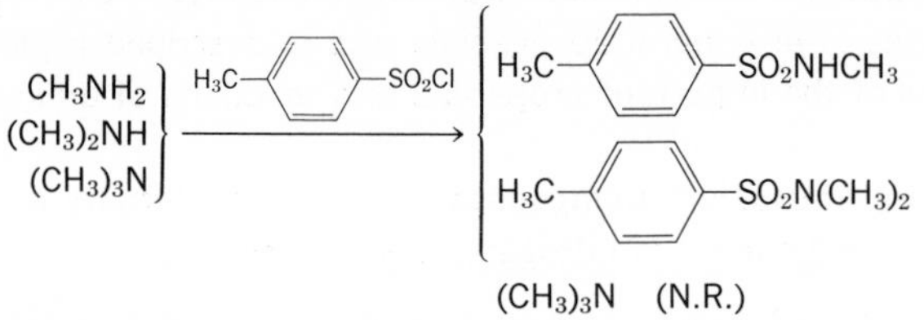

CH3NH2
(CH3)2NH
(CH3)3N
H3C—SO2Cl
H3C—SO2NHCH3
H3C—SO2N(CH3)2
(CH3)3N (N.R.)

The tertiary amine, having no replaceable hydrogen, does not react with acid chlorides or anhydrides to form amides (Sec. 13.15). The sulfonamides are unable to bind a proton and are acidic in nature (see above). It is now only necessary to add dilute acid to form a salt of the tertiary amine.

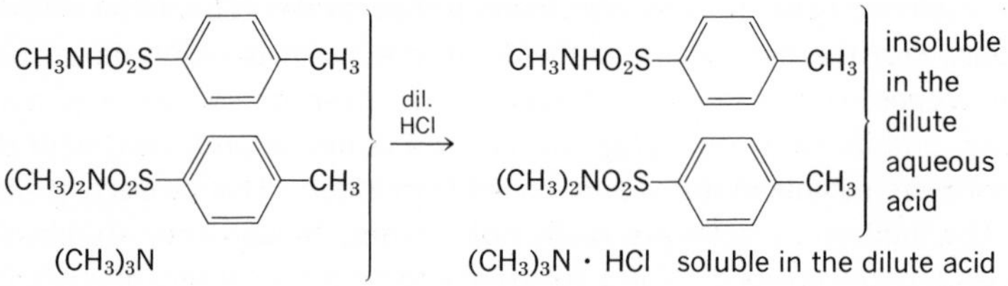

CH3NHO2S—CH3
(CH3)2NO2S—CH3
(CH3)3N
dil. HCl
CH3NHO2S—CH3
(CH3)2NO2S—CH3
insoluble in the dilute aqueous acid
(CH3)3N · HCl soluble in the dilute acid

As usual, the ionized salt has widely different properties from the covalent amides and is readily separated, for example, by water extraction. The addition of a strong base to its water solution liberates the amine.

$$(CH_3)_3N{\cdot}HBr + KOH \longrightarrow (CH_3)_3N + KBr + H_2O$$

This particular phase of the separation can be accomplished equally well with other acid chlorides or anhydrides, acetyl chloride, for example. In the separation of primary and secondary amines in the next step, given below, the sulfonic acid derivative is required, since carboxylic amides are not sufficiently strong acids.

In the separation of the remaining two amines advantage is taken of the fact that the *N*-methyl-*p*-toluenesulfonamide from the primary amine is acidic (Table 14.3).

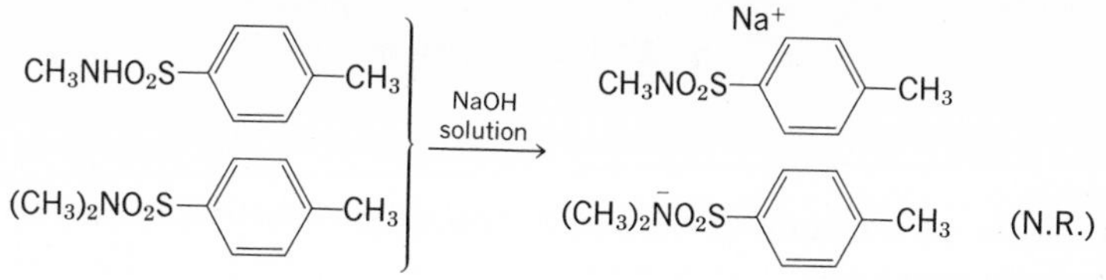
CH3NHO2S— —CH3
(CH3)2NO2S— —CH3
NaOH
solution
Na+
CH3NO2S— —CH3
(CH3)2NO2S— —CH3
(N.R.)

As before, the ionized salt is separated from the covalent amide (which has no ionizable H on the N and, therefore, does not react) by ether-water partition. The amines are finally regenerated by hydrolysis (prolonged boiling).

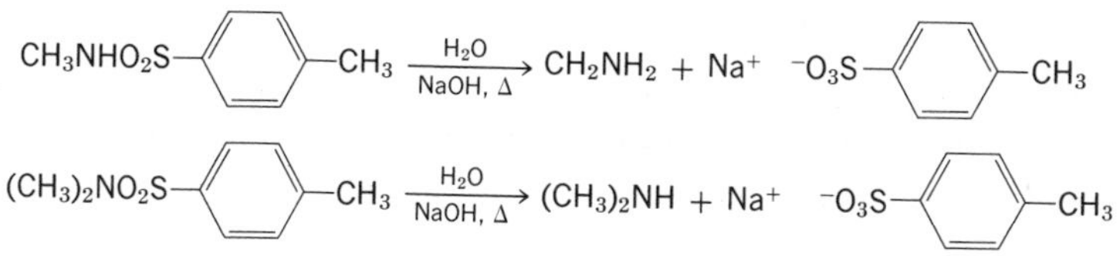
CH3NHO2S— —CH3
H2O
NaOH, Δ
CH2NH2 + Na+ −O3S— —CH3
(CH3)2NO2S— —CH3
H2O
NaOH, Δ
(CH3)2NH + Na+ −O3S— —CH3

All the ether-water separations based on the formation and properties of salts are carried out promptly, and at room temperature. Under these conditions sulfonic acid amides and esters are not appreciably hydrolyzed. Thus, they can be separated from acids and bases.

It must be understood, of course, that this tedious process need not be carried out in every case where a mixture of amines is encountered. When the boiling points of primary and secondary amines lie far enough apart, a careful distillation may be used. Methods such as the above are most commonly used to remove small amounts of more-or-less substituted amines from samples that may be predominantly one type.

14.7 Pure Primary Amines The use of a large excess of ammonia favors the formation of primary amines, in some cases to 60 to 70 percent yield, and with the higher amines the method is often practical. As usual, an aromatic halide is too unreactive to give amines by this method except under drastic conditions, and the aromatic amines are, as a rule, prepared by other methods (see below). There are methods available for the synthesis of amines, particularly primary amines, free from the side products encountered above. Some of these have been discussed already.

The Hofmann rearrangement yields pure primary amines as the final product, and the reaction is of a general nature (Sec. 13.17).

$$CH_3CH_2CONH_2 \xrightarrow[H_2O,\ \Delta]{NaOBr} CH_3CH_2NH_2$$

The reduction of certain types of nitrogen compounds leads to the formation of primary amines relatively free of side products. One of the most versatile of these methods involves the reduction of amides by $LiAlH_4$.

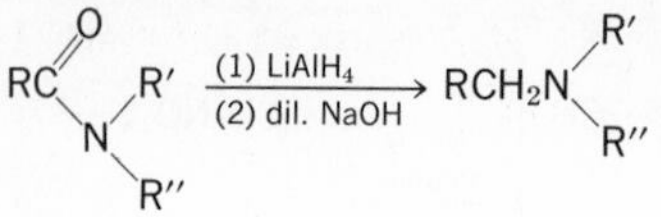
RC
O
R′
N
R″
(1) LiAlH4
(2) dil. NaOH
RCH2N
R′
R″

Since a variety of amides can be readily made, the reaction is clearly useful. Analogous reductions are:

$$CH_3CH_2C{\equiv}N \xrightarrow[\text{or } LiAlH_4]{H_2/\text{cat.}} CH_3CH_2CH_2NH_2$$

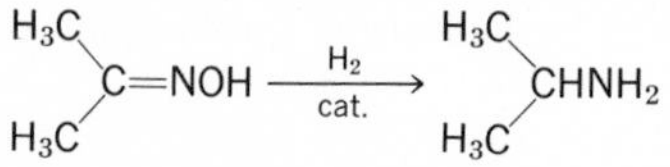
H3C
C=NOH
H3C
H2
cat.
H3C
CHNH2
H3C

The reduction of nitro compounds may be carried out simply by heating with iron or tin and acid; the method is of major importance in the aromatic series, aniline being made commercially in this way.

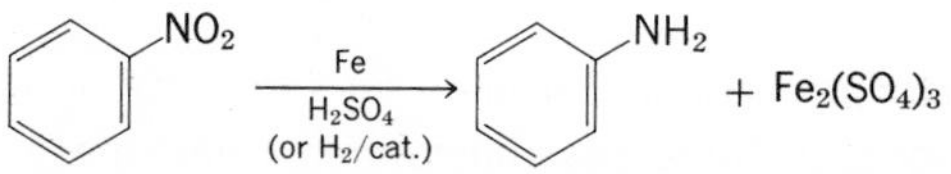
NO2
Fe
H2SO4
(or H2/cat.)
NH2
+ Fe2(SO4)3

Aniline is also made commercially by direct reaction of chlorobenzene with ammonia at high temperature and pressure. The conditions of the reaction are similar to those required for the hydrolysis of chlorobenzene in the commercial production of phenol.

14.8 Gabriel Synthesis One of the most ingenious and convenient methods for the synthesis of pure primary amines is the phthalimide method of Gabriel.

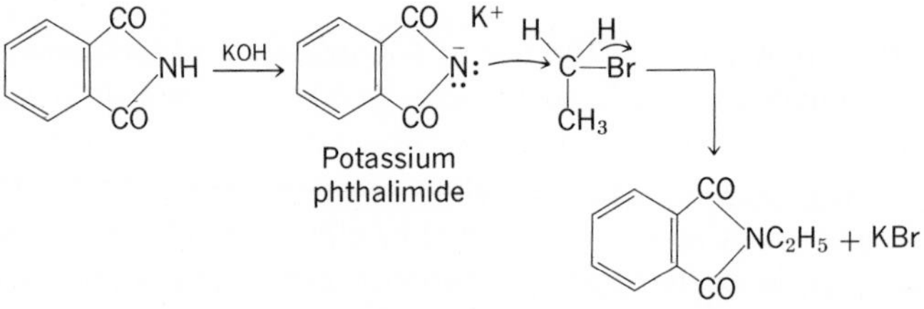
CO
NH
CO
KOH
CO
K+
N:
CO
Potassium
phthalimide
H
H
C—Br
CH3
CO
NC2H5 + KBr
CO

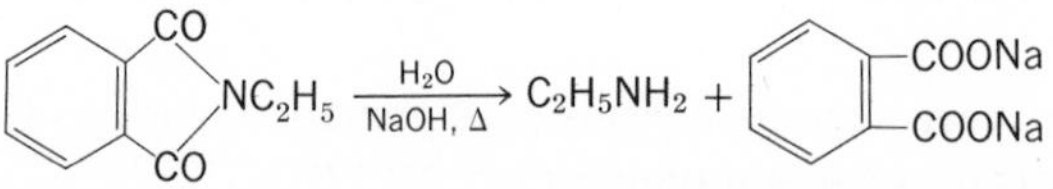
CO
NC2H5
CO
H2O
NaOH, Δ
C2H5NH2 +
COONa
COONa

Phthalimide is a weak acid; its potassium salt can be formed, and this in turn reacts with alkyl halides like all salts of weak acids (nucleophilic agents). Hydrolysis of the resulting substituted amide yields a pure primary amine and allows easy recovery of the phthalic acid as its water-soluble sodium salt.

More complex and less general methods are used for the preparation of pure secondary and tertiary amines. The quaternary salts, the final product of alkylation reactions, are prepared by direct alkylation of ammonia or amines using a large excess of halide, thus forcing the reaction to completion.

14.9 Salt Formation The most characteristic property of the amines as a class is their ability to form salts with acids. The application of this property to amine separations has been pointed out. Water-insoluble salts of complex acids sometimes crystallize well and are useful in characterizing amines. Salts of chloroplatinic acid, H_2PtCl_6, are often used. A number of acidic organic substances serve a similar purpose. Among the most common of these is picric acid.

$$RNH_2 + H_2PtCl_6 \longrightarrow RNH_2 \cdot H_2PtCl_6$$

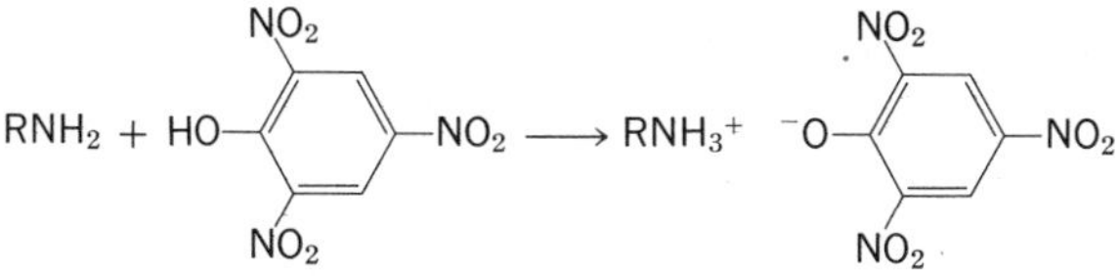

RNH2 + HO
NO2
NO2
NO2
RNH3+ −O
NO2
NO2
NO2

Picric acid   An amine picrate

The salts of picric acid, called *picrates,* are solids which, as a rule, can be readily purified by crystallization; as with all such salts the base is recovered by treating with alkali.

14.10 Reactions of Aliphatic Amines with Nitrous Acid With nitrous acid, HNO_2, reactions other than that of salt formation are observed. Here the different classes of amines behave differently, and for this reason nitrous acid is often used to distinguish between them.

All primary amines, and in fact all $—NH_2$ compounds including amides, react with HNO_2 at ordinary temperatures to evolve nitrogen.

$$CH_3CH_2NH_2 + HONO \longrightarrow (CH_3CH_2OH) + N_2\uparrow + HOH$$

The yields of hydroxy compounds are usually very low in the aliphatic series, and the reaction is of no value in the preparative sense. An explanation for this fact is forthcoming when the mechanism of the reaction is examined. When nitrous acid reacts with a primary amine, several nonisolable intermediates are formed.

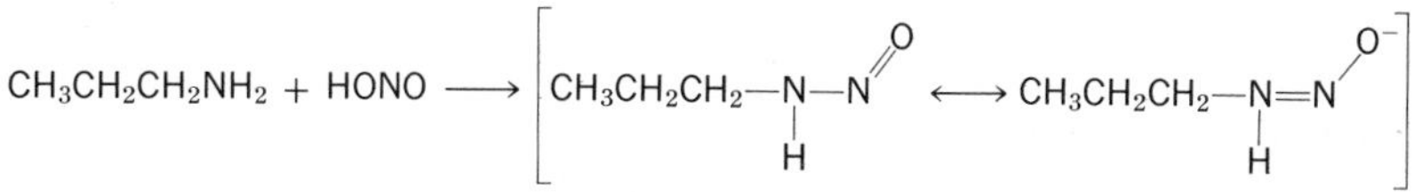

CH3CH2CH2NH2 + HONO
CH3CH2CH2—N—N
O
H
CH3CH2CH2—N=N
O−
H

A nitrosamine
(amide of nitrous acid)

The nitrosamine cannot be isolated if the nitrogen bears a proton, and the reaction proceeds via the loss of H_2O to yield a very unstable diazonium ion.

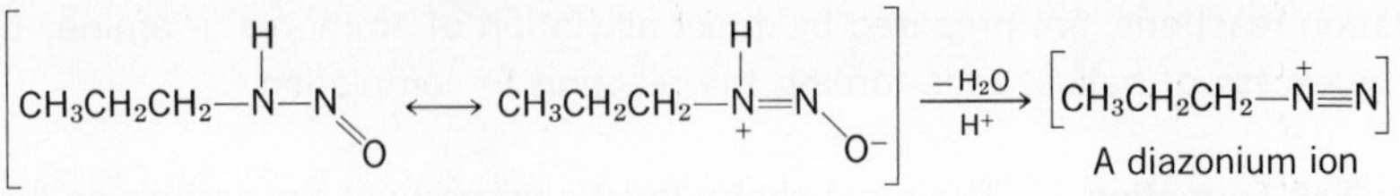

$$\left[CH_3CH_2CH_2-\overset{H}{\underset{}{N}}-N\overset{}{\underset{O}{\diagdown}} \quad \longleftrightarrow \quad CH_3CH_2CH_2-\overset{H}{\underset{+}{N}}=N\overset{}{\underset{O^-}{\diagup}}\right] \xrightarrow[H^+]{-H_2O} \left[CH_3CH_2CH_2-\overset{+}{N}\equiv N\right]$$

A diazonium ion

The instability of the diazonium ion is due in large part to its ability to form the very stable nitrogen molecule (N_2). The energy released on the formation of molecular nitrogen is sufficient to make the formation of the very unstable primary carbonium ion come within reach.

$$\left[CH_3CH_2CH_2-\overset{+}{N}\equiv N\right] \longrightarrow \quad N\equiv N:\uparrow \quad + \left[CH_3CH_2CH_2{}^+\right]$$

Very stable

As a consequence, the deamination reaction leads in all cases to very S_N1-like intermediates which have a strong tendency to rearrange, eliminate, etc. In the case of n-propylamine a variety of products are obtained.[1]

$$CH_3CH_2CH_2-\overset{+}{N}\equiv N \longrightarrow N_2 + \left[CH_3CH_2CH_2{}^+\right] \xrightarrow{H_2O} CH_3CH_2CH_2OH + H^+$$

$$\xrightarrow{-H_2O} CH_3CH=CH_2 + H_3O^+$$

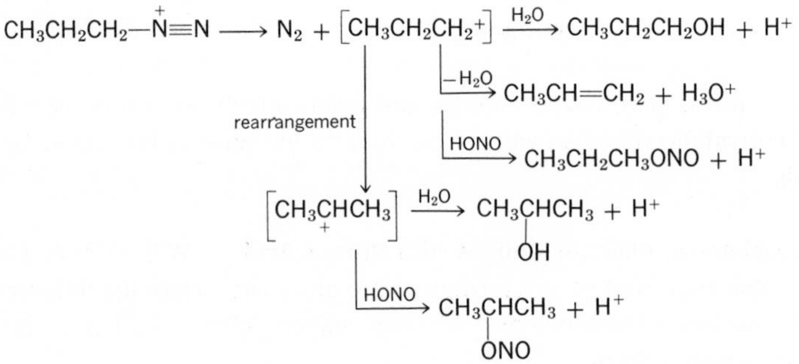

rearrangement

$$\xrightarrow{HONO} CH_3CH_2CH_3ONO + H^+$$

$$\left[CH_3\overset{+}{C}HCH_3\right] \xrightarrow{H_2O} CH_3\underset{OH}{CH}CH_3 + H^+$$

$$\xrightarrow{HONO} CH_3\underset{ONO}{CH}CH_3 + H^+$$

The products formed from this reaction are isopropyl alcohol, propylene, n-propyl alcohol, cyclopropane, nitrate esters, and various others depending on the medium. The reaction does not find great synthetic utility because of the extensive rearrangement which often accompanies it.

In this reaction the evolution of nitrogen is quantitative. It is the basis of the widely used Van Slyke amino-nitrogen analysis which is specific for the —NH_2 group (primary aliphatic and aromatic amines, and even amides).

With aromatic amines the diazonium salts are stable enough to be obtained in solution or even isolated in some cases, although when dry they are often explosive.

Secondary amines react with nitrous acid in a manner similar to that of primary amines, but since they lack a second hydrogen on the nitrogen, the nitrosoamine does not go on to a diazonium ion.

$$\underset{H_3C}{\overset{H_3C}{\diagdown}}NH + HONO \longrightarrow \underset{H_3C}{\overset{H_3C}{\diagdown}}N-N\overset{O}{\diagup} \longleftrightarrow \underset{H_3C}{\overset{H_3C}{\diagdown}}\overset{+}{N}=N\overset{O^-}{\diagup} \xrightarrow{/\!\!/} \underset{H_3C}{\overset{H_3C}{\diagdown}}\overset{+}{N}\equiv N$$

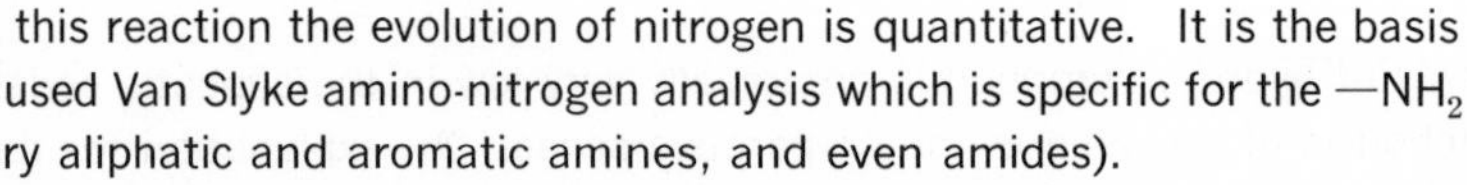

[1] Other products, such as cyclopropane, are also formed in this reaction, but this and other mechanistically complex products shall not be considered here.

These nitrosamines are isolable, and being amides of nitrous acid, are not basic. The secondary amine can be regenerated by hydrolysis just as with other amides. The formation of nitrosamines is specific for aliphatic or aromatic secondary amines.

Tertiary alkylamines are inert to the action of nitrous acid except for the expected salt formation, and these are recovered unchanged from nitrous acid solutions by addition of alkali.

$$(CH_3)_3N + HONO \longrightarrow (CH_3)_3\overset{+}{N}H \quad ONO^-$$

Tertiary aromatic amines are substituted in the ring to give C-nitroso compounds in much the same way nitric acid acts to produce nitro compounds.

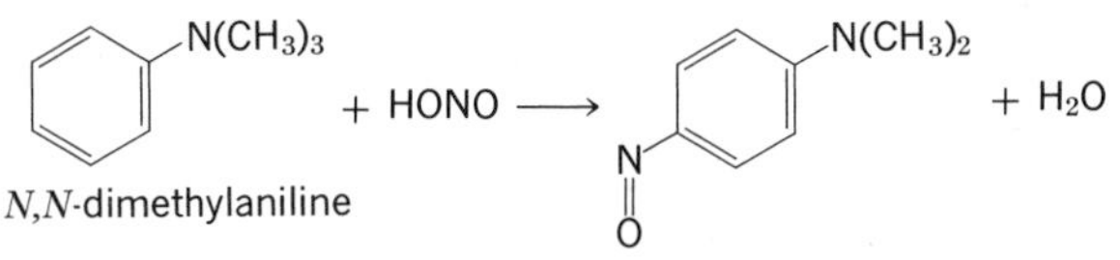
N(CH3)3
+ HONO ⟶
N(CH3)2
+ H2O
N
O

N,N-dimethylaniline

These often exhibit a green or blue-green color which can be used to help differentiate tertiary aromatic amines from aliphatic amines.

14.11 Carbylamine Reaction A reaction that has all but been replaced by spectroscopic and other methods, but which is of some historical interest, is the *carbylamine* test.

$$C_2H_5NH_2 + CHCl_3 \xrightarrow{NaOH} C_2H_5N{\equiv}C + NaCl + H_2O$$

A carbylamine
(ethyl isocyanide)

The carbylamines or isocyanides are neutral substances characterized by a particularly unpleasant sweetish odor. A whiff of the carbylamine reaction mixture is usually enough to make identification of a primary amine possible without the necessity of isolating products. Unfortunately false positive tests are given by other amines containing a little primary amine, and false negations by nonvolatile primary amines. A trace of isocyanide is usually formed in the reaction of cyanide ion with alkyl halides which accounts for the odor of some cyanide preparations.

14.12 Oxidation Reactions Although aliphatic amines can be oxidized, the reaction, in general, is of little practical utility because of the variety of products obtained. With aromatic amines the oxidation is more facile and invariably involves extensive destruction of the aromatic ring. The situation here closely parallels that observed in the case of phenols in that the oxidation products commonly undergo extensive degradations. A freshly distilled sample of pure aniline in contact with air soon turns dark orange in color owing to the formation of complex oxidation products. The sensitivity to oxidizing agents is utilized in the preparation of photographic developers (reducing agents) which are often complex hydroxy and aminobenzenes.

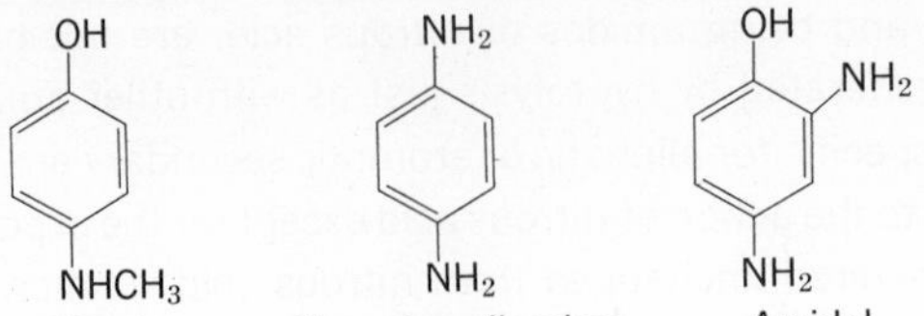

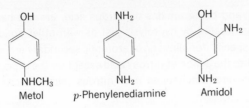

Metol | *p*-Phenylenediamine | Amidol

Amines, like the phenols, are more susceptible to oxidation in alkaline medium than in acid. This is due to the conversion of the amine by acid to its ammonium salt. The unpaired electrons on nitrogen can no longer become involved with the aromatic ring and the electron-rich ring of the free amine is made electron poor.

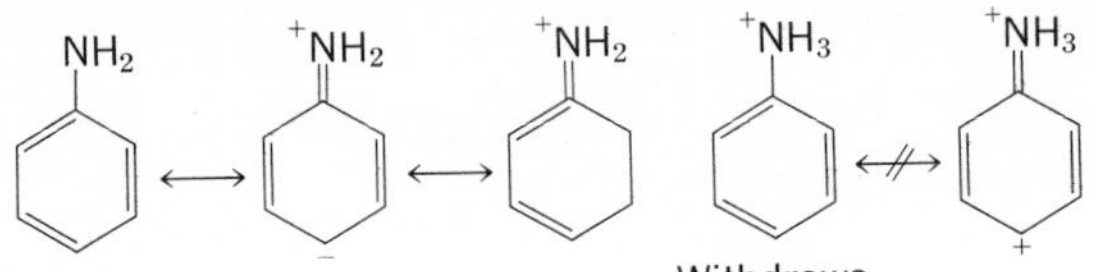

Increases electron density of ring | Withdraws electrons inductively

Since oxidizing agents are electrophilic[1] by nature, both the ring and the nitrogen function represent preferred sites of attack for oxidizing agents.

In performing synthetic operations on aromatic amines under oxidizing conditions (as with most electrophilic reagents) it is, therefore, necessary to *protect* the amino group. This is usually done by converting the amine to an amide with a convenient acid derivative. This process reduces the availability of electrons to the ring so that oxidation is less prevalent, but does not destroy the *o,p* directing effect of the nitrogen.

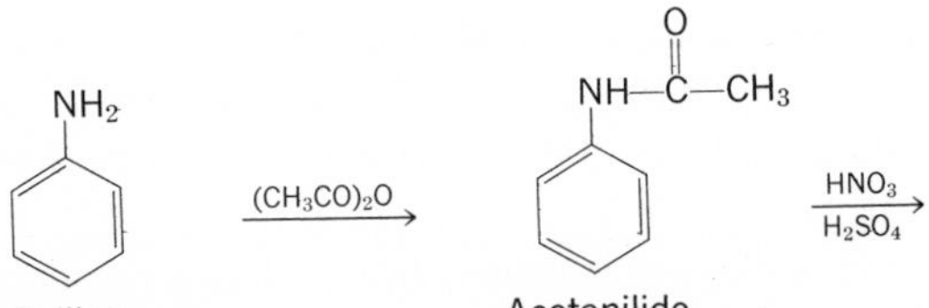

Aniline (strong *o,p* director but easily oxidized) | Acetanilide (*o,p* director; not easily oxidized)

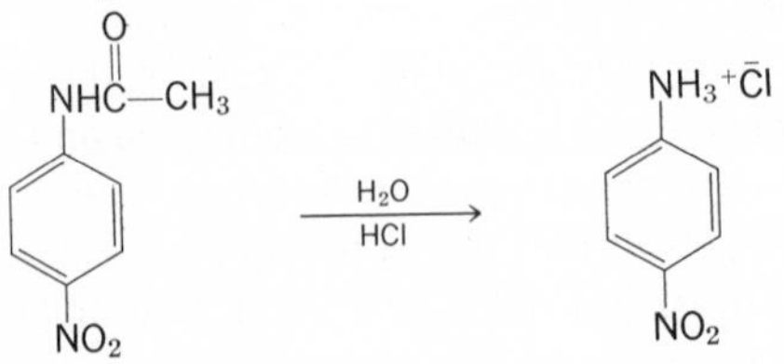

[1]Oxidation may always be thought of as the transfer of electrons from a reducing agent to an oxidizing agent. This may be an actual transfer or, as defined previously (Sec. 13.2), an effective transfer, as in converting C—H to C—Cl where the pair of bonding electrons is more strongly held by the more electronegative —Cl atom.

In the hydrolysis step here, HCl rather than KOH is used to catalyze the removal of the protecting acetyl group in order to minimize oxidation which occurs more readily with the free base (that is, in alkaline solution) than with its salts. *p*-Nitroaniline is, of course, obtained by an ionic reaction on addition of alkali to the final product.

The direct reaction of aniline with nitric-sulfuric acid is unsatisfactory both because of oxidation and because the amine is converted to the strongly deactivated ammonium salt.

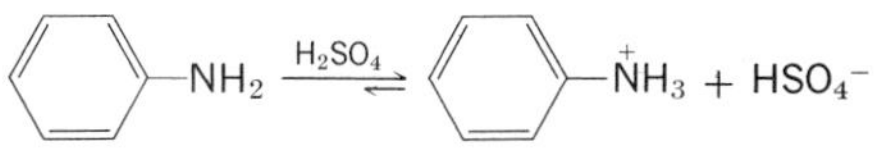

14.13 Halogenation The reaction of aniline with halogens takes an expected course (compare phenols, Sec. 8.24).

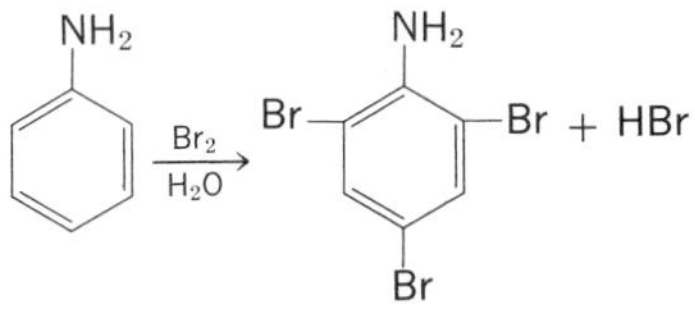

This substitution is uncontrollable; 2,4,6-tribromoaniline is obtained as the only product in practically quantitative yields. By protecting the amine group, partially substituted anilines can be prepared.

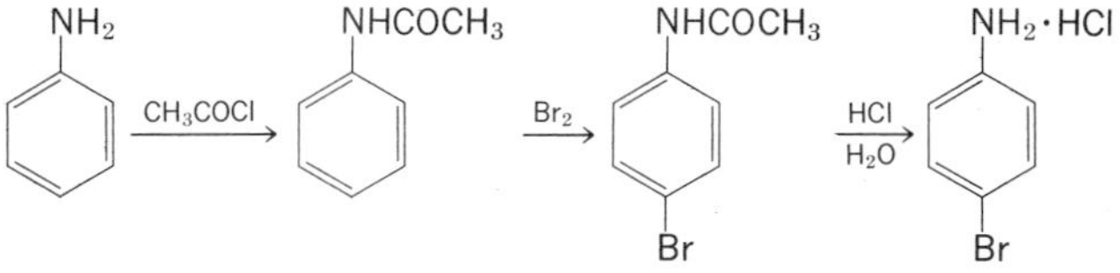

What has been done here and in the nitration reaction above is to decrease the activating effect of the amino group by substituting it with an electron-attracting group (an acyl group). This has the effect of decreasing the ability of the amino group to donate electrons to the ring and so decreasing the contribution of the resonance forms which would enhance ring activation. Obviously an alkyl group, being an electron-repelling group, would not be a very effective "protecting group." It also suffers from the fact that removal of alkyl groups is not as straightforward as that of acyl groups.

14.14 Sulfonation Sulfonation of the ring can be accomplished with aniline without "protection" because sulfuric acid is not a particularly effective oxidizing agent. The reaction is not carried out in the presence of excess sulfuric acid since the aniline would be converted to its strongly deactivated salt, but rather only one equivalent of sulfuric acid is used. The salt is then heated, and since water is driven off, the equilibrium ultimately favors the sulfonation product.

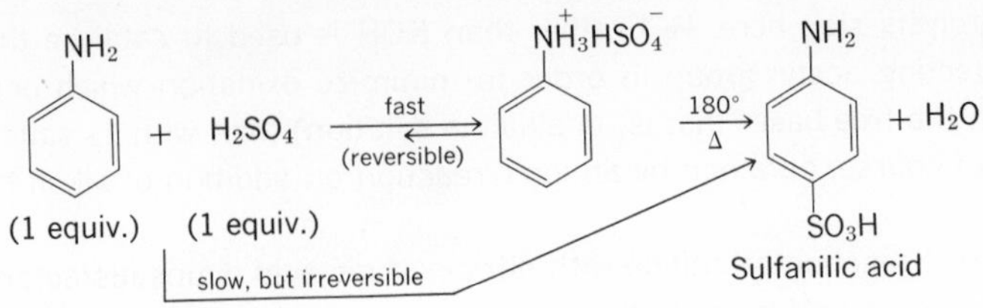
NH2
+ H2SO4
fast
(reversible)
NH3HSO4
180°
Δ
NH2
+ H2O
(1 equiv.)
(1 equiv.)
slow, but irreversible
SO3H
Sulfanilic acid

14.15 Exhaustive Methylation A reaction which has been of great value in the proof of structure of amino compounds, particularly the alkaloids, is known as exhaustive methylation. This sequence represents an important phase of the chemistry of quaternary salts and bases.

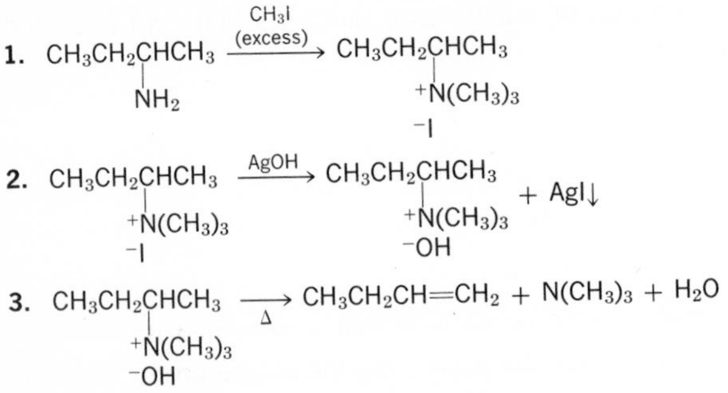
1. CH3CH2CHCH3
NH2
CH3I
(excess)
CH3CH2CHCH3
+N(CH3)3
-I
2. CH3CH2CHCH3
+N(CH3)3
-I
AgOH
CH3CH2CHCH3
+N(CH3)3
-OH
+ AgI↓
3. CH3CH2CHCH3
+N(CH3)3
-OH
Δ
CH3CH2CH=CH2 + N(CH3)3 + H2O

Reaction 1 illustrates the general method used in the preparation of quaternary salts; it may be applied for this purpose to any amine and reactive halide, but three methyl groups on nitrogen are required for Reaction 3.

Reaction 2 calls attention to the chief characteristic of the quaternary salt, *complete ionization*, and to the strength of the base from which it is derived. Since quaternary hydroxides, the free bases, are about as strong as sodium hydroxide, the addition of aqueous alkali is without effect on them. By addition of AgOH ($Ag_2O + H_2O$), the insoluble silver halide is precipitated, leaving only OH^- in solution with the R_4N^+ ion. On evaporation of such a solution, the water-soluble quaternary base is obtained in solid form.

In Reaction 3 the thermal decomposition of a quaternary base is shown. When the three alkyl groups are methyl groups, the decomposition usually follows the course shown with the ultimate formation of an alkene. The removal of a hydrogen from the chain with the $N(CH_3)_3$ reverses the usual rule (Chap. 3), that is, hydrogen is removed predominantly from that carbon having the most hydrogens. The nitrogen is completely removed from the molecule leaving an alkene. From the structure of the alkene it is often possible to deduce the structure of the original amine. The reaction proceeds via an E_2 elimination rather than an E_1 so that carbonium ions and their concurrent rearrangements are not involved. In general, the competing displacement reaction (S_N2) is not important unless no β hydrogens are available for elimination.

Some quaternary salts are found in nature among the water-soluble constituents of plants and animals. Choline in muscle and betaine, an "inner salt" from sugar beets, are examples.

$$\underset{\text{Choline}}{(CH_3)_3N^+CH_2CH_2OH\ OH^-} \qquad \underset{\text{Betaine}}{(CH_3)_3\overset{+}{N}—CH_2COO^-}$$

14.16 Ylide Formation: The Wittig Reaction If a quaternary salt which cannot give an elimination is treated with a very strong base, such as butyllithium, it is possible to remove an α hydrogen to form a species known as an *ylide*.[1]

$$(C_6H_5)_3\overset{+}{N}—CH_3I^- \xrightarrow{C_4H_9^-Li^+} \underset{\text{An ylide}}{[(C_6H_5)_3\overset{+}{N}—CH_2^-]} + C_4H_{10}\uparrow + LiI$$

In the case of nitrogen these ylides are very unstable and of limited interest. However, with the element phosphorus, which lies just below nitrogen in the periodic table, much more stable and synthetically useful ylides are formed.

$$\underset{\text{Triphenyl phosphine}}{(C_6H_5)_3P:} + CH_3I \longrightarrow \underset{\text{Triphenylmethyl phosphonium iodide}}{(C_6H_5)_3\overset{+}{P}—CH_3\ I^-}$$

$$(C_6H_5)_3\overset{+}{P}—CH_3\ I^- + C_4H_9^-Li^+ \longrightarrow (C_6H_5)_3\overset{+}{P}—CH_2^- + C_4H_{10}\uparrow + LiI$$

The phosphorus ylide can be stabilized by resonance involving *d* orbitals (not possible with nitrogen), and of course phosphorus is less electronegative than nitrogen.

$$(C_6H_5)_3\overset{+}{P}—CH_2^- \longleftrightarrow (C_6H_5)_3P{=}CH_2$$

The highly polar phosphorus ylides add to carbonyl groups of aldehydes and ketones to form new carbon-carbon bonds plus triphenylphosphine oxide.

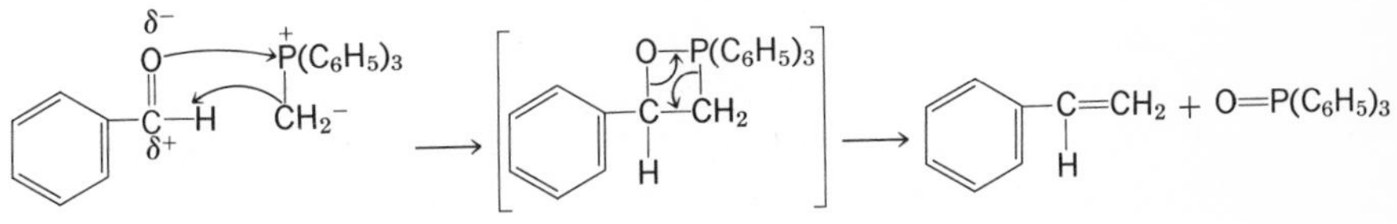
δ−
O
P(C6H5)3
C—H
CH2−
δ+
O—P(C6H5)3
C—CH2
H
C=CH2 + O=P(C6H5)3
H

The Wittig reaction, named after its discoverer, is general for the preparation of a variety of alkenes from aldehydes and ketones.

[1] Pronounced ĭll-ĭd.

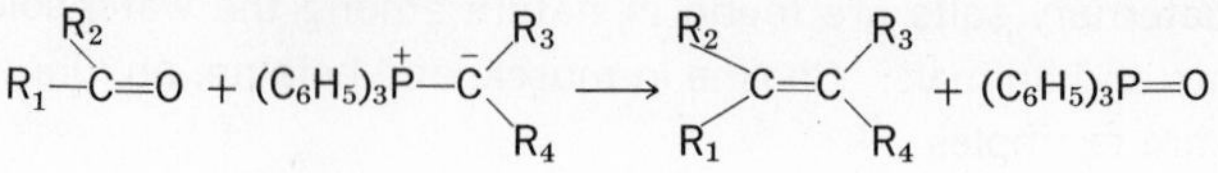

$$R_1{-}C{=}O + (C_6H_5)_3\overset{+}{P}{-}\overset{-}{C}\diagdown_{R_4}^{R_3} \longrightarrow \overset{R_2}{\underset{R_1}{\diagup}}C{=}C\overset{R_3}{\underset{R_4}{\diagdown}} + (C_6H_5)_3P{=}O$$

Here the R's may be the same or different; the chief restriction is that no other electrophilic groups such as H^+ or $C{=}O$ may be present.

DIAZONIUM SALTS

14.17 Diazotization and Coupling   In the case of most aliphatic compounds, diazonium salts are transient intermediates of limited synthetic utility. With aromatic compounds, however, diazonium salts are stable enough to be formed at low temperatures and, in some cases, even isolated, although they are still very reactive. Aromatic diazonium salts find widespread use in synthetic reactions. They are prepared from aromatic amines and nitrous acid in the presence of mineral acid and in fairly dilute solution. The reaction is called "diazotization."

$$\xrightarrow[\text{low temp., } H_2O]{NaNO_2, \text{ HCl}}$$

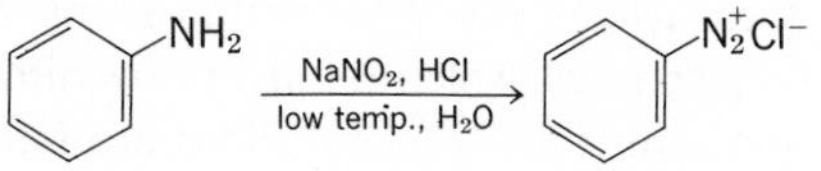

Benzene diazonium chloride

The structure of diazonium salts will be considered later. Some of the wide variety of reactions undergone by this group are indicated below to illustrate its importance in aromatic syntheses.

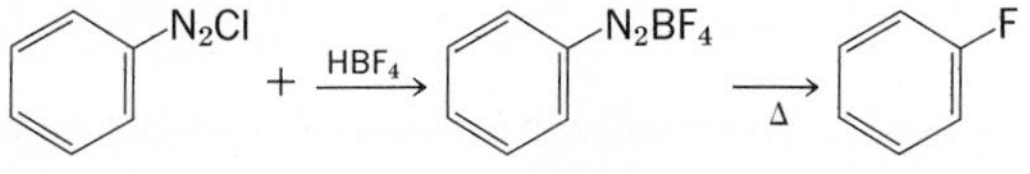

Isolated

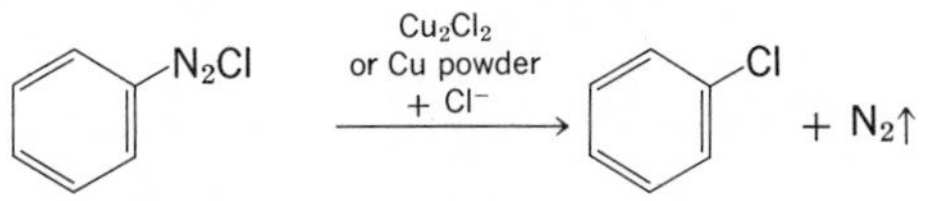

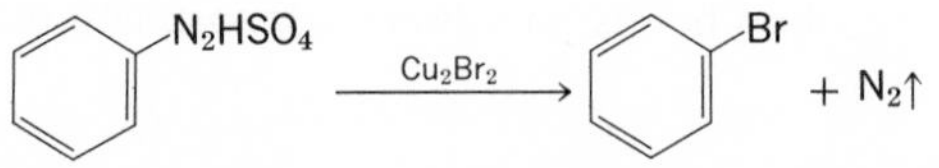

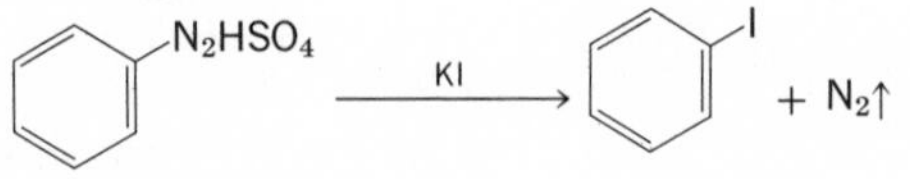

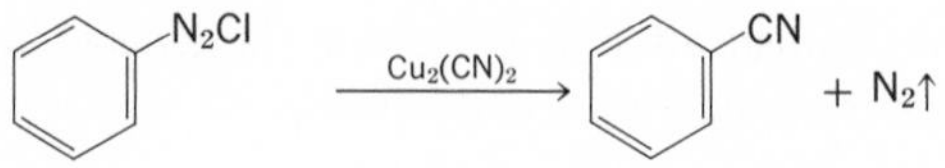

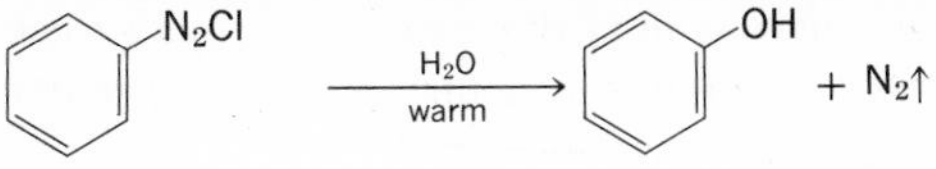
N2Cl
H2O
warm
OH
+ N2↑

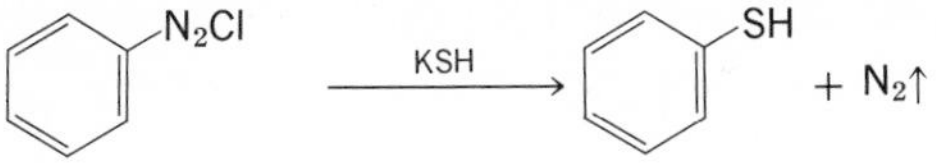
N2Cl
KSH
SH
+ N2↑

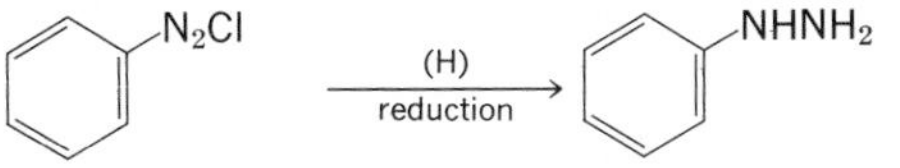
N2Cl
(H)
reduction
NHNH2

Phenylhydrazine

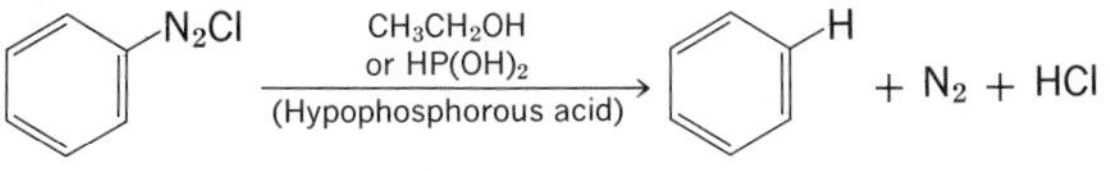
N2Cl
CH3CH2OH
or HP(OH)2
(Hypophosphorous acid)
H
+ N2 + HCl

The application of some of these reactions, such as the reaction with KI (almost the only general method for the preparation of aromatic iodides), will be evident at once. Because of the ease of substitution in the benzene series, the replacement of the diazonium group by halogen is useful only in special cases where directive influences are unfavorable for normal substitution. For example, the synthesis of *m*-chlorobromobenzene can be accomplished as follows:

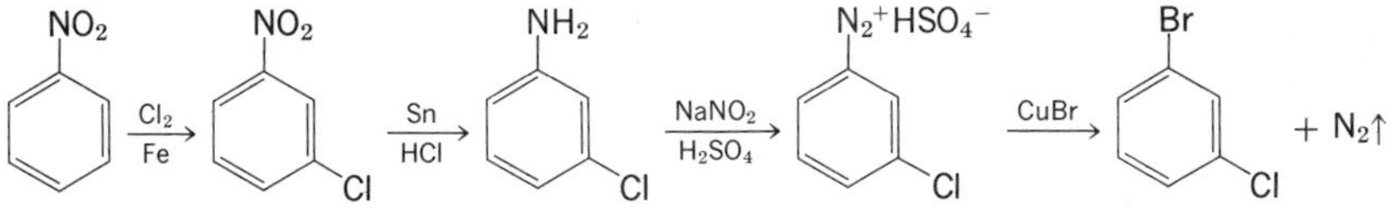
NO2
Cl2
Fe
NO2
Cl
Sn
HCl
NH2
Cl
NaNO2
H2SO4
N2+HSO4−
Cl
CuBr
Br
Cl
+ N2↑

In addition to the reactions given above, diazonium salts undergo an important and characteristic reaction known as the *coupling reaction*. When aromatic compounds having a reactive benzene ring, that is, containing an amino or hydroxy group, are added to neutral solutions of diazonium salts, the following reaction takes place:

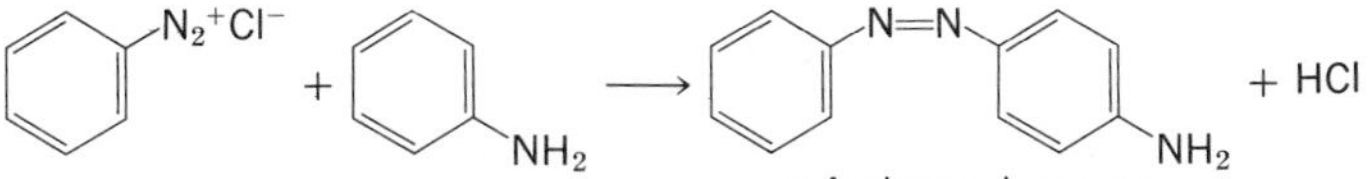
N2+Cl−
+
NH2
N=N
NH2
+ HCl

p-Aminoazobenzene

This would be an important side reaction during the diazotization of aniline if the reaction were not carried out in the presence of excess mineral acid so as to keep the concentration of free amine low.

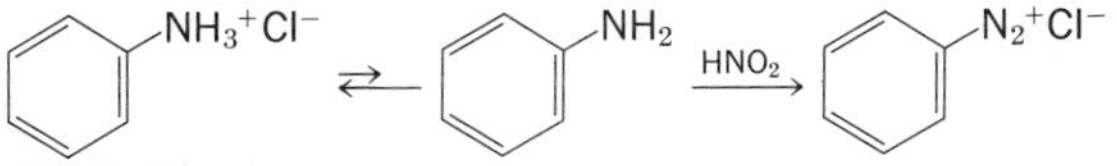
NH3+Cl−
NH2
HNO2
N2+Cl−

Most of the aniline is in this unreactive form

low conc.

Coupling will not take place with m directors or even with weak o,p directing groups, but only with phenols and amines in neutral or alkaline solution. It always occurs in the para position unless this position is already occupied, in which case the ortho compound results.

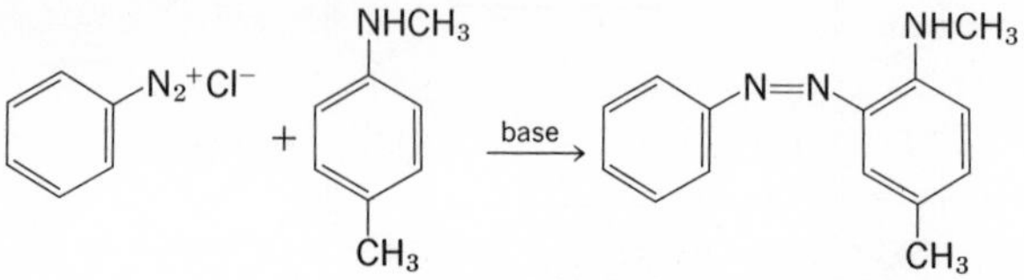

NHCH3
N2+Cl−
+
CH3
base
N=N
NHCH3
CH3

2-Methylamino-5-methylazobenzene

All azo compounds, such as those given above, are also more or less colored. By varying the groups on the two rings, substances with a variety of colors can be produced and a large number of dyes, known as *azo dyes,* are prepared by this

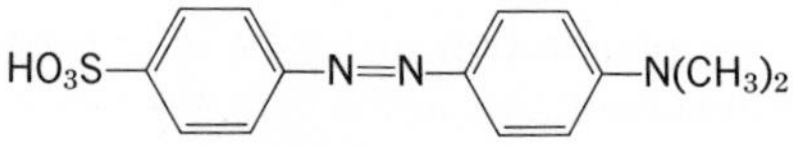

HO3S
N=N
N(CH3)2

Methyl orange

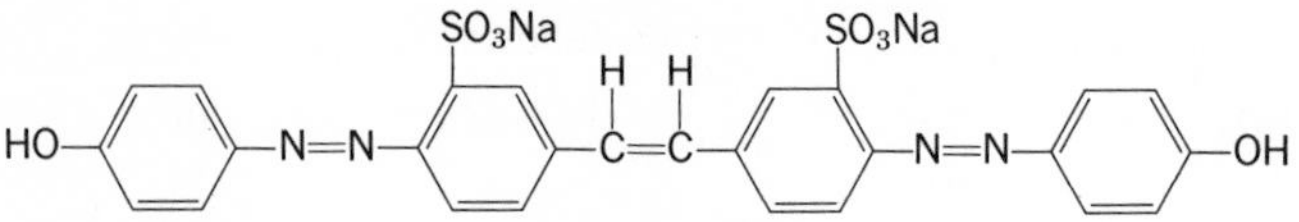

SO3Na
SO3Na
H H
HO
N=N
C=C
N=N
OH

Brilliant yellow

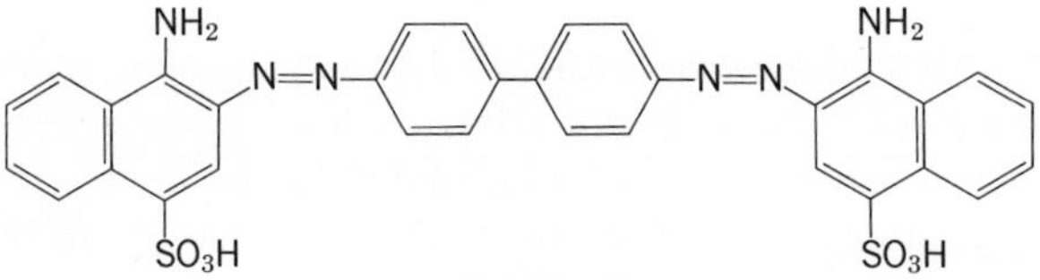

NH2
N=N
N=N
NH2
SO3H
SO3H

Congo red

reaction. Examples of a few commercial azo dyes are given above (note the presence of sulfonic acid groups to improve water solubility and aid in fixing the dyestuff to the fabric).

Azobenzene is taken as the parent substance for naming these compounds.

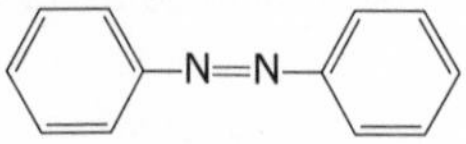

N=N

Azobenzene

The point of attachment of the nitrogen to each ring is always taken as the reference point in numbering substituents. Thus *p*-aminoazobenzene is also 4-aminoazobenzene. The substance below

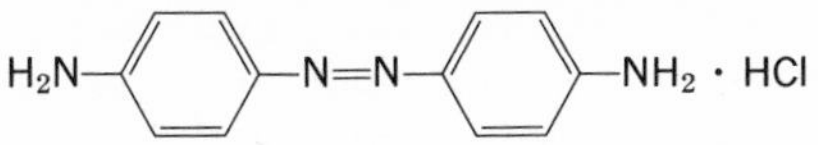

would be *p,p′*-diaminoazobenzene monohydrochloride, or 4,4′-diaminoazobenzene monohydrochloride. The prefix *azo-* is used when both the double-bonded nitrogens are attached to carbon; when one is attached to carbon and one to some other atom, the suffix is *diazo-*, for example, benzenediazohydroxide, $C_6H_5N{=}NOH$; and when one of the nitrogens is in the quaternary state, the ending *-onium* has to be used: $C_6H_5\overset{+}{N}{\equiv}N$, benzenediazonium ion.

Because of the mild conditions (low temperature, dilute neutral solution) under which coupling takes place, the reaction has been of value in "labeling" proteins in immunological research. Such azoproteins may form antibodies specific for the particular group coupled to them. Much valuable information on the nature of antitoxins, etc., has been accumulated by the use of this technique.

14.18 Structure of the Diazonium Group In general, the structure of the diazonium group is best represented as the ionic structure I*a* or more precisely as the resonance hybrid of I*a*, I*b*, I*c*, and I*d*.

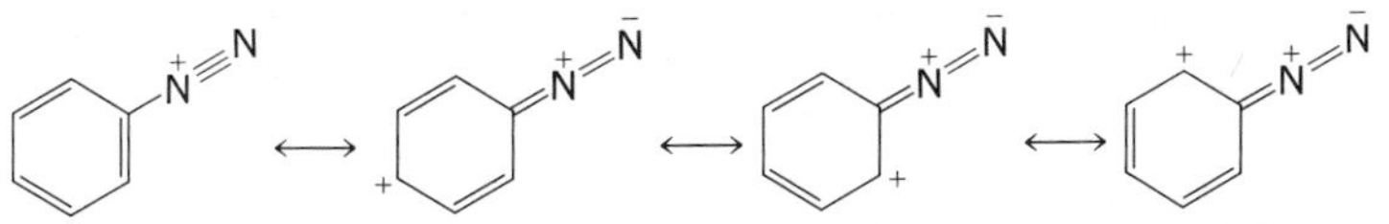

I*a* I*b* I*c* I*d*

Although structures I*b*, I*c*, and I*d* are not of particularly low energy, their contribution is to bring the very high energy of the aliphatic diazonium ion down to the energy level observed for aromatic diazonium ions.

Certain more covalent forms of the diazonium ion are known, such as diazohydroxides, but most of the reactions are best considered through the diazonium form.

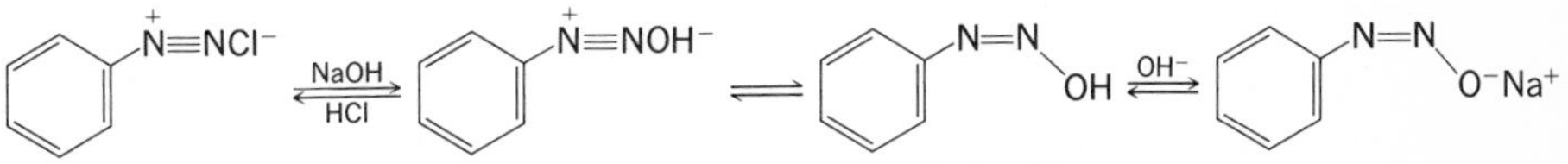

Diazonium hydroxide Diazohydroxide

The actual mechanism of replacement of the diazonium group is far more complex than the simple formulations would imply. It appears to involve free radicals (catalyzed by such species as Cu^+) rather than the ionic decomposition observed with aliphatic diazonium compounds.

The diazonium ion displays weak electrophilic character with highly activated aromatic rings as evidenced by the coupling reaction.[1]

[1] Note that coupling occurs with the formally uncharged terminal nitrogen. The other nitrogen cannot form an additional covalent bond without expanding its octet; the only process that can remove the formal charge is the recapture of a pair of bonding electrons by attack at the other nitrogen.

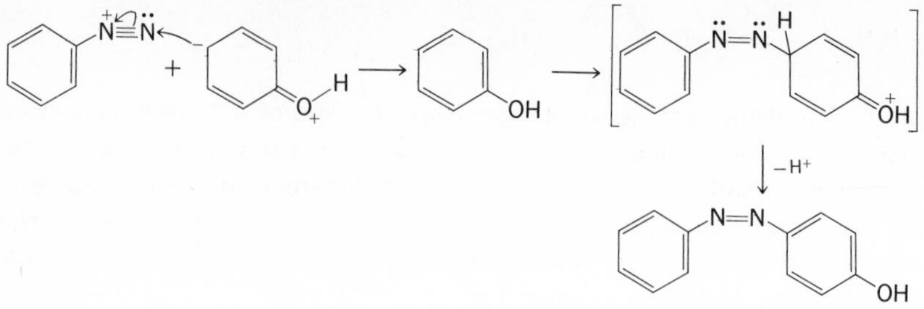

If the aromatic ring is not especially reactive, one of the other reactions, such as reaction with solvent, occurs.

14.19 Spectroscopic Properties Amines possess characteristic infrared N—H stretching bands in the region 3500 to 3300 cm^{-1} in dilute solution with the appearance of broad bands due to H bonding in more concentrated solution. Since alcohols have a hydrogen-bonded band in the 3500 to 3300 cm^{-1} region (the free OH band comes at 3700 to 3500 cm^{-1}), the presence of an amine cannot always be unequivocally established if oxygen is present in the molecule. The nmr of amines is strongly affected by hydrogen bonding and the N—H proton may be observed anywhere from $\delta = 0.6$ to 5.0. In hydroxylic solvents it may not be observed due to exchange. One ready test for the presence of amino hydrogen is given by shaking a solution in CCl_4 with D_2O. The NH is replaced by ND and the peaks shown previously by the NH disappear on such treatment. Other exchangeable groups, of course, also show such behavior.

Tertiary amines do not exhibit the above phenomena and are often difficult to demonstrate by spectroscopic means.

The amine function in itself does not give rise to absorption in the ultraviolet, but when conjugated with unsaturated systems it often gives rise to long wavelength shifts of absorption maxima.

Functions such as diazonium ions and azo compounds have characteristic uv and infrared maxima, but are of more specialized interest and will not be discussed here.

OUTLINE OF AMINE CHEMISTRY

Preparation

1. Reaction of alkyl or side-chain halides with ammonia, alkyl, or aryl amines.

$$RX + NH_3 \longrightarrow RNH_3X(RNH_2 \cdot HX) \xrightarrow{NaOH} RNH_2 + NaX$$

Compare

$$HX + NH_3 \longrightarrow NH_4X \xrightarrow{NaOH} NH_3 + NaX$$

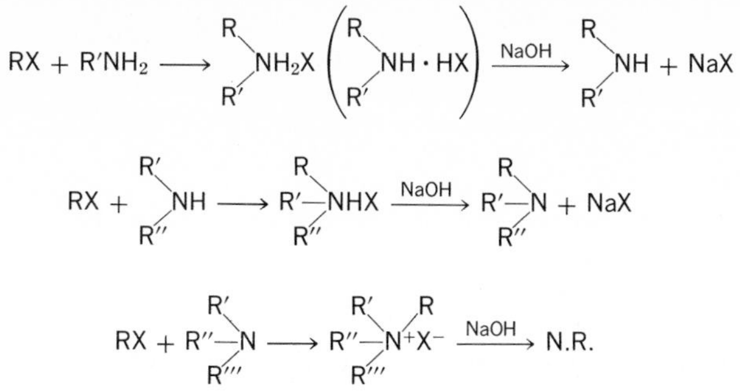

$$RX + R'NH_2 \longrightarrow \underset{R'}{\overset{R}{>}}NH_2X \left(\underset{R'}{\overset{R}{>}}NH \cdot HX \right) \xrightarrow{NaOH} \underset{R'}{\overset{R}{>}}NH + NaX$$

$$RX + \underset{R''}{\overset{R'}{>}}NH \longrightarrow \underset{R''}{\overset{R}{R'-}}NHX \xrightarrow{NaOH} \underset{R''}{\overset{R}{R'-}}N + NaX$$

$$RX + \underset{R'''}{\overset{R'}{R''-}}N \longrightarrow \underset{R'''}{\overset{R'}{R''-}}N^+_R X^- \xrightarrow{NaOH} N.R.$$

These reactions cannot be controlled. Thus, the reaction of C_2H_5Br and NH_3 gives a mixture of $C_2H_5NH_3Br$, $(C_2H_5)_2NH_2Br$, $(C_2H_5)_3NHBr$, and $(C_2H_5)_4NBr$. NOTE: ϕX cannot be used in these reactions.

2. Hofmann reaction.

$$RCONH_2 \xrightarrow[NaOH]{Br_2} RCONHBr \xrightarrow{NaOH} RN{=}C{=}O \xrightarrow{NaOH} RNH_2 + Na_2CO_3$$

An acid amide   An isocyanate

3. Reduction methods.

   a. $RNO_2 \xrightarrow[\substack{Sn + HCl, \\ Fe + HCl, \\ H_2 + Pt, etc.}]{(H)} RNH_2$

   b. $\underset{R'}{\overset{R}{>}}C{=}NOH \xrightarrow[\substack{or \\ LiAlH_4}]{(H)} \underset{R'}{\overset{R}{>}}CHNH_2$

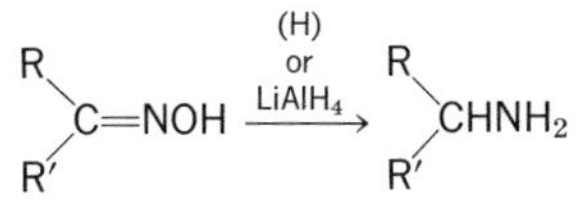

     An oxime

   c. $RC{\equiv}N \xrightarrow[\substack{or \\ LiAlH_4}]{(H)} RCH_2NH_2$

   d. $RCONR'R'' \xrightarrow{LiAlH_4} RCH_2NR'R''$

    $R{-}$ cannot be $H{-}$, $C_6H_5{-}$, or $\overset{R'}{\underset{R'}{R'C{-}}}$ for reaction *d*.

4. Phthalimide synthesis of pure primary amines.

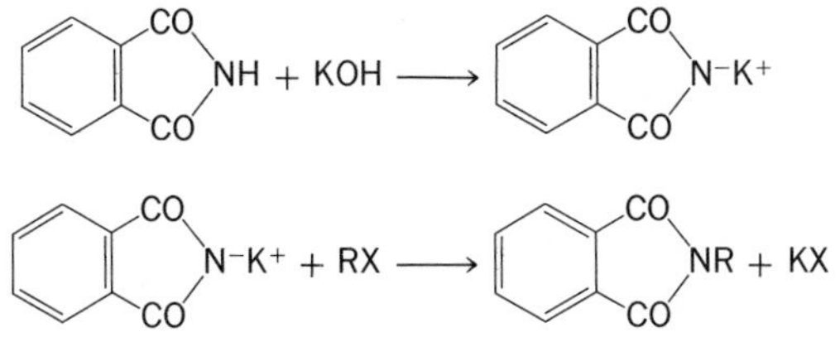

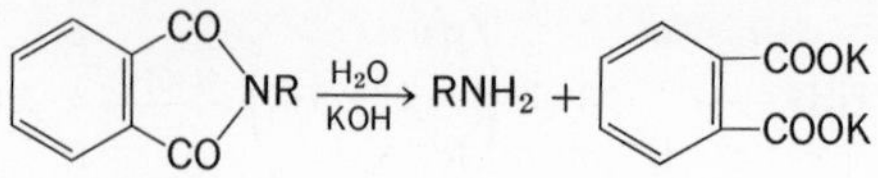

Properties

1. Salt formation. Amines, both aliphatic and aromatic, are very similar to ammonia in their basic properties.

$$RNH_2 + HX \longrightarrow RNH_3X$$
$$RNH_3X + NaOH \longrightarrow RNH_2 + H_2O + NaX$$

2. Carbylamine reaction.

$$RNH_2 + CHCl_3 + KOH \longrightarrow \underset{\substack{\text{An isocyanide}\\ \text{or carbylamine}\\ \text{(stink!)}}}{RN{\equiv}C} + KCl + H_2O$$

$$\left.\begin{matrix} R_2NH \\ R_3N \end{matrix}\right\} + CHCl_3 + KOH \longrightarrow \text{N.R. (no carbylamine)}$$

3. $RNH_2 + HNO_2 \longrightarrow ROH + N_2\uparrow + H_2O$ (not of preparative value)

$$\phi NH_2 + HNO_2 + HCl \xrightarrow[\text{cold}]{} \phi N_2Cl + H_2O \xrightarrow[\Delta]{} \phi OH + N_2\uparrow + HCl$$

$$R_2NH + HNO_2 \longrightarrow \underset{\text{A nitrosamine}}{R_2N{-}N{=}O} + H_2O$$

$R_3N + HNO_2 \longrightarrow$ N.R. (except salt formation or C—nitrosation of aromatic amines)

4. $RNH_2 + R'COCl \longrightarrow RNH{-}\underset{\underset{O}{\|}}{C}R' + HCl$

$$R_2NH + R'COCl \longrightarrow R_2N{-}\underset{\underset{O}{\|}}{C}R' + HCl$$

5. In preparing nitrated aromatic amines, it is usually necessary to "protect" the amino group before the nitration to minimize oxidative destruction.

$$C_6H_5NH_2 \xrightarrow{(CH_3CO)_2O} C_6H_5NHCOCH_3 \xrightarrow[H_2SO_4]{HNO_3}$$

$$p\text{-}C_6H_4\begin{matrix} NHCOCH_3 \\ NO_2 \end{matrix} \xrightarrow[HCl]{H_2O} p\text{-}C_6H_4\begin{matrix} NH_2 \\ NO_2 \end{matrix} + CH_3COOH$$

6. 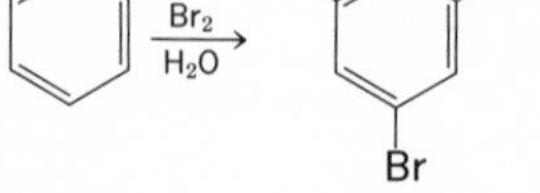

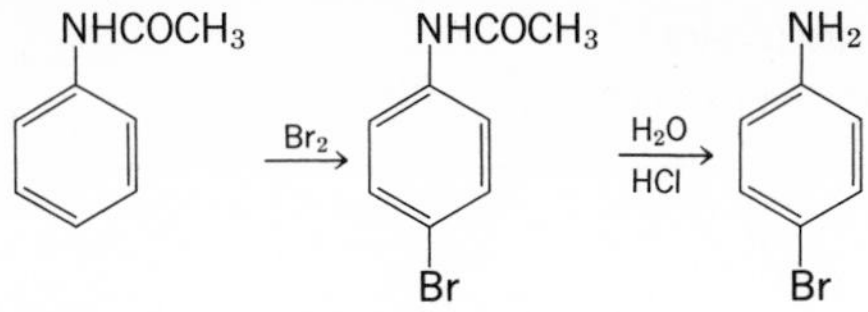
NHCOCH3
Br2
NHCOCH3
Br
H2O
HCl
NH2
Br

For reactions with halides, see Preparation.

QUATERNARY SALTS AND BASES

Preparation

$$R_3N + RX \longrightarrow R_4N^+X^-$$

Properties

1. In general, the quaternary salts behave as the salts of very strong bases. Thus, treatment of R_4NX with aqueous NaOH simply gives a solution containing R_4N^+, X^-, Na^+, and OH^-.
2. $R_4NX + AgOH \longrightarrow R_4NOH + AgX\downarrow$
3. $RCH_2CH_2N^+(CH_3)_3OH^- \xrightarrow[\Delta]{} RCH{=}CH_2 + (CH_3)_3N + H_2O$

   See Sec. 14.15.
4. Ylide formation. $CH_3N^+\phi_3I^- + C_4H_9Li \longrightarrow \bar{C}H_2N^+\phi_3 + LiI$

DIAZONIUM SALTS

Preparation

$$\phi NH_2 \xrightarrow[\substack{\text{HCl or}\\ H_2SO_4\\ \text{low temp.}}]{NaNO_2} \phi N_2Cl(\text{or } \phi N_2HSO_4) + NaCl\ (\text{or } NaHSO_4)$$

$$\phi{-}N{=}N{-}Cl \rightleftharpoons (\phi{-}N{\equiv}N)^+ + Cl^-$$

Properties

The properties of diazonium sulfates, nitrates, etc., are the same as those of the chlorides.

1. $\phi N_2Cl \xrightarrow[\Delta]{H_2O} \phi OH + N_2\uparrow + HCl$
2. Sandmeyer reaction.
   a. $\phi N_2Cl + Cu_2Cl_2 \longrightarrow \phi Cl + N_2\uparrow + Cu_2Cl_2$
   b. $\phi N_2Cl + Cu_2Br_2 \longrightarrow \phi Br + N_2\uparrow + Cu_2Cl_2$
   c. $\phi N_2Cl + Cu_2(CN)_2 \longrightarrow \phi CN + N_2\uparrow + Cu_2Cl_2$
3. $\phi N_2Cl + KI \longrightarrow \phi I + N_2\uparrow + KCl$

4. $\phi N_2Cl \xrightarrow{(H)} \phi NHNH_2 \cdot HCl$

A phenylhydrazine hydrochloride

5. $\phi N_2Cl \xrightarrow[\text{or } HP(OH)_2]{CH_3CH_2OH} \phi H$

6. $\phi N_2BF_4 \longrightarrow \phi F$

7. Coupling reactions: widely used in the preparation of azo dyes.

a. 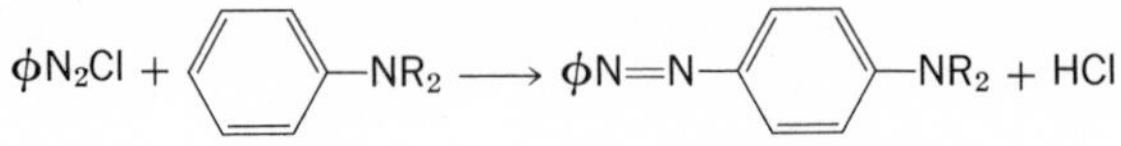

A *p*-dialkylaminoazo compound

b. 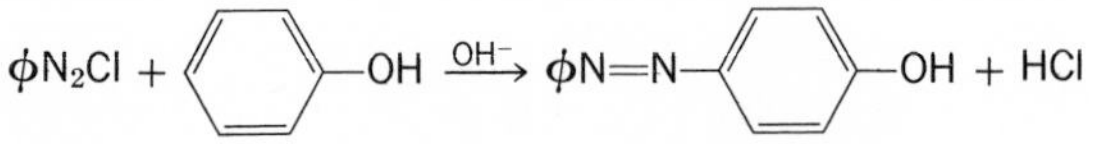

A *p*-hydroxyazo compound

With very few exceptions, OH, NH_2, NHR, or NR_2 groups must be present on the ring to permit this reaction with diazonium salts.

c. 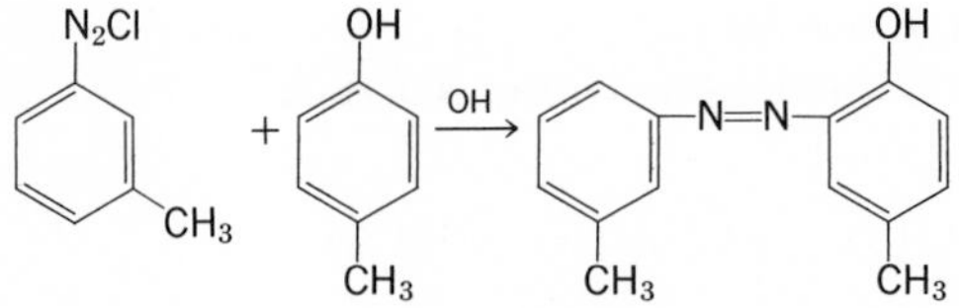

2-Hydroxy-5,3′-dimethylazobenzene

PHOSPHORUS DERIVATIVES

Preparation

Many of the reactions of phosphorus are similar to those of nitrogen except that phosphorus is less basic.

1. Alkylation

$$\underset{\text{Phosphine}}{PH_3} + RX \longrightarrow RPH_2 \longrightarrow R_2PH \longrightarrow R_3P \longrightarrow \underset{\text{Phosphonium salt}}{R_4P^+}$$

2. Wittig Reaction

$$\phi_3P + R_1\overset{X}{\overset{|}{C}}H{-}R_2 \longrightarrow \phi_3\overset{+}{P}CR_2X^- \text{(with H and } R_1\text{ on C)} \xrightarrow{C_4H_9Li} \underset{\text{Wittig reagent}}{\phi_3\overset{+}{P}{-}\overset{-}{C}R_1R_2}$$

$$\phi_3\overset{+}{P}{-}\overset{-}{C}R_1R_2 + R_3R_4C{=}O \longrightarrow R_3R_4C{=}CR_1R_2 + \phi_3P{-}O$$

CHAPTER 14 EXERCISES

1. In each of the following series arrange the bases in the expected order of decreasing base strength.

★*a.* $H_3C-\underset{\displaystyle H}{\underset{|}{N}}-CH_3$ $\quad H_5C_6-NH_2 \quad NH_3$

b. 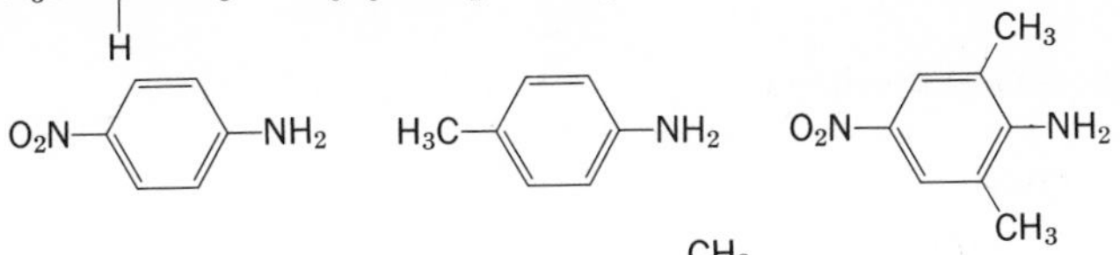

★*c.* 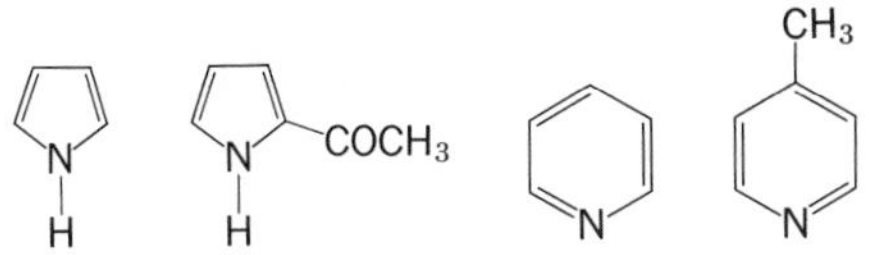

d. Aniline, ethylamine, nitrobenzene

★*e.* Phenol, acetamide, ammonia

f. Acetamide, phthalimide, phthalic acid

★*g.* Trimethylamine, sodium hydroxide, water

2. Show how you could distinguish between the following pairs of structures both by chemical and spectral (ir or nmr) means.

a. Butylamine and dibutylamine

b. $(CH_3)_3NHCl$ and $(CH_3)_4NCl$

c. Aniline and cyclohexylamine

d. *p*-Ethylaniline and $C_2H_5CONHC_6H_5$

e. $C_2H_3CH{=}NOH$ and $CH_3CONHOH$

f. $CH_3CHOH-CH_2NH_2$ and $(CH_3)_2COHNH_2$

g. CH_3COONH_4 and CH_3CONH_2

3. Show by equations how the following compounds could be synthesized from simple, available materials.

★*a.* 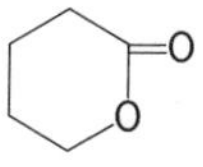

b. 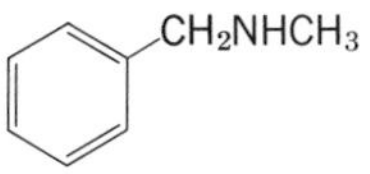

★*c.* 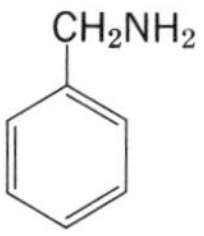

d. 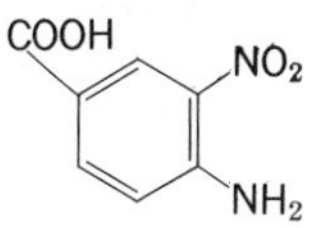

★*e.* $C_6H_5CONHCH_2COOH$

f.

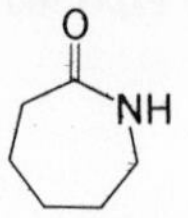

O
NH

(Seven-membered ring)

g.

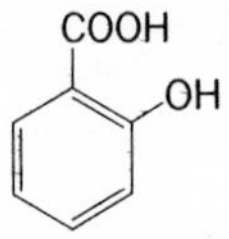

COOH
OH

★*h.*

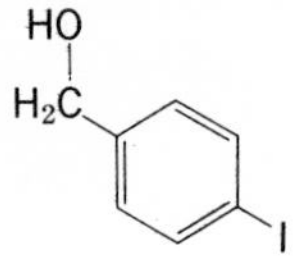

HO
H₂C
I

4. Convert:

a. Acetonitrile ⟶ diethylamine

b. Toluene ⟶ $C_6H_5CH_2NHCOC_6H_5$

c. Acetone ⟶ CH_3CN

d. 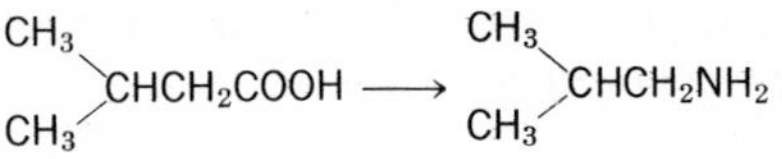

CH₃
CHCH₂COOH ⟶
CH₃
CH₃
CHCH₂NH₂
CH₃

e. Butyric acid ⟶ $CH_3CH_2CH_2NHCOCH_2CH_2CH_3$

5. How would you separate the following mixtures? Use methods involving chemical transformations to avoid reliance on unpredictable physical properties.

a. $(C_2H_5)_3N$ and hexanol

b. *p*-Nitroaniline and 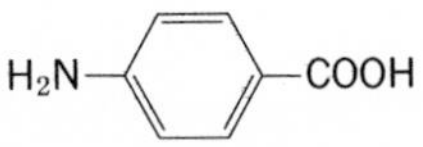

H₂N
COOH

c. C_2H_5—NH_2 and *N*-methylaniline

d. $CH_3CONHC_6H_5$ and 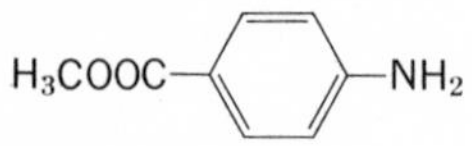

H₃COOC
NH₂

e. $CH_3CH_2CH_2CN$ and $CH_3CH_2CH_2CONH_2$

f. 

6. 

has a high melting point and is insoluble in ether. The infrared spectrum differs from that expected for a sulfonic acid. Explain.

7. Show how you would carry out the following conversions, using any compound of three carbon atoms or less (in addition to those shown).

★*a.* 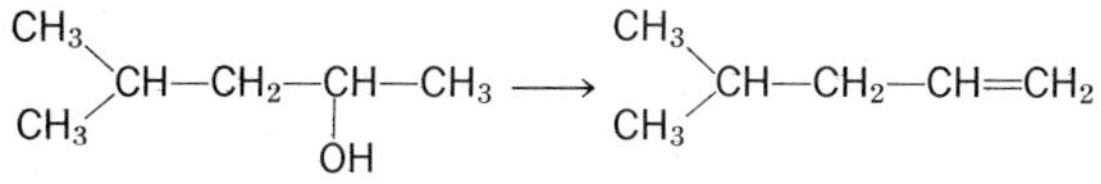

b. 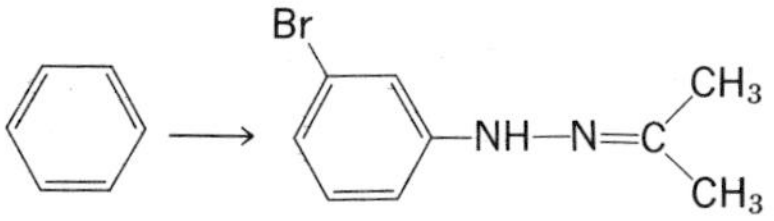

★*c.* 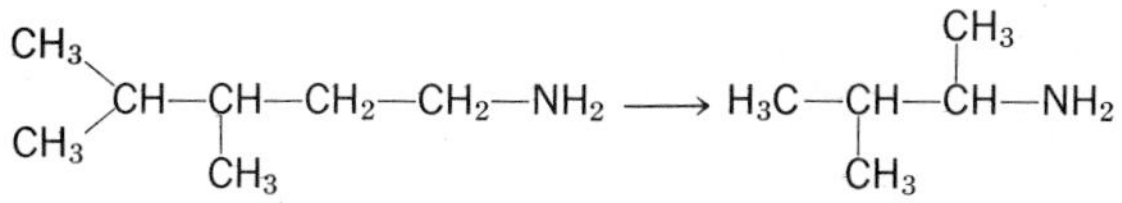

d. $CH_3-CH_2-CH_2-CH_2-Br \longrightarrow CH_3-CH_2-CH_2-CH_2NH_2$ (pure)

★*e.* 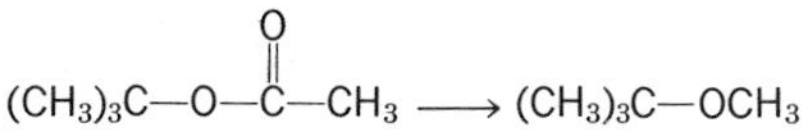

8. Account for the following. Use equations showing mechanisms where appropriate.

a. On adding one mole of KOH to a mixture of

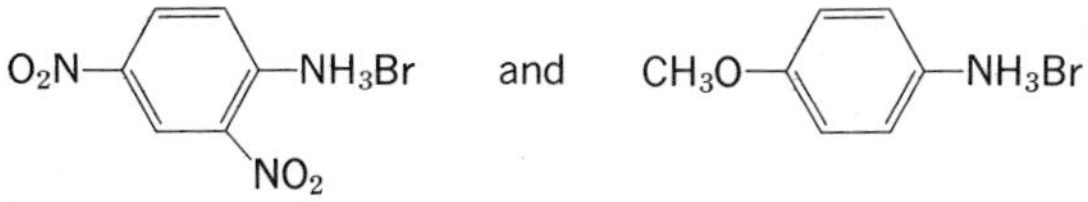

one of the substances is rendered insoluble in water, the other remaining in solution.

b. The two carbonyl groups of

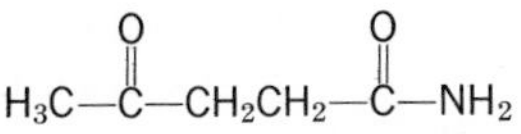

absorb at different frequencies in the infrared.

★9. A certain compound X, $C_{10}H_{13}N$, behaves as follows:

a. $X + NaNO_2 + HCl \longrightarrow$ X dissolves and can be recovered unchanged by adding KOH

b. $X \xrightarrow[\text{(excess)}]{CH_3I} \xrightarrow[H_2O]{Ag_2O} C_{11}H_{17}NO \xrightarrow[\Delta]{} (Y)C_{11}H_{15}N$

c. $Y \xrightarrow{KMnO_4} C_{10}H_{13}O_2N \xrightarrow{PCl_5} \xrightarrow{NH_3} \xrightarrow{NaOBr} (Z)C_9H_{14}N_2$

d. 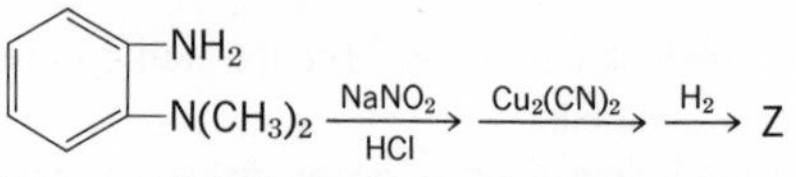

NH2
N(CH3)2
NaNO2
HCl
Cu2(CN)2
H2
Z

Interpret briefly each reaction and deduce the formula of X.

10. Fill in the structures

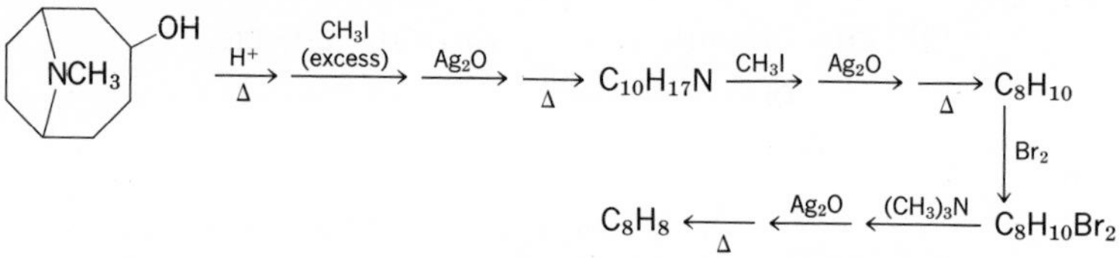

OH
NCH3
H+
Δ
CH3I
(excess)
Ag2O
Δ
C10H17N
CH3I
Ag2O
Δ
C8H10
Br2
C8H10Br2
(CH3)3N
Ag2O
Δ
C8H8

★11. A synthetic morphine substitute, "Methodone," $C_{21}H_{27}ON$, gives the following reactions:

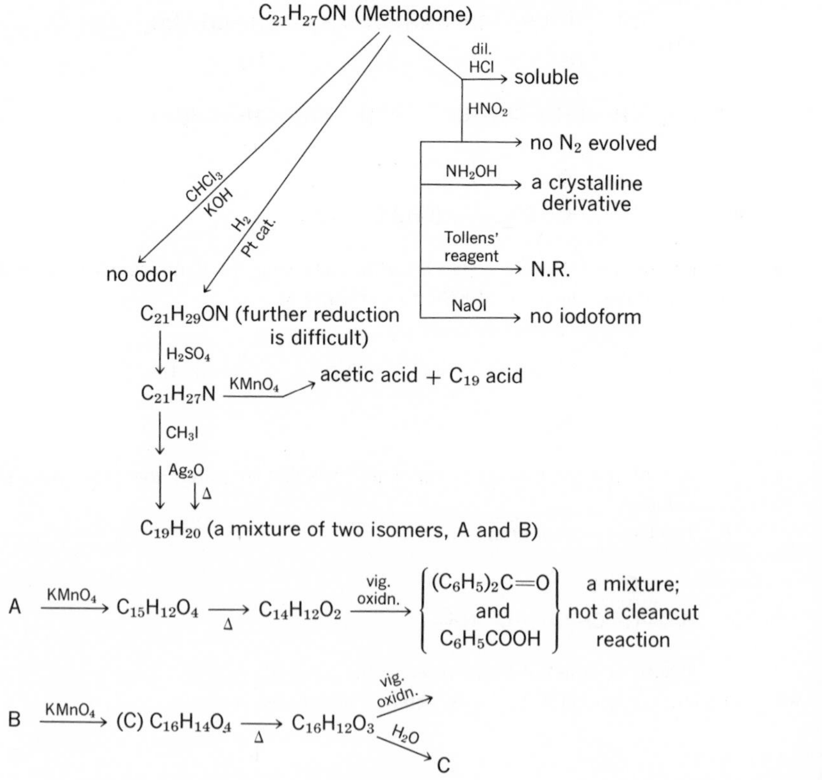

C21H27ON (Methodone)
dil. HCl
soluble
HNO2
no N2 evolved
NH2OH
a crystalline derivative
Tollens' reagent
N.R.
NaOI
no iodoform
CHCl3
KOH
no odor
H2
Pt cat.
C21H29ON (further reduction is difficult)
H2SO4
C21H27N
KMnO4
acetic acid + C19 acid
CH3I
Ag2O
Δ
C19H20 (a mixture of two isomers, A and B)
A
KMnO4
C15H12O4
Δ
C14H12O2
vig. oxidn.
(C6H5)2C=O
and
C6H5COOH
a mixture; not a cleancut reaction
B
KMnO4
(C) C16H14O4
Δ
C16H12O3
vig. oxidn.
H2O
C

Deduce the structure of Methodone. Assume no rearrangements in the above reactions.

12. $A = C_{13}H_{20}ClN$ {freely soluble in water; neutral; no carbylamine reaction; $\lambda = 6.15\mu$ among others}

↓ mild oxidation at low temp.

$B = C_{11}H_{16}O_2ClN$ {freely soluble in water; acidic; ir $\lambda = 6.15\mu$ replaced by $\lambda = 5.92\mu$.}

\+

$C = C_2H_4O_2$ (neutralization equivalent = 60)

B ↓ aq. NaOH

$D = C_{11}H_{14}O_2NNa$ (freely soluble in water; basic)

↓ (1) CH_3I (2) Ag_2O/H_2O, heat

$E = C_9H_8O_2$ + trimethylamine ($\lambda = 6.13\mu$)

↓ $KMnO_4$

$F = C_8H_6O_4$ (neutralization equivalent = 83)

↓ heating

$G = C_8H_4O_2$

Establish structures for all lettered compounds.

13. Indicate the ir absorption(s) you would expect to find in each product in the following sequences:

a. $CH_3Br \xrightarrow{KCN} A \xrightarrow{H_2O} C_2H_5ON \xrightarrow{CH_3OH} C \xrightarrow{LiAlH_4} D$

b. $CH_2CH(OH){-}CH_2CHO \xrightarrow[\Delta]{} A \xrightarrow{\text{Tollens' reagent}} C \xrightarrow{PCl_5} D \xrightarrow{NH_3} E \xrightarrow{NaOBr} F$

14. Show by equations a rational mechanism for the following reactions:

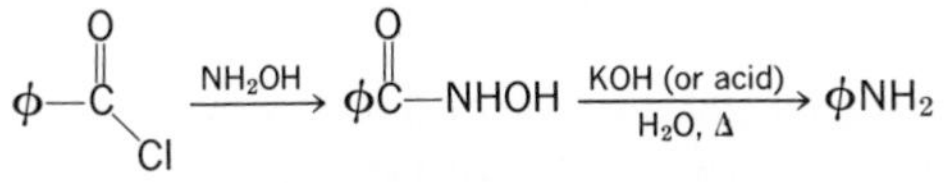

15. Why does the oxidation of *m*-diaminobenzene, which occurs very readily, not yield any *m*-benzoquinone? The para compound is readily converted to *p*-benzoquinone.

★16. Which of the following compounds would you expect to show absorption in the visible region, i.e., be colored?

a. 

b. 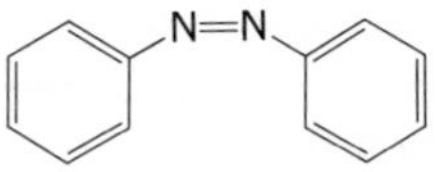

c. 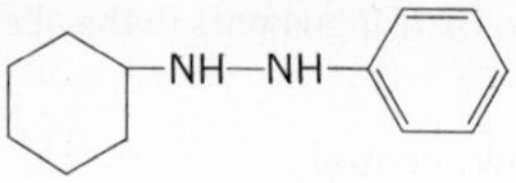
NH—NH

d. 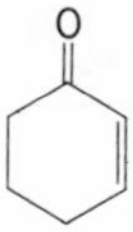
O

e. 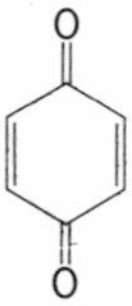
O
O

f. $(C_6H_5)_3N$

17. Arrange the compounds in the following groups in order of decreasing basicity.

a. Aniline, ethylamine, nitrobenzene

b. Phenol, acetamide, ammonia

c. Acetamide, phthalimide, phthalic acid

d. Trimethylamine, sodium hydroxide, water

18. Show how you could separate a mixture of:

a. Aniline, triethylamine, acetanilide, phenol

b. Methylethylamine, *p*-nitro-*N*-methylamine, tetramethylammonium chloride

c. Acetamide, phthalimide, ammonium phthalate

d. 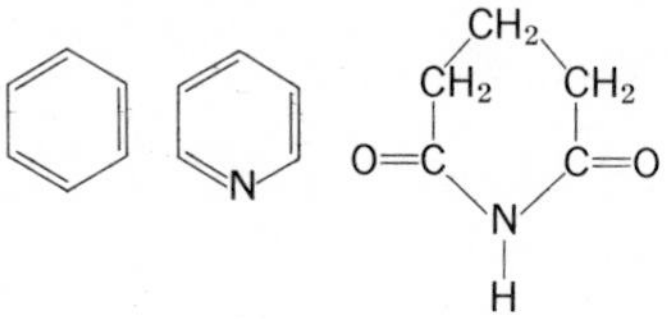
N
CH2
CH2
CH2
O=C
C=O
N
H

19. What products might be expected from the reaction of 2-methyl-1-aminopropane with nitrous acid, assuming the reaction to go by way of a carbonium ion intermediary? Explain.

★20. It is observed that both aromatic amines and phenols are more easily oxidized in basic solution than in acid. How is this explained?

21. Oxidation of a derivative of nicotine yields hygrinic acid, $C_6H_{11}O_2N$. Twice-repeated, exhaustive methylation of this substance gives the acid $C_5H_6O_2$. Catalytic hydrogenation of this acid shows it to contain either two double bonds or one triple bond, and yields *n*-valeric acid. Give the structure of hygrinic acid and show your reasoning clearly. Write equations for the reactions.

★22. The alkaloid piperine, $C_{17}H_{19}NO_3$, is the substance chiefly responsible for the sharp taste of pepper. Hydrolysis of piperine by acids gives piperic acid and a nitrogenous compound, $C_5H_{11}N$. Exhaustive methylation of the latter compound gives an unsaturated amine, which on repetition of the exhaustive methylation procedure gives a diolefin containing five carbons which is oxidizable to malonic acid. Derive the structure of the nitrogenous compounds. Show your reasoning clearly.

Piperic acid has the formula $C_{12}H_{10}O_4$. Titration shows its equivalent weight to be about 220. It adds 205 ml (S.T.P.) of hydrogen per gram on catalytic hydrogenation. It can be ozonized to piperonal

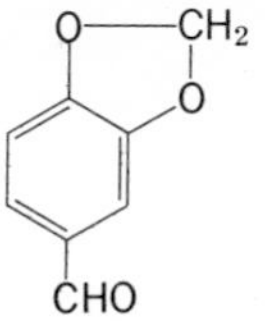

Piperonal
(heliotropine)

and oxalic acid can be isolated from its oxidation. Deduce the structure of piperine. Show your reasoning clearly.

23. Show by equations how exhaustive methylation could be used to distinguish:

a. Cyclohexylamine and 2-aminohexane

b. 1-Aminobutane and 1-dimethylaminobutane

24. Show by equations how the following conversions could best be accomplished:

a. *o*-Nitroaniline to *o*-nitrobenzoic acid

b. Benzene to *m*-dibromobenzene

c. Aniline to *m*-aminobenzoic acid

d. Aniline to *p*-dimethylaminoazobenzene

e. Nitrobenzene to *p*-dibromobenzene

f. Toluene to 3,5-dibromo-4-cyanotoluene

g. Aniline to *p*-hydroxyazobenzene

h. Benzene to *p*-hydroxy-*p*′-chloroazobenzene

25. Tyramine, a naturally occurring pressor amine, has the structure

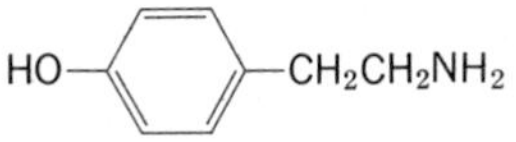

How could you prove this structure by chemical reactions?

★26. Compound X is basic, reacts with 2 moles of acetic anhydride, hydrolyzes to *p*-aminobenzoic acid and 2-aminoethanol. Its infrared spectrum shows a maximum at 1626 cm^{-1} which is absent from both hydrolysis products. What is the structure of X?

27. Neurine, a toxic substance formed in decaying animal tissue, has the structure

$$(CH_3)_3\overset{+}{N}{-}CH{=}CH_2 \quad OH^-$$

Devise a proof of this structure. Synthesize the compound from easily available starting materials.

★28. Chloromycetin has the formula $C_{11}H_{12}O_5N_2Cl_2$, fails to evolve a gas with nitrous acid, and does not react with Tollens' reagent. It has ir maxima at 1600, 1660, and 3200 cm^{-1}, but none at 1720 to 1740 or 2650 to 2880 cm^{-1}. Hydrolysis of chloromycetin forms two compounds: A, $C_2H_2O_3$, and B, $C_9H_{12}O_4N_2$. A has ir maxima at 1720 and 2800 cm^{-1} plus an nmr peak at $\delta = 9.5$ which does not exchange with D_2O, and it dissolves in $NaHCO_3$ with the evolution of a gas. B contains no carbonyl groups, gives an unpleasant smelling product with $CHCl_3$ and NaOH, and forms a triacetyl derivative with acetyl chloride. Careful oxidation of B gives a keto acid, $C_9H_8O_5N_2$, that evolves a gas with HNO_2; more vigorous oxidation produces *p*-nitrobenzoic acid. Deduce a formula for chloromycetin and show your reasoning.

29. Show how you would carry out the resolution of the following racemic compounds into their optical enantiomorphs. Assume you have available an optically active tertiary amine

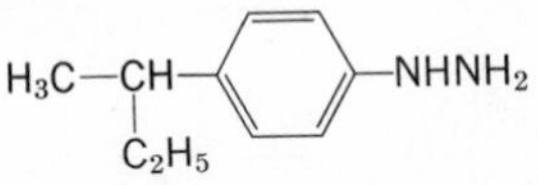

H3C—CH—
C2H5
—NHNH2

and optically active lactic acid.

a. 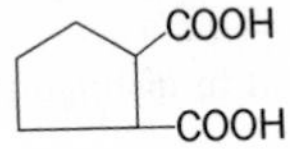

COOH
COOH

(Would this have to be cis or trans for resolution?)

b. 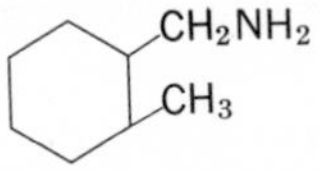

CH2NH2
CH3

c. 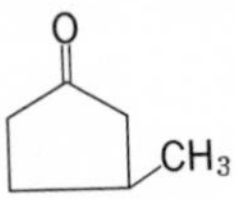

O
CH3

d. 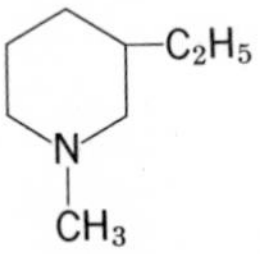

C2H5
N
CH3

e. $CH_3CH(C_2H_5)—CONH_2$

HINT: Try hydrolysis and regenerate.

★30. Suggest a reasonable formula for each of the lettered compounds which fits all the data given. Identify the absorption bands.

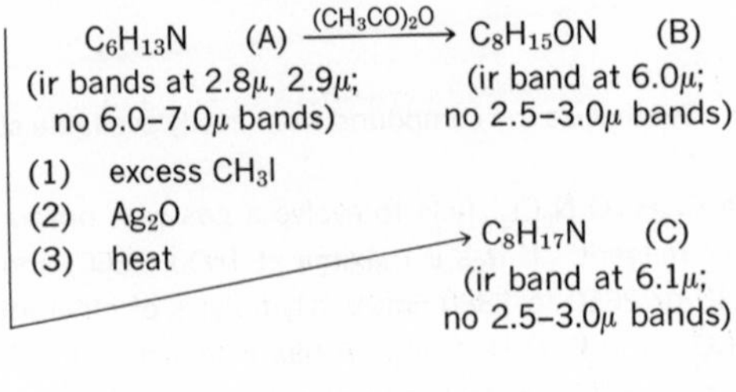

C6H13N (A)
(ir bands at 2.8μ, 2.9μ;
no 6.0–7.0μ bands)
(CH3CO)2O
C8H15ON (B)
(ir band at 6.0μ;
no 2.5–3.0μ bands)
(1) excess CH3I
(2) Ag2O
(3) heat
C8H17N (C)
(ir band at 6.1μ;
no 2.5–3.0μ bands)

15 SULFUR COMPOUNDS

15.1 Nomenclature Sulfur, being of the same group in the Periodic Table as oxygen, is capable of forming compounds in which sulfur atoms replace oxygen atoms. These are usually referred to as *thio* compounds. In the Geneva nomenclature, *thiol* is used to indicate an —SH group.

CH_3SH

Methanethiol
(methyl mercaptan)

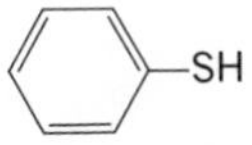

Thiophenol
(mercaptobenzene)

$C_2H_5—S—CH_3$

Methyl ethyl thioether
(methyl ethyl sulfide)

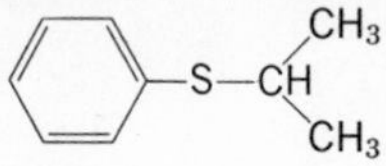

Isopropyl phenyl sulfide

$H_3C—S—S—CH_3$

Methyl disulfide

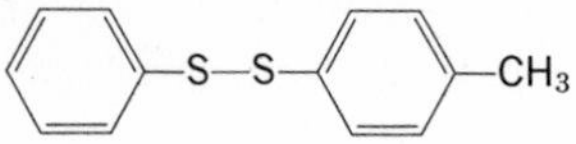

Phenyl *p*-tolyl disulfide

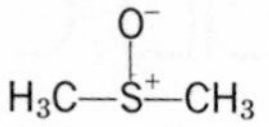

Methyl sulfoxide

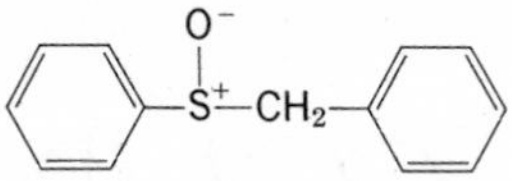

Benzyl phenyl sulfoxide

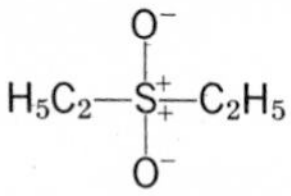

Ethyl sulfone

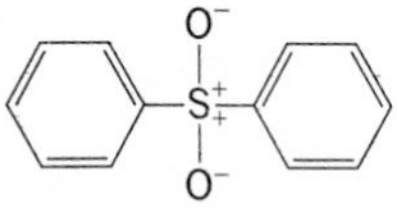

Phenyl sulfone

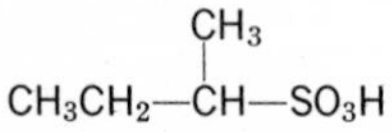

1-Methylpropanesulfonic acid

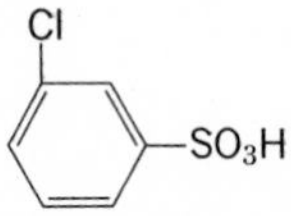

m-Chlorobenzenesulfonic acid

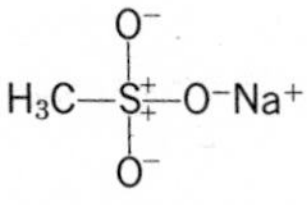

Sodium methanesulfonate

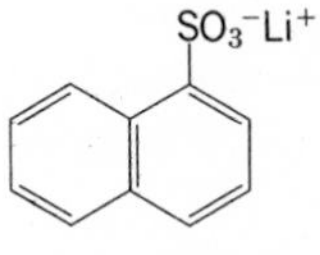

Lithium α-naphthalene-
sulfonate

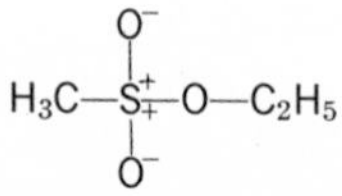

Ethyl methanesulfonate

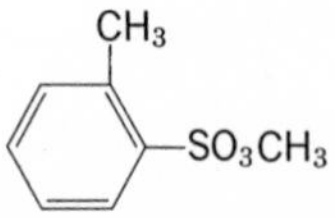

Methyl *o*-toluenesulfonate

$CH_3CH_2CH_2CH_2SO_2Cl$

Butanesulfonyl chloride

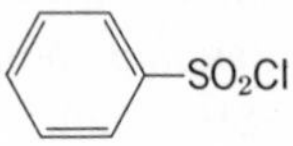

Benzenesulfonyl chloride

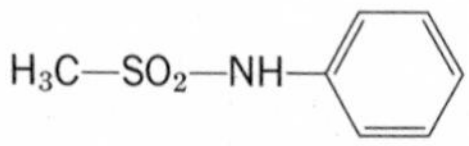

N-Phenylmethanesulfonamide

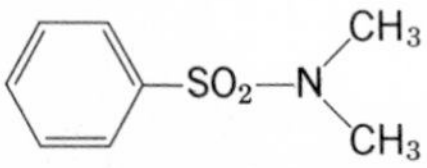

N,N-Dimethylbenzenesulfonamide

$(CH_3)_3S^+I^-$
Trimethylsulfonium iodide

These substances bear a marked formal resemblance to the oxygen analogs, but differ in a number of ways. The chief difference arises from the greater tendency of the sulfur atom to form coordinate covalences as compared to oxygen. This arises from the greater distance of valence electrons from the nucleus, and the lower electronegativity of sulfur.

15.2 Properties and Reactions A great many sulfur compounds analogous to oxygen compounds, such as thiol esters $\left(\mathrm{R\overset{O}{\overset{\|}{C}}{-}SR}\right)$, thio esters $\left(\mathrm{R\overset{S}{\overset{\|}{C}}{-}OR}\right)$, dithio acids $\left(\mathrm{R\overset{S}{\overset{\|}{C}}{-}SH}\right)$, thio aldehydes $\left(\mathrm{R\overset{S}{\overset{\|}{C}}{-}H}\right)$, thio ketones $\left(\mathrm{R\overset{S}{\overset{\|}{C}}R}\right)$, etc., are known, but many of these are of limited importance at present. On the other hand, the thiols, thio ethers, and their oxidation products are of considerable importance.

The aliphatic thio alcohols, often called "mercaptans," a name derived from the ability of these compounds to form insoluble mercury derivatives, are usually prepared synthetically. Only a few representatives of this series are found in nature. *n*-Butylmercaptan is the odorous constituent of the secretion of the skunk; this peculiarly disagreeable type of odor characterizes the lower members of the series. The odor of ethyl mercaptan is said to be detectable at a dilution of 1 part mercaptan to 50,000,000,000 parts air.[1]

Mercaptans may be synthesized by a displacement reaction from an alkyl halide and sodium or potassium hydrogen sulfide. Since the hydrogen sulfide ion is a better nucleophile and a poorer base than hydroxide ion, elimination is less important than in the alcohol synthesis (Sec. 8.4).

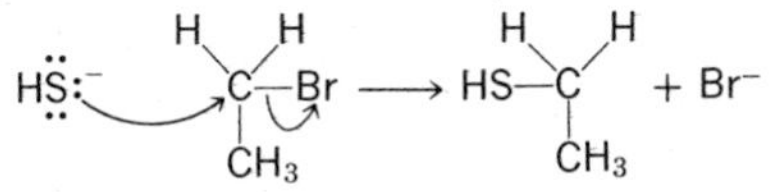

H
H
HS:⁻
C
Br
CH3
HS
C
H
H
CH3
+ Br⁻

One of the outstanding differences between the thiols and alcohols lies in the greater acidity of the former. Bases react with thiols to form salts which are not excessively hydrolyzed in aqueous solution.

$$C_2H_5SH + NaOH \rightleftharpoons C_2H_5S^- \,{}^+Na + HOH$$

$$C_2H_5OH + NaOH \rightleftharpoons C_2H_5O^- \,{}^+Na + HOH$$

This stability is due to the fact that the thiols, in contrast to the alcohols, are stronger acids than water.

[1] Some mercaptans are used because of their odor. Mercaptan by-products from the refining of petroleum are added to household gas to warn of leaks or escaping gas. They are also used in industrial warning signals of various kinds.

A practical application of the acidity of the thiols is found in the refining of petroleum. Small amounts of thiols are often found in crude petroleums and are undesirable because of their odor and the corrosive nature of their combustion products (SO_2 and SO_3). They may be removed from petroleum (and other mixtures) by converting them to their water-soluble, nonvolatile salts.

Although the S—H bond is more completely ionized than the O—H bond, the grouping as a whole shows much less tendency to form hydrogen bridges, or bonds. The ability to form hydrogen bonds is a function both of the electronegativity *and* the radius of the atom to which H is attached. Only O—H, N—H, and F—H are important hydrogen bonders. In C—H, the C is not electronegative enough; in Cl—H, the radius of Cl is too large. Both effects may operate with S—H to some extent. The low boiling points of SH compounds, relative to their oxygen analogs, indicate a lesser degree of association in the former (see Table 15.1).

TABLE 15.1
BOILING POINTS OF SOME MERCAPTANS AND ALCOHOLS

Mercaptan	*Boiling point, °C*	*Alcohol*	*Boiling point, °C*
CH_3SH	5.8	CH_3OH	64.7
C_2H_5SH	36	C_2H_5OH	78.3
n-C_4H_9SH	98	*n*-C_4H_9OH	117.7
HSH	−61.8	HOH	100

Mercaptans may be separated from mixtures and analyzed in the form of their salts with heavy metals. (Recall that HSH forms precipitates with mercury, silver, and lead ions.) The mercury and silver salts have been particularly useful because of their insolubility in water. All these salts are decomposed readily by strong acids to regenerate the original mercaptan.

$$C_2H_5SH + HgO \longrightarrow (C_2H_5S)_2Hg$$

$$(C_2H_5S)_2Hg \xrightarrow[H_2SO_4]{\text{dil.}} C_2H_5SH + HgSO_4$$

Thiols react with organic acids in a manner analogous to the esterifications of alcohols, although the greater acidity of the S—H tends to make the formation of esters more difficult than with alcohols.

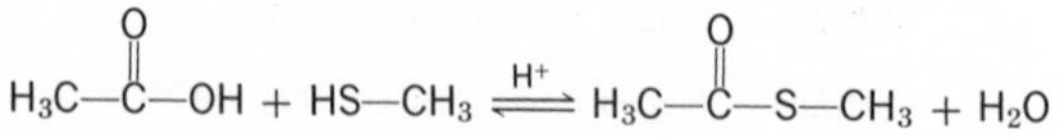

Similarly, with ketones and aldehydes, thioketals and thioacetals are formed (compare Chap. 11).

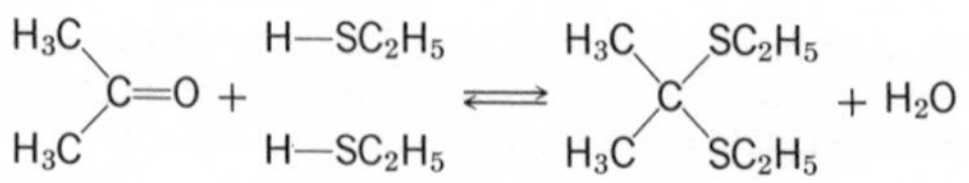

Thioethers, often called "sulfides," are prepared by a modified Williamson synthesis (displacement).

$$CH_3CH_2Br + Na^+ \, {}^-SCH_3 \longrightarrow CH_3CH_2SCH_3 + Na^+Br^-$$

These compounds often have rather sweet odors (when pure), although compounds like allylsulfide $(H_2C{=}CH{-}CH_2)_2S$ are active ingredients in the odors of onion and garlic.

The methyl-thioether ($S{-}CH_3$) group plays an important role in animal metabolism through the reactions of the amino acid methionine (Sec. 19.1).

The tendency of sulfur to form bonds with its extra electron pairs extends to reaction with alkyl halides.

$$(CH_3)_2S + CH_3Br \longrightarrow (CH_3)_2S^+{-}CH_3 \;\; Br^-$$

Trimethylsulfonium bromide

The sulf*onium* halides are almost completely ionized in solution and are comparable with quaternary amm*onium* halides in their properties. Like the quaternary ammonium compounds, they are salts of strong bases, the sulfonium hydroxides.

$$(CH_3)_2S^+(C_2H_5) \;\; OH^- \xrightarrow[\Delta]{} (CH_3)_2S + H_2C{=}CH_2 + H_2O$$

$$(CH_3)_3N^+(C_2H_5) \;\; OH^- \xrightarrow[\Delta]{} (CH_3)_3N + H_2C{=}CH_2 + H_2O$$

15.3 Oxidation Reactions The behavior of thiols toward oxidizing agents is different from that of the oxygen analogs in that these reagents react with the sulfur atom itself, the carbon chain remaining intact.

$$CH_3CH_2SH \xrightarrow{\text{mild oxidizing agent such as air or } Br_2 + H_2O \text{ or } FeCl_3} CH_3CH_2S{-}SCH_2CH_3$$

Diethyl disulfide

Disulfides can be easily reduced to regenerate the mercaptans.

$$CH_3CH_2S{-}SCH_2CH_3 \xrightarrow{\text{mild reducing agent}} 2CH_3CH_2SH$$

This mobile oxidation-reduction system is utilized in the respiration of living cells. Here the more complex amino acids cysteine and cystine are involved.

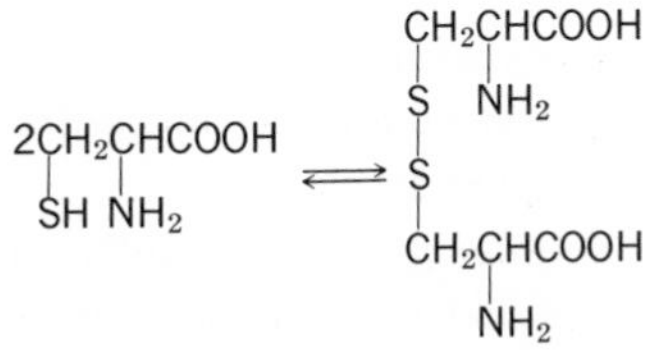

Cysteine Cystine

A more vigorous oxidation decomposes the disulfide and leads to the formation of a sulfonic acid.

$$C_2H_5SH \longrightarrow C_2H_5S{-}SC_2H_5 \longrightarrow C_2H_5\overset{+}{\underset{+}{S}}(\rightarrow O^-)_2{-}OH$$

Sulfonic acids, like the carboxylic acids and sulfuric acid, resist further action of oxidizing agents. The direct reduction of sulfonic acids to thiols is impractical, although the chlorides and esters can be reduced (compare carboxylic acids and derivatives).

The sulfonic acids, final product of thiol oxidation, may be regarded as analogous to substituted sulfuric acids,

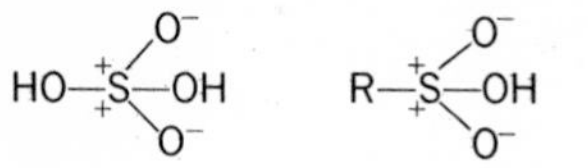

Sulfuric acid A sulfonic acid

Like sulfuric acid, they are very extensively ionized and, as a rule, readily soluble in water. Of increasing importance in the aliphatic series, they are of considerable industrial significance in the aromatic series. Here they are always prepared by direct sulfonation, a reaction that proceeds slowly even at elevated temperatures. The sulfuric acid used in direct sulfonation is often maintained at the highest concentration possible by dissolving SO_3 in it to make the so-called "fuming" sulfuric acid.

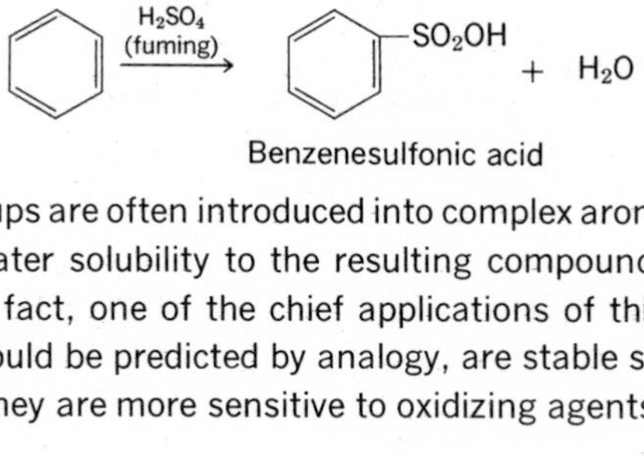
H2SO4
(fuming)
SO2OH
+ H2O

Benzenesulfonic acid

Sulfonic acid groups are often introduced into complex aromatic structures, such as dyes, to impart water solubility to the resulting compounds; this increasing of water solubility is, in fact, one of the chief applications of this grouping.

Thioethers, as would be predicted by analogy, are stable substances, especially in alkaline medium; they are more sensitive to oxidizing agents, however, than their oxygen analogs.

$$CH_3CH_2SCH_2CH_3 \xrightarrow[H_2O_2,\ 25°]{(O)} CH_3CH_2\overset{O^-}{\overset{|}{\overset{+}{S}}}CH_2CH_3 \xrightarrow[H_2O_2,\ 100°]{H^+} CH_3CH_2\overset{O^-}{\underset{O^-}{\overset{+}{\underset{+}{S}}}}CH_2CH_3$$

Diethyl sulfoxide Diethyl sulfone

We find here again the tendency of the oxidizing agent to attack the sulfur atom preferentially. The sulfones and sulfoxides, in contrast to the analogous ether peroxides (see Sec. 10.7), are quite stable.

Again, as in Sec. 15.2, an analogy between sulfur and nitrogen chemistry may be pointed out. Tertiary amines, like the thioethers, are attacked by oxidizing agents at the unshared pairs of electrons to form stable products.

$$(CH_3)_3N{:} \xrightarrow{\text{oxidation}} (CH_3)_3N^{+}{-}O^{-}$$

An amine oxide

$$(CH_3)_2\ddot{S}{:} \xrightarrow{\text{oxidation}} (CH_3)_2S^{+}{-}O^{-}$$

A sulfoxide

An interesting practical application of the reactions discussed above is found in the commercial synthesis of the hypnotic drug Sulfonal.

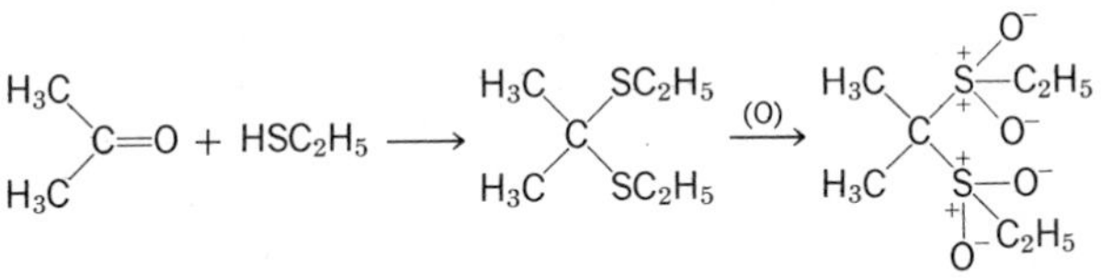

Sulfonal[1]

Sulfoxides, in particular dimethylsulfoxide, and, to a lesser extent, sulfones have been found to be exceptional solvents for displacement reactions by negatively charged ions. Thus, displacement by NaCN, potassium phthalimide, and a number of other ions occur many orders of magnitude faster in dimethylsulfoxide (DMSO) than in alcohol. This unique solvent action is attributed to a specific solvation of the cation, leaving the anion more free to do its displacing.

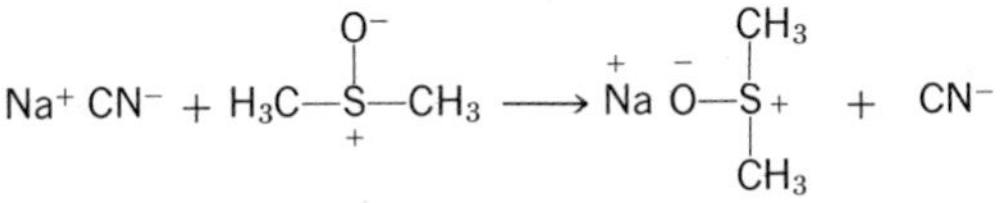

15.4 Sulfonic Acid Derivatives With PCl_5, sulfonic acids yield sulfonyl chlorides. These are able to react with alcohols and amines to form esters and amides in the same manner as carboxylic acid chlorides.

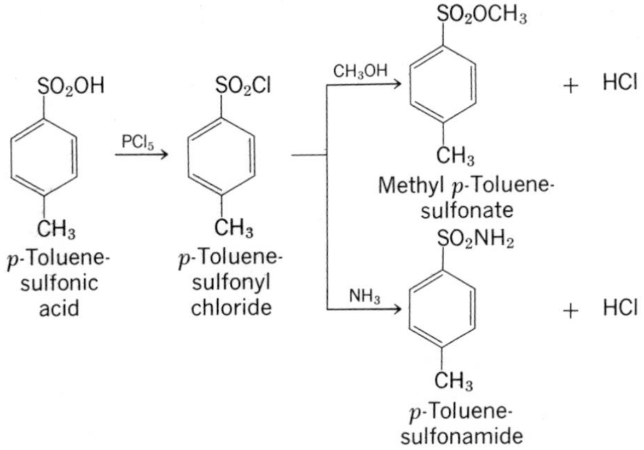
p-Toluene-
sulfonic
acid
p-Toluene-
sulfonyl
chloride
Methyl p-Toluene-
sulfonate
p-Toluene-
sulfonamide

---

[1]Sulfonal is a trade-marked name for diethylsulfondimethylmethane. It has sedative and hypnotic properties. A sedative is an agent that calms, or allays, irritability; a hypnotic is an agent that produces sleep.

All these derivatives may be hydrolyzed to the parent sulfonic acids. The hydrolysis of sulfonamides takes place much more slowly than that of the corresponding carboxylic acid derivatives and requires long boiling with strong acid.

Sulfonic acid esters have been invaluable in the study of reaction mechanisms. Esters of *p*-toluenesulfonic acid are formed from *p*-toluenesulfonyl chloride in a basic solvent such as pyridine.

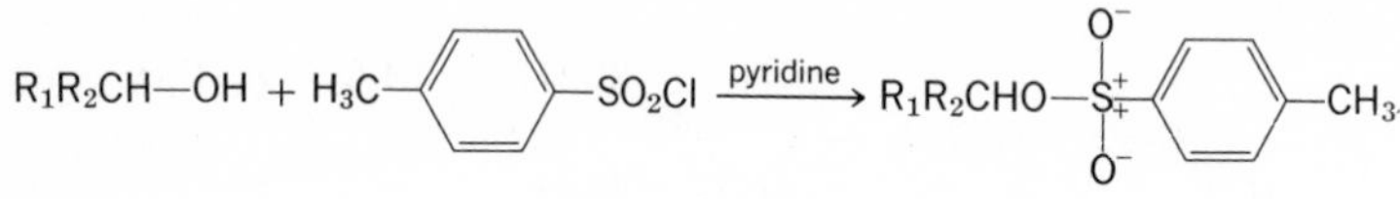

R1R2CH—OH + H3C
SO2Cl
pyridine
R1R2CHO
CH3

The group

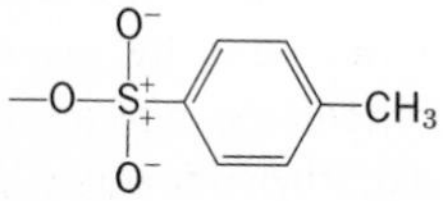

CH3

behaves very much like a bromide group in reactions such as displacement reactions and is nicknamed the *tosyl* group (for *p*-TOlueneSulfonYL).

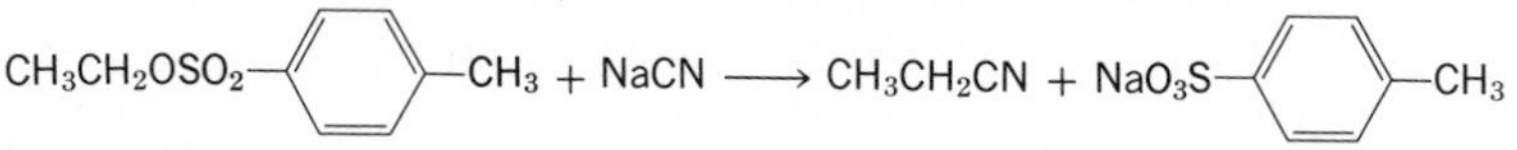

CH3CH2OSO2
CH3 + NaCN
CH3CH2CN + NaO3S
CH3

The advantage of this group for mechanistic studies lies in the fact that the formation of toluenesulfonate esters involves displacement on sulfur so that the C—O bond is not broken, whereas the formation of a halide necessarily breaks the C—O bond and may involve inversion, retention, or racemization, depending on the reaction mechanism.

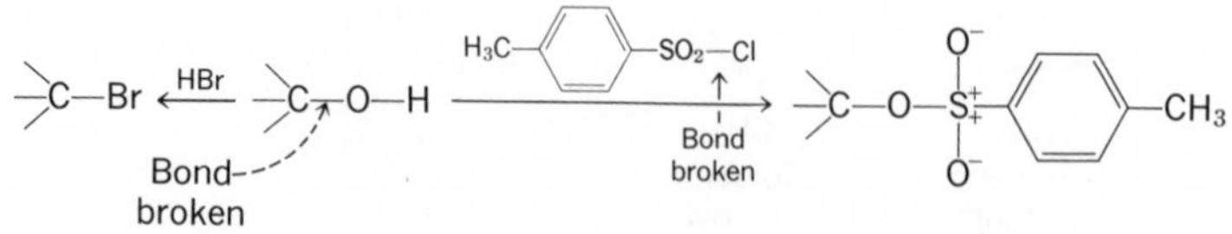

C—Br
HBr
C—O—H
Bond
broken
H3C
SO2—Cl
Bond
broken
CH3

The use of a *p*-toluenesulfonate ester in the following sequence helped establish the fact that inversion occurs in the S_N2 reaction (Sec. 9.6). In this sequence an alcohol of the *S* configuration is converted by three consecutive steps into the same alcohol of opposite (*R*) configuration of almost identical optical purity (as shown by the magnitude of the rotation). Such a process is called a "Walden cycle."

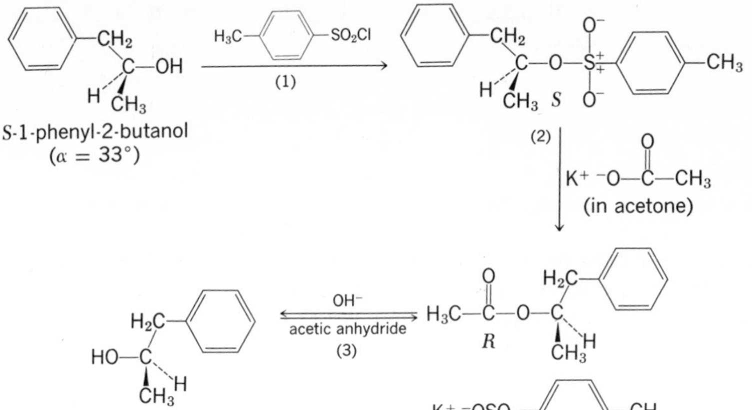
CH2
C—OH
H
CH3
H3C
SO2Cl
(1)
CH2
C—O—S
O−
CH3
S
(2)
K+ −O—C—CH3
(in acetone)
O
H2C
H3C—C—O—C
R
CH3
H
K+ −OSO2
CH3
OH−
acetic anhydride
(3)
H2C
HO—C
H
CH3
S-1-phenyl-2-butanol
($\alpha = 33°$)

R-1-phenyl-2-butanol
($\alpha = -32°$)

In Step 1 the C—O bond is not involved so that inversion of configuration cannot have occurred. In Step 3, ample evidence has been derived (Secs. 8.5 and 13.3) to show that cleavage of esters by base occurs at the carbonyl oxygen and not at the alcoholic carbon-oxygen bond. This is confirmed by the re-esterification of the *R* alcohol with acetic anhydride to give the same (*R*) acetate from which it was formed. It is thus clear that an essentially complete inversion must have occurred in Step 2. This is the type of behavior that we have come to expect for the S_N2 reaction, and it is experiments such as the one above that have led to the establishment of this mechanism.

Amides and substituted amides of simple sulfonic acids have acquired considerable importance in medicine. When treated with chlorine water, the so-called "chloramine" antiseptics are formed.

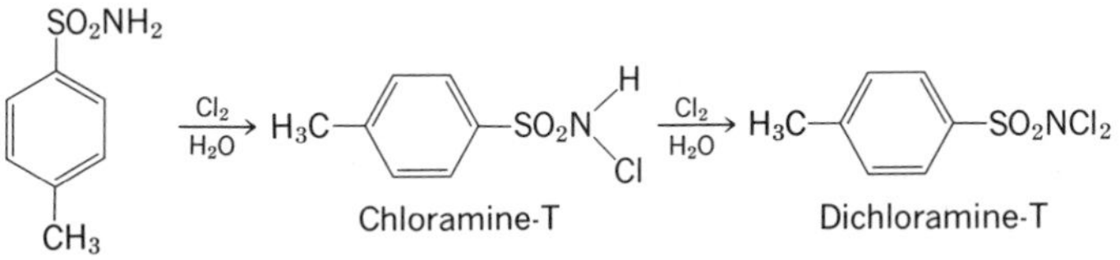
SO2NH2
CH3
Cl2
H2O
H3C
SO2N
H
Cl
Cl2
H2O
H3C
SO2NCl2
Chloramine-T

Dichloramine-T

These compounds owe their effectiveness to their slow hydrolysis, through which HOCl, a powerful germicide, is formed.

15.5 Sulfanilamides The most successful of the now well-known *sulfa* drugs have been amides and substituted amides of sulfanilic acid.

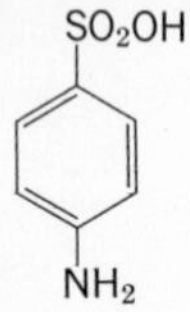
Sulfanilic acid

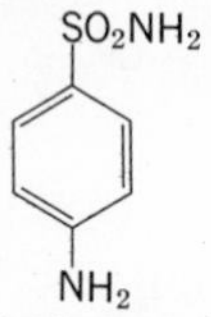
Sulfanilamide

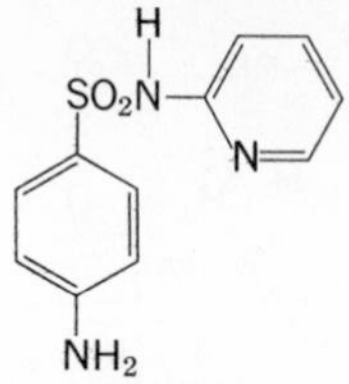
Sulfapyridine

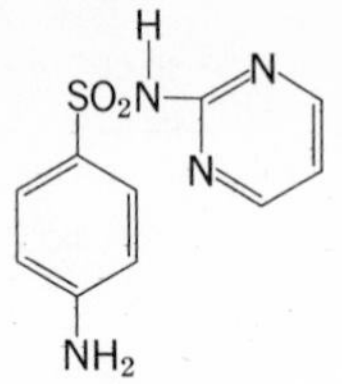
Sulfadiazine

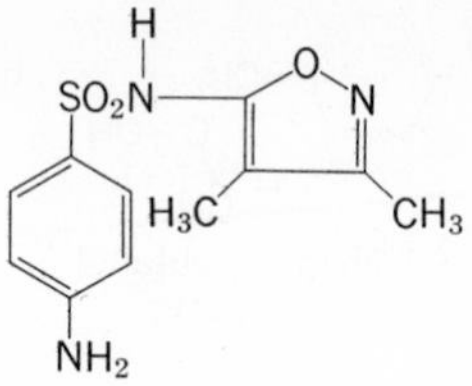
Sulfisoxazole

Because of the presence of an amino group in sulfanilic acid, it is not possible to prepare the acid chloride or amide directly from this acid. The amino group is, therefore, protected by acetylation (Secs. 14.12 and 14.13).

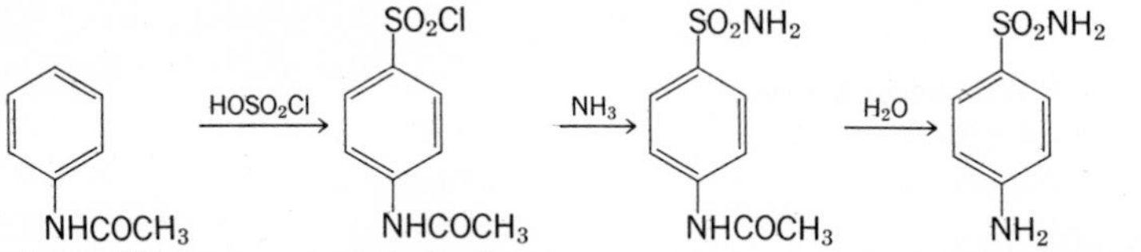
Acetanilide — Acetylsulfanyl chloride — Sulfanilamide

Two steps are combined into one in this commercial synthesis by using chlorosulfonic acid instead of sulfuric acid, thus obtaining the sulfonyl chloride group on the ring directly. In the last step, advantage is taken of the greater ease of hydrolysis of carboxylic acid amides over that of the corresponding sulfonic acid amides.

15.6 Saccharin Another interesting medical application of simple aromatic sulfonic acid derivatives is found in saccharin.

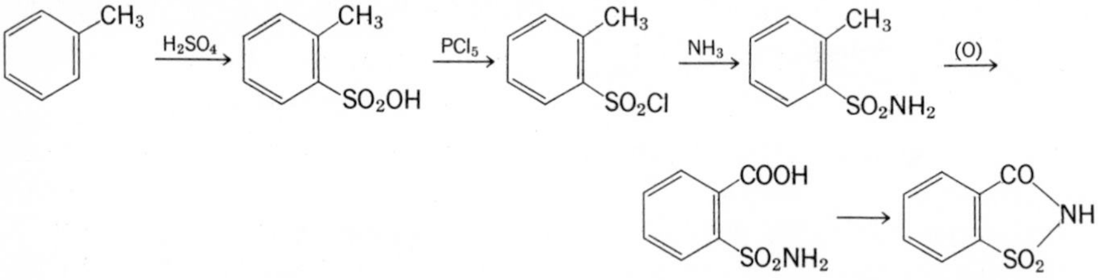
Saccharin

The extraordinary sweetness of saccharin, approximately 550 times that of sugar, makes it a suitable sweetening agent for diabetics or others unable to utilize natural sugars. The difference in structure from sugar (Chap. 21) makes saccharin unable to replace sugar in the metabolic process, and so avoids the difficulties which make sugars harmful in diabetes.

In the commercial synthesis outlined above, it is noted that the oxidation of the methyl group is accomplished without hydrolyzing the sulfonamide group. This is another example of the relative stability of sulfonic acid derivatives. Since saccharin is relatively insoluble in water, it is converted to its sodium salt for use;

the combined effect of carbonyl and 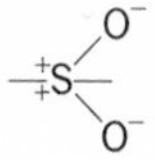 groups on the imide nitrogen makes it a weak acid (compare phthalimide).

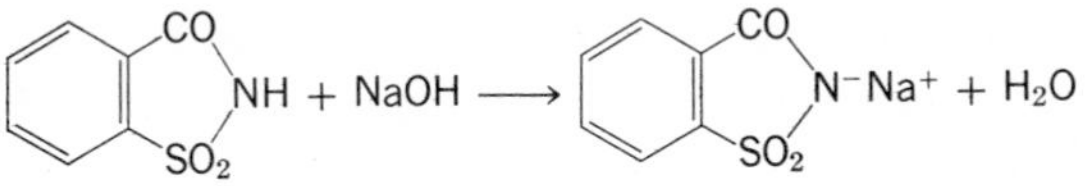

15.7 Chemical Structure and Physiological Action The interest in the role of sulfur compounds in biochemistry stems from the discovery of a number of important vitamins and drugs containing this element. The "B vitamins," biotin and thiamin, the penicillins, the sulfa drugs, sulfoxone (used in treatment of leprosy), and a number of enzymes contain sulfur in various linkages. The discussion of all these cannot profitably be undertaken in an elementary text.

What can be done without going into great detail is to sketch the method of approach which has initiated some of these investigations and brought them to issue. In the first place, an observation is made which leads to an idea. Many observations are made without consequent ideas, and these may be useful as a body of data, but the body is not informed with life until an idea has developed. This can be illustrated by an example. Many histologists had observed that cells of plants and animals could be stained rather specifically with certain dyestuffs; cells of a given kind, or specific parts of cells, could be stained while contiguous matter remained unaffected; bacteria could be stained without staining the tissue of the host. It remained for Paul Ehrlich to develop the idea that on the basis of analogous chemical specificity, it might be possible to prepare substances which would attack bacteria and other parasites in animals without appreciably injuring the animal.

But the observation and the idea are usually not sufficient, because while often many previous observations may have been made, they were not made with the idea in mind: they were empirical observations. What is usually needed is further observation, but observation guided by a hypothesis based on the idea. This is experimentation in the best meaning of the word. It meant synthesizing hundreds of compounds and testing them against parasites, such as harmful bacteria, always looking for the *chemical structure,* or group, which might carry a specific action against some organism. This idea, and the hypothesis based upon it, that certain molecules might be devised (by the combination of suitable groups) which would have a specific effect on definite living cells, was embodied in the term *chemotherapy,* coined by Ehrlich. Chemotherapeutic investigations, a study on the borderline between chemistry and pharmacology, have not been limited to antibacterial substances, but cover all biologically active substances (vitamins, hormones, etc.).

The method of experimentation which would follow from such an idea may also be illustrated. Among the substances reported of use as a chemotherapeutic agent was a dyestuff, Prontosil, reported by Domagk (in the early 1930s) as effective against a particularly virulent streptococcus. This substance has the formula

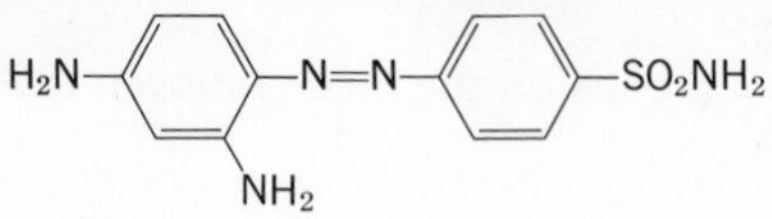

2,4-Diaminoazobenzene-4′-sulfonamide

A systematic examination of this structure in the spirit of Ehrlich's idea led to the recognition several years later by Tréfouëls and coworkers that the active part of the molecule was the grouping

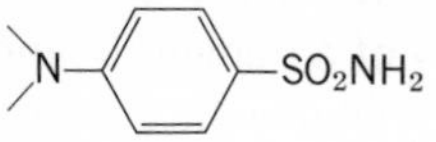

It was shown that sulfanilamide, *p*-aminobenzenesulfonamide,

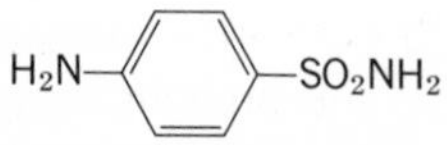

was therapeutically active in septicemias, streptococcal pneumonia, peritonitis, and a number of other diseases. The corresponding ortho and meta compounds were inactive! There were many diseases in which sulfanilamide was inactive (such as tuberculosis, measles, and most virus diseases), and it became a prime subject of research to synthesize analogs and variations on the sulfanilamide theme in an endeavor to determine the effectiveness of substituent groups and to obtain improved therapeutic power and versatility of action. Activity in this field was so great that by the end of 1944 some 3,380 sulfanilamide derivatives had been prepared, and counting related compounds not derivatives of sulfanilamide, a grand total of over 5,400 compounds. The most useful of these, as regards relatively high therapeutic effectiveness combined with relatively low toxicity, were described in Sec. 15.5. Each of these compounds seems to have usefulness in a specific field.

An investigation of this kind is not closed with the discovery of specific chemotherapeutic agents. There still remains the question of mechanism of action, the problem of how the substance exerts its effect. The solution of this problem is essential to the design of better agents, and it bears also on the biochemistry of the living organisms, the host and parasite. Of the thousands of sulfanilamide derivatives which have been prepared, relatively few are of practical value at present. This underlines the need for a more detailed knowledge of the chemical reactions of living cells. It is such knowledge which would give better direction to future research in chemotherapy.

Some insight into this problem has been gained in the field from which the above examples were taken, through the observation that sulfanilamide is antagonistic to the compound *p*-aminobenzoic acid. This substance appears to be essential to the well-being of certain bacteria (a kind of "vitamin" for them). It is required in some set of reactions in the bacterial cell, perhaps as part of an enzyme.[1]

If sulfanilamide is present also and available to the cell (that is, present in the medium in which the cell finds itself), it is picked up by the cell, replacing, because it is similar to it, the *p*-aminobenzoic acid. Whereupon it interferes with the *normal* metabolic processes of the cell by failing to perform the function of *p*-aminobenzoic acid. Under these conditions, the cell is more vulnerable to the defense mechanism of the host than it would be in a normal state. The effect of the sulfanilamide in these cases can be overcome by increasing the relative proportion of *p*-aminobenzoic acid present, and this suggests a mass-action effect. Whether the ability of sulfanilamide to substitute for *p*-aminobenzoic acid is due to a similarity in size and shape between the two,

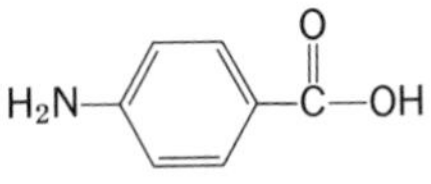

H₂N
C—OH
O

p-aminobenzoic acid
(PAB)

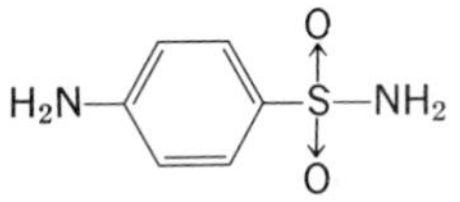

H₂N
S—NH₂
O
O

or to similarities in degree of dissociation, or solubility, is still under investigation. Some reflection will show, however, that the problem can be far more complex than has been indicated here.

In recent years the sulfa drugs have been largely supplanted by the more effective and generally less toxic *antibiotics*. The latter substances are materials synthesized by various molds to protect themselves from bacteria. They are substances of greatly varying structure which generally interfere with bacterial protein synthesis (Sec. 22.10).

15.8 Spectroscopic Properties Mercaptans exhibit characteristic SH absorption in the infrared at 2550 to 2600 cm^{-1}. These absorptions are weak, but since little else absorbs in this region, they are often distinguishable. Sulfides show no unique peaks, but sulfoxides absorb at 1030 to 1070 cm^{-1}, and sulfones at 1120 to 1160

[1] A type of enzyme for which this requirement appears has for part of its structure one of the *folic acids*. The structure of one of these, below, shows the *p*-aminobenzoic acid group, indicated by dotted lines, as part of the molecule.

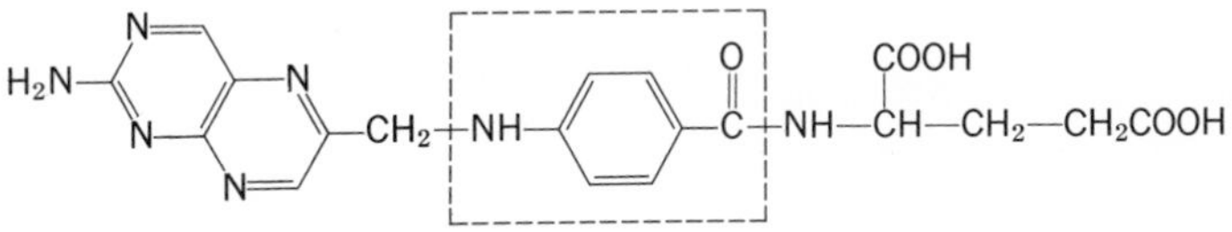

H₂N
N
N
N
N
CH₂
NH
C
O
NH—CH—CH₂—CH₂COOH
COOH

The folic acids are members of the B group of vitamins, and are essential to the life of many organisms. Some microorganisms, however, do not require folid acid for life, and upon them sulfanilamide is without effect.

and 1300 to 1350 cm^{-1}. The nmr of the SH group shows up at $\delta = 1.2$ to 1.6 ppm (exchangeable). Since sulfones and sulfoxides possess positively charged sulfur atoms, the alkyl substituents are shifted downfield.

Sulfonamides show infrared absorption of 1300 to 1370 and 1140 to 1180 cm^{-1}, and sulfonic acid esters and sulfate esters show bands at 1145 to 1230 and 1330 to 1440 cm^{-1}.

OUTLINE OF SULFUR-COMPOUND CHEMISTRY

MERCAPTANS (THIOLS)

Preparation

$$RX + KSH \longrightarrow KX + RSH$$

Properties

1. $RSH + NaOH \xrightarrow{H_2O} RSNa + H_2O$

   $RSH + HgO \longrightarrow (RS)_2Hg\downarrow + H_2O$

   $RSH + HgCl_2 \longrightarrow RSHgCl\downarrow + HCl$

2. Oxidation.

   a. $RSH + \xrightarrow[HNO_3]{(O)} \underset{\text{A sulfonic acid}}{RSO_3H}$

   b. $RSH \xrightarrow[I_2]{NaOH} \underset{\text{A disulfide}}{RSSR}$

   This oxidation takes place with mild oxidizing agents and is reversed by mild reducing agents.

THIOETHERS (SULFIDES)

Preparation

1. $2RX + K_2S \longrightarrow RSR + 2KX$
2. $RSNa + XR' \longrightarrow RSR' + NaX$

Properties

1. $RSR + \begin{Bmatrix} NaOH \\ HgO \end{Bmatrix} \longrightarrow$ N.R.
2. Oxidation.

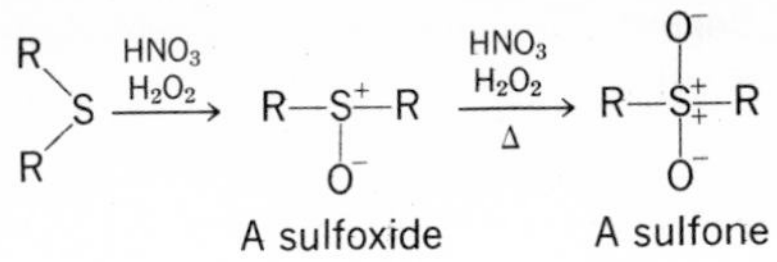
HNO3
H2O2
HNO3
H2O2
Δ
A sulfoxide
A sulfone

SULFONIC ACIDS

Preparation

1. Oxidation of mercaptans.

$$RSH \xrightarrow{(O)} RSO_3H$$

2. Direct sulfonation of *aromatic* compounds.

Properties

1. Sulfonic acids are strong acids, and resemble sulfuric acid in their salt-forming properties.

2. $\phi\ SO_3H$ (or the Na salt) $\xrightarrow[\Delta \text{ fuse}]{NaOH} \phi\ ONa$

   $\phi\ ONa \xrightarrow{H^+} \phi\ OH$

3. Formation of aromatic cyanides.

$$\phi\ SO_3Na \xrightarrow[\Delta \text{ fuse}]{NaCN} \phi\ CN$$

4. Formation of sulfonyl chlorides.

$$RSO_3H \text{ (or the Na salt) } \xrightarrow{PCl_5} RSO_2Cl$$

   Sulfonylchlorides react with water, alcohols, and amines in the same way as carboxylic acid chlorides do.

5. $\phi\ SO_3H \xrightarrow{\text{superheated steam}} \phi\ H$

6. $ArSO_2Cl + ROH \xrightarrow{\text{pyridine}} R{-}OSO_2Ar \xrightarrow{X^-} RX + {}^-O_3SAr$

   An alkane sulfonate

CHAPTER 15 EXERCISES

1. Compare the mechanism for hydrolysis of a sulfonic acid ester with that of a carboxylic acid ester.
2. Suggest a method for separating the butyl mercaptan from the fats found in the secretion of the skunk.

★3. Show how you could distinguish by chemical and/or spectral methods between:

★*a.* CH_3CH_2SH and C_2H_5OH

b. CH_3SCH_3 and $CH_3\overset{+}{S}(O^-)CH_3$

★*c.* $HSCH_2CH_2OCH_3$, $HOCH_2CH_2SCH_3$, and $HOCH_2CH_2SSCH_3$

d. $C_6H_5SO_3H$ and C_6H_5COOH

★*e.* 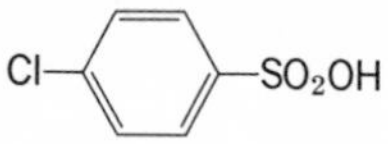 and 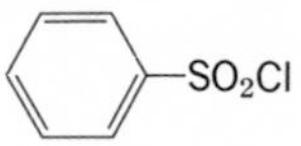

f. 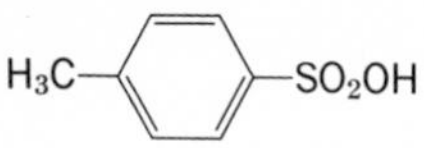 and 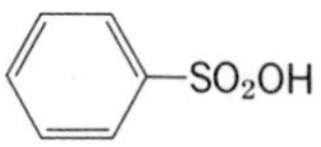

4. Indicate by equations how you could accomplish the following conversions:

a. Ethyl alcohol to sulfonal

b. *n*-Propanol to *n*-propyl mercaptan

c. *p*-Toluenesulfonic acid to toluene

d. *p*-Toluenesulfonamide to toluene (one step)

e. Ethanol to 2-propanesulfonic acid

f. Methanol to diethyl disulfide

g. Methanol to methyl ethyl sulfide

5. Show the expected stereochemistry of the products of the following reactions

★*a.* 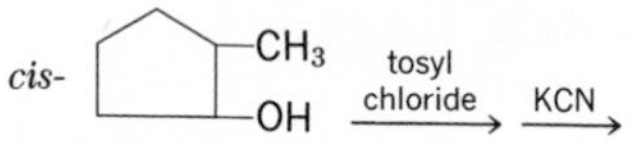

b. $R\text{—}CH_3CH(OTosyl)\text{—}C_6H_5 \xrightarrow{KBr}$

★*c.* 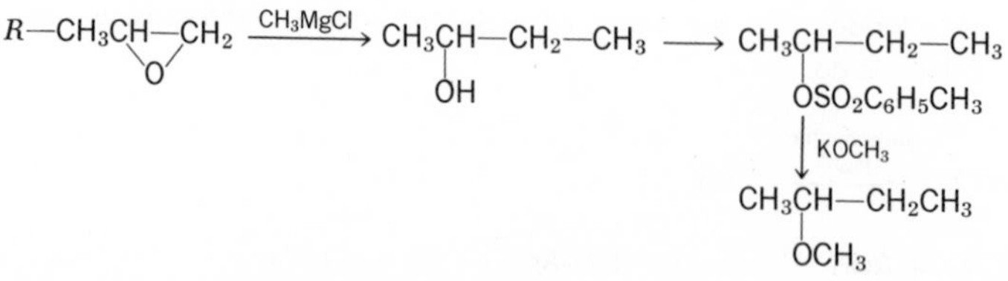

6. Compound A, $C_6H_5O_2NS$, was soluble in NaOH solution though insoluble in water. Oxidation of A followed by fusion with NaOH gave a product (B) that contained no sulfur. B was reduced, diazotized, and coupled with phenol to give *p,p'*-dihydroxyazobenzene. Deduce structures for A and B and show your reasoning.

7. Sulfonium compounds (Sec. 15.2) are analogous to ammonium compounds in most respects. Write equations for the exhaustive methylation (Sec. 14.15) of 1-butanethiol.

8. Show how you would synthesize the following compounds from benzene and organic intermediates containing no more than three carbon atoms.

★*a.* 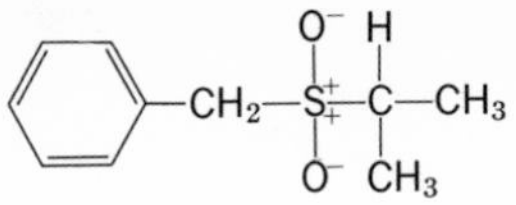

b. 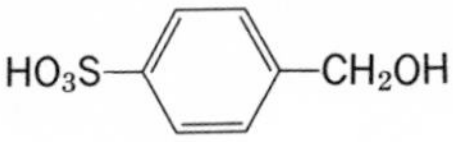

★*c.* 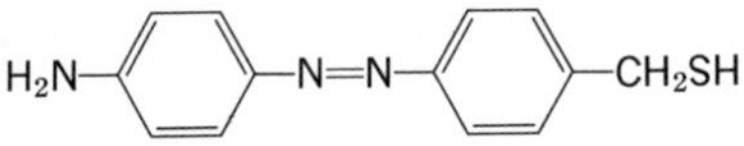

d. 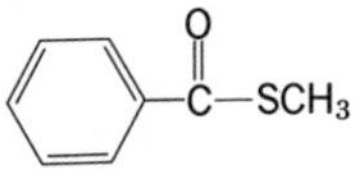

★*e.* HO—CH₂—CH₂—S—S—CH₂—CH₂OH

f. 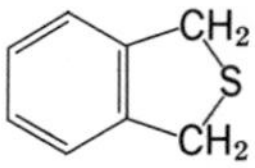

16

CARBONIC ACID DERIVATIVES

16.1 Carbonyl Chloride Carbonic acid is a dibasic acid containing only one carbon atom, and so occupies a unique position in the carboxylic acid series. Although the acid itself is unstable,[1] many of its derivatives are known and are of great importance. In this chapter will be considered those derivatives of carbonic acid which correspond to the carboxylic acid derivatives already dealt with (Chap. 13): the acid chlorides, esters, and amides. The anhydride (CO_2) and the salts of carbonic acid will not be considered here.

The acid chloride of carbonic acid is *phosgene,* or *carbonyl chloride.* It cannot,

[1] The free acid has recently been isolated as the etherate.

of course, be prepared from the very unstable free acid. The commercial method of synthesis is from chlorine and carbon monoxide.[1]

$$Cl_2 + CO \longrightarrow O=C\begin{smallmatrix} Cl \\ \\ Cl \end{smallmatrix}$$

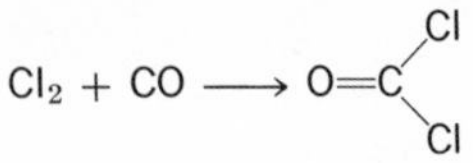

Phosgene

Phosgene is a volatile (bp—8.3°), poisonous substance. It has been used as a war gas. It is said to have an odor resembling new-mown hay.

Phosgene behaves chemically like an acyl chloride. It can be hydrolyzed, alcoholyzed, and ammonolyzed, and the reactions follow the courses to be expected of an acid chloride. However, since phosgene is a diacid chloride, certain additional reactions may be observed. A number of the reactions are shown below. All are vigorous reactions, a behavior understandable on the basis of the reactive carbonyl group present, the group being made more electrophilic by the two electronegative Cl substituents.

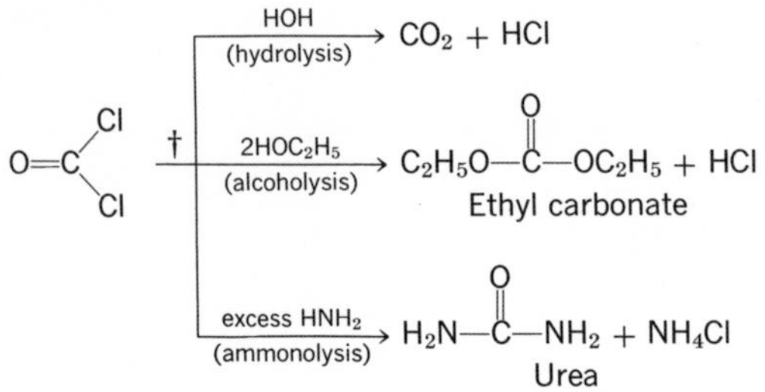

$$O=C\begin{smallmatrix} Cl \\ \\ Cl \end{smallmatrix} \dagger$$

$\xrightarrow[\text{(hydrolysis)}]{HOH} CO_2 + HCl$

$\xrightarrow[\text{(alcoholysis)}]{2HOC_2H_5} C_2H_5O-\overset{O}{\underset{}{C}}-OC_2H_5 + HCl$
Ethyl carbonate

$\xrightarrow[\text{(ammonolysis)}]{\text{excess }HNH_2} H_2N-\overset{O}{\underset{}{C}}-NH_2 + NH_4Cl$
Urea

Some of these reactions may be interrupted at an intermediate stage.

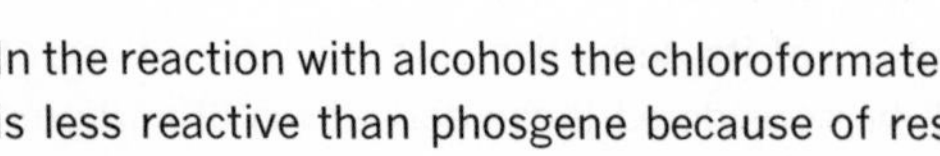

$$Cl-\overset{O}{\underset{}{C}}-Cl + CH_3CH_2OH \longrightarrow CH_3CH_2O-\overset{O}{\underset{}{C}}-Cl$$
Ethyl chloroformate

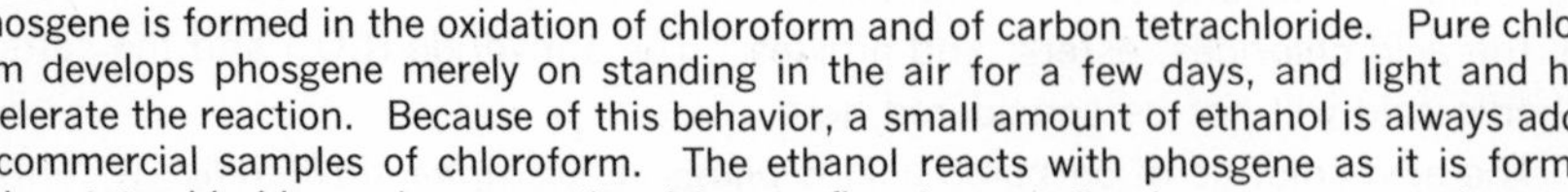

$$Cl-\overset{O}{\underset{}{C}}-Cl + C_6H_5NH_2 \longrightarrow \left[C_6H_5\overset{H}{\underset{}{N}}-\overset{O}{\underset{}{C}}-Cl \right] \xrightarrow{C_6H_5-NH_2} C_6H_5N=C=O + C_6H_5NH_3^+Cl^-$$
Phenyl isocyanate

In the reaction with alcohols the chloroformate (sometimes called a chlorocarbonate) is less reactive than phosgene because of resonance stabilization

---

[1] Phosgene is formed in the oxidation of chloroform and of carbon tetrachloride. Pure chloroform develops phosgene merely on standing in the air for a few days, and light and heat accelerate the reaction. Because of this behavior, a small amount of ethanol is always added to commercial samples of chloroform. The ethanol reacts with phosgene as it is formed. Carbon tetrachloride used as an extinguisher on fires in a confined space may produce toxic amounts of phosgene by a partial hydrolysis.

† A tertiary amine is usually added to absorb the HCl formed.

thus the reaction can be stopped at this stage by the use of a limited amount of alcohol.

$$CH_3CH_2O\overset{\overset{O}{\|}}{C}-Cl \longleftrightarrow CH_3CH_2-\overset{+}{O}=\overset{\overset{O^-}{|}}{C}-Cl$$

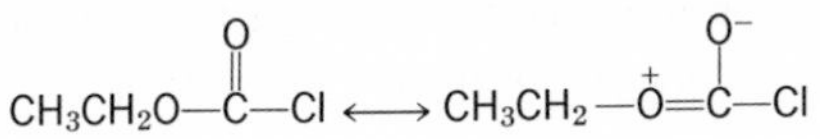

In the reaction with primary amines, the expected intermediate product, $C_6H_5NHCOCl$, still possesses a hydrogen atom adjacent to the carbonyl and readily loses HCl to give the isocyanate. In the presence of up to two moles of amine, only isocyanate and amine hydrochloride are formed. If more than two equivalents of amine are used, the isocyanate reacts further to give a substituted urea.

$$C_6H_5N=C=O + C_6H_5NH_2 \longleftrightarrow C_6H_5NH\overset{\overset{O}{\|}}{C}NHC_6H_5$$

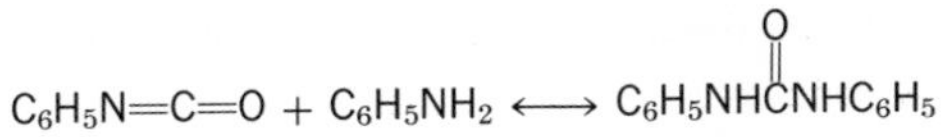

N,N'-diphenyl urea

Both chloroformates and isocyanates will react with alcohols and amines, and mixed products may often be made by different routes.

$$CH_3CH_2O\overset{\overset{O}{\|}}{C}-Cl + C_6H_5NH_2 \longrightarrow CH_3CH_2O-\overset{\overset{O}{\|}}{C}NHC_6H_5 \longleftarrow CH_3CH_2OH + C_6H_5N=C=O$$

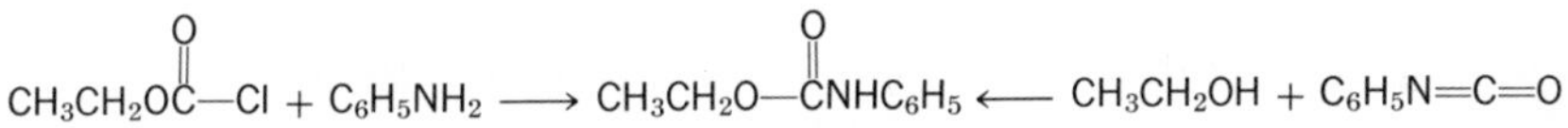

A urethane

When a derivative would be formed in which a hydrogen is attached to oxygen or nitrogen, the intermediate chloro product may decompose to carbon dioxide.

$$H_2O + Cl-\overset{\overset{O}{\|}}{C}-Cl \xrightarrow{-HCl} \left[HO-\overset{\overset{O}{\|}}{C}-Cl\right] \longrightarrow O=C=O\uparrow + HCl$$

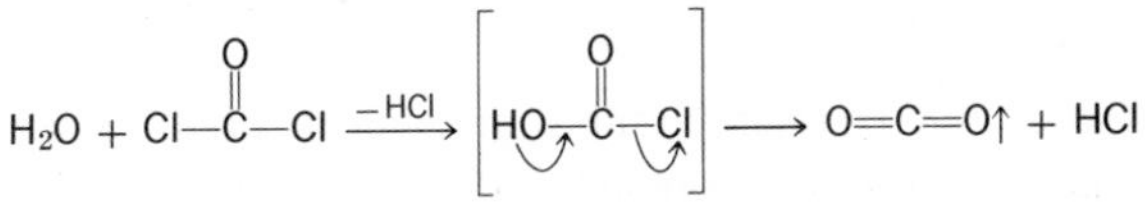

16.2 Urea   The most important of the simple derivatives of carbonic acid is the diamide, urea. This may be prepared by all the standard methods given in Chap. 13 for amides. An important commercial synthesis is based on the reaction of ammonia and carbon dioxide.

$$2NH_3 + CO_2 \rightleftharpoons H_2N\overset{\overset{O}{\|}}{C}-O-NH_4^+ \rightleftharpoons H_2N-\overset{\overset{O}{\|}}{C}-NH_2 + H_2O$$

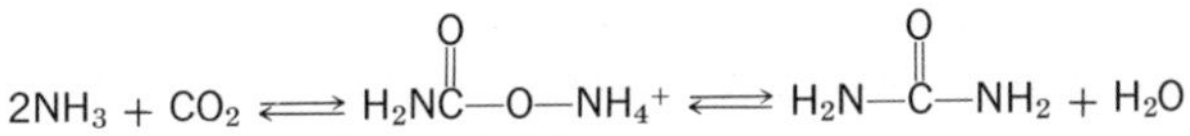

Ammonium
carbamate

The reaction is carried out under pressure. Upon cooling and releasing the pressure, any ammonium carbamate present decomposes into carbon dioxide and ammonia, which may be recovered and recycled. The initial step is the familiar carbonyl addition.

$$H_3N:\overset{\overset{O}{\|}}{C} \rightleftharpoons H_2N-\overset{\overset{OH}{|}}{C} \xrightarrow{NH_3} H_2N-\overset{\overset{O^-NH_4^+}{|}}{C}$$

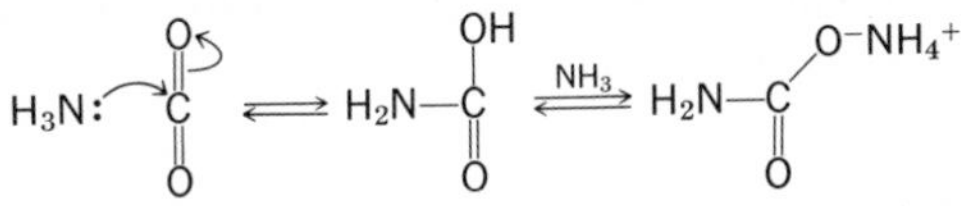

The properties of urea, as is often the case with low-molecular-weight compounds, differ in some respects from those of the higher molecular weight analogs. While other amides are either neutral or weakly acidic, urea is somewhat basic. The single carbonyl group is not able to suppress the basic properties of the two amino groups completely. Urea will form salts with strong acids, for example, nitric acid.

$$NH_2\overset{\overset{O}{\|}}{C}NH_2 + HNO_3 \longrightarrow NH_2\overset{\overset{O}{\|}}{C}NH_3NO_3$$

Urea nitrate happens to be relatively insoluble in strong nitric acid solution; advantage is taken of this fact in the isolation of urea from urine.

In either acid or basic medium, urea and its substitution products are fairly readily hydrolyzed.

$$NH_2CONH_2 \xrightarrow[H_2O]{NaOH} 2NH_3 + Na_2CO_3$$

$$NH_2CONH_2 \xrightarrow[H_2O]{HCl} 2NH_4Cl + CO_2$$

$$CH_3NHCONHC_2H_5 \xrightarrow[H_2O]{NaOH} CH_3NH_2 + C_2H_5NH_2 + Na_2CO_3$$

Nitrous acid, as with all $—NH_2$ compounds, liberates free nitrogen from urea according to the equation:

$$NH_2CONH_2 + 2HNO_2 \longrightarrow 2N_2\uparrow + CO_2 + 3H_2O$$

Only in the case of the Hofmann rearrangement of urea do we encounter an unexpected product. The reaction results in the formation of free nitrogen since the expected hydrazine, $H_2N—NH_2$, is oxidized to N_2 by the reagent.

$$NH_2CONH_2 + NaOBr \longrightarrow N_2\uparrow + CO_2 + H_2O + NaBr$$

A measurement of the amount of nitrogen evolved in this manner is used in the clinical estimation of urea in various body fluids.

At elevated temperatures, urea, having the properties of both an amine and carbonyl compound, reacts with itself to form an amide; the product of this reaction is called *biuret*, and it forms colored copper salts in the *biuret test* for complex amides.

$$NH_2CONH_2 \longrightarrow \underset{\text{Biuret}}{NH_2CONH—CONH_2} \xrightarrow[\text{alkali}]{Cu^{++}} \text{pink color}$$

Not only urea but also many proteins and peptides (Chap. 22) give the biuret test; the development of the color apparently depends on the presence of several amide groups in the same molecule.

16.3 Ureides Acting as an amine, urea will react with derivatives of acids to form ureides.

$$CH_3\overset{O}{\overset{\|}{C}}Cl + NH_2CONH_2 \longrightarrow CH_3\overset{O}{\overset{\|}{C}}NHCONH_2$$

Ureide of acetic acid

Those ureides which are formed from dicarboxylic acids are of considerable biochemical significance. Barbituric acid, for example, is prepared as shown:

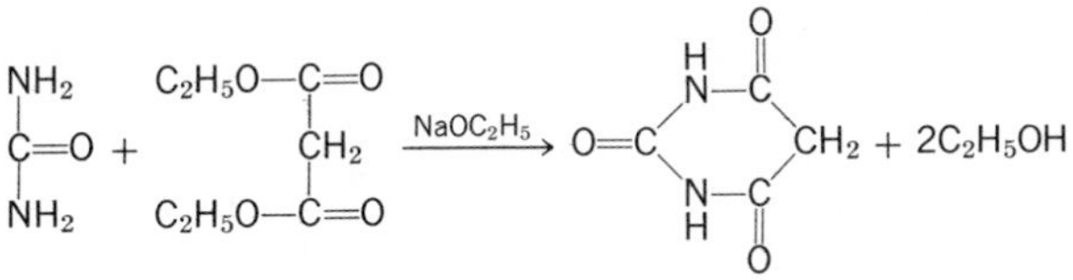

Barbituric acid
(a cyclic ureide)

When this type of reaction is carried out with suitably substituted malonic acid derivatives, members of the important group of hypnotics known as the *barbiturates* are produced.

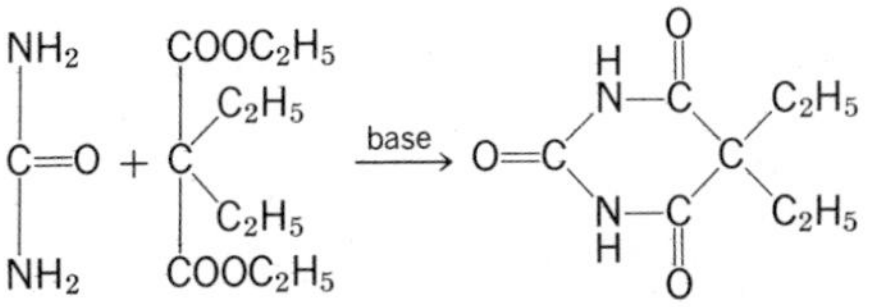

Veronal[1]
(5,5-diethylbarbituric acid)

Cyclic amides of similar structure are found to play an important role in the metabolism in living tissues; some of the vitamins, the nucleic acids, and some enzymes are known to contain cyclic ureide types of rings. These are sometimes rather complex, as in the case of uric acid.

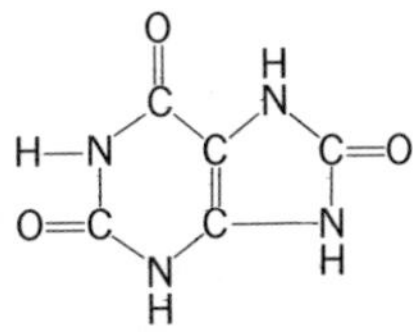

Uric acid

Two large general groups of these heterocyclic structures related to urea are recognized: the pyrimidines (or 1,3-diazines), and the purines. The ring skeletons are given below.

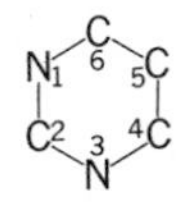

Pyrimidine skeleton

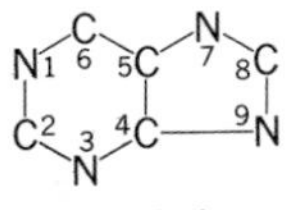

Purine skeleton

[1] Veronal is a typical barbiturate of the type used as hypnotics.

Although none of the cyclic ureide structures shown contain explicit acid groups, they are acid in nature and are named as acids.

The acidity arises from the possibility for stabilization of the anion which results on loss of a proton.

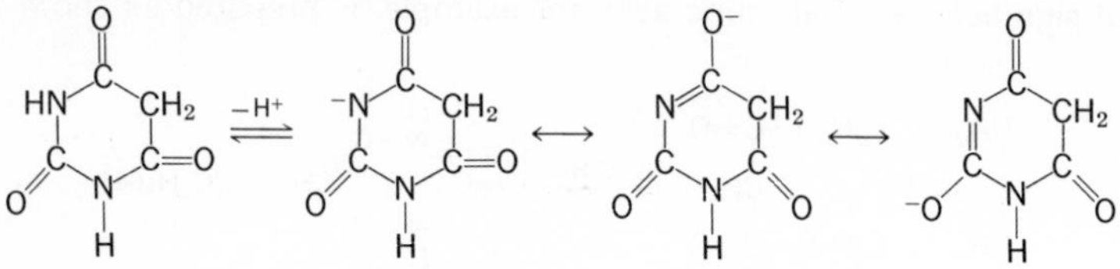

The actual structure of the barbituric acid may be that shown or an enol form (analogous to phenol) such as:

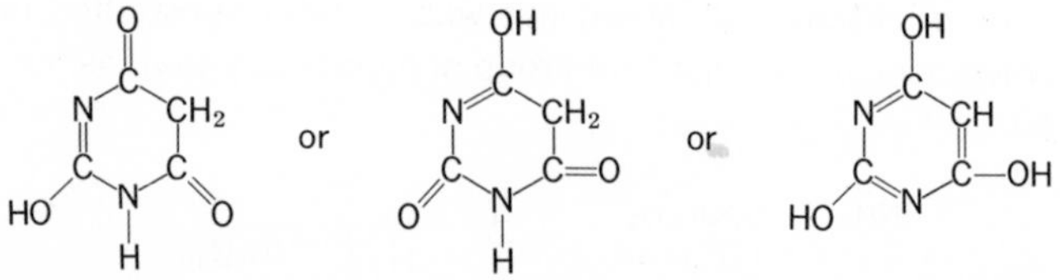

It is sometimes difficult to ascertain the correct structure of the acid form because of the variety of possibilities, although methods such as nmr have resolved a number of these cases. The important feature is that the acidity of these heterocyclic structures is due to resonance stabilization of the anion.

16.4 Isocyanates It was mentioned previously (Sec. 16.1) that isocyanates were formed from primary amines and phosgene. They may also be formed by the Curtius reaction (Sec. 13.16).

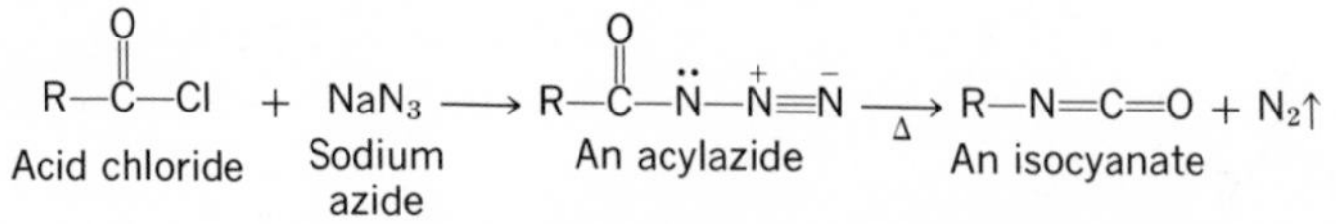
Acid chloride
Sodium azide
An acylazide
An isocyanate

The reactions of isocyanates with alcohols and amines proceed via carbonyl addition reactions, which often do not require catalysis because of the resonance stabilization of the adduct.

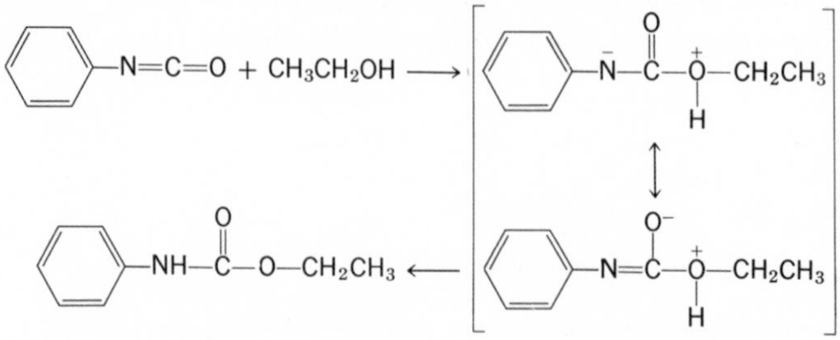

The proton could combine with either oxygen or nitrogen to give the final products I or II below:

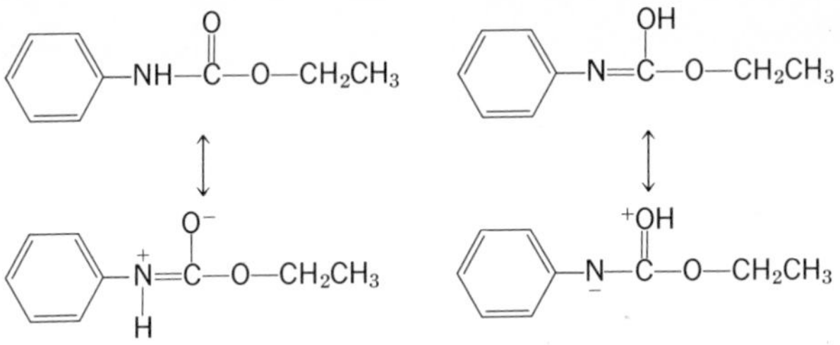

I II

However, in the charge-separated contributing structure, II has the + charge on oxygen while I has it on nitrogen, and vice versa for the negative charges. Since oxygen is more electronegative than nitrogen, there will be an energetic preference for I. Isocyanates, as mentioned, react with amines to give substituted ureas in a similar fashion.

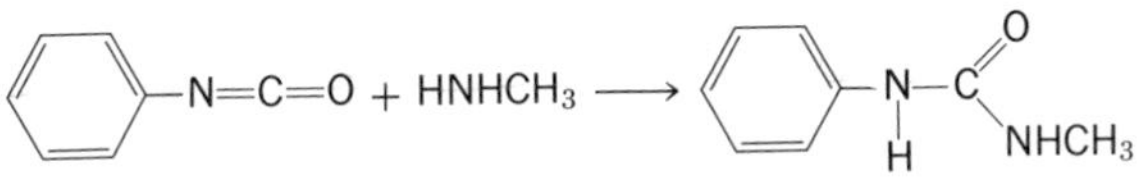

N,N′-Phenylmethylurea

Urethanes and substituted ureas are often stable crystalline substances well suited to the characterization of amines and alcohols. Because the reactions are rapid and almost quantitative, phenyl isocyanate is often used as a reagent in the identification of alcohols and amines.

Isothiocyanates behave in a manner quite analogous to the isocyanates. They are prepared from halides and KNCS, or from carbylamines and sulfur. Certain members of this series are found in nature; for example, allyl isothiocyanate, $H_2C{=}CH{-}CH_2NCS$, occurs in mustard seed. Isothiocyanates are often referred to as the *mustard oils.*

If calcium cyanamide is treated with acid, cyanamide is formed.

$$\text{CaNCN} \xrightarrow{H^+} H_2N{-}C{\equiv}N \rightleftharpoons [H{-}N{=}C{=}N{-}H]$$

This reaction involves the formation of a weak acid from its salt through the action of a stronger acid. The hydrogens on nitrogen, here, are more acid than those in NH_3 for the reasons already discussed in connection with amides.

It is apparent from the formula given above for cyanamide that it is the analog, in the nitrogen series of compounds, of O═C═O in the oxygen series. There is known a whole series of organic compounds which may formally be considered as being derived from ammonia, much as the alcohols, carbonyl compounds, ethers, and carboxylic acids are derived (formally) from water. The extensive discussion of these compounds cannot be undertaken in an elementary course, though some members of the series are dealt with. Thus the primary amines are "nitrogen analogs" of the alcohols, and the tertiary amines of the ethers. Some of these nitrogen compounds are of great importance industrially, and many of them are encountered in living organisms.

If cyanamide is treated with water, it yields urea; if it is treated with ammonia, it yields the all-nitrogen analog of urea, guanidine.

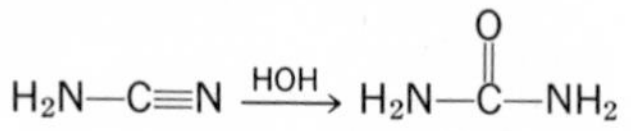

$$H_2N-C\equiv N \xrightarrow{HOH} \underset{\text{Urea}}{H_2N-\overset{\displaystyle O}{\overset{\|}{C}}-NH_2}$$

$$H_2N-C\equiv N \xrightarrow{HNH_2} \underset{\text{Guanidine}}{H_2N-\overset{\displaystyle NH}{\overset{\|}{C}}-NH_2}$$

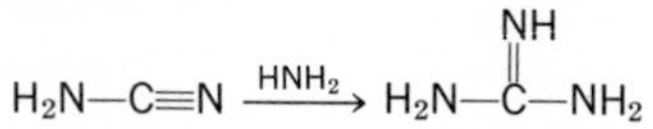

The guanidine grouping is found in many substances of biological importance (for example, arginine, Sec. 19.1).

16.5 Spectroscopic Properties   The infrared spectra of carbonates, chloro-formates, ureas, and urethanes are quite similar to those of other esters, acid chlorides, and amides, respectively. Isocyanates have characteristic maxima in the region 2100 to 2200 cm^{-1}, a region in which few other groups absorb. The nmr spectra are not unique and the uv is not of much utility.

OUTLINE OF CHEMISTRY OF CARBONIC ACID DERIVATIVES

PHOSGENE (CARBONIC ACID CHLORIDE)

Preparation

$$CO + Cl_2 \xrightarrow{\Delta} COCl_2$$

Properties

1. $COCl_2 + H_2O \longrightarrow CO_2 + 2HCl$
2. $COCl_2 + ROH \longrightarrow ROCOCl + HCl$
    $ROCOCl + ROH \longrightarrow (RO)_2CO + HCl$
       A dialkyl
       carbonate
3. $COCl_2 + NH_3 \longrightarrow CO(NH_2)_2 + HCl\ (NH_4Cl)$

UREA (CARBONIC ACID DIAMIDE)

Preparation

1. $COCl_2 + NH_3 \longrightarrow NH_2CONH_2 + NH_4Cl$
2. $CO(OR)_2 + 2NH_3 \longrightarrow NH_2CONH_2 + 2ROH$
3. $(NH_4)_2CO_3 \xrightarrow{\Delta} NH_2COONH_4 \xrightarrow{\Delta} NH_2CONH_2 + H_2O$

4. $CaC_2 + N_2 \xrightarrow[\Delta\Delta]{} CaNCN \xrightarrow[H_2SO_4]{H_2O} NH_2CONH_2 + CaSO_4$

Calcium cyanamide

Properties

1. $NH_2CONH_2 \xrightarrow{NaOH} NH_3 + Na_2CO_3$
2. $NH_2CONH_2 \xrightarrow{HNO_2} N_2\uparrow + CO_2 + H_2O$
3. $NH_2CONH_2 \xrightarrow{\text{conc. } HNO_3} NH_2CONH_2 \cdot HNO_3$
4. $NH_2CONH_2 + 3NaOBr \longrightarrow N_2\uparrow + CO_2 + 2H_2O + 3NaBr$

ISOCYANATES

Preparation

1. $RN{\equiv}C \xrightarrow[HgO]{(O)} RN{=}C{=}O$
2. $RI + AgNCO \longrightarrow RN{=}C{=}O + AgI$

   NOTE: $\phi I + AgNCO \longrightarrow$ N.R.
3. $\phi NH_2 + COCl_2 \longrightarrow \phi N{=}C{=}O + HCl$
4. $R{-}CON_3 \xrightarrow[\Delta]{} R{-}N{=}C{=}O + N_2$

Properties

1. $RN{=}C{=}O + NH_3 \longrightarrow RNH{-}CO{-}NH_2$
2. $RNCO + R'NH_2 \longrightarrow RNHCONHR'$
3. $RNCO + R'OH \longrightarrow RNHCOOR'$

   A urethane
4. $RNCO + H_2O \longrightarrow CO_2 + RNH_2 \xrightarrow{\text{unchanged } RNCO} RNHCONHR$
5. $RNCO \xrightarrow[H_2O]{NaOH} RNH_2 + Na_2CO_3$

CHAPTER 16 EXERCISES

1. Compare the relative acidity or basicity of:

a. Urea and the reaction product of $C_6H_5N{=}C{=}O + NH_3$

b. Barbituric acid and 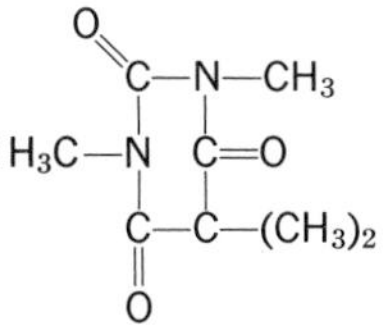

2. Formulate the reactions and show mechanisms for:

a. $CH_3N{=}C{=}O + H_2O \longrightarrow C_3H_8ON_2$

b. Diethyloxalate + thiourea $\longrightarrow$

c. Uric acid + hydrolysis $\longrightarrow$

d. $C_6H_5N{=}C{=}S + C_2H_5OH \longrightarrow$

e. Diethylcarbonate + 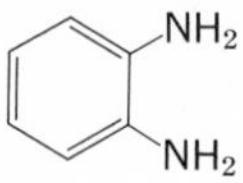 $\longrightarrow$

f. $C_6H_5N{=}C{=}O + CH_3MgX \longrightarrow \xrightarrow{H_2O}$

g. $C_6H_5NH_2 + COCl_2 \longrightarrow C_7H_5ONC$

3. Show how the following could be prepared from simple materials.

★*a.* $C_6H_5NHCONHCH_3$

b. 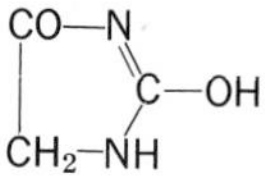

★*c.* 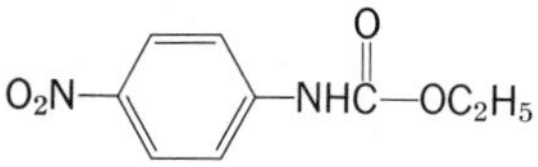

4. How could you distinguish by chemical and/or spectral means?

a. 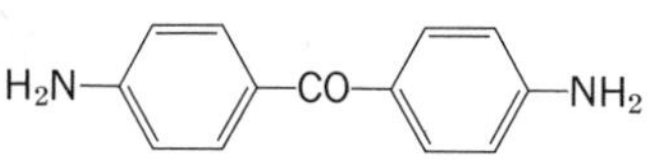 and 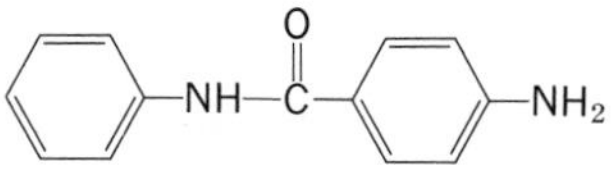

b. 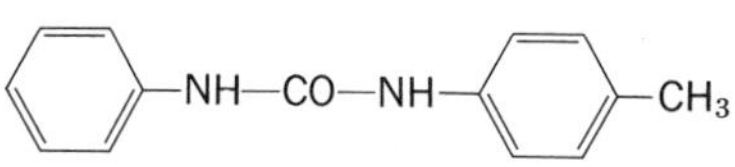 and 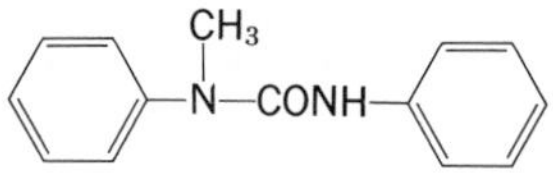

c. $NH_2COOC_2H_5$ and $C_2H_5CONH_2$

d. $CH_3CH(OH){-}CH_2{-}COOC_2H_2$ and $CH_3CH_2CH(OH)COOC_2H_5$

e. $C_6H_5CH{=}CHCH_2COOH$ and $C_6H_5CH{=}CHCOOH$

f. 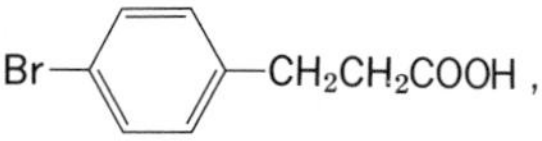, 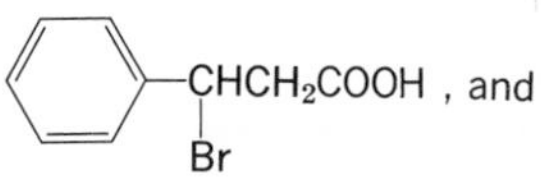, and 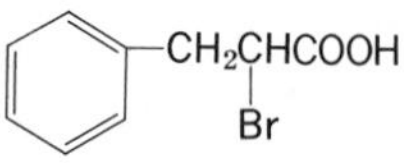

★**5.** A local anesthetic—more active and less toxic than novocaine—has the following structure. Show how it might be synthesized.

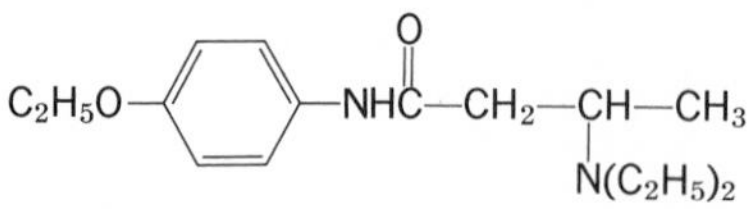

6. $C_6H_5CONH_2 + Br_2$ + base in CH_3OH solution (no H_2O) gives a C_8 product. Predict the structures of the C_8 product and outline a mechanism for the reaction.

★7. Write equations for the reactions by which barbituric acid could be prepared, starting with carbon as the only source of this element, using whatever other reagents are needed.

8. Show how you could distinguish by chemical and/or spectral means:

a. $C_2H_5NHCONH_2$, $(CH_3)_2NCONH_2$, $C_2H_5NHCONHC_2H_5$

b. Barbituric acid and

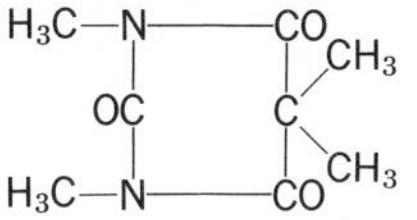

c. C_6H_5NCO, C_6H_5NCS, $C_6H_5CONH_2$

d. $NH_2COOC_4H_9$ and $NH_2CH_2CH_2OCOOC_2H_5$

e.

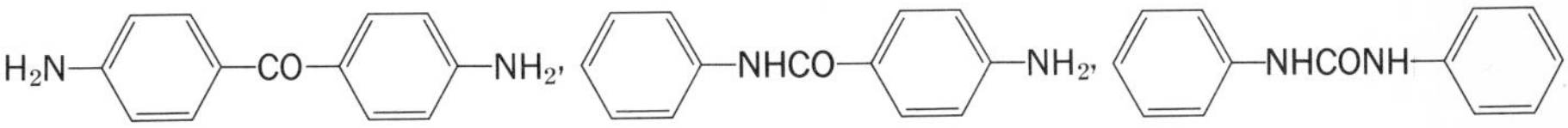

★9. Show, by means of equations, how the following compounds may be prepared: (*a*) Aponal, $CH_3CH_2C(CH_3)_2OCONH_2$; (*b*) Voluntal, $CCl_3CH_2OCONH_2$. NOTE: These compounds are sedatives.

10. Predict the relative positions of the characteristic carbonyl group bands in the infrared region for the following pairs of compounds:

a. $H_2N\text{—}\overset{\text{O}}{\overset{\|}{C}}\text{—}OC_2H_5$ and $CH_3\overset{\text{O}}{\overset{\|}{C}}\text{—}CH_2CH_2NH_2$

b. $NH_2CH_2\overset{\text{O}}{\overset{\|}{C}}\text{—}CH_2NH_2$ and $CH_3NH\underset{\text{O}}{\underset{\|}{C}}NHCH_3$

c. $C_6H_5\underset{\text{O}}{\underset{\|}{C}}NHCH_3$ and $C_6H_5CH_2\underset{\text{O}}{\underset{\|}{C}}\text{—}NH_2$

★11. Would pressure have any effect on the position of the equilibrium in the reaction leading to the formation of urea from NH_3? Explain.

12. The ultraviolet absorption spectra of phenol and

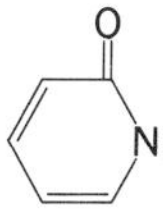

NH in basic solution are much alike. Explain.

★13. Account for the fact that on exposure to moist air, phenyl isocyanate, C_6H_5NCO, forms *N,N′*-diphenylurea.

$$\phi N{=}C{=}O + \text{small amounts of } H_2O \longrightarrow \phi NH\text{—}\overset{\text{O}}{\overset{\|}{C}}\text{—}NH\phi + CO_2$$

17

SUBSTITUTED ACIDS: HALOGEN, HYDROXY, AND OLEFINIC CARBOXYLIC ACIDS

17.1 Introduction; Nomenclature In the study of organic chemistry, a great many compounds are met which contain two or more different reactive groups in the same molecule; a few have been discussed already. The question naturally arises "Will these groups act independently, or will one exert some significant influence on the chemical behavior of the other?" A great deal of the work in the study of advanced organic chemistry is devoted to answering this question and dealing with the situations which arise in such molecules. The question has been partially answered in previous chapters. The influence of one reactive group on a similar or a different one has been encountered in the phenomena of conjugation (Sec. 6.4), orienting effect of substituents on the benzene ring (Sec. 6.12), and influence of substituents on the strength of acids and on the basicity of nitrogen compounds (Secs. 12.2 and 14.3). Certain answers have been derived for this question, and a number of them have considerable generality, so that it is possible to do a small amount of prediction (the crucial test of any generalization). It has been shown, for example, that resonance has a stabilizing effect on the groups involved in the resonance, and the conditions under which resonance can occur have been partially set forth. It has been shown that there are groups which attract electrons (for reasons which we have not generalized any further) and that such groups exert a marked and predictable influence on the release of, or the binding of, protons by neighboring atoms such as O and N which have pairs of unshared electrons. It has been observed that the influence of such groups diminishes rapidly with distance from the atom to be influenced.

In this and the succeeding three chapters, the study of the influence upon each other of dissimilar functional groups in the same molecule will be continued. The substituted aliphatic carboxylic acids, being reactive substances rather frequently encountered, serve well for this further study. Many of these substances occur naturally and have been given trivial names. Those used in this text are the ones usually found in the chemical literature. Systematic names may be derived for these compounds on the basis of the rules already studied. In addition, there is a group of names in which Greek letters are used to indicate the position of substituents relative to the carboxyl group of the fatty acid. A few examples will serve to illustrate these methods of nomenclature (see Table 17.1).

The halogen-substituted acids, which will be considered first, are not widely distributed[1] in nature and are obtained by synthetic methods which, for the most

TABLE 17.1
SYSTEMS OF NOMENCLATURE OF SUBSTITUTED ACIDS

Structure	*Trivial name*	*Systematic name*	*Greek-letter name*
$CH_3CH(OH)COOH$	Lactic acid	2-Hydroxypropanoic acid	α-Hydroxypropionic acid
$CH_3CH{=}CHCOOH$	Crotonic acid	2-Butenoic acid	
$CH_2(Cl)CH_2CH_2CH_2CH_2COOH$	. . .	6-Chlorohexanoic acid	ω-Chlorocaproic acid[a]

[a] The Greek letter *omega* (ω) is often used in organic-chemical nomenclature to mean "on the far end." This acid has a chlorine substituted on the end carbon farthest from the carboxylic acid group. Strictly speaking, this is an *epsilon*-chloro acid (ε-chloro). The device of *omega* saves remembering the entire Greek alphabet. The letters beyond *delta* are seldom used.

[1] One naturally occurring chlorinated acid derivative is the antibiotic Chloromycetin.

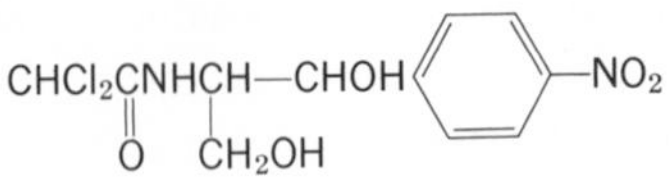

CHCl2CNHCH—CHOH
O
CH2OH
NO2

part, have already been discussed. Most useful are direct substitution and addition reactions with suitably constituted starting materials.

17.2 Preparation of Halogen Acids Substitution on the α carbon atom by halogen proceeds more easily in the case of acid derivatives than with the free acids themselves. The electron-attracting effect of the carbonyl group is apparently more effective in these substances than in the acids. For this reason, the halogenation of the α carbon atom of fatty acids is carried out as shown in the equations below. This is known as the *Hell-Volhard-Zelinsky* (HVZ) reaction (Sec. 12.14).

$$CH_3CH_2COOH \xrightarrow[PBr_3]{Br_2} CH_3\underset{\underset{\displaystyle Br}{|}}{C}HCOBr$$

A mild hydrolysis removes the halogen atom from the carbonyl carbon, leaving the α halogen untouched.

$$CH_3\underset{\underset{\displaystyle Br}{|}}{C}HCOBr \xrightarrow{H_2O} CH_3\underset{\underset{\displaystyle Br}{|}}{C}HCOOH + HBr$$

It is helpful to regard this as another example of the general rule that addition reactions are rapid compared to displacements. The acyl-halide hydrolysis is primarily an addition at the carbonyl group (Sec. 13.3); removal of the α halogen, on the other hand, involves a displacement and is slower.

In the addition of unsymmetrical addenda, such as HCl, to the double bond of unsaturated acids in which the double bond is situated in the α,β position, it is found that Markownikoff's rule no longer appears to hold; instead the major reaction goes as follows:

$$H_2C{=}CHCOOH + HCl \longrightarrow \underset{\underset{\displaystyle Cl}{|}}{C}H_2{-}CH_2COOH$$

This can be readily understood if one remembers that the carboxyl group is stabilized by resonance of the type

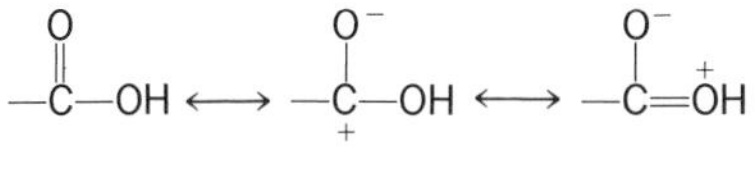

so that the carbon atom of the carboxyl group bears appreciable positive charge. Since HCl adds by initial addition of H^+, the two alternative carbonium ions are:

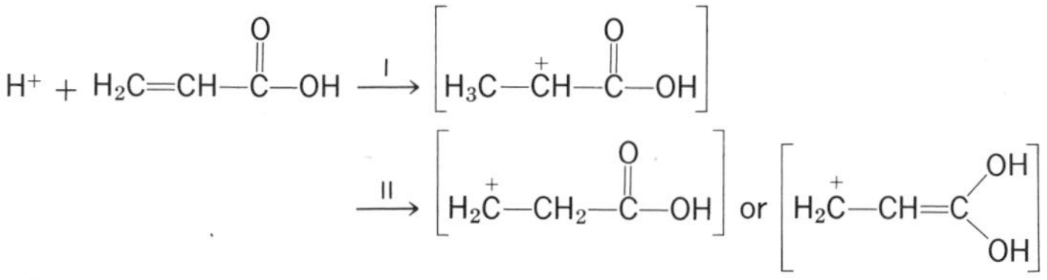

Combination with chloride ion gives the product. Path I would put the positive charge on the adduct directly adjacent to the positively polarized carbon of the carbonyl group. Therefore, Path II is followed as the "lesser of two evils." The addition is much slower than that to simple alkenes.

If Markownikoff's rule is formulated in more mechanistic terms such that "the positively charged addend adds so as to place the newly generated positive charge in the energetically most favorable location," then many apparent deviations from Markownikoff's rule fall in line. Thus, HX[1] will add to any alkene to give

$$\mathrm{C{=}C{-}Y + HX \longrightarrow X{-}C{-}C(H){-}Y} \qquad \text{if Y is electron withdrawing}$$

and

$$\mathrm{C{=}C{-}Z + HX \longrightarrow H{-}C{-}C(X)(Z)} \qquad \text{if Z is electron donating}$$

Although nucleophilic reagents do not add readily to double bonds, suitably conjugated double bonds (with strongly electron-withdrawing groups) add nucleophiles readily.

$$\mathrm{:NH_3 + H_2C{=}CH{-}C{\equiv}N \longrightarrow H_2N{-}CH_2{-}CH_2{-}CN}$$

Note that a nucleophile adds so as to best stabilize the negative charge generated. Thus

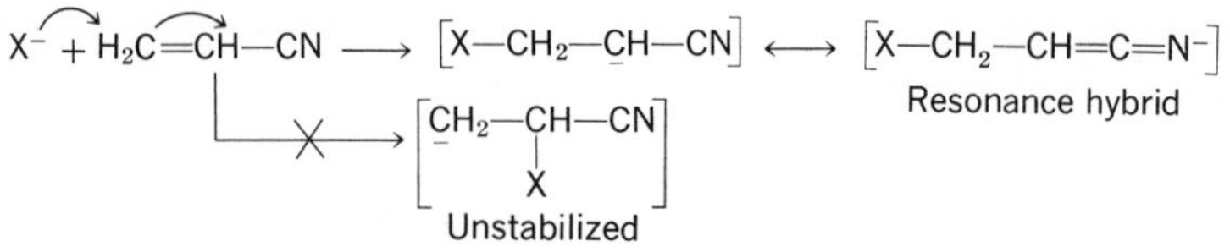

X⁻ + H₂C=CH—CN
[X—CH₂—CH—CN] ⟷ [X—CH₂—CH=C=N⁻]
Resonance hybrid
[CH₂—CH—CN, X]
Unstabilized

As the distance between the carboxyl group and the double bond increases, this effect diminishes; electrical effects (inductive effects) do not generally extend beyond three CH_2 groups. For example, oleic acid yields a mixture of products with HBr.

$$\mathrm{CH_3(CH_2)_7CH{=}CH(CH_2)_7COOH \xrightarrow{HBr} CH_3(CH_2)_7\underset{\displaystyle Br}{\underset{|}{C}}H{-}CH_2(CH_2)_7COOH}$$

$$\mathrm{+\ CH_3(CH_2)_7CH_2{-}\underset{\displaystyle Br}{\underset{|}{C}}H(CH_2)_7COOH}$$

When hydroxy acids are available from natural or other sources, the hydroxyl groups can be substituted by halogen in the usual manner.

---

[1] This will necessarily be true only for ionic additions. It will, however, generally be true if X^- rather than H^+ adds first.

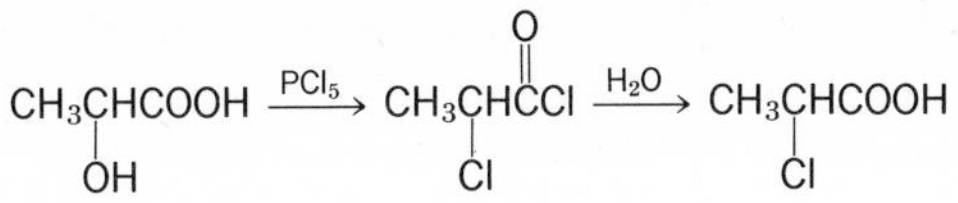

$$CH_3\underset{\underset{OH}{|}}{CH}COOH \xrightarrow{PCl_5} CH_3\underset{\underset{Cl}{|}}{CH}\overset{\overset{O}{\|}}{C}Cl \xrightarrow{H_2O} CH_3\underset{\underset{Cl}{|}}{CH}COOH$$

Lactic acid

As previously noted, we find that a halogen *directly linked* to a carbonyl group (an acyl halide) is far more easily hydrolyzed than a halogen on a saturated chain.

17.3 Preparation of Hydroxy Acids   Hydrolysis of any halide normally leads to the production of the corresponding hydroxy compound. A comparison of the hydrolysis of a series of halogen acids with aqueous alkali is instructive.

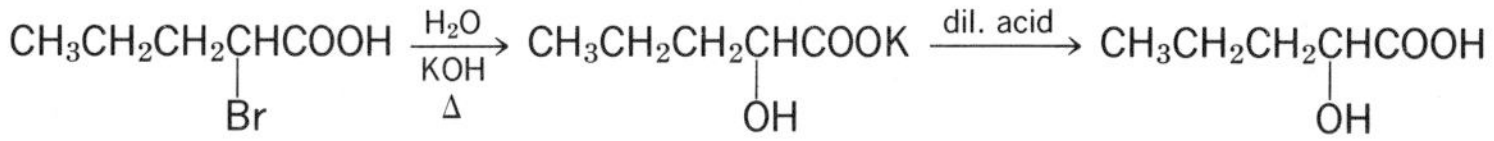

$$CH_3CH_2CH_2\underset{\underset{Br}{|}}{CH}COOH \xrightarrow[\text{KOH}]{H_2O} CH_3CH_2CH_2\underset{\underset{OH}{|}}{CH}COOK \xrightarrow{\text{dil. acid}} CH_3CH_2CH_2\underset{\underset{OH}{|}}{CH}COOH$$

α-Bromovaleric acid

$$CH_3CH_2\underset{\underset{Br}{|}}{CH}CH_2COOH \xrightarrow[\text{KOH}]{H_2O} CH_3CH_2CH{=}CHCOOK \xrightarrow{\text{dil. acid}} CH_3CH_2CH{=}CHCOOH$$

β-Bromovaleric acid

$$CH_3\underset{\underset{Br}{|}}{CH}CH_2CH_2COOH \xrightarrow[\underset{\Delta}{\text{KOH}}]{H_2O} CH_3\underset{\underset{OH}{|}}{CH}CH_2CH_2COOK \underset{\longleftarrow}{\xrightarrow{\text{dil. acid}}} CH_3\underset{\underset{O}{|}}{CH}CH_2CH_2C{=}O$$

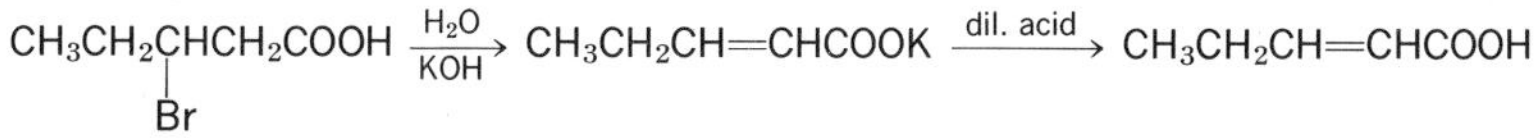

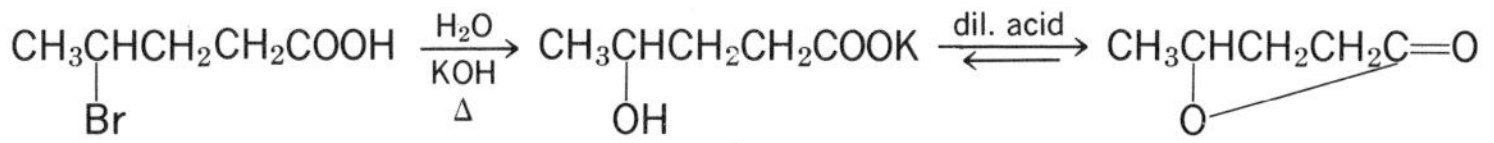

γ-Bromovaleric acid   A lactone
                                                   (γ-valerolactone)

The student should realize that the long bonds drawn in the lactone formula are merely for convenience and are not meant to imply an unusual type of bond. This formula may be redrawn to present a better picture of the lactone structure as follows:

$$\begin{array}{c} H_2C{-\!\!-\!\!-}CH_2 \\ | \qquad\quad | \\ H_3C{-}HC \qquad C{=}O \\ \phantom{H_3C{-}HC}\diagdown_O\diagup \end{array}$$

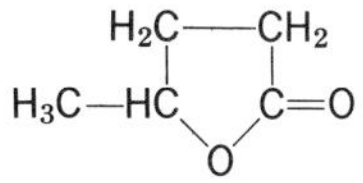

It will be observed that the only new reactions involved here are the loss of HBr from the β-bromovaleric acid and the formation of a lactone *on acidifying* the salt of the γ-valeric acid. For reasons that are discussed in Sec. 11.12, this loss of HX from β-halogen-substituted acids is extraordinarily easy. Only under very mild conditions can β-hydroxy acids be obtained.[1]

Although in the case of γ-bromovaleric acid the hydrolysis appears to proceed normally, it is probably assisted by an intramolecular displacement

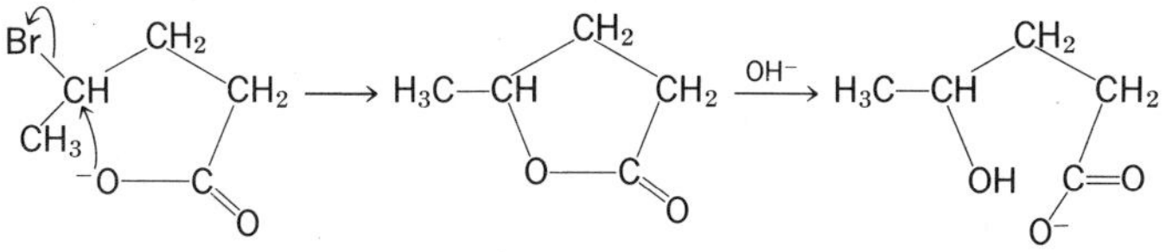

---
[1] Moist silver oxide is a reagent that leads to a minimum of unsaturated-acid formation.

since the reaction occurs hundreds of times faster than with bromo acids where the bromine is farther out in the chain. This is one example of a neighboring-group effect (Sec. 20.6).

The formation of the salt of the hydroxy acid is comparable to the hydrolysis of an ordinary ester and the equilibrium is displaced to the side of the salt. On acidification, however, the lactone is re-formed; the hydroxy acid is not isolated in many instances when a five- or six-membered lactone can be formed.

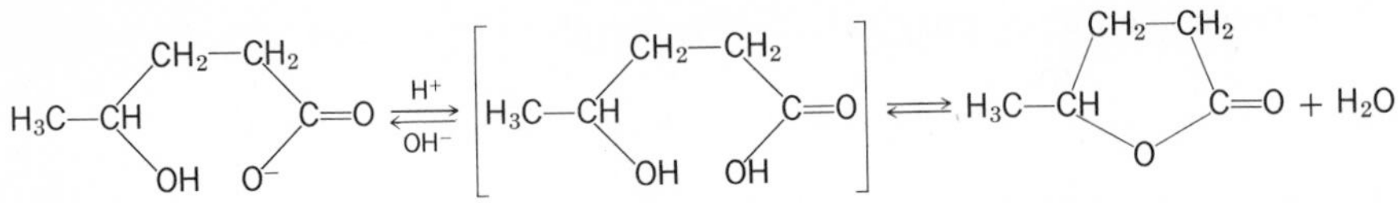

This is a simple esterification reaction, but since it is an *intra*molecular reaction the effective concentrations of acid and alcohol are much higher than with *inter*molecular esterification, and the equilibrium is displaced far on the side of the lactone. When a halogen (or for that matter any other) group is more than three CH_2 groups away, interaction with a carboxyl group is negligible under ordinary conditions.

It is interesting that lactones derived from beta-substituted acids are known. They are usually prepared indirectly and are very reactive substances, as might be expected because of the ring size. In the reactions of beta-substituted acids, described above, they do not appear in significant amounts.

Because of the versatility of the aliphatic hydroxyl group as a starting material for organic syntheses, the hydroxy acids are particularly useful. A number of them, often with very complex structures, are found in nature, where they seem to play important roles. Some interesting ones are listed below.

TABLE 17.2
SOME IMPORTANT SUBSTITUTED ACIDS

Name	*Structure*	*Source*
Malic acid	HOOCCH(OH)CH₂COOH	Sour apples, many berries, etc.
Tartaric acid	HOOCCHOHCHOHCOOH	Grapes and many plants
Citric acid	HOOCCH₂C(OH)(COOH)CH₂COOH	Citrus fruit and many organisms

17.4 Amino Acids

Like other halides, halogen acids react with ammonia and amines; this property is of value in the synthesis of amino acids, particularly the natural α-amino acids.

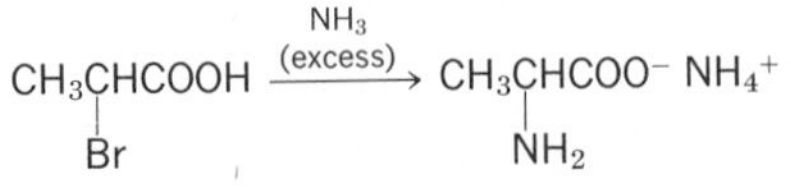

Here, as would be expected, the formation of secondary and tertiary amines is a side reaction. By use of a large excess of ammonia these side reactions become of minor importance in this case. β-Halogen acids with ammonia and amines behave as with other alkaline reagents.

$$\mathrm{CH_3\underset{\underset{Cl}{|}}{C}HCH_2COOH} \xrightarrow{\mathrm{NH_3}} \mathrm{CH_3CH{=}CHCOONH_4 + NH_4Cl}$$

Halogen atoms more remote from the carboxyl group are replaced normally by ammonia to form mixtures of amino acids. γ- and δ-amino acids tend, like the corresponding hydroxy acids, to form cyclic amides on heating. These are called *lactams.*

$$\mathrm{CH_3\underset{\underset{Br}{|}}{C}HCH_2CH_2COOH} \xrightarrow{\mathrm{NH_3}} \mathrm{CH_3\underset{\underset{NH_2}{|}}{C}HCH_2CH_2COO^-\ NH_4^+} \xrightarrow{\mathrm{acid}} \mathrm{CH_3\underset{\underset{HN}{|}}{C}HCH_2CH_2C{=}O}$$

A lactam

It is found that those ring systems containing nitrogen and other atoms, *heterocyclic* ring systems, are most easily formed when the ring contains five or six atoms in all. Lactones or lactams beyond the δ position (six-ring) are difficult to obtain but, when formed, resemble other esters or amides in stability. The situation is analogous to that found with the cycloalkanes (Sec. 4.3).

17.5 Cyanohydrin Synthesis Aldehydes, both aromatic and aliphatic, are often used in the preparation of hydroxy acids (see preceding pages for methods from halogen acids). The addition of HCN to aldehydes leads directly to α-hydroxy nitriles (cyanohydrins) and is one of the most useful approaches to the corresponding acids.

$$\mathrm{C_6H_5CH{=}O + HCN \longrightarrow C_6H_5\underset{\underset{OH}{|}}{C}H{-}C{\equiv}N \xrightarrow[\mathrm{acid}]{\mathrm{H_2O}} C_6H_5\underset{\underset{OH}{|}}{C}H{-}\overset{\overset{O}{\|}}{C}{-}OH}$$

Mandelic acid

17.6 Reformatsky Reaction Because of the rapid reaction between Grignard reagents and acids or acid derivatives, this otherwise valuable synthetic tool is of little use in the preparation of substituted acids. It is found, however, that the zinc analog of the Grignard reagent, readily prepared from *α-halogen* esters, adds smoothly to carbonyl compounds. The organozinc reagent is prepared in the presence of the aldehyde or ketone with which it is expected to react, in order to minimize reaction with the ester carbonyl. The synthesis is referred to as the *Reformatsky reaction.*

$$\mathrm{(CH_3)_2C{=}O + Br\underset{\underset{CH_3}{|}}{C}HCOOCH_3 \xrightarrow{Zn} (CH_3)_2\underset{\underset{OZnBr}{|}}{C}{-}\overset{\overset{CH_3}{|}}{C}HCOOCH_3}$$

The zinc bromide salt, analogous to the familiar magnesium bromide intermediate of the Grignard synthesis, is readily decomposed by water to yield the free hydroxy ester.

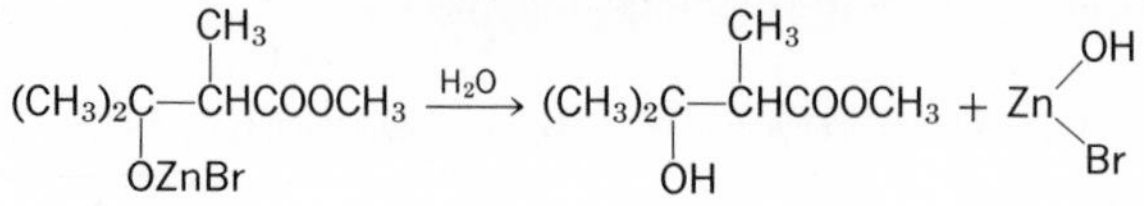
CH3
(CH3)2C—CHCOOCH3
OZnBr
H2O
CH3
(CH3)2C—CHCOOCH3 + Zn
OH
OH
Br

The Reformatsky reaction depends for its success on the fact that the intermediate halogen zinc alkyls react at a greater rate with aldehydes and ketones than with esters. It is of interest to note that it is the esters of α-halogen acids only which undergo the Reformatsky reaction (since the negative charge can be stabilized by resonance).

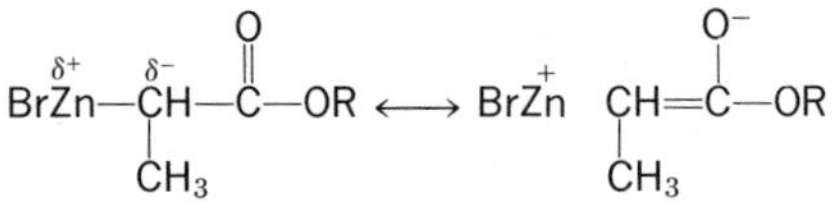
δ+ δ−
BrZn—CH—C—OR
O
CH3
BrZn CH=C—OR
O−
CH3

The free acids cannot be used in this reaction; acids, even such weak acids as water and ammonia, will decompose the Reformatsky, as they will the Grignard, reagent. Also it will be recalled that attempts to form the reagent with free acids will lead instead to a different reaction, the α-halogen acids being rather strong acids in most cases.

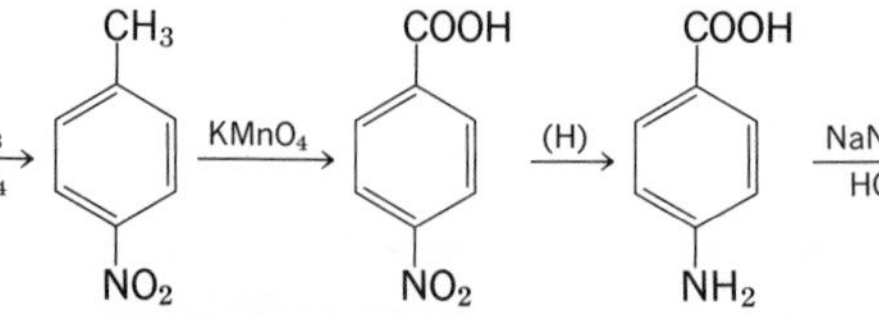
2CH2COOH + Zn
Cl
CH2COO
Cl
2
Zn + H2↑

17.7 Preparation of Phenolic Acids Phenolic acids, some of which are of industrial and medicinal importance, may be prepared by the use of reactions already studied.

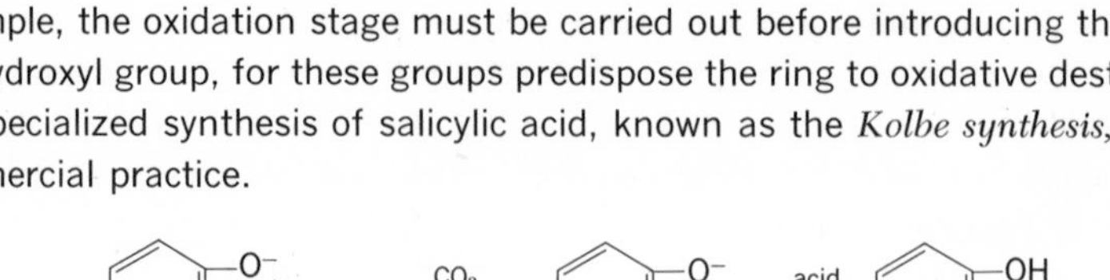
CH3
HNO3
H2SO4
CH3
NO2
KMnO4
COOH
NO2
(H)
COOH
NH2
NaNO2
HCl
COOH
OH

In reaction sequences of this sort, the order of operations is often important. Here, for example, the oxidation stage must be carried out before introducing the amino or the hydroxyl group, for these groups predispose the ring to oxidative destruction.

A specialized synthesis of salicylic acid, known as the *Kolbe synthesis,* is used in commercial practice.

O−
CO2
pressure, Δ
O−
COO−
acid
OH
COOH

Phenolate ion
(from sodium phenolate)

Salicylic acid

At the elevated temperature of the reaction the phenolate ion apparently has enough nucleophilic character at the ortho position to combine with the carbonyl group of CO_2 in the manner already observed with other reagents. This reaction and others of a similar sort that will not be discussed are limited to the phenols; benzene itself, for example, does not add to CO_2 in this manner. We have here another example of the activating effect of the phenol group on the benzene ring (Sec. 8.24).

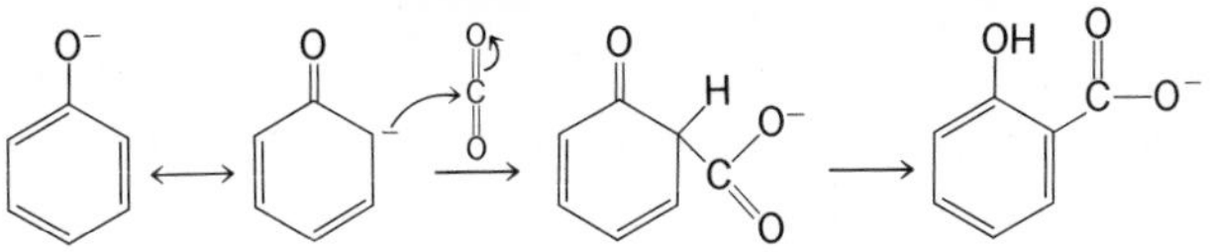

Here the course of the reaction is complicated by the counter ion. With Na^+, the major product is the *ortho* with very little *para* being found. If the potassium salt is used, the *para* product is favored. Salicylic acid is an ingredient of many medicinals and is itself an effective germicide. Acetylsalicylic acid is the familiar aspirin, and methyl salicylate is found in oil of wintergreen.

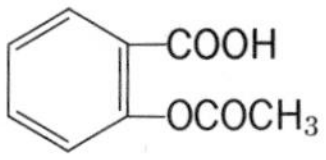

Aspirin

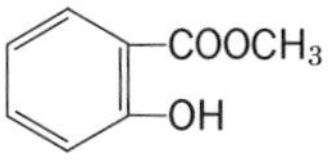

Methyl salicylate
(in oil of wintergreen)

The properties of phenolic acids can generally be predicted on the basis of the known properties of phenols and of acids; the two groups have little effect on each other.

17.8 Properties of Aliphatic Hydroxy Acids In many respects, the aliphatic hydroxy acids exhibit more or less independently the reactions of both the hydroxyl and the carboxyl group. In certain instances, the same reagents may affect both groups and act to limit the generality of this statement. For example, the acid chlorides of the hydroxy acids are unknown; attempts to prepare them always involve the hydroxyl group as well as the carboxyl group.

$$\underset{\displaystyle OH}{CH_3\overset{}{C}HCOOH} + PCl_5 \longrightarrow \underset{\displaystyle Cl}{CH_3CHCOCl}$$

On the other hand, reagents toward which either group is inert may react normally with the other.

$$\underset{\text{Lactic acid}}{\underset{\displaystyle OH}{CH_3CHCOOH}} \xrightarrow{(O)} \underset{\text{Pyruvic acid}}{\underset{\displaystyle O}{\overset{}{CH_3\overset{}{C}COOH}}}$$

$$\underset{\displaystyle OH}{CH_3CHCOOH} + NaOH \xrightarrow{H_2O} \underset{\displaystyle OH}{CH_3CHCOONa} \xrightarrow{CH_3I} \underset{\displaystyle OH}{CH_3CHCOOCH_3}$$

$$\underset{\displaystyle OH}{CH_3CHCOOH} + CH_3COCl \longrightarrow \underset{\displaystyle OCOCH_3}{CH_3CHCOOH}$$

17.9 Decarboxylation Those acids in which the carboxyl and hydroxyl groups are located close to one another in the molecule give characteristic reactions with hot dilute sulfuric acid. The α-hydroxy acids decompose with loss of a single carbon atom.

$$CH_3\underset{\underset{OH}{|}}{CH}-COOH \xrightarrow[\Delta]{\text{dil.} \atop H_2SO_4} CH_3CH{=}O + CO + H_2O$$

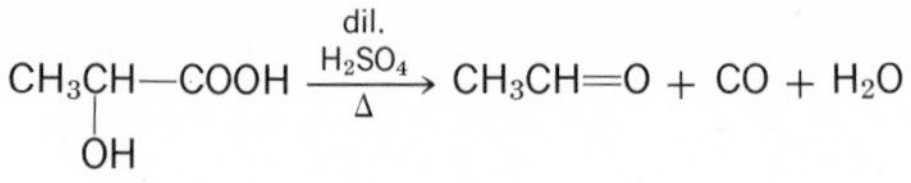

Presumably the reaction proceeds by a dehydration mechanism.[1]

$$CH_3\underset{\underset{OH}{|}}{CH}-\overset{\overset{O}{\|}}{C}-OH \xrightarrow{H^+} \left[CH_3\underset{\underset{OH}{|}}{CH}-\overset{\overset{O}{\|}}{\underset{\underset{H}{|}}{C}}-\overset{+}{O}H \right] \longrightarrow \left[CH_3\underset{\underset{OH}{|}}{CH}-\overset{\overset{O}{\|}}{C}{}^+ \right]$$

$$\downarrow$$

$$CH_3\underset{\underset{O}{\|}}{CH} + H^+ + CO$$

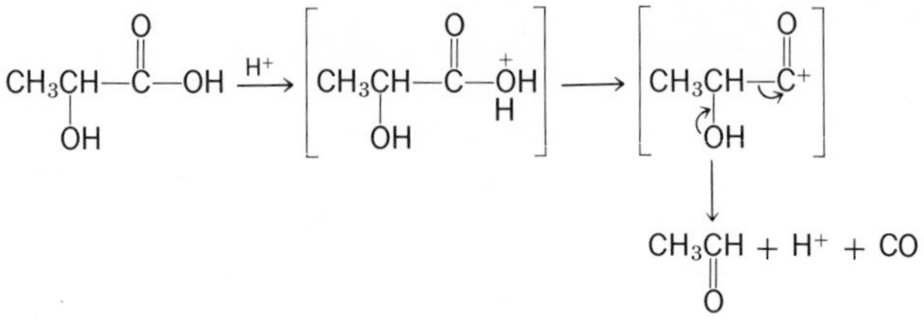

This reaction of α-hydroxy acids is reminiscent of a number of other decompositions of carbon compounds bearing two functional groups on the same carbon atom. It has been noted that in the case of compounds of type

$$-\underset{\underset{X}{|}}{\overset{\overset{|}{}}{C}}-OH$$

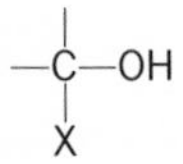

where X may be OH, OR, OCOR, NH_2, NHR, NR_2, SH, Cl, Br, or I, the substances usually are so unstable as to defy isolation. In some cases where X is strongly electronegative, as in CN, COOH, or SO_3H, the compounds can be isolated but are readily decomposed. In all cases, the decomposition product is a carbonyl compound,[2] as in the present case of the α-hydroxy acids. This decomposition has been utilized as a general method for decreasing the length of a carbon chain by one carbon atom.

$$CH_3CH_2COOH \xrightarrow{Br_2} CH_3\underset{\underset{Br}{|}}{CH}COOH \xrightarrow{H_2O} CH_3\underset{\underset{OH}{|}}{CH}COOH \xrightarrow{H_2SO_4} CH_3CH{=}O \xrightarrow{(O)} CH_3COOH$$

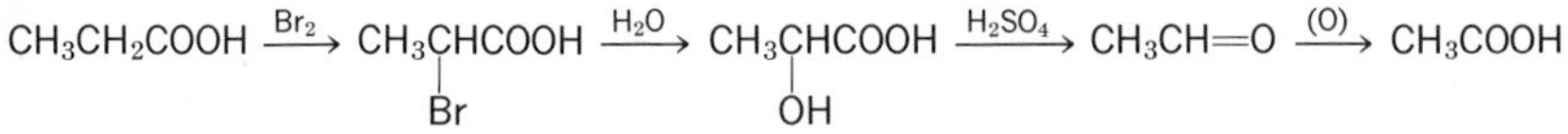

[1] It should be noted that this is not the preferred locus of protonation of an α-hydroxy acid. Either

$$CH_3\underset{\underset{+OH_2}{|}}{CH}-COOH \quad \text{or} \quad CH_3\underset{\underset{OH}{|}}{CH}-\overset{\overset{O}{\|}}{\underset{\underset{+}{OH}}{C}}-OH$$

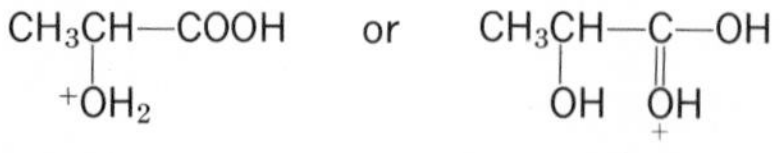

should be preferred. However, since neither of these leads to a competing reaction product, they may be ignored as blind alleys. All that is required is that the above method of protonation occurs some of the time, since the subsequent steps are irreversible at ordinary pressures (CO escapes).

[2] As was pointed out in Sec. 13.2 such substances are all of the same oxidation level with respect to the carbon atom in question.

17.10 Dehydration Reactions A side reaction in the above series results from the catalysis of esterification by sulfuric acid. It is observed with α-hydroxy acids and involves the formation of a cyclic ester (note the size of the ring).

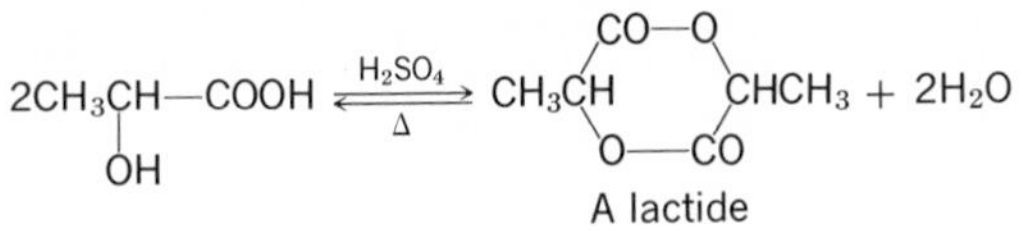

A lactide

By a suitable choice of conditions, either lactide formation or degradation can be made to predominate.

β-Hydroxy acids under the same conditions lose the elements of water in an almost spontaneous reaction. The reasons for this easy loss of water are the same as those discussed in Sec. 11.12 in the case of aldol.

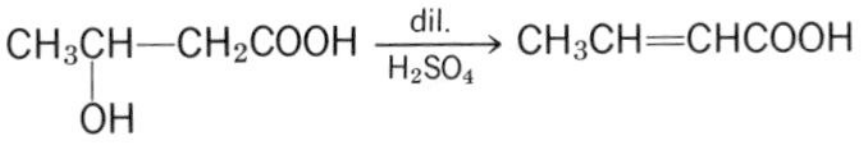

As already pointed out, γ- and δ-hydroxy acids under these ester-forming conditions form lactones.

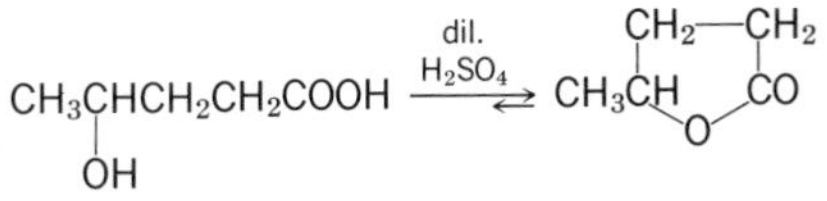

We have, then, in dilute sulfuric acid a useful test reagent for identifying the position of the hydroxyl group in aliphatic hydroxy acids, in that the above reactions seem quite general and apply even to rather complex cases. The most rapid and characteristic reaction of the three is that of the β hydroxyl group (and α hydrogen atom) to yield a double bond. In cases where alternative reactions are possible, unsaturated-acid formation often predominates.

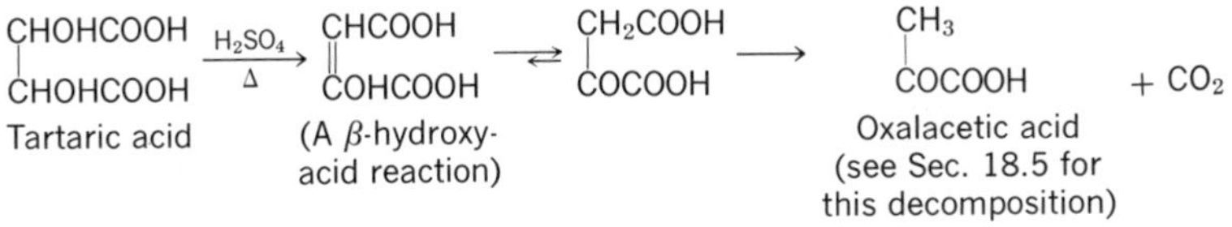

Tartaric acid
(A β-hydroxy-acid reaction)
Oxalacetic acid (see Sec. 18.5 for this decomposition)

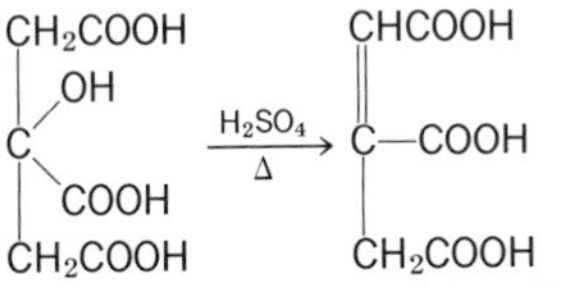

Citric acid　　Aconitic acid

In the examples given above of the natural acids citric and tartaric (often called *fruit acids*), it will be observed that the primary reaction in each case is a typical β-hydroxy-acid reaction although the substances can also be regarded as α-hydroxy acids. The dehydration products frequently undergo further degradation in cases

TABLE 17.3
PROPERTIES OF MALEIC AND FUMARIC ACIDS

Acid	*Melting point,* °C	*Specific gravity*	*Solubility,* *g/100 g water*	*Heat of combustion,* *kcal/mole*
Maleic	130	1.609	79 (at 25°)	327.1
Fumaric	287	1.635	0.7 (at 17°)	320.2

of this sort, as is shown above. Very similar reaction sequences are found to take place in living tissues.[1]

17.11 Maleic and Fumaric Acids Malic acid, found naturally in fruits, undergoes a simple dehydration. Instead of a single unsaturated acid, however, there are found two different isomeric acids, maleic and fumaric acids.

Both acids contain one carbon-to-carbon double bond and two carboxyl groups; on hydrogenation of either acid, two atoms of hydrogen are taken up with the formation of succinic acid as the only product. Only one form of succinic acid is known.

$$\begin{matrix} CHOHCOOH \\ | \\ CH_2COOH \end{matrix} \xrightarrow{-H_2O} \left\{ \begin{matrix} \text{maleic acid} \\ \text{fumaric acid} \end{matrix} \right\} \xrightarrow{H_2} \begin{matrix} CH_2COOH \\ | \\ CH_2COOH \end{matrix}$$

Succinic acid

It is evident that we are dealing with geometrical *isomers* of the type discussed in Sec. 7.2. When maleic acid is heated it loses water to form a (subliming) cyclic anhydride. Fumaric acid reacts slowly and only at very high temperature to form some maleic anhydride. This is sufficient for us to deduce that maleic acid is the *cis* form and fumaric is the *trans*.

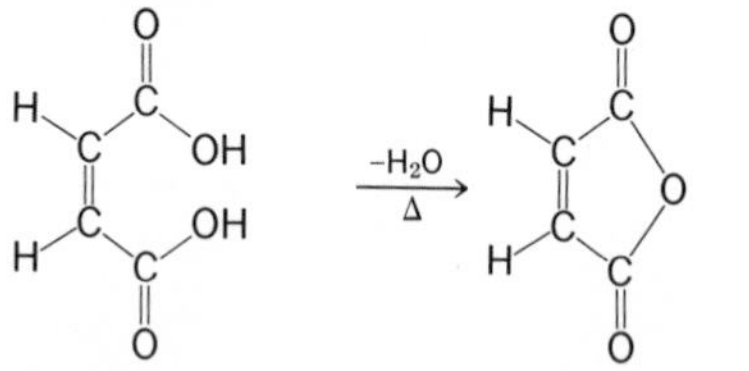

Maleic acid   Maleic anhydride

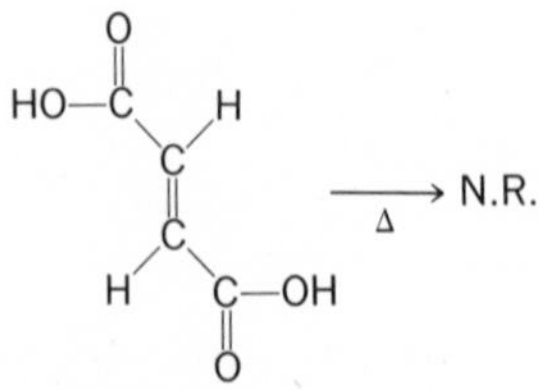

Fumaric acid

[1] All these acids seem to be intermediates in the carbohydrate metabolism of plants and animals.

With fumaric acid the carboxyl groups cannot come close enough together to interact, and since rotation about the double bond is energetically difficult, the formation of a cyclic anhydride is not readily observed.

17.12 Structure and Physical Properties Some interesting correlations between structure and physical properties are found with maleic and fumaric acids (see Table 17.3). The two carboxyl groups, being close together in maleic acid, can interact with each other, a fact which may be related to the observed higher dissociation constant of maleic, as compared to fumaric, acid. But the bulky carboxyl groups may be rather crowded when on the same side of the double bond, and this may increase the strain on the bond, accounting at least in part for the higher heat of combustion of maleic acid, evidence that it contains more available energy than fumaric. This crowding and lack of symmetry of the molecule may contribute to less harmonious packing in the crystal, and so account for the fact that the melting point of maleic is substantially lower than that of fumaric acid. Also, intramolecular interaction of the two carboxyl groups in maleic acid would diminish intermolecular interactions, which might bind neighboring molecules and would be the only kind of interaction possible to fumaric acid. This intermolecular interaction combined with the better packing, which would come with its greater symmetry, would adequately account for the greater density (specific gravity) and higher melting point of fumaric acid, and also for its lesser solubility in water. In general, those isomers with higher melting points, indicating, among other things, better bound crystals, are also less soluble. That this is not an isolated example of such correlations may be seen from an examination of the physical properties of the two 1,2-dichloroethylenes (see Table 17.4). Here again the more symmetrical trans form melts higher than the cis. By actual measurement the halogen atoms are found to be farther apart in the trans form. The halogen atoms, which are electronegative with respect to the carbon atoms to which they are attached, are opposed across the trans molecule and yield a zero dipole moment, indicating electrical symmetry. In the cis form, they are on the same side, and the molecule is definitely dissymmetric.

17.13 Cis-Trans Interconversion An equilibrium is set up between cis and trans forms when, under the influence of heat or reagents of various kinds, the double

TABLE 17.4
SOME PHYSICAL PROPERTIES OF THE 1,2-DICHLOROETHYLENES

Substance	*Melting point,* °C	*Distance* Cl *to* Cl, Å[a]	*Dipole moment*[b]
Cis form	−80.5	3.6	1.89
Trans form	−50	4.1	0

[a] This distance was measured by electron diffraction: electrons are scattered in the same manner as x-rays by the centers of atoms.
[b] The dipole moment is measured in *Debye units;* it is a measure of electrical symmetry of the molecule; the more symmetry, the lower the dipole moment.

bond is weakened to the point where rotation becomes possible. Usually the more stable form predominates in this equilibrium mixture. Maleic acid on moderate heating rearranges to fumaric, but if either isomer is heated very hot, so that the chemical reaction of anhydride formation can ensue, then the equilibrium is shifted as the molecule is fixed in the cis form by anhydride formation. If energy in the form of ultraviolet light is added to fumaric acid, some of it can be converted to the form that is richer in energy, maleic acid.

The most convincing proof of distinction between cis-trans isomers may be obtained in those cases where two groups in the same molecule are capable of reacting with one another, as for instance in the case of maleic and fumaric acids. Anhydride formation is often used, but, in principle, any reaction will do. For example,

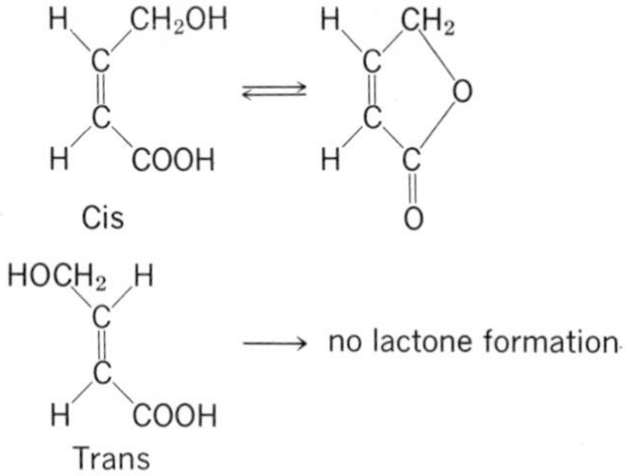
Cis
no lactone formation
Trans

The real distinction lies in the greater distance between the two groups.

The student should realize that the formulas as written above for the cis and trans forms bear such a close correspondence to the actual shapes of the molecules that predictions about the number of isomers and their reactions can be made from them.

Aside from physical and chemical differences between geometrical isomers (stability, ionization, rate of reaction), there are biochemical differences in some cases. Fumaric acid is an intermediate in metabolism, while maleic acid is not found in nature and is a toxic substance. *cis*-Cinnamic acid has plant-hormone activity; *trans*-cinnamic has none.

The number of geometrical isomers possible in a substance such as vitamin A, which has many double bonds, indicates the diversity which this phenomenon permits to some naturally occurring substances.

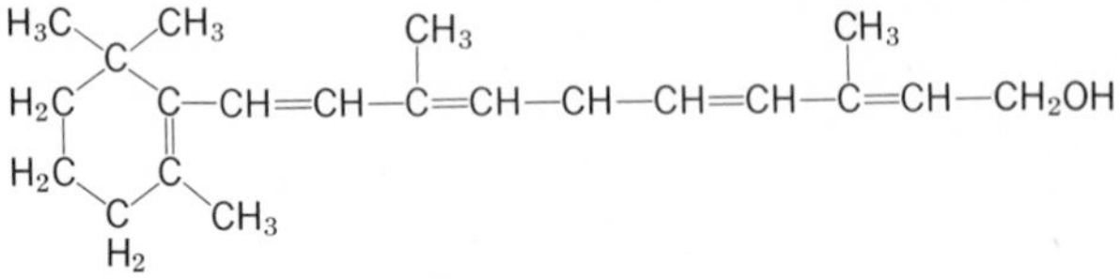

Vitamin A[1]

[1]Compare this structure with that of β-carotene, Sec. 3.20, which is also vitamin A active. Even such a complicated substance as this has been commercially synthesized.

There is evidence for the occurrence in nature of several of the possible isomers in the case of carotene (see Sec. 3.20).

17.14 Reactions of Unsaturated Acids Except for the phenomenon of geometrical isomerism, unsaturated acids for the most part exhibit the combination of alkene and acid properties that would be expected from an inspection of their formulas.

$$\underset{\text{Crotonic acid}}{CH_3CH{=}CHCOOH} \xrightarrow{Br_2} CH_3\underset{Br}{\underset{|}{C}}H{-}\underset{Br}{\underset{|}{C}}HCOOH$$

$$CH_3CH{=}CHCOOH \underset{H^+}{\overset{C_2H_5OH}{\rightleftharpoons}} CH_3CH{=}CHCOOC_2H_5$$

$$CH_3CH{=}CHCOOH \xrightarrow{KMnO_4} CH_3COOH + [HOOCCOOH]$$

$$CH_3CH{=}CHCOOH \xrightarrow{SOCl_2} CH_3CH{=}CHCOCl$$

$$CH_3CH{=}CHCOOH \xrightarrow{HBr} CH_3\underset{Br}{\underset{|}{C}}H{-}CH_2COOH$$

The α,β types are easily available from β-hydroxy acids or from Perkin or aldol condensation products. The α,β (conjugated) types are more stable than other forms. At elevated temperatures, particularly in contact with strong acids, rearrangements often occur, actually shifting the double bond into this preferred position.

$$H_2C{=}CH{-}CH_2COOH \underset{\Delta}{\overset{acid}{\rightleftharpoons}} H_3C{-}CH{=}CHCOOH$$

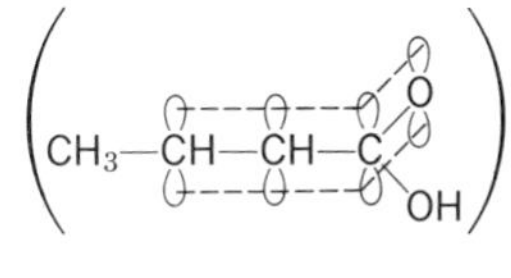
CH3—CH—CH—C
O
OH

Aside from the α,β-unsaturated acids, only a few isolated types of unsaturated acids are of importance at present.

Oleic acid has already been referred to. Some other analogous unsaturated acids occur as glycerides in many oils. For example,

$$CH_3(CH_2)_4CH{=}CH{-}CH_2{-}CH{=}CH(CH_2)_7COOH$$

Linoleic acid
(linseed oil, soybean oil)

$$CH_3(CH_2)_5\overset{OH}{\overset{|}{C}}H{-}CH_2CH{=}CH(CH_2)_7COOH$$

Ricinoleic acid
(castor oil)

There is evidence that oils containing unsaturated acids of this sort play an important role, and are required, in the diet of mammals (vitamin F).

17.15 Spectroscopic Properties The infrared and nuclear magnetic resonance spectra of substituted acids do not differ appreciably from what is expected for the various substituents separately. Not unexpected deviations, such as more extensive hydrogen bonding in hydroxy acids and small shifts in the carbonyl frequencies of conjugated unsaturated acids, are observed.

In certain instances spectroscopic methods have proved themselves invaluable. For example, both the infrared and nmr spectra of maleic acid are different from those of fumaric acid, and it is possible by correlating such differences to assign cis and trans structures to a large number of alkenes from the spectra (especially nmr) alone. Such considerations are beyond the scope of this book, however, and will not be dealt with here.

OUTLINE OF CHEMISTRY OF HALOGEN, HYDROXY, AND UNSATURATED ACIDS

HALOGEN ACIDS

Preparation

1. 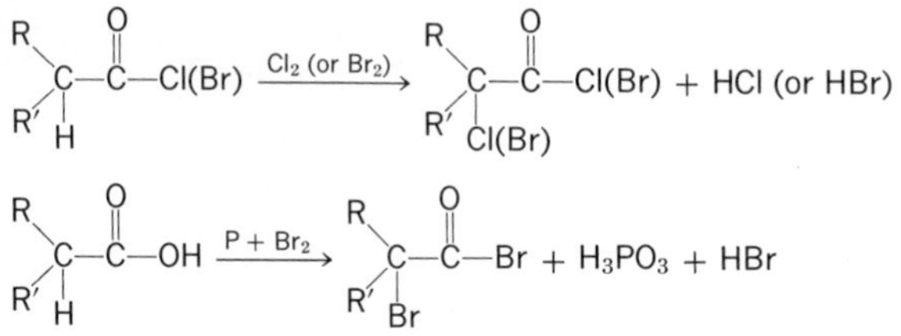

R
O
C—C—Cl(Br)
R′
H
Cl₂ (or Br₂)
R
O
C—C—Cl(Br) + HCl (or HBr)
R′
Cl(Br)
R
O
C—C—OH
R′
H
P + Br₂
R
O
C—C—Br + H₃PO₃ + HBr
R′
Br

2. $CH_3CH_2CHOHCOOH \xrightarrow{PCl_5} CH_3CH_2CHClCOCl \xrightarrow{H_2O} CH_3CH_2CHClCOOH$

3. $H_2C{=}CH{-}COOH + HCl \longrightarrow CH_2Cl{-}CH_2{-}COOH$

Properties

1. Halogen acids have all the ordinary properties of carboxylic acids and of alkyl halides.
2. Reaction with alkali.
   a. α-Halogen acids.

   $$CH_3CH_2CHXCOOH \xrightarrow{KOH} CH_3CH_2CHOHCOOK \xrightarrow[HCl]{dil.} CH_3CH_2CHOHCOOH$$

   b. β-Halogen acids.

   $$CH_3CHXCH_2COOH \xrightarrow{KOH} CH_3CH{=}CHCOOK \xrightarrow[HCl]{dil.} CH_3CH{=}CHCOOH$$

   c. γ-Halogen acids.

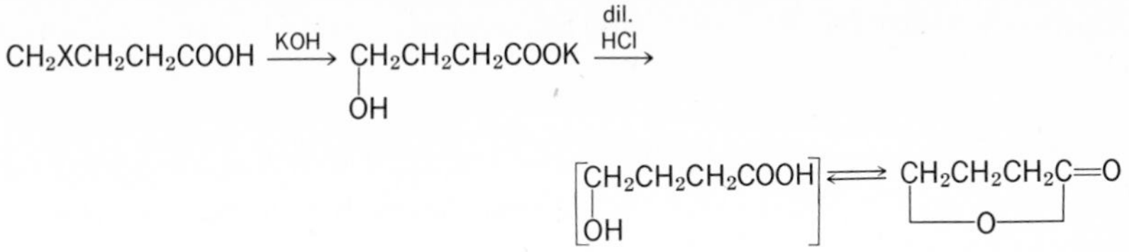

$$CH_2XCH_2CH_2COOH \xrightarrow{KOH} \underset{\underset{OH}{|}}{CH_2CH_2CH_2COOK} \xrightarrow{\overset{dil.}{HCl}}$$

$$\left[\underset{\underset{OH}{|}}{CH_2CH_2CH_2COOH} \right] \rightleftharpoons CH_2CH_2CH_2C{=}O$$

HYDROXY ACIDS

A. Aliphatic and side-chain hydroxy acids (that is, alcoholic hydroxyl group).

Preparation

1. $RCHXCOOH \xrightarrow[\overset{aq.}{NaOH}]{} RCHOHCOONa \xrightarrow[\overset{dil.}{HCl}]{} RCHOHCOOH + NaCl$

2. For α-hydroxy acids only.

$$\underset{R'}{\overset{R}{>}}C{=}O + HCN \longrightarrow \underset{R'}{\overset{R}{>}}\underset{CN}{\overset{OH}{C}} \xrightarrow[HCl]{H_2O} \underset{R'}{\overset{R}{>}}\underset{COOH}{\overset{OH}{C}} + NH_4Cl$$

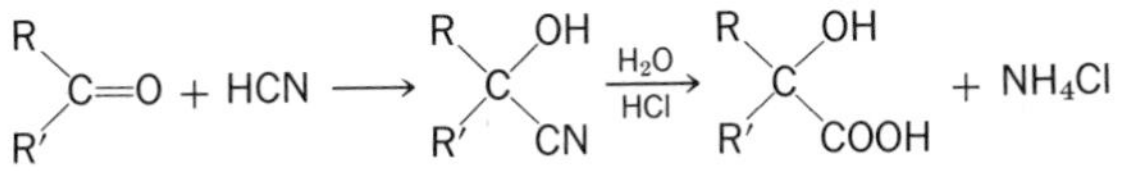

An aldehyde or ketone, aliphatic or aromatic, may be used here.

3. The Reformatsky reaction yields β-hydroxy acids only.

$$RCHO + Br{-}CH_2COOC_2H_5 \xrightarrow[ether]{Zn} \underset{}{\overset{OZnBr}{R\dot{C}HCH_2COOC_2H_5}} \xrightarrow{H_2O}$$

$$RCHOHCH_2COOC_2H_5 + Zn(OH)Br$$

Properties

The following properties, unless otherwise indicated, hold for α-, β-, and γ-hydroxy acids.

1. Aliphatic hydroxy acids have most of the usual properties of acids. Notable exceptions are that they do not form stable anhydrides or acid chlorides.

2. $R{-}CHOH{-}COOH \xrightarrow{(O)} R{-}CO{-}COOH$

3. $R{-}CHOH{-}COOH \xrightarrow{PCl_5} R{-}CHCl{-}COCl \xrightarrow{H_2O} R{-}CHCl{-}COOH$

4. $R{-}CHOH{-}COOH \xrightarrow{R'COCl} \underset{\underset{O-CO-R'}{|}}{R{-}CH{-}COOH} + HCl$

5. *a.* α-Hydroxy acids only.

$$\underset{R'\ OH}{\overset{R}{>}}C{-}COOH \xrightarrow[\Delta]{H_2SO_4} \underset{R'}{\overset{R}{>}}C{=}O + HCOOH$$

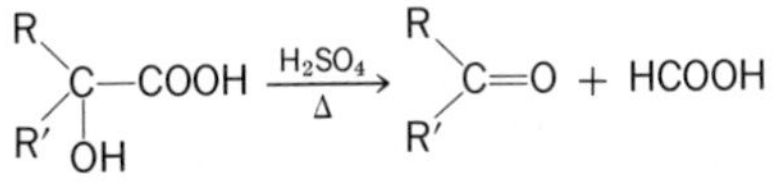

Either R or R′, or both, may be hydrogen.

b. β-Hydroxy acids.

$$RCHOH—CH_2COOH \xrightleftharpoons[\Delta]{H_2SO_4} RCH{=}CHCOOH + H_2O$$

An α,β-unsaturated acid

c. γ-Hydroxy acids.

$$\underset{\underset{OH}{|}}{R}CHCH_2CH_2COOH \xrightleftharpoons{dil.\ H_2SO_4} RCHCH_2CH_2 + H_2O$$
$$\underset{O{\longrightarrow}C{=}O}{}$$

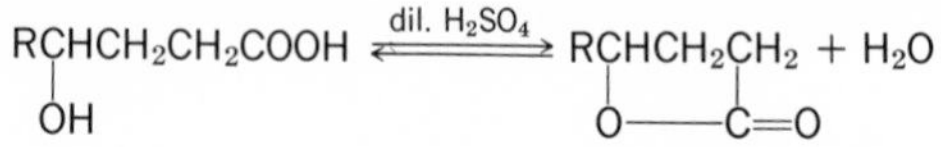

A lactone

6.

$$2RCHOHCOOH \xrightleftharpoons[\Delta]{H_2SO_4}$$

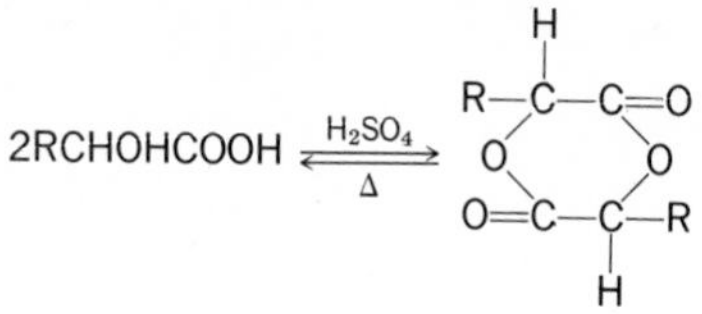

A lactide

B. Aliphatic polybasic hydroxy acids.
   1. Malic acid (hydroxysuccinic acid).

Properties

a.
$$\underset{\underset{CH_2COOH}{|}}{CHOHCOOH} \xrightarrow{\Delta} \underset{\underset{HCCOOH}{\|}}{HCCOOH} + \text{related compounds}$$

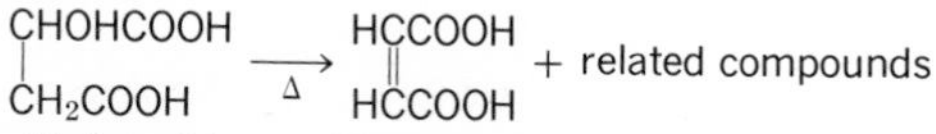

Malic acid Maleic acid

b.
$$\underset{\underset{CH_2COOH}{|}}{CHOHCOOH} \xrightarrow{(O)} \underset{\underset{CH_2COOH}{|}}{COCOOH}$$

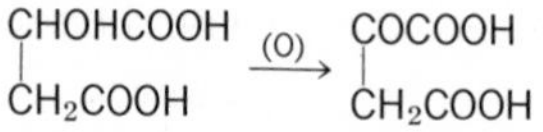

Oxalacetic acid

2. Tartaric acid (dihydroxysuccinic acid).

Preparation

a. By-product in wine fermentation.

b.
$$\underset{\underset{CHCOOH}{\|}}{CHCOOH} \xrightarrow[KMnO_4]{dil.} \underset{\underset{CHOHCOOH}{|}}{CHOHCOOH}$$

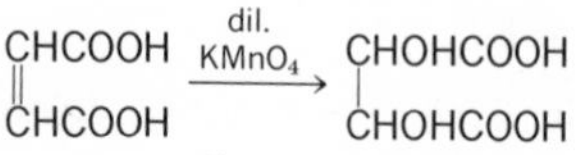

Properties

a.
$$\underset{\underset{CHOHCOOH}{|}}{CHOHCOOH} \xrightarrow[\Delta]{H_2SO_4} \underset{\underset{COCOOH}{|}}{CH_3} + CO_2 + H_2O$$

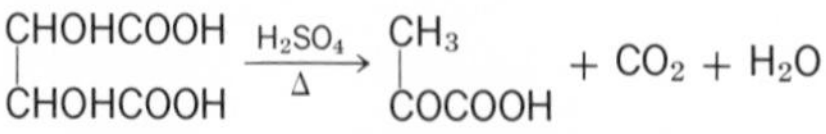

Pyruvic acid

C. Aromatic hydroxy acids (phenolic hydroxyl group).

Preparation

1. 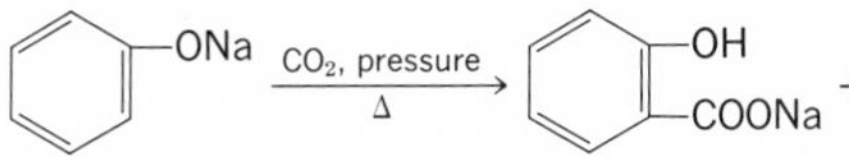 + a little para compound

Sodium salicylate

2. $$H_4C_6(COOH)(NH_2) \xrightarrow[(b)\ \Delta]{(a)\ \text{diazotize}} H_4C_6(COOH)(OH)$$

(*o*, *m*, or *p*) (*o*, *m*, or *p*)

Properties

The phenolic acids have, in general, all the important properties of phenols and acids.

UNSATURATED ACIDS

Attention will be restricted to α,β-unsaturated acids.

Preparation

1. $CH_2OHCH_2COOH \xrightarrow[\Delta]{} H_2C{=}CHCOOH$

   α-Hydroxy acids do not yield unsaturated acids except in special cases.
2. Mild oxidation of α,β-unsaturated aldehydes.

$$CH_3CH{=}CHCHO \xrightarrow{Ag_2O} CH_3CH{=}CHCOOH$$

Crotonaldehyde (from aldol condensation of acetaldehyde)

3. 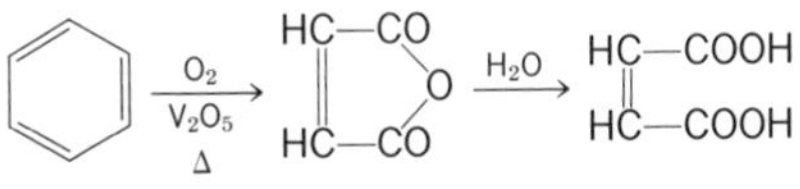

   Compare with the oxidation of naphthalene to phthalic anhydride.

4. 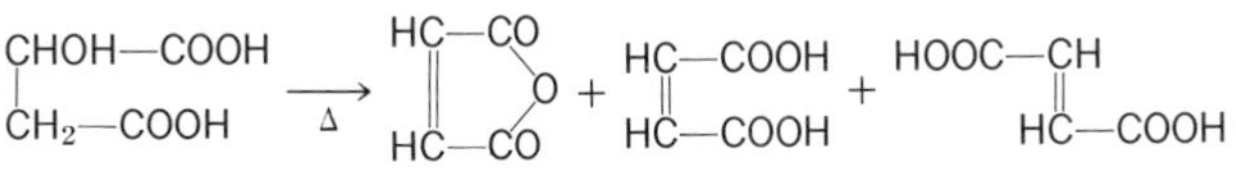

Properties

1. Unsaturated acids form salts, esters, acid chlorides, and anhydrides in the normal manner.
2. $H_2C{=}CHCOOH \xrightarrow[\text{cat.}]{H_2} CH_3CH_2COOH$
3. $H_2C{=}CH{-}COOH \xrightarrow{X_2} CH_2X{-}CHX{-}COOH$

4. $CH_3CH{=}CHCOOH \xrightarrow[\text{KMnO}_4]{\text{conc.}} CH_3COOH + \begin{matrix}COOH \\ | \\ COOH\end{matrix}$ (or $CO_2 + H_2O$)

5. $H_2C{=}CHCOOH \xrightarrow{\text{HX}} \underset{X}{CH_2CH_2COOH}$

6. $\begin{matrix}HC{-}COOH \\ \| \\ HC{-}COOH\end{matrix} \xrightarrow{\Delta} \begin{matrix}HC{-}CO \\ \| \quad \searrow O \\ HC{-}CO \end{matrix} + H_2O$ 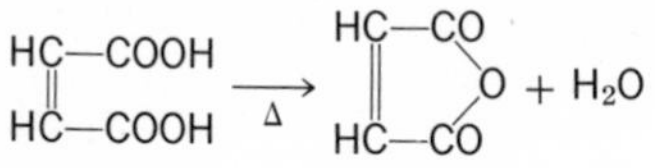

Fumaric acid forms no anhydride of its own.

CHAPTER 17   EXERCISES

1. Write equations for the reactions of α-bromopropionic acid and p-bromobenzoic acid with:
   a. Cold dil. NaOH
   b. Hot dil. NaOH
   c. Excess NH_4OH
   d. KCN (after forming salt of acid)
   e. Absolute ethanol and dry HCl
   f. Thionyl chloride
   g. Methylmagnesium bromide

2. Show by equations how lactic acid could be converted to each of the following compounds.
   a. n-Butyl lactate
   b. Methyl-α-methoxypropionate
   c. Acetic acid
   d. Methyl puruvate
   e. α-Aminopropionic acid (alanine)
   f. Lactamide
   g. Ethyl-α-acetoxypropionate [$CH_3CH(OCOCH_3)COOC_2H_5$]
   h. Propylene glycol (1,2-propanediol)

3. Show by equations how the Reformatsky reaction could be used to prepare the following compounds from starting materials having a smaller number of carbon atoms.

   a. $\underset{\quad\;\; OH}{C_6H_5CH}{-}CH_2{-}\underset{OH}{CH_2}$ 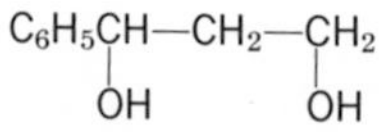

   b. $\begin{matrix}H_3C \\ \\ H_3C\end{matrix}\!\!\underset{}{\overset{}{C}}{=}C\!\!\begin{matrix}CH_3 \\ \\ COOH\end{matrix}$ 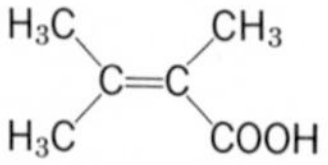

   c. $\begin{matrix}H_5C_6 \\ \\ H_5C_6\end{matrix}\!\!\underset{}{C}\!\!\begin{matrix}{-}CH_2COOH \\ \\ OCH_3\end{matrix}$ 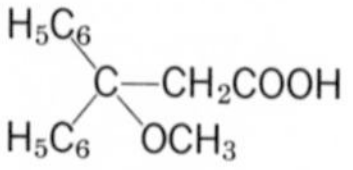

   d. $CH_3CH{-}\underset{}{CH}{-}CH_2OH$ 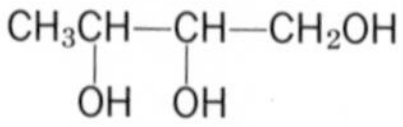
      $\quad\;\;\; \underset{OH}{}\;\; \underset{OH}{}$

e.

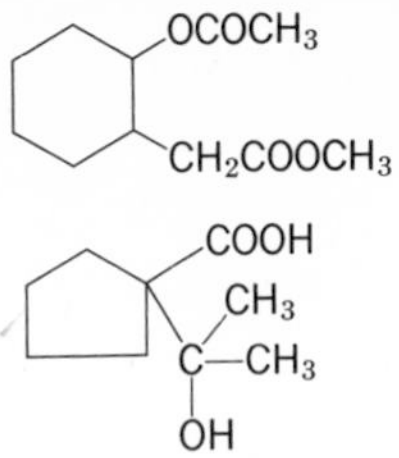

4. Explain in each of the syntheses in Exercise 3 why the use of other methods (Grignard reactions, etc.) is less desirable than the Reformatsky reaction.

★5. Write all the stereoisomers of the following molecules and state their optical properties.

★*a.* HOOC—CH(CH$_3$)—CH=CH—CH(CH$_3$)—COOH

b. H$_2$C=CH—CH=C=CH—CH=CH$_2$

★*c.*

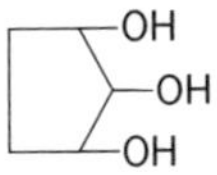

d. HOOC—CH=CH—CH=CH—COOH

6. Show how you could distinguish:

a.

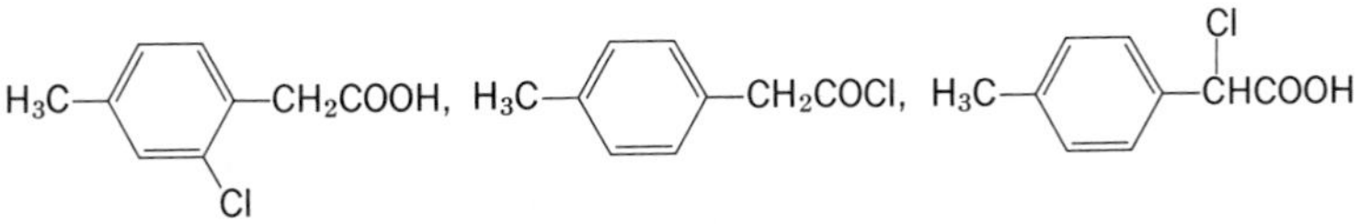

b.

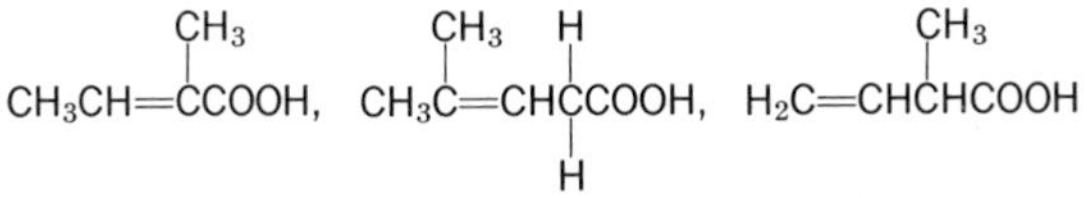

★7. Show all the cis-trans isomers possible for the following compounds and show how you could distinguish between any two of them.

★*a.*

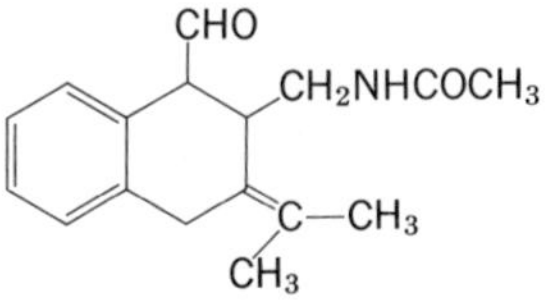

b.

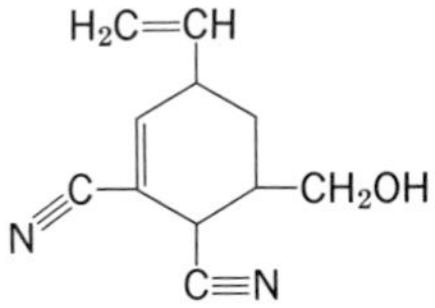

c.

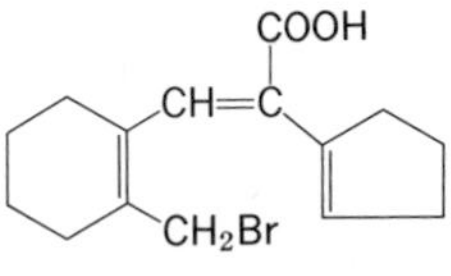

★*d.* 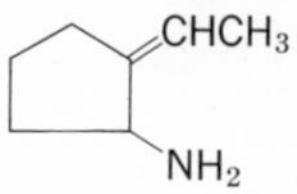

8. In the following reaction sequence show clearly all the stereoisomers which would be produced at each step. Mark them *R*, *S*, meso, trans, etc., and show any *R,S* pairs.

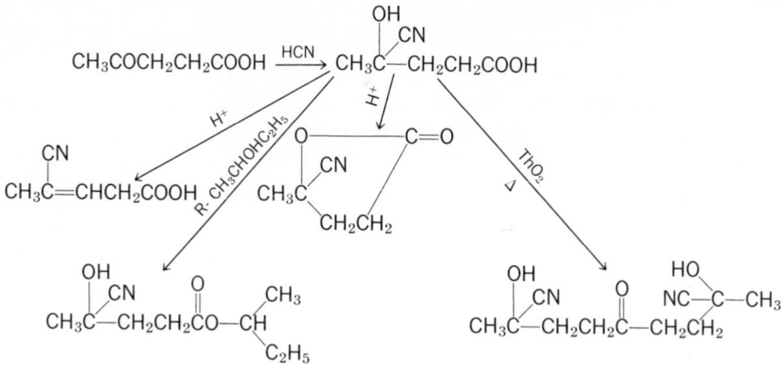

CH3COCH2CH2COOH
HCN
OH
CN
CH3C—CH2CH2COOH
H+
CH3C=CHCH2COOH
R. CH3CHOHC2H5
O
C=O
CH3C
CH2CH2
ThO2
Δ
CH3C—CH2CH2CO—CH
CH3
C2H5
HO
NC—C—CH3
CH3C—CH2CH2C—CH2CH2

★9. *Cis* and *trans* 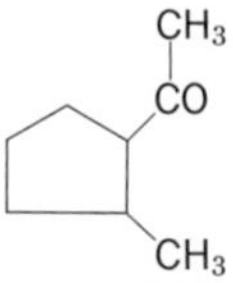 are inter-converted (come to equilibrium) in basic medium. Explain.

10. Explain the failure to obtain cis and trans forms of cyclopentene in view of the existence of *cis*- and *trans*-cyclooctadecene.

★11. How many geometrical isomers of vitamin A are theoretically capable of existence? Explain.

12. A certain compound A, $C_{10}H_{11}O_2Br$, does not react with $NaHCO_3$ solution but reacts slowly on heating in NaOH solution to form a water-soluble substance B, $C_3H_5O_3Na$, and an insoluble oil C, C_7H_8O. C reacts with methylmagnesium iodide to yield 1 mole of methane per mole of C; on oxidation it forms benzoic acid. B, when heated with acid, forms some acetaldehyde and some $C_6H_8O_4$. What is the structure of A? Show your reasoning.

★13. The plastic Lucite® is manufactured by polymerizing methacrylic acid esters.

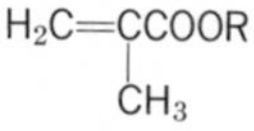

Assuming that the polymerization occurs by a free-radical mechanism, suggest a probable (repeating) structure for the polymer. The transparency of Lucite® is due in large part to its lack of crystallinity. What stereochemical factors might tend to reduce the ability of such a polymer to crystallize?

14. Two acids, $C_6H_{12}O_3$, both react with acetyl chloride but give no iodoform with NaOH and iodine. On heating with acid one forms a single substance, $C_6H_{10}O_2$, while the other gives two substances with the same formula, $C_6H_{10}O_2$. There are no other products in either case. Deduce structures for the acids that fit these data.

15. The infrared spectra of *cis-* and *trans*-1,2-cyclopentanediols differ from each other in the following way. In one case there appears a band in the 2 to 3-μ region that is absent in the other; this band does not change in relative intensity on dilution of the diol with an inert solvent such as $CHCl_3$. Explain. (HINT: Consider association.)

18

SUBSTITUTED ACIDS: KETO ACIDS AND MALONIC ESTER

18.1 Claisen Condensation Aliphatic keto and aldehydo acids are known in which a carbonyl group is found in various positions on the chain relative to the carboxyl. As in the case of other substituted acids in the previous chapter, only α-, β-, and γ-keto (or aldehydo) acids are of interest here. They may be prepared by oxidation of the corresponding hydroxy acids or by oxidation of appropriate unsaturated acids when these are available. More important by far as sources of keto and other substituted acids are the *acetoacetic-ester* and *malonic ester syntheses* to be described in this section.

When an ester, for example, ethyl acetate, is heated with a sufficiently strong base, such as sodium ethoxide, reaction takes place with the eventual formation of a β-keto ester.

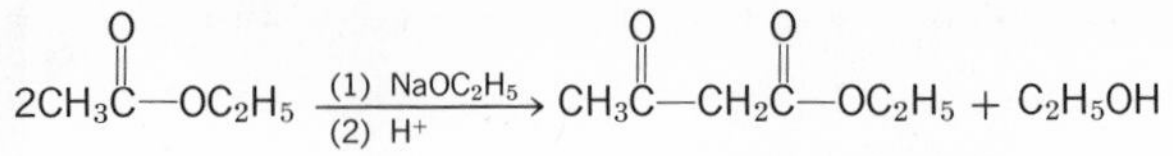

$$2CH_3\overset{O}{\overset{\|}{C}}-OC_2H_5 \xrightarrow[\text{(2) H}^+]{\text{(1) NaOC}_2\text{H}_5} CH_3\overset{O}{\overset{\|}{C}}-CH_2\overset{O}{\overset{\|}{C}}-OC_2H_5 + C_2H_5OH$$

acetoacetic ester
(ethyl acetoacetate)

The reaction, known as the Claisen condensation, is a complex one, and the properties of the product are so unusual that the reaction attracted a great deal of attention in the early days of organic chemistry. A modern view of the reaction mechanism is as follows:

$$CH_3\overset{O}{\overset{\|}{C}}-OC_2H_5 + \bar{O}C_2H_5 \rightleftharpoons \left[CH_2\overset{O}{\overset{\|}{C}}-OC_2H_5 \longrightarrow H_2C=\overset{O^-}{\overset{|}{C}}-OC_2H_5\right] + HOC_2H_5$$

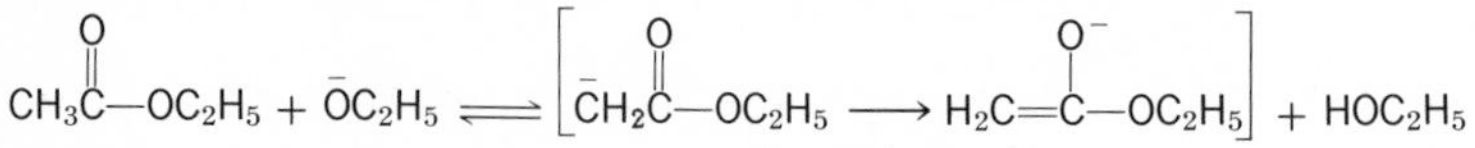

An enolate anion

The enolate anion is formed in very low concentration and is not isolated; in fact, the most rapid reaction is addition of ethoxide to the carbonyl.

$$CH_3\overset{O}{\overset{\|}{C}}-OC_2H_5 + \bar{O}C_2H_5 \rightleftharpoons CH_3\overset{O^-}{\underset{\overset{|}{O}C_2H_5}{\overset{|}{C}}}-OC_2H_5$$

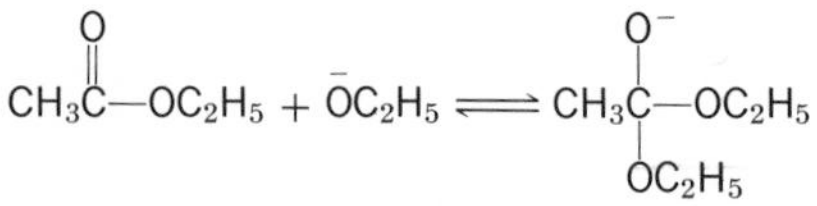

This reaction, however, cannot lead to product and, accordingly, may be ignored here.

The enolate ion formation requires an α-hydrogen atom and, of course, does not proceed in its absence. Like ethoxide and other nucleophilic reagents, the enolate ion can add to the carbonyl group of another molecule.

$$\left[\begin{array}{c}\bar{C}H_2\overset{O}{\overset{\|}{C}}OC_2H_5 \\ \updownarrow \\ H_2C=\overset{O}{\overset{|}{C}}-OC_2H_5\end{array}\right] + CH_3\overset{O}{\overset{\|}{C}}-OC_2H_5 \rightleftharpoons \left[C_2H_5-O\overset{O}{\overset{\|}{C}}-CH_2-\overset{O^-}{\underset{\overset{|}{C}H_3}{\overset{|}{C}}}-OC_2H_5\right]$$

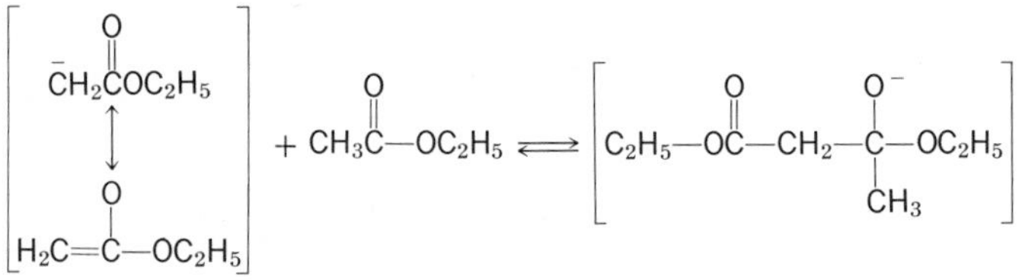

The initial adduct can now eliminate an ethoxide ion.

$$\left[C_2H_5-O\overset{O}{\overset{\|}{C}}-CH_2-\overset{O^-}{\underset{\overset{|}{C}H_3}{\overset{|}{C}}}OC_2H_5\right] \rightleftharpoons C_2H_5O\overset{O}{\overset{\|}{C}}-CH_2-\overset{O}{\overset{\|}{C}}-CH_3 + \bar{O}C_2H_5$$

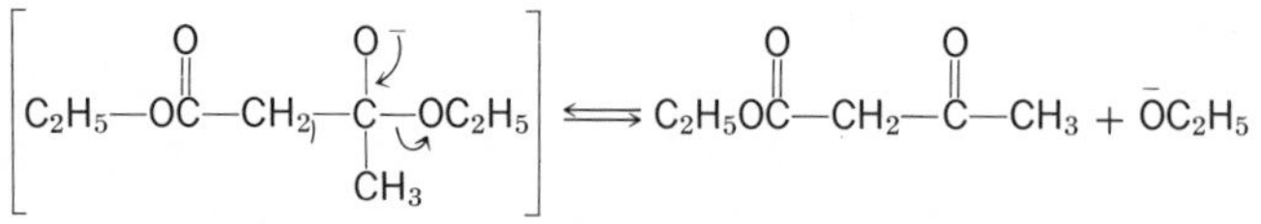

All of the above steps are reversible and the equilibrium would not favor the condensed product were it not for the next step.

$$C_2H_5O\overset{O}{\overset{\|}{C}}-CH_2-\overset{O}{\overset{\|}{C}}CH_3 + \bar{O}Et \rightleftharpoons \left[C_2H_5O-\overset{O}{\overset{\|}{C}}-CH-\overset{O}{\overset{\|}{C}}-CH_3\right] + HOEt$$

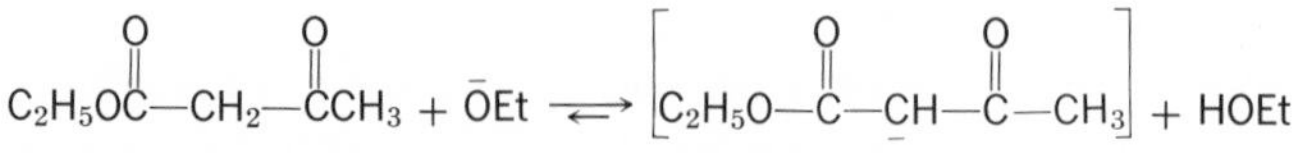

The anion above is so strongly resonance stabilized

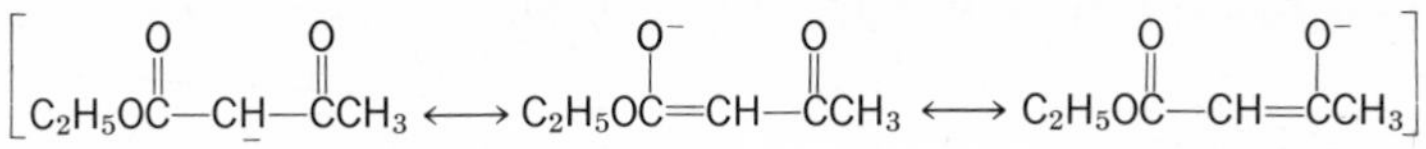

that it is formed in high concentration and tends to drive the reaction toward the condensed product. (Recall the base-catalyzed hydrolysis of esters in Sec. 13.6.)

The final product is thus a sodium salt of the enol form of a β-keto ester, in this case of acetoacetic ester. The addition of dilute acid regenerates the free keto ester and this is usually extracted with ether and purified.

The reaction is generally applicable to esters having *two* α hydrogen atoms, the same limitations on mixed condensations applying here as in the case of the aldol condensation. Condensation always involves the α-carbon only. A few examples will illustrate the application of the reaction to the formation of substituted β-keto esters.

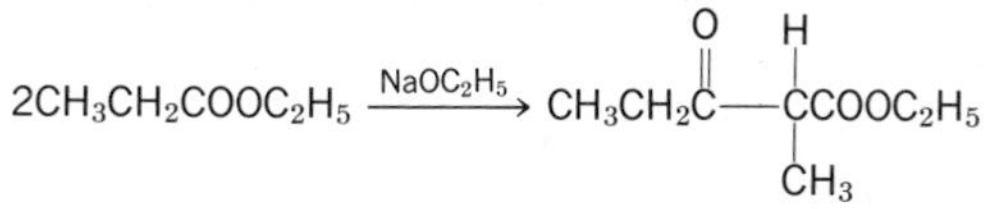

Final product after acidification

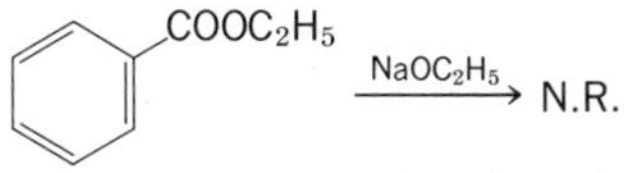

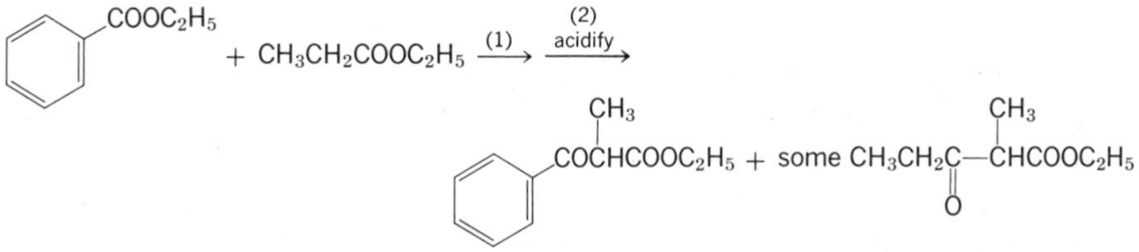

In the case of a mixed condensation, such as that given above between ethyl benzoate and ethyl propionate, a reasonable yield of the desired product is obtained only if one of the esters has *no* α hydrogen atoms. This diminishes the number of possible condensation products from four to two and simplifies the isolation of the desired keto ester (compare Sec. 11.12).

Ethoxide-catalyzed condensations do not succeed when the ester has only one α-hydrogen. The final step, formation of an enolate anion of the β-keto ester, does not occur with ethoxide ion because the product has no strongly acidic α hydrogen atom situated between two carbonyl groups. Exceedingly strong bases serve to form an anion elsewhere in the molecule and allow the reaction to proceed to the β-keto ester stage.

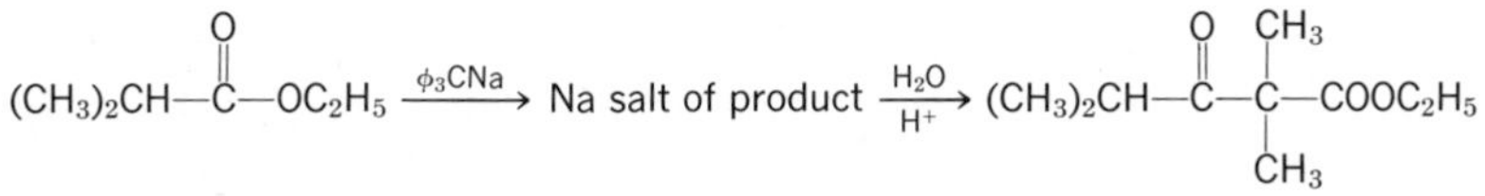

18.2 Reactions of Acetoacetic Ester The chemical properties of acetoacetic ester and its homologs and analogs differ in many respects from those of other esters. The following chart gives some of the reactions of acetoacetic ester:

$CH_3COCH_2COOC_2H_5$ +

$HCN \longrightarrow$	addition reaction	(18.1)
$NH_2OH \longrightarrow$	oxime formation	(18.2)
$H_2 \xrightarrow{\text{cat.}}$	$CH_3CHOHCH_2COOC_2H_5$	(18.3)
$CH_3COCl \longrightarrow$	an acetyl derivative	(18.4)
$FeCl_3 \longrightarrow$	colored iron salt	(18.5)
$Br_2 \longrightarrow$	rapid decolorization	(18.6)

18.3 Keto-Enol Tautomerism The first three reactions are those to be expected of a compound of the structure shown above; reactions (18.4) to (18.6) are not usually given by simple ketones, by esters, or by keto esters in general, except the β-keto type. These latter reactions are those of an unsaturated hydroxy compound with weakly acidic properties, and they resemble the reactions of phenols. The phenomenon observed above in which a *single* compound exhibits two sets of reactions, indicative of two different structural formulas, is known as *tautomerism*. The apparently pure liquid acetoacetic ester has been found to be an equilibrium mixture of two forms, a *keto* and an *enol* form.

$$\underset{\textit{Keto}\text{ form}}{CH_3COCH_2COOC_2H_5} \rightleftharpoons \underset{\textit{Enol}\text{ form}}{CH_3\overset{\overset{\large OH}{|}}{C}{=}CHCOOC_2H_5}$$

Any chemical reagent that reacts with one form will, of course, upset this equilibrium in the direction of that form. Since equilibrium is rapidly established at ordinary temperatures, the substance appears to react as *though* it were either 100 percent keto or 100 percent enol, depending on the reagent used. Although equilibrium is established rapidly under ordinary conditions, by careful exclusion of alkali (which catalyzes the tautomeric shift) and operation at a low temperature (liquid nitrogen), the keto and enol forms in this case have actually been isolated. On warming them again to room temperature both forms soon revert to the equilibrium mixture (about 7 percent enol). *The student is reminded that this is not a case of resonance.* An actual shift of a hydrogen atom is involved here as well as a shift of electrons.

18.4 Structure and Enolization In ordinary ketones the percentage of enol at equilibrium is negligibly small. In the case of β-keto esters and β-diketones several factors combine to make this percentage significant. These factors are: (1) inductive destabilization of the β-diketo form which places the two positively charged carbon atoms near one another, (2) resonance stabilization of the enol via a conjugated system, and (3) intramolecular hydrogen bonding in the enol.

Compounds containing cyclic hydrogen bridges are often called "chelated" (pronounced kē′ lāted) compounds. Chelation has the effect of tying up the OH group within the molecule, thus decreasing the interaction of the molecule with its neighbors. The word chelation is derived from the Greek word *chelos* for *claw*, and describes the situation in which an atom is bound by two or more parts of the same molecule. Either hydrogen atoms (as with β-keto esters and β-diketones) or metal ions may be involved. For example, copper forms the chelate:

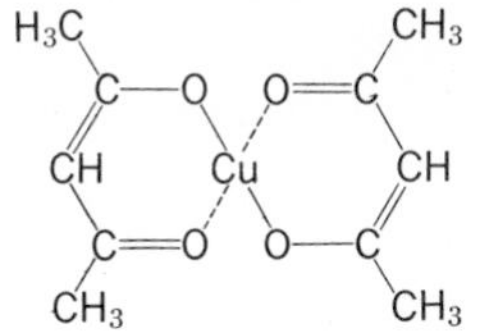

which is a volatile, organic-solvent-soluble substance. The chelated form of the enol of acetoacetic ester may be drawn

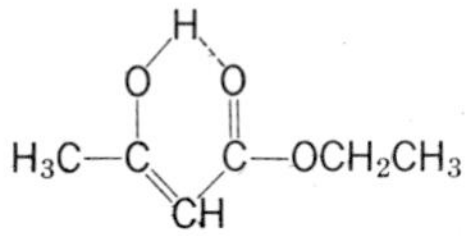

The formation of enols generally requires all three of the above factors for appreciable stability, except where Factor 2 is enhanced by other features such as incorporation into an aromatic ring. If a few percent or more of enol is present, a complex color reaction is often observed on treatment with $FeCl_3$. This has been used as a crude criterion of enol formation as illustrated in Table 18.1.

TABLE 18.1
ENOLIZATION OF SOME KETONES

Keto formula	*Enol formula*	$FeCl_3$ *test*
$CH_3C(=O)CH_2C(=O)CH_3$	$CH_3C(O{-}H\cdots O{=})\!=\!CHCCH_3$	+
$CH_3C(=O)CH_2CH_2C(=O)CH_3$	$CH_3C(OH){=}CHCH_2C(=O)CH_3$[a]	−
$C_6H_5C(=O)CH(CH_3)CN$	$C_6H_5C(OH){=}C(CH_3)CN$	+
$CH_3C(=O){-}C(CH_3)_2COOC_2H_5$	$H_2C{=}C(OH){-}C(CH_3)_2COOC_2H_5$	−
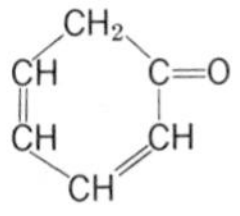	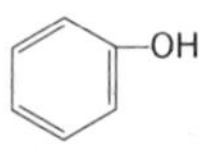 (fully conjugated enol in this form)	+

[a] Dienols are not easily formed.

In the absence of acids or bases, enols can be readily detected by nmr if sufficient (5 percent or more) enol is present. The hydrogens of the central carbon of the keto form fall in the region $\delta = 3.15$ to 3.65 ppm, while in the enol this carbon atom is an alkene carbon and its hydrogen falls at about 5.5 ppm. More significantly, the chelated —OH hydrogen falls in the region of $\delta = 11$ to 16 ppm. This region, which falls outside the range of almost all other hydrogens, is indicative of the highly deshielded environment of the chelated hydrogen and makes its detection and identification much simpler.

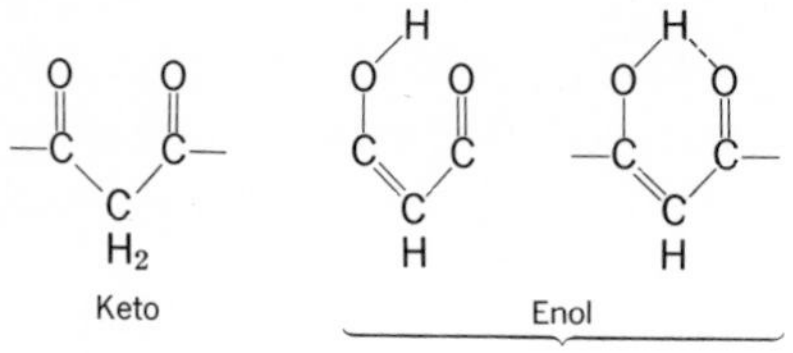
Keto
Enol

$\delta = 3.15$–3.65 ppm $\delta = 5.5$ ppm $\delta = 11$–16 ppm (chelated)

In cases where the OH is not chelated (as with phenols) the OH appears at higher field (more shielded) and its position varies markedly with concentration (due to intermolecular H bonding).

If acid or base are present, a rapid equilibration causes an averaging of the keto and enol hydrogens and the results are less readily interpreted.

Evidence for the presence of enol may also be obtained from the infrared in which a peak at 1540 to 1690 cm^{-1}, attributed to the chelated carbonyl, is seen in addition to the normal ester and/or keto carbonyl frequencies.

18.5 Reactions of β-Keto Esters Acetoacetic ester is an acid with an acidity (in water) almost equal to that of phenol, and like phenol, it can be converted to its anion in sodium hydroxide solution. Unlike phenol it can also undergo hydrolysis of the ester function in aqueous base and, consequently, is customarily converted to its salt in anhydrous solvents such as ethanol.

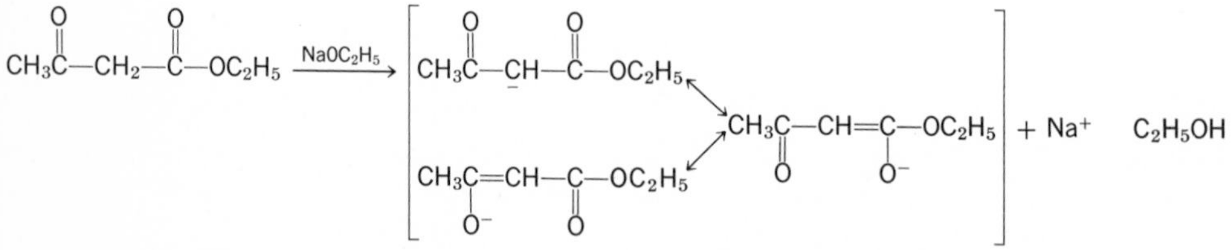

This ion, called an "enolate" ion, is a resonance hybrid, not a tautomer, as is the enol. This enolate ion is a good nucleophilic displacing group and, if it is treated with an alkyl halide which is susceptible to S_N2 (not S_N1) displacement, attack (almost invariably by the carbon atom of the enolate) ensues.

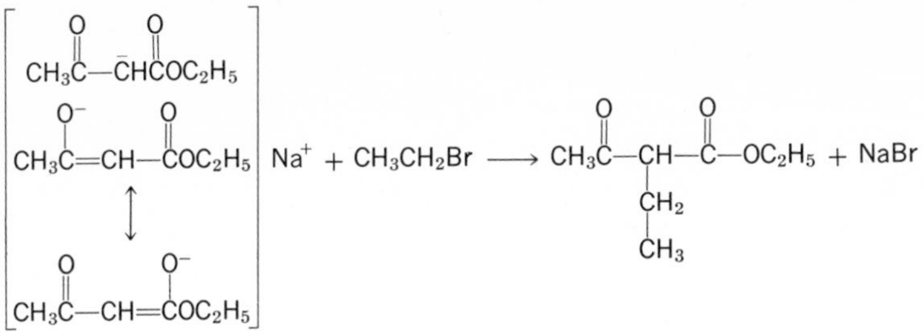

The importance of the synthesis lies in the fact that we have here another general method of forming new carbon-to-carbon bonds, limited only to neutral (containing no acidic groups) primary or secondary alkyl halides.

Moreover, the process can be repeated to introduce still a second group on the α-carbon, if two α-hydrogens were originally present.

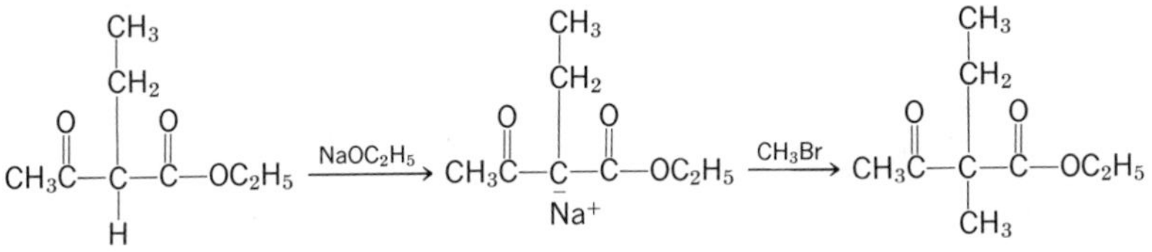

No further groups can be introduced since there are no more readily replaceable α hydrogen atoms in the molecule acidic enough to enolize appreciably under these conditions.

Not only do β-keto esters and other 1,3-dicarbonyl compounds undergo this *alkylation* reaction on the α carbon atom, but they are hydrolyzed in an interesting manner. Two types of cleavage can be effected, the result depending on the conditions of reaction. Yields of 100 percent are never obtained; both types of cleavage always occur, but either one can be made to predominate as shown.

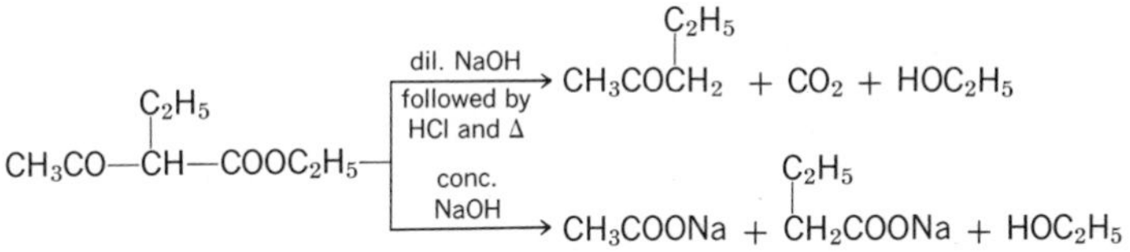

Attempts to obtain the uncleaved hydrolysis product—acetoacetic acid—are ordinarily unsuccessful owing to the instability of β-keto acids.

$$CH_3COCH_2COOC_2H_5 \xrightarrow[\text{alkali}]{\text{dil.}} CH_3COCH_2COONa \xrightarrow[\text{acid}]{\text{dil.}}$$

$$[CH_3COCH_2COOH] \xrightarrow[\text{decarboxylation}]{\text{spontaneous}} CH_3COCH_3 + CO_2$$

The cleavage in strong alkali is formally the reverse of the Claisen condensation of esters.

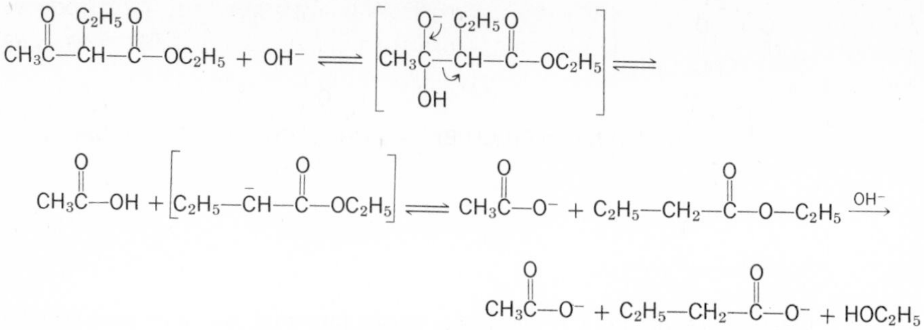

This reaction is "driven" in the direction of free acids by the stability of the carboxylate ion.

The decarboxylation of β-keto acids is occasioned by the formation of the resonance-stabilized enolate anion and by the irreversible loss of carbon dioxide.

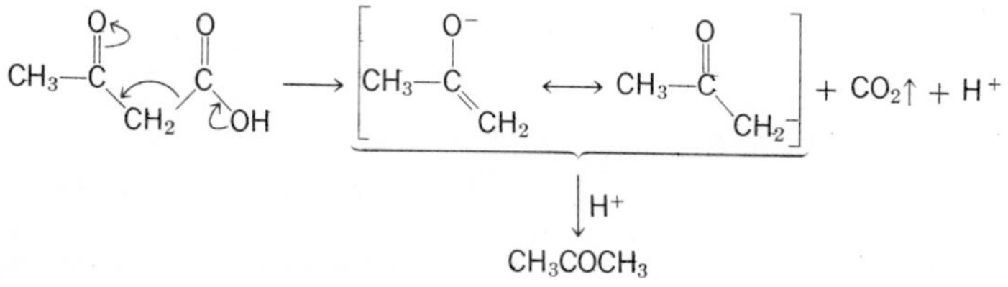

The application of these decompositions to a large number of variously substituted β-keto esters has led to the synthesis of both ketones and acids of the types

where R, R_1, and R_2 may be any neutral alkyl or substituted primary or secondary alkyl group or hydrogen. When two alkyl groups are present on the α carbon atom of acetoacetic ester, the hydrolysis to yield ketones is difficult, the main product being a dialkylacetic acid. The synthesis of a γ-keto acid by this method is illustrative of its versatility.

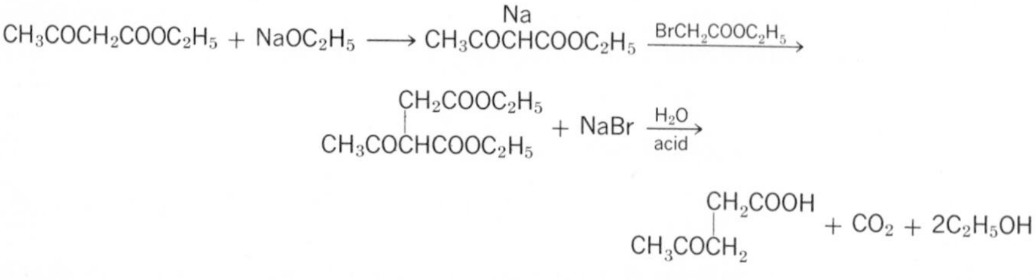

With the keto group in the γ position, the free acid is stable.

As in the above example, even carboxyl-containing groups can be introduced into acetoacetic ester if their acidic properties are suppressed by esterification. Not only acetoacetic ester but other β-keto esters are useful in synthesis.

$$CH_3CH_2COOC_2H_5 + C_6H_5COOC_2H_5 \xrightarrow{NaOC_2H_5} \xrightarrow{\text{acidify}} C_6H_5CO\underset{}{\overset{CH_3}{\overset{|}{C}}}HCOOC_2H_5 \xrightarrow{NaOC_2H_5} \xrightarrow{CH_3I} C_6H_5CO\underset{\underset{CH_3}{|}}{\overset{CH_3}{\overset{|}{C}}}COOC_2H_5$$

18.6 Malonic-ester Syntheses Another 1,3-dicarbonyl compound in common use for syntheses of the acetoacetic-ester type is the dicarboxylic acid ester, *malonic ester*.

Malonic ester itself is prepared from chloracetic acid on a commercial scale.

$$ClCH_2COO^- \xrightarrow{CN^-} CNCH_2COO^- \xrightarrow[H_2SO_4]{C_2H_5OH} CH_2(COOC_2H_5)_2$$

The simultaneous hydrolysis and esterification of the nitrile group is often used to save one step in a synthesis of this sort. Because malonic ester is easily available and its alkylation product can lead only to acids, forming no ketones as a side reaction (as in acetoacetic ester), it is the reagent of choice for the preparation of substituted acetic acids.

$$C_2H_5O{-}\overset{O}{\overset{\|}{C}}CH_2\overset{O}{\overset{\|}{C}}{-}OC_2H_5 \qquad \text{malonic ester}$$

$$CH_3{-}\overset{O}{\overset{\|}{C}}CH_2\overset{O}{\overset{\|}{C}}{-}OC_2H_5 \qquad \text{acetoacetic ester}$$

The close similarity to acetoacetic ester in structure, as shown above, is borne out in the reactions of malonic ester with sodium and alkyl halides. Although it does not contain enough free enol at room temperature to give a ferric chloride test, it readily forms a sodium salt in alcohol solution.

$$C_2H_5O\overset{O}{\overset{\|}{C}}{-}CH_2\overset{O}{\overset{\|}{C}}OC_2H_5 \xrightarrow{Na^+\ ^-OC_2H_5} \left[C_2H_5O\overset{O}{\overset{\|}{C}}{-}\bar{C}H\overset{O}{\overset{\|}{C}}OC_2H_5 \longleftrightarrow C_2H_5O\overset{O^-}{\overset{|}{C}}{=}CH\overset{O}{\overset{\|}{C}}OC_2H_5\right]$$

$$C_2H_5O\overset{O}{\overset{\|}{C}}{-}\bar{C}H\overset{O}{\overset{\|}{C}}OC_2H_5 + BrC_4H_9 \longrightarrow C_4H_9CH(COOC_2H_5)_2 + Br^-$$

The structure of the sodium salt (anion) is similar to that in acetoacetic ester.

Hydrolysis of malonic ester and its alkylated product is normal. The resulting malonic acid, however, is easily decarboxylated by heat to form a substituted acetic acid

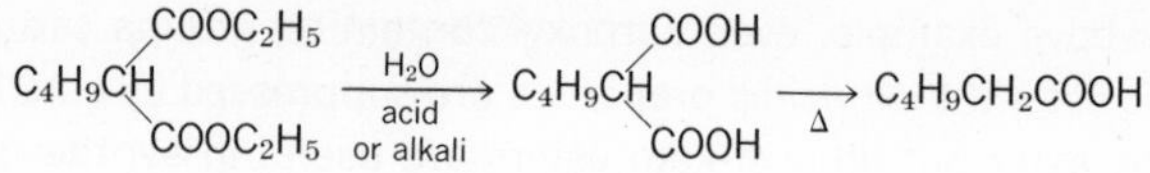
COOC2H5
C4H9CH
COOC2H5
H2O
acid
or alkali
C4H9CH
COOH
COOH
Δ
C4H9CH2COOH

By the malonic-ester synthesis just described, compounds of the general formulas

R
CHCOOH
R1
and
R
COOH
C
R1
COOH

can be prepared. R and R_1 have the same limitations as in acetoacetic-ester alkylations. The second R group is always introduced in a separate step prior to hydrolysis.

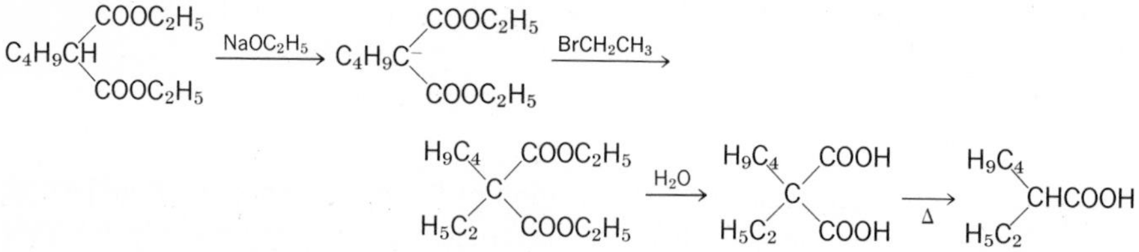
COOC2H5
C4H9CH
COOC2H5
NaOC2H5
C4H9C
COOC2H5
COOC2H5
BrCH2CH3
H9C4
COOC2H5
C
H5C2
COOC2H5
H2O
H9C4
COOH
C
H5C2
COOH
Δ
H9C4
CHCOOH
H5C2

Although beyond the scope of this book, many applications of both malonic and acetoacetic esters have been made in the synthesis of complex, naturally occurring heterocycles such as the ureides. Alkylations of this type supplemented by the now-familiar Grignard reagents make possible many of the involved synthetic operations often needed in biochemical work.

It is noteworthy that both these syntheses are actually merely applications of the principle already studied in Chap. 11: the electron-attracting power of the carbonyl group. Here we find two such groups attached to a single carbon, and there results a much greater tendency to release a proton from that carbon than from other saturated carbon atoms—hence the reaction with strong bases to form salts. The fact that the anions so produced are stabilized by resonance is another contributing factor here. Like the nucleophilic anions of all weak acids, they react with alkyl halides by a displacement mechanism. Subsequent cleavages depend on the crowding together on one carbon atom of two active groups, a phenomenon already observed in other cases.

18.7 α-Keto Acids Although of less general importance, keto acids and esters with the carbonyl group in other than the β position are also known. The γ-, δ-, and other keto acids in which the carbonyl group is more remote from the carboxyl exhibit the expected combination of ketone and acid properties more or less independently. One γ-keto acid, levulinic acid, is a commercial product; others are prepared by the acetoacetic-ester synthesis.

One α-keto acid, pyruvic acid, is of sufficient biochemical importance to deserve mention here.

$$CH_3CHOHCOOH \xrightarrow{(O)} CH_3COCOOH$$

Lactic acid Pyruvic acid

Presumably because of the proximity of the carbonyl groups, pyruvic acid is a very reactive substance; in a manner resembling the α-hydroxy acids, it decomposes when heated with sulfuric acid.

$$CH_3COCOOH \xrightarrow{H_2SO_4} CH_3COOH + CO$$

This reaction is a general one for α-keto acids and serves to distinguish them from other classes of keto acids.

18.8 Summary of Condensation Reactions It is instructive at this point to compare the three condensation reactions most often used in the synthesis of complex structures. These are the aldol, Reformatsky, and Claisen reactions; all of them follow the same pattern of addition to a carbonyl group of an anion with its charge on an *alpha* carbon atom. They may all be represented by the same general equation.

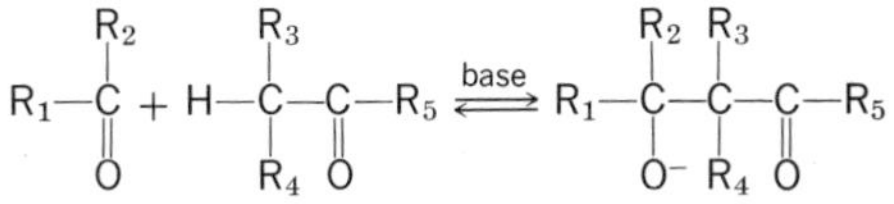

For each reaction there are certain specific limitations and applications:

Aldol. All R groups may be hydrogen or alkyl or aryl. Because mixtures are otherwise produced, the reaction is usually restricted to condensation between identical molecules and yields are often unsatisfactory unless $R_1 = H$ (Sec. 11.12).

Reformatsky. H must be ZnBr (from the halide and Zn) and R_5 must be alkoxy (OC_2H_5, etc.); other R groups may be alkyl or aryl or hydrogen. The reaction is generally useful for condensing two different molecules.

Claisen. R_2 and R_5 must be alkoxy and R_3 must be hydrogen; other R groups may be alkyl or aryl or hydrogen. In this reaction R_2 is eliminated to produce the final product, a β-keto ester. The reaction is often restricted to condensations between identical molecules (Sec. 18.1). It is useful largely because of the variety of synthetic applications of the products (Sec. 18.5).

18.9 Spectroscopic Properties β-Keto esters and β-diketones show an additional peak in the ir at about 1650 cm^{-1} due to the chelated form of the enol. In the nmr both the keto and enol forms will show up uniquely. In the absence of catalysis it is often possible to obtain an accurate measure of the percentage of enol present (provided it is present in greater than about 5 percent).

OUTLINE OF KETO-ACID CHEMISTRY

α-KETO ACIDS

Preparation

$$CH_3CHOHCOOH \xrightarrow{(O)} CH_3COCOOH$$

Properties

1. Pyruvic acid has the usual properties of a carboxylic acid and of a ketone.
2. $CH_3COCOOH \xrightarrow[\Delta]{H_2SO_4} CH_3COOH + CO$

β-KETO ACIDS

Acetoacetic acid and substituted acetoacetic acids are of importance.

Preparation

The Claisen condensation.

$$2CH_3COOC_2H_5 \xrightarrow{C_2H_5ONa} CH_3COCH(Na)COOC_2H_5 \xrightarrow{H^+} CH_3COCH_2COOCH_5$$

$$CH_3COCH(Na)COOC_2H_5 \xrightarrow{RX} CH_3CO\underset{\substack{| \\ R}}{C}HCOOC_2H_5$$

Properties

$$CH_3COCH_2COOC_2H_5 \xrightarrow{\substack{\text{conc.} \\ NaOH}} 2CH_3COONa + C_2H_5OH$$

$$CH_3COCH_2COOC_2H_5 \xrightarrow{\substack{\text{dil.} \\ NaOH}} CH_3COCH_2COONa + C_2H_5OH$$

$$CH_3COCH_2COONa \xrightarrow{\substack{\text{dil.} \\ HCl}} [CH_3COCH_2COOH] \xrightarrow[\Delta]{} CH_3COCH_3 + CO_2$$

OUTLINE OF MALONIC ESTER CHEMISTRY

1. Preparation of malonic ester.

$$Cl{-}CH_2COONa \xrightarrow{NaCN} NC{-}CH_2COONa \xrightarrow[H_2SO_4]{C_2H_5OH} CH_2(COOC_2H_5)_2 + (NH_4)_2SO_4 + Na_2SO_4$$

2. Preparation of substituted malonic esters.

$$CH_2(COOC_2H_5)_2 \xrightarrow{NaOC_2H_5} NaCH(COOC_2H_5)_2 \xrightarrow{RX} RCH(COOC_2H_5)_2$$

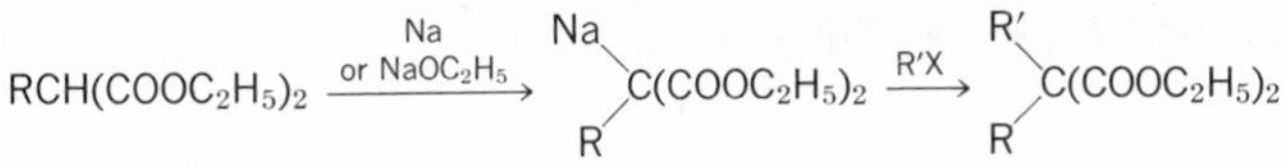

$$RCH(COOC_2H_5)_2 \xrightarrow[\text{or NaOC}_2H_5]{Na} \overset{Na}{\underset{R}{C(COOC_2H_5)_2}} \xrightarrow{R'X} \overset{R'}{\underset{R}{C(COOC_2H_5)_2}}$$

In these reactions nuclear aromatic halides cannot be used.

$$\overset{R}{\underset{R'}{C}}\overset{COOC_2H_5}{\underset{COOC_2H_5}{}} \xrightarrow[H_2O]{NaOH} \overset{R}{\underset{R'}{C}}\overset{COONa}{\underset{COONa}{}} \xrightarrow[H_2O]{H_2SO_4} \overset{R}{\underset{R'}{C}}\overset{COOH}{\underset{COOH}{}} \xrightarrow{\Delta} \overset{R}{\underset{R'}{CHCOOH}} + CO_2$$

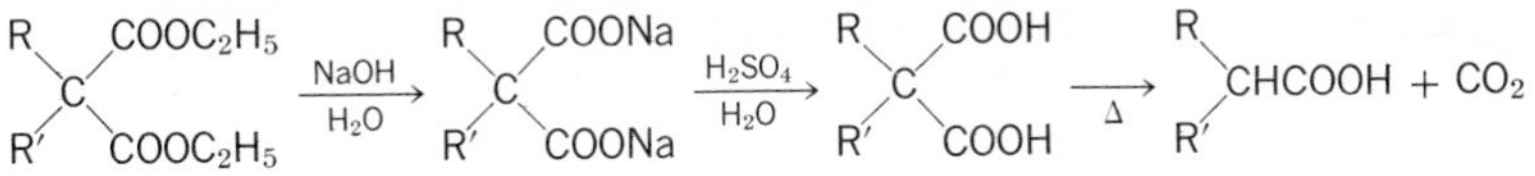

Spectroscopic Properties of β-Dicarbonyl Compounds

	ir, cm^{-1}	nmr, δ (ppm)
keto form	1705–1725 (ketone C=O) 1740–1755 (ester C=O)	6.15–6.65 [(CO)$_2$CH$_2$]
enol form	1540–1690 (chelated C=O) (very intense)	11–16 (chelated OH) 5.5 (C=C—H)

CHAPTER 18   EXERCISES

★1. How could β-aldehydo esters be prepared in general? For example, write the equations for the preparation of

$$\underset{\underset{O}{\overset{\|}{}}\ \underset{CH_3}{|}}{CHCHCOOC_2H_5}$$

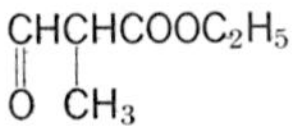

2. Write the formulas of three substances, other than acetoacetic ester or its homologs, that would be expected to exist in tautomeric forms. What structural changes in the molecules would you make to prevent the possibility of tautomerism?

★3. Account for the fact that a trace of base rapidly brings either the pure keto or the pure enol form to an equilibrium mixture.

4. Starting with no carbon compounds, except methyl acetate and products prepared from this, show how you could synthesize the following. Use no Grignard reagents.

a. $(C_2H_5)_2CHCOOH$

b. 3-Methylbutane-2-thiol

c. $(CH_3)_2CH\overset{\overset{CONH_2}{|}}{C}\underset{\underset{CONH_2}{|}}{H}$ 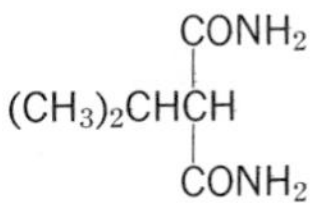

d. $CH_2(COOH)CH(COOH)CH_2COOH$

e. $H_3C—\overset{\overset{CO—NH}{|}}{C}H\quad \overset{\overset{}{|}}{C}O$ 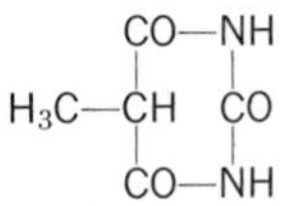
$\qquad\qquad\ \ \underset{CO—NH}{|}$

★**5.** Explain why the alkylation of acetoacetic acid or malonic ester with *t*-butylbromide is unsuccessful.

6. In the decarboxylation of a β-keto ester in the presence of Br_2, a monobromo ketone is formed. Is this consistent with the mechanism proposed for this reaction? Explain. (Assume no direct reaction between Br_2 and the ketone under these conditions.)

7. Write the equations for the following reactions:

★*a.* $CH_3COO(CH_2)_4COOCH_3 + NaOC_2H_5 \longrightarrow$

b. 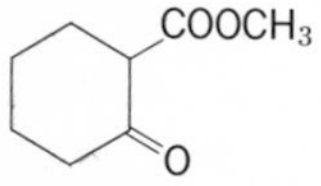 $+ NaOCH_3 + CH_3I \longrightarrow$

★*c.* 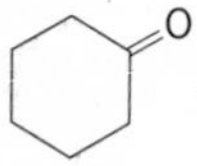 $+ CH_3COOC_2H_5 + NaOC_2H_5 \longrightarrow$

d. $HCOOC_2H_5 + CH_3CH_2COOC_2H_5 + NaOC_2H_5 \longrightarrow$

★*e.* $CH_3CH{=}CH_2 + NaCH(COOC_2H_5)_2 \longrightarrow$

f. $BrCH_2\overbrace{CH{-}CH_2}^{O} + NaCH(COOC_2H_5)_2 \longrightarrow$

★*g.* $C_6H_5{-}COCH_3 + C_6H_5COOC_2H_5 \xrightarrow{NaOC_2H_5}$

8. Show clearly how you could distinguish by chemical and/or spectral methods among:

a. $CH_3CH_2CH_2CH_2COOCH_3$, $CH_3CH(COOCH_3)_2$, $CH_3COCH_2CH_2COOCH_3$

b. $CH_3CH_2COCOOCH_3$, $CH_3CH_2CHOHCOOCH_3$, $\underset{\underset{O}{\|}}{C}HCH_2CH_2COOCH_3$

c. 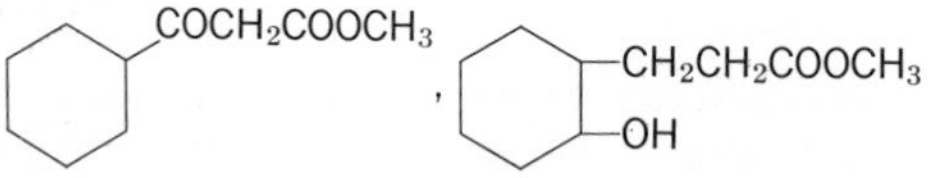

d. $CH_3\underset{CH_3}{\overset{CH_3}{C}}CH_2COCOOCH_3$, $CH_3CO\underset{CH_3}{\overset{CH_3}{C}}COOCH_3$, $CH_3\underset{CH_3}{\overset{CH_3}{C}}COCH_2COOCH_3$

9. Write equations for a practical synthesis of Amytal (below). What would be the products of a complete hydrolysis of Amytal?

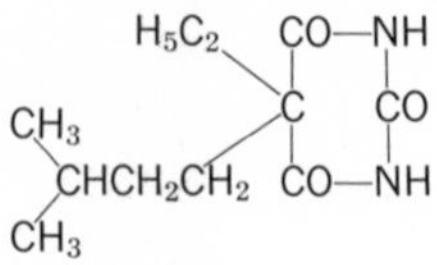

10. Account for the fact that the following reaction does *not* occur as shown. What would the product(s) be?

$$CH_3COCH_2COOC_2H_5 + BrCH_2COOC_2H_5 \xrightarrow{Zn} \underset{\underset{CH_2-COOC_2H_5}{|}}{\overset{\overset{O\;ZnBr}{||}}{CH_3C-CH_2COOC_2H_5}}$$

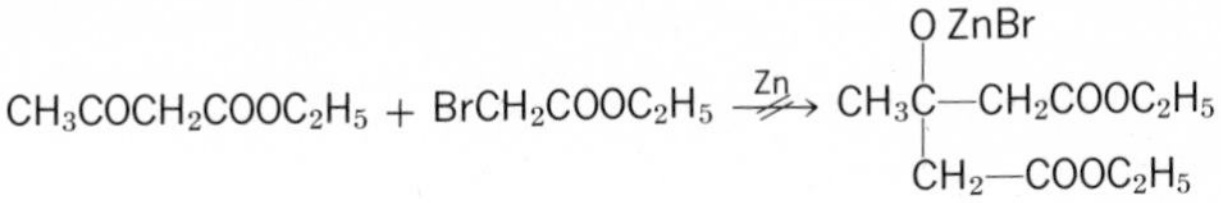

★11. Suggest a mechanism to account for the fact that ethyl carbonate, labeled with isotopic carbon, reacts with unlabelled malonic ester thus:

$$\underset{C_2H_5O\quad OC_2H_5}{\overset{O}{\overset{||}{C^{14}}}} + CH_2(COOC_2H_5)_2 \xrightarrow[\text{(trace)}]{NaOC_2H_5} C_2H_5OOC^{14}CH_2C^{14}OOC_2H_5$$

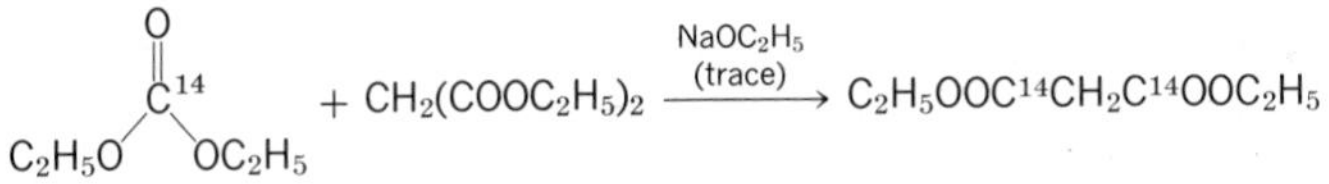

12. Explain why the resolution of optically inactive methyl acetoacetic ester presents difficulties.

★13. Show how the following compounds could be prepared from simpler substances without the use of Grignard reagents or Reformatsky syntheses.

★*a.* $\underset{\underset{O}{\underset{\diagdown\diagup}{CH_2\quad C=O}}}{\overset{CH_2-CH_2}{|\qquad}}$ 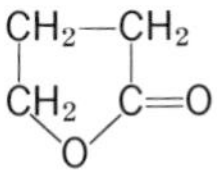

b. $C_6H_5CH_2COCH_2C_6H_5$

★*c.* $\underset{NH_2\;\;CH_2OH}{\overset{\;\;\;\;\;|\quad\;\;|}{CH_3CH_2CH_2CH-CHCH_2CH_3}}$ 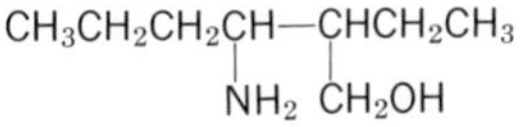

d. $\underset{\underset{COOCH_3}{|}}{\overset{\overset{CH_2-CH_2}{|\qquad\diagdown}}{\underset{CH_2-CH}{\;\;\;\;\;\;\;\;\;\;\;C=O}}}$ 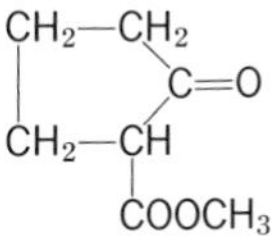

★*e.* $\underset{H_5C_2}{\overset{H_3C}{\diagdown}}CH-CH_2S-S-CH_2CH\underset{C_2H_5}{\overset{CH_3}{\diagup}}$ 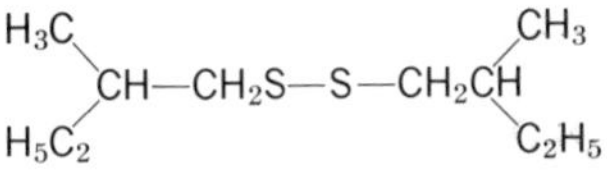

14. A certain substance A reacts as follows:

$$A + FeCl_3 \longrightarrow color$$
$$A + Br_2 \longrightarrow \text{an addition product}$$
$$A + NH_2OH \longrightarrow \text{a crystalline derivative}$$
$$A + Na \longrightarrow \text{a gas evolved}$$
$$A + NaHCO_3 \longrightarrow \text{no visible reaction}$$
$$A + NaOH + H_2O \longrightarrow \text{propionic acid and phenol}$$

Deduce the structure of A from these data. Write equations for the synthesis of A from propionic acid and phenol.

★15. How would you account for the fact that the ester

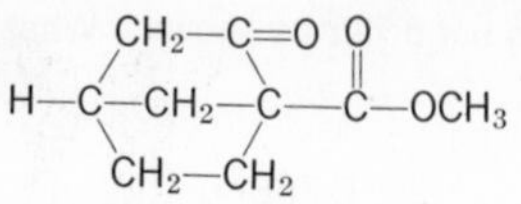

does not give a color with $FeCl_3$? The corresponding free acid, unlike most β-keto acids, does not lose CO_2 readily. Explain.

16. Show clearly by equations why the Claisen condensation stops at the "first stage," i.e., why little $(CH_3CO)_2CHCOOC_2H_5$ is formed from ethyl acetate.

★17. Cis and trans isomers would be predicted for the following compounds. Explain why these are difficult to obtain.

a.

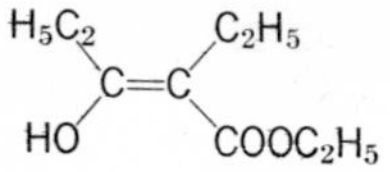

b.

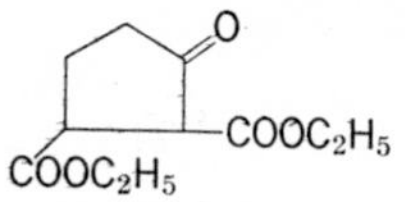

c.

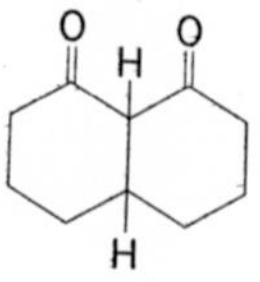

19

SUBSTITUTED ACIDS: AMINO ACIDS

19.1 Nomenclature The amino acids are closely related, both chemically and biochemically, to the hydroxy and keto acids, and because of their important role as building units for the very large molecules of proteins, the *alpha*-amino acids are of greatest interest. In fact, the term "amino acid" is often interpreted to imply α-amino acid. Acids other than those with the amino group in the α position will be considered in this chapter only when significant differences in reactions or properties occur.

Because of their natural origin, trivial names are used almost exclusively for these substances. Table 19.1 contains the names and structural formulas of those amino acids found in proteins. The names underlined are those most often encountered and should be memorized, with their formulas, by the student.

TABLE 19.1
SOME AMINO ACIDS

Glycine	CH_2COOH NH_2
Alanine	$CH_3CHCOOH$ NH_2
Valine[a]	H $(CH_3)_2CH—CCOOH$ NH_2
Leucine[a]	$(CH_3)_2CHCH_2CHCOOH$ NH_2
Isoleucine[a]	CH_3 $CH_3CH_2CHCHCOOH$ NH_2
Norleucine	$CH_3CH_2CH_2CH_2CHCOOH$ NH_2
Serine	$CH_2—CH—COOH$ OH NH_2
Threonine[a]	$CH_3CH—CHCOOH$ OH NH_2
Cysteine	$CH_2—CHCOOH$ SH NH_2
Cystine	$CH_2CHCOOH$ S NH_2 S $CH_2CHCOOH$ NH_2
Methionine[a]	$CH_2CH_2CHCOOH$ S NH_2 CH_3
Phenylalanine[a]	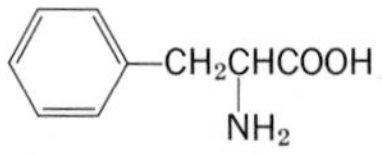
Tyrosine	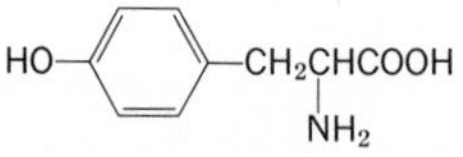
Diiodotyrosine	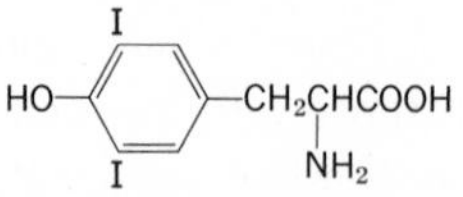
Thyroxine	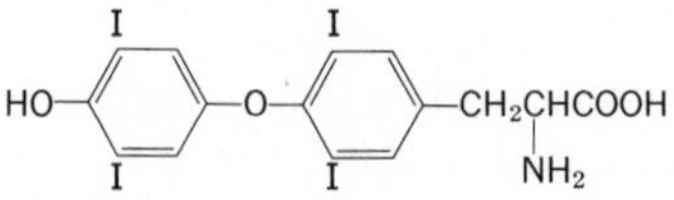

TABLE 19.1 *(continued)*

Tryptophan[a]	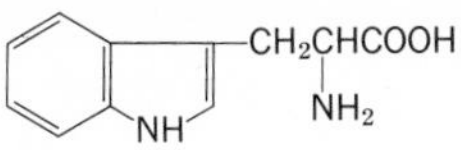
Proline	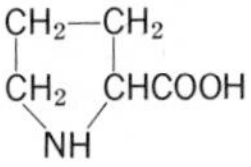
Hydroxyproline	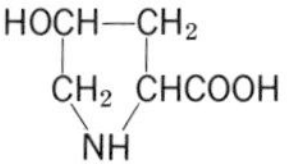
Aspartic acid	NH_2 CHCOOH CH_2COOH
Glutamic acid	$HOOCCH(NH_2)CH_2CH_2COOH$
Arginine[b]	$NH_2CNH(CH_2)_3CHCOOH$ NH $\quad$ NH_2
Lysine[a]	$NH_2(CH_2)_4CHCOOH$ NH_2
Histidine[b]	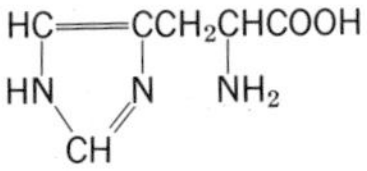

[a]These amino acids have been found to be required in the diet of man because they are not synthesized by the body.
[b]These two amino acids are semiessential, in that synthesis by the body is sometimes inadequate.

19.2 Naturally Occurring Amino Acids With a very few exceptions all amino acids found in nature are of the same configuration (see Chap. 7), that is, the *S* (formerly called L) configuration. The corresponding enantiomeric *R* (or D) amino acids are either totally useless to living organisms or must be converted by some chemical process into the *S* configuration before they can be used. (Glycine has no asymmetric carbon atom of course.) Consequently, the amino acids obtained from natural sources are usually optically active.

The most important sources of amino acids are proteins, which are polymers comprised of many different amino acids linked by amide (sometimes called "peptide") linkages into long chains. Hydrolysis of proteins yields gross mixtures of amino acids from which certain of the amino acids can be isolated in quantity by the use of specialized techniques.

19.3 Synthesis of Amino Acids There are several general synthetic methods for preparing pure (racemic) α-amino acids.

Halogen and hydroxy acids are converted to amino acids by methods already studied. The interrelations are shown as follows:

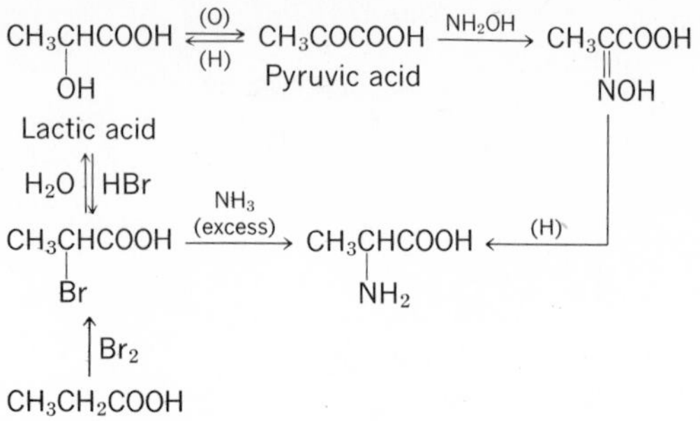
CH3CHCOOH
OH
Lactic acid
(O)
(H)
CH3COCOOH
Pyruvic acid
NH2OH
CH3CCOOH
NOH
H2O
HBr
CH3CHCOOH
Br
NH3
(excess)
CH3CHCOOH
NH2
(H)
Br2
CH3CH2COOH

When coupled with a malonic-ester synthesis of acids, these reactions make many of the amino acids available;[1] the synthesis of phenylalanine below illustrates a typical amino-acid synthesis.

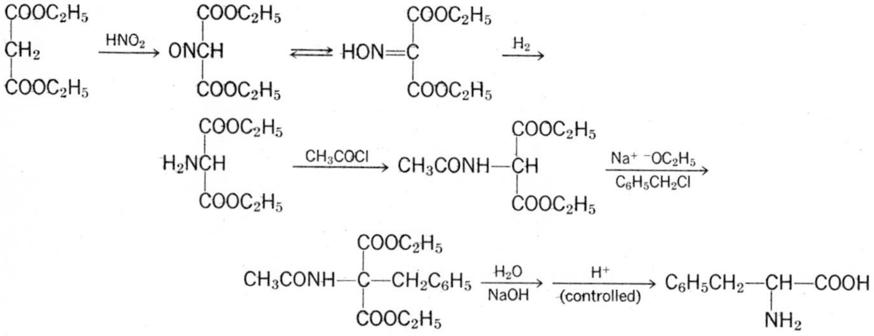
COOC2H5
CH2
COOC2H5
HNO2
ONCH
HON=C
H2
H2NCH
CH3COCl
CH3CONH—CH
Na+ −OC2H5
C6H5CH2Cl
CH3CONH—C—CH2C6H5
H2O
NaOH
H+
(controlled)
C6H5CH2—CH—COOH
NH2

Methods depending on the reaction of NH_3 with halogen acids yield some side products (Sec. 14.4), but by the use of a large excess of NH_3 these are minimized.

The use of the phthalimide method effectively prevents these side reactions and generally gives good yields.

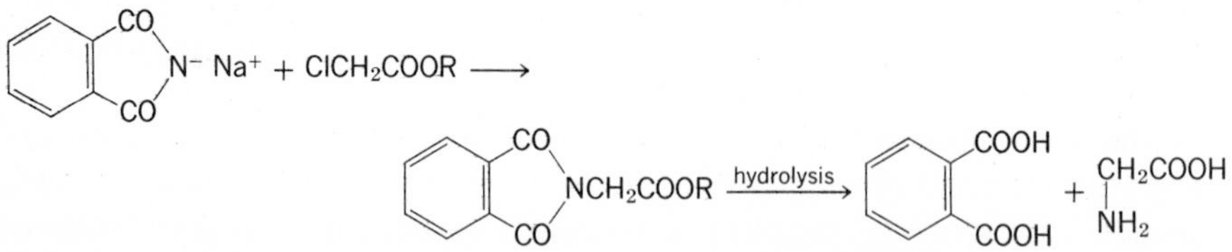
CO
N− Na+ + ClCH2COOR
CO
NCH2COOR
hydrolysis
COOH
COOH
CH2COOH
NH2

Halogen esters with the halogen in other than the α position react in the same manner. The β-halogenated acids lose HX as already observed.

A modification of the aldehyde and ketone addition reactions known as the *Strecker synthesis* is particularly convenient for α-amino acids.

$$CH_3CHO + NH_3 \longrightarrow [CH_3CH{=}NH] + H_2O$$

[1]The first step is a reaction which appears to involve the addition of NO^+ (from HNO_2) to the enol form of malonic ester.

$$[CH_3CH{=}NH] + HCN \xrightarrow{\Delta} CH_3CH(CN)(NH_2) \xrightarrow[\text{hydrolysis}]{H_2O,\ \text{acid}} CH_3CH(COOH)(NH_2)$$

Both of these steps can be accomplished at once by the use of NH_4CN. Aromatic as well as aliphatic aldehydes may be used in this reaction with success.

The only amino acids which cannot be made by these general syntheses are those in which the amino group is directly attached to an aromatic ring. These are often prepared by reduction of corresponding nitro compounds or by the rearrangement of amides. The marked resistance of the COOH group to reduction makes the former a practical process. The synthesis of the physiologically important *p*-aminobenzoic acid is carried out as shown below.

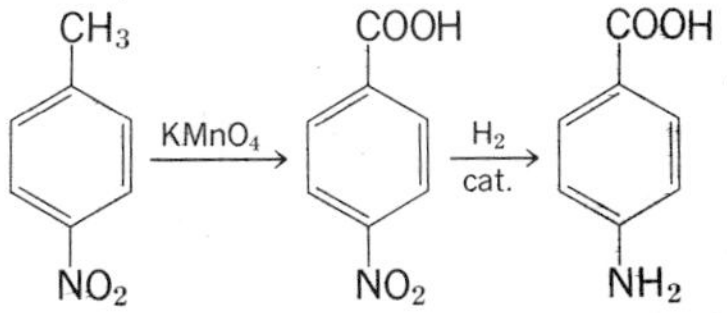
CH3
KMnO4
NO2
COOH
H2
cat.
NO2
COOH
NH2

p-Aminobenzoic acid
(PAB)

19.4 Physical Properties A comparison of the physical properties of amino acids with those of other substituted acids of comparable molecular size is instructive (Table 19.2). In all their physical properties amino acids are seen to resemble electrovalent compounds rather than the covalent type represented by the usual formulation

$$CH_3CH(NH_2){-}COOH$$

A review of the properties of the carboxyl and amino groups will remind the student that the coexistence of these two groups in the same molecule would lead to some interaction. The carboxylic acid group is a proton donor and the amino group a

TABLE 19.2
SOME PHYSICAL PROPERTIES

Formula	*Melting point,* °C	*Boiling point,* °C	*Solubility, water*	*Solubility, ether and other organic solvents*
$CH_3CH(Cl)COOH$	Liquid	186	Soluble	Soluble
$CH_3CH(OH)COOH$	26	Decomposes	Soluble	Soluble
$CH_3CH(NH_2)COOH$	295	Decomposes	Soluble	Insoluble
CH_3COONa	324		Soluble	Insoluble

proton acceptor. There is a great deal of evidence pointing to an ionic structure for amino acids both in the solid state and in solution. The carboxylic acid group donates its proton to the amino group to form a salt of the structure

$$CH_3\underset{\underset{\displaystyle NH_3{}^+}{|}}{CH}\text{—}COO^-$$

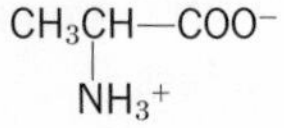

This is the *zwitterion* structure of the molecule. It may be looked upon as an internal salt analogous to

$$R\text{—}NH_2 + R'\text{—}\overset{\displaystyle O}{\overset{\|}{C}}\text{—}OH \rightleftharpoons R\text{—}\overset{+}{N}H_3 + R'\text{—}\overset{\displaystyle O}{\overset{\|}{C}}\text{—}O^-$$

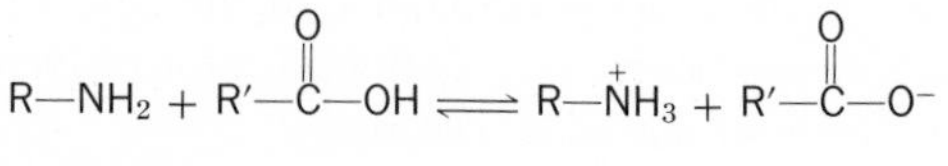

in which R and R' are part of the same molecule. Zwitterions can form from α-, β-, γ-, etc., amino acids since the electrovalent bond has no fixed length or direction.

The zwitterion formulation agrees not only with the observed physical properties, but also with the chemical properties of the amino acids. Being amphoteric substances, they form salts with both acids and bases. Both the acidity and the basicity of amino acids is much lower than one would anticipate from the simple (non-zwitterion) formula. Thus the ionization constant for an amino acid is much lower

$$H_3C\text{—}\overset{\displaystyle H}{\underset{\displaystyle NH_2}{\overset{|}{\underset{|}{C}}}}\text{—}\overset{\displaystyle O}{\overset{\|}{C}}OH + H_2O \rightleftharpoons H_3C\text{—}\overset{\displaystyle H}{\underset{\displaystyle NH_2}{\overset{|}{\underset{|}{C}}}}\text{—}\overset{\displaystyle O}{\overset{\|}{C}}\text{—}O^- + H_3O^+$$

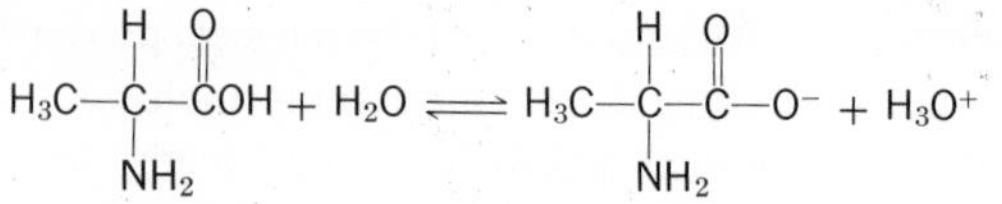

than that for ordinary carboxylic acids, and is similar to that observed for compounds of the type RNH_3^+. The above equation is more properly written

$$H_3C\text{—}\overset{\displaystyle H}{\underset{\displaystyle NH_3{}^+}{\overset{|}{\underset{|}{C}}}}\text{—}\overset{\displaystyle O}{\overset{\|}{C}}\text{—}O^- \quad + H_2O \rightleftharpoons H_3C\text{—}\overset{\displaystyle H}{\underset{\displaystyle NH_2}{\overset{|}{\underset{|}{C}}}}\text{—}\overset{\displaystyle O}{\overset{\|}{C}}\text{—}O^- + H_3O^+$$

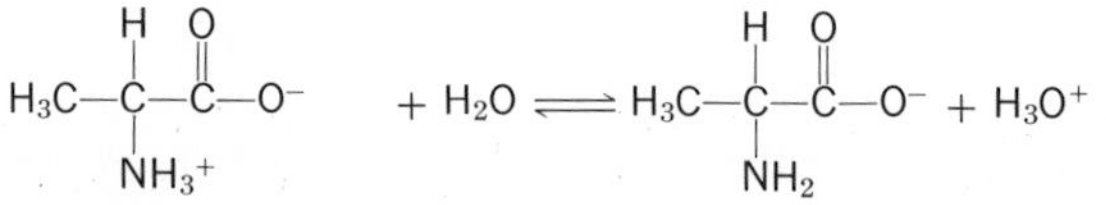

Comparatively weak acid

Similar considerations hold for the basic ionization constant, the reaction

$$H_3C\text{—}\overset{\displaystyle H}{\underset{\displaystyle NH_3{}^+}{\overset{|}{\underset{|}{C}}}}\text{—}\overset{\displaystyle O}{\overset{\|}{C}}\text{—}O^- \quad + H_2O \rightleftharpoons H_3C\text{—}\overset{\displaystyle H}{\underset{\displaystyle \overset{NH_3}{+}}{\overset{|}{\underset{|}{C}}}}\text{—}\overset{\displaystyle O}{\overset{\|}{C}}\text{—}OH + OH^-$$

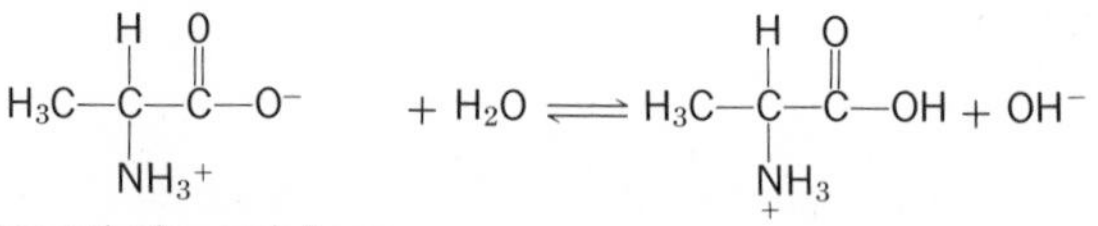

Comparatively weak base

being favored over

$$H_3C\text{—}\overset{\displaystyle H}{\underset{\displaystyle NH_2}{\overset{|}{\underset{|}{C}}}}\text{—}\overset{\displaystyle O}{\overset{\|}{C}}\text{—}OH \quad + H_2O \rightleftharpoons H_3C\text{—}\overset{\displaystyle H}{\underset{\displaystyle \overset{NH_3}{+}}{\overset{|}{\underset{|}{C}}}}\text{—}\overset{\displaystyle O}{\overset{\|}{C}}\text{—}OH + OH^-$$

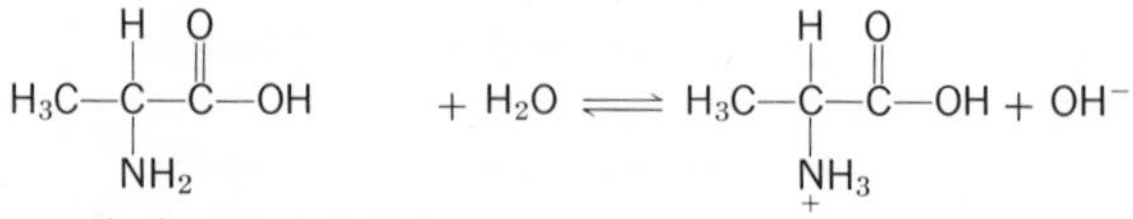

Comparatively strong base

That is, the amino acid behaves more like the weak base $RCOO^-$ than the more basic RNH_2.

19.5 Isoelectric Point The amphoteric nature of amino acids introduces certain other features. The equilibrium in water may be summarized as:

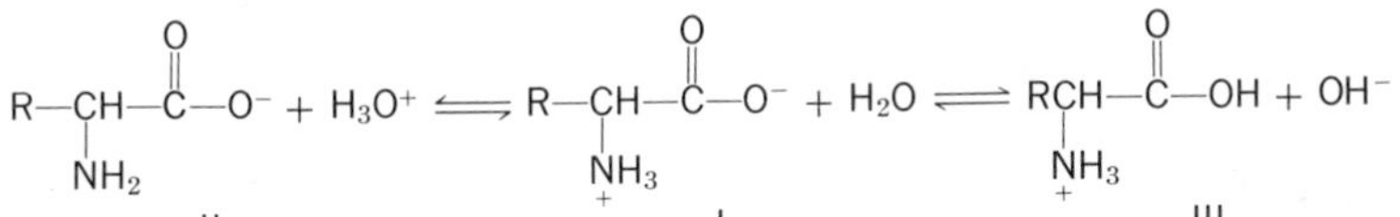

II I III

In alkaline solution (at high pH) the equilibria are shifted to the left and most of the molecules exist in the form II. In acid solution the reverse is true and most of the molecules exist in the form III. At some intermediate pH, not necessarily a neutral solution, most of the molecules are in the form of the electrically balanced zwitterion I, the remainder being equally divided between II and III. This point, reached by adjusting the pH, is known as the *isoelectric point*. At this point, if positive and negative electrodes are placed in the solution, there will be no net migration toward either electrode. Isoelectric points are always expressed in terms of pH and are characteristic for particular amphoteric substances, depending on their relative strengths as acids and bases.

When an amino acid such as alanine is dissolved in water, the majority of the molecules will be in solution as the zwitterion. By the process of hydrolysis (characteristic of salts of weak acids and bases—not the hydrolysis as described for esters, etc.) the equilibrium described above will be set up. Since OH^- and H_3O^+ are also in equilibrium, however, it is evident that these equilibria are not independent, but are related by the ionization constant of water:

$$K_w = (OH^-)(H_3O^+) = 10^{-14}$$

Because the carboxyl group is slightly stronger as an acid than the amino group is as a base, the above equilibria will be shifted slightly toward the left and the concentration of H_3O^+ and $RCHNH_2—COO^-$ will be slightly greater than that of (OH^-) and $RCHNH_3^+—COOH$. This means that the solution will be slightly acidic. In order to drive the solution toward its isoelectric point where the concentration of II equals that of III, it is necessary to add acid so as to displace the equilibria toward the right. Alanine thus has an isoelectric point at pH 6.

With basic amino acids, such as lysine, which have two amino groups per molecule the equilibrium is driven toward the right since the carboxyl group is "outnumbered." In order to make the number of negatively charged ions equal the number of positively charged it is necessary to add excess OH^- to drive the

TABLE 19.3
ISOELECTRIC POINTS OF SOME AMINO ACIDS

Amino acid	*Isoelectric* pH	*Amino acid*	*Isoelectric* pH
Glycine	6.0	Lysine	9.7
Alanine	6.0	Proline	6.3
Leucine	6.0	Tyrosine	5.7
Glutamic acid	3.2		

equilibrium back toward the left. For acidic amino acids, such as glutamic acid, the reverse is true.

A few examples of amino-acid isoelectric points are given in Table 19.3.

At their isoelectric points amino acids and proteins will not migrate in an electric field, will have a minimum solubility, and can usually be obtained in the free state by concentration of the solution. At other pH values they are more soluble and crystallize out on evaporation only as salts containing other ions (Na^+ or Cl^-, for example). In the isolation of amino acids, proteins, or amphoteric substances in general from synthetic reaction mixtures or from natural sources, a knowledge of the isoelectric points involved and adjustment of the pH of the solution play an important role.

19.6 Reactions In order to obtain typical carboxyl-group reactions from amino acids (esterification, for example), it is necessary to add a source of protons, that is, a strong acid, to shift from the zwitterion to the COOH form.

$$\underset{\displaystyle NH_3^+}{CH_3CHCOO^-} \xrightarrow[\text{dry}]{HCl} \underset{\displaystyle NH_3^+Cl^-}{CH_3CHCOOH} \xrightleftharpoons{HOC_2H_5} \underset{\displaystyle NH_3^+Cl^-}{CH_3CHCOOC_2H_5}$$

For this reason, a large excess of hydrogen chloride is used in the preparation of amino-acid esters. To obtain the free amino ester, it is only necessary to add cautiously the calculated amount of cold dilute alkali.

$$\underset{\displaystyle NH_3^+Cl^-}{CH_3CHCOOC_2H_5} \xrightarrow[\text{(dil., cold)}]{NaOH} \underset{\displaystyle NH_2}{CH_3CHCOOC_2H_5} + NaCl + H_2O$$

Esters of α-amino acids, being no longer zwitterions, are usually water-insoluble liquids that can be distilled in vacuum; in other words, they behave as covalent compounds with the reactions to be expected of primary amines and of esters. Since one of the common reactions of esters is that with amines to form amides, amino-acid esters are not particularly stable, but may react

$$R-\underset{\displaystyle NH_2}{\overset{\displaystyle H}{C}}-\overset{\displaystyle O}{\overset{\|}{C}}-OEt + R-\underset{\displaystyle NH_2}{\overset{\displaystyle H}{C}}-\overset{\displaystyle O}{\overset{\|}{C}}-OEt \longrightarrow R-\underset{\displaystyle NH_2}{\overset{\displaystyle H}{C}}-\overset{\displaystyle O}{\overset{\|}{C}}-NH-\overset{\displaystyle R}{CH}-\overset{\displaystyle O}{\overset{\|}{C}}-OEt \longrightarrow \text{etc.}$$

They may, however, be stored indefinitely as their hydrochlorides.

In order to obtain the typical amino-group reactions, for example, amide formation, it is necessary, because of the zwitterion structure of the free acid, to work in alkaline medium.

$$\underset{\displaystyle \mathrm{NH_3^+}}{\overset{}{\mathrm{CH_2COO^-}}} \xrightarrow{\mathrm{NaOH}} \underset{\displaystyle \mathrm{NH_2}}{\mathrm{CH_2COO^-Na^+}} \xrightarrow{\mathrm{C_6H_5COCl}} \mathrm{C_6H_5CONHCH_2COO^-Na^+}$$

To obtain the acylated amino acid from its salt, dilute acid is added.

$$\underset{\displaystyle \mathrm{NHCOC_6H_5}}{\mathrm{CH_2COO^-Na^+}} \xrightarrow[\text{(cold)}]{\text{dil., HCl}} \underset{\displaystyle \underset{\text{Hippuric acid}^1}{\mathrm{NHCOC_6H_5}}}{\mathrm{CH_2COOH}}$$

Acylated amino acids exhibit the normal carboxyl-group reactions and the solubility analogous to other substituted acids. They will decompose $NaHCO_3$, while the simple amino acids themselves generally do not (unless more carboxyl than amino groups are present).

$$\mathrm{CH_3}\underset{\displaystyle \mathrm{NH_3^+}}{\mathrm{C}}\mathrm{HCOO^-} + \mathrm{NaHCO_3} \longrightarrow \text{N.R.}$$

$$\mathrm{CH_3}\underset{\displaystyle \mathrm{NHCOCH_3}}{\mathrm{C}}\mathrm{HCOOH} + \mathrm{NaHCO_3} \longrightarrow \mathrm{CH_3}\underset{\displaystyle \mathrm{NHCOCH_3}}{\mathrm{C}}\mathrm{HCOONa} + \mathrm{CO_2}\uparrow + \mathrm{H_2O}$$

Because of the weakly acidic character of most amino acids, neutralization equivalents cannot be determined in the ordinary way; it is first necessary to block off the basic amino group. The most convenient way of accomplishing this is by the use of formaldehyde. This reagent reacts with a water solution of an amino acid to form a mixture of products (not isolated in practice) in which the basic properties of the amino group are depressed.

$$\mathrm{CH_2O} + \mathrm{H_3\overset{+}{N}CH_2COO^-} \longrightarrow \mathrm{H_2C{=}NCH_2COOH} + \mathrm{H_2O}$$
$$\mathrm{H_2C{=}NCH_2COOH} + \mathrm{NaOH} \longrightarrow \mathrm{H_2C{=}NCH_2COONa} + \mathrm{H_2O}$$

Titrations carried out in this way are known as *formol titrations* and have been used in amino-acid identification and analysis.

Like other ammonium salts, amino acids form amides when heated.

$$\underset{\text{Ammonium acetate}}{\mathrm{CH_3COO^-NH_4^+}} \xrightarrow[\Delta]{} \underset{\text{Acetamide}}{\mathrm{CH_3\overset{\displaystyle O}{\overset{\|}{C}}NH_2}}$$

[1] Hippuric acid is produced from benzoic acid in the diet. As the name suggests, it is found in the urine of horses.

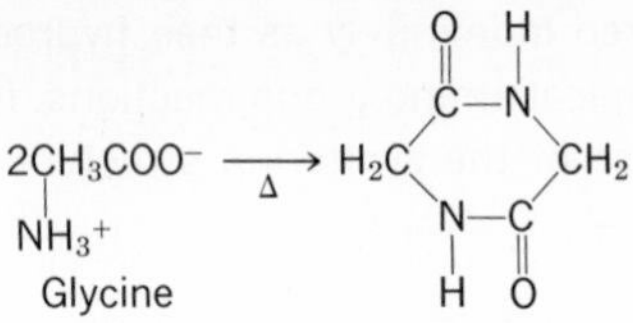

Glycine

(a cyclic diamide)
Diketopiperazine

Esters likewise form amides when heated with amines, and amino esters react in a similar way, sometimes forming polypeptides (amino acid polymers, see below).

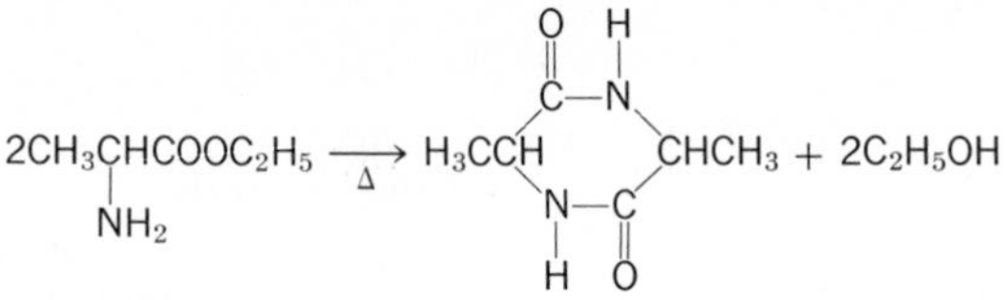

Cyclic amides of this type (compare lactides, Sec. 17.10) are called "diketopiperazines." This is a poor name, of course, because no keto group is present. As they are readily formed when α-amino acids or esters are subjected to high temperatures, they are found among the products of protein hydrolyses which involve long heating; they are believed, however, to be artifacts and not to occur in nature or to represent any part of protein structure. γ-Amino acids on heating form *lactams* (compare lactones).

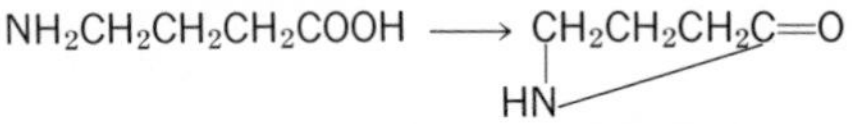

γ-Butyrolactam

A most important type of linkage in α-amino-acid chemistry is that between two amino-acid molecules, an amide linkage, formed from the carboxyl group of one molecule and the α-amino group of another.

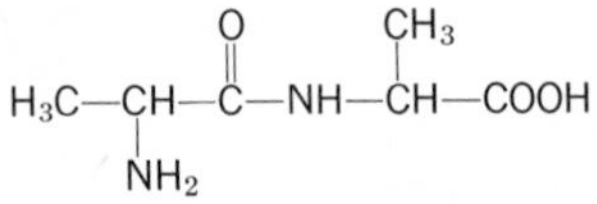

Such linkages are known as *peptide* linkages and constitute the major type of linkage found in proteins. This represents another name for *amide* linkages and the term "peptide linkage" would be superfluous were it not for its widespread use.

19.7 Spectroscopic Properties of Amino Acids Primary amino acids possess rather characteristic infrared spectra due to their zwitterionic structures. They exhibit an N—H stretching band at 3030 to 3130 cm^{-1}, N—H bending bands at 1610 to 1660 cm^{-1}, and bands at 1485 to 1550 cm^{-1} plus the COO^- band at 1560 to 1600 cm^{-1}. The nmr is less clearcut due to the rapid exchange of COOH and NH protons which tends to "wash out" their spectra.

OUTLINE OF AMINO-ACID CHEMISTRY

Preparation

1. Hydrolysis of proteins by means of acids, alkalis, or enzymes.

$$\underset{R'}{\overset{R}{>}}C=O \xrightarrow[\Delta]{NH_4CN} \underset{R'}{\overset{R}{>}}\underset{CN}{\overset{NH_2}{C}} \xrightarrow[HCl]{conc.} \underset{R'}{\overset{R}{>}}\underset{COOH}{\overset{NH_2 \cdot HCl}{C}}$$

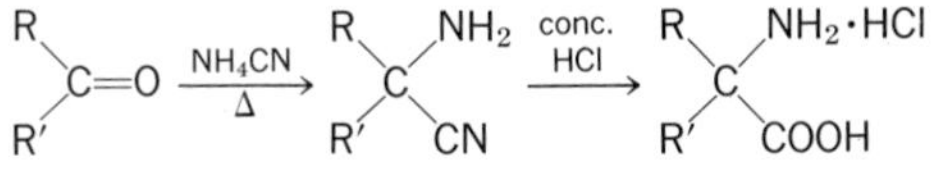

An aldehyde or ketone, aliphatic or aromatic, may be used here.

3. α-Halogenated acid with NH_3.

$$RCHXCOOH \xrightarrow{NH_3} \underset{NH_2}{RCHCOONH_4}$$

Since for a great many uses optically active (S) amino acids are desired, those obtained from natural products are most useful.

Properties

1.
$$\underset{NH_2}{RCHCOOH} \begin{cases} \xrightarrow{HCl} \underset{NH_2 \cdot HCl}{RCHCOOH} \\ \xrightarrow{NaOH} \underset{NH_2}{RCHCOONa} \end{cases}$$

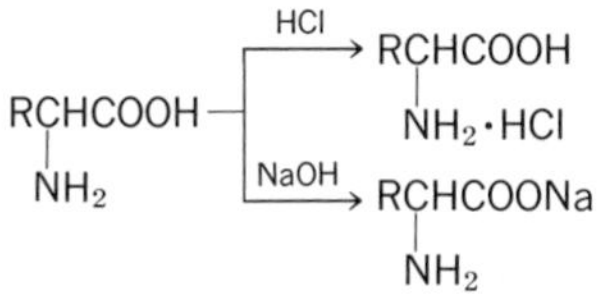

In most of their reactions α-amino acids behave as though they have the zwitterion structure $\underset{NH_3^+}{RCHCOO^-}$.

2. $\underset{NH_2}{RCHCOOH} \xrightarrow{R'COCl} \underset{NHCOR'}{RCHCOOH} + HCl$

This reaction does not proceed well in the absence of base such as NaOH.

3. $\underset{NH_2}{RCHCOOH} \xrightarrow{HNO_2} N_2\uparrow$ (quantitative)

4. $\underset{NH_2}{RCHCOOH} \xrightarrow{R'CHO} \underset{N=CHR'}{RCHCOOH}$

This reaction is used in an analytical reaction known as the formol titration of amino acids (Sec. 19.6).

5. $\underset{\overset{|}{NH_2}}{RCHCOOH} \xrightleftharpoons{\overset{C_2H_5OH \;+\; HCl(excess)}{}} \underset{\overset{|}{NH_2 \cdot HCl}}{RCHCOOC_2H_5} \xrightarrow[Na_2CO_3]{H_2O} \underset{\overset{|}{NH_2}}{RCHCOOC_2H_5}$

6. $R—\underset{\overset{|}{NH_2}}{CH}—CO—OC_2H_5 \xrightarrow{\Delta}$ 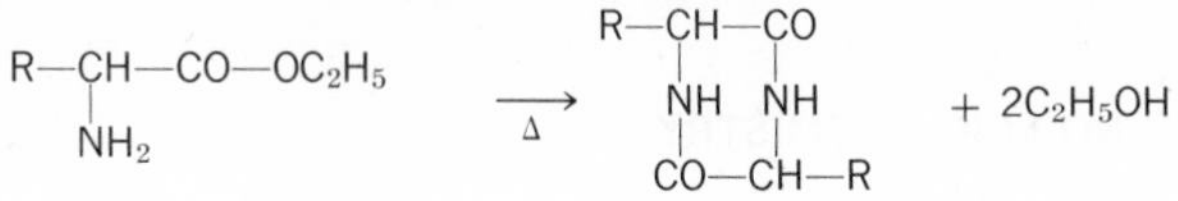
$\begin{array}{c} R—CH—CO \\ \;\;|\quad\quad| \\ NH\quad NH \\ \;\;|\quad\quad| \\ CO—CH—R \end{array} \;+\; 2C_2H_5OH$

A diketopiperazine

CHAPTER 19   EXERCISES

1. Make a list of the heterocyclic (Sec. 19.1) amino acids. Which of these would be the strongest base? Explain.

2. The protein from hair is found to contain considerable amounts of sulfur. Which amino acids are probably present? How could you distinguish between them?

3. Why, in the synthesis of glycine from bromoacetic acid, is a large excess of ammonia used?

4. Write equations for the synthesis of leucine, using (*a*) acetoacetic ester and (*b*) the Strecker synthesis.

5. Many living organisms have enzyme systems capable of decarboxylating and deaminating ($NH_2 \longrightarrow OH$) amino acids. What has this to do with the occurrence of amyl alcohols in the fermentation of corn and other materials?

6. Although most amino acids do not liberate CO_2 from solutions of $NaHCO_3$, glutamic acid does. Explain this fact.

7. Predict the isoelectric points, if any, of the following substances. Your answer should be "pH 8 to 11," "pH 3 to 6," or "not amphoteric."

★ *a.* Cysteine

  b. Thyroxine

★ *c.* Hippuric acid (*N*-benzoylglycine)

  d. $CH_3\underset{\overset{|}{NH_2}}{CH}SO_3H$

★ *e.* $CH_3\underset{\underset{(CH_3)_3}{\overset{|}{N^+}}}{CH}COO^-$

  f. Histidine

★ *g.* $CH_3\underset{\overset{|}{CH_3}}{CH}COOCH_3$

  h. $\begin{array}{c} \quad\quad CONH \\ H_3CCH\quad\; HCCH_3 \\ \quad\quad NHCO \end{array}$ 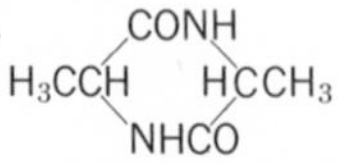

★ *i.* β- and γ-Aminobutyric acids

8. Suppose the isoelectric pH of a certain amino acid is 8. Will a solution of the pure amino acid in pure water have a pH greater or less than 7, and greater or less than or equal to 8? Can you generalize this result and define within limits the pH of a solution of a pure ampholyte in pure water?

9. In solution at pH 6.0, toward which pole will the following migrate? The isoelectric point of alanine is at pH 6.0.

★*a.*

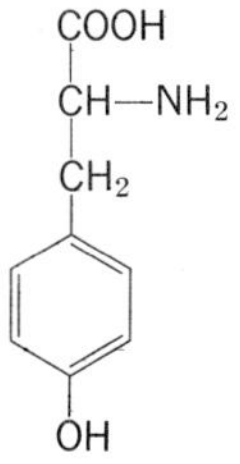

★*b.*

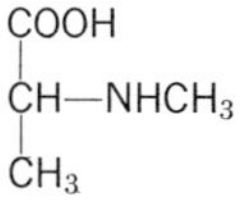

★*c.*

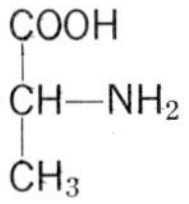

★*d.*

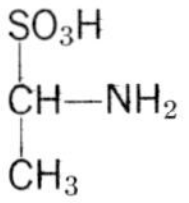

10. Distinguish by chemical tests:
   a. Alanine, propionamide, ammonium propionate, lactic acid amide
   b. Phenylalanine, tyrosine, leucine
   c.

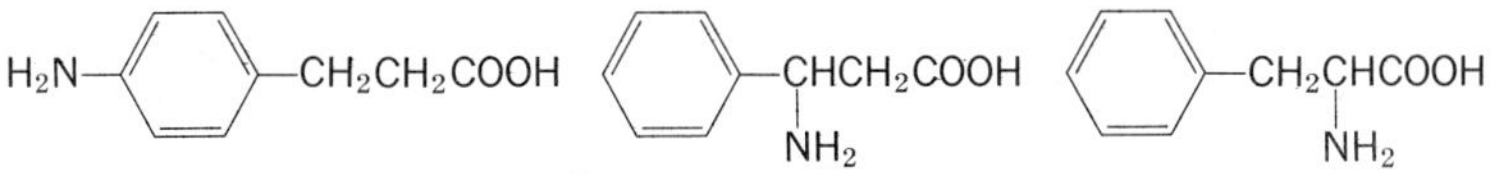

   d. $C_2H_5NHCOOCH_3$

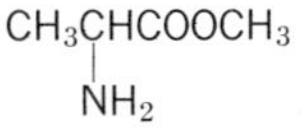

★**11.** Explain why the separation of a mixture of three amino acids (choose any three) is more difficult than the separation of a mixture of phenol, benzene, and benzoic acid.

12. Account for the fact that the characteristic 1724 cm^{-1} absorption band present in lactic acid is not present (is shifted) in alanine.

★**13.** Which of the natural amino acids would be expected to show appreciable absorption in the ultraviolet?

14. Explain how the use of infrared spectra would aid in following the course of a Strecker synthesis.

★**15.** Natural amino acids are of the *S* configuration. Draw the structure of the diketopiperazine from *S* alanine. Is it optically active? Does an inactive diketopiperazine exist?

16. Explain briefly the following:

a. Lysine

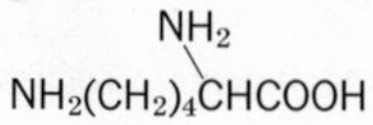

NH2
NH2(CH2)4CHCOOH

in solution at pH 3 migrates toward the positive (or negative, which?) pole in an electrophoresis experiment.

b.

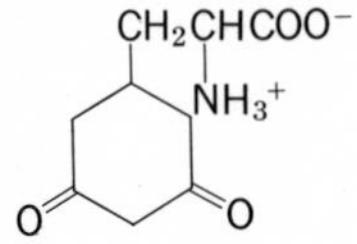

CH2CHCOO−
NH3+
O
O

has an isoelectric point at a higher (or lower, which?) pH than that of alanine.

20

MOLECULES WITH MORE THAN ONE FUNCTIONAL GROUP

20.1 Interaction of Functional Groups Organic chemistry is in many ways the chemistry of systems containing more than one reaction site or functional group. Any transformation directed toward one group may involve other functional groups in conflicting, competitive, or cooperative roles.

20.2 Conflicting Functional Groups We have already observed a number of instances in which two different functional groups cannot exist in the same molecule, that is, they are incompatible. In essence, any two groups which are capable of reacting with one another are usually not compatible substituents. This is most obvious when the two groups can react with one another within the molecule.

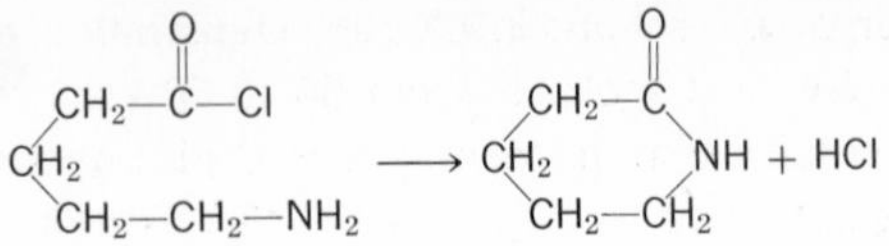

Not stable

Less obvious, but equally important, is the case of molecules where such direct action is not possible. Thus, in principle, the Grignard reagent formed from *p*-bromobenzaldehyde cannot react with itself because the two groups cannot get close to each other. However, it is very important to remember that any real system involves

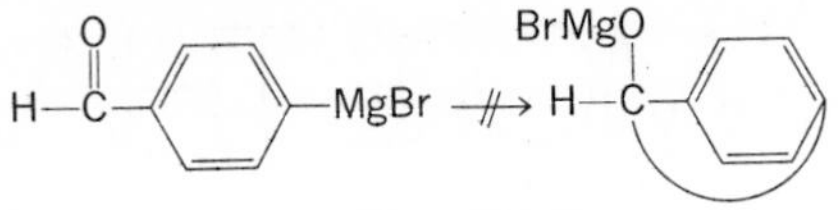

large numbers of molecules. For this reason the above molecule would be unstable except in extremely dilute solution[1] due to the reaction

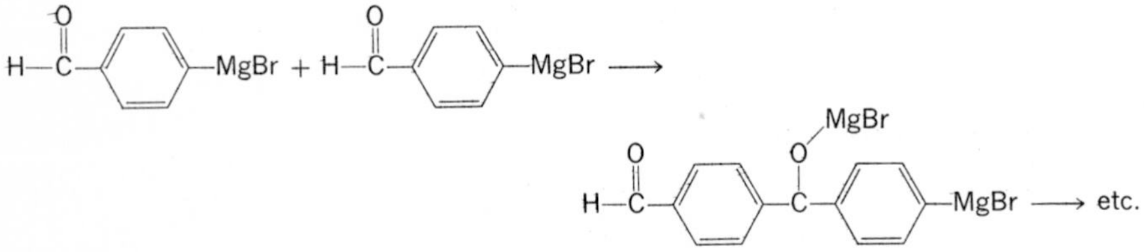

In this case the Grignard reagent would not be formed in any amount because the products from the reaction would coat the magnesium and make it unreactive.

Incompatibility can be of a more subtle nature, as in the case of amino-acid esters. Since amines react slowly with esters at ordinary temperatures to form amides, these esters cannot be kept for long periods of time except at very low temperatures. The hydrochlorides of amino-acid esters, on the other hand, are stable indefinitely.

$$H_2N\!-\!\underset{\displaystyle R}{\underset{|}{CH}}\!-\!CO_2R' + H_2N\!-\!\underset{\displaystyle R}{\underset{|}{CH}}\!-\!CO_2R' \xrightarrow{\text{slowly}} H_2N\!-\!\underset{\displaystyle R}{\underset{|}{CH}}\!-\!CONH\!-\!\underset{\displaystyle R}{\underset{|}{CH}}\!-\!CO_2R' + HOR', \text{ etc.}$$

$$Cl^-H_3\overset{+}{N}\!-\!\underset{\displaystyle R}{\underset{|}{CH}}\!-\!CO_2R' \longrightarrow \text{N.R.}$$

20.3 Competition More common, and accordingly, more serious, than the direct conflict of functional groups is competition of different groups during reaction with a given reagent. Many functional groups react with many reagents at not dissimilar rates so that it is impossible, or at least very difficult, to perform an operation on a selected functional group.

[1] In most instances the concentrations which would be required are so low that the reagent would be destroyed by trace impurities in the solvent.

Most common organic reactions involve key steps with activation energies (Sec. 3.8) ranging from a few kcal/mole to about 50 to 60 kcal/mole. Since so many transformations must fall within this energy range, it is not surprising that competition between various pathways occurs so commonly. Fortunately for the practicing organic chemist (and very unfortunately for the predictions of the theoretical chemist) a difference in activation energies of 1.4 kcal/mole can make a difference of about 10 in the relative rates of two reactions, that is, one reaction will produce over 90 percent of the product. With a difference of 2.8 kcal/mole it will be favored by a factor of 100, or more than 99 percent! The present state of chemical theory is such that it is not usually possible to predict energy differences to anywhere near this order of precision. It is thus the task of the organic chemist to find reactions and conditions such that the desired pathways are favored, a task which is still largely empirical.

In many cases it is possible to arrange the sequence of steps of a multistep transformation so as to minimize competition. We have seen instances of this technique in aromatic substitution sequences (Sec. 6.21). Thus, the transformation

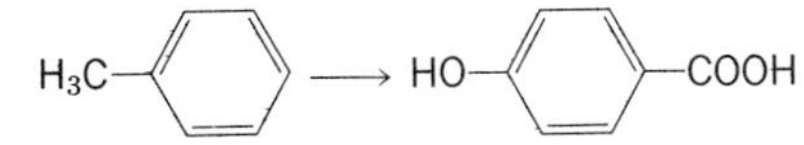
H3C
HO
COOH

cannot be accomplished by the following sequence:

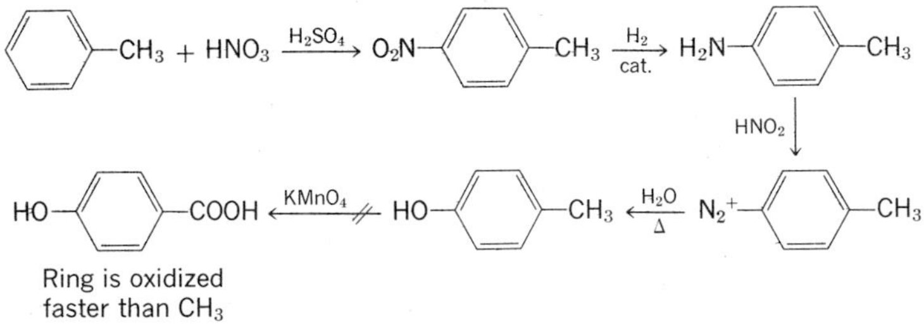
CH3 + HNO3
H2SO4
O2N
CH3
H2
cat.
H2N
CH3
HNO2
HO
COOH
KMnO4
HO
CH3
H2O
Δ
N2+
CH3
Ring is oxidized
faster than CH3

In the above sequence the hydroxyl group activates the aromatic ring toward oxidation so that it is more readily oxidized than the methyl group. By varying the sequence of the same steps, however, the conversion can be effected according to the scheme:

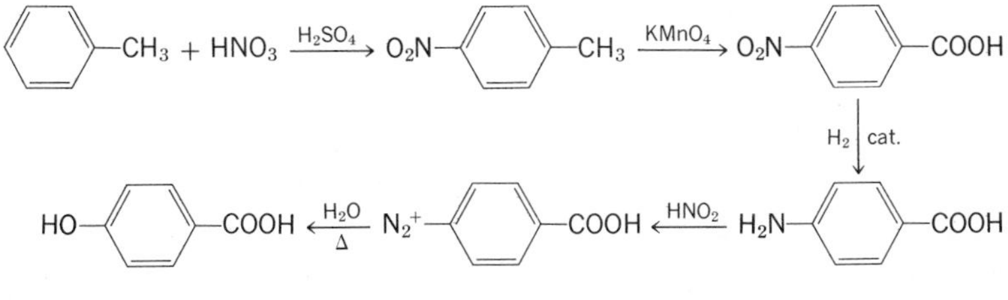
CH3 + HNO3
H2SO4
O2N
CH3
KMnO4
O2N
COOH
H2
cat.
HO
COOH
H2O
Δ
N2+
COOH
HNO2
H2N
COOH

In this sequence oxidation of the nitro-substituted ring is difficult because the nitro group withdraws electrons from the ring. The reduction step does not cause trouble because of the well-known resistance of the carboxyl group to reduction (Sec. 12.3).

It is often not possible to find such clearcut differences in reactivity. The organic chemist is continually striving to find reagents which are increasingly more selective in their action. An example of this is given by the methods of oxidation and reduction which have been developed to permit a variety of reactions to be carried out competitively.[1] In Table 20.1 are shown a number of functional groups and a collection of reducing agents. In Table 20.2 is shown a similar series for oxidizing agents. Entries marked with a plus indicate that the functional group is reduced (oxidized) by that reagent, with a minus that is not reduced, and no mark means that it is not clearly defined.

In addition to selective reductions, selective hydrolyses (Sec. 9.5), selective substitutions (Sec. 9.5), as well as many other selective reactions have been worked out.

20.4 Protection of Functional Groups From Table 20.1 it can be seen that highly reactive groups like aldehydes and ketones can be reduced in the presence of the less reactive esters, but that the converse is not true. In order to carry out a conversion on a relatively unreactive functional group in the presence of a more reactive group, it is necessary to "protect" the latter group by converting it to some derivative which is less reactive in the desired transformation, but from which the original group can be readily regenerated after the transformation is completed.

One of the simplest forms of protection is accomplished by a change in pH.

TABLE 20.1
SELECTIVE REDUCTIONS OF VARIOUS FUNCTIONAL GROUPS

Group	*Reagent*				
	$LiAlH_4$	$NaBH_4$	H_2/Pt (1 atm)	H_2/Ni (high pressure)	$H_2/Zn—Cr$ oxide
R—CHO[a]	+	+	+	+	+
R_2—CO[a]	+	+		+	+
R—CO_2R'[a]	+	−	−	+	+
R—COOH[a]	+	−	−	+	−
R—CN[b]	+	−		+	
ϕ—CH_2—O—R[c]	−	−	+	+	+
C═C[d]	−	−	+	+	−
C≡C[d,e]	−	−	+	+	−
Aromatic rings[d]	−	−	−	+	−

[a] Reduction product is the alcohol.
[b] Reduction product is the primary amine.
[c] Reduction product is toluene and the alcohol ROH.
[d] Reduction product is the saturated hydrocarbon if sufficient hydrogen is used.
[e] May be interrupted at the alkene stage if the amount of hydrogen is limited.

[1] It is possible to show but a few of the more general methods which have been developed.

TABLE 20.2
SELECTIVE OXIDATIONS OF VARIOUS FUNCTIONAL GROUPS

Group	*Reagents*					
	Ag^+ *in* NH_4OH	$KMNO_4$ *(dil.)*	$KMNO_4$ *(conc.)*[a]	$KMNO_4$ *(hot, conc.)*	H_2CrO_4, H_2SO_4	CrO_3, *pyridine*
RCHO	+	+	+	+	+[b]	−
R_2CO	−	−	−	+	−	−
RCH_2OH	−	−	+	+	+[b]	+
R_2CHOH	−	−	+	+	+	+
R_3COH	−	−	−	+	−	−
RCH═CHR	−	+[c]	+[d]	+	−	−
RC≡CR	−	+	+	+	−	−
R—O—R	−	−	−	+	−	−
$R—CH(OR)_2$[e]	−	−	−	+	[e]	−
Ar—CH_2—R	−	−	−	+	·	·

[a] Hot $KMNO_4$ will destroy most aliphatic chains regardless of functional group.
[b] The aldehyde may be prepared from the alcohol with this reagent if the aldehyde is sufficiently volatile so that it can be distilled off as formed.
[c] Yields the 1,2-diol.
[d] Cleaves the molecule at this point to give ketones or acids.
[e] In acid solution this compound will be converted to the aldehyde, which will usually then be oxidized.

For example, if the diazotization of aromatic amines were carried out in neutral solution, the unreacted amine would couple with the diazonium ion formed in the reaction (Sec. 14.17).

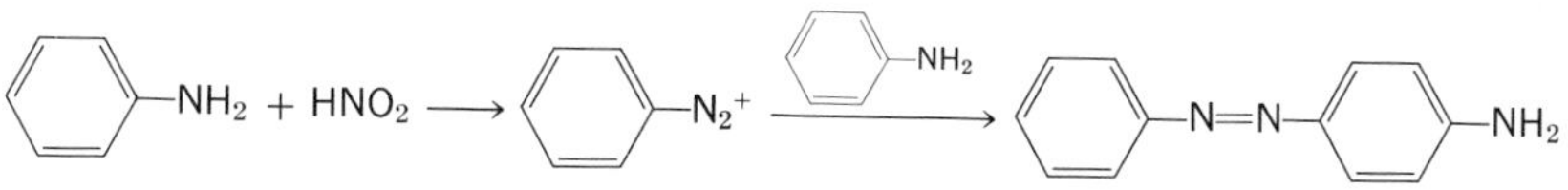

NH2 + HNO2
N2+
NH2
N=N
NH2

In the presence of fairly strong acid, most of the amine is in the form of the unreactive ammonium salt and is not available for coupling (which requires the free amine). Since the reaction of the small amount of free amine available from the equilibrium

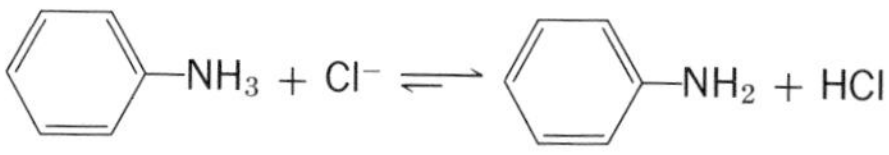

NH3 + Cl−
NH2 + HCl

is more rapid with nitrous acid than with diazonium salt, the reaction between diazonium salt and free amine is effectively prohibited. After all of the amine (and ammonium salt) has been converted to the diazonium salt, the pH may be raised if desired since there is no more possibility of its coupling.

For many compounds, simple pH changes are not enough to provide the necessary protection, and more firmly bound protecting groups must be used. For example, an aldehydic acid may be converted to the aldehydic alcohol by the following sequence:

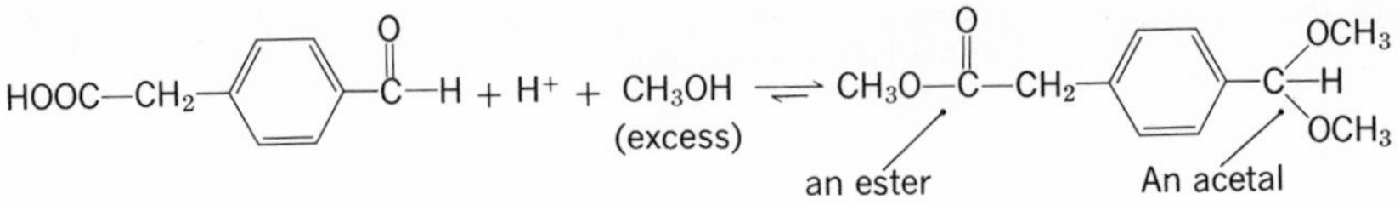

(excess)
an ester
An acetal

Reaction of carboxylic acids with alcohols yields esters which are readily reduced with lithium aluminum hydride. Aldehydes yield acetals under similar conditions, and acetals are, like ethers, stable to alkaline reagents such as lithium aluminum hydride. Thus, reduction of the above compound yields

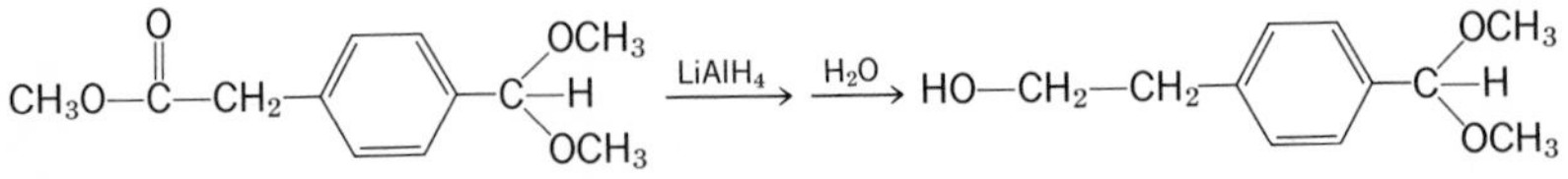

The aldehyde can be regenerated with dilute acid.

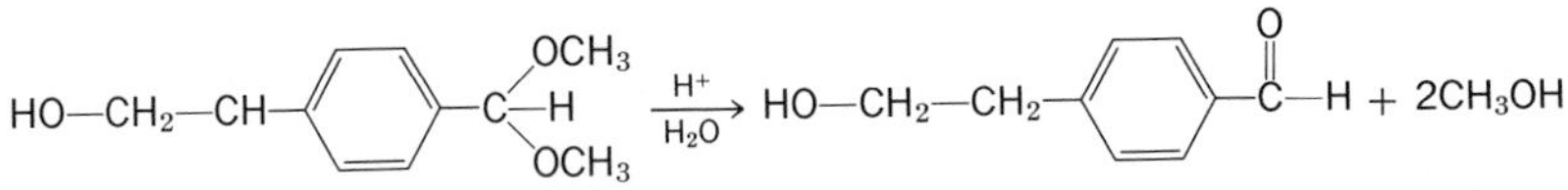

Any attempt at direct reduction leads to preferential reduction of the aldehyde function, or, at best, simultaneous reduction of both functional groups.

A variety of other protecting groups have been devised, a few examples of which are given in Table 20.3.

TABLE 20.3
REPRESENTATIVE PROTECTING GROUPS

Functional group	*Protecting group*	*Provides protection against*	*Removed by*
Aldehyde, ketone	acetal, ketal	basic or neutral reagents, oxidation, reduction	dilute acids
Amines	$R-\overset{O}{\overset{\Vert}{C}}-$ (acyl)[a]	oxidation, alkylation	acid hydrolysis
Alkenes	bromine	oxidation	zinc dust
Alcohols	benzyl ethers	acylation, alkylation, oxidation	hydrogen/platinum[b]

[a] See Sec. 22.10 for more specialized examples of this type of group as applied to protein synthesis.
[b] See Footnote *c*, Table 20.1.

20.5 Cooperation When two functional groups lie in certain relationships with respect to one another, reactions can occur that would not be observed with either functional group alone. Most commonly this situation arises when one of the functional groups is unsaturated (multiply bonded) and lies adjacent to the other functional group (separated by a single bond). Such systems are said to be *con-*

jugated. Some common examples are:

$$C=C-C=C \quad C=C-\overset{\overset{\displaystyle O}{\|}}{C}- \quad C\equiv C-\overset{\overset{\displaystyle O}{\|}}{C}-$$

$$[C=C-C^+] \quad C-C=C-\bigcirc \quad \left[\bigcirc-CH_2{}^+\right]$$

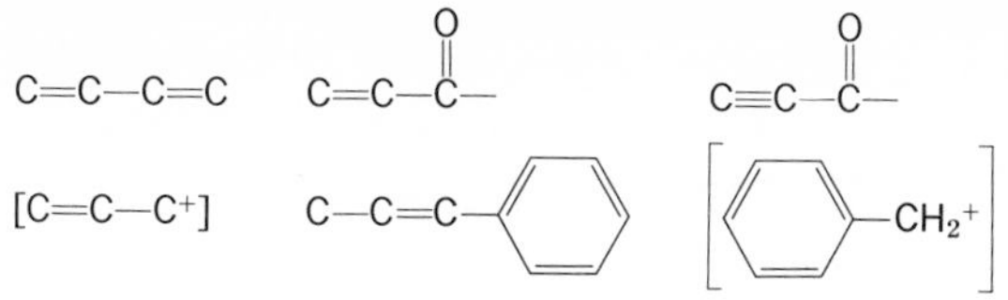

The addition of various reagents to conjugated dienes and carbonyl compounds frequently occurs by a 1,4-process (Sec. 6.4).

$$C=C-C=C + Br_2 \longrightarrow Br-C-C=C-C-Br + Br-\overset{\overset{\displaystyle |}{C}}{\underset{\overset{\displaystyle |}{Br}}{C}}-C-C=C$$
$$\text{1,4-Addition}$$

$$C=C-\overset{\overset{\displaystyle |}{C}}{\underset{\displaystyle |}{}}=O + R_2NH \longrightarrow R_2N-C-C-\overset{\overset{\displaystyle |}{C}}{\underset{\displaystyle |}{}}=O$$

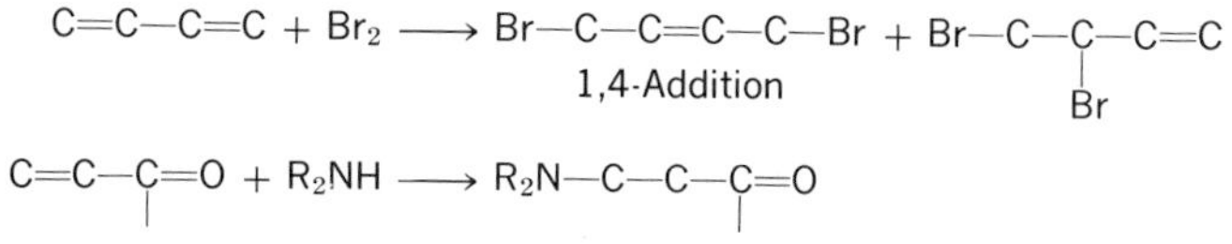

$$C=C + R_2NH \longrightarrow \text{N.R.}$$

The reactions of conjugated systems so frequently involve conjugated ions or radicals that it is appropriate to consider all of them together. The addition of bromine to butadiene occurs by way of a conjugated carbonium ion as follows:

$$C=C-C=C + [Br^+] \longrightarrow [Br-C-C=C-\overset{+}{C} \longleftrightarrow Br-C-C-\overset{+}{C}=C]$$

The ion is stabilized by resonance, yielding an ion with positive character at two sites. Subsequent reaction with Br$^-$ can (and does) occur at either location.

Nucleophilic addition to alkenes is ordinarily a poor reaction which occurs only with very strongly nucleophilic reagents such as alkyllithium compounds. In most reactions alkenes themselves act as nucleophiles. Carbonyl groups, on the other hand, undergo nucleophilic additions readily, although the first-formed products are often unstable due to the presence of two electronegative groups on carbon.

$$R_2NH + H_3C-\overset{\overset{\displaystyle O}{\|}}{C}-CH_3 \rightleftharpoons \left[H_3C-\overset{\overset{\displaystyle O^-}{|}}{\underset{\overset{\displaystyle |}{\underset{\displaystyle R_2}{+NH}}}{C}}-CH_3 \right] \rightleftharpoons H_3C-\overset{\overset{\displaystyle OH}{|}}{\underset{\overset{\displaystyle |}{NR_2}}{C}}-CH_3$$
$$\text{Unstable}$$

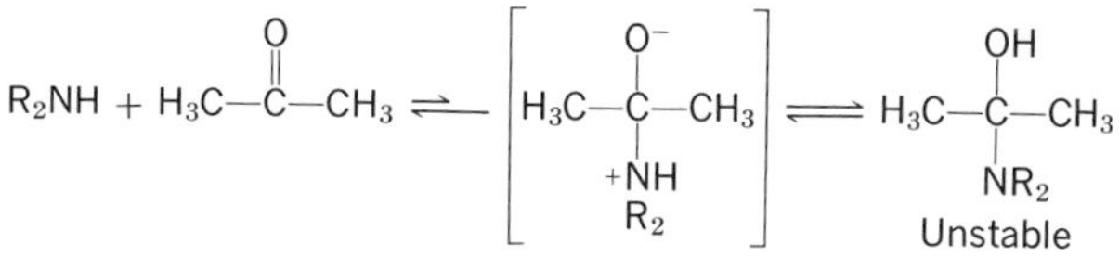

When double bonds are conjugated with carbonyl compounds, however, a new mode of nucleophilic reaction becomes possible.

$$R_2\overset{..}{N}H + H_2C=CH-\overset{\overset{\displaystyle O}{\|}}{C}-CH_3 \rightleftharpoons \left[R_2\overset{+}{\underset{\displaystyle H}{N}}-CH_2-CH=\overset{\overset{\displaystyle O^-}{|}}{C}-CH_3 \right.$$

$$R_2N-CH_2-CH=\overset{\overset{\displaystyle OH}{|}}{C}-CH_3 \rightleftharpoons \left. R_2\overset{+}{\underset{\displaystyle H}{N}}-CH_2-CH-\overset{\overset{\displaystyle O}{\|}}{C}-CH_3 \right]$$

$$R_2N-CH_2-CH_2-\overset{\overset{\displaystyle O}{\|}}{C}-CH_3$$
$$\text{Stable product}$$

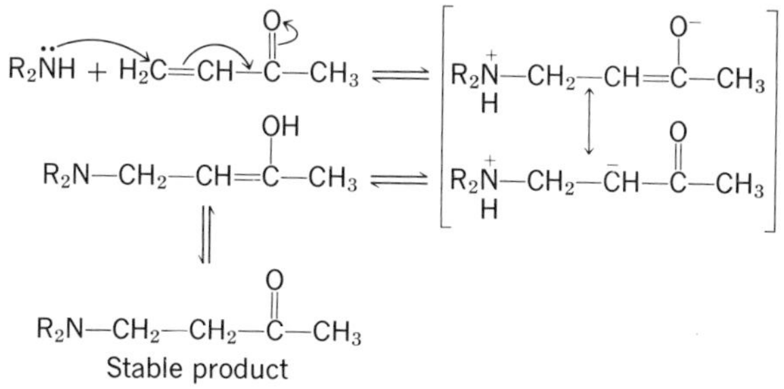

This type of nucleophilic addition, which may be viewed as a 1,4-addition (Sec. 6.4) to an unsaturated carbonyl compound, is called a "Michael reaction."

In general, reactive intermediates such as carbonium ions, carbanions, and free radicals are more stable when generated adjacent to an unsaturated group. Most often this stabilization arises from resonance interaction. Some common examples are:

$$\left[C{=}C{-}\overset{+}{C} \longleftrightarrow \overset{+}{C}{-}C{=}C\right] \qquad \left[C{=}C{-}\dot{C} \longleftrightarrow \dot{C}{-}C{=}C\right]$$

$$\left[C{=}C{-}\overset{-}{C} \longleftrightarrow \overset{-}{C}{-}C{=}C\right]$$

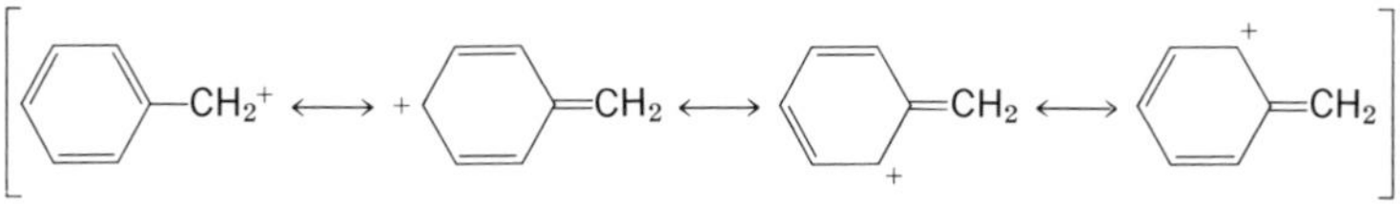

(Similarly for the radical and carbanion)

(Similarly for radical) (Carbonium ion not stabilized)

Exceptions occur when "stabilization" would violate one of the resonance rules (Sec. 1.7), such as placing a positive charge on an electronegative atom without addition of extra bonds (the case above) or where the required structure would exceed the stable valence shell of a first-row element (as below).

Groups other than double bonds can stabilize intermediates if they have unshared pairs of electrons.

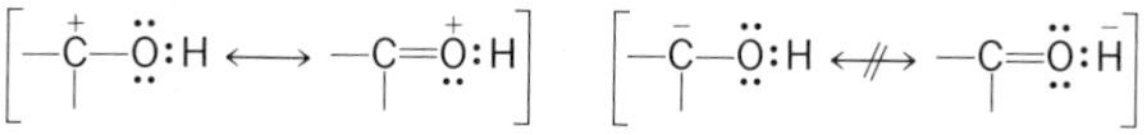

Is stabilized Is not stabilized

Another type of cooperative behavior of a very special type is found in the Diels-Alder reaction. This reaction occurs between a conjugated diene and an unsaturated molecule. Two new single bonds are formed to give a six-membered ring. All of the original multiple bonds disappear, but a new double bond is formed between the two originally conjugated bonds.

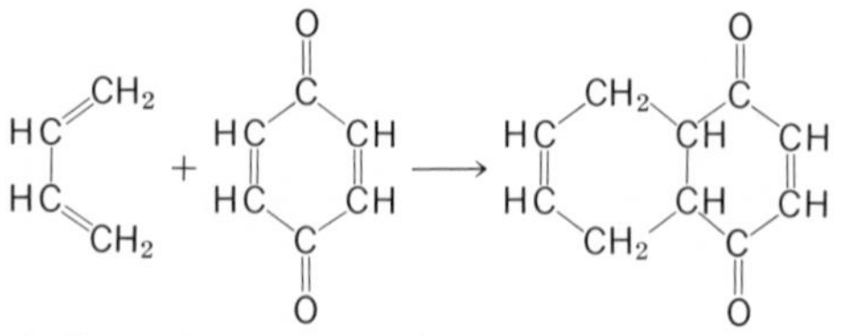

A diene An unsaturated group

The Diels-Alder reaction has been carefully studied and many important features of it have been discovered.

1. The reaction takes place between a conjugated diene, which must be able to assume a *cis*-like configuration, and another, multiply bonded substance such as an alkene or an alkyne. The latter substance is called a "dienophile."[1]

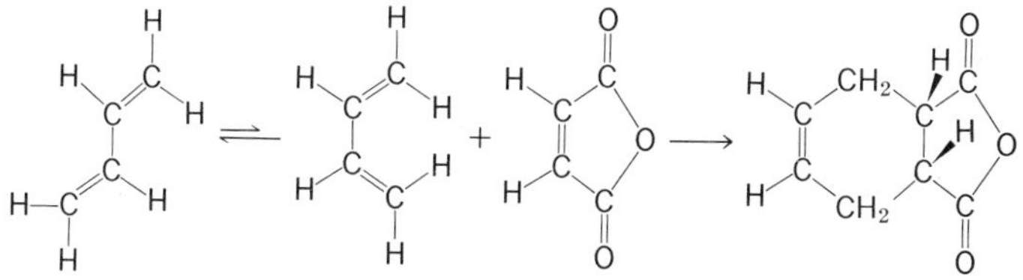

A *trans*-like conformer (unreactive) | A *cis*-like conformer (reactive) | A dienophile | A Diels-Alder adduct

2. The dienophile will be more reactive if it is substituted with electron-withdrawing groups such as $—CO_2R$, —CN, etc. With less-reactive dienophiles the reaction will only proceed at high temperatures and pressures. Very reactive dienophiles, such as tetracyanoethylene (TCNE) and acetylenedicarboxylic ester, will often react readily at room temperature. Less-reactive ones such as maleic anhydride require some heating.
3. The reaction is reversible, but it is usually possible to run it under conditions where the reverse reaction is so slow as to be negligible. The first-formed product is not always the most stable, so that prolonged or excessive heating can isomerize the product. At high enough temperatures the reverse reaction is favored.

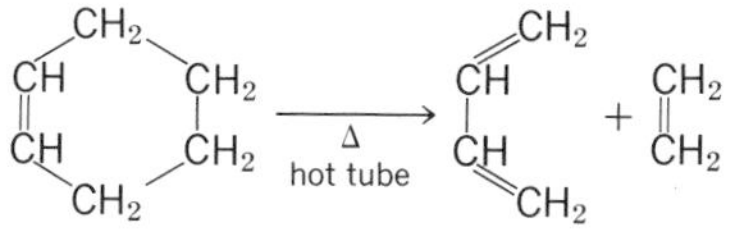

4. The initially formed product (usually the major product) is formed in a very specific fashion: (*a*) Both new bonds are formed on the *same side* of the *cis*-like diene; (*b*) both new bonds to the dienophile are *cis*.

a.

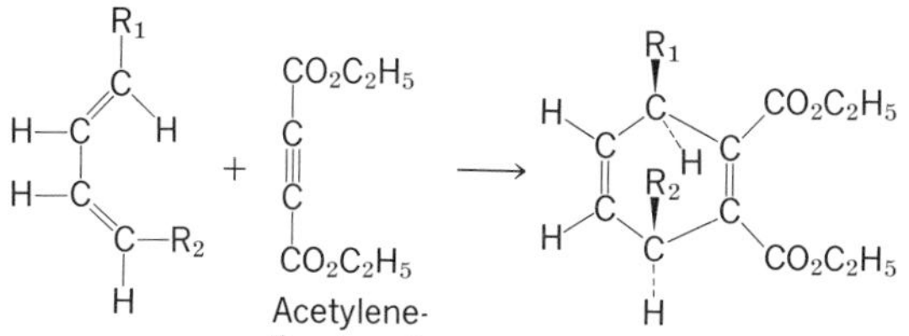

Acetylene-dicarboxylic ester

[1] Dienophile means diene liking, analogous to nucleophile (nucleus liking) and electrophile (electron liking).

b.

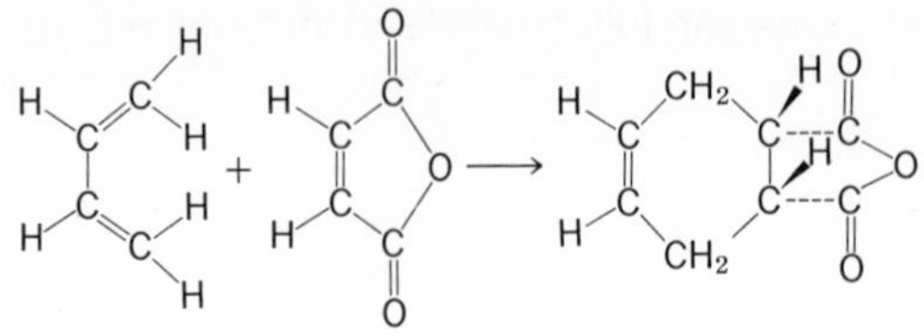

Maleic anhydride

20.6 Neighboring-group Participation The interaction of functional groups in adjacent (conjugated) situations does not usually occur when these groups are separated by one or more saturated carbon atoms. However, a number of situations do arise in which groups can interact at greater distances. This type of interaction is generally called "neighboring-group participation."

The most common instances of such interactions are 1,3-interactions. These can best be described by consideration of specific examples. A classic case is the hydrolysis of *R*-2-bromopropionic acid under various conditions.

1. 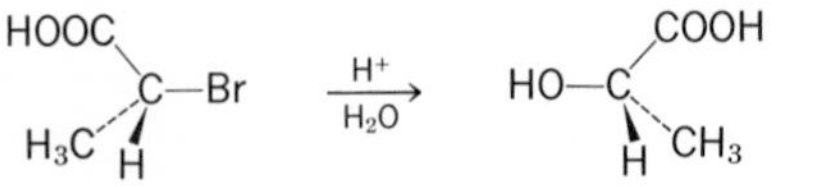

R-2-bromopropionic acid

S-2-hydroxypropionic acid

2. 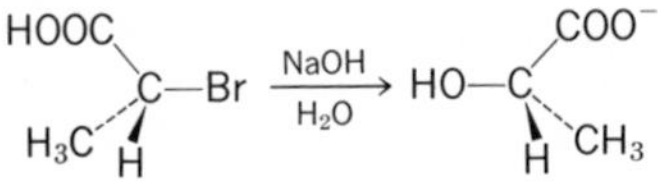

R *S*

3. 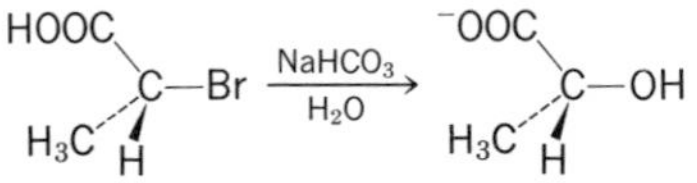

R *R*

Reactions 1 and 2 are believed to involve normal S_N2 displacements (Sec. 9.6) by H_2O and HO^-, respectively, yielding an inverted product. Reaction 3, however, occurs with net *retention!* The most reasonable mechanistic explanation for this behavior is that *two* S_N2 displacements have occurred, each with inversion so that the overall result is two inversions, which is indistinguishable from net retention.

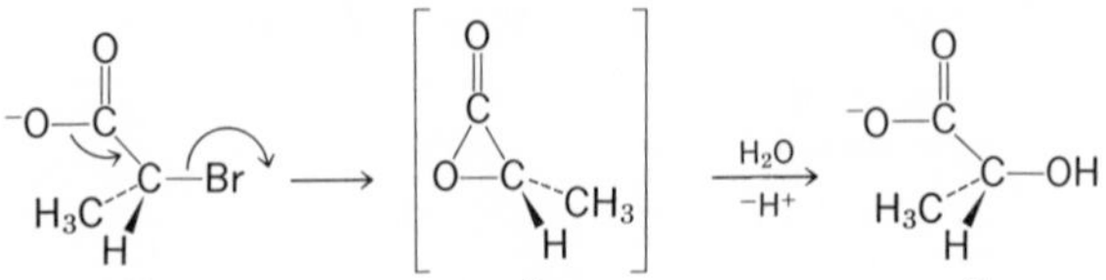

R *S* *R*

A highly strained α-lactone

This behavior occurs only in bicarbonate solution because it is only at this pH that one has a high concentration of the —COO⁻ group but little HO⁻, so that the S_N2 displacement reaction by the latter is less important.

The hydrolysis of mustard gas is another example of an assisted reaction. Mustard gas, $(ClCH_2CH_2)_2S$, hydrolyzes about 10^7 times as fast as ethyl chloride. This has been shown to involve internal displacement by the sulfur atom.

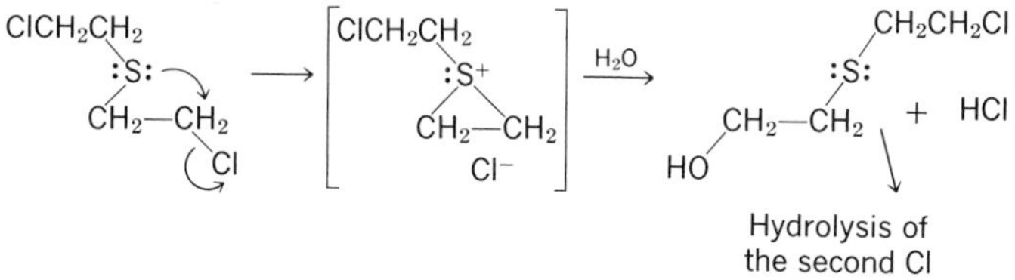
H₂O
+ HCl
Hydrolysis of the second Cl

Similar behavior is shown by neighboring nitrogen, oxygen, and even halogen atoms, although the importance of such assistance varys considerably. However, even when rate differences between assisted and unassisted (normal) reactions are small, control of stereochemistry often indicates that an assisted reaction is occurring. For example, the treatment of 2-bromo-3-hydroxybutane with HBr provides the following results:

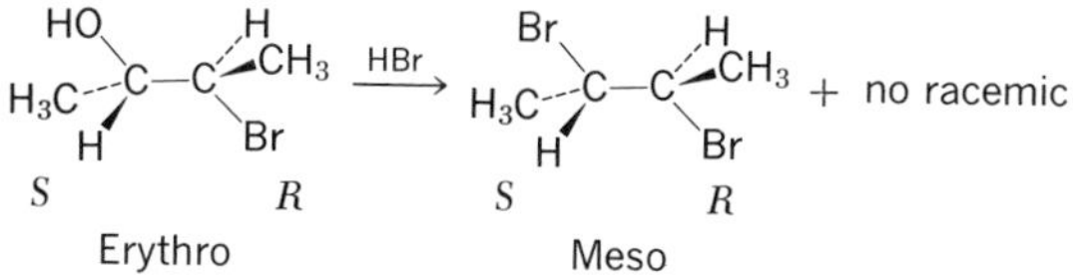
HBr
+ no racemic
S R S R
Erythro
Meso

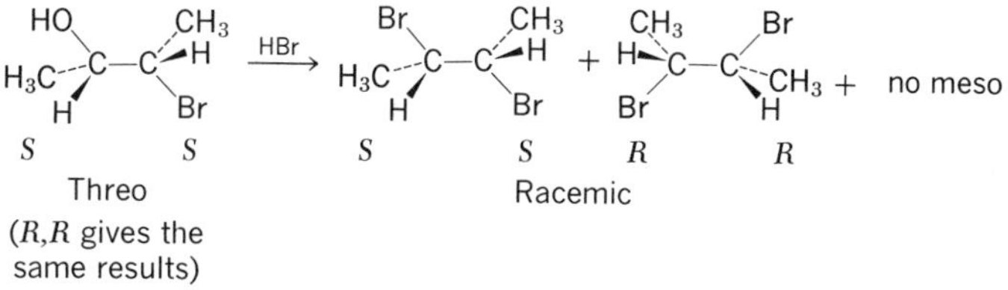
HBr
+ no meso
S S S S R R
Threo
(R,R gives the same results)
Racemic

These observations can best be explained by participation of the neighboring bromine as shown for the erythro case.

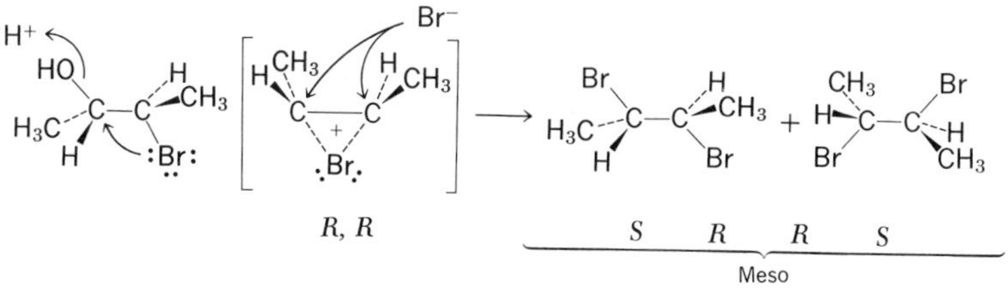
R, R
S R R S
Meso

Participation by atoms such as oxygen, bromine, nitrogen, and sulfur involves unshared electrons on these atoms, most commonly *p*-orbital electrons. It is also possible for the p orbitals in π bonds to participate to yield intermediates which resemble allylic ions in aliphatic systems and benzylic ions in aromatic systems. Such ions are called "homoallylic" ions in the former case and "phenonium ions" in the latter.

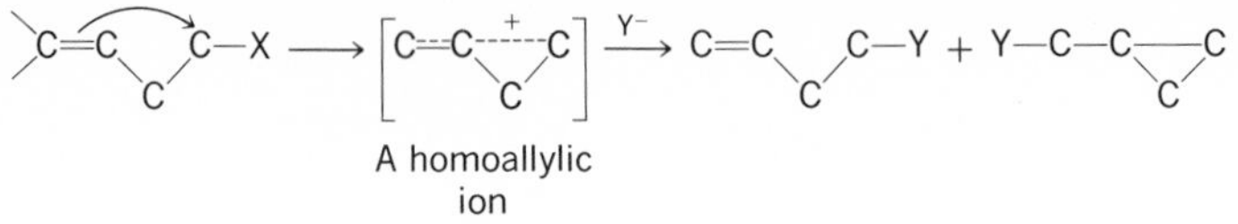

A homoallylic ion

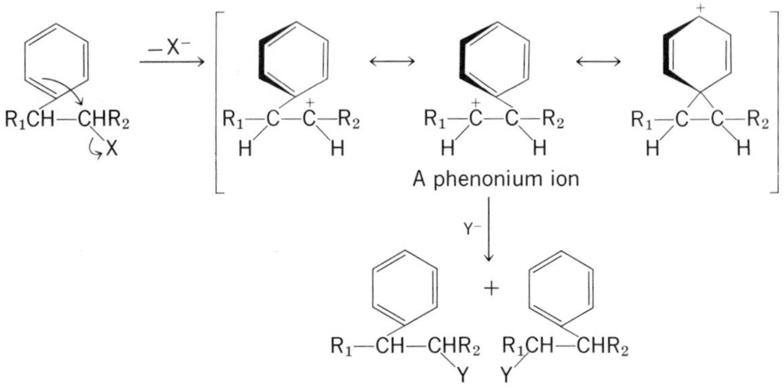

A phenonium ion

Participation by π electrons which are not directly conjugated, that is, the type involved in the systems above, is usually less effective than that in the conjugated systems because the π system and the carbonium ion center are farther apart. Due to the greater flexibility of these systems, the probability of the two positions coming together in just the right way to interact is also much lower. As a result of this, the effect of such interactions is more often seen as a control of stereochemistry (which requires a relatively small interaction) rather than a large effect on the rate of reaction. Exceptions occur when the rest of the molecule holds the π bond and the carbonium ion center in the right geometry, and in highly ionizing solvents where such participation becomes more important. In many reactions, particularly nucleophilic displacements (Sec. 9.6) under S_N2 conditions, such participation is unimportant.

Interactions at distances greater than 1,3 are generally not important, with the exception of 1,5 and 1,6 interactions where there is a reasonably high probability of the two ends coming together (Sec. 17.3). For example, the reaction of certain acetyl derivatives has been shown to involve 1,5 participation.

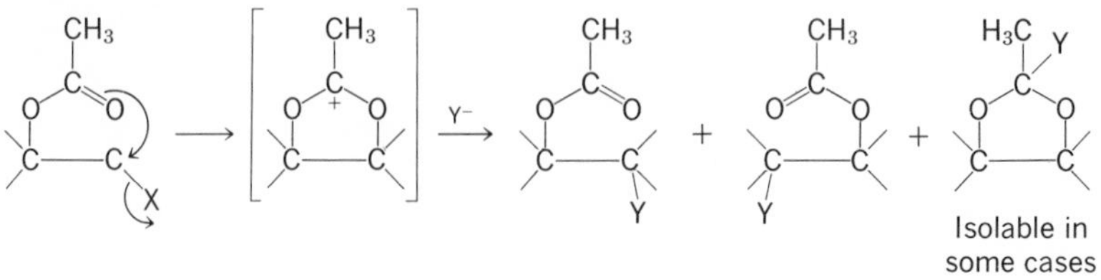

Isolable in some cases

In essence, cyclic anhydride, lactone, and related compound formation involve neighboring-group reactions in which stable cyclic products result. For example,

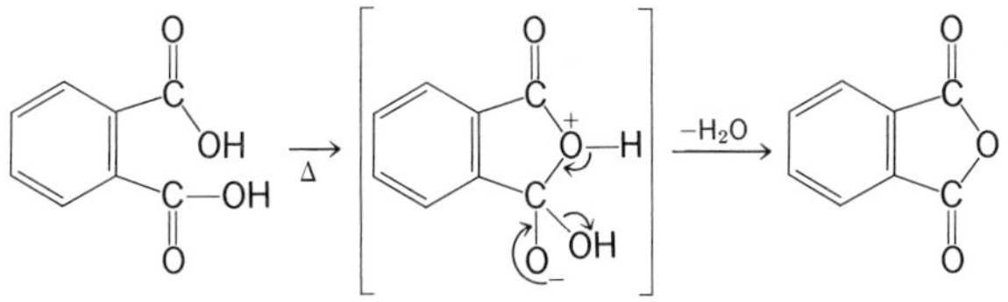

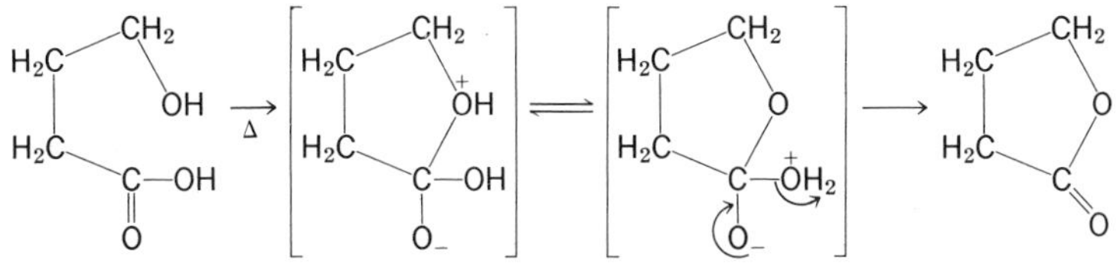

Such groups as vinyl and phenyl also participate in 1,5 and 1,6 interactions, but are subject to somewhat greater restrictions than the corresponding 1,3 interactions discussed above.

Neighboring-group reaction assumes a more important role in biochemical systems. In many molecules, especially ones such as enzymes, certain groups are held in specific relationships to other parts of the system. It is then possible for highly specific reactions, similar to the ones described above, to take place rapidly at low temperatures in which one particular site is especially suited to react and does so to the exclusion of other equally reactive, but not suitably situated, locations.

CHAPTER 20 EXERCISES

★1. Which, if any, of the following would be expected to have a reasonable half-life in solution? Explain.

★ *a.* CH_3COCH_2MgBr

b. $HOOCCH_2CH_2NH_2$

★ *c.* $HOOCCH_2—CH(CH_3)—OH$

d. HO—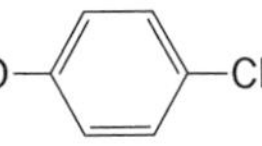—CH_2COOH

★ *e.* $HOCH_2CH_2CHO$

f. $(CH_3)_3N^+—CH_2CH_2COOH$ Cl^-

★ *g.* $CH_3C(=O)—CH^-—C(=O)CH_3$ Na^+

h. $CH_3\overset{O}{\overset{\|}{C}}-CH_2\underset{Zn^+Br}{\overset{-}{C}H}-\overset{O}{\overset{\|}{C}}-OCH_3$

★ *i.* $MgClCH_2CH_2\underset{\underset{O}{\|}}{C}OCH_3$

2. Show how the following transformations, involving selective reactions, might be accomplished.

a. $CH_3CH{=}CH-\overset{O}{\overset{\|}{C}}-CH_3 \longrightarrow CH_3CH{=}CH-\underset{OCH_3}{\underset{|}{C}H}-CH_3$

b. 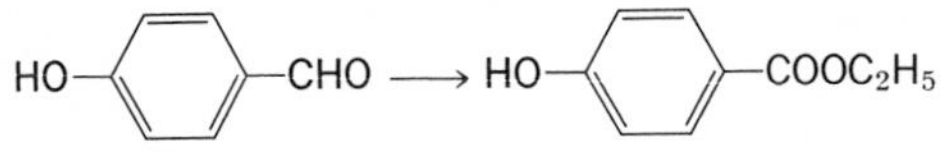

c. $CH_3\overset{O}{\overset{\|}{C}}-CH_2CH_2\overset{O}{\overset{\|}{C}}-OC_2H_5 \longrightarrow CH_3\underset{OH}{\underset{|}{C}H}-CH_2CH_2COOC_2H_5$

d. 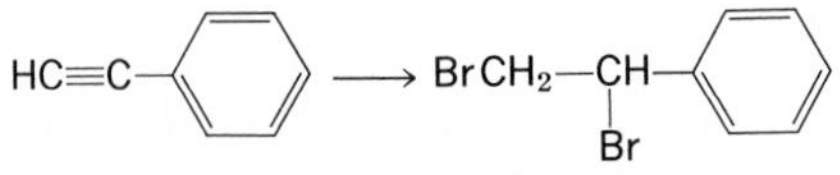

e. 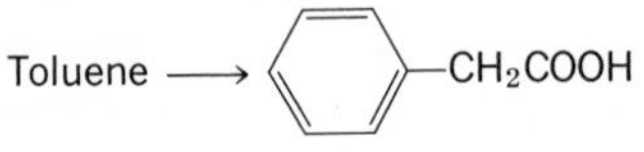

f. 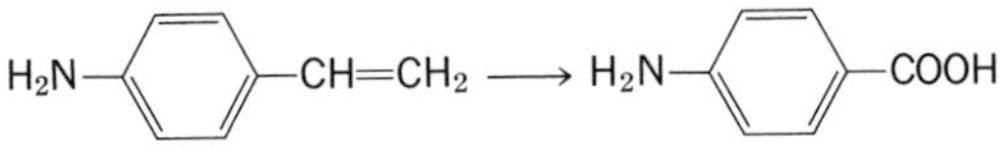

g. 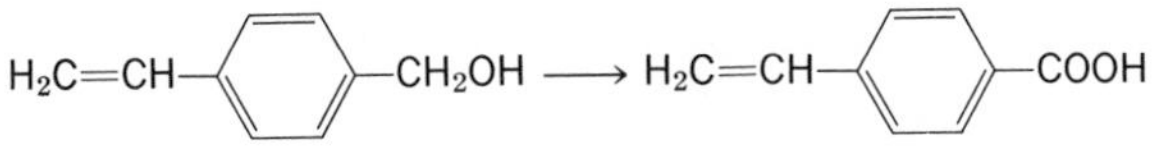

h. $CH_2OH-CH_2CH_2\overset{O}{\overset{\|}{C}}H \longrightarrow HOOC-CH_2CH_2\overset{O}{\overset{\|}{C}}H$

i. 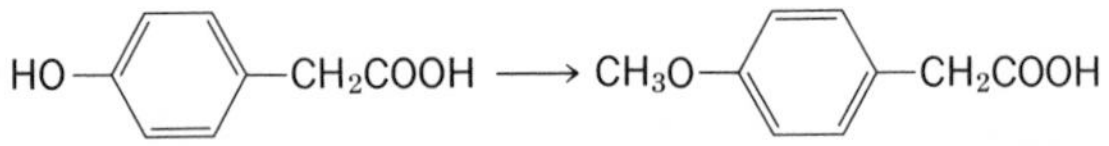

★**3.** Account for the fact that the reaction product isolated from

$$C_6H_5CH{=}CH\overset{O}{\overset{\|}{C}}-C_6H_5 + C_6H_5MgBr$$

contains some ketonic material (ir spectra) even when excess of Grignard is present. Predict the structure of the ketone.

4. Predict products and indicate mechanisms for the following reactions. Show expected stereochemistry where pertinent.

a. $C_6H_5CH{=}CH{-}COOC_2H_5 + NaCH(COOC_2H_5)_2$

b. 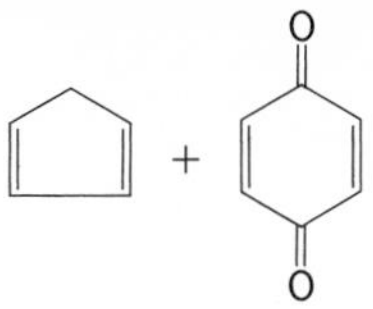

c. $H_2C{=}CH{-}CN + CH_3NH_2$

d. 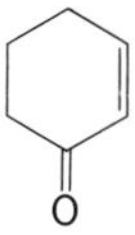 + butadiene

e. $C_6H_5{-}CH{=}CH{-}COOH + HBr$

f. $CH_3COOC{\equiv}CCOOCH_3 + C_2H_5CH{=}CH{-}CH{=}CH{-}CH_3$

★5. Assuming a neighboring-group effect, show products and their stereochemistry in the following reactions.

★ *a.* $CH_3C{\equiv}C{-}CH_3 \xrightarrow[\text{}]{H_2 \text{ (1 mole)}} \xrightarrow[\text{(trans addn.)}]{HOBr} \xrightarrow{HBr}$

b. *R,R*- $CH_3{-}\underset{Br}{CH}{-}\underset{Br}{CH}{-}CH_3 + H_2O \longrightarrow C_4H_9OBr$

★ *c.* 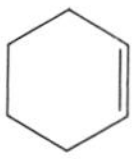 $+ Br_2 \longrightarrow \xrightarrow{\text{dil. base}} C_6H_{11}OBr$

d. *S*- $CH_3\underset{Cl}{CH}{-}COOH + NH_3$

★ *e.* 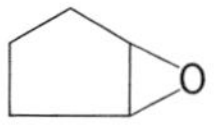 $\xrightarrow{HBr \text{ (1 mole)}} \xrightarrow{H_2O} C_5H_{10}O_2$

f. $(ClCH_2CH_2)_2NH + H_2O$

★ *g.* *R,R*- $CH_3\overset{NH_2}{CH}{-}\overset{OH}{C}HCH_3 + HNO_2 \longrightarrow C_4H_{10}O_2$

h. *cis* 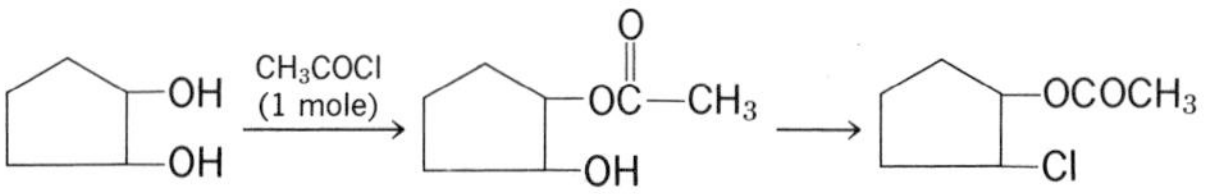

6. Account for the fact that the reaction of

R,S- $CH_3\overset{Br}{CH}{-}\overset{OH}{C}HCH_3 + HBr$

proceeds by S_N1 kinetics, yet yields only one stereoisomer (which?) of 2,3-dibromobutane.

21

CARBOHYDRATES: THE SUGARS

21.1 Definition and Nomenclature The carbohydrates, a class of substances which includes cane sugar, cotton, cellulose, rayon, starch, blood sugar, and derivatives of glycerol, occupy an important place in organic chemistry. They are of major importance industrially, since the production and distribution of some of them has influenced world diet, economy, and politics, and they are of great theoretical interest because they illustrate that small changes in a general molecular pattern can produce an extraordinary variety of effects.

In the early history of carbohydrate chemistry, all the known sugars gave analytical results indicative of the same empirical formula, $(CH_2O)n$, hence the name *carbohydrates*, or *hydrates of carbon*. Since this time the definition of carbohydrates has been extended to include compounds of different empirical formulas. The naturally occurring polyhydroxy aldehydes and ketones and their derivatives and isomers are now all regarded as carbohydrates.

The carbohydrates are usually divided into three categories:

1. *Monosaccharides.* Carbohydrates that cannot be further degraded by hydrolysis; the ultimate unit. Example: glucose, $C_6H_{12}O_6$.
2. *Disaccharides, trisaccharides, etc.* Carbohydrates that are decomposed by hydrolysis into two, three, etc., molecules of monosaccharides. Example: sucrose, $C_{12}H_{22}O_{11}$.
3. *Polysaccharides.* Carbohydrates of large molecular weight that are decomposed by hydrolysis into di- and trisaccharides and eventually into monosaccharides. Example: cellulose, $(C_6H_{10}O_5)n$.

In this chapter we shall confine ourselves to a study of mono- and disaccharides, deferring until Chap. 22 the consideration of the high-molecular-weight polysaccharides.

21.2 Glucose Analysis and molecular-weight determinations of glucose, the most widely occurring and well-known monosaccharide, correspond to the formula $C_6H_{12}O_6$. Glucose gives a positive test with Tollens' reagent and reacts with acetic anhydride to form a pentaacetate. These reactions are those of a pentahydroxy aldehyde. Kiliani by a simple process proved glucose to be a straight-chain compound with a terminal aldehyde group.

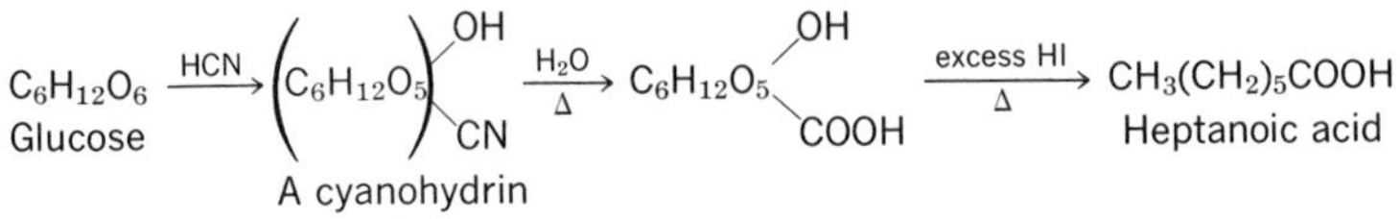

C6H12O6
Glucose
HCN
C6H12O5
OH
CN
A cyanohydrin
H2O
Δ
C6H12O5
OH
COOH
excess HI
Δ
CH3(CH2)5COOH
Heptanoic acid

The reaction with hydriodic acid has the effect of converting all hydroxyl groups to hydrogen by way of the iodides (see Sec. 8.25). Because of the well-known instability of compounds having two hydroxyl groups on a single carbon atom, the most reasonable structure to be deduced for glucose from this evidence is

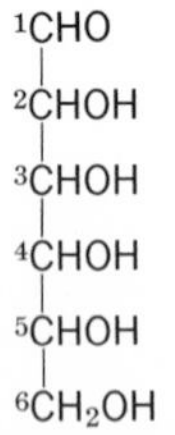

1CHO
2CHOH
3CHOH
4CHOH
5CHOH
6CH2OH

Glucose
(an aldohexose)

Carbohydrates are given the suffix *-ose.* Sugars such as glucose, with an aldehyde group, constitute the class of *aldoses,* and those with six carbon atoms are *aldohexoses.* If instead of an aldehyde group a keto group is present in the sugar, then it is called a *ketose.* Fructose, a sugar found in many fruits and in honey, is a typical ketose. Most natural ketoses have the keto group on carbon number 2.

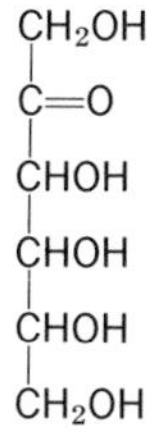

Fructose
(a ketohexose)

When treated by Kiliani's method, ketoses yield branched-chain acids.

While the structure of glucose is for the most part well represented by the formula above, it will be noted that this structure contains four asymmetric carbon atoms. According to stereochemical theory, there should be 2^4 or 16 stereoisomeric aldohexoses of which only one is natural, dextrorotatory glucose. Corresponding to this is the enantiomeric levorotatory (unnatural) glucose. All of the possible diastereomers have now been prepared and characterized; each enantiomeric pair has a special name (always ending in -ose): glucose, mannose, galactose, etc. These aldohexoses all have the same *structure* but differ in *configuration,* that is, in the arrangement of the groups in space. The establishment of these configurations by Emil Fischer is one of the masterpieces of classical organic chemistry.

Before going into the matter of configurations, it is convenient to discuss certain typical reactions and properties of the carbohydrates in general.

21.3 Reactions of Monosaccharides Not all carbohydrates have the sweet taste of familiar sugars, but all are exceedingly soluble in water and insoluble in ether and most other organic solvents; they are solids with relatively low vapor pressures and cannot be distilled. This apparent exception to the general solubility behavior of covalent compounds is evidently due to extensive hydrogen-bridge formation between the —OH groups of the sugars and the water. The great water solubility, while useful at times, is more often a disadvantage in laboratory manipulations. Most carbohydrates do not crystallize readily on evaporation of their solutions, tending to form syrups instead. The purification of many sugars by crystallization is made still more difficult by the close similarity in properties of the diastereomer mixtures often encountered. In fact, little progress could be made in this field until the discovery by Emil Fischer of the action of phenylhydrazine on carbohydrates. The overall reaction of this reagent with an aldohexose is given.

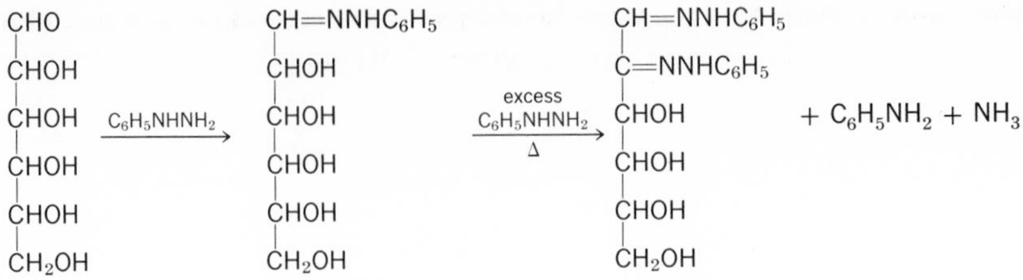

A phenylhydrazone An osazone

In the presence of excess phenylhydrazine, carbon 2 of an aldose (or carbon 1 of a 2-ketose) is apparently oxidized to a carbonyl group, which in turn reacts with the reagent. Osazones are easily recrystallized, water-insoluble derivatives which have been of great value in characterizing carbohydrates. The intermediate phenylhydrazones can, as a rule, be isolated but are of less general importance than the osazones. The mechanism of the reaction is complex and will not be elaborated here. There seems to be no tendency for carbons other than 1 and 2 to react.

Hydrolysis of an osazone does not regenerate the original material but yields an osone.

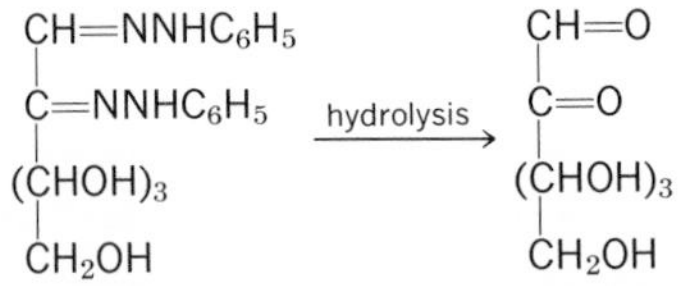

An osazone An osone

It is important to note that the osazone (and the osone) of an aldose has one less asymmetric carbon atom than the sugar from which it is derived; Fischer made use of this fact in establishing the configuration of glucose.

Ketoses react in an analogous way with phenylhydrazine. Fructose, for example, yields the same osazone as glucose, a fact which gives information about the relative configurations of these two sugars.

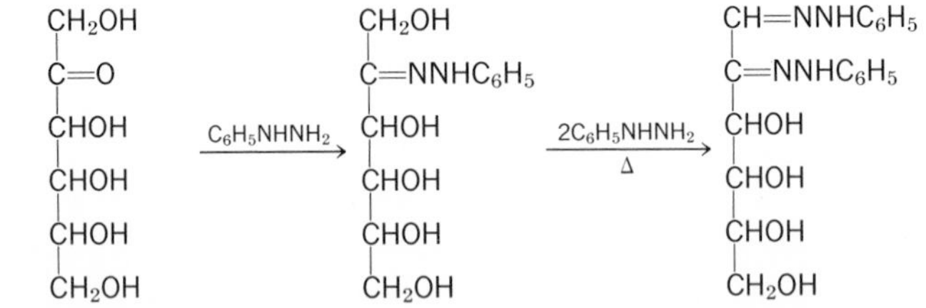

Fructose (a ketohexose) Glucosazone (an osazone)

21.4 Extension of Chain In addition to the reaction with phenylhydrazine, Fischer used two other general reactions in working out the configuration of glucose: the extension of the aldose chain and oxidation of aldoses to dibasic acids. By the one method it is possible to proceed, for example, from an aldopentose to an aldohexose.

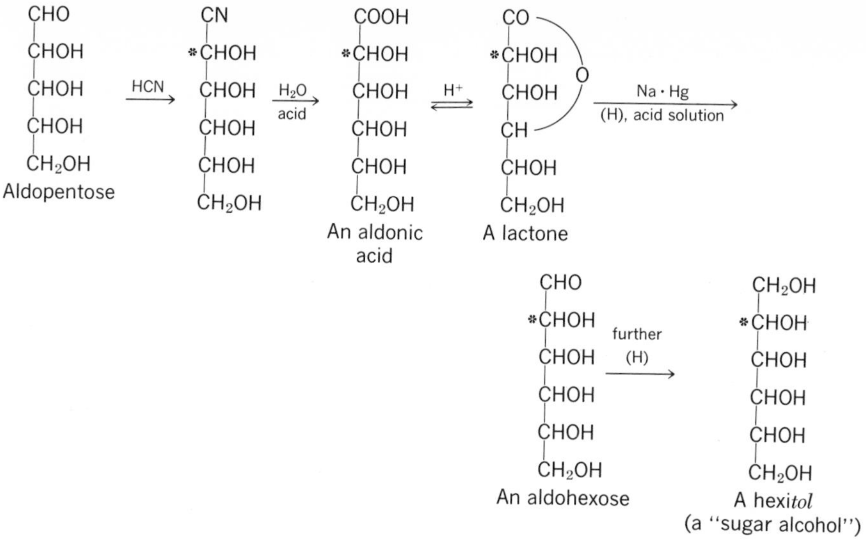

CHO
CHOH
CHOH
CHOH
CH₂OH
Aldopentose
HCN
CN
*CHOH
CHOH
CHOH
CHOH
CH₂OH
H₂O
acid
COOH
*CHOH
CHOH
CHOH
CHOH
CH₂OH
An aldonic acid
H⁺
CO
*CHOH
CHOH
CH
CHOH
CH₂OH
O
A lactone
Na · Hg
(H), acid solution
CHO
*CHOH
CHOH
CHOH
CHOH
CH₂OH
An aldohexose
further
(H)
CH₂OH
*CHOH
CHOH
CHOH
CHOH
CH₂OH
A hexi*tol*
(a "sugar alcohol")

Note that there is nothing unusual about any of these steps. The first two steps are general reactions of aldehydes and cyanohydrins; the third step is characteristic of γ-hydroxy acids; the reduction with sodium amalgam (although not described by us before) is a general reaction of lactones. There are, however, two important points in this synthesis. In the addition of HCN, a new asymmetric carbon atom (the one with an asterisk) is produced; since all natural aldoses are optically active, the introduction of a new asymmetric carbon atom produces *two diastereomers.* Diastereomers in such cases do not form in exactly equal amounts, and, as always, they have different physical properties. In practice, two aldohexoses (and hexonic acids) are obtained from a pure aldopentose by this process.

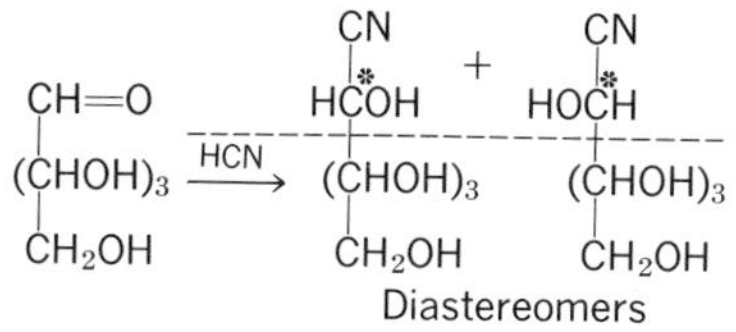

CH=O
(CHOH)₃
CH₂OH
HCN
CN
HCOH
(CHOH)₃
CH₂OH
+
CN
HOCH
(CHOH)₃
CH₂OH
Diastereomers

Note that the configuration below the dashed line is not altered by the reaction. In the sodium amalgam (Na · Hg) step, it is necessary to form the lactone before carrying out the reduction; aldonic acids, like other carboxylic acids, resist the action of most reducing agents, although their esters (here a lactone or inner ester) can be readily reduced. Lactone formation is usually effected by treating the acid in solution with a strong acid such as HCl and recrystallizing the resulting lactone. The reduction process must be carefully controlled to minimize reduction of the aldehyde group. The reducing power of the aldoses toward Fehling solution and the changing optical rotation are useful guides here in following the course of the reduction.

21.5 Shortening of Chain   The converse of this process, the degradation of aldoses, is readily accomplished by a modification of an α-hydroxyacid reaction.

$$RCHCOOH \xrightarrow[\Delta]{H_2SO_4} RCHO$$
$$| $$
$$OH$$

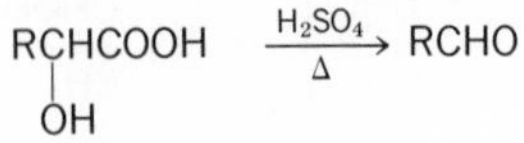

A typical
α-hydroxy acid

This reaction, as it stands, is not used with sugar acids, because hot dilute sulfuric acid causes extensive decomposition of the very sensitive carbohydrate molecules. Instead, a very mild reaction using hydrogen peroxide is employed, the aldose being first converted to the corresponding acid with a mild oxidizing agent.   The overall process is known as the *Ruff degradation*.

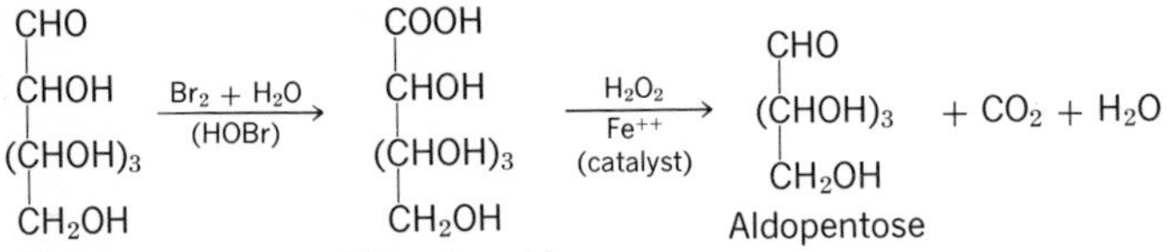

$$
\begin{array}{ccccc}
CHO & & COOH & & CHO \\
| & & | & & | \\
CHOH & \xrightarrow[(HOBr)]{Br_2 + H_2O} & CHOH & \xrightarrow[(catalyst)]{H_2O_2 \atop Fe^{++}} & (CHOH)_3 \quad + CO_2 + H_2O \\
| & & | & & | \\
(CHOH)_3 & & (CHOH)_3 & & CH_2OH \\
| & & | & & \\
CH_2OH & & CH_2OH & & \text{Aldopentose} \\
\text{Glucose} & & \text{Gluconic acid} & &
\end{array}
$$

The *-onic* suffix is used for acids formed by the mild oxidation of aldoses.   Ketoses do not react under these conditions.
  More vigorous oxidizing agents yield a different result.

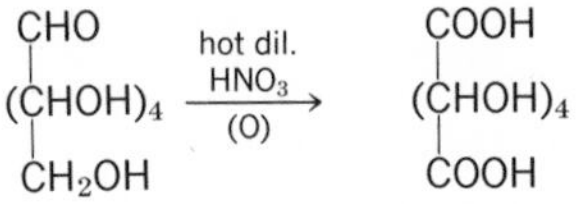

$$
\begin{array}{ccc}
CHO & \text{hot dil.} & COOH \\
| & \xrightarrow[(O)]{HNO_3} & | \\
(CHOH)_4 & & (CHOH)_4 \\
| & & | \\
CH_2OH & & COOH \\
& & \text{A saccharic acid}
\end{array}
$$

Here the greater rate of oxidation of primary alcohols as compared to secondary is emphasized.   The products are called *saccharic acids* and because of their potential symmetry are useful in correlating configurations in this series.

21.6 Proof of Configuration of Carbohydrates   The methods and reasoning used in establishing the configurations of the carbohydrates are most easily understood by starting with the simplest sugar[1] containing an asymmetric carbon atom.

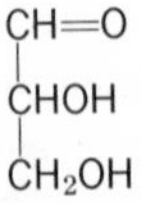

$$
\begin{array}{c}
CH{=}O \\
| \\
CHOH \\
| \\
CH_2OH \\
\text{Glyceraldehyde}
\end{array}
$$

Only two forms of glyceraldehyde are possible: an R and an S form (see Sec.

---

[1] Fischer's proof was accomplished in a slightly different manner using only those sugars known to him at the time; the reactions and type of reasoning, however, are the same as those used here.

7.8). The dextrorotatory (+) form of glyceraldehyde (see Sec. 7.4) has been found to have the *R* configuration[1] as shown below, where the "bow ties" represent groups coming above the plane of the page and the dashed lines represent those projected below the page. It is necessary to specify both the configuration (*R* or *S*) and the observed rotation (+ or −) for each species.

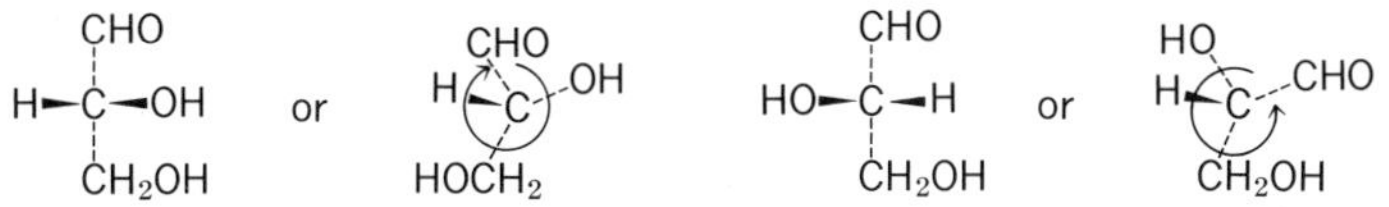

R(+)-Glyceraldehyde   *S*(−)-Glyceraldehyde

Observed rotation depends upon the configuration and the nature of the molecule, as well as on the solvent, temperature, and wavelength of light. Opposite configurations of a molecule will always give equal and opposite rotations under the same conditions, but the direction of the observed rotation in the first place will depend on the groups present about the asymmetric carbon. An example may clarify this:

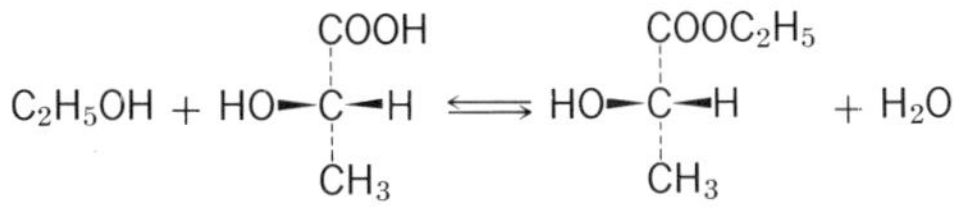

S(+)-Lactic acid   *S*(−)-Ethyl lactate

The conversion of an acid to an ester, a reaction which cannot change the *configuration* of the molecule, has changed its observed rotation. The *R* and *S* convention is a means of keeping track of the *structural configurational* relationships of molecules, a relationship that is more fundamental than observed rotation.

21.7 Fischer projection Since it is tedious to draw out the three-dimensional (bow-tie) formulas for all the asymmetric carbon atoms of the various carbohydrate molecules, it is customary to use a formalized projection, called the "Fischer projection." In this projection it is always assumed that the vertical bonds are *below* the plane of the page and the horizontal bonds are above the plane. Thus, for *R*(+)-glyceraldehyde

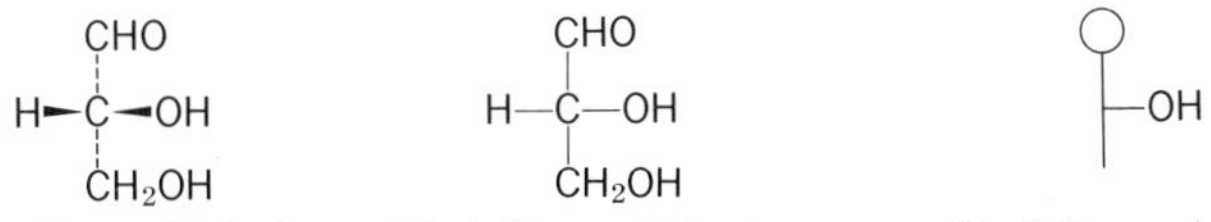

R(+)-Glyceraldehyde (three-dimensional representation)   *R*(+)-Glyceraldehyde (Fischer projection)   *R*(+)-Glyceraldehyde (formalized Fischer projection)

[1]In the older literature *R*(+)-glyceraldehyde was called D(+)-glyceraldehyde, and *S*(−)-glyceraldehyde was called L(−)-glyceraldehyde. We shall use the newer *R* and *S* for all cases of individual atom configuration, reserving the terms D and L for certain molecular configurations as outlined in Sec. 21.8.

The commonly used Fischer projection in the center represents the three-dimensional model shown on the left. The formalized projection on the right is sometimes used for even greater simplicity. Once the projection is made it must be used with certain restrictions. It can be rotated within the plane of the paper, but it cannot be flipped over.

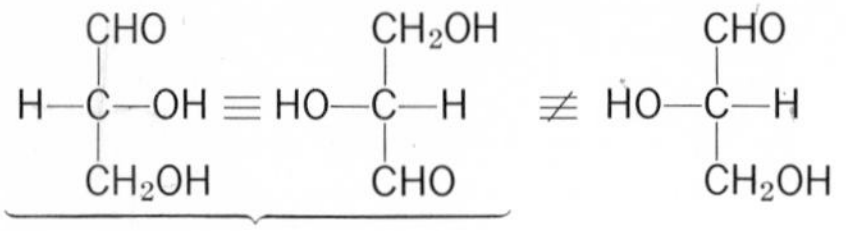

Equivalent — *S*(−)-Glyceraldehyde

The reason for this rule will become apparent as more examples are encountered; that is, only certain orientations are acceptable for projection and once the projection is made it is necessary to take account of these rules. Thus, the projection of *S*(−)-glyceraldehyde can be made.

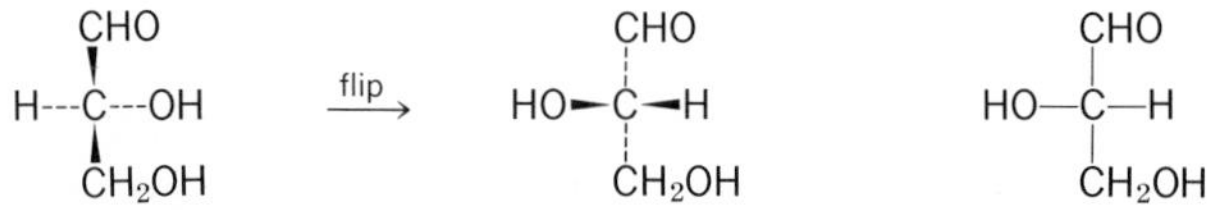
flip

S(−)-Glyceraldehyde (do not project in this orientation) — *S*(−)-Glyceraldehyde (in proper orientation for projection) — *S*(−)-Glyceraldehyde (Fischer projection)

Flipping the projection over would not give *S*(−)-glyceraldehyde because the rules require vertical bonds to be down, and horizontal to be up.

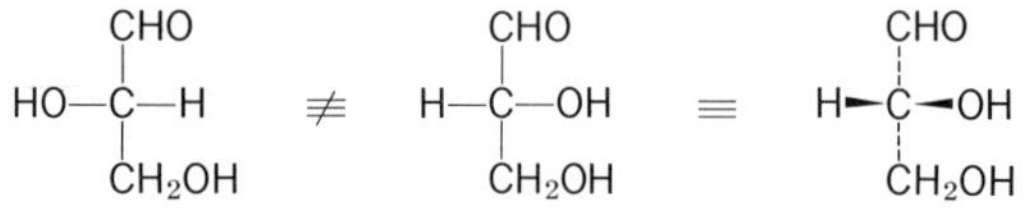

S(−)-Glyceraldehyde — "Flipped-over projection" — *R*(+)-Glyceraldehyde

Thus, flipping the projection over would amount to an inversion of configuration.

The same rules apply for multiple optically active centers

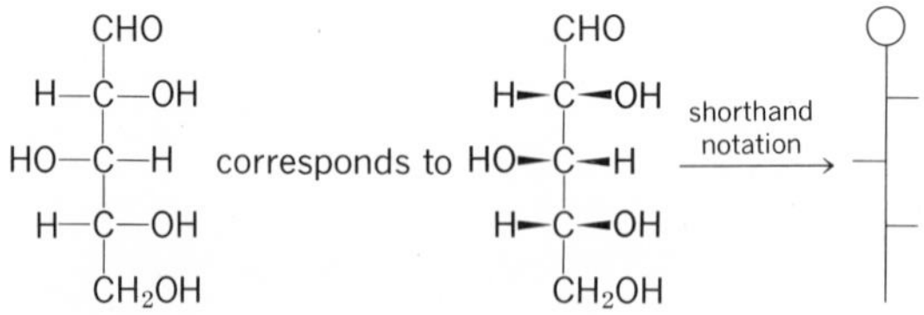
corresponds to
shorthand notation

21.8 Relationship of the Configurations of Other Sugars to *R*(+)-Glyceraldehyde When *R*(+)-glyceraldehyde is converted to the next higher aldose by the HCN method, two diastereomers, D(−)-erythrose and D(−)-threose are obtained.

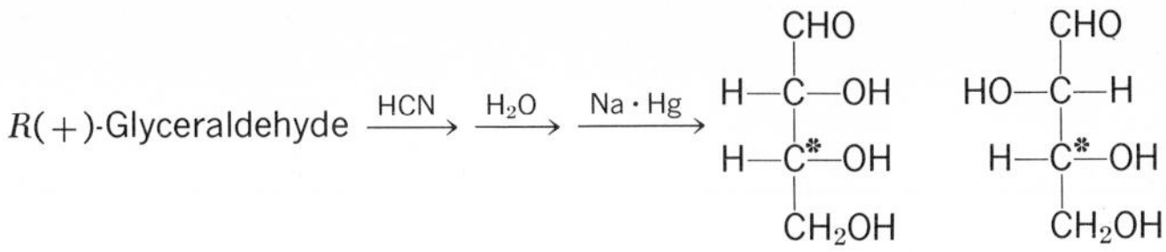
R(+)-Glyceraldehyde
HCN
H₂O
Na · Hg

The configurations are represented here in the Fischer projection. The starred carbon is the original *R*(+)-glyceraldehyde asymmetric carbon atom. Both D-erythrose and D-threose must have the *R*-configuration of this carbon atom; they can differ only in the configuration of the other asymmetric carbon atom. We shall use the term "D" to indicate *that a given carbohydrate has the R configuration of the carbon atom adjacent to the* CH_2OH *group,* and the term "L" to indicate that the corresponding carbon atom has the *S* configuration.[1] The configurations of the other carbon atoms are not specified by this "D" and "L" terminology. (They are, however, determined by the prefix D or L and the name of the carbohydrate in question as we shall see below.[2])

The assignment of the correct structures to D-erythrose and D-threose to the actual substances can be done as follows: When oxidized with hot dilute nitric acid, D-threose yields an optically active tartaric acid, while D-erythrose yields optically inactive *meso*-tartaric acid. This establishes the following configurations:

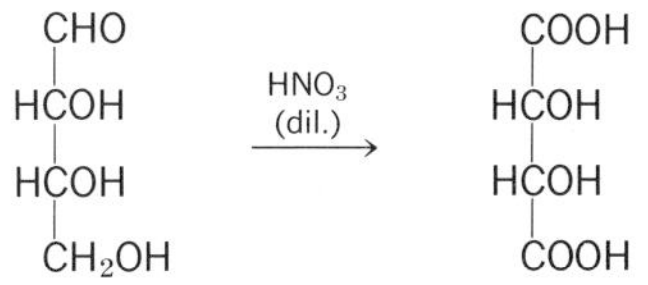

D(−)-Erythrose *meso*-Tartaric acid (inactive)

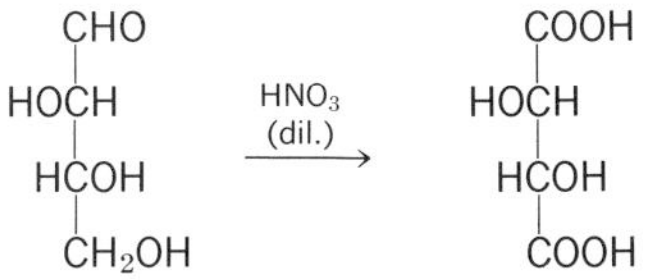

D(−)-Threose D(−)-Tartaric acid

It is evident that this conclusion could have been reached equally well by reduction of erythrose and threose to the corresponding alcohols (tetritols); here again erythrose yields an optically inactive, nonresolvable (meso) tetritol and threose an active one of the D series.

[1]This usage of "D" and "L" differs from the older usage where these terms were used to specify the relative configurations of all optically active molecules. In deference to the large body of literature in the carbohydrate field which uses these terms we shall retain them for this purpose. In other instances we will use the more precise *R* and *S* terminology, referring to the older terminology only where necessary for clarity.

[2]It is important to note that the terms *R* and *S* refer to the configurations of each individual asymmetric carbon atom. The terms D and L have been used both for individual asymmetric atoms and for entire overall molecular configurations. We shall restrict the use of D and L to molecular configurations of molecules having two or more asymmetric carbon atoms.

This illustrates another instance of the power of symmetry considerations in deciphering the structures of molecules. By conversion of the two different groups, —CHO and —CH_2OH, into two identical —COOH groups, the plane of symmetry (Sec. 7.7) of the meso form indicates that the two carbon atoms must have opposite configurations (Secs. 7.7 and 7.8). Since the —CH_2OH group is increased in priority (Sec. 7.8) during the oxidation, the mesoform can only have come from the erythrose, in which both carbon atoms had the same configuration. As will become evident, almost all of the carbohydrate structures have been determined by the use of such symmetry considerations.

Four pentoses of the D series are theoretically possible, and four are known. Their names and some reactions are given below:

$$\text{D}(-)\text{-Erythrose} \xrightarrow{\text{HCN method}} \begin{matrix}\text{D}(-)\text{-arabinose} \\ \text{D}(-)\text{-ribose}\end{matrix} \xrightarrow{C_6H_5NHNH_2} \text{the same osazone}$$

$$\text{D}(-)\text{-Threose} \xrightarrow{\text{HCN method}} \begin{matrix}\text{D}(+)\text{-xylose} \\ \text{D}(-)\text{-lyxose}\end{matrix} \xrightarrow{C_6H_5NHNH_2} \text{the same osazone}$$

Arabinose and ribose are seen, by the above data, to be identical in all respects save for the configuration of carbon atom 2. Xylose and lyxose also bear this relationship to each other. Diastereomeric pairs of sugars bearing this relationship to one another are known as *epimers.*

Nitric acid oxidation of D-arabinose leads to an optically active dibasic acid, while D-ribose in the same reaction forms an inactive acid that cannot be resolved: a meso form. Note that this meso form has a plane of symmetry indicated by the dashed line.

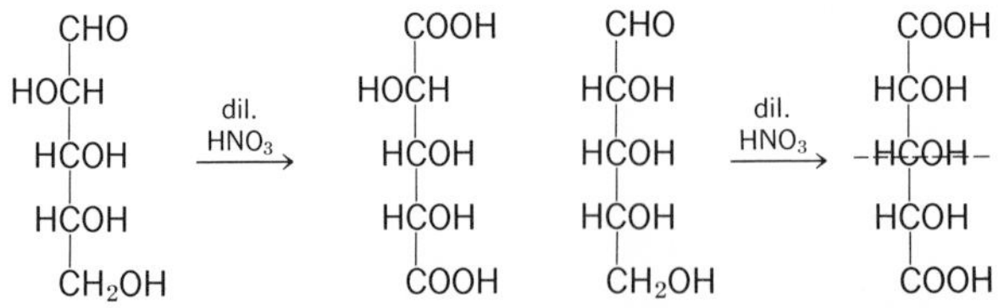

CHO
HOCH
HCOH
HCOH
CH_2OH
dil. HNO_3
COOH
HOCH
HCOH
HCOH
COOH
CHO
HCOH
HCOH
HCOH
CH_2OH
dil. HNO_3
COOH
HCOH
HCOH
HCOH
COOH

D-Arabinose | Optically active | D-Ribose | A meso form

By exactly analogous experiments and arguments, the configurations of xylose and lyxose were established and the work extended to include the aldohexoses. Table 21.1 gives the configurations of the D-series monosaccharides through the hexoses. In this table the vertical lines represent the carbon chains, with a carbon atom understood at each intersection. At the lower end is understood the —CH_2OH group; at the upper end the circle signifies an aldehyde group, and the horizontal lines indicate hydroxyl groups, the configurations of the asymmetric carbon atoms being written in the conventional manner.

Those stereoisomers commonly found in natural sources are underlined in the

TABLE 21.1
THE D SERIES OF ALDOSES

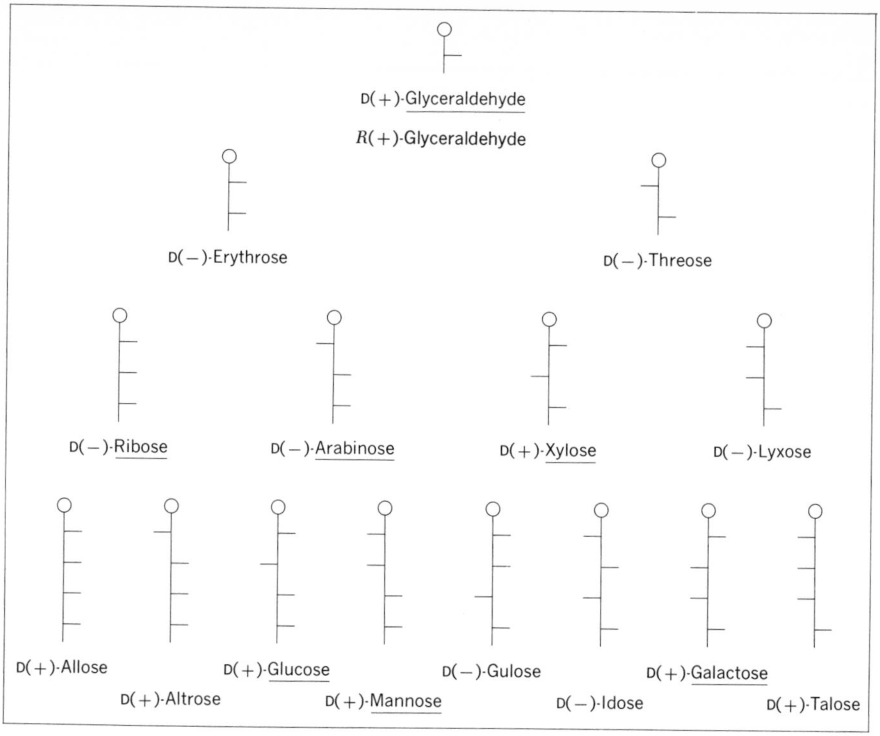
D(+)-Glyceraldehyde
R(+)-Glyceraldehyde
D(−)-Erythrose
D(−)-Threose
D(−)-Ribose
D(−)-Arabinose
D(+)-Xylose
D(−)-Lyxose
D(+)-Allose
D(+)-Altrose
D(+)-Glucose
D(+)-Mannose
D(−)-Gulose
D(−)-Idose
D(+)-Galactose
D(+)-Talose

table. There is a corresponding L series of aldoses represented by the mirror images of the structures in Table 21.1.[1]

Since ketoses react with phenylhydrazine to form osazones, they are easily related to the known aldoses by this means. For example, fructose and glucose yield the same osazone (see Sec. 21.3); hence, the configurations of these two sugars are the same at carbons 3, 4, and 5.

[1]Note that L(−)-glucose differs in the configuration of every carbon atom from D(+)-glucose. This can be seen from the formulas and full configurational names (Sec. 7.8).

mirror images

D(+)-Glucose
(2-*R*, 3-*S*, 4-*R*, 5-*R*, 6, -penta-hydroxyhexanal)

L(−)-Glucose
(2-*S*, 3-*R*, 4-*S*, 5-*S*, 6, -penta-hydroxyhexanal)

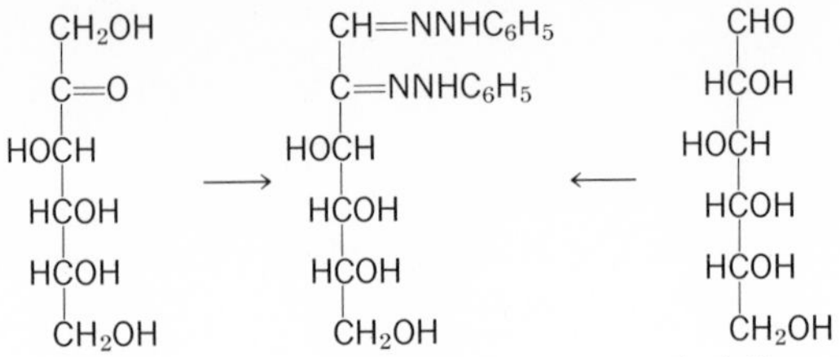

D(−)-Fructose — D-Glucosazone or D-Fructosazone (identical) — D(+)-Glucose

21.9 Hydroxyl-group Reactions

21.9 Hydroxyl-group Reactions The hydroxyl groups of the carbohydrates react for the most part as they do in other molecules. With acid derivatives esters are formed, and ethers can also be prepared.

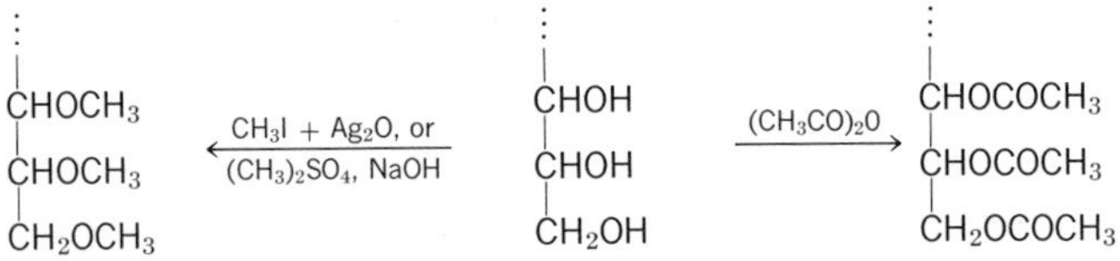
CH3I + Ag2O, or
(CH3)2SO4, NaOH
(CH3CO)2O

Methyl ether[1] — Part of a carbohydrate chain — Acetate

It is not generally possible to attack one of the alcoholic hydroxyls without also attacking all other available ones; the direct formation of partially acetylated or methylated sugars by ordinary methods is unsuccessful.

21.10 Reactions with Acid and Alkali All those sugars which are oxidized by Fehling solution are referred to as *reducing sugars*. They are very unstable toward alkaline reagents of all sorts. Deep-seated degradations and rearrangements take place with glucose in sodium hydroxide solution; the detailed mechanism of these reactions is unknown but parallels interestingly the course of the breakdown of glucose in nature.

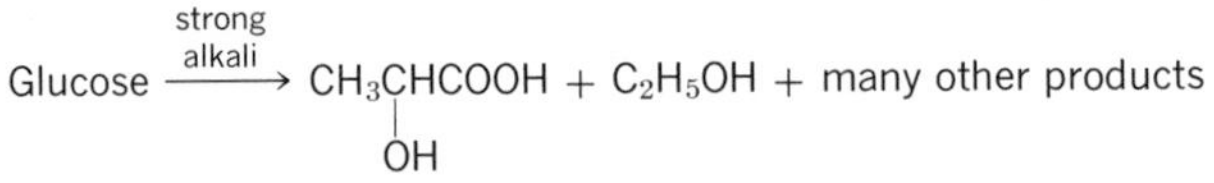
Glucose
strong
alkali
CH3CHCOOH + C2H5OH + many other products
OH

Hot concentrated acids also lead to the breakdown of glucose and its analogs, although they are much more stable in acid than in alkaline medium. Pentoses under these conditions yield furfural. Hexoses yield a furfural derivative and levulinic acid by complex dehydration reactions.

[1] Dimethyl sulfate is often used in place of the more expensive methyl iodide in the formation of methyl ethers. This is a modified Williamson synthesis—a displacement reaction.

$$RONa + (CH_3)_2SO_4 \longrightarrow ROCH_3 + NaCH_3SO_4$$

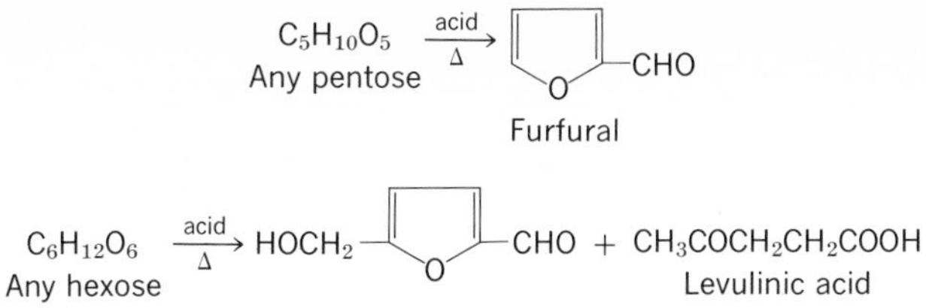
$C_5H_{10}O_5$
Any pentose
acid
Δ
CHO
O
Furfural
$C_6H_{12}O_6$
Any hexose
acid
Δ
$HOCH_2$
O
CHO + $CH_3COCH_2CH_2COOH$
Levulinic acid

The formation of furfural is the basis of some of the qualitative tests for carbohydrates; the furfural is made to condense with various aromatic compounds to produce characteristic colors.

Furfural is manufactured commercially from crude, naturally occurring, pentose-containing materials such as straw, cornstalks, and oat hulls by the reaction shown above.

Very dilute alkali converts glucose, through a reaction which appears to depend on an enolization of the carbonyl group, into an equilibrium mixture of glucose, mannose, and fructose (three of the most abundant naturally occurring hexoses).

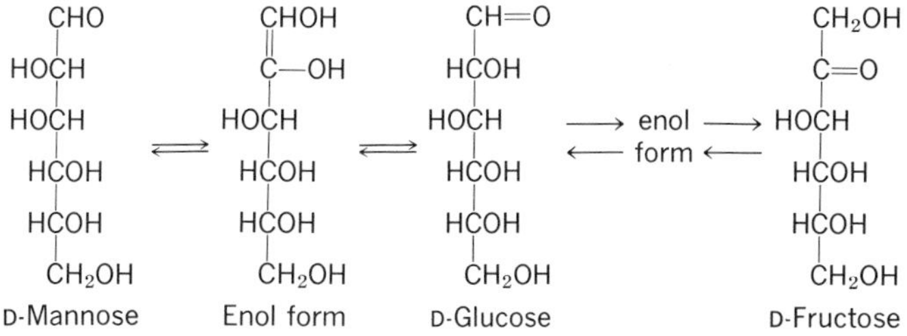
CHO
HOCH
HOCH
HCOH
HCOH
CH_2OH
D-Mannose
CHOH
C—OH
HOCH
HCOH
HCOH
CH_2OH
Enol form
CH=O
HCOH
HOCH
HCOH
HCOH
CH_2OH
D-Glucose
enol
form
CH_2OH
C=O
HOCH
HCOH
HCOH
CH_2OH
D-Fructose

The enol form has one less asymmetric carbon than glucose and, hence, leads to the formation of the diastereomeric mannose, or to the ketose (fructose) by tautomeric shift. The fact that fructose, a ketone, is able to reduce Fehling solution is due in part to the rapid establishment of this equilibrium and the further degradations due to the alkaline reaction medium. It should be pointed out that although Fehling solution is used in the analysis of natural products for reducing sugars, the reaction is not a simple stoichiometric aldehyde oxidation; different sugars yield different amounts of Cu_2O as a result of further degradations, and the analysis must be calibrated for each sugar.

21.11 Synthesis The laboratory synthesis of a carbohydrate such as glucose from simple materials presents an imposing problem, owing largely to the number of asymmetric carbon atoms involved. Since any ordinary synthesis without the use of optically active reagents gives rise to a mixture of all possible stereoisomers, a "glucose synthesis," from formaldehyde, for instance, yields a complex mixture. Merely treating formaldehyde with alkali produces (aside from polymerization) the following equilibria:

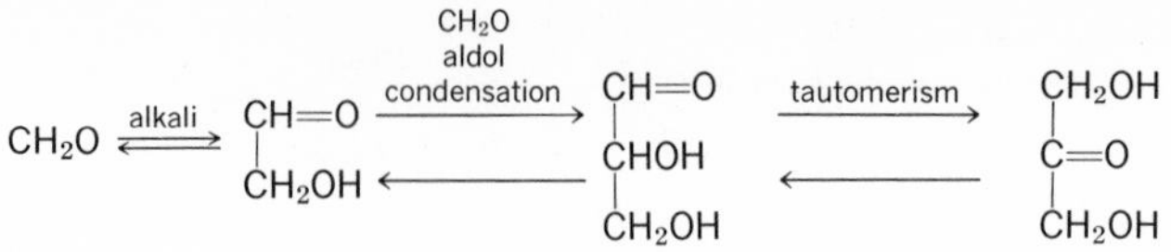
CH2O
aldol
condensation
alkali
tautomerism
CH2O
CH=O
CH2OH
CH=O
CHOH
CH2OH
CH2OH
C=O
CH2OH

Racemic Glyceraldehyde Dihydroxyacetone

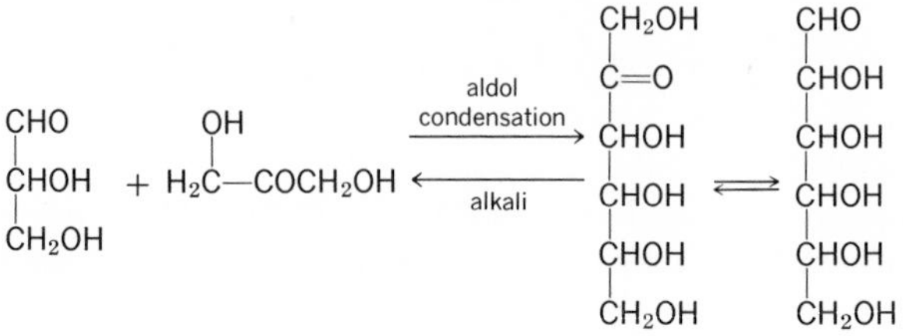
CHO
CHOH
CH2OH
OH
+ H2C—COCH2OH
aldol
condensation
alkali
CH2OH
C=O
CHOH
CHOH
CHOH
CH2OH
CHO
CHOH
CHOH
CHOH
CHOH
CH2OH

A mixture of hexoses

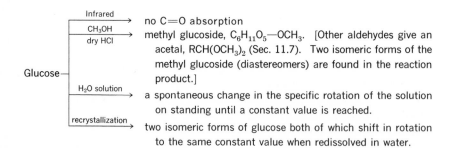

It may be seen in the above equations that the application of known aldehyde condensation reactions and ketose-aldose interconversions leads eventually to a hexose. Emil Fischer actually carried out such a synthesis and, without isolating all of the intermediates shown, was able to obtain minute amounts of D-glucose. The reaction sequence as shown, although of no practical laboratory value, is a close approximation to that followed in nature.

21.12 Acetal Formation There are a number of observations that do not fit in well with the structural formula developed in Sec. 21.2 for glucose. Some of these are given below.

Glucose —
- Infrared → no C=O absorption
- CH_3OH, dry HCl → methyl glucoside, $C_6H_{11}O_5—OCH_3$. [Other aldehydes give an acetal, $RCH(OCH_3)_2$ (Sec. 11.7). Two isomeric forms of the methyl glucoside (diastereomers) are found in the reaction product.]
- H_2O solution → a spontaneous change in the specific rotation of the solution on standing until a constant value is reached.
- recrystallization → two isomeric forms of glucose both of which shift in rotation to the same constant value when redissolved in water.

This information is interpreted in terms of a hemiacetal (Sec. 11.7) formula for glucose. Alcohols react with glucose to form stable substances with the properties of acetals. These are called "glucosides," and involve only one molecule of alcohol. It is evident that one of the hydroxyls of glucose itself is able to take part in this acetal formation.

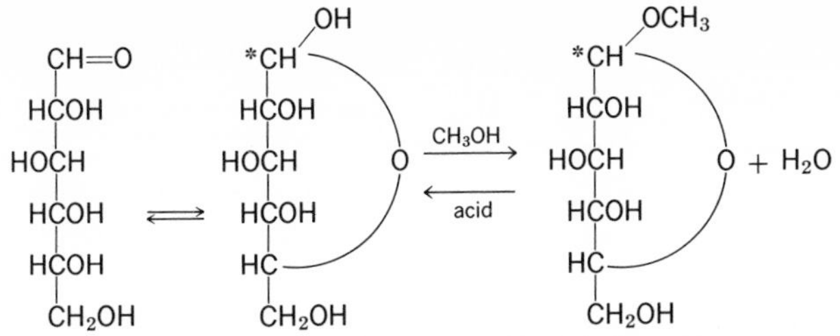

D(+)-glucose (aldehyde form) — A hemiacetal — Methyl glucoside (an acetal)

The powerful electrophilic attraction of the carbonyl group for hydroxy compounds has already been noted. In glucose and other hexoses, the hydroxyl groups on carbons 4 and 5 can approach close to the carbonyl group (strain theory), and this facilitates reaction.

Long bonds such as those shown above are merely artifacts of the Fischer projection (Sec. 21.7). A more accurate representation of the above reaction is given by changing from the Fischer projection back to the three-dimensional representation.

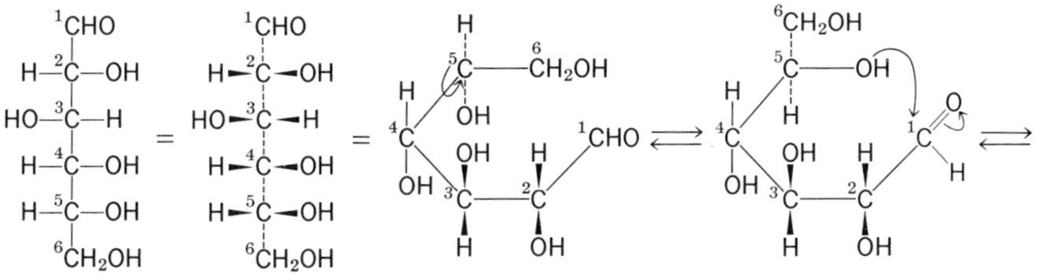

I*a* Fischer projection — I*b* Three-dimensional form — I*c* — I*d*

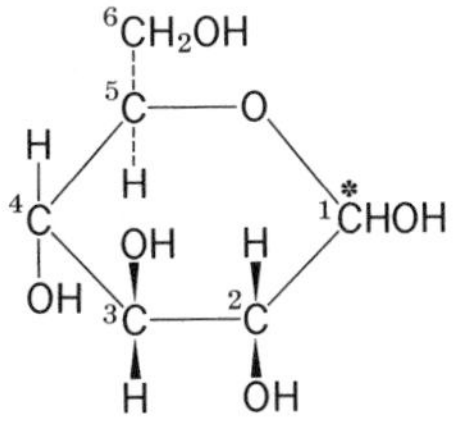

I*e*

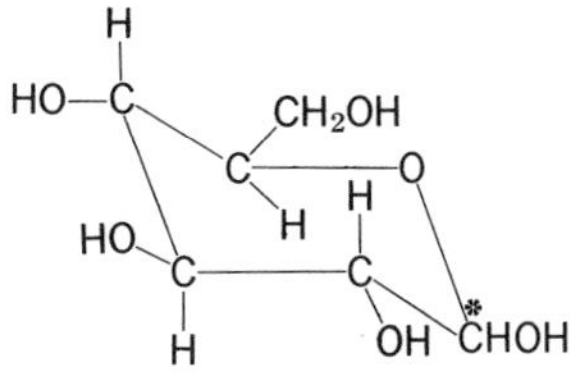

I*f*

The Fischer projection, I*a*, is a projection of the three-dimensional formula, I*b*, in which the bow ties represent bonds coming up, and the dotted lines represent bonds extending below the carbon atom to which they are attached. This means that the molecule I*b* is really curved around as shown (rotated by 90°) in I*c*. Now, in order to form a six-membered ring (as shown, closure between atoms 1 and 6 would yield an unfavorable seven-membered ring), the rotation about the bond between atoms 4 and 5 (indicated by the curved arrow) is performed, yielding I*d*. Ring closure to yield the hemiacetal, I*e*, gives the six-membered cyclic structure in which all bonds are of normal length. This can be drawn in the puckered conformational representation, I*f*, but for comparison of cis-trans isomers, etc., the flat structures are often more useful.

In the above hemiacetal formula it is seen that the carbon atom 1, the previously "aldehydic" carbon (starred), becomes an asymmetric carbon and, therefore, can exist in two configurations. This accounts for the existence of two forms of glucose and two methyl glucosides; these are called "alpha" and "beta" forms.

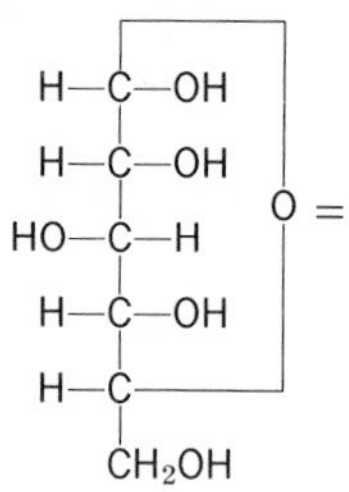
H—C—OH
H—C—OH
HO—C—H
H—C—OH
H—C
CH2OH
O =
alpha-D-Glucose

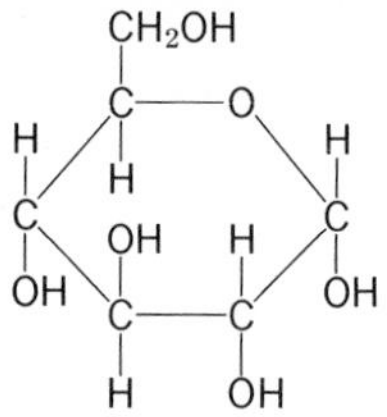
CH2OH
C—O
H
H
C
OH
OH
H
C—C
H
OH
H
C
OH
alpha-D-Glucose

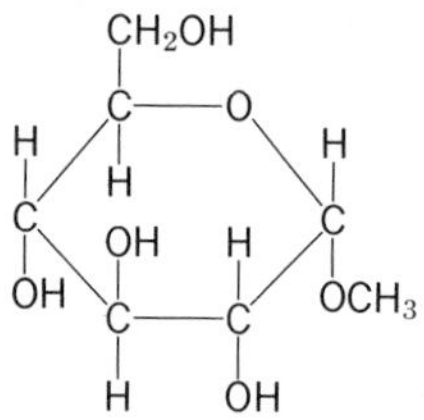
CH2OH
C—O
H
H
C
OH
OH
H
C—C
H
OH
H
C
OCH3
alpha-methyl-D-glucoside

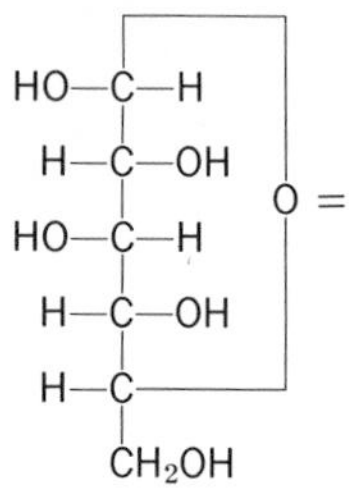
HO—C—H
H—C—OH
HO—C—H
H—C—OH
H—C
CH2OH
O =
beta-D-Glucose

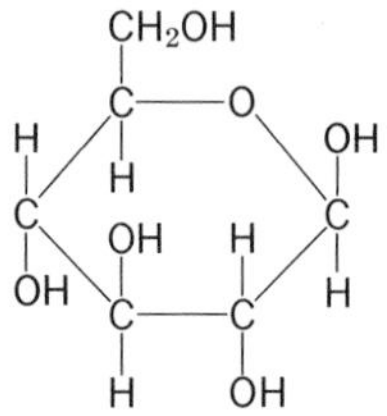
CH2OH
C—O
H
H
C
OH
OH
H
C—C
H
OH
OH
C
H
beta-D-Glucose

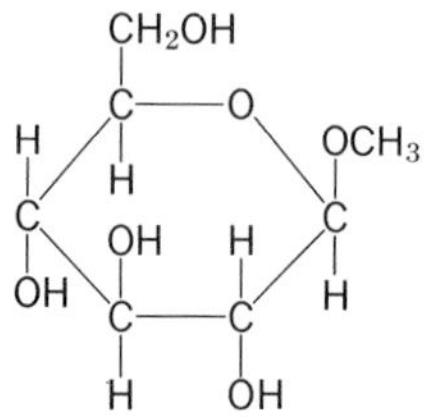
CH2OH
C—O
H
H
C
OH
OH
H
C—C
H
OH
OCH3
C
H
beta-methyl-D-glucoside

21.13 Mutarotation The α and β forms of glycosides[1] are distinguished by the nature of the enzymes which will catalyze their hydrolysis. Most naturally occurring glycosides are of the β configuration and are hydrolyzed specifically by the enzyme *emulsin*. α-Glycosides are specifically cleaved by *maltase*, an enzyme from malt. The proof of configuration of the reducing carbon atom in free sugars and glycosides is beyond the scope of this work; the formulas as written are the presently accepted ones. In the crystalline state, α- and β-glucose are stable individuals with characteristic melting points. When either form is dissolved in water, the following equilibria are set up slowly:

[1] The term *glycoside* is a general one referring to the acetal formed between any carbohydrate and another hydroxyl compound. Specific compounds are named by the sugars from which they are derived: glucoside from glucose, fructoside from fructose, etc.

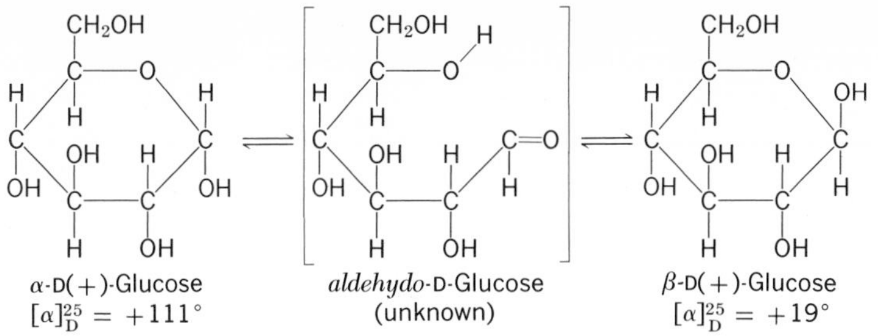
α-D(+)-Glucose
$[\alpha]_D^{25} = +111°$
aldehydo-D-Glucose
(unknown)
β-D(+)-Glucose
$[\alpha]_D^{25} = +19°$

Establishment of this equilibrium mixture, which has a specific rotation of +53°, accounts for the observed spontaneous shift in rotation of freshly prepared (α- or β-) glucose solutions. This phenomenon, which is observed with all reducing sugars, is known as *mutarotation.* Methyl glucoside (and of course glycosides of other alcohols and other carbohydrates in general) does not undergo mutarotation; the full acetal structure, in contrast to the hemiacetal free sugar, is stable except to acid hydrolysis.

With the exception of a few specialized cases, carbohydrates appear to exist almost entirely in the hemiacetal or lact*ol* (not lactone) form. In the formulas above, glucose is represented as having a six-membered lactol ring. From the strain theory we would expect to encounter also five-membered rings, and this expectation is realized in certain glycosides. In general, the five-membered rings are less common and less stable than the larger ring in this series. The nomenclature of these lactol forms is derived from the heterocycles.

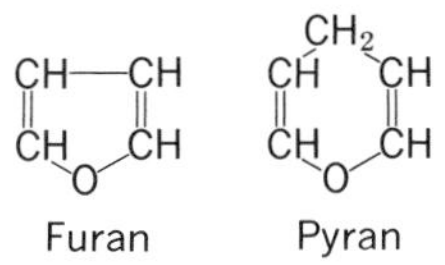
Furan
Pyran

In the lactol rings, of course, there are no double bonds.

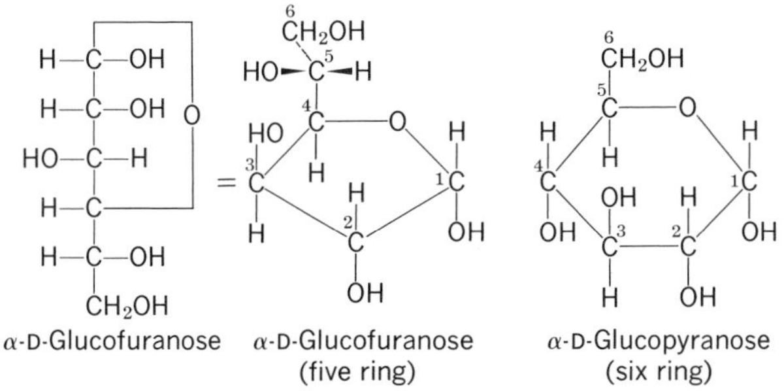
α-D-Glucofuranose
α-D-Glucofuranose
(five ring)
α-D-Glucopyranose
(six ring)

Not only the aldoses but also the ketoses form hemiacetals and exist predominantly in the lactol form. Thus the structure of fructose is represented:

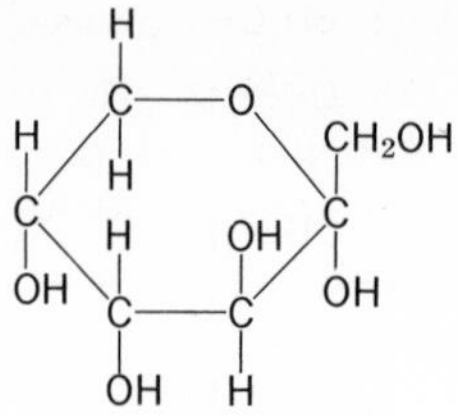

α-D-Fructopyranose
(a hemiketal)

Like the aldoses, ketoses readily form glycosides and undergo mutarotation; they can exist both in pyranose forms as shown above and as furanose derivatives. The best chemical method of distinguishing a ketose from an aldose is oxidation with bromine water; the ketose is unchanged by this reagent.

21.14 Size of Ring The establishment of ring size in simple and complex carbohydrates has been carried out by methylation. The equations illustrate the reactions used to prove the presence of the pyranose ring in glucose. The term

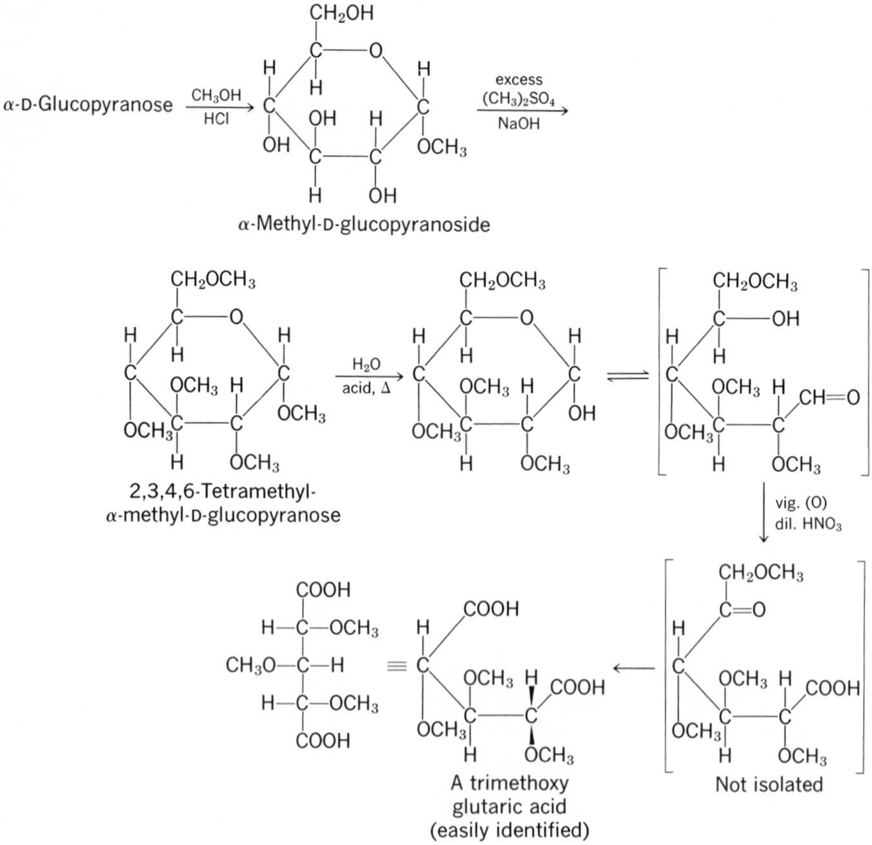
α-D-Glucopyranose
CH₃OH
HCl
excess
(CH₃)₂SO₄
NaOH
α-Methyl-D-glucopyranoside
H₂O
acid, Δ
2,3,4,6-Tetramethyl-
α-methyl-D-glucopyranose
vig. (O)
dil. HNO₃
Not isolated
A trimethoxy
glutaric acid
(easily identified)

methylation refers in the sugar series only to replacement of alcoholic hydrogen with methyl; the products are called "methylated sugar derivatives."[1]

Note that the first step is acid-catalyzed acetal (glucoside) formation. This glucoside is stable to base and is not changed by the methylation with methylsulfate in the second step. However, the third step involves hydrolysis with dilute acid and, while far too mild to cleave ordinary ether linkages, easily reopens this acetal. This is exactly analogous to the cases we have seen previously.

$$R—O—CH_3 \xrightarrow{\text{dil. } H^+} \text{N.R.}$$

$$\text{An ether} \xrightarrow[\text{reflux}]{\text{HI}} \text{cleavage to RI, ROH}$$

$$R—CH(OCH_3)_2 \xrightarrow[\Delta]{\text{dil. } H^+} RCHO + 2CH_3OH$$

An acetal

The final vigorous oxidation attacks the free (or available from the hemiacetal) aldehyde function and the unmethylated OH (which was previously involved in the hemiacetal). Cleavage can occur on either side of the carbonyl, but as long as some of the product of cleavage shown is isolated it is sufficient to prove the methylation pattern.

21.15 Periodic-acid Method A more direct method for the determination of ring size in carbohydrate derivatives is based on the discovery of the use of periodic acid as an oxidant (Sec. 8.26). It will be recalled that this reagent is specific for adjacent free OH groups. The application of this oxidation to pyranosides and furanosides is shown.

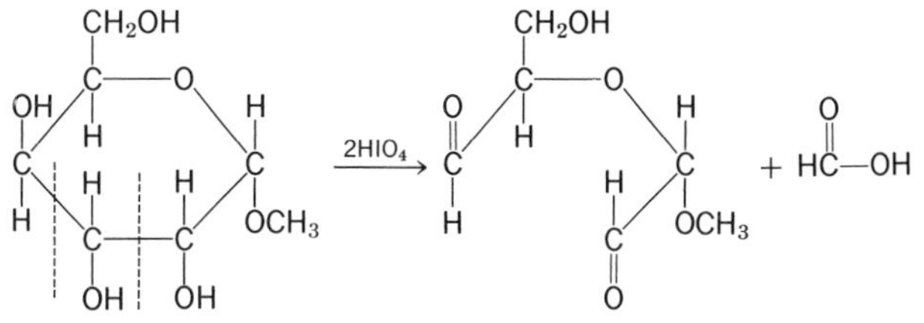

CH₂OH
OH
H
OCH₃
2HIO₄
O
+ HC—OH

A pyranoside

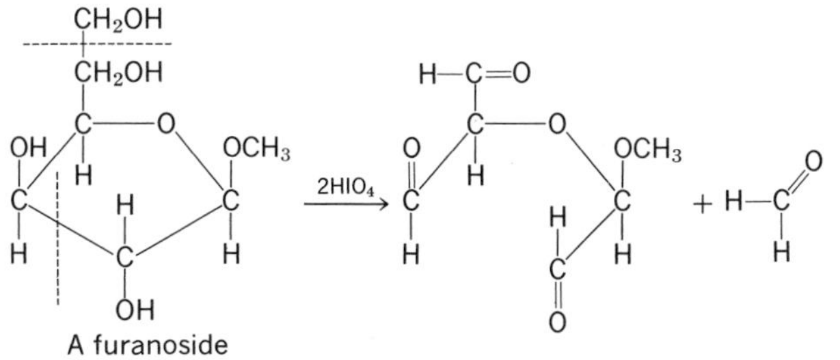

CH₂OH
OH
OCH₃
2HIO₄
H—C=O
+ H—C

A furanoside

[1]The notation O-methyl is used to distinguish a methyl on oxygen from a methyl on carbon (C-methyl).

It is seen that both pyranosides and furanosides consume 2 moles of periodic acid; the other products, HCOOH and CH_2O, are different in the two cases and allow a distinction to be made. This same reagent has proved of great value in other carbohydrate structural problems as well. For example, the rather difficult distinction between 3-O-methyl and 4-O-methyl pyranoside is simplified by the fact that only one will react with periodic acid.

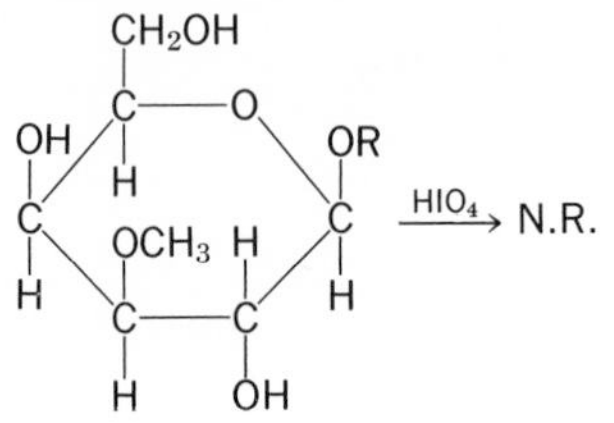

CH₂OH
OH
OR
H
C
OCH₃ H
HIO₄
N.R.
H
OH

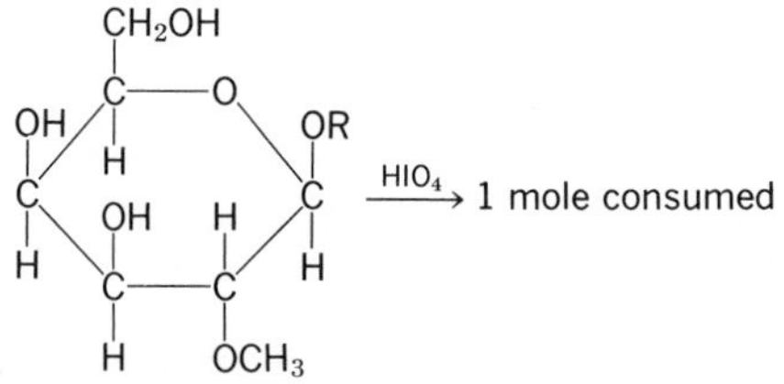

CH₂OH
OH
OR
H
OH
HIO₄
1 mole consumed
OCH₃

Periodic-acid oxidation also serves to correlate configurations about the glycosidic carbon atom. Inspection of the reactions reveals that all α-D-methylpyranosides (from glucose, mannose, etc.) give the same oxidation product since the product contains only the original asymmetric centers at C_1 and at C_5.

21.16 Disaccharides Disaccharides, and also higher saccharides, are composed of monosaccharides linked together by glycoside linkages. In these compounds, the alcohol portion of the glycoside consists of one of the available hydroxyl groups of another monosaccharide. The naturally occurring disaccharides vary in a number of respects: the type of linkage (α and β), the point of linkage to the "alcohol" sugar, the size of the lactol rings, and the constituent monosaccharides.

Lactose, a natural disaccharide found in milk, has the following structure:

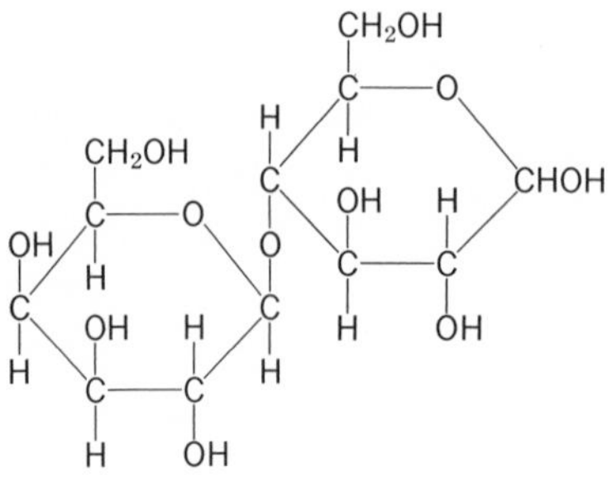

CH₂OH
CH₂OH
CHOH
OH
OH H
H
OH

The proof of the position of the linkage between the galactose and glucose portions in this molecule was obtained by a methylation process followed by hydrolysis.

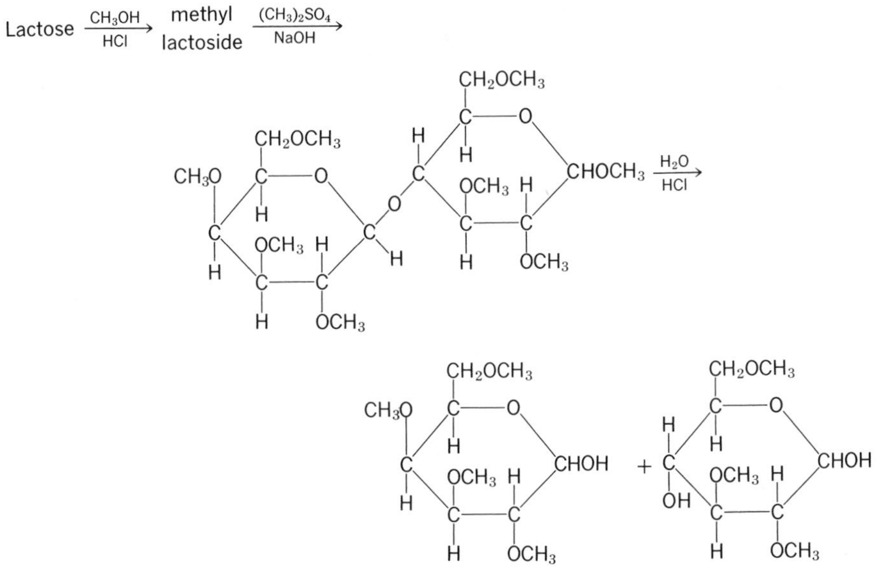
Lactose
CH3OH
HCl
methyl lactoside
(CH3)2SO4
NaOH
H2O
HCl

2,3,4,6-Tetramethylgalactose   2,3,6-Trimethylglucose

All those hydroxyl groups which did not combine with the methylating agent prior to hydrolysis must have been involved in either a *lactol ring* or a *glycosidic linkage*, since these are both types of bonds that are broken on acid hydrolysis while the normal methyl-ether groups would be expected to be stable. The data obtained in the equation above are in accord with the proposed structural formula.

21.17 Sucrose Sucrose, or cane sugar, is also a disaccharide. Hydrolysis of sucrose by acids or enzyme preparations yields a mixture of glucose and fructose (the so-called *invert sugar*).[1] Sucrose itself does not reduce Fehling solution, although both the constituent monosaccharides do; this indicates that the glucose and fructose molecules must be so linked together that their reducing groups are *both* masked by glycoside linkages. This situation is possible only with a formula of the type shown on page 468. Methylation and hydrolysis of sucrose by a procedure exactly analogous to that used in the case of lactose furnish evidence for the existence of the furanose ring in the fructose portion of the molecule as shown.

A number of other di- and trisaccharides are known, both natural products and synthetic ones. In all of them the general types of linkage between the constituent monosaccharides are the same as those found in lactose and sucrose, and the

[1]This mixture is called invert sugar because during the hydrolysis of sucrose the observed direction of optical rotation of the mixture changes from dextro to levo.

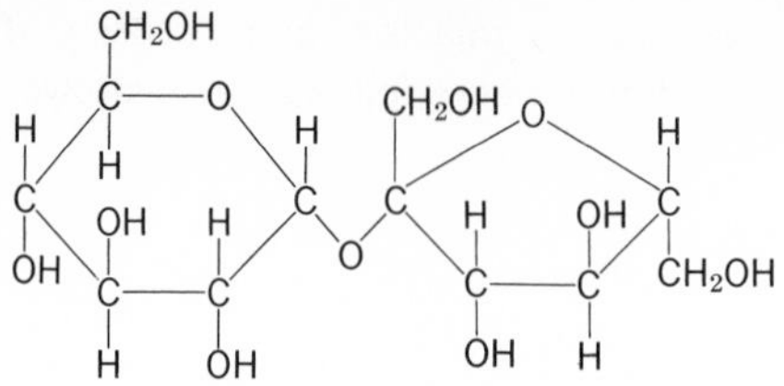
CH2OH
C
O
H
H
CH2OH
O
H
H
C
OH
H
C
C
H
OH
C
OH
C
C
O
C
C
CH2OH
H
OH
OH
H
The structure of sucrose

methods used in the proof of their structures are also the same. These substances will, therefore, not be considered in detail here.

21.18 Polysaccharides Substances like starch and cellulose (see Chap. 22) are made up in the same general way, by consecutive linking of monosaccharide molecules through glycoside linkages. Even in the case of these very large and sometimes complex molecules, the method of methylation and hydrolysis described above has been very useful in determining structure. It is important to understand the simple principle used in these studies: (1) all the available free hydroxyl groups are protected by stable methyl-ether groups; (2) examination of the hydrolysis products of methylated sugars reveals a number of hydroxyl groups that were not methylated or had lost their methyl groups on hydrolysis; (3) these, then, must be the groups involved in the bonds between the monosaccharides and in the lactol rings.

Information of this sort makes it possible to draw significant conclusions about the structure of even complex sugars.

21.19 Spectroscopic Properties At a first level of approximation carbohydrates do not provide much information via spectroscopy since they are all stereoisomers. Within a given series, however, the nmr in particular may provide valuable configurational relationships. These, however, are beyond the scope of this book.

OUTLINE OF CARBOHYDRATE CHEMISTRY

Properties

Most of the reactions will be illustrated for the case of an aldohexose. Similar reactions are given by other monosaccharides, *including ketoses* (compare throughout with aldehyde and ketone reactions, Chap. 11).

1.

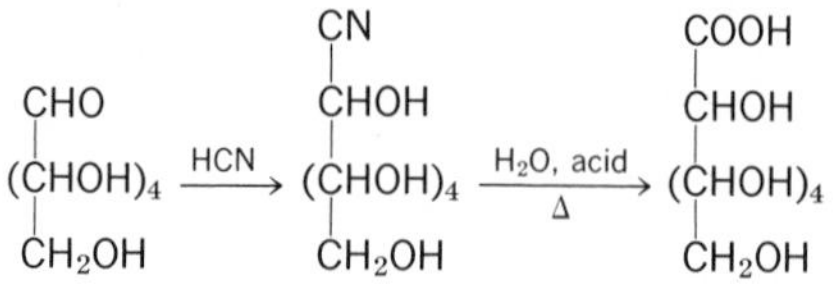
CN
COOH
CHO
CHOH
CHOH
(CHOH)4
HCN
(CHOH)4
H2O, acid
Δ
(CHOH)4
CH2OH
CH2OH
CH2OH

2. 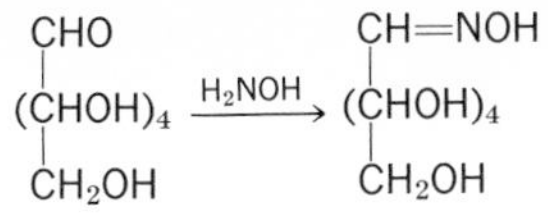

CHO
(CHOH)4
CH2OH
H2NOH
CH=NOH
(CHOH)4
CH2OH

3. 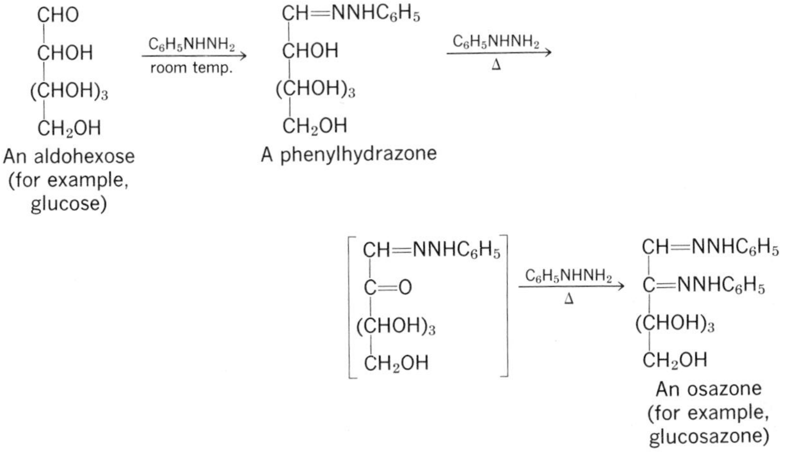

C6H5NHNH2
room temp.
C6H5NHNH2
Δ
An aldohexose (for example, glucose)
A phenylhydrazone
An osazone (for example, glucosazone)

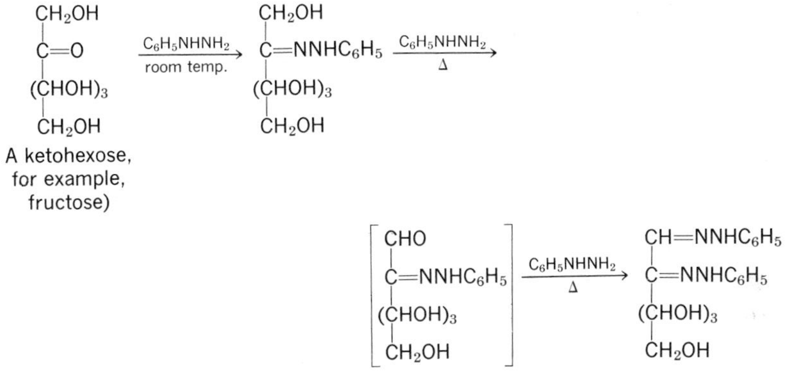

C6H5NHNH2
room temp.
C6H5NHNH2
Δ
A ketohexose, for example, fructose)

4. Oxidation.

a. With Fehling solution. (Sugars which give this test are known as reducing sugars.) Ketoses also give this reaction.

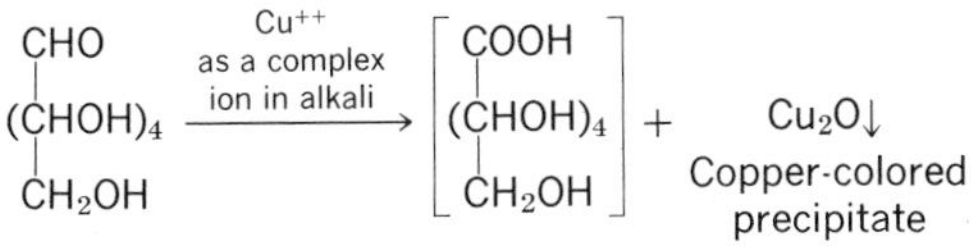

Cu++ as a complex ion in alkali
Cu2O↓
Copper-colored precipitate

b. With Tollens' reagent.

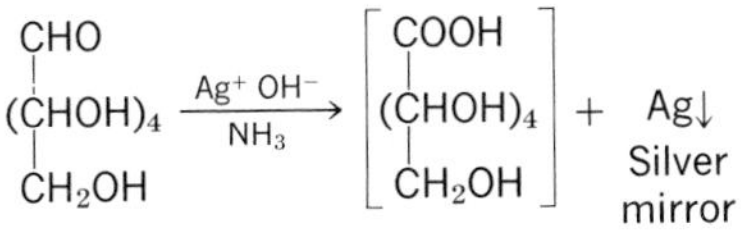

Ag+ OH−
NH3
Ag↓
Silver mirror

Ketoses are also oxidized by Tollens' reagent.

c. With bromine water.

$$\underset{\substack{\text{CHO} \\ | \\ (\text{CHOH})_4 \\ | \\ \text{CH}_2\text{OH}}}{} \xrightarrow[\text{H}_2\text{O}]{\text{Br}_2} \underset{\substack{\text{COOH} \\ | \\ (\text{CHOH})_4 \\ | \\ \text{CH}_2\text{OH}}}{}$$

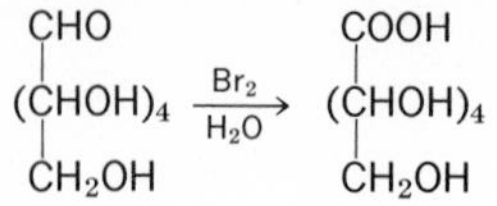

An aldonic
acid

Ketoses are not affected by this reagent. In this case, in contrast to *a* and *b* above, the aldonic acid is obtained in excellent yield.

d. With nitric acid.

$$\underset{\substack{\text{CHO} \\ | \\ (\text{CHOH})_4 \\ | \\ \text{CH}_2\text{OH}}}{} \xrightarrow{\text{HNO}_3} \underset{\substack{\text{COOH} \\ | \\ (\text{CHOH})_4 \\ | \\ \text{COOH}}}{}$$

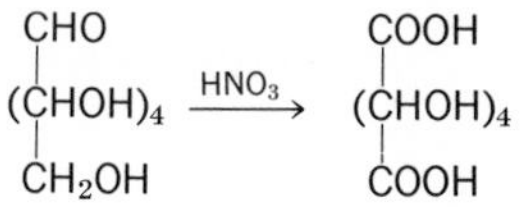

A saccharic
acid

e. With periodic acid. See Sec. 21.15.

5. Breakdown of carbohydrates by strong acids and alkalis.

$$\text{C}_5\text{H}_{10}\text{O}_5 \xrightarrow{\text{conc. HCl}}$$
Pentoses

$$\underset{\substack{\text{CH}\!-\!\text{CH} \\ | \quad\quad | \\ \text{CH} \quad \text{C}\!-\!\text{CHO} \\ \diagdown\ \text{O}\ \diagup}}{}$$

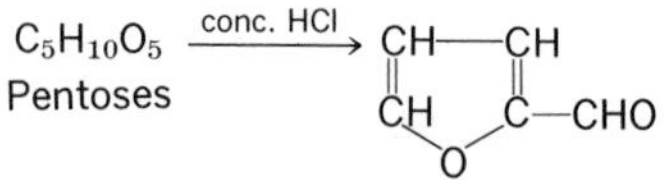

Furfural

$$\text{C}_6\text{H}_{12}\text{O}_6 \xrightarrow{\text{conc. HCl}} \text{CH}_3\text{COCH}_2\text{CH}_2\text{COOH} + \text{HCOOH}$$
Hexoses Levulinic acid

$$\text{Polysaccharides} \xrightarrow{\text{dil. HCl}} \text{monosaccharides} \xrightarrow{\text{conc. HCl}}$$

above products

Reducing sugars $\xrightarrow{\text{NaOH}}$ complicate breakdown and atmospheric oxidation. Non-reducing sugars show a marked stability to alkali.

6. In almost all other cases the sugars appear to react in the lactol (hemiacetal) form.

$$\underset{\substack{\diagup\text{CHOH} \\ | \\ \text{O}\ (\text{CHOH})_3 \\ | \\ \diagdown\text{CH} \\ | \\ \text{CH}_2\text{OH}}}{} + \text{ROH} \underset[\substack{\text{H}_2\text{O, trace} \\ \text{HCl}}]{\overset{\substack{\text{trace of HCl} \\ \text{(anhyd.)}}}{\rightleftharpoons}} \underset{\substack{\diagup\text{CHOR} \\ | \\ \text{O}\ (\text{CHOH})_3 \\ | \\ \diagdown\text{CH} \\ | \\ \text{CH}_2\text{OH}}}{} \xrightarrow{\substack{\text{dil.} \\ \text{NaOH}}} \text{N.R.}$$

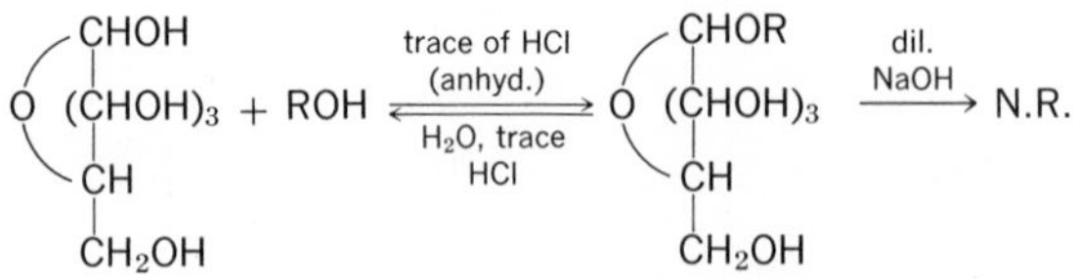

A glycoside
(nonreducing)

Ketoses and reducing polysaccharides react similarly.

7. 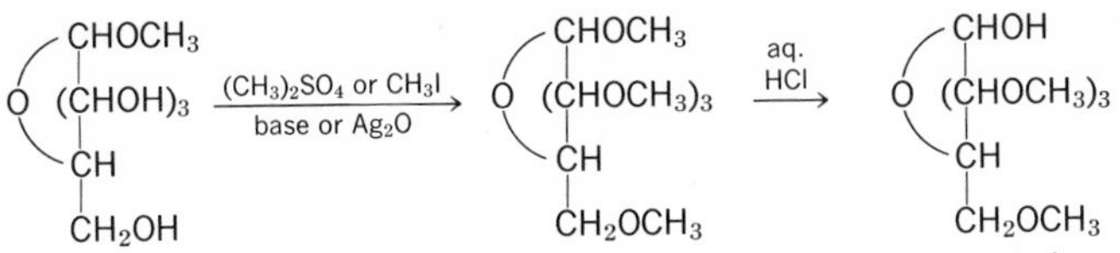

$$\underset{\text{CH}_2\text{OH}}{\overset{\text{CHOCH}_3}{\text{O}\;(\text{CHOH})_3}} \xrightarrow[\text{base or Ag}_2\text{O}]{(\text{CH}_3)_2\text{SO}_4 \text{ or CH}_3\text{I}} \underset{\text{CH}_2\text{OCH}_3}{\overset{\text{CHOCH}_3}{\text{O}\;(\text{CHOCH}_3)_3}} \xrightarrow[\text{HCl}]{\text{aq.}} \underset{\text{CH}_2\text{OCH}_3}{\overset{\text{CHOH}}{\text{O}\;(\text{CHOCH}_3)_3}}$$

A tetramethyl-methyl glycoside (nonreducing)   A tetramethyl sugar (reducing)

Ketoses and polysaccharides react similarly, all available hydroxyl groups being methylated.

8. 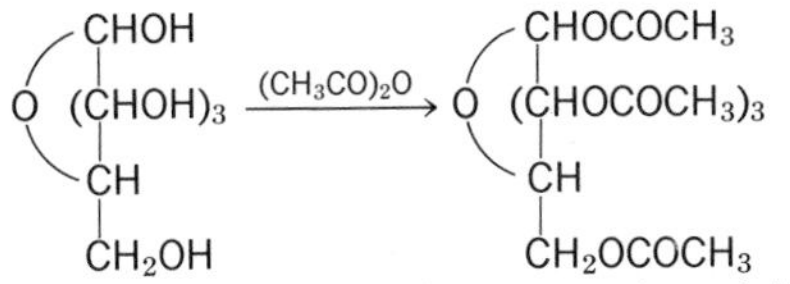

$$\underset{\text{CH}_2\text{OH}}{\overset{\text{CHOH}}{\text{O}\;(\text{CHOH})_3}} \xrightarrow{(\text{CH}_3\text{CO})_2\text{O}} \underset{\text{CH}_2\text{OCOCH}_3}{\overset{\text{CHOCOCH}_3}{\text{O}\;(\text{CHOCOCH}_3)_3}}$$

A sugar pentaacetate

9. To increase the carbon chain of an aldose by one carbon atom:

$$\underset{\text{CH}_2\text{OH}}{\overset{\text{CHO}}{(\text{CHOH})_3}} \xrightarrow{\text{HCN}} \underset{\substack{\text{CHOH}\\(\text{CHOH})_3\\\text{CH}_2\text{OH}}}{\overset{\text{CN}}{}} \xrightarrow[\text{H}_2\text{O}]{\text{HCl}} \underset{\substack{\text{CHOH}\\(\text{CHOH})_3\\\text{CH}_2\text{OH}}}{\overset{\text{COOH}}{}} \rightleftharpoons \underset{\substack{\text{CH}\\\text{CHOH}\\\text{CH}_2\text{OH}}}{\overset{\text{C}=\text{O}}{\text{O}\;(\text{CHOH})_2}} \xrightarrow{(\text{H})} \underset{\substack{\text{CHOH}\\(\text{CHOH})_3\\\text{CH}_2\text{OH}}}{\overset{\text{CHO}}{}}$$

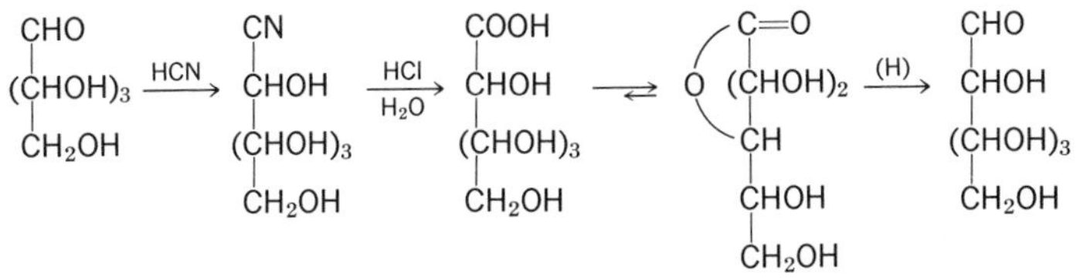

γ-Lactone (an ester; therefore, can be easily reduced)

10. To decrease the carbon chain of an aldose by one carbon atom:

$$\underset{\text{CH}_2\text{OH}}{\overset{\substack{\text{CHO}\\\text{CHOH}\\(\text{CHOH})_3}}{}} \xrightarrow[\text{H}_2\text{O}]{\text{Br}_2} \underset{\text{CH}_2\text{OH}}{\overset{\substack{\text{COOH}\\\text{CHOH}\\(\text{CHOH})_3}}{}} \xrightarrow[\text{(Fe}^{++}\text{ catalyst)}]{\text{H}_2\text{O}_2} \underset{\text{CH}_2\text{OH}}{\overset{\substack{\text{CHO}\\(\text{CHOH})_3}}{}} + CO_2 + H_2O$$

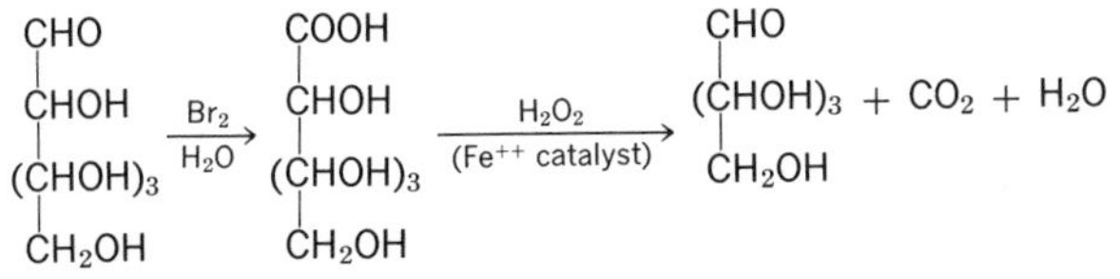

Compare the reactions of α-hydroxy acids.

CHAPTER 21   EXERCISES

1. In the preceding outline no "preparations" are given for carbohydrates. Why is this? What is the source of carbohydrate materials?
2. Write equations illustrative of the proof of structure of fructose by the method of Kiliani (Sec. 21.2).

3. Indicate a process which could be used for resolving D,L-glucose into its enantiomorphous components.
4. Indicate by equations how the following conversions could be accomplished:
   a. D-Glucose to pentaacetylglucose
   b. D-Fructose to γ-hydroxyvaleric acid
   c. D-Mannose to D-fructose
   d. D-Glucose to D-arabonic acid
   e. Starch to D-glucose
   f. D-Glucose to a ketohexose
   g. Glucose to an aldoheptose
   h. Glucose to an aldopentose

★5. Write the structure of the aldopentose which can be degraded to D-tartaric acid and which gives an optically active trihydroxyglutaric acid on oxidation.

6. Write the configuration of *all* the products obtained from D-arabinose by the following series of reactions:

$$\text{D-Arabinose} \xrightarrow{\text{HCN}} \text{cyanohydrin} \xrightarrow[\text{HCl}]{H_2O} \text{monobasic acid} \longrightarrow \text{lactone} \xrightarrow{\text{(H)}}$$

$$\text{hexitol} \xrightarrow{\text{(O)}} \text{monobasic acid} \longrightarrow \text{lactone} \xrightarrow{\text{(H)}} \text{aldohexose}$$

★7. A, B, and C represent three aldohexoses. Of these A and B yield the same hexahydric alcohol on reduction, but different phenylosazones, while B and C give the same phenylosazone but different alcohols. Indicate the configurational relationships of A, B, and C.

8. Write structural formulas for the isomeric 2-ketopentoses of the D series.

9. Given D(−)-erythrose as a reference substance of known configuration, show how it would be possible to prove the configuration of the trihydroxyglutaric acid shown below:

```
 COOH
  |
HCOH
  |
HCOH
  |
HCOH
  |
 COOH
```

10. Given D(−)-threose as a reference substance of known configuration, show how you could prove the configuration of D(+)-xylose.

11. Why does the mutarotation of D-glucose lead to a product which is optically active, even though the analogous process of racemization usually leads to an inactive product?

12. How many ketopentoses having a furanose ring are possible? How many 2-ketohexoses having a pyranose ring? How many aldoheptoses having a pyranose ring?

13. Glucosamine, $C_6H_{13}O_5N$, gives N_2 with HNO_2; with phenylhydrazine it gives the same osazone as D-glucose; it can be oxidized to a monobasic acid, glucosaminic acid, $C_6H_{13}O_6N$; it shows mutarotation in solution. Give a probable structure for glucosamine, and indicate clearly what elements of doubt there are in the structure you propose. Indicate a method for synthesizing glucosaminic acid, starting with a readily available substance.

★**14.** A certain disaccharide is hydrolyzed by dilute acid to give D-glucose and D-arabinose. The disaccharide reduces Fehling solution. If it is oxidized to a monobasic acid and then hydrolyzed, glucose and arabonic acid are formed. Methylation of the disaccharide followed by hydrolysis gives 2,3,4,6-tetramethylglucose and 2,3-dimethylarabinose. Give the probable structure of the disaccharide. What points concerning the structure are uncertain, and what additional information could be found to settle these points?

15. Partial acetylation of methyl-D-mannopyranoside gives a small yield of monoacetate. This compound reacts with 1 mole of HIO_4 per mole of monoacetate. Draw the structure of the monoacetate showing all configurations.

16. Show by equations that α-methyl-D-glucopyranoside and α-methyl-D-galactopyranoside give the same products with HIO_4.

17. Draw out the conformations of the two glycosides of Exercise 16.

18. A certain nonreducing disaccharide X yields glucose and arabinose on hydrolysis. Reaction of X with HIO_4 shows that 3 moles of oxidant are consumed per mole of disaccharide and one mole of CH_2O produced. Show by formulas possible structures for X.

19. Sucrose has the configuration shown below between the glucose portion and the fructose portion.

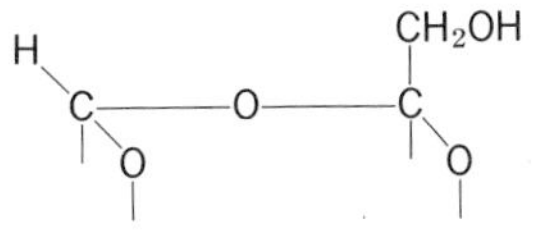

1-α-D-Glucopyranosyl-β-D-Fructofuranoside

Explain the full meaning of the two parts of the name, and explain why sucrose is not a reducing sugar.

20. Maltose has been proved to have the structure and configuration shown in the formula below. Would you expect maltose to be a reducing sugar? Explain. Write the equations for the complete methylation and hydrolysis of maltose.

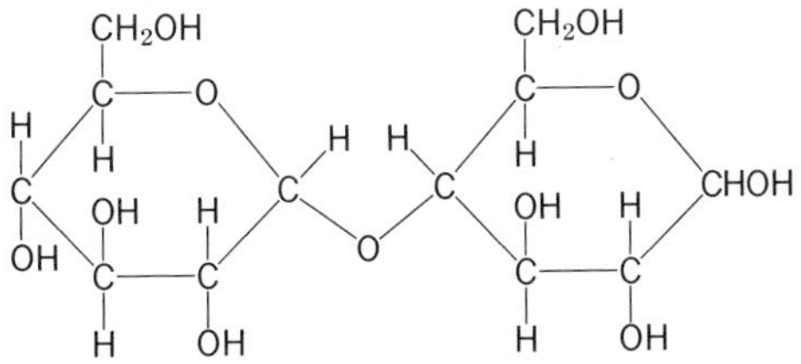

21. A certain disaccharide reduces Fehling solution and on hydrolysis forms a mixture of galactose and fructose. Complete methylation followed by hydrolysis yields a mixture of 2,3,5,6-tetramethylgalactose and 1,4,6-trimethylfructose. Deduce a structure for the disaccharide. Are there any points of structure still undetermined by this evidence?

22. A naturally occurring aldopentose, A, $C_5H_{10}O_5$, gives on nitric-acid oxidation a dibasic acid, $C_5H_8O_7$, which is optically active. When A is treated with bromine water and then with H_2O_2 and a ferrous-salt catalyst, there is obtained a compound B, $C_4H_8O_4$. Upon nitric acid oxidation of B there is formed *meso*tartaric acid. Write equations for these reactions, and write a structural formula for A that shows the configurations of the asymmetric atoms in the conventional manner.

23. Show how you could distinguish:

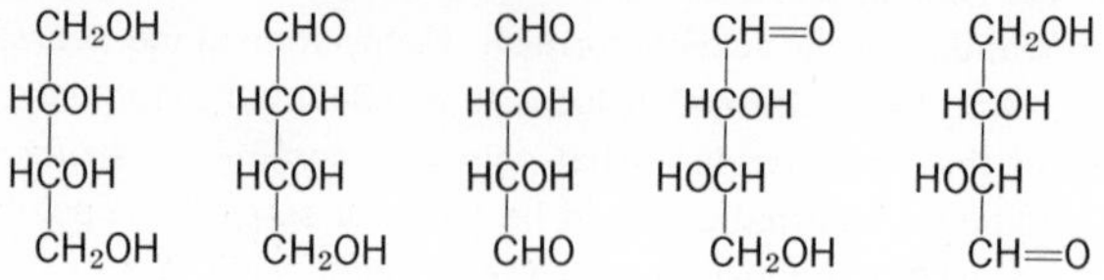

CH2OH
HCOH
HCOH
CH2OH
CHO
HCOH
HCOH
CH2OH
CHO
HCOH
HCOH
CHO
CH=O
HCOH
HOCH
CH2OH
CH2OH
HCOH
HOCH
CH=O

★24. Compounds A, B, C, D, and E have the following properties:

a. Compound A, $C_{10}H_{20}O_6$, + acetic anhydride $\longrightarrow$ $C_{18}H_?O_?$

b. A is stable to alkali and gives a negative Fehling test

c. A + hot acid $\longrightarrow$ B, $C_5H_{10}O_5$, + C, $C_5H_{12}O_2$

d. B gives a positive Fehling test. B + CH_3COCl $\longrightarrow$ $C_{13}H_?O_?$

e. C + CH_3MgBr $\longrightarrow$ 2 moles of CH_4 per molecule of C

f. C + excess $KMnO_4$ $\longrightarrow$ D, $C_5H_8O_4$ $\longrightarrow$ $C_4H_8O_2$

g. B is optically active. B + hot dil. HNO_3 $\longrightarrow$ E, $C_5H_8O_7$

h. E is optically inactive and cannot be resolved

Interpret clearly each of the above statements. Deduce formulas and configurations for lettered compounds.

★25. A trisaccharide A, $C_{18}H_{32}O_{16}$, has the following properties:

a. A is nonreducing and hydrolysis yields D-galactose and D,L-glucose.

b. Partial hydrolysis of A yields D-galactose + B, a nonreducing disaccharide.

c. $A + (CH_3)_2SO_4 + NaOH \longrightarrow$ undecamethyl-A (C_{29}) $\xrightarrow[\text{acid}]{H_2O}$

a tetramethyl D-galactose +
a tetramethyl L-glucose +
2,4,6-trimethyl D-glucose

d. Destructive oxidation of the tetramethyl D-galactose with dil. HNO_3 $\longrightarrow$ L-dimethyltartaric acid and no larger-molecular-weight products.

e. Destructive oxidation of the tetramethyl L-glucose $\longrightarrow$ a *meso*-trimethoxyglutaric acid.

f. A is cleaved completely by maltose, and emulsin cleaves it to L-glucose + a new reducing disaccharide.

Deduce structures and show reasoning. Is B optically active? Are any configurational features left unspecified?

22

MACROMOLECULES: POLYSACCHARIDES, PROTEINS, AND PLASTICS

22.1 Introduction This chapter deals with an important group of substances of very high molecular weight. Their chemical unreactivity, physical strength, and insolubility are the basis of many of their industrial applications. In this group are found materials like cotton, wool, silk, wood, rubber, and plastics. Some of these substances are spoken of as *high polymers* because they are made up of large numbers of small molecules condensed or added together. Their properties are closely related to their molecular size and structure and it, therefore, seems appropriate to emphasize this in the title of this chapter. The relations of some physical properties to molecular size and structure will first be considered, and then three classes of macromolecules will be described.

As molecular size increases, certain properties change with it. For example, in any homologous series boiling point rises with molecular weight, and so does melting point (Fig. 12-2). Solubility usually decreases, though there may be some short-range variations observed. The examinations of homologous series made so far have not been extended to very large molecules. The macromolecules with which this chapter is concerned have molecular weights of the order of 10,000 to several million, so that to extend the generalizations made on simple homologous series to these substances would be to carry them rather to extremes. In such a process it would not be surprising to find new phenomena turning up and to find that the generalizations previously laid down are inapplicable.

In most of the homologous series dealt with so far, those member substances with more than about 20 carbon atoms have been certain to be solids at room temperature. It is not surprising, therefore, that substances with 2,000 carbon atoms are all solids. Nor is it surprising that they are nonvolatile. But interesting new phenomena appear when some other properties are examined, and here the large molecules have to be classified on the basis of their structure:

1. *Linear polymers* are those which consist of chains of atoms. Each chain may be extremely long, and may be somewhat branched, but is not linked laterally to other long chains. These are sometimes called *one-dimensional* because their length may be very great but their cross section relatively small.
2. *Network, or sheet, polymers* are those in which long chains of atoms are linked laterally

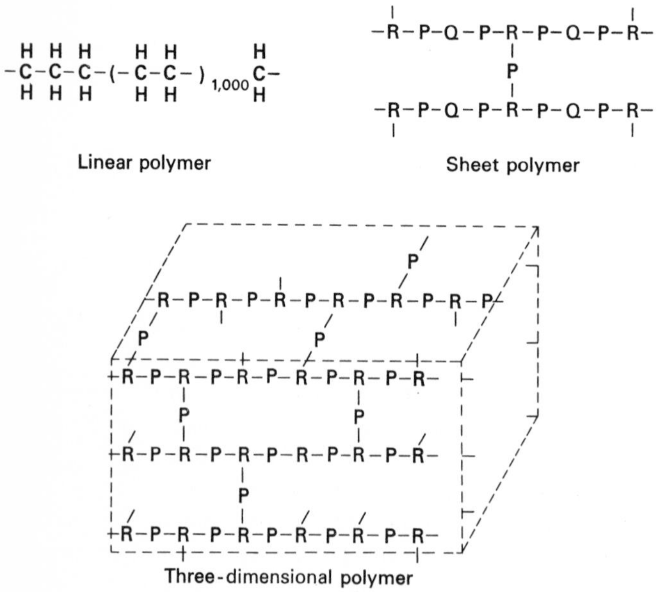
Linear polymer
Sheet polymer
Three-dimensional polymer

figure 22.1
Types of polymers. Diagramatic.

to other chains to form networks. These molecules are essentially *two-dimensional* in that they may have great length and breadth, and yet not be very thick.

3. *Three-dimensional polymers* are those in which chains are crosslinked into nets, which are crosslinked in the direction of their thickness to other nets, so that a whole large mass may be a single molecule.

The different kinds of polymers can be prepared in the laboratory, and representatives of these different kinds can be found in nature. Even when pure, these materials may be mixtures of molecules with different chain lengths and different places and degrees of branching.

22.2 Physical Properties: Solubility When the property of solubility is considered, it can be understood that the linear polymers may be soluble in some solvents, particularly in hot solvents; for if the chains are not fastened together along their lengths to other chains, then solvent molecules can get in between them and joggle them apart until, if there is enough solvent, the chains may move about quite freely. But the process of solution is interesting. As solvent molecules work their way in between the long, thread-like molecules, the substance swells; it, in effect, is dissolving the solvent at this point. After enough solvent gets into the mass, it may form a viscous solution which may have properties of tackiness [if the chemical constitution is of the right sort (see below)], and as more solvent enters the mass, it may dissolve—first to a thick, viscous liquid, then to a thinner one, with more solvent. If such a liquid is forced through a small tube, it is found (by optical means which cannot be discussed here) that the shearing action of flow causes the molecules to line up in the direction of flow, side by side along their lengths, and the liquid then becomes partially oriented and shows some properties usually associated only with crystals. The ease with which a linear polymer dissolves depends to some extent on its chemical makeup, and this will be discussed further below.

Sheet polymers behave in an analogous manner, but when they swell, it is chiefly in the direction of their thickness; the solvent molecules pry the sheets apart, as it were. Many network polymers do not dissolve but only swell. The chief examples of sheet polymers are found among inorganic substances. It must be apparent at this point that three-dimensional polymers will not dissolve. If the spaces between the rings of which they are composed are large enough, solvent molecules may work their way into the structure, and some polymers imbibe large quantities of solvent without swelling much. The three-dimensional polymer can be broken down only by breaking chemical bonds.

If *melting* is thought of as analogous to solution (which is legitimate because both involve a setting free of the molecules so that they move in a disorganized way), then it is easy to understand from the discussion above that linear polymers may melt (at an elevated temperature) and network polymers may melt, or soften, but three-dimensional polymers do not melt. They are "infusible," and if the temperature is raised sufficiently high, they may decompose and burn, like most compounds of carbon.

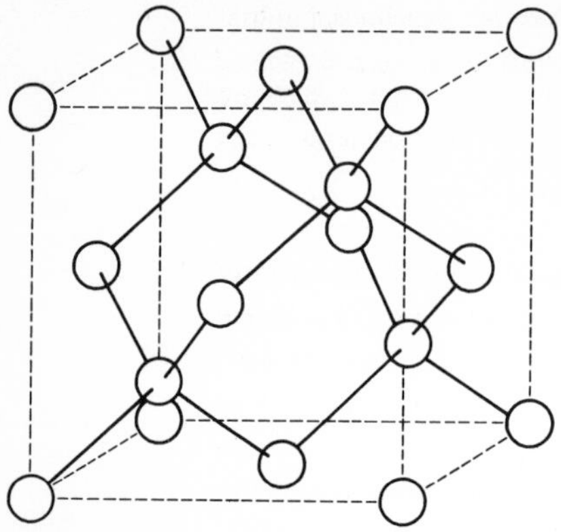

figure 22.2
Three-dimensional diamond structure of carbon. *(From Linus Pauling, "The Nature of the Chemical Bond," 3d ed., Cornell University Press, 1960.)*

22.3 Strength In principle, an object made from a three-dimensional polymer consists of one huge molecule. If the network is regular and the linkages are numerous, such an object can be very strong, as it will be necessary to break a large number of chemical bonds in order to break such an object apart. The diamond structure of carbon represents such a network in which carbon atoms are linked by four sp^3 bonds as indicated in Fig. 22-2. The hardness of diamond is a consequence of such an arrangement.

In contrast to the diamond structure lies that of vulcanized rubber. As illustrated schematically in Fig. 22-3, rubber consists of long floppy hydrocarbon molecules tied together in a few places by the vulcanizing agent. When rubber is stretched, the molecules will extend to the point where they are more or less straightened out and then will resist further extension. If greater stress is applied, chemical bonds will be broken. Unlike diamond, however, only a small number of bonds must be broken because of the small number and random location of the crosslinks.

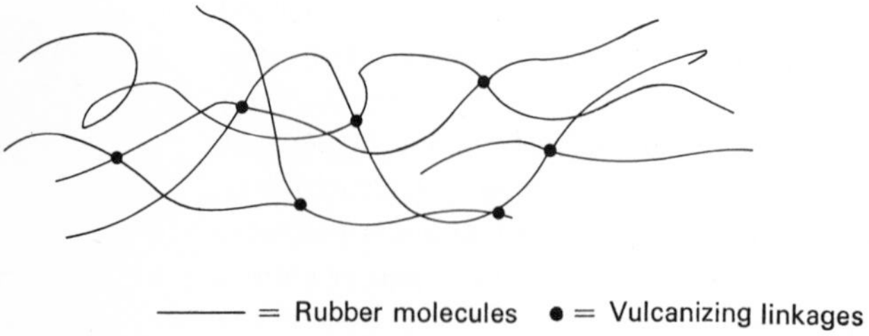
——— = Rubber molecules
• = Vulcanizing linkages

figure 22.3
Schematic model of vulcanized rubber.

Three-dimensional polymers will have in common the properties of insolubility, and infusibility. They can vary in hardness and strength all the way from diamond to very soft, comparatively weak materials, depending on the number and regularity of the bonds tying the network together (called "crosslinks"). Highly crosslinked polymers will be impermeable, while lightly crosslinked species like rubber will swell when immersed in solvents, but will not actually dissolve in them. On the other hand, highly crosslinked materials, although strong, will tend to be brittle, that is, they will resist a deformation with all their strength, but when this limit is exceeded, they will shatter (even diamond has this property).

The very properties that give three-dimensional polymers their useful characteristics also strongly limit their utility. Objects prepared from them must be shaped during the polymerization; they cannot be readily modified later. It is difficult or impossible to form them into films or fibers since they cannot be melted or dissolved. It is for these reasons that linear, noncrosslinked polymers occupy a more important position than the three-dimensional polymers, in spite of the inherently weaker intermolecular forces. Such molecules can be separated by thermal agitation to give melts, or by solvent molecules to give solutions, which are much like melts, and solutions of smaller molecules. Because of the greater chain lengths, however, they will have grossly higher viscosities than small molecules because the chains will continually get entangled with one another.

The question naturally arises, "To what do these linear polymers owe their mechanical strength?" In the molecular weight range commonly encountered—10,000 to 10,000,000—one can estimate that the longest molecule, fully extended, would be no more than 0.1 mm long. It is thus readily apparent that chemical bonds need not be broken to break objects made of such materials. In fact, it is the so-called weak forces that hold these molecules together, the van der Waals', dipole-dipole, and hydrogen-bond forces. With long polymer chains there are so many such interactions per molecule that the net force between molecules approaches a good percentage of that exerted by an actual chemical bond. When the comparatively strong hydrogen bonds are involved, molecular weights of 10,000 provide sufficient strength; with van der Waals' forces alone, molecular weights must be on the order of 100,000 to 10,000,000.

Because the intermolecular forces are of the nondirectional, time-dependent type, linear polymers can yield under stress without shattering (in addition to being soluble and fusible). When a fiber made of such material is stretched under certain conditions the molecules can actually line up parallel with one another to give what is termed an *oriented* fiber which has its greatest strength in the linear direction. The wide variety of possibilities open to linear polymers has tended to put them in a preeminent position in the polymer field.

The *chemical inertness* of many macromolecular substances which may contain ordinarily quite reactive groups (glycoside linkages in cellulose, double bonds in rubber) is related to the difficulty of bringing reagents into effective contact with the groups owing to the insolubility or rigidity of the molecules. Often also there are too few groups present per molecule to have an appreciable effect on the behavior of the molecule. Cellulose, for example, has barely detectable reducing power although it contains one reducing group per molecule (molecular weight up to 300,000).

The *purification* of macromolecules, whether natural or synthetic, is a problem of greatest difficulty. Even those substances which are reasonably soluble in appropriate solvents seldom crystallize and have no characteristic melting points or other simple criteria of purity. The only purification techniques generally applicable are arbitrary fractional precipitations from different solvents; these can sometimes be modified, as in the case of proteins, by adjustment of pH with respect to the isoelectric point or by adding large amounts of neutral salts such as $(NH_4)_2SO_4$ to salt out the large molecules, literally crowding them from solution. Except in cases where there may be some particular feature, such as the iron in hemoglobin (a protein) or the physiological activity associated with protein hormones and enzymes, there is no convenient criterion of purity to control these precipitations except complex analyses for constituents and determination of molecular weights. Consequently, the term pure, as applied to cellulose or rubber or a protein, must be accepted with reserve until the criteria of purity have been examined.

22.4 Determination of Molecular Weight The usual methods for the determination of molecular weights are of practically no value with macromolecules. It is often impossible to get enough molecules of such substances in solution to produce a measurable effect on freezing point or boiling point. Often, true solutions cannot be obtained at all or only with considerable degradation of the molecule. Most proteins (egg white, for example) are *denatured*, or coagulated, at elevated temperature, and other macromolecules like starch tend to form suspensions of particles, or jellies.

Synthetic polymers and many natural polymers have no unique molecular weight but often occur over quite a range of molecular weights. The "average" molecular weights which are determined for these molecules are sensitive to the method of determination.

Several physical methods are now available that lead to reliable average-molecular-weight values for macromolecules. The pressure developed across a semipermeable membrane, such as cellophane, between a solution containing a substance unable to pass through the membrane and the pure solvent is called *osmotic pressure*. Although the accurate measurement of osmotic pressure presents a considerable number of experimental difficulties, it was one of the early methods used in protein chemistry to gain information about the molecular weights of proteins in solution. Since the osmotic pressure developed by a solution is a function of the molecular weight of the solute, molecular weights can be determined in this way.

In a similar manner, the viscosity of a solution of a macromolecular substance is related to the molecular weight of the solute. This is an easily measurable property of solutions and is widely used in the plastics industry to measure and help control the average molecular weight of the products of synthetic operations. Although the method is useful in a relative sense, the theoretical aspects of the problem of calculating absolute molecular weights from viscosities leave much to be desired (see above for the effect of flow on the structure of solutions of polymers).

Perhaps the most reliable method is the centrifugal one. Solutions of macromolecules are spun in an ultracentrifuge to very high speeds. This applies a tremendous centrifugal force to the molecules—a force up to 400,000 times that of gravity, depending on the speed—which tends to force them toward the outer edge of the spinning centrifuge. The solutions are in small chambers in a rotor which is spun. Under the influence of the tremendous gravitational force, the large molecules tend to settle out of solution toward the circumference of the rotor. The rate of settling may be used to calculate the molecular weight, the calculations depending on some knowledge of molecular shape (long and thin, globular, etc.) usually derivable from x-ray or diffusion studies.

Another, more recent method is that of gel-permeation chromatography. This is a type of chromatography (Sec. 22.8) in which a solution of the polymer is passed down a column of a lightly crosslinked polymer. The crosslinked polymer is insoluble in the solvent but is considerably swollen by the solvent to give a loose network with solvent-filled voids. Such a network is called a "gel," from which the technique gets its name. The voids are of the order of molecular dimensions so that the molecules of the soluble polymer permeate through the voids in the gel. The rate of passage of the polymer molecules down the column is a function of the molecular weight, the larger molecules being held up in the voids more effectively than the smaller ones. By measuring the amount of material passing through such a column as a function of time one can obtain a measure of the molecular-weight distributions in a given sample of polymer.

All these methods do not agree perfectly with each other in every case, but they serve to establish the approximate molecular weights beyond reasonable doubt. They all depend on getting the material into solution. This is often a difficult thing to accomplish without degradation of macromolecules. Sometimes the more extensive the "purification," the lower the molecular-weight values become; the molecules are degraded during the "purification." An example of this is found in cellulose.

For purposes of discussion, the macromolecules are here divided into three classes: polysaccharides, proteins, and synthetics.

POLYSACCHARIDES

While there are many representatives of this group, some of great physiological interest, we shall confine our discussion to the three best-known polysaccharides: cellulose, starch, and glycogen. Much less is definitely known about the other, more complex, members of this class, but the methods of investigation used with them are on the whole the same as those used in cellulose research.

Cellulose is an important structural material in the plant kingdom. The cell walls of plants are made up in large part of cellulose; cotton is almost pure cellulose, and wood, while it contains lignin (which is not a carbohydrate), contains much cellulose. *Starch* and *glycogen* are not structural materials. They will disperse in water and they serve as reserve food supplies for plants and animal cells, respectively.

All three polysaccharides yield exclusively D-glucose on complete hydrolysis; partial hydrolysis yields cellobiose from cellulose, and maltose from starch and glycogen.

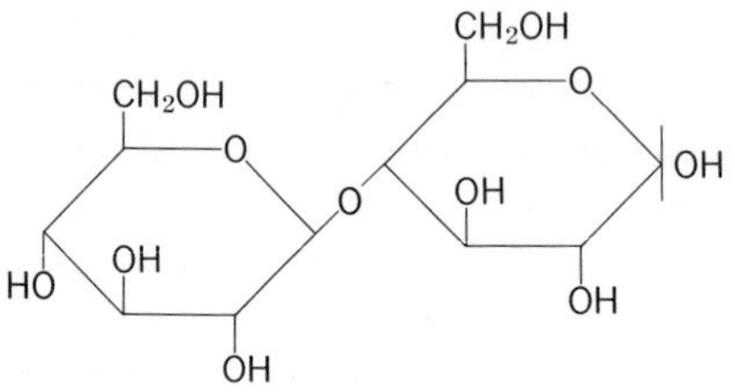

Cellobiose

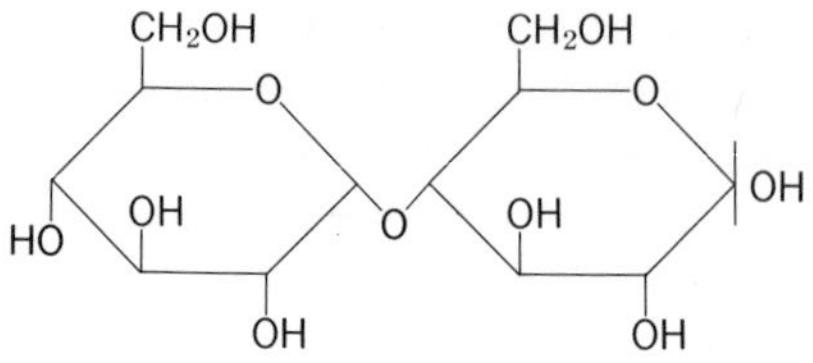

Maltose

It will be noted that in these fragments the glucose units are linked together from carbon 1 to carbon 4 in a β-glucoside linkage in cellobiose and an α-glucoside linkage in maltose.

22.5 Structure Determination Studies by the methylation method have yielded valuable information on the structure of these polysaccharides; the following formula for cellulose has been deduced in part from these experiments:

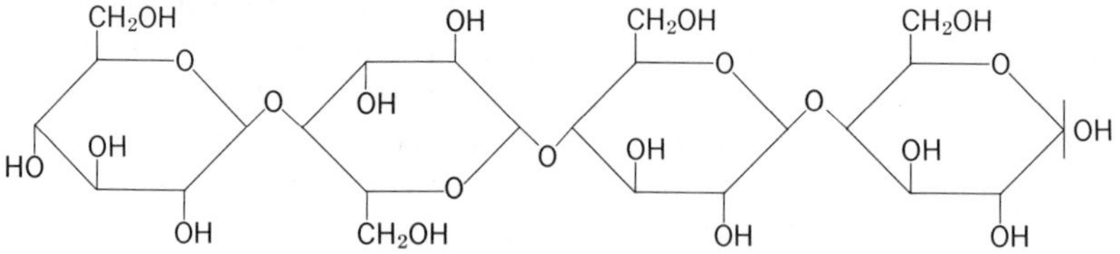

I II III IV

cellulose chain

Methylation of cellulose, although difficult, can be carried practically to completion with the formation of a methylated product containing an average of three methoxyl groups per glucose unit; each of the central units in the chain above has three free hydroxyls. A hydrolysis of methylated cellulose yields chiefly 2,3,6-trimethylglucose and a trace (0.5 percent, approximately) of 2,3,4,6-tetramethylglucose.

A consideration of the cellulose formula above reveals that the terminal glucose unit at the left, I, has four free alcoholic hydroxyls and should form a tetramethyl derivative. The other end of the chain also has a glucose unit with four free hydroxyl groups, but one of these is a glucosidic hydroxyl, and it forms a methylglucoside which is hydrolyzed during the cleavage of the cellulose chain.

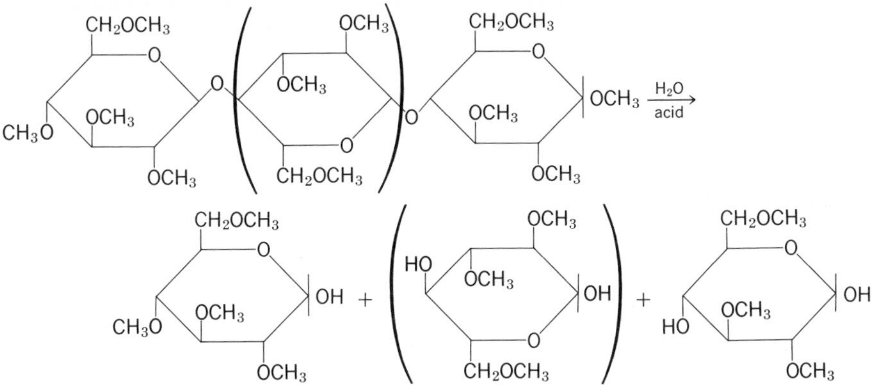

H_2O
acid

2,3,4,6-Tetramethylglucose 2,3,6-Trimethylglucose (x molecules) 2,3,6-Trimethylglucose

From the amount of 2,3,4,6-tetramethylglucose found, one can deduce a chain length for cellulose of 100 to 200 glucose units, or a molecular weight of about 17,000 to 34,000. Cellulose purified under milder conditions, however, yields less 2,3,4,6-tetramethylglucose, and it is believed that this value for the molecular weight is a minimum one. The ultracentrifuge gives much higher values than this chemical or *end-group* method; results of the order of 2,000 glucose units are obtained.

Similar studies on starch and glycogen reveal differences from cellulose in addition to the α,β difference found in the partial hydrolysis products. Both starch and glycogen, in contrast to cellulose, give intense colors, blue and red, respectively, with iodine in aqueous solution. This complex reaction is often employed as a specific test for these polysaccharides.

Methylation of both starch and glycogen can be carried practically to completion, and hydrolysis of the methylated derivatives, following the procedure outlined above for cellulose, yields mostly 2,3,6-trimethylglucose as before. Much larger amounts of 2,3,4,6-tetramethylglucose (the end group) are found, together with considerable 2,3-dimethylglucose. The quantitative estimation of the partially methylated sugars in such a mixture was a difficult task until the development of chromatography (Sec. 22.8), but evidence has been accumulated to indicate a chain length of 12 or 18 glucose units for glycogen and 24 units for starch. That is, there is found one end group for every 24 glucose molecules in the hydrolysis product of methylated starch.

Molecular weights as determined by the ultracentrifuge or other physical methods are always found to be much higher in starch and glycogen samples than the end-group method indicates. Values of the order of 3,000 to 5,000 glucose units are regularly obtained. These conflicting results are explained by a structure such as that given below:

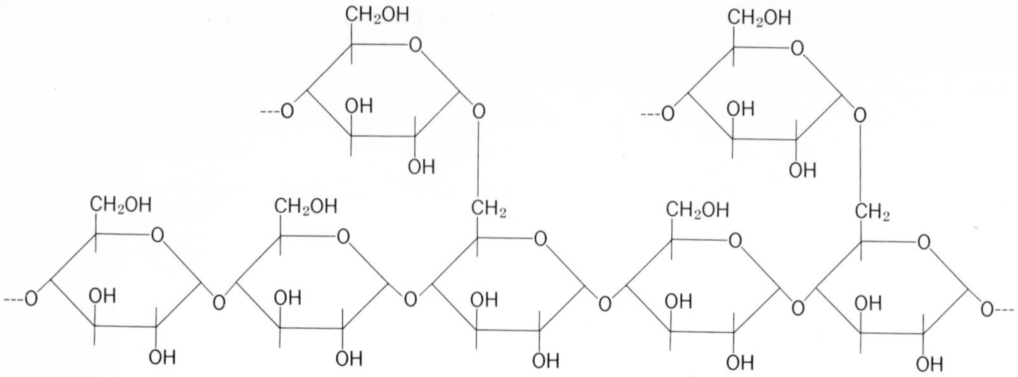

Section of a branched-chain polysaccharide

The occurrence of dimethylglucose in the hydrolysis products of fully methylated starch and glycogen is due to the branched-chain structure. Note that the glucose units at the "stem" ends of the branches have only two free hydroxyls, and these give rise to the dimethylglucose found on hydrolysis.

The question might be raised now as to the relation between the structure and physical properties of these molecules. Cellulose is made up of linear molecules but is insoluble and fiber-like, whereas starch and glycogen are much branched and somewhat soluble in water. Examination of cellulose by means of x-ray diffraction shows that it has an orderly structure with a repeated pattern. The chains of glucose units, some 200 or more to a chain, appear to lie side by side, bound in bundles of ten or so chains, and oriented parallel to the axis of the cellulose fiber. These chains are bound by hydrogen bridges it is thought, and this crosslinking accounts for the insolubility and toughness of the cellulose. The hydrogen bonds can be broken by very polar substances, as will be shown below, whereupon the cellulose becomes soluble. Starch and glycogen, on the other hand, are found to be amorphous. The molecules do not pack as well, and so any crosslinking by hydrogen

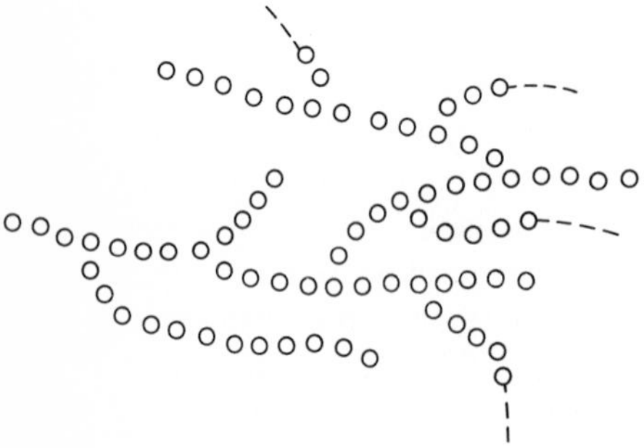

figure 22.4
Schematic diagram of a section of a glycogen molecule.

bridges is of less importance; water molecules can penetrate more readily between the chains and cause swelling to occur, and gummy, pasty hydrated masses can form. These, since they are composed of molecules with very polar groups, can wet and adhere to polar surfaces. The nonoriented starch and glycogen have little structural value to the plant and animal, but their solubility makes them useful as convenient reservoirs for glucose: places in which the very soluble glucose can be tied up with a hydrolyzable link until it is needed in the economy of the living organism.

22.6 Conformation It is interesting to note that the β linkage enables the glucose molecules in cellulose to be linked in a conformationally stable (all large groups in the least-crowded positions) linear array as follows:

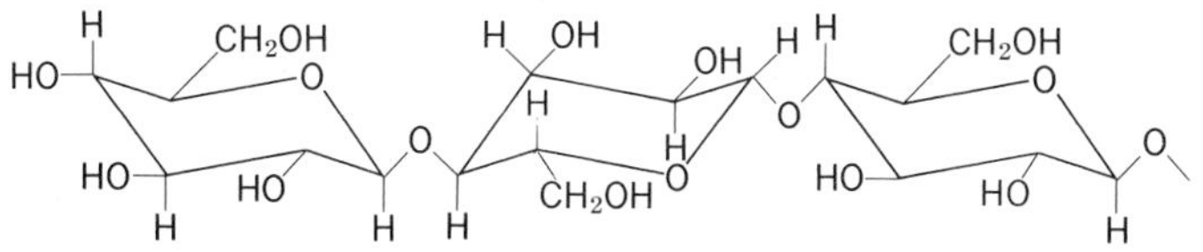

This can readily yield linear molecules which are structural materials. The α linkage, however, does not allow such an array to be formed even if less favorable conformations are found. The tendency is to form a helix or coil. This simple difference may well be sufficient to make cellulose a structural material (by efficient packing and hydrogen bonding) and to prevent starch and glycogen from doing the same.

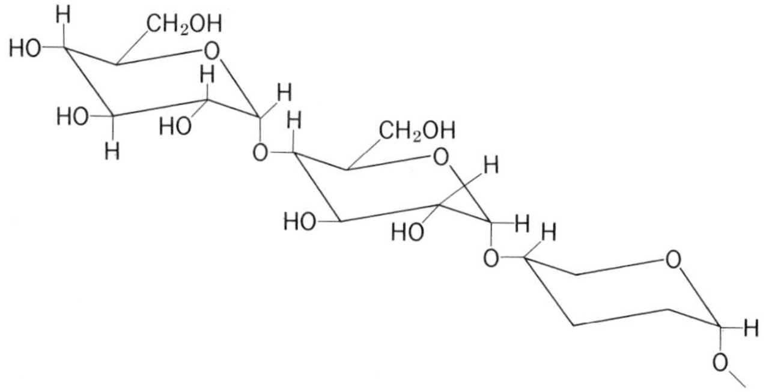

PROTEINS

In the animal kingdom much of the structural material, aside from bones, is protein in nature. In plants it was found to be cellulosic in nature. There is a great chemical difference between proteins and polysaccharides, but they have some points in common. Both are macromolecules made up of a number of smaller units connected into long chains by regularly repeating linkages. In the polysaccharides, it is the glycosidic linkage which is repeated; in proteins, it is the amide or peptide linkage. Some proteins seem to be long-chain molecules; others seem to be branched and crosslinked. Perhaps the most significant point of difference between these two classes is the number and variety of building units, amino acids, found in proteins as compared to the relatively few monosaccharide types involved in polysaccharides. This together with the lower stability of proteins is one of the main reasons why our detailed knowledge of protein molecules has been developed more slowly than that of cellulose and starch.

Proteins are of two main types:[1] the *simple proteins,* yielding exclusively amino acids on hydrolysis, and the *conjugated* proteins, which yield amino acids and some other compound of different structure. This "foreign" substance is called a *prosthetic group* and may be an iron compound, as in hemoglobin, or a vitamin, or any one of a great variety of substances.

22.7 Structure All available evidence indicates that the amide bond, called a peptide linkage, is the only type of chemical bond holding the amino-acid units together in simple proteins. A gradual hydrolysis of such a protein accompanied by an analysis for amino groups (Van Slyke, Sec. 14.10) and carboxyl groups (formol titration, Sec. 19.6) shows that for each amino group set free during hydrolysis, a carboxyl group also appears. We have, therefore, the following picture of a protein chain in which each cleavage would liberate one NH_2 per COOH.

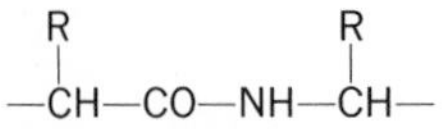

The peptide linkage

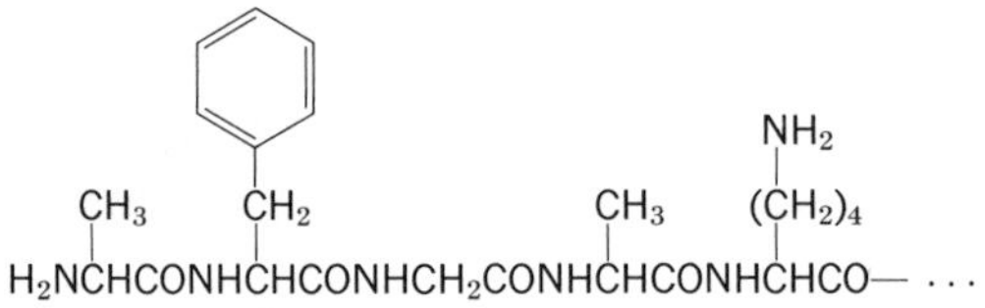

Section of a protein chain

Here the groups attached to the main chain may be any one of the twenty or so groups of the amino acids known to be produced by protein hydrolysis (Sec. 19.1) arranged in any order.[2] The number of combinations possible in a chain of the size of a typical protein is ample to account for the vast number of individual proteins known to exist. In addition, the possibilities of interaction between R groups (especially such groups as aspartic acid, lysine, cysteine) allow for a still greater number of modifications in structure and shape of protein molecules by crosslinking and cyclizing the chains.

[1] Proteins are further subdivided, on the basis of their solubilities, into albumins, globulins, etc. This classification results from differences in secondary structure which are beyond the scope of this book.

[2] This is a problem as yet unsolved: how much order (regular repeated structure) there is, but see Sec. 22.9.

The observed amphoteric character of proteins is largely due to the acidic (COOH, SH, phenolic OH) and basic (NH_2, $NH{-}\underset{\underset{NH}{\|}}{C}{-}NH_2$) groups present in these R groups. Certain proteins contain relatively large amounts of lysine and arginine, and, consequently, are predominantly basic in behavior (isoelectric point above pH 7), while others are more acidic. All proteins, however, seem to be amphoteric and to exhibit a characteristic isoelectric point.

In addition to the above, or primary, structure most proteins have weak secondary interactions brought about by hydrogen bonds and other interactions. Many of the important biological functions of proteins are vitally dependent on this secondary structure. As this secondary interaction is very easily destroyed, by heat, pH, salt concentration, and many other changes, it is much more difficult to study, and only relatively recently has much progress been made in this area.

Considerable gains have been made in the study of primary protein structure and we shall devote our attention to this aspect. One of the most basic problems here is the great similarity of the chemical and physical properties of amino acids, proteins, and fragments of proteins. (The latter consist of two or more amino acids linked together and are called "peptides.") Many very clever and sophisticated techniques have been devised in order to separate and analyze the various species, and it is necessary to have some understanding of these techniques in order to proceed. By far the most used and most generally applicable technique is that of chromatography.

22.8 Chromatography[1] This represents a class of techniques with certain basic features in common. A chromatographic system consists of two phases—a stationary phase and a moving phase. The stationary phase almost invariably involves or utilizes a polymer, either organic or inorganic, which may be packed in a column, spread in a sheet, or cut into strips. The moving phase is most often a liquid or gas and is immiscible with (does not mix with) the solid phase. In operation, the liquid (or gaseous) phase containing the mixture to be analyzed (usually in one small portion or slug) is applied at one point to the stationary phase (e.g., the top of a column or sheet) and more liquid is passed through the column or down the sheet. As the moving phase progresses the components of the mixture interact with the stationary phase. This interaction is usually slightly different for different compounds, and as the moving phase progresses along the stationary phase, it is repeated many times. The more strongly a substance interacts with the stationary phase, the more slowly it will move with respect to the rest of the moving phase, and, in time, a separation will occur. This is illustrated schematically for a column in Fig. 22-5.

The variables of time, amount, and activity of stationary phase, and liquid phase, temperature, etc., may be standardized.

[1] So-called because one variation of this technique was used to separate colored pigments from plant extracts.

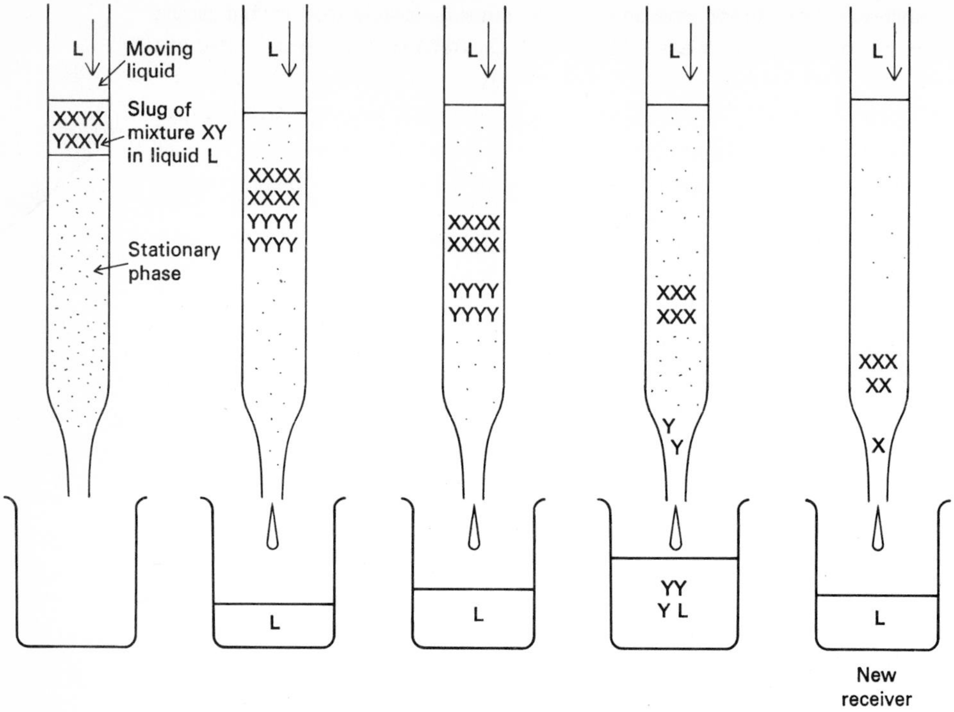
L
Moving liquid
XXYX
YXXY
Slug of mixture XY in liquid L
Stationary phase
XXXX
XXXX
YYYY
YYYY
L
XXXX
XXXX
YYYY
YYYY
L
XXX
XXX
Y
Y
YY
Y L
XXX
XX
X
L
New receiver

figure 22.5
Schematic chromatographic separation of mixture XY.

The amount of material required can, in some instances, be on the order of micrograms or it may be considerably larger (up to many grams). The advantage of the technique lies in the continuous repetition of the separation as the material passes down the column so that even very minute differences in properties are sufficient to effect a separation.

The types of interaction most commonly utilized are those of adsorption, ion exchange, and solvent partition. Inorganic polymers such as alumina and silica gel (aluminum and silicon oxides) have on their surfaces many polar groups such as —O—Al—O⁻ (Al bearing two further bonds, one above and one below), Si—OH, Al—O^-Na^+, etc., which exert very strong short-range forces on other polar groups. As a solution of a substance passes by such a column, the substance is alternately attracted to these polar groups (adsorbed) and redissolved in the solvent (desorbed). One may also prepare inorganic or (more commonly) organic polymers with a large number of salt-like groups on the surface (—C(=O)—O^-, —$N(CH_3)_3^+$) in which the group is bound to the polymer and cannot move away. Passage of ionic substances through such a column gives rise to salt formation and exchange of ionic partners, i.e., ion-exchange chromatography. Finally, a polymer may have a liquid phase so strongly bound that it cannot move so that when a second immiscible liquid passes by, the substance to be separated is alternately dissolved in the stationary and the moving liquid. Cellulose is a polymer which holds a number of water molecules in this fashion so that the technique of paper chromatography is one of liquid-liquid partition chromatography.

So many variations of the above techniques are now in common use that it is impossible to discuss them in any detail. Currently, column, thin-layer (a thin layer of an adsorbent on a glass sheet), paper, gas-liquid, and ion-exchange chromatography are indispensable analytical and preparative tools. With the amino acids, paper, ion-exchange, and thin-layer chromatography have been developed to the point where complete separation of all the known amino acids can be effected readily and even quantitatively.

TABLE 22.1
APPROXIMATE AMINO ACID COMPOSITION OF SOME PROTEINS, PERCENT

	Protein			
	Gelatin	*Hemoglobin*[a]	*Egg albumin*	*Insulin*
	Molecular weight			
Amino acid	*Up to 70,000*	*68,000*	*40,000–46,000*	*46,000*
Arginine	7.6	3.5	5.7	3
Histidine	1	7.6	1.8	4
Lysine	4.3	8	4.5	2
Tyrosine	0.2	3	4.2	12
Tryptophan	0	1–2	1.4	
Phenylalanine	2	6.7	6	1
Cystine	0.1	0.4	1.7	12
Methionine	0.8	0.5	5.0	
Serine	3.3	5.2	7.6	4
Threonine	1.5	6.8	3–4	3
Leucine	3.7	16.6	9.4	30
Isoleucine	1.7	1.5		
Valine	2.5	8.2	6.8	
Glutamic acid	5.4	5.7	16.3	21
Aspartic acid	8.5	8.1	8.2	
Glycine	23.6	. . .	1.9	
Alanine	10	7–8	7.4	
Proline	15.3	2	4–5	10
Hydroxyproline	13			

[a]On the basis of its iron content, a "minimum molecular weight" can be calculated for hemoglobin. The molecular weight given above is a multiple of this (see Exercise 12).

22.9 Determination of the Primary Structure of Proteins The determination of the primary structure of a protein, assuming it can be obtained in a pure form, is accomplished by determining the composition and amino acid sequence. The composition is obtained by hydrolyzing the protein into its constituent amino acids and then determining the kind and relative amounts of amino acids present, using the above chromatographic techniques.

The sequence is determined in stages—first the terminal amino acid is labeled, often by treatment with dinitrofluorobenzene.

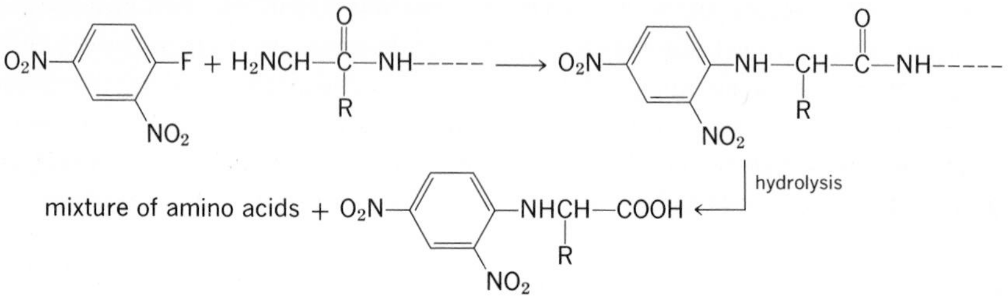

The labeled amino acid can be identified and, as it is the only monobasic acid with a dinitrophenyl group, it can be assigned the terminal position. Next a sample of the protein is partially hydrolyzed to give a mixture of peptides. These can be separated and labeled with dinitrofluorobenzene; further hydrolysis gives each terminal amino acid. Each peptide can be further degraded and labeled. Finally, one will cut these down to a series of known di- and tripeptides. By fitting these overlapping pieces together the entire sequence can ultimately be determined. Needless to say, it is a very tedious and time-consuming task.

Sanger applied this general method to the crystalline protein insulin and established its structure, at least as far as the amino-acid sequence is concerned. Some of the peptides isolated are shown below, with the sequence deduced from a knowledge of their structures.

Peptides Isolated

Phenyl alanine—valine
valine—aspartic acid
aspartic acid—glutamic acid
glutamic acid—histidine—leucine
valine—aspartic acid—glutamic acid

Reconstruction of a Portion of Insulin Chain

Phenyl alanine—valine—aspartic acid—glutamic acid—histidine · · ·

In Table 22.1 the amino-acid compositions and molecular weights of a few typical proteins are given. It is obvious from this table that a protein, which might contain several hundred amino-acid residues representing twenty or so amino acids, can be a very complex substance indeed. One of the most important present-day problems of physiological chemistry is how such complex molecules as these can be duplicated over and over by a cell and copied from generation to generation.

22.10 Synthesis of Proteins Once the sequence of a protein is known, the protein can, in principle, be synthesized. In practice this is an extremely difficult procedure. In the first place, even if each step went in 90 percent yield adding only 20 amino acids, in sequence this would result in a yield of $(100)\,(0.9)^{20}$ percent or 12.5 percent. Added to this is the difficulty that amino acids are bifunctional so that one end must be protected, and the other end activated so as to ensure that the right end of amino acid A reacts with the other end of amino acid B and not with another molecule of itself. A number of N-protecting groups have been devised which are stable enough to carry through the necessary reaction, but which can be readily removed when desired. Two of these are the carbobenzoxy group and the phthalimido group.

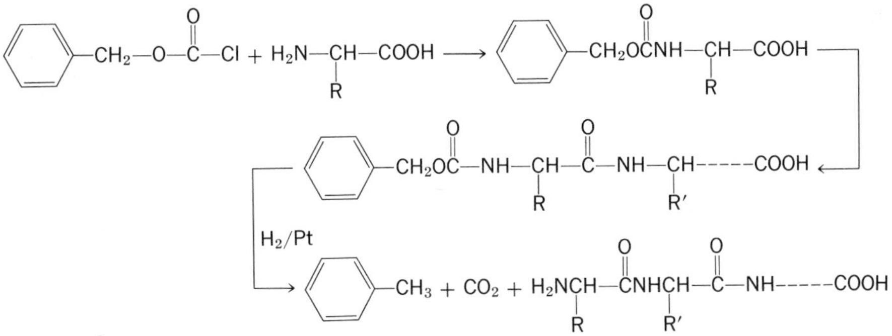
H2/Pt

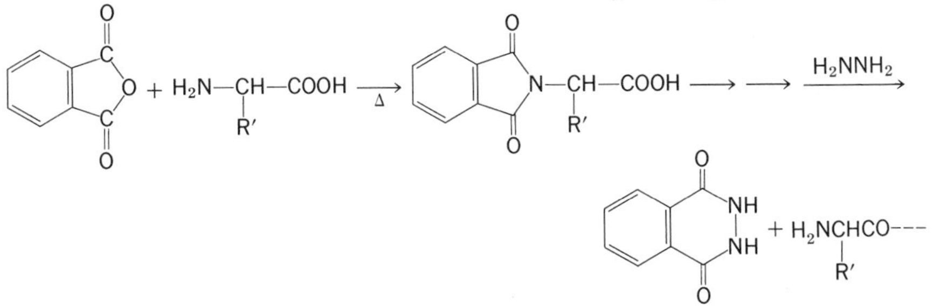
Δ
H2NNH2

The former utilizes the tendency of benzyl alcohol derivatives to hydrogenolyze to toluene and the second the ease of reaction of hydrazine with phthalimides.

Such methods have helped in the synthesis, but other problems remain. One of the major problems is the similarity of starting materials and products and the difficulty of separating them if reaction is not quantitative (which it seldom is).

Many clever devices have been used to minimize this problem. For example, the protecting group can be a polymer molecule, such that:

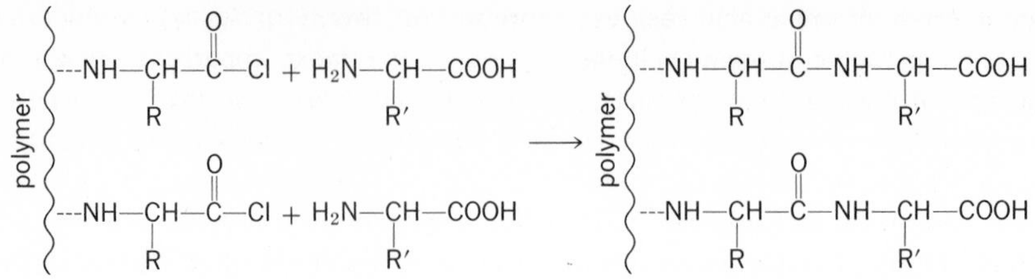

$$\underset{\text{polymer}}{\Big\{} \cdots\text{NH}-\underset{\underset{\text{R}}{|}}{\text{CH}}-\overset{\overset{\displaystyle\text{O}}{\|}}{\text{C}}-\text{Cl} + \text{H}_2\text{N}-\underset{\underset{\text{R}'}{|}}{\text{CH}}-\text{COOH}$$

$$\underset{\text{polymer}}{\Big\{} \cdots\text{NH}-\underset{\underset{\text{R}}{|}}{\text{CH}}-\overset{\overset{\displaystyle\text{O}}{\|}}{\text{C}}-\text{Cl} + \text{H}_2\text{N}-\underset{\underset{\text{R}'}{|}}{\text{CH}}-\text{COOH}$$

$$\longrightarrow$$

$$\underset{\text{polymer}}{\Big\{} \cdots\text{NH}-\underset{\underset{\text{R}}{|}}{\text{CH}}-\overset{\overset{\displaystyle\text{O}}{\|}}{\text{C}}-\text{NH}-\underset{\underset{\text{R}'}{|}}{\text{CH}}-\text{COOH}$$

$$\underset{\text{polymer}}{\Big\{} \cdots\text{NH}-\underset{\underset{\text{R}}{|}}{\text{CH}}-\overset{\overset{\displaystyle\text{O}}{\|}}{\text{C}}-\text{NH}-\underset{\underset{\text{R}'}{|}}{\text{CH}}-\text{COOH}$$

Reaction of the attached amino acid with a large excess of $\text{H}_2\text{N}-\text{CHR}'-\text{COOH}$ and then filtration of the polymer removes the product mechanically from the other reactants. The still-attached dipeptide can be converted to an active derivative (e.g., an acid chloride, etc.) and the process repeated. When enough units have been added, the peptide can be released from the polymer. This and related methods have enabled chemists to synthesize moderately large peptides, and still larger ones are being produced as time goes on.

22.11 Peptide Hormones   Utilizing the same kinds of techniques and reasoning described for the insulin structural problem, the structures of some peptide hormones have been determined. These somewhat simpler protein-like molecules have been more amenable to synthesis than macromolecules, and several structures have indeed been confirmed by total synthesis. One of these, oxytocin, a hormone isolated from the posterior pituitary gland, is shown.

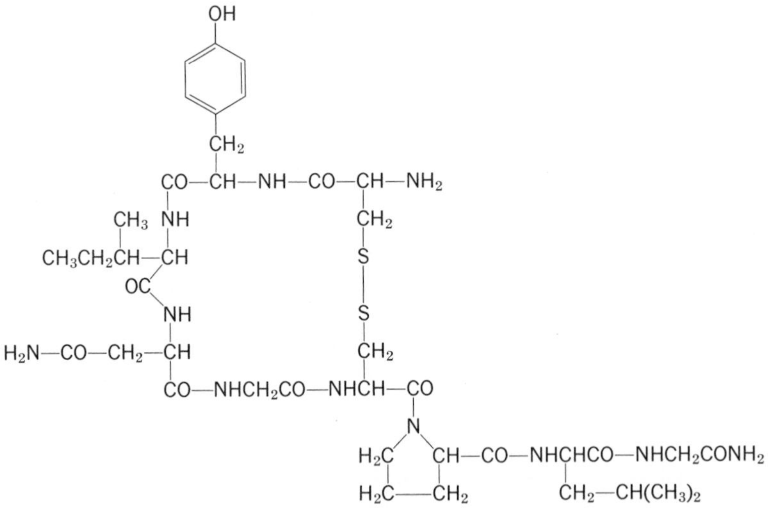

Oxytocin

22.12 Secondary Structure of Proteins   Because proteins have a wealth of carbonyl and NH groups in addition to free carboxyl and amino groups (from dibasic and diamino acids), it is not surprising that strong hydrogen bonds can be formed. This is manifested in the water solubility of many proteins. In recent years Pauling has shown that *intramolecular* hydrogen bonds are of extreme importance to many proteins. For instance, α-keratin occurs in the form of an α-helix with 3:6 amino-acid residues per turn. A helix is like a coil spring  or a spiral staircase. The advantage in forming such an arrangement comes from the maximum opportunity for the amide groups of one amino acid to hydrogen bond with the carbonyl of another, plus the fact that all of the bulky α-substituents end up on the outside of the helix, out of each other's way. (Because all natural amino acids are of the S configuration, the helix is a right-hand helix.)[1]

Other such interactions give protein molecules rod-like, globular, or flat shapes. Many of the very highly specific properties of proteins depend on which amino acid resides end up near each other in space. It is supposed that changes in this secondary structure will cause profound changes in these relationships, and, consequently, will impair or alter the biological function of the molecule. Such changes are part of the process called "denaturation" of proteins.

SYNTHETIC MACROMOLECULES—PLASTICS

Because the natural macromolecules frequently have some inconvenient or undesirable properties, much effort has been expended, particularly in recent years, in the development of synthetics; these have come to be called plastics and synthetic resins. The problem has been attacked from two angles: modification of natural macromolecules and synthesis from inexpensive materials of other new types, often modeled on the natural ones.

Both polysaccharides and proteins, but particularly the former, can be converted to derivatives with a marked change in physical properties, especially in solubility. The esters and ethers of cellulose, for example, are soluble in organic solvents and can be dissolved and reprecipitated in a different physical form: as a film, by evaporation (lacquers and enamels), or as a thread, by extrusion in a fine stream into a different medium. When the —OH groups are covered up, the hydrogen-bridging is eliminated, so that the molecules become insoluble in water. Nitrates, acetates, and methyl and ethyl ethers of cellulose all find commercial application under names such as Celluloid (nitrate), and as photographic film.

These derivatives are prepared by the usual reactions.

ACETYLATION OF AN ALCOHOL

$$ROH + (CH_3CO)_2O \longrightarrow ROCOCH_3 + CH_3COOH$$

[1] It is apparent from the structure that an R amino acid will not fit into such a helix without badly distorting it. Thus, one can understand one reason why unnatural (R-series) amino acids cannot usually be utilized by living substances.

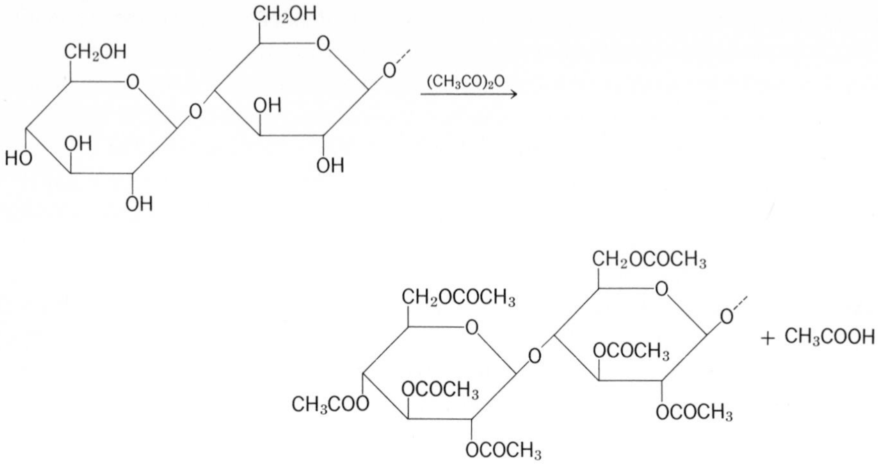
CH2OH
OH
HO
OH
O
(CH3CO)2O
CH2OCOCH3
OCOCH3
CH3COO
+ CH3COOH

Cellulose acetate

A major part of the rayon and Cellophane produced derives from the easily hydrolyzed xanthate of cellulose.

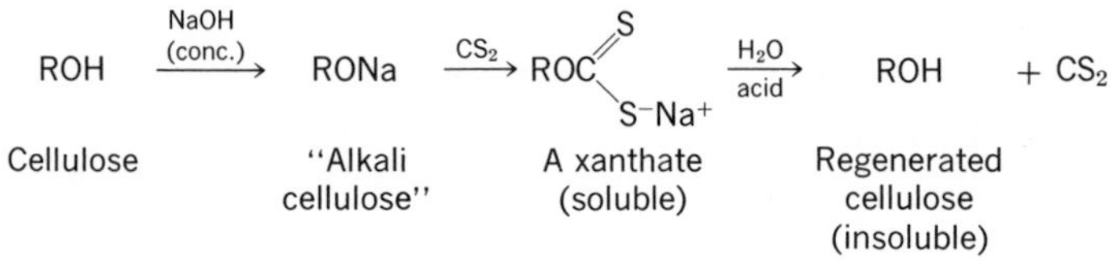
ROH
NaOH (conc.)
RONa
CS2
ROC
S
S⁻Na⁺
H2O acid
ROH
+ CS2
Cellulose
"Alkali cellulose"
A xanthate (soluble)
Regenerated cellulose (insoluble)

When the xanthate is extruded in solution into a bath of dilute acid as a sheet or thin stream, it is rapidly hydrolyzed, regenerating a slightly degraded form of cellulose as a sheet (Cellophane) or fiber (rayon). This is known as the *viscose process*.

Some efforts have been made to use native proteins also for fibers by precipitating them as a fiber and hardening (denaturing) with formaldehyde.

22.13 Totally Synthetic Polymers The world of synthetic polymers has grown almost beyond the imagination. From essentially nothing a little over 30 years ago the world production of synthetic polymers climbed to well over 10 billion lb by 1966. There are as many subclasses and types of synthetic polymers as there are uses. The types are most conveniently classified as thermoplastic or thermosetting. *Thermoplastic* polymers are linear, branched, or very lightly crosslinked polymers that soften on heating and harden on cooling reversibly so that they can be molded readily and yet retain their shape at ordinary temperatures. *Thermosetting* polymers are materials that on mixing or heating irreversibly crosslink so that they become infusible and insoluble without degradation.

22.14 Thermoplastic Polymers   These are of two types—addition polymers and condensation polymers. Addition polymers are usually formed by the addition of alkenes to one another without the loss or gain of any small molecules.

$$n \; \underset{H}{\overset{H}{}}C{=}C\underset{H}{\overset{H}{}} \longrightarrow \left(\overset{H}{\underset{H}{}}C{-}C\overset{H}{\underset{H}{}} \right)_n$$

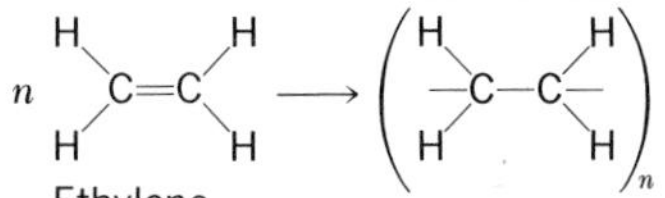

Ethylene   Polyethylene

Condensation polymers are formed by condensation of two groups with the loss of water or other small molecules.

$$n\mathrm{H_2N(CH_2)_6NH_2} \; + \; n\mathrm{HO\overset{O}{\overset{\|}{C}}{-}(CH_2)_4\overset{O}{\overset{\|}{C}}{-}OH} \xrightarrow{\Delta} \mathrm{H{-}(NH(CH_2)_6NH\overset{O}{\overset{\|}{C}}(CH_2)_4\overset{O}{\overset{\|}{C}}{)_n}OH} + 2n\mathrm{H_2O}$$

Hexamethylene diamine   Adipic acid   Nylon

In terms of volume, the most important polymers are the addition polymers: polyethylene, polypropylene, polystyrene, and polyvinyl chloride.

$$n\mathrm{H_2C{=}CH_2} \longrightarrow \left(\overset{H}{\underset{H}{}}C{-}C\overset{H}{\underset{H}{}} \right)_n$$

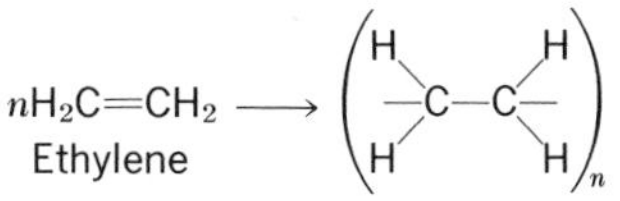

Ethylene   Polyethylene

$$n\mathrm{CH_2{=}\underset{CH_3}{CH}} \longrightarrow \left(\overset{H}{\underset{H}{}}C{-}C\overset{H}{\underset{CH_3}{}} \right)_n$$

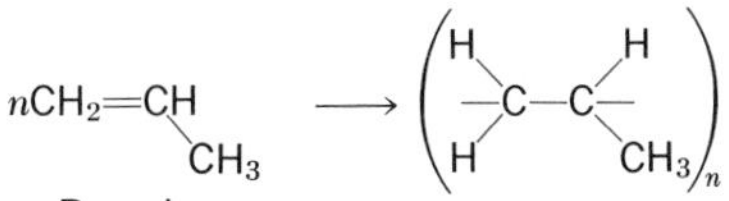

Propylene   Polypropylene

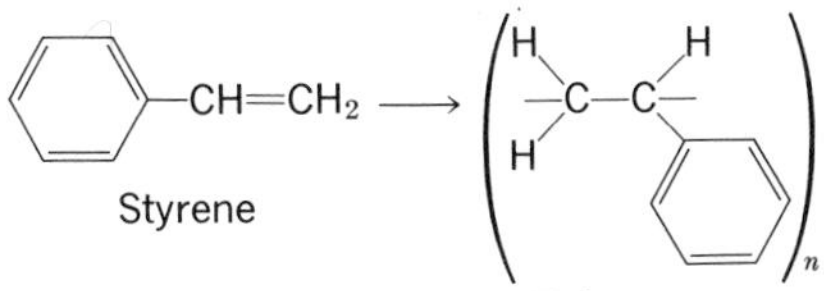

Styrene

Polystyrene

$$\mathrm{H_2C{=}C\overset{H}{\underset{Cl}{}}} \longrightarrow \left(\overset{H}{\underset{H}{}}C{-}C\overset{H}{\underset{Cl}{}} \right)_n$$

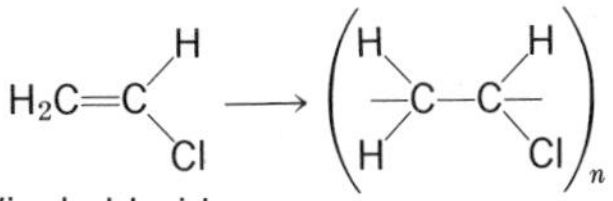

Vinyl chloride   Polyvinyl chloride
("vinyl")

The most common method of forming such polymers involves free-radical catalysis, illustrated for polystyrene (see Sec. 3.12).

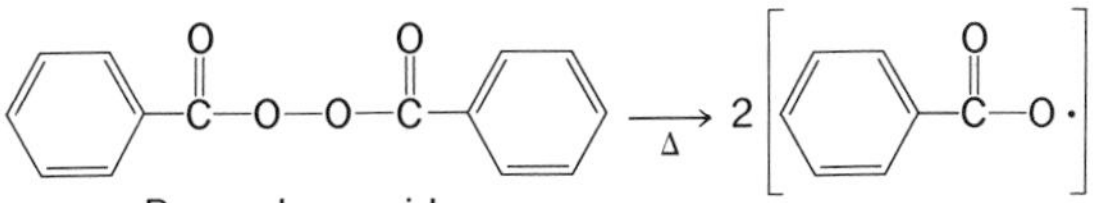

Benzoyl peroxide
(initiator)

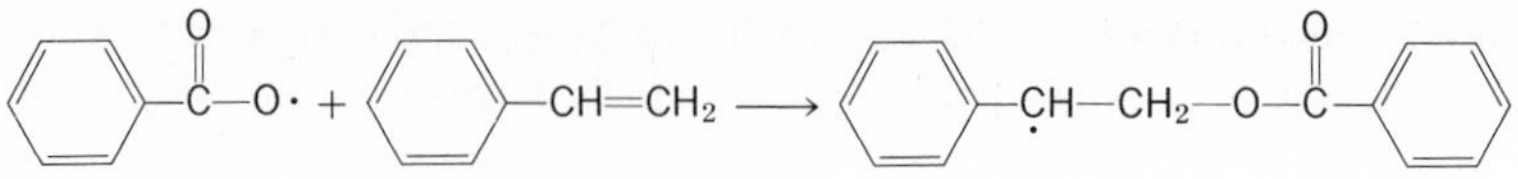

(initiation)

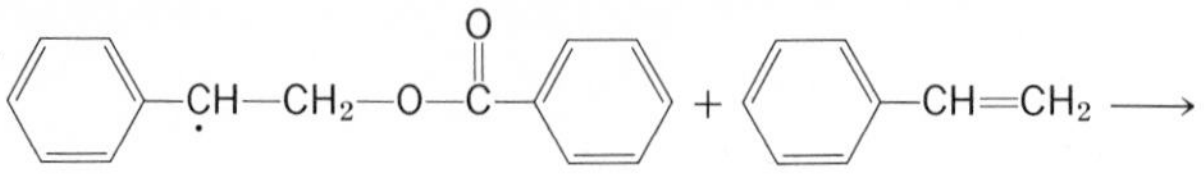

(propagation)

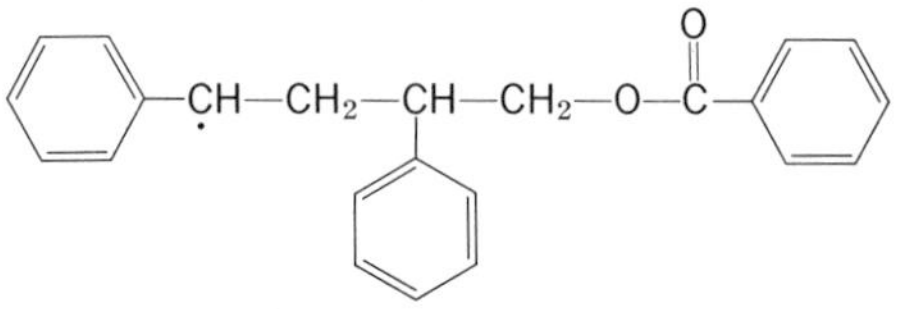

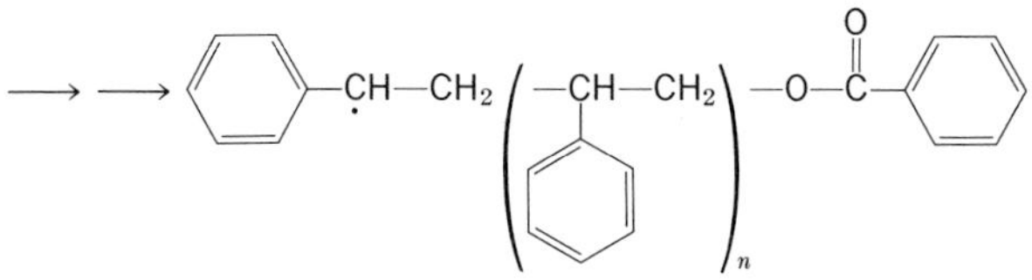

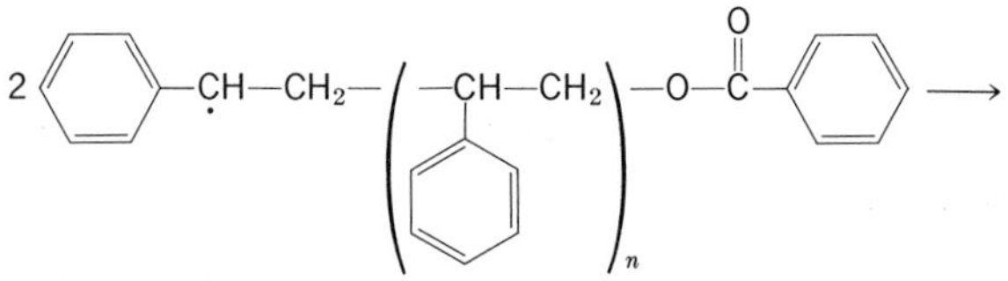

(termination)

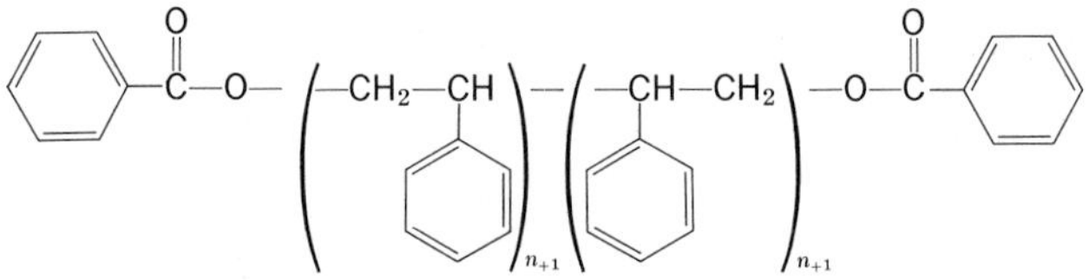

This is an example of a chain reaction in which a few free radicals are formed; these add to styrene molecules which, in turn, add to other styrene molecules until long polymer radicals form. Eventually two polymer radicals (or a polymer radical and some other radical such as an initiator radical) meet and combine, thus ending the chain reaction for these radicals. New radicals are continually being formed by the initiator and the process starts all over again until most of the styrene is consumed. Styrene is called a "monomer" and is converted to a polymer.

With ethylene, propylene, and certain other hydrocarbons another type of catalysis involving a transition metal and an aluminum alkyl is very important in that it gives very regular chains with no branching. Such polymers are more compact and are usually called "high-density polymers." The mechanism of this process is quite complex and will not be considered here.

Other addition polymers of considerable importance are polymethylmethacrylate

(Lucite®, Plexiglass®), polytetrafluoroethylene (Teflon®), polyformaldehyde (Delrin®), and polyacrylonitrile (Orlon®, Acrylan®).[1]

$$H_2C{=}C\underset{\underset{\displaystyle CH_3}{|}}{}\overset{\overset{\displaystyle O}{\|}}{C}{-}OCH_3 \longrightarrow \left(-CH-\underset{\underset{\displaystyle CH_3}{|}}{\overset{\overset{\displaystyle CO_2CH_3}{|}}{C}}- \right)_n$$

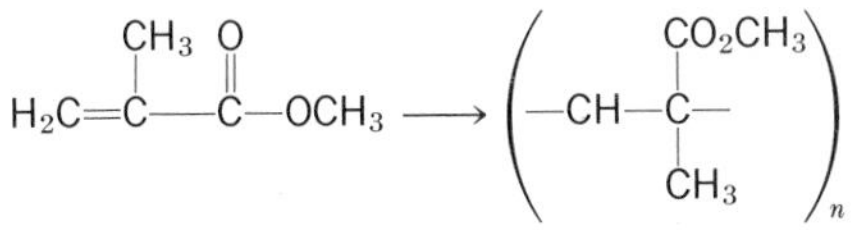

Methylmethacrylate   Lucite® or Plexiglass®

$$\underset{F}{\overset{F}{}}C{=}C\underset{F}{\overset{F}{}} \longrightarrow \left(-\underset{\underset{\displaystyle F}{|}}{\overset{\overset{\displaystyle F}{|}}{C}}-\underset{\underset{\displaystyle F}{|}}{\overset{\overset{\displaystyle F}{|}}{C}}- \right)_n$$

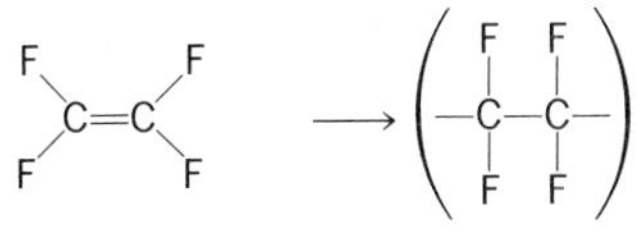

Tetrafluoroethylene   Teflon®

$$\underset{H}{\overset{H}{}}C{=}O \longrightarrow \left(-O-\underset{\underset{\displaystyle H}{|}}{\overset{\overset{\displaystyle H}{|}}{C}}- \right)_n$$

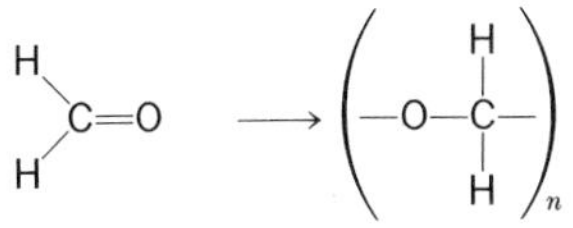

Formaldehyde   Delrin®

$$H_2C{=}C\overset{\overset{\displaystyle H}{}}{\underset{\underset{\displaystyle CN}{}}{}} \longrightarrow \left(CH_2{-}\underset{\underset{\displaystyle CN}{|}}{CH} \right)_n$$

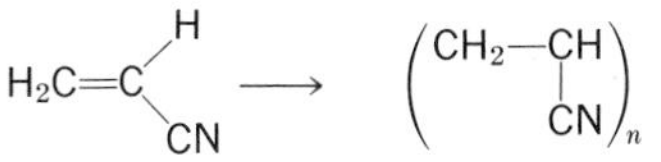

Acrylonitrite   Orlon® or Acrylan®

The condensation polymers of great commercial importance include nylon (a polyamide not unlike proteins) (Sec. 22.7).

Dacron® or Mylar® are formed from terephthalic acid and ethyleneglycol.

$$HO\overset{\overset{\displaystyle O}{\|}}{C}{-}\langle\bigcirc\rangle{-}\overset{\overset{\displaystyle O}{\|}}{C}{-}OH + HOCH_2CH_2OH \xrightarrow{\Delta} HO\overset{\overset{\displaystyle O}{\|}}{C}\left({-}\langle\bigcirc\rangle{-}\overset{\overset{\displaystyle O}{\|}}{C}{-}OCH_2CH_2{-}O\overset{\overset{\displaystyle O}{\|}}{C}{-} \right)_n$$

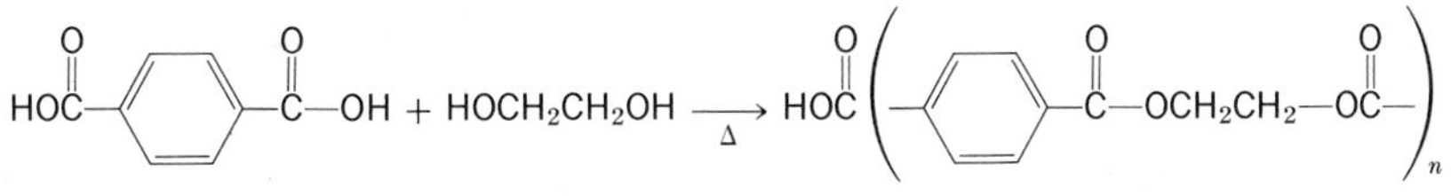

Dacron® or Mylar®

The thermoplastic polymers may be either plastics (polyethylene, polypropylene, polystyrene, Lucite®, Teflon®, Delrin®, or Mylar®) or they may be artificial fibers (nylon, Dacron®, Orlon®, or polypropylene).

22.15 Thermosetting Polymers   The thermosetting polymers are not fiber materials as they are usually too crosslinked to be drawn into fibers. Some of the more important ones are phenol-formaldehyde (Bakelite®), melamine, and epoxy.

---

[1] The symbol ® indicates that these names are registered trade marks and are the property of the company registering them. Since the more important plastics are commonly known by these names we use them here in preference to the generic names such as "polymethyl-methacrylate," "polytetrafluoroethylene," etc.

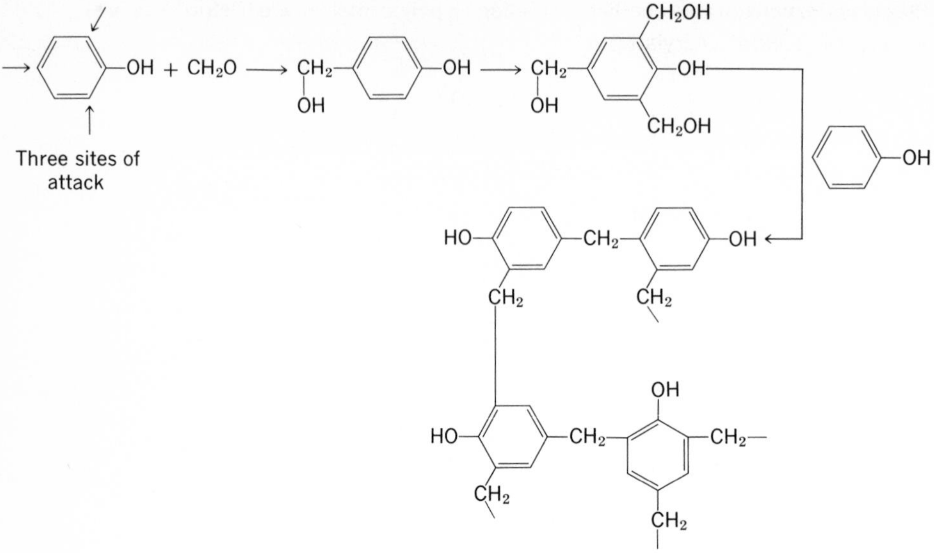
Three sites of attack

Bakelite®
(crosslinked)

Melamine-formaldehyde polymers give a very hard, tough material used extensively for counter tops, etc. The crosslinked structure involved is much like that involved in phenol-formaldehyde polymers.

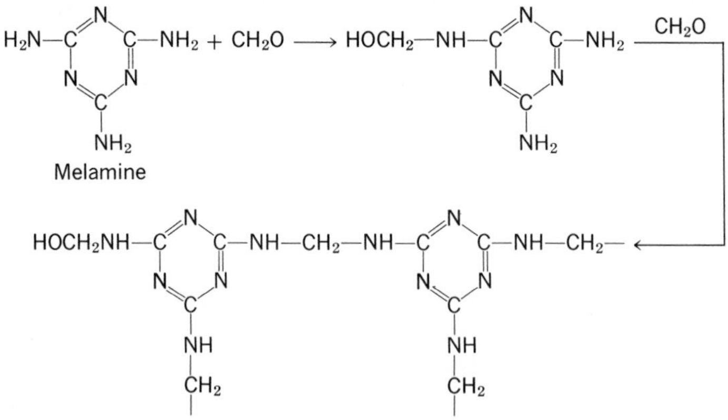
Melamine

Epoxy polymers utilize the reaction of an epoxide with (usually) an amine.

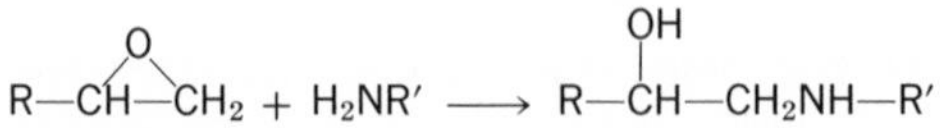

By utilization of amine-epoxide combinations such as the ones shown, highly crosslinked structures can be formed.

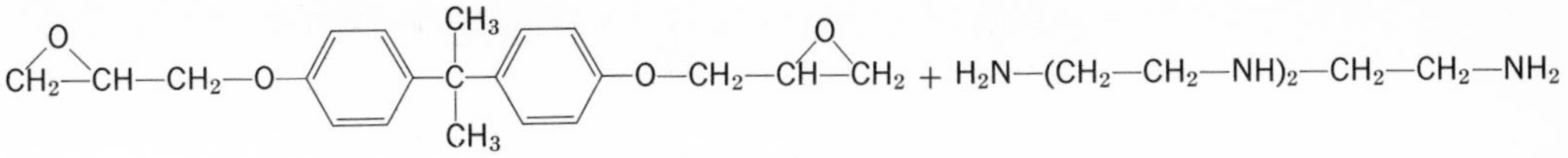
CH3
O
O
CH2—CH—CH2—O—
C
—O—CH2—CH—CH2 + H2N—(CH2—CH2—NH)2—CH2—CH2—NH2
CH3

Urethane polymers use diisocyanates and polyhydroxy compounds, the linking reaction being the reaction of an isocyanate with an alcohol (Sec. 16.2).

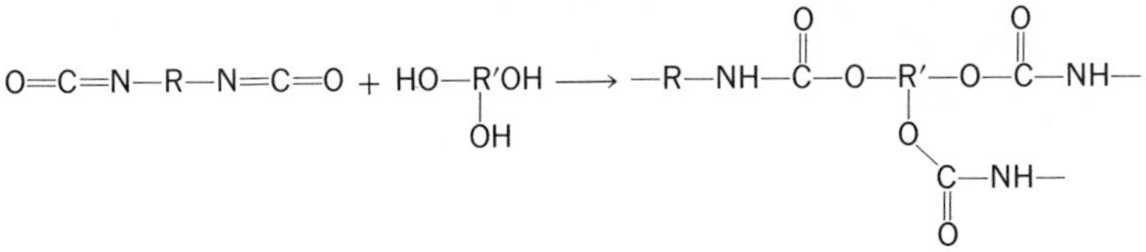
O=C=N—R—N=C=O + HO—R′OH ⟶ —R—NH—C—O—R′—O—C—NH—
OH
C—NH—

A polyurethane

By inclusion of difunctional groups it is possible to prepare thermosetting polymers from molecules like styrene that would ordinarily give linear polymers; control over the degree of crosslinking may be had by varying the ratio of difunctional compound to monomer.

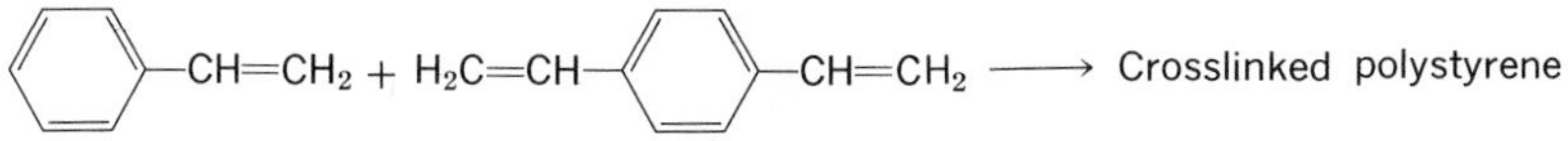
—CH=CH2 + H2C=CH—
—CH=CH2 ⟶ Crosslinked polystyrene

Few percent

In addition to the above types of polymers it is possible to prepare graft polymers (in which shorter chains of one type of polymer are attached at intervals to the linear chain of a different polymer). Natural polymers, such as rubber and leather, have either been duplicated or closely simulated by man. In many cases the synthetic polymer has properties superior to the natural material (e.g., neoprene and Corfam®).

ANSWERS TO EXERCISES

INTRODUCTION

Answers to selected exercises are included as an aid to study. An effort has been made to answer typical types of questions and to give more detailed answers as each new type is first encountered.

CHAPTER 1

3. Polar bonds are those between elements of different electronegativity. Polar molecules:

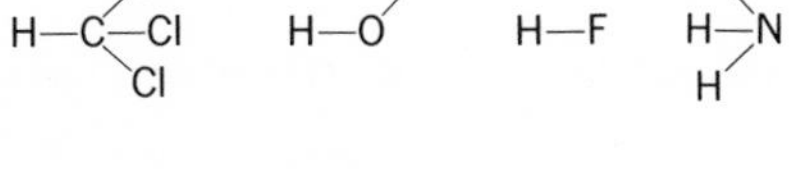

Nonpolar molecule:

O=C=O

Polar bonds:

$\overset{+}{C}-\overset{-}{Cl}$ $\overset{+}{H}-\overset{-}{O}$ $\overset{+}{H}-\overset{-}{F}$ $\overset{+}{H}-\overset{-}{N}$ $\overset{-}{O}=\overset{+}{C}$

The two polar C=O bonds in CO_2 oppose each other yielding a molecule which has no net polarity.

5. Electronegativity, in general, increases along each row (with increasing completeness of valence shell) in the periodic table. Also, elements in the first row are generally more electronegative than in the second. The more electronegative component is shown in boldface:

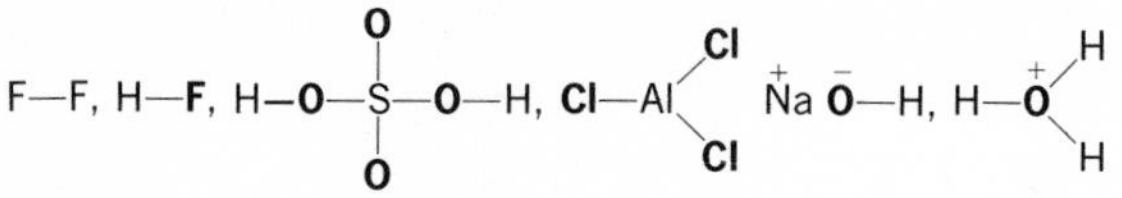

7.

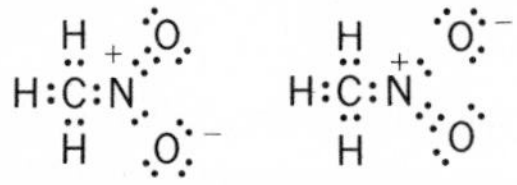

Yes. Resonance stabilization of the nitro group is important in this molecule.

9. Yes. Because of the tetrahedral structure of carbon, the carbon-bromine dipoles do not cancel.

11.

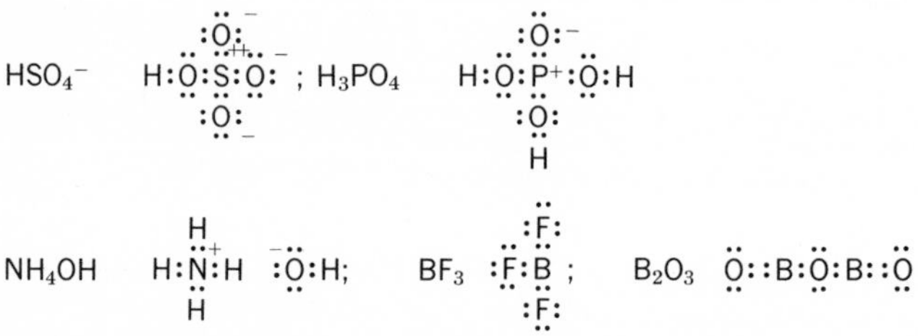

13. Ammonia, NH_3, exists in a nearly tetrahedral geometry. This can result from three sp^3-hybrid orbitals yielding N—H bonds with the fourth orbital containing an unshared pair of electrons.

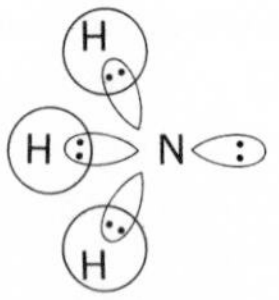

15.

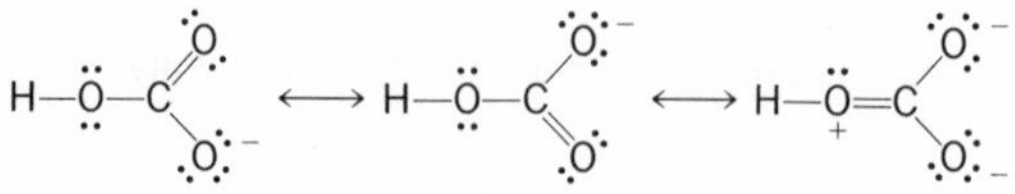

No. Two of the C—O bonds are identical, the third (C—O—H) C—O bond would be longer because of less C=O character (lesser importance of the third resonance structure due to unfavorable charge separations).

17. H:C:::C:H or H—C≡C—H

Due to the high concentration of electrons in the vicinity of the carbon atoms (as they are not shared with other elements), it will be a good nucleophile compared to CH_4.

19. The molecule

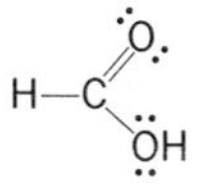

is resonance stabilized, the major contributing structures to the hybrid being:

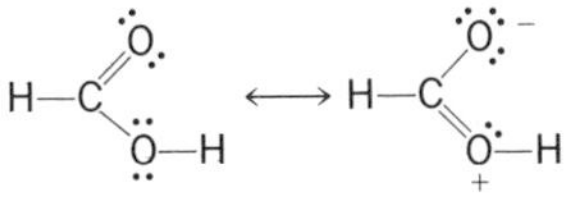

with the former structure being most important because of the more favorable charge distribution. The ion H—CO_2^- is greatly stabilized by resonance because both contributing structures are identical.

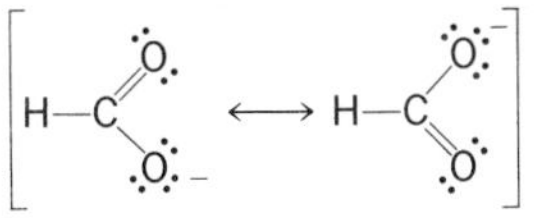

This means that the ion gains more from resonance stabilization than the molecule from which it is formed. The effect of this is to lower the energy of the ionization.

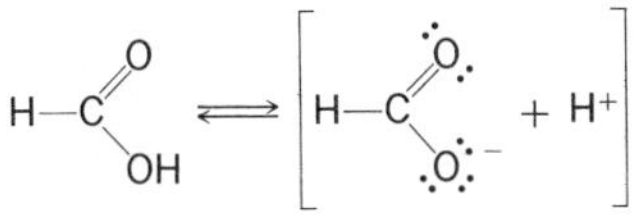

In the case of H—CH_2—OH $\longrightarrow$ [H—CH_2—O^- + H^+] there is no change in resonance stabilization on going to the ion so that the net ionization energy is larger and it is more difficult to form the ion.

23. H_2C=CH_2, but only at very low wavelengths (high energies), below 2000 Å.

25. Lewis acids: BCl_3, CH_3^+.
Lewis bases: H_2O, NH_3, ^-OH, and KBr (very weak).

29. No. Electrons and bonds cannot be rearranged in the ion without shifting atoms as well.

CHAPTER 2

1. Identical: *a* and *n*, *b* and *j*, *c* and *l*, *d* and *h*, *f* and *m*, *g* and *i*, *k* and *o*.
Isomers: *c* or *l* and *e*, and *k* or *o*. *f* or *m* and *g* or *i*. Members of same homologous series: none.

3\. *a.*

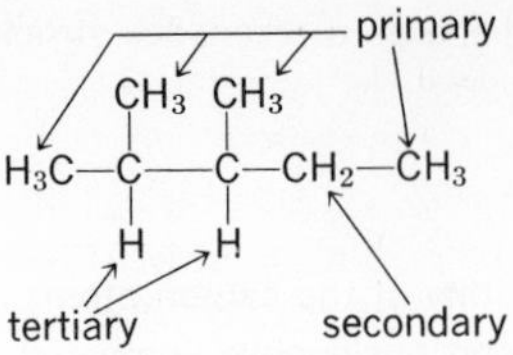

primary
tertiary
secondary

2,3-Dimethylpentane

nmr: Three different primary hydrogen peaks, one secondary hydrogen peak, two tertiary hydrogen peaks.

c.

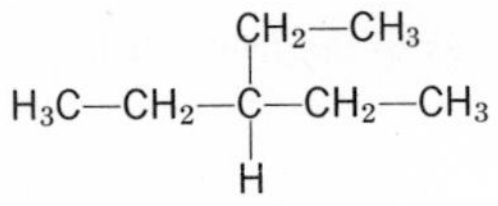

or

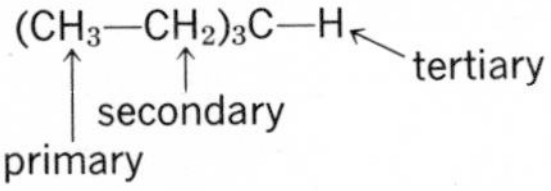

tertiary
secondary
primary

3-Ethylpentane

nmr: Three peaks, one for primary, one for secondary, and one for tertiary.

5\. Heptane. (Consider the boiling points of propane and $C_{20}H_{22}$ relative to the problem of containing the first, and of vaporizing the second in the engine.)

7\. The most convenient would be nmr since one molecule has a tertiary hydrogen and the other has no absorption in that region. Infrared could also be used if the pure compounds were available as there will be differences in the fingerprint region of their infrared spectra.

9\. One monobromo derivative. Four dibromo derivatives (two cis and two trans).

11\. *a.* 2-Methylbutane
c. 2,4-Dimethylpentane
e. 2-Methyl-3-ethylpentane (not 3-ethyl-4-methylpentane)
g. 3,3-Diethylpentane

13\. If the reaction were carried out at about 25°C the polychlorination products CH_2Cl_2, $CHCl_3$, and CCl_4 are all liquids, and as such occupy much less volume than they would as gases. The volume decrease would be one indication of reaction. This will not work for hexane and bromine at ordinary temperatures. The evolution or production of acid, HBr or HCl, would also indicate reaction. How would you test for this?

15\.

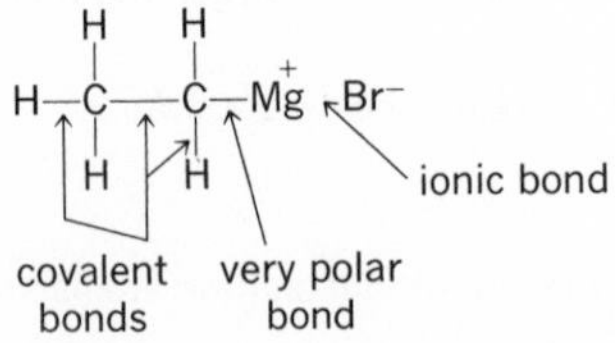

ionic bond
covalent bonds
very polar bond

$[H_3C{-}CH_2^-]$ having a negative charge would be very nucleophilic.

17\. No. Approximately 20.

19\. Bonding between carbon and sulfur utilizes the $3p$ orbitals of sulfur, whereas that between carbon and oxygen utilizes the $2p$ orbitals of oxygen. Since the radius of sulfur atoms is larger than that of oxygen due to the filled L shell of sulfur, and since $3p$ orbitals extend (on the average) farther from the nucleus than do $2p$ orbitals, the point of maximum bond strength occurs at a greater internuclear distance for sulfur than for oxygen. Similar considerations apply for C—F, C—Cl, C—Br, and C—I.

21. It is a property of a tetrahedron that the vector sum of four equal vectors directed in tetrahedral directions (toward the apices of a tetrahedron with the carbon atom at the center) will add up to zero. Thus the four equal C—Cl dipoles of CCl_4 just cancel. With $CHCl_3$, this cancellation will not occur and there will be a net dipole of about 1.9 units directed along the C—H bond axis. (If the hydrogen were a chlorine, a dipole of 2.3 units would be needed to just cancel out the resultant from the other three dipoles, as in CCl_4.)

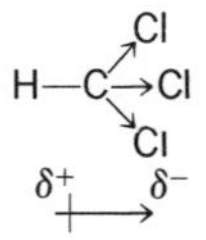

Cl
H—C→Cl
Cl
δ+
δ−

22. They would all give one peak in the nmr. They would, however, have quite different boiling points.

A chain of 59 carbon atoms would have primary (CH_3-type) hydrogens amounting to 5 percent of the total hydrogen.

23. The bulk of the three methyl groups causes a crowding (steric effect) which can be relieved by distortion of the bond angles. This angle distortion is energetically unfavorable in itself, so that a compromise is reached between methyl-group crowding and bond-angle distortion.

25. $CH_3Cl + Mg \xrightarrow{\text{ether}} CH_3MgCl \xrightarrow{D_2O} CH_3D + MgClOD$

27. *a.* See Exercise 3*b*.
b. $(CH_3)_2CH—CH(CH_3)_2$
c. $(CH_3)_2C{=}C(CH_3)_2$

CHAPTER 3

1. $H_2C{=}CH—CH_2—CH_2—CH_3$
1-Pentene

$H_2C{=}CH—C(CH_3)—CH_3$ (CH₃ on the third carbon)
3-Methylbutene-1

$H_3C—CH{=}CH—CH_2—CH_3$
2-Pentene
(cis and trans)

$H_3C—CH{=}C(CH_3)—CH_3$
2-Methylbutene-2

$H_3C—CH_2—C(CH_3){=}CH_2$
2-Methylbutene-1

There are only three isomeric pentanes.

3. *a.* $(CH_3)_2C{=}CH—CH_2CH_3$
c. $H_2C{=}CH—CH_2—CH{=}CH_2$

e. $H_2C{=}CH{-}CH(CH_3)_2$

g. 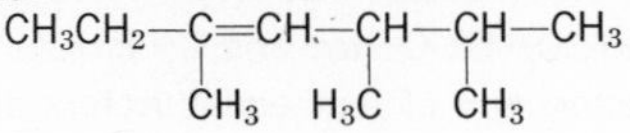

i. $CH_3CH{=}CH{-}CH_2CH_3$

5. *a.* nmr: One isomer would have CH_3 absorption of 33 percent of the total hydrogen, split into three peaks by the CH_2, while the other would have 67 percent CH_3, split into a doublet.

c. nmr: One isomer gives a single, large CH_3 peak and no other except the OH hydrogen.

e. Chemical: One isomer would give only $(CH_3)_2CO$ and the other would give that plus CH_3CH_2COOH.

nmr: One isomer has only a single type of CH_3 group and no other hydrogens.

7. Geranial:

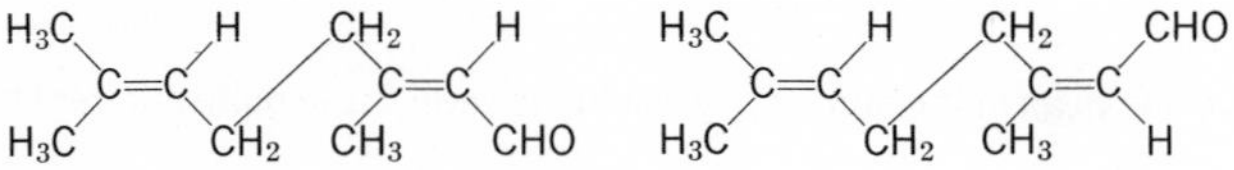

Linalool—only one cis-trans isomer.

9. Dehydration is a poor method of structure proof because of the possibility of rearrangement. Thus, structure I could easily rearrange so as to give the same major alkene product as II.

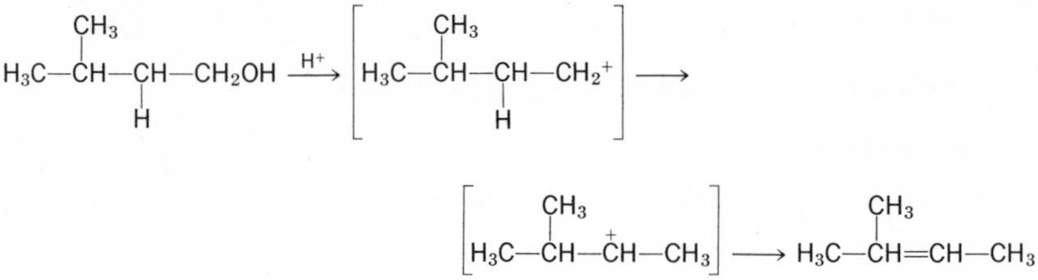

11. *a.* 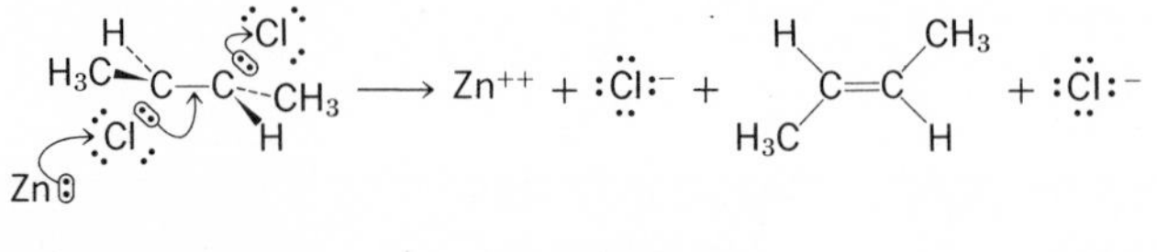

c. $Cl_2 \xrightarrow[\text{light}]{} 2[Cl\cdot]$

$[Cl\cdot] + (CH_3)_3CH \longrightarrow [(CH_3)_3C\cdot] + HCl$

$[(CH_3)_3C\cdot] + Cl_2 \longrightarrow (CH_3)_3CCl + [Cl\cdot]$

$[Cl\cdot] + (CH_3)_3CH \longrightarrow$ etc.

e. $RO{-}OR \xrightarrow{\text{spontaneously}} 2[RO\cdot]$

A peroxide

$[RO\cdot] + HBr \longrightarrow ROH + [Br\cdot]$

$[Br\cdot] + (CH_3)_2C{=}CH_2 \longrightarrow [(CH_3)_2\dot{C}{-}CH_2Br]$

$[(CH_3)_2\dot{C}{-}CH_2Br] + HBr \longrightarrow (CH_3)_2CH{-}CH_2Br + [Br\cdot]$

$[Br\cdot] + (CH_3)_2C{=}CH_2 \longrightarrow$ etc.

13. Oxidizing agents are agents which tend to remove electrons from other species. They are accordingly electrophilic.

15. Shaking gasoline with concentrated sulfuric acid will convert the olefins (alkenes) to carbonium ions and thence to water-soluble, gasoline-insoluble products. Shaking with alkaline permanganate solution is not too efficient for removal of olefins as they are not very water-soluble and complete removal is difficult. The decoloration of the permanganate solution does constitute a good test for the presence of alkenes, however.

17. *a.*

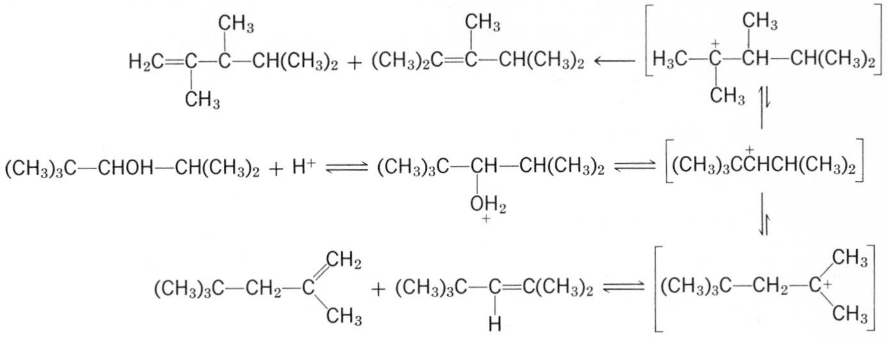

plus alcohols, etc., corresponding to each carbonium ion.

c.

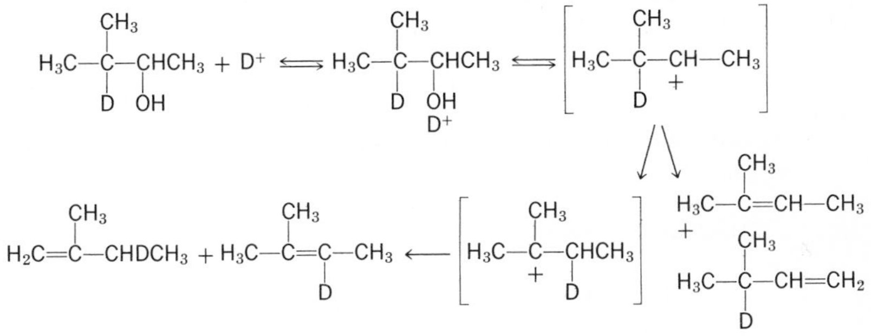

19. *a.* Alkaline $KMnO_4$ or Br_2.

c. Addition of CH_3CH_2OH. (One would yield *n*-butane, a gas, the other would not react.)

21. Compound *b* would react much more rapidly, as the intermediate chlorocarbonium ion would be tertiary rather than secondary as from *a*.

23.

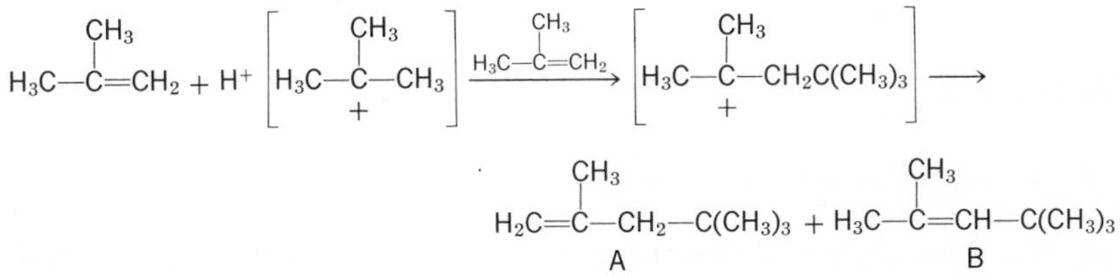

Oxidation of A with permanganate would give CO_2 and $CH_3COCH_2C(CH_3)_3$, while B would give $(CH_3)_2CO$ and $(CH_3)_3C—COOH$.

The nmr of B would show two sets of CH_3 groups in the ratio of 2:3, and only one alkene hydrogen, while A would have CH_3 groups in the ratio of 1:3 as well as secondary hydrogens and two alkene hydrogens.

25. *a.* 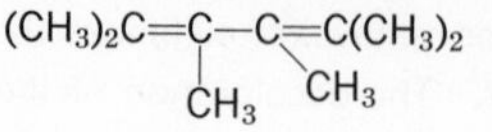

b. 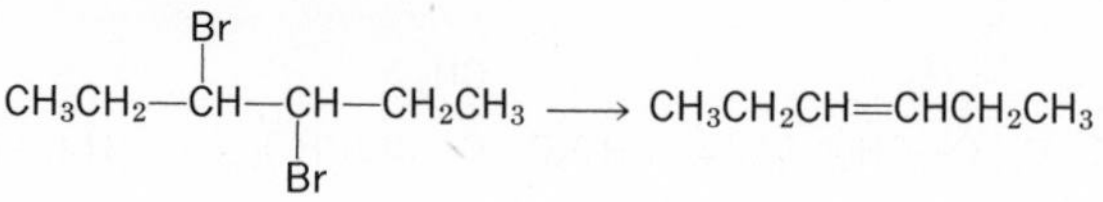

27. *a.* 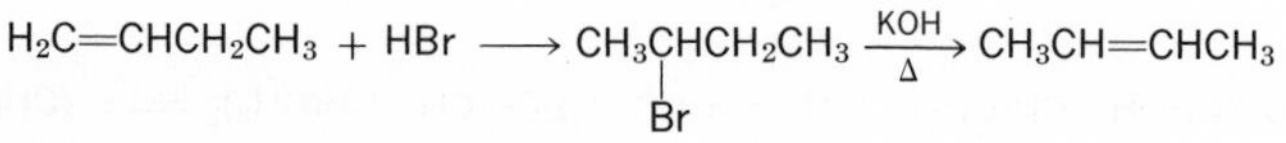

c. $CH_3CHBrCHBrCH_2CH_2CH_3 + \text{zinc} \longrightarrow CH_3CH{=}CHCH_2CH_2CH_3 \xrightarrow{H_2/Pt} CH_3(CH_2)_4CH_3$

e. 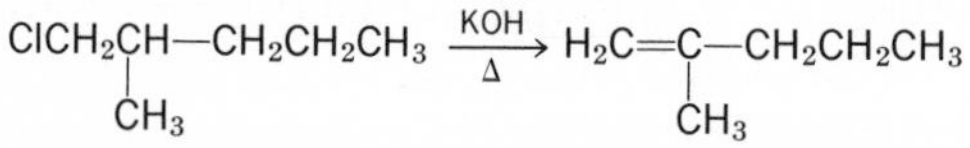

g. $H_2C{=}CHCH_2CH_3 + HBr \xrightarrow{\text{peroxides}} Br{-}CH_2CH_2CH_2CH_3$

29. I would be more stable as it would be resonance-stabilized.

$$[CH_3\bar{C}H{-}CH{=}CHCH_3 \longleftrightarrow CH_3CH{=}CH{-}\bar{C}HCH_3]$$

The first carbonium ion is both tertiary and resonance-stabilized while the second is only tertiary.

31. *a.* 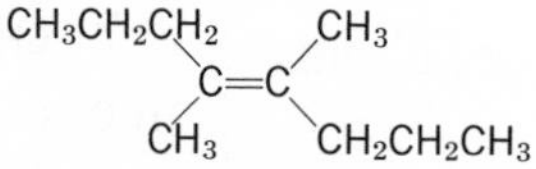

c. 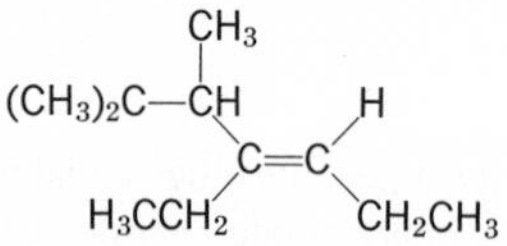

Use of cis and trans here is ambiguous unless specific groups are specified.

CHAPTER 4

1. *a.* 1-Ethyl-2-methylcyclobutane.
b. Cycloheptane.

3. The probability of two atoms, five carbons away from each other (chain ends), meeting each other in the normal course of molecular motion is very high; that for two groups twenty carbons away is less than that for collision with atoms of many other molecules.

5. 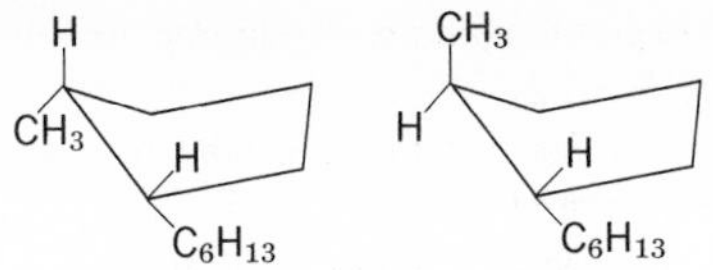

(cis) (trans)

1-Methyl-3-hexylcyclopentane

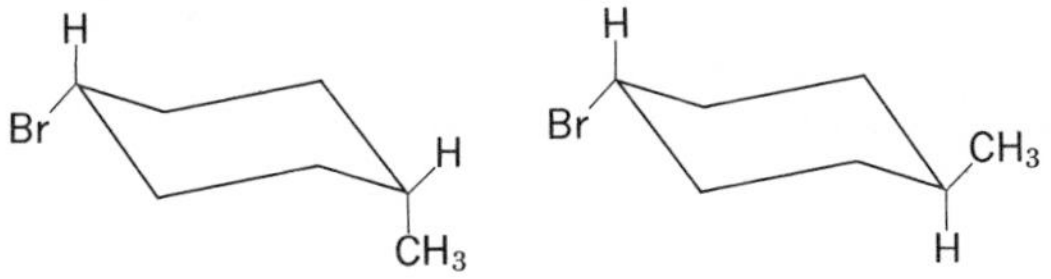

(cis) (trans)

1-Bromo-4-methylcyclohexane

7. No. (HINT: Consider the problems of putting a trans double bond in a five-membered ring.)

9. *b.* A penicillin.

d. Pyrethrin.

17. *a.* 1; *b.* 1 and 0; *c.* 2; *d.* 1 and 0; *e.* 1.

19. At high temperatures the cyclohexane rings are flipping rapidly between the two favored chair conformations. As the temperature is lowered, the flipping becomes slow enough so that the axial and equatorial fluorines are in different magnetic environments since they are no longer averaged out by the flipping process.

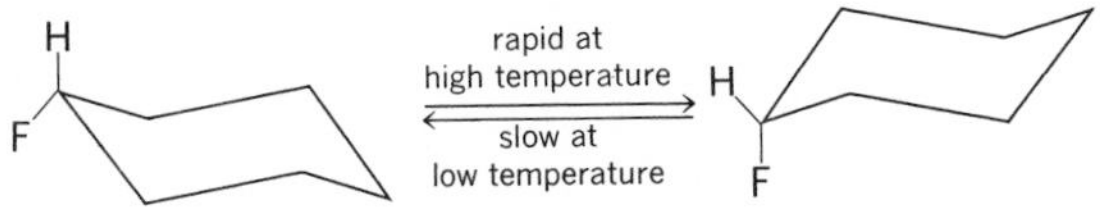
rapid at
high temperature
slow at
low temperature

Equatorial F Axial F

CHAPTER 5

1. *a.* $CH_3C{\equiv}CCH_3 + 2KMnO_4 \longrightarrow 2MnO_2 + 2CH_3COO^-K^+$

c. $CH_3C{\equiv}CH \xrightarrow[\text{(excess)}]{\text{HCl}} CH_3C(Cl)_2CH_3$

e. $3HC{\equiv}CH + 8KMnO_4 + 4H_2O \longrightarrow 6CO_2 + 8MnO_2 + 8KOH$

g. $KC{\equiv}CH + ICH_2CH_3 \longrightarrow CH_3CH_2{-}C{\equiv}CH + KI$

3. One could bubble the contaminated sample through dilute H_2SO_4 containing $HgSO_4$ and then through a column packed with KOH.

5. *a.* 1-Butene will have a characteristic C=C in the infrared.

1-Butyne will have a characteristic C≡C in the infrared and will react with alkaline Cu^+ and Ag^+.

2-butyne will have only one peak in the nmr. (Its infrared C≡C will be very weak or missing because of its symmetry.)

Butane will have no unsaturation in the infrared or nmr, and will not react with $KMnO_4$.

c. Propane and cyclopropane will not decolorize permanganate. Cyclopropane will have only one peak in the nmr at very high field ($\delta = 0.2$).
Propyne will have the characteristic acetylene infrared and will exhibit two peaks in the nmr in the ratio 3:1. Cyclopropene will have two regions in the ratio 1:1.

7. No. The sigma bond is, like all sigma bonds, formed by overlap along the axis of the two carbon atoms. The two π bonds are indistinguishable.

9. For vinyl acetate (using BF_3 as the Lewis acid):

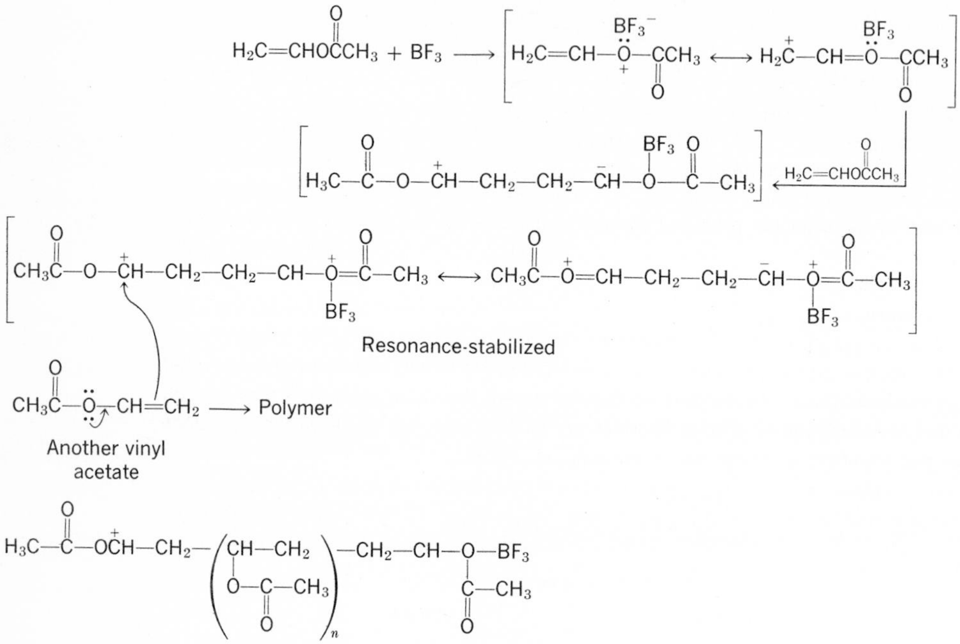
Resonance-stabilized
Polymer
Another vinyl acetate

11.

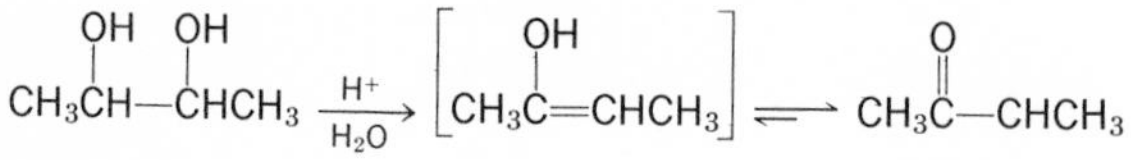

CHAPTER 6

1. If *o*-xylene had Structure I it could give only glyoxal and methylglyoxal by the cleavages shown (dashed lines). If Structure II were the structure, only glyoxal and diacetyl would be formed. The observation of all three products means that either a mixture of I and II, a rapidly equilibrating mixture of I and II, or some species intermediate between I and II best represents *o*-xylene.

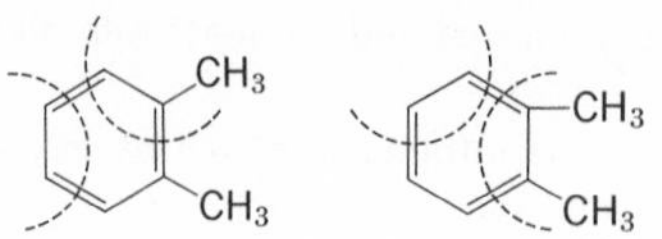

I II

3. The reactions of benzene are those of substitution rather than addition. The formula shown should undergo addition very readily, while the Kekulé formula for benzene, in which substitution preserves the conjugated system while addition destroys it, provides a logical preference for substitution. Also there are only two trisubstitution products possible.

5. *a.* *m* Directing and deactivating.

b. *o, p* Directing and activating (an unshared electron pair is available).

c. *o, p* Directing and activating (same reasons as *b*).

d. *m* Directing and deactivating (unshared pair is no longer available and a positive charge is present).

7. *a.* This is true for displacement reactions involving either S_N2 or S_N1 attack (see Chap. 9). Nucleophilic substitution on sp^2-hybridized carbon atoms is poor because the backside is crowded by the two other coplanar groups (in this case the ring), and attack from above or below is bad because it destroys the aromatic system. S_N1 substitution is poor because the carbonium ion center is in the plane of the ring and cannot get assistance from the π electrons, and it cannot become planar.

Certain other reactions, such as the formation of Grignard reagents, occur readily with bromobenzene, presumably by another mechanism.

c. Chlorine, being considerably more electronegative than carbon, is electron-withdrawing due to the C—Cl dipole. Three chlorines are sufficiently electron-withdrawing to make a negative charge on carbon moderately stable.

d. Elimination to form an alkene predominates with tertiary halides and bases such as the acetylide ion.

9. One might be tempted to say that both systems yield the ethyl cation, $CH_3CH_2^+$. In view of the reluctance of primary systems to form carbonium ions, however, it is more likely that two similar, strongly polarized (carbonium-like) reagents are formed, and that they both attack benzene as electrophilic reagents.

$$H_2C{=}CH_2 + H_2SO_4 \longrightarrow \overset{\delta^+}{CH_3CH_2}\text{—}\overset{\delta^-}{OSO_3H} \quad (x)$$

$$CH_3CH_2Cl + AlCl_3 \longrightarrow \overset{\delta^+}{CH_3CH_2}\text{—}\overset{\delta^-}{ClAlCl_3} \quad (x)$$

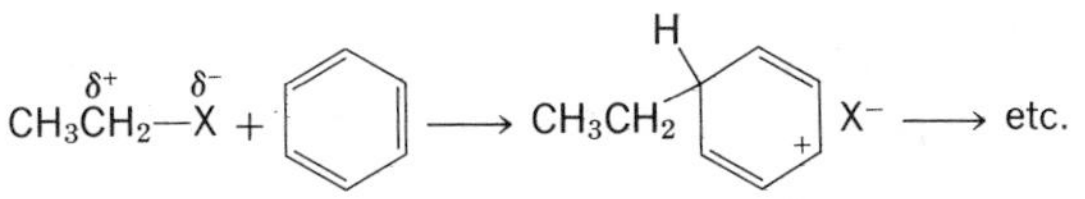

δ+ δ−
CH3CH2—X +
CH3CH2
H
+
X−
etc.

11. *a.* $CH_3CH_2OH + HCl \longrightarrow CH_3CH_2Cl + H_2O$

$$CH_3CH_2Cl + KOH \longrightarrow H_2C{=}CH_2 \xrightarrow{Cl_2} ClCH_2CH_2Cl$$

$$ClCH_2CH_2Cl + KOH \longrightarrow HC{\equiv}CH \xrightarrow[\text{liq. }NH_3]{Na} NaC{\equiv}CH$$

$$NaC{\equiv}CH + CH_3CH_2Cl \longrightarrow CH_3CH_2C{\equiv}CH \xrightarrow[Hg^{++}]{H_2SO_4\text{—}H_2O} CH_3CH_2\overset{\overset{\displaystyle O}{\|}}{C}CH_3$$

c. 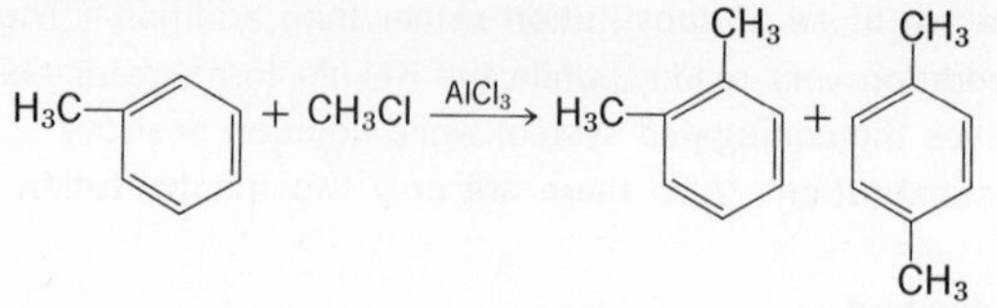

Separate isomers (by distillation—poor with this mixture—or by fractional crystallization). Or carry through next step and then separate.

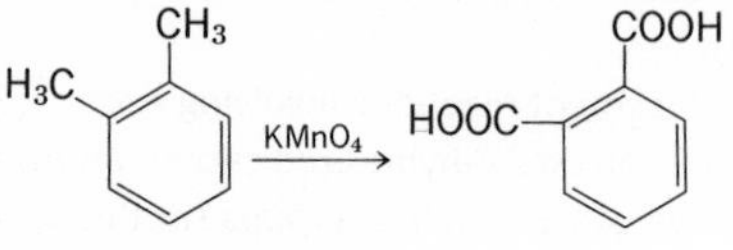

e. $C^\circ + CaO \xrightarrow[\Delta]{} CaC_2 \xrightarrow{H_2O} HC{\equiv}CH \xrightarrow[Hg^{++}]{H_2O}$

$CH_3{-}CH{=}O \xrightarrow{H_2} CH_3CH_2OH$

(NOTE: Syntheses not unlike that above are used for the preparation of isotopically labeled compounds, e.g., from $^{14}CO_2$.)

g. 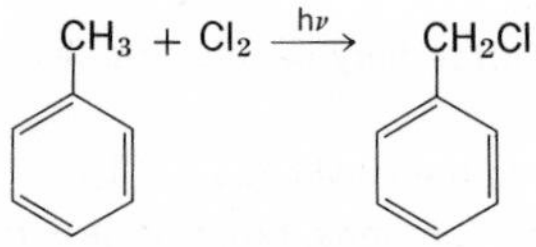

$HC{\equiv}CH \xrightarrow[\text{liq. } NH_3]{+Na^0} NaC{\equiv}CH$

(From carbon via syntheses *e* and *a*, if necessary)

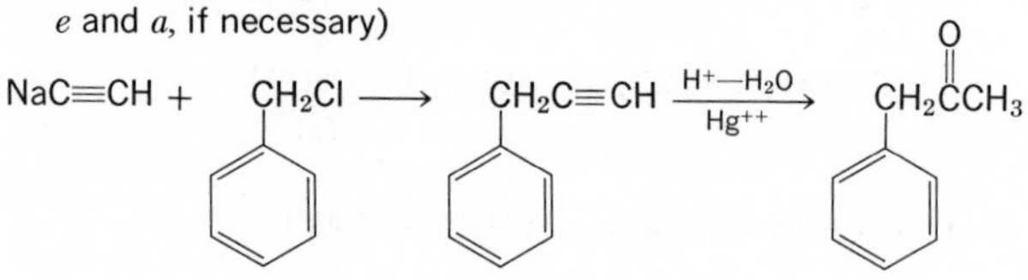

13. This reaction does not proceed in the absence of a trace of moisture or of CH_3CH_2Cl. With water a small amount of HCl is formed which adds to give a trace of CH_3CH_2Cl. Then a small amount (catalytic amount) of

$$\overset{\delta^+}{CH_3CH_2}{-}\overset{\delta^-}{ClAlCl_3}$$

is formed. As it is strongly polarized it can add to an alkene as follows:

$$\overset{\delta^+}{CH_3CH_2}{-}\overset{\delta^-}{ClAlCl_3} + H_2C{=}CH_2 \longrightarrow [CH_3CH_2CH_2CH_2^+]\ AlCl_4^-$$

$$[CH_3CH_2(CH_2CH_2)nCH_2CH_2^+]AlCl_4^- \xleftarrow{nH_2C{=}CH_2}$$

$$\downarrow -H^+$$

$$CH_3CH_2(CH_2CH_2)nCH{=}CH_2$$

Needless to say, this mode of polymerization proceeds very poorly with ethylene, as primary carbonium ion-like intermediates would be involved. It does proceed quite readily with higher alkenes.

15. The nitro group is m directing and strongly deactivating. However, in order for it to exert this deactivating influence in the adjacent ring, structures such as I, which destroy the aromatic resonance of both rings (as compared with II), must be invoked. Thus the influence of the nitro group is much weaker in the adjacent ring and products III and IV are formed in greatest quantity.

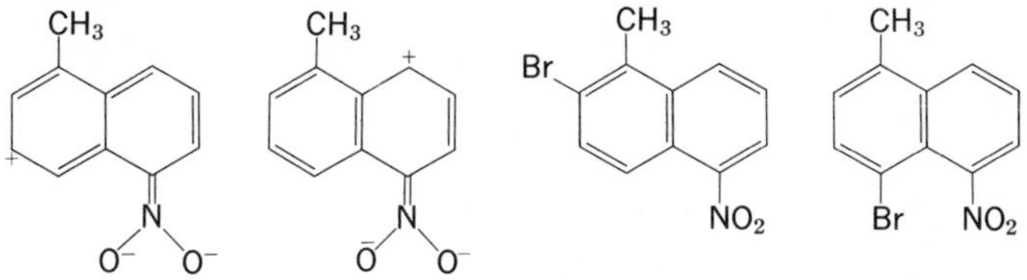

CH3
N
O⁻ O⁻
CH3
Br
NO2
Br NO2

I II III IV

17. *a.* The ion A is a very highly resonance-stabilized ion, while the parent molecule is an essentially unstabilized triene.

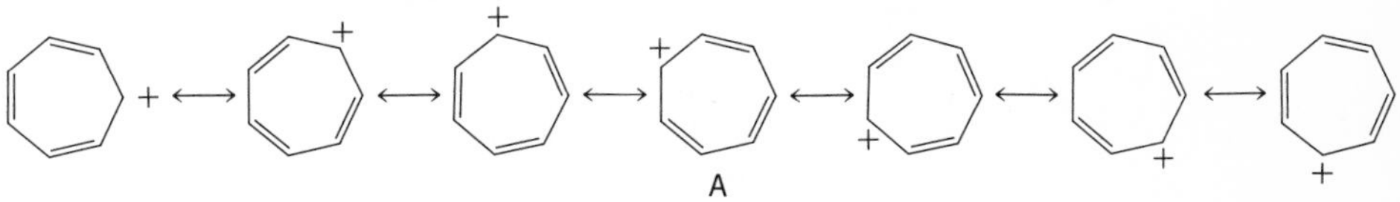

A

b. Groups which are activating toward electrophilic substitution are also activated toward attack by other electrophilic reagents, especially oxidizing agents. The converse is true for deactivating groups. Thus the ring with the OH substituent is destroyed.

19. C_9H_8. Unsaturation number $= 9 + 1 - 8/2 = 6$ (benzene $= 4$).

a. Terminal alkyne—this accounts for unsaturation over that of benzene.

b. Vigorous oxidation leaves eight carbon atoms—two substituents yielding two COOH groups.

c. Only the p isomer will yield a single monitro derivative.

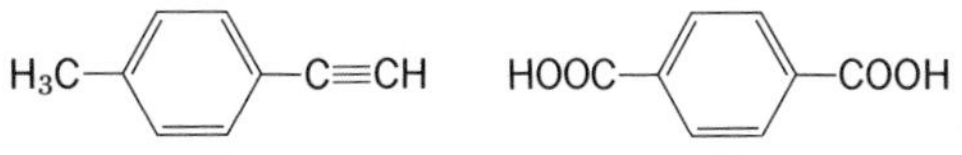

H3C
C≡CH
HOOC
COOH

A B

21. X has $N_u = 6$.

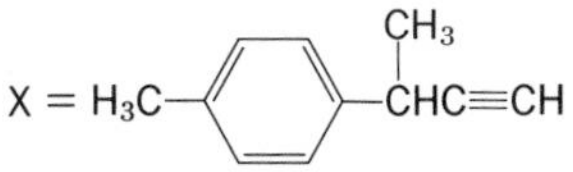

X = H3C
CH3
CHC≡CH

23.

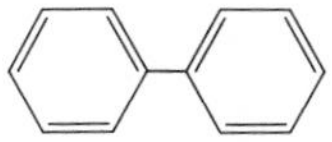

25. Fluorine is very electronegative and the trifluoromethyl group is accordingly very strongly electron-withdrawing. Since the sulfonyl group (SO_2) bears a formally electropositive sulfur, the overall group is very strongly electron-withdrawing indeed. This should accordingly be *m* directing and strongly deactivating.

27. See Exercise 25. $CF_3CH_2CH_2Cl$. (A primary carbonium ion-like species is better than $CF_3—C^+$.)

29. *a.* 3. (A triplet for the CH_3, a quartet for the CH_2, and a rather broad mixture for the aromatic H's.)

c. 2. (The two NH hydrogens and the aromatic H's.)

e. 2. (The aromatic hydrogens and the acidic hydrogen.)

CHAPTER 7

1. *a.*

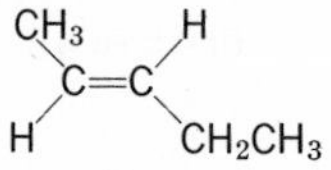

Trans

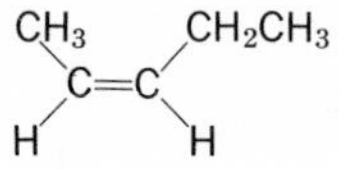

Cis

c.

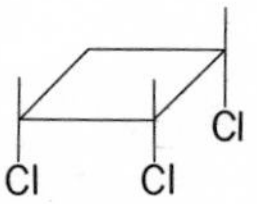

cis, cis

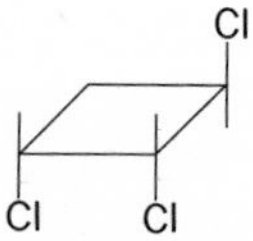

cis-trans
(or trans-cis)

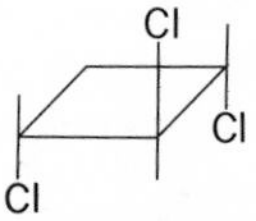

trans-trans

e.

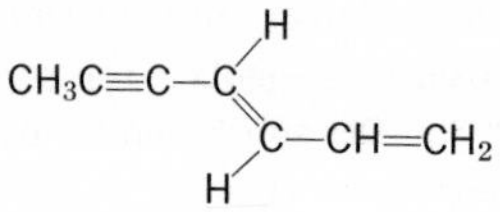

Trans

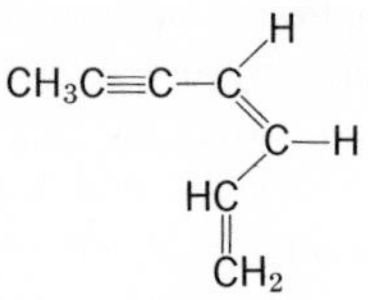

Cis

g.

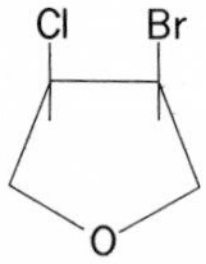

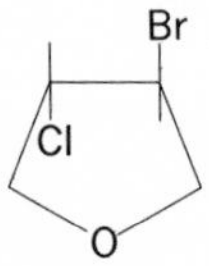

5. *a.*

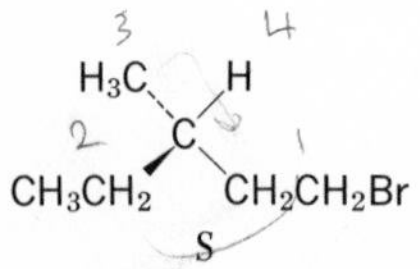

S

b. Is inactive.

c.

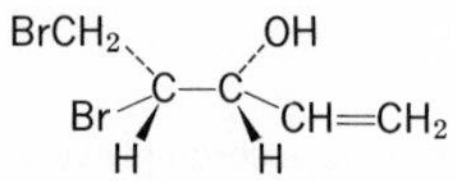

S, R

d. 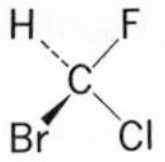

S

e. Is inactive.

7. No. Both enantiomers would give salts which were still enantiomers and not diastereomers.

9. The displacement reaction does not change the configuration of the optically active center. Since the new substituent, —C≡C—, still has the highest sequence of any of the substituents, each half of the molecule will still be *R*. In order for a meso product to be formed, one-half of the molecule must be the mirror image of the other half, that is, one-half must be *R* and the other *S*. The product will thus be optically active.

11. *a.* 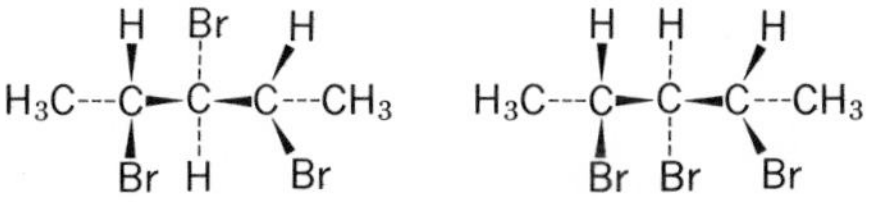

Meso

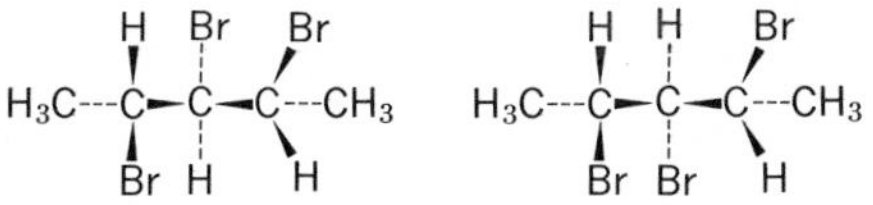

Active

b. 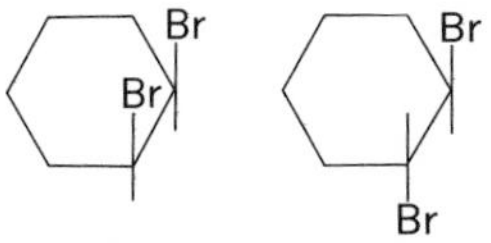

Cis (meso) — Trans (active)

c. 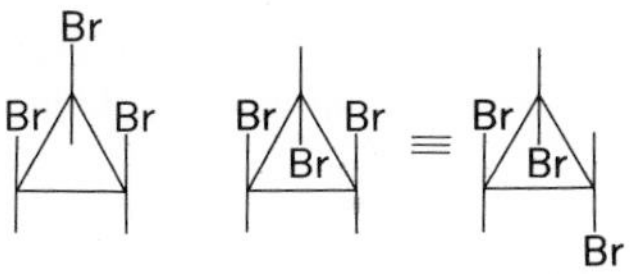

Cis — Trans (both inactive)

13. 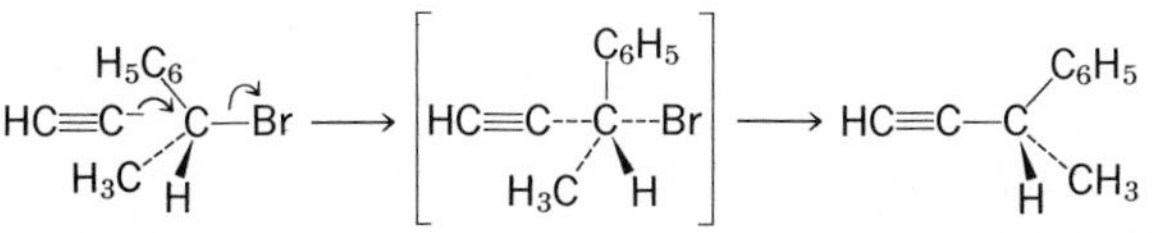

R — *R*

Displacement occurs from backside with inversion of configuration. Because the HC≡C— group is of lower priority than C_6H_5 (phenyl), the absolute configuration is still called *R* even though an inversion has occurred. When priorities of groupings are changed in a reaction, the change (or lack of it) from R to S does not necessarily mean inversion (or retention).

15. Cyclopentene and dil. $KMnO_4$ yield 1,2-dihydroxycyclopentane. The cis isomer will be meso and not resolvable (see Exercise 11*b*), whereas the trans isomer will be active. Therefore, both hydroxy groups must have been added from the same side of the ring (cis addition).

CHAPTER 8

1. *a.* $H_2C{=}CH_2 + KMnO_4$ (dil.) $\longrightarrow HOCH_2{-}CH_2OH$

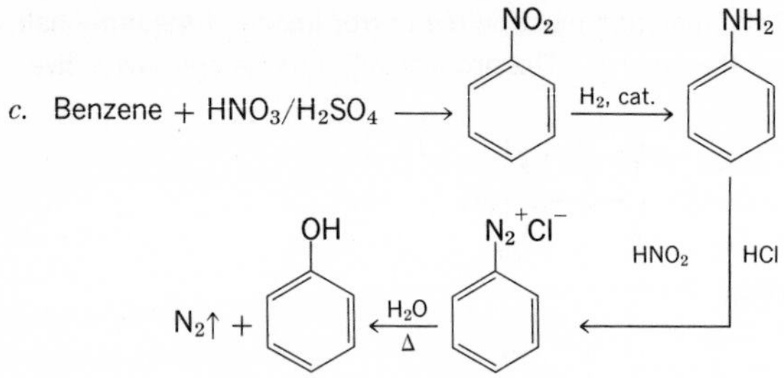

c. Benzene + HNO_3/H_2SO_4
NO_2
NH_2
H_2, cat.
HNO$_2$
HCl
OH
$N_2^+Cl^-$
$N_2\uparrow$ +
H_2O
Δ

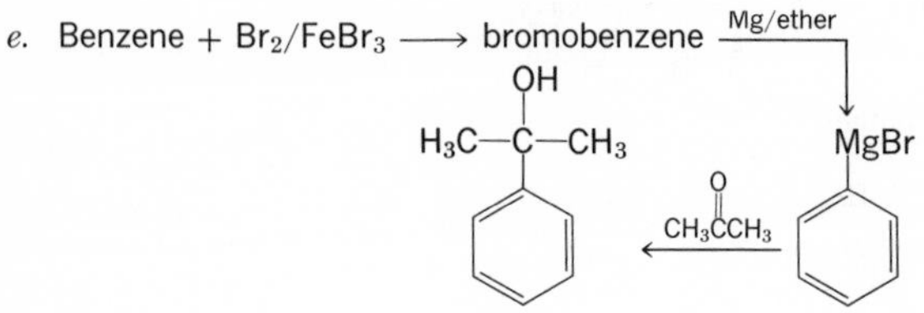

e. Benzene + $Br_2/FeBr_3$ ⟶ bromobenzene
Mg/ether
MgBr
OH
$H_3C{-}C{-}CH_3$
O
CH_3CCH_3

3. H_3O^+. Stronger base. One can compare the equilibrium conversions of the type:

$$B\!: + HX \rightleftharpoons BH^+ + X^-$$

The stronger the base, the more the equilibrium will be shifted toward the right.

5. The electron density on the aromatic ring is increased due to contributions of the following structures to the resonance hybrid:

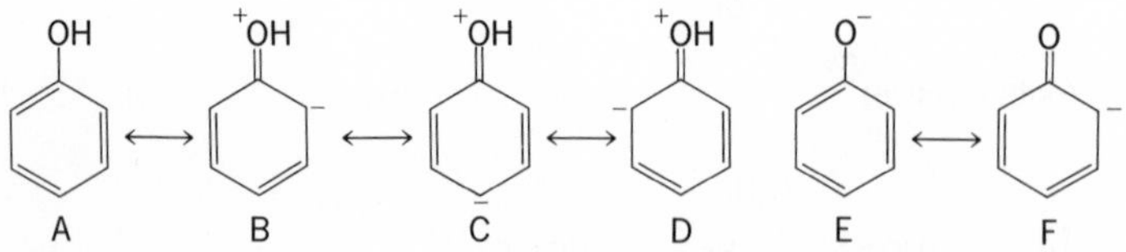

OH
$\overset{+}{O}H$
$\overset{+}{O}H$
$\overset{+}{O}H$
O^-
O
A
B
C
D
E
F

Note that structures B, C, and D have a positive charge on an electronegative atom (in itself energetically poor) but they compensate for this in part by having an extra bond to the oxygen.

The properties of the OH group are affected, especially insofar as the ion derived by loss of a proton, E, is highly resonance-stabilized by resonance structures of type F, so that the acidity of these alcohols (phenols) is increased over that of regular alcohols.

7. Order of increasing solubility from left to right:

a. C_4H_9Br, C_4H_9OH, CH_3OH. (Like dissolves like, and the larger the percentage of solubilizing groups—water-like—the more soluble a molecule will be.)

c. Sodium phenolate, phenol, 1-decanol (its sodium salt would be completely hydrolyzed in water), benzene. [Salts, if not hydrolyzed, are usually strongly solvated (hydrated) in water, and this makes them more soluble—the other compounds follow the rule set forth in *a.*]

9. *a.* Nmr would show differences, for example in the CH_3—splittings and intensity. One molecule would show a CH_3 as a triplet in the normal CH_3 region; another would show two CH_3 absorptions, one a triplet, and the other a doublet; the other would show a very intense singlet in the CH_3 region. (It is left for the student to decide which is which.)

Oxidation of one of these would give a ketone, the other two acids. The two which gave acids could also be converted to *p*-toluenesulfonate esters—one of these would readily undergo nucleophilic substitution, the other not.

c. Nmr will show CH_3 groups that are triplets in one case and a doublet in the other. Periodic acid will cleave the *cis*-1,2-diol, but not the 1,3-diol.

f. One compound will have no absorption in the CH_3 region of the nmr, the other will have two different (probably overlapping) triplets.

Dehydration of both alcohols followed by permanganate oxidation will give a mixture of three monocarboxylic acids in one case and a single, dibasic acid in the other case.

12. There are twelve different tautomers possible for this molecule, involving differences in placement of oxygen-bound hydrogen.

16. 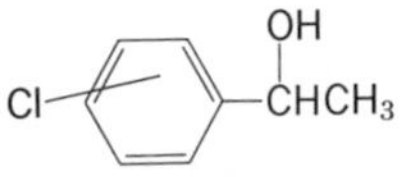
OH
Cl
CHCH3

o, *m*, or *p*

18. *a.* $C_6H_5OH > CH_3C_6H_4OH$. (CH_3 group is electron-donating, tending to destabilize a negative charge on the ring.)

c. $CCl_3CH_2OH > CH_3CH_2OH$. Cl is electron-withdrawing, except in situations where strong, positive charge is present.

e. 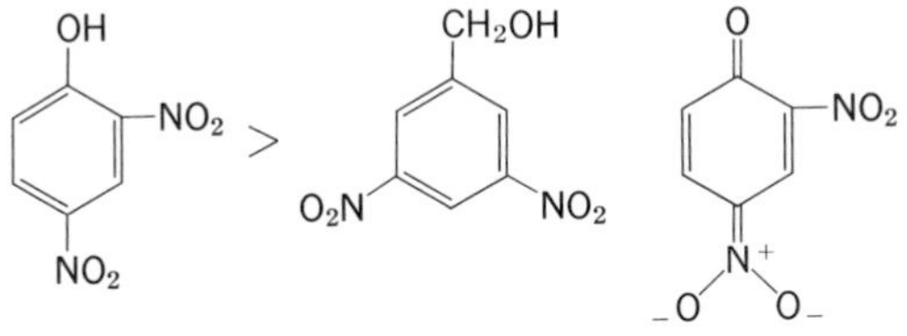
OH
NO2
NO2
>
CH2OH
O2N
NO2
O
NO2
N+
O
O

I

The dinitrophenylate ion can be stabilized by resonance (contributors like I), whereas the CH_2 group "insulates" the O^- from direct resonance interaction, and only a weaker inductive effect is left.

20. Yes. *S*. The usual esterification reaction does not affect the alcohol O—C bond and, therefore, cannot change the configuration.

22. No. The products would be aldehydes without asymmetric carbon atoms.

24. *a.*

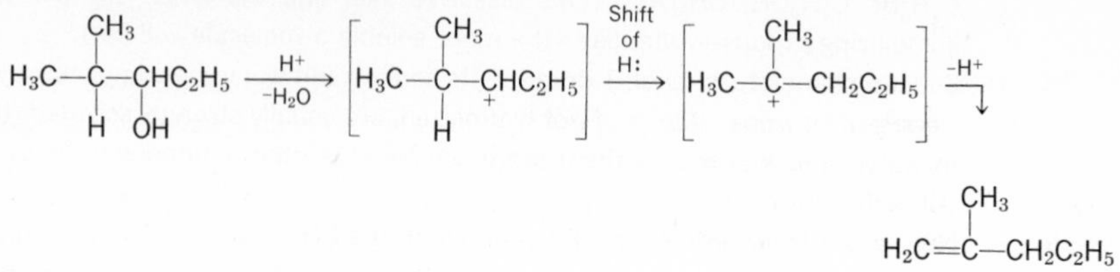
Shift
of
H:
-H2O
-H+

Major product most likely is

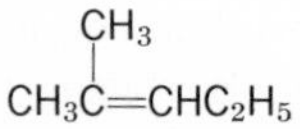

c. HX (X = Cl^-, OH^-, etc.) + $AlCl_3$ ⇌ $H^+AlCl_3X^-$

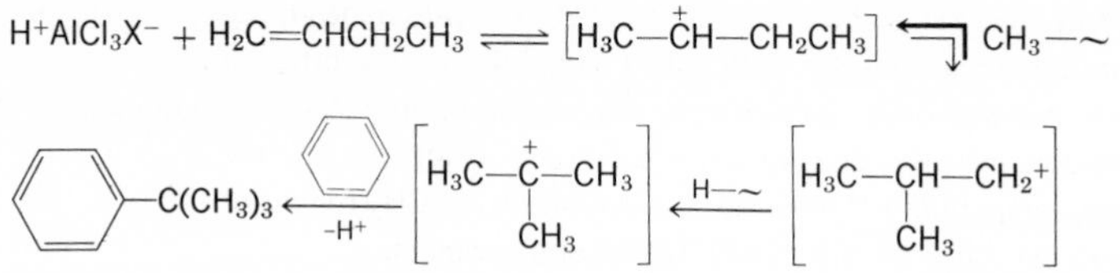

Above rearrangement would be extremely unfavorable as it involves primary carbonium ion (or more likely a similar, high-energy pathway) and, in fact, the major product would be

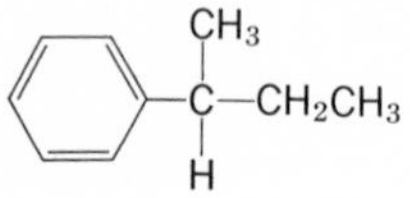

26. *a.* All $C_7H_{14}O$ alcohols (alcohols with one double bond or ring) except those with the OH group on the double bond (vinyl alcohols) which are unstable.

c. Cyclic ethers, ethers with rings not containing the oxygen —O—R, and ketones.

28. The compound X, $(CH_3CH_2)_3C—OH$, would give only the products indicated. Other isomers would give additional products.

30.

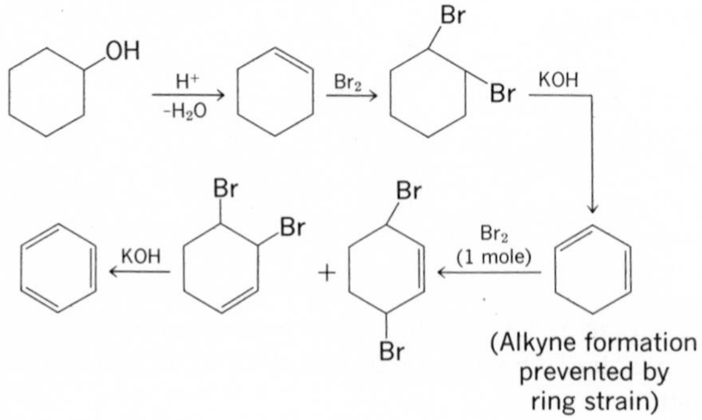
H+
-H2O
Br2
KOH
KOH
Br2
(1 mole)
(Alkyne formation
prevented by
ring strain)

Cyclohexanol is made from benzene or materials easily converted to benzene.

32. —(CH—CH$_2$)$_n$—.
with C$_2$H$_5$ attached to the CH

34. C_6H_5—HC=CH—CH_2CH_3. The double bond is conjugated with the aromatic ring which is an energetically favorable product, and which has an energetically lower transition state for its formation.

36. *a.* $(CH_3)_3C$—OH will have a singlet for the CH_3 groups at $\delta = 1.2$. *n*-Butyl alcohol will have a triplet at about 0.9 plus multiplets at 1.2–1.4 and a triplet or multiplet at 7.6 ppm.

In both compounds the position and, in the latter the splitting pattern, will depend on purity and concentration.

c. CH_3CH_2OH (in the presence of a trace of acid to exchange the OH) will show a triplet at $\delta = 1.2$ and a quartet at $\delta = 3.55$ ppm.
$CH_3CH_2CH_2OH$ under similar conditions will show a triplet at $\delta = 1.0$–1.1 ppm, and a triplet at about 3.55 ppm plus a multiplet at about 1.4 ppm.

e. Quinone will have hydrogens in the alkene 4.4–7.6 region and no exchangeable hydrogens, while hydroquinone will be in the aromatic 6–9 ppm region with two exchangeable (e.g., with D_2O) hydrogens at a variable location.

CHAPTER 9

1. *a.* $CH_3CH(Br)$—CH_3 + NaCN ⟶ $CH_3CH(CN)$—CH_3 + NaBr

c.

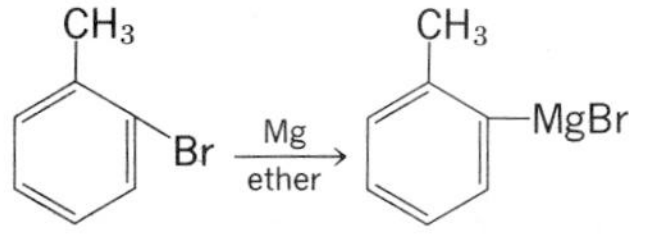
CH3
Br
Mg
ether
CH3
MgBr

e. CH_3CH_2—I + H_2O + NaOH ⟶ CH_3CH_2OH + some H_2C=CH_2

g. $CH_3CH(OH)CH_2CH_3$ + $SOCl_2$ ⟶ $CH_3CH(Cl)CH_2CH_3$ + SO_2 + HCl

i. H_2C=C(CH_3)—HC=CH_2 + HCl ⟶ H_3C—CCl(CH_3)HC=CH_2 + H_3C—C(CH_3)=CH—CH_2Cl

Major products

3. *a.*

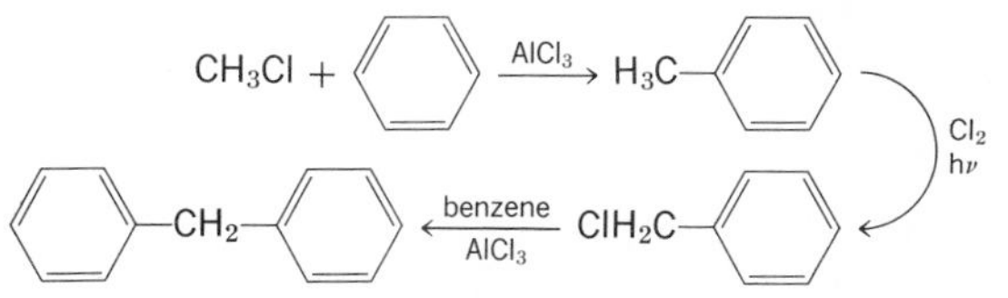
CH3Cl +
AlCl3
H3C
Cl2
hν
CH2
benzene
AlCl3
ClH2C

c.

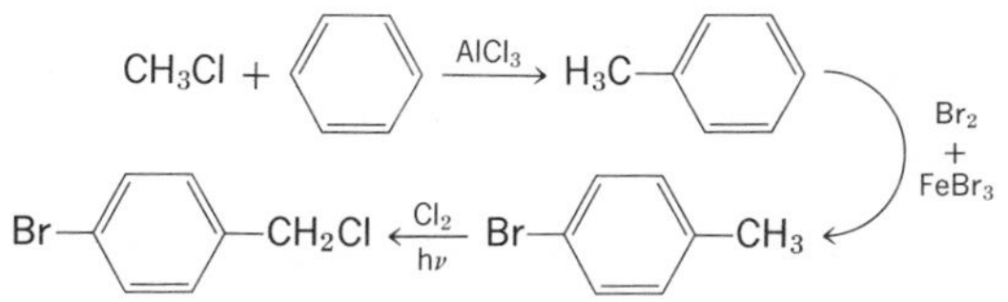
CH3Cl +
AlCl3
H3C
Br2
+
FeBr3
Br
CH2Cl
Cl2
hν
Br
CH3

e. $H_2C{=}CH_2 + HBr \longrightarrow CH_3CH_2Br$

$$HC{\equiv}CH + Na^0 \longrightarrow HC{\equiv}C^-Na^+ \xrightarrow{CH_3CH_2Br} CH_3CH_2C{\equiv}CH$$

$$\downarrow H_2/\text{cat.}$$

$$CH_3CH_2CH_2CH_2Br \xleftarrow[\text{peroxide}]{HBr} CH_3CH_2HC{=}CH_2$$

g. $C_6H_6 \xrightarrow[AlCl_3]{C_2H_5Cl} C_6H_5C_2H_5 \xrightarrow{Cl} C_6H_5CCl_2CH_3 \longrightarrow$

$$C_6H_5C{\equiv}CH \longrightarrow C_6H_5C{\equiv}CNa \xrightarrow{CH_3Cl} C_6H_5C{\equiv}C{-}CH_3$$

5. *a.* The first compound has a chlorine that is unreactive to $AgNO_3$ or base. The second and third can most easily be distinguished by nmr, as one will show a doublet at about $\delta = 1.6$ for the CH_3 (shifted down by the Cl) and a quartet at about 5 for the CH (shifted by both the phenyl and the Cl); the other will have two triplets at about 3.8 and 3.2 ppm.

c. One isomer will eliminate readily with base, the other will be very resistant toward elimination.

6. $CH_3CH_2CH_2CH_2Br + KOH \longrightarrow CH_3CH_2CH_2CH_2OH + KBr$
$CH_3CH_2CH_2CH_2OH + KOH \rightleftharpoons CH_3CH_2CH_2CH_2O^- + HOH$
$CH_3CH_2CH_2CH_2O^- + CH_3CH_2CH_2CH_2{-}Br \longrightarrow CH_3CH_2CH_2CH_2{-}O{-}CH_2CH_2CH_2CH_3$

8. Replacing a hydrogen on the alkene $CH_3HC{=}CH_2$ by an electron-withdrawing chlorine, as in $CH_3ClC{=}CH_2$, lowers the stability of the carbonium ion

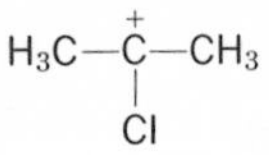

which would be formed on addition of H^+ to the corresponding alkene, enough so as to make it less reactive than the alkyne.

10. 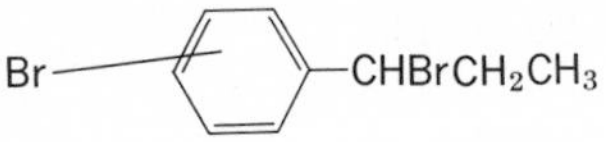

12. First, 2-methyl-1-chlorobutane is a primary halide with little tendency toward competing S_N1 reaction.
Second, the backside approach in 2-methyl-1-chlorobutane is a little less hindered than that of 2-methyl-2-chlorobutane:

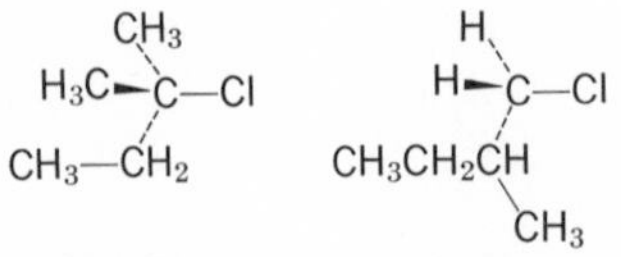

2-Methyl-2-chlorobutane　　2-Methyl-1-chlorobutane

14. *a.* $ICH_2{-}C{\equiv}C{-}C{\equiv}C{-}CH_2I$
b. CF_3CH_2OH
c. $(CH_3)_2CHCH_2Br$

16. Elimination via a carbonium ion to an alkene and readdition to the same (racemic) alcohol could occur under acidic but not basic conditions. Both the carbonium ion and the alkene are symmetrical, so recombination of the carbonium ion with H_2O also racemizes.

$$CH_3(C_6H_5)COH{-}CH_2CH_3 + H^+ \rightleftharpoons CH_3(C_6H_5)C{=}CHCH_3 + H_2O$$

CHAPTER 10

1. Nitrate ion is a poor nucleophile, but a good oxidant. Therefore, the cleavage would be slow and the phenol produced would be oxidized by the nitric acid and/or nitrated.

3. Yes. The same tendencies toward elimination apply as in other carbonium-ion reactions, especially with ethers of secondary and tertiary alcohols. However, the high nucleophilicity of I^- helps reduce elimination somewhat.

5. With a free carbonium-ion intermediate, both cis and trans isomers would be expected to give the same mixture of diastereomeric $CH_3CHClCHOHCH_3$ products.

7. *a.* Shaking with aqueous NaOH solution would convert the phenol into the water-soluble phenoxide (phenylate) ion. Separation of the aqueous and nonaqueous phases and acidification of the NaOH solution would regenerate the free phenol.

   b. Ethanol is water-soluble while diethyl ether is poorly soluble in water. However, ethanol would increase the solubility of diethyl ether in water, and, furthermore, ethanol is difficult to separate from water. Reaction with Na will yield a nonvolatile, ether-insoluble salt, C_2H_5ONa.

9. *a.* $CH_3CH_2CH_2CH_2OH + Na^0 \longrightarrow CH_3CH_2CH_2CH_2O^-Na^+ + \frac{1}{2}H_2$

   $$CH_3CH_2CH_2CH_2O^-Na^+ \xrightarrow{CH_3Br} CH_3CH_2CH_2CH_2OCH_3$$

   c. $C_6H_5OC_2H_5 + HI \longrightarrow C_6H_5OH + CH_3CH_2I$

   $C_6H_5OH \xrightarrow[\Delta]{\text{Zn dust}} \text{benzene}$

   e. $C_2H_5OC_2H_5 + HI \longrightarrow C_2H_5I + C_2H_5OH$

   $C_2H_5OH + \text{chromic acid} \xrightarrow[\Delta]{} CH_3CHO$

   $C_2H_5I + Mg \xrightarrow{\text{ether}} C_2H_5MgI \xrightarrow{CH_3CHO} \xrightarrow{H^+} C_2H_5CH(OH)CH_3$

11. The Williamson ether synthesis involves an S_N2 displacement. These just do not occur with tertiary halides.

    Tertiary alcohols are usually unstable enough in the presence of strong acid that they proceed to the alkene, and very little of the alcohol survives long enough to trap the carbonium ion and give the symmetrical ether via the classical dehydration synthesis.

13. $HOH_2C{-}CH(OCH_3){-}CH_2{-}OH$

    1,3-Dihydroxy-2-methoxypropane

15. $CH_3OCH_2CH_2CH(CH_3){-}OH$ or $HOCH_2CH_2CH(CH_3){-}OCH_3$

17. Answer = *a*. Displacements on tertiary halides are essentially nonexistent.

CHAPTER 11

1. 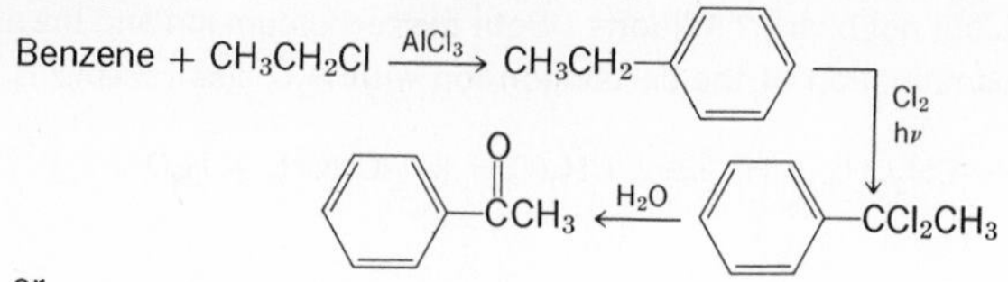

or

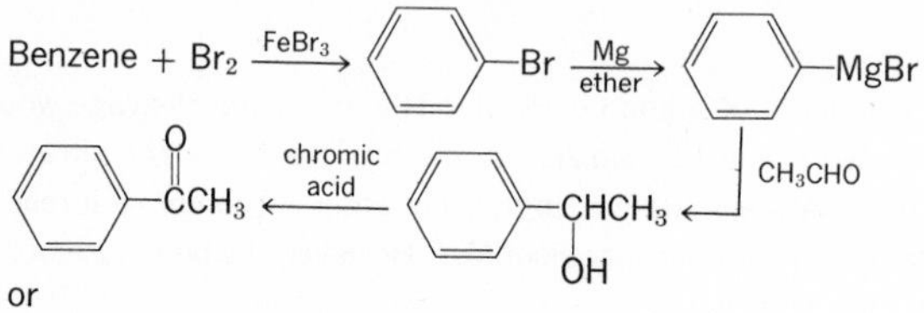

or

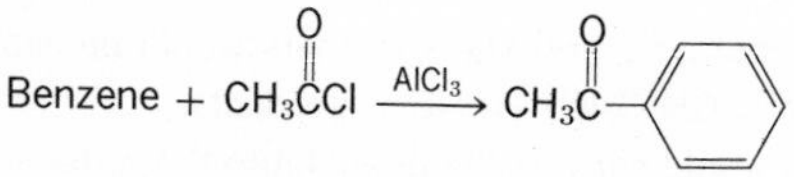

The third method is by far the easiest.

3. Allyl alcohol and propionaldehyde are easily oxidized by $KMnO_4$ while acetone is not. Propionaldehyde will react with carbonyl reagents such as 2,4-dinitrophenylhydrazine while allyl alcohol will not (if pure). Allyl alcohol, on the other hand, will react exothermally with acetyl chloride while propionaldehyde will not.

The infrared spectra of these compounds will show considerable differences. Allyl alcohol will have broad OH bands in the 3350–3360 cm^{-1} region (as well as some in the 3500–3700 range), plus C=C at 1640–1660 and 3000–3030 cm^{-1}. Propionaldehyde will have the aldehyde C=O at 1740 and a CH at 2720 cm^{-1}. Acetone will have a band at 1710 cm^{-1} with no band at 2720 cm^{-1}.

4. *a.* $CH_3\overset{O}{\overset{\|}{C}}{-}CH_3 \xrightarrow{H} CH_3\overset{OH}{\overset{|}{C}}HCH_3 \xrightarrow{H^+} CH_3CH{=}CH_2 \xrightarrow{O_3} CH_3\overset{O}{\overset{\|}{C}}H$

c. $CH_3\overset{O}{\overset{\|}{C}}{-}CH_3 \xrightarrow{PCl_5} CH_3CCl_2CH_3$ or from $CH_3C{\equiv}CH$ as in *e.*

e. $CH_3\overset{O}{\overset{\|}{C}}{-}CH_3 \xrightarrow{\text{as in } a} CH_3CH{=}CH_2 \xrightarrow{Cl_2} CH_3\underset{Cl}{\underset{|}{C}}H{-}\underset{Cl}{\underset{|}{C}}H_2 \xrightarrow{OH^-} CH_3C{\equiv}CH$

g. From *i*-propyl alcohol or propene via *a*, above.

i. $CH_3\overset{O}{\overset{\|}{C}}{-}CH_3 + CH_3MgBr \longrightarrow \xrightarrow{H_2O} CH_3\overset{OH}{\underset{CH_3}{\overset{|}{\underset{|}{C}}}}{-}CH_3 \xrightarrow{H^+} \xrightarrow[Pt]{H_2} CH_3\underset{CH_3}{\underset{|}{C}}H{-}CH_3$

5. *a.* $CH_3COCH_3 + LiAlH_4 \longrightarrow CH_3CHOHCH_3 \xrightarrow[\Delta]{H^+} CH_3HC{=}CH_2 \xrightarrow{O_3} CH_3CHO + OCH_2$

$CH_3COCH_3 + I_2 + OH^- \longrightarrow CHI_3 + CH_3COOH \xrightarrow{LiAlH_4} CH_3CH_2OH \xrightarrow{\text{chromic acid}} CH_3CHO$

c. $CH_3COCH_3 + PCl_5 \longrightarrow CH_3CCl_2CH_3$

e. Step *c* above, then $\xrightarrow[\Delta]{KOH} CH_3C{\equiv}CH$

g. $CH_3COCH_3 + H_2/cat. \longrightarrow CH_3CHOHCH_3 \xrightarrow{HCl} CH_3CHClCH_3$

or $CH_3CHOHCH_3 + H^+ \xrightarrow[\Delta]{} CH_3HC{=}CH_2 \xrightarrow{HCl} CH_3CHClCH_3$

i. $CH_3COCH_3 + CH_3MgCl \longrightarrow (CH_3)_3COH \xrightarrow{H^+} (CH_3)_2C{=}CH_2$

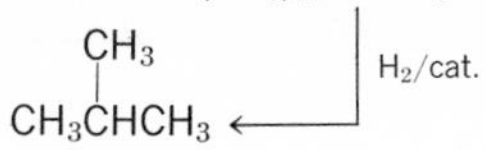

CH3
CH3CHCH3
H2/cat.

7. X =

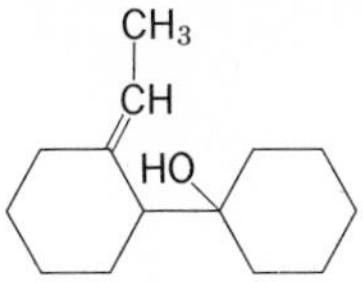

CH3
CH
HO

Y =

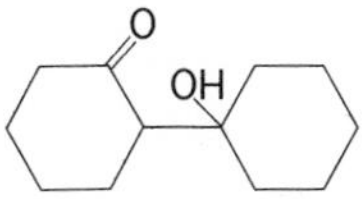

O
OH

Z =

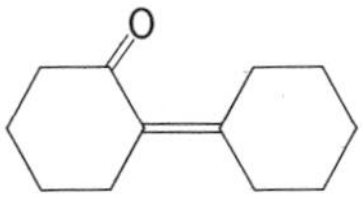

O

9. Q =

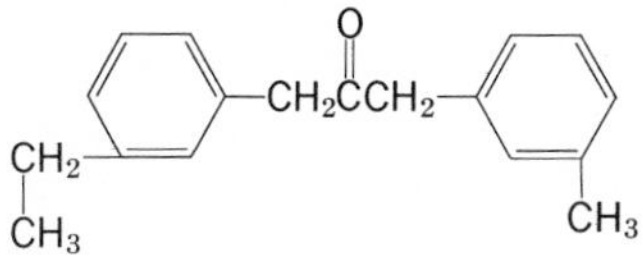

O
CH2CCH2
CH2
CH3
CH3

S =

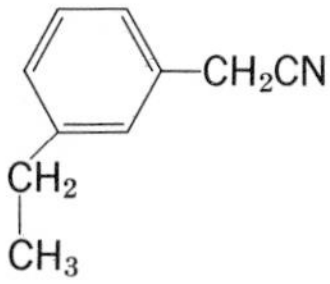

CH2CN
CH2
CH3

T =

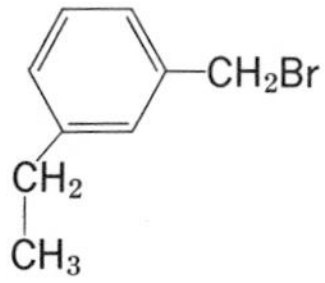

CH2Br
CH2
CH3

11. $CH_3\overset{O}{\overset{\|}{C}}CH_2CH_2CHO \xrightarrow{(O)} CH_3\overset{O}{\overset{\|}{C}}CH_2CH_2COOH$

13. *a.*

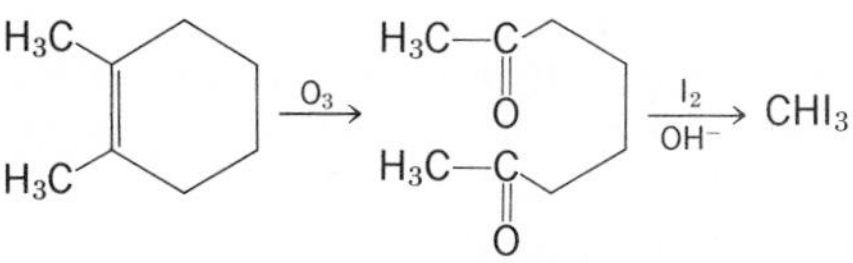

H3C
H3C
O3
H3C—C
O
H3C—C
O
I2
OH−
CHI3

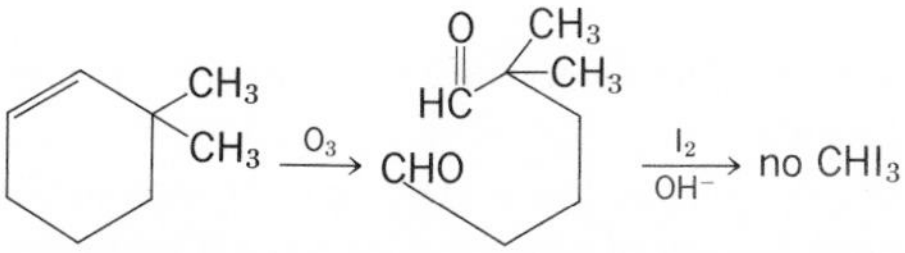

CH3
CH3
O3
O
HC
CH3
CH3
CHO
I2
OH−
no CHI3

In the nmr the first compound would show a singlet of intensity 6 at $\delta = 1.7$ and no hydrogens in the 4.5–7.6 region. The second compound would show a singlet of intensity 6 at $\delta = 0.9$, plus two protons in the 4.5–7.6 range.

c. One compound would not give the iodoform test. (Which?) One of the other two would oxidize with dilute permanganate while the other would not.
In the nmr one compound would have no aliphatic hydrogens, one would have aliphatic and aromatic but no alkene hydrogens, and the third would have all three.

15. *a.*

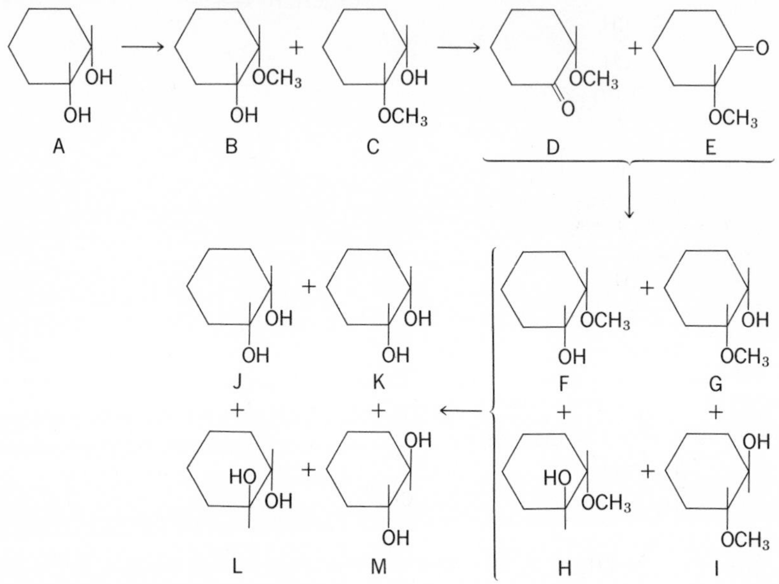
OH
OH
A
OCH3
OH
B
OH
OCH3
C
OCH3
O
D
O
OCH3
E
OH
OH
J
OH
OH
K
OCH3
OH
F
OH
OCH3
G
HO
OH
L
OH
OH
M
HO
OCH3
H
OH
OCH3
I

b. B and C are enantiomers and cis.
D and E are enantiomers.
F and G are enantiomers and cis.
H and I are enantiomers and trans.
F and G are both diastereomers of H and I.
J and K are identical (meso), cis, and also identical to A.
L and M are enantiomers and trans.
J and K and A are diastereomers of L and M.
B and C, D and E, F and G, H and I, and L and M are all racemic mixtures.

17. *a.* Propanol reacts with acetyl chloride (heat) or sodium metal (hydrogen evolution), but not with phenylhydrazine.

Propanal does not react with acetyl chloride, does not readily evolve hydrogen with sodium, and does form a phenylhydrazone. It is readily oxidized by Fehling's solution.

Propanone behaves similarly to propanal above except that it does not react with Fehling's solution.

c. Ethylene oxide will not form a crystalline derivative with phenylhydrazine, and will not readily reduce Fehling's solution.

19. The uncatalyzed reaction between acetone and hydroxylamine is too slow to be practical. At low H^+ concentration the carbonyl group of acetone does not get protonated so that nucleophilic addition of hydroxylamine is too slow. At high H^+ concentration the acetone is adequately protonated, but the amine function of the hydroxylamine is completely protonated and it loses its nucleophilic character.

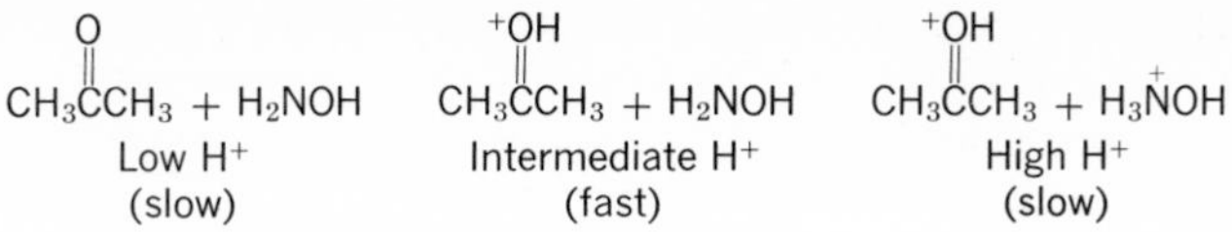
Low H+
(slow)
Intermediate H+
(fast)
High H+
(slow)

21.

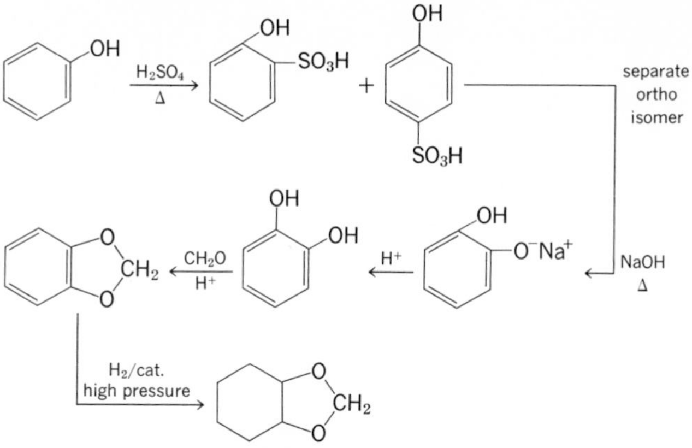
H2SO4
Δ
separate
ortho
isomer
NaOH
Δ
H+
CH2O
H+
H2/cat.
high pressure

23. Lack of discrimination between two such different (in reactivity) reagents is good evidence for a prior, slow (rate-determining) step, followed by a fast and irreversible reaction with the halogen. This is most likely the enolate-formation reaction.

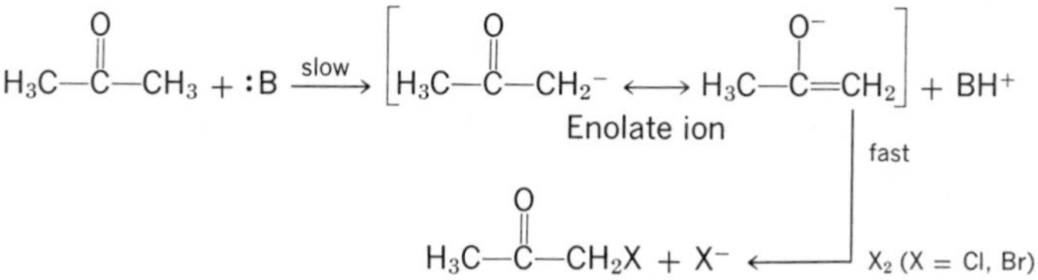
slow
Enolate ion
fast
X2 (X = Cl, Br)

25. CCl_3CHO cannot condense in aldol fashion as it has no α hydrogens.

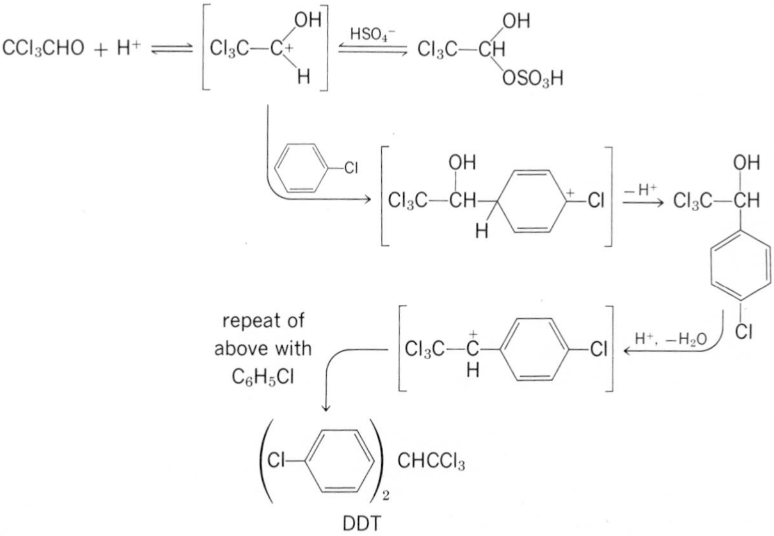
repeat of
above with
C6H5Cl
DDT

CHAPTER 12

5. The stronger acid in each case is:

a. C_6H_5OH.

c. $NO_2C_6H_4OH$.

e. Both are very similar as they will form the same ion.

g. 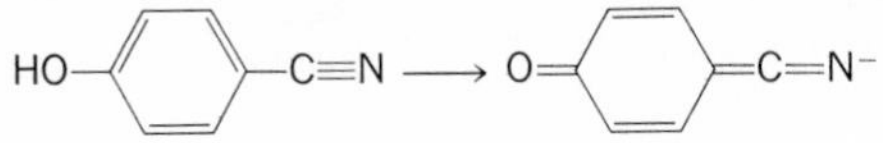

HO
C≡N
O
=C=N⁻

7. *a.* 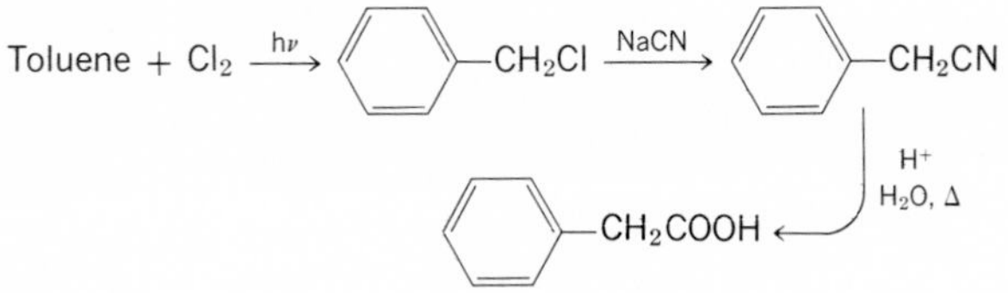

Toluene + Cl2
hν
CH2Cl
NaCN
CH2CN
H+
H2O, Δ
CH2COOH

c. 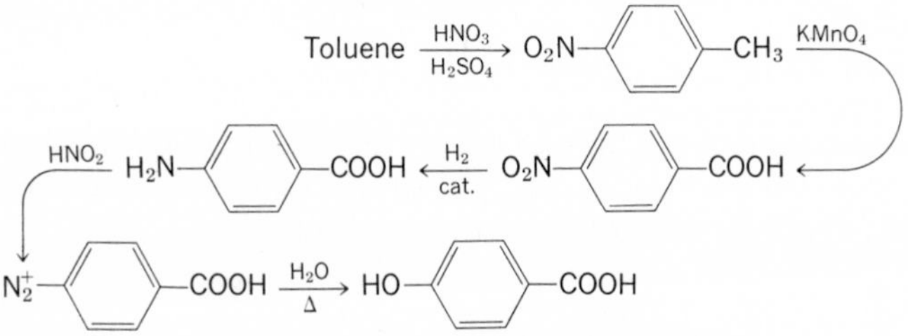

Toluene
HNO3
H2SO4
O2N
CH3
KMnO4
COOH
H2
cat.
H2N
HNO2
N2+
H2O
Δ
HO

e. $\text{Propene} + \text{HCl} \longrightarrow CH_3CHClCH_3 \xrightarrow[\text{ether}]{\text{Mg}} (CH_3)_2CHMgCl$

$$(CH_3)_2CHMgCl \xrightarrow{CO_2,\ H^+} (CH_3)_2CHCOOH$$

g. 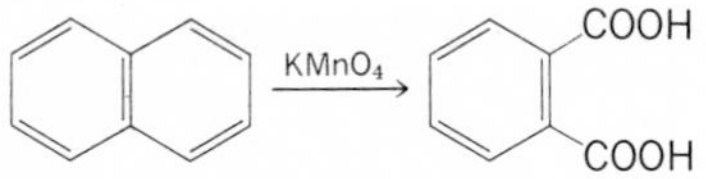

KMnO4
COOH
COOH

i. 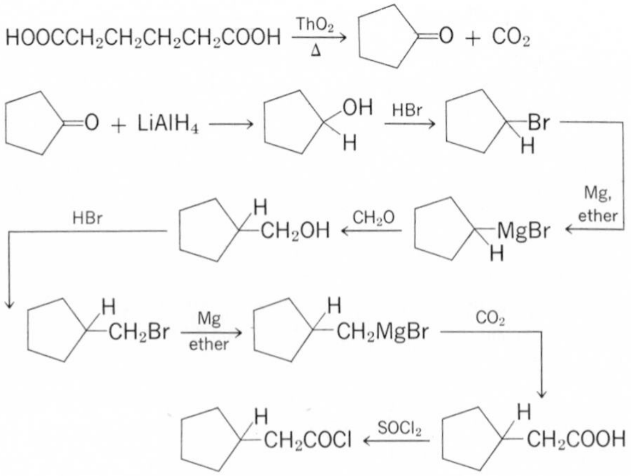

HOOCCH2CH2CH2CH2COOH
ThO2
Δ
O + CO2
O + LiAlH4
OH
H
HBr
Br
H
Mg,
ether
MgBr
H
CH2O
H
CH2OH
HBr
H
CH2Br
Mg
ether
H
CH2MgBr
CO2
H
CH2COOH
SOCl2
H
CH2COCl

9. $HOOCCHOHCH_2COOH$, $HOCH_2CH(COOH)_2$, $CH_3COH(COOH)_2$

11. $CH_3CHOHCH_2CH{=}CH_2$ or $H_2C{=}C(CH_3)CH_2CH_2OH$

13.

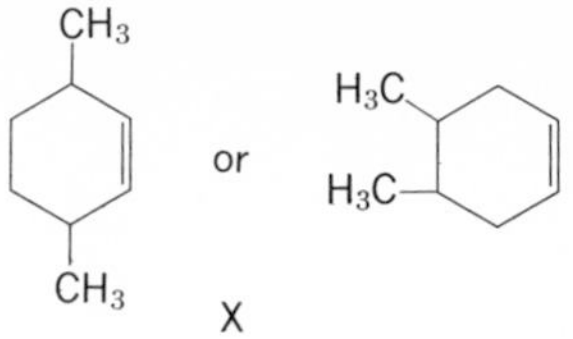

X

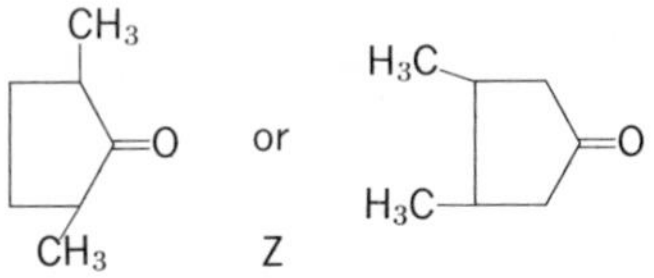

Z

(NOTE: Five-membered ketones absorb in the infrared at 1740–1750 cm^{-1}.)

15. *a.* $C_6H_5COOH + LiAlH_4 \longrightarrow C_6H_5CH_2OH \xrightarrow{HBr} C_6H_5CH_2Br$

$$C_6H_5CH_2Br \xrightarrow{Mg,\ ether} C_6H_5CH_2MgBr$$

$$C_6H_5CH_3 \xleftarrow{H_2O} C_6H_5CH_2MgBr$$

c. $CH_3CH_2COOH + LiAlH_4 \xrightarrow{H^+} CH_3CH_2CH_2OH$

$$CH_3CH_2CH_2OH + HBr \longrightarrow CH_3CH_2CH_2Br \xrightarrow[ether]{Mg} CH_3CH_2CH_2MgBr$$

$$CH_3CH_2CH_2MgBr \xrightarrow{CO_2} CH_3CH_2CH_2COOH$$

$$CH_3CH_2CH_2COOH + CH_3CH_2CH_2OH \overset{H^+}{\rightleftharpoons} CH_3CH_2CH_2\overset{O}{\overset{\|}{C}}{-}O{-}CH_2CH_2CH_3$$

e.

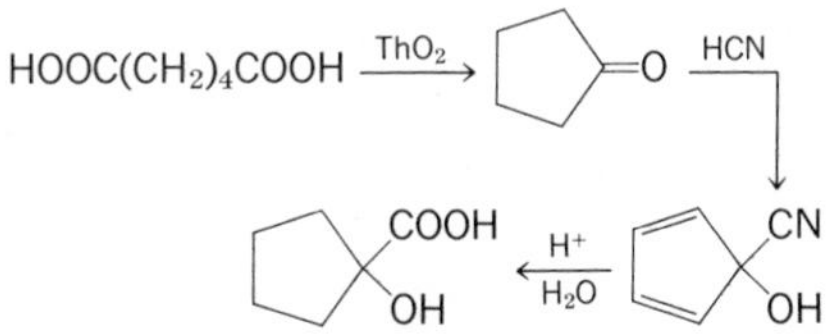

g.

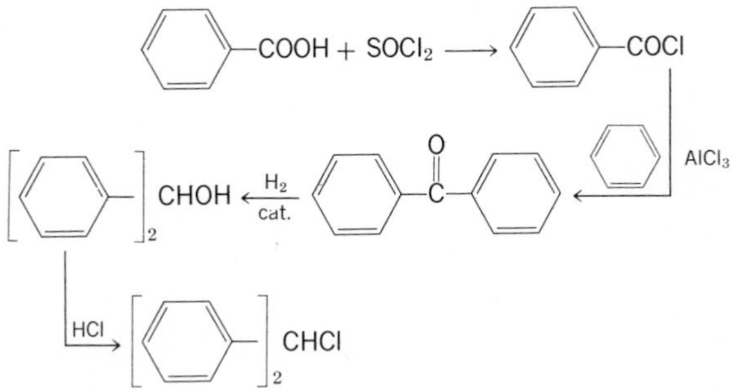

17. There are three isomers possible. The cis (meso), and the two enantiomeric (*R,R*, and *S,S*) trans. The cis isomer will form a cyclic anhydride easily, while the trans would require highly strained rings in order to do so.

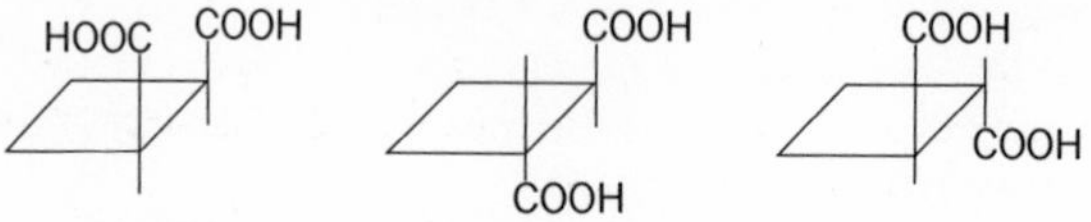

HOOC
COOH
COOH
COOH
COOH
COOH

meso (cis) *R,R* (trans) *S,S* (trans)

18. The αH is on the asymmetric center in one case and is reversibly removed in base.

19. $CH_3CH_2OCH_2COOH$

CHAPTER 13

1. *a.*

$$CH_3\overset{O}{\overset{\|}{C}}-OCH_3 + H_2NC_2H_5 \rightleftharpoons \left[H_3C-\overset{O^-}{\underset{\underset{H_5C_2}{\overset{|}{N}H_2^+}}{\overset{|}{\underset{|}{C}}}}-OCH_3 \right] \rightleftharpoons \left[H_3C-\overset{OH}{\underset{\underset{H_5C_2}{\overset{|}{N}H}}{\overset{|}{\underset{|}{C}}}}-OCH_3 \right] \rightleftharpoons \left[H_3C-\overset{O-H}{\underset{\underset{H_5C_2}{\overset{|}{N}H}}{\overset{|}{\underset{|}{C}}}}-\overset{O}{\underset{}{}} \right]$$

$$CH_3\overset{O}{\overset{\|}{C}}NHC_2H_5 + HOCH_3 \leftarrow$$

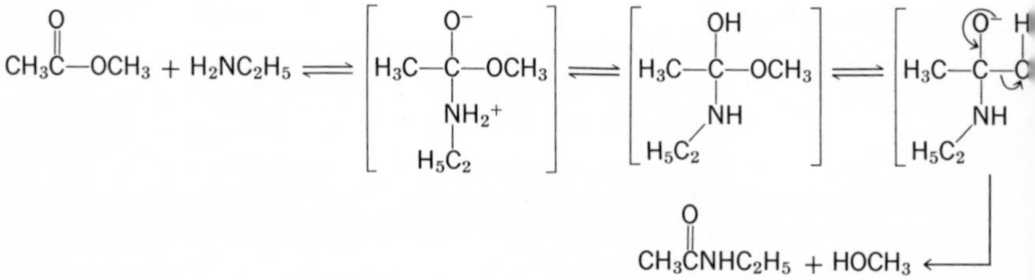

c. Acetic anhydride + C_6H_5—$NHCH_3$ ⟶ $CH_3C(=O)$—$N(CH_3)$—C_6H_5 + $HOC(=O)$—CH_3

Mechanism similar to *a.*

e.

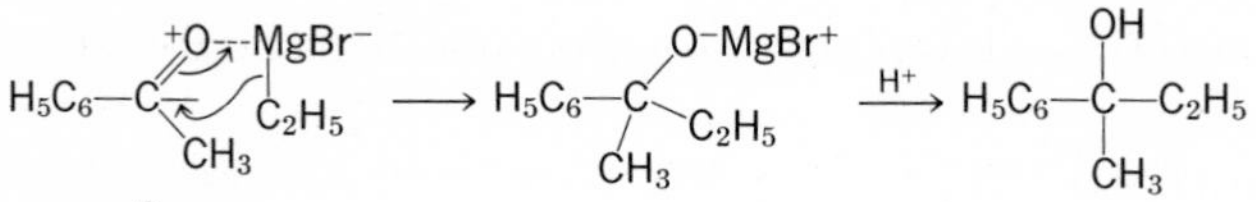

g. H_5C_6—$C(=O)$—NH_2 + HOOC—C_6H_5

Mechanism similar to *a.*

3. A = $CH_3CH_2OC(=O)CH_3$

Ethyl acetate

B = $CH_3CH_2CH_2COOH$

Butyric acid

C = $CH_3OCH_2CH_2CHO$

3-Methoxypropanal

D =

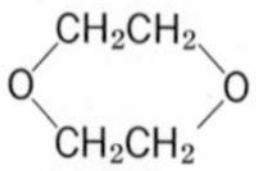

Dioxane

(1,4-dioaycyclohexane)

Chemical tests: B would dissolve in $NaHCO_3$ solution with the evolution of carbon dioxide.
C would form a phenylhydrazone and reduce Fehling's solution.
A would react with boiling NaOH to dissolve, and would not come out again on cooling and acidification since both fragments are too soluble in water.
D would not react with any of the above.

Spectroscopic differences: Infrared: A would have absorption at 1740 and 1245 cm^{-1}.
B would have a broad band at 1710 cm^{-1} and another strong band at 1250 cm^{-1}.
C would have maxima at 1730 and 2720 cm^{-1}.
D would have no maxima in the carbonyl or OH region, but would have a strong band at 1100 cm^{-1}.

4. *a.* $CH_3COCl > CH_3COI$

5. The OH region of the spectra of these compounds would differ in that A2 has more opportunity to hydrogen bond internally.

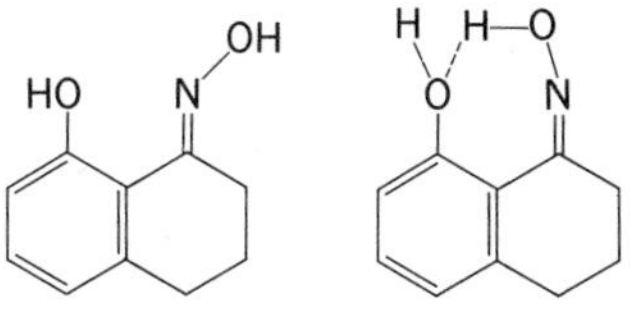

A1 A2

7. *a.* One isomer would have a CH_3 split into a triplet, with relative intensity-3. The other would have CH_3 as a doublet with relative intensity-6. The other hydrogens would be correspondingly split.

c. The first compound would have a sharp CH_3 triplet with a weak (and spread out by splitting with the D) multiplet superimposed. The CH_2 would be complex. The second compound would have a more intense but broader triplet in the same region. The CH_2 would be a sharp pentuplet.

e. The alkene hydrogens would be split by about 14 cycles in the trans compound and by about 9 cycles in the cis, if it were not for symmetry. It would be necessary to make one end different (e.g., by converting it to an ester) in order to see this.

9. The reaction of a ketal such as $(CH_3)_2C(OCH_3)_2$ with water is an equilibrium reaction in which the equilibrium is displaced very far to the right as written below:

$$(CH_3)_2C(OCH_3)_2 + H_2O \rightleftharpoons (CH_3)_2CO + 2CH_3OH$$

Since the esterification of methanol and acetic acid is an equilibrium reaction in which water is produced, the effect of putting the ketal in the mixture is to consume the water produced as indicated above, and thus shift the equilibrium to the right.

$$CH_3COOH + HOCH_3 \rightleftharpoons CH_3COOCH_3 + H_2O \text{ (removed as indicated above)}$$

11. *a.*

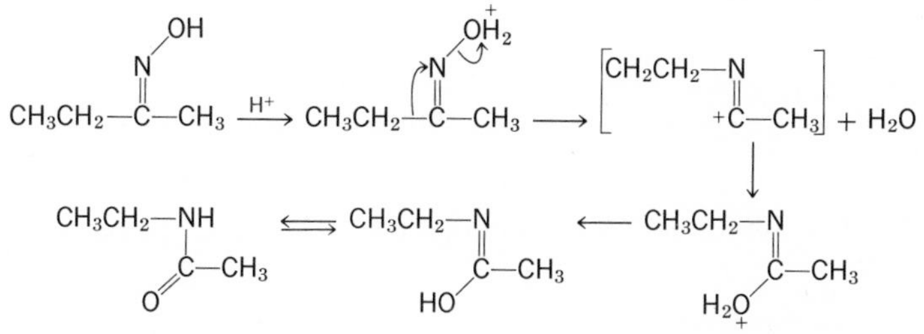

c. $C_6H_5NH_2$

e.

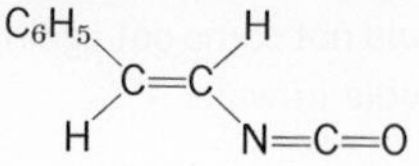

13. *a.*

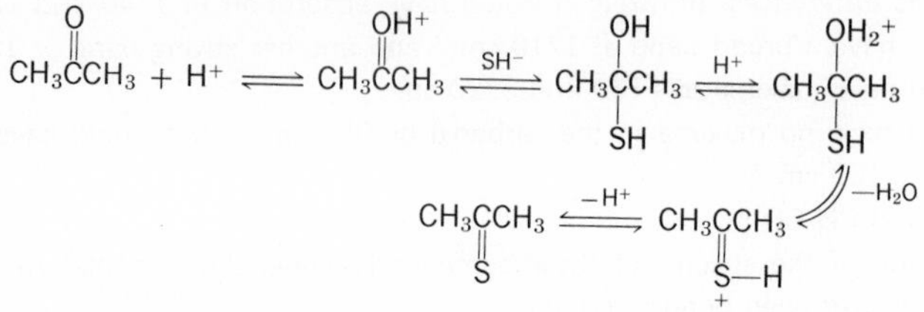

c.

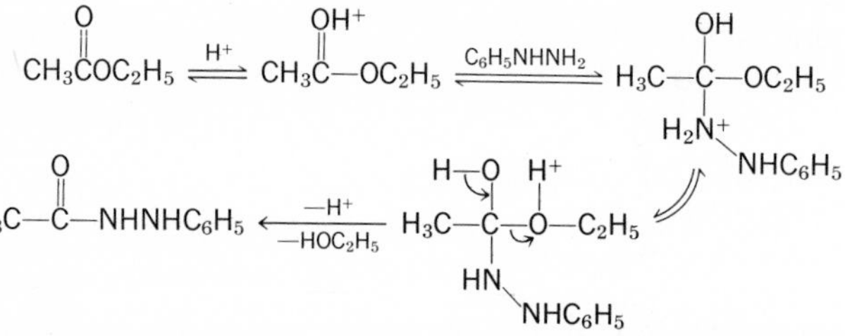

e.

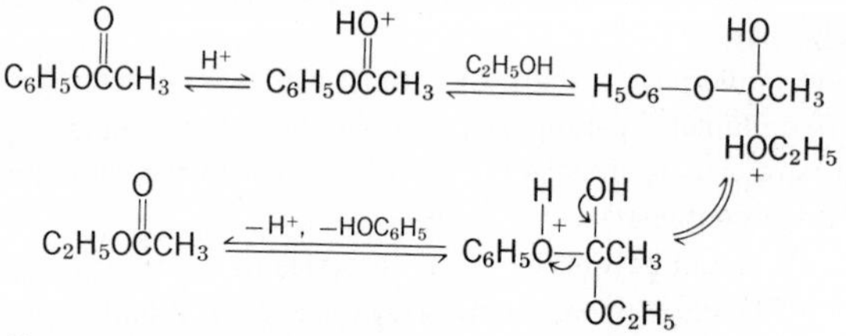

Alcohol esters are more stable than phenol esters.

g.

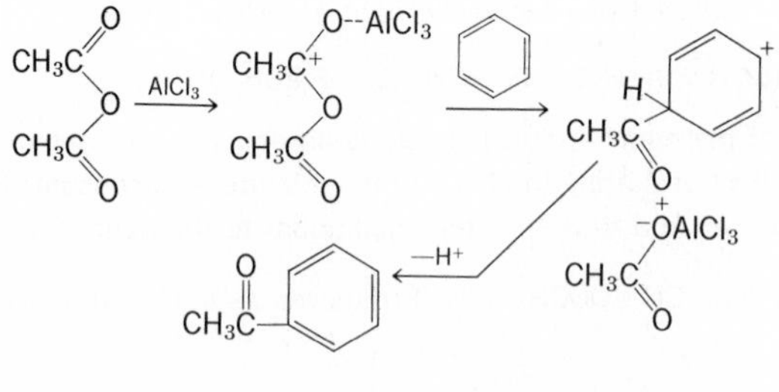

15. *b.*

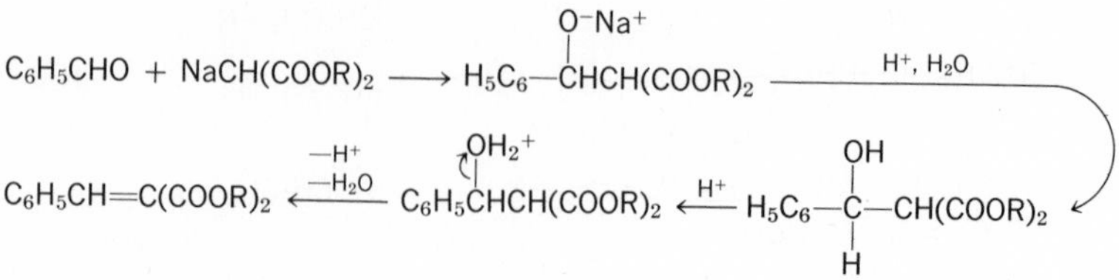

d. $CH_3COCH_3 + NaC{\equiv}CH \longrightarrow (CH_3)_2C(O^-){-}C{\equiv}CH$

f. $CH_3C{\equiv}N + H^+ \rightleftharpoons CH_3C{\equiv}\overset{+}{N}H \xrightarrow{CH_3OH} CH_3C(=NH){-}\overset{+}{O}H(CH_3) \longrightarrow CH_3C(=NH_2^+){-}OCH_3$

17.

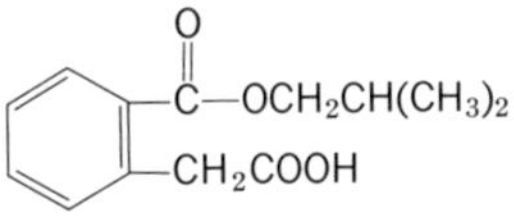

A

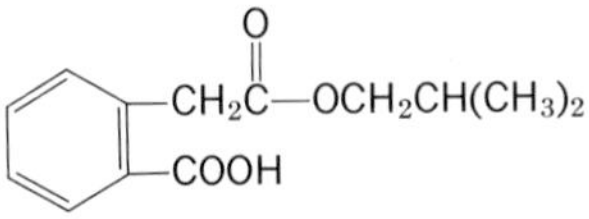

(would have ir at 1740 not 1720 cm^{-1})

$HOCH_2CH(CH_3)_2$
B

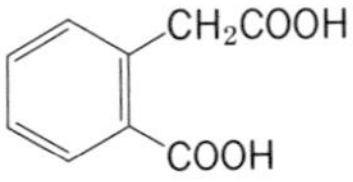

C

$OCHCH(CH_3)_2$
D

19. Saponification equivalent is 185.

a. A 50% (by weight) hydrocarbon impurity would double it.

b. A 50% C_{10} carboxylic acid would change it to 181.

21. The preparation of nitriles from halides involves an S_N2 displacement reaction by cyanide ion. Tertiary halides are extremely poor in S_N2 displacements.

23. The second compound has a double bond conjugated with the carbonyl group of the ester function, thus making a better chromaphore in the ultraviolet. The infrared spectra would differ in that the carbonyl absorption of conjugated esters is shifted down about 15 to 20 cm^{-1} and the C—C absorption would also be shifted down by about 40 cm^{-1}. The nmr would differ in the splitting of the CH_3 groups and in the positions of the alkene hydrogens among other things.

CHAPTER 14

1. In decreasing base strength:

a. $(CH_3)_2NH$, NH_3, $C_6H_5NH_2$.

c.

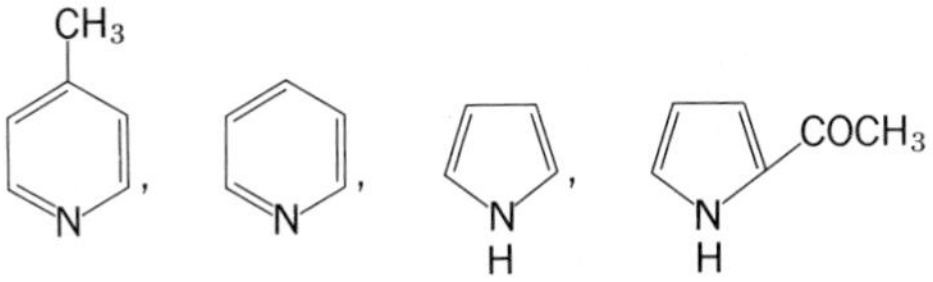

e. Ammonia, acetamide, phenol.

g. Sodium hydroxide, trimethylamine, water.

3. *a.*

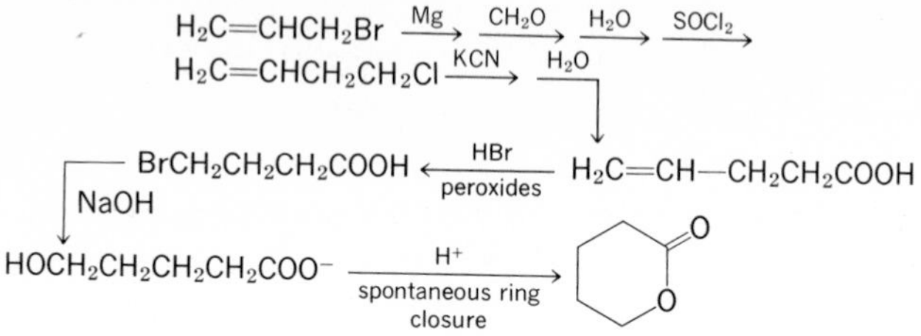

c.

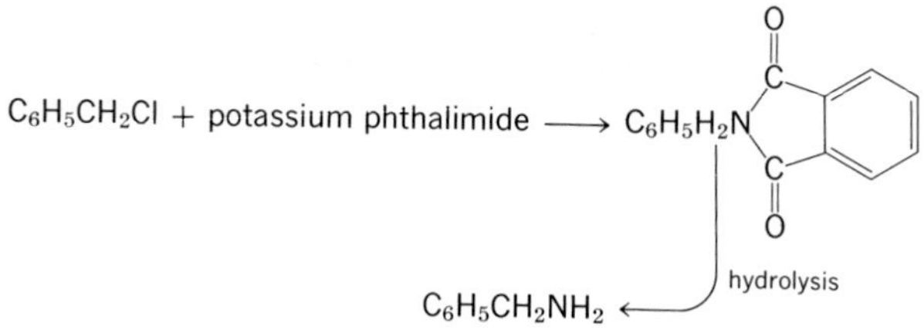

e. Benzoic acid + $SOCl_2 \longrightarrow C_6H_5COCl + SO_2\uparrow$

$$C_6H_5COCl + \underset{\text{Sodium salt of glycine}}{H_2NCH_2COONa} \xrightarrow[\text{two-phase reaction}]{H_2O} C_6H_5CONHCH_2COONa \xrightarrow{HCl} C_6H_5CONHCH_2COOH$$

h.

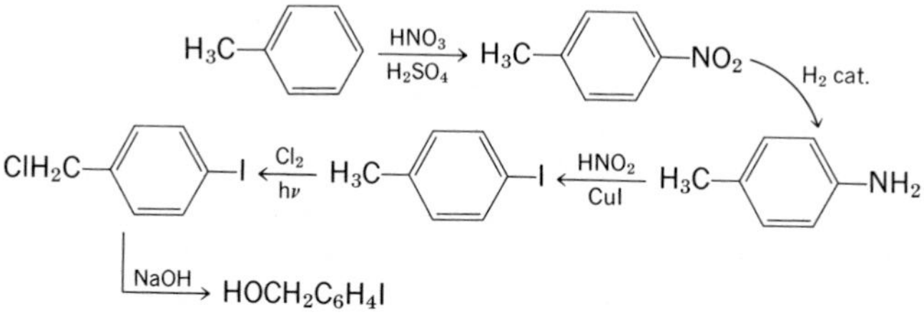

7. *a.* $ROH + HBr \longrightarrow RBr \xrightarrow{(CH_3)\ N} RN(CH_3)_3I$

$$RN(CH_3)_3I \xrightarrow{Ag_2O,\ H_2O} RN(CH_3)_3{}^+OH^- \xrightarrow{\Delta} \text{product}$$

c. $RNH_2 \xrightarrow{CH_3I} \xrightarrow{Ag_2O} \xrightarrow[\Delta]{} \xrightarrow{KMnO_4} \xrightarrow{SOCl_2} \xrightarrow{NH_3} \xrightarrow{NaOBr} \text{product}$

e. $ROAc + NaOCH_3 \longrightarrow RO^- \xrightarrow{CH_3I} ROCH_3$

9. X=

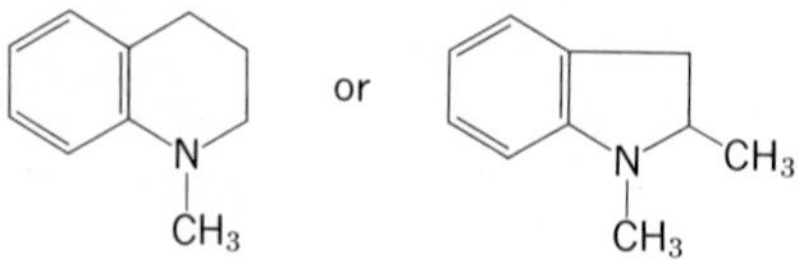

11. 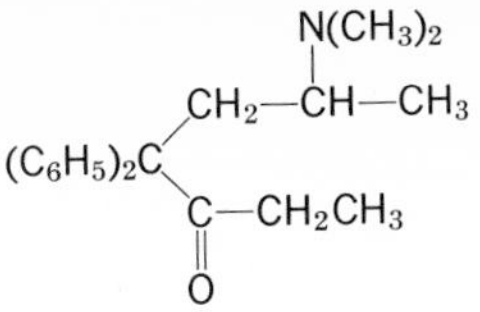

16. *b* and *e*.

20. The NH_2 group and the ^-O group are much more electron-donating than the corresponding NH_3^+ group and HO group which would be present in acid solution.

22. 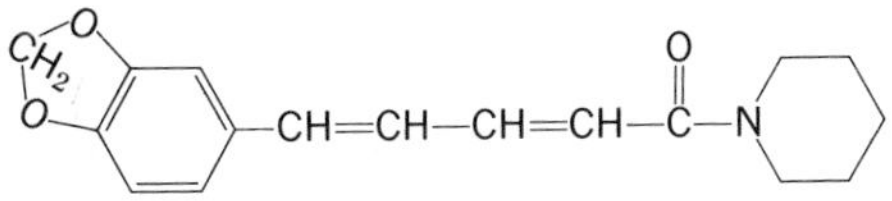

26. 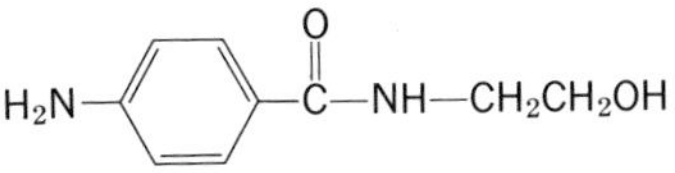

28. 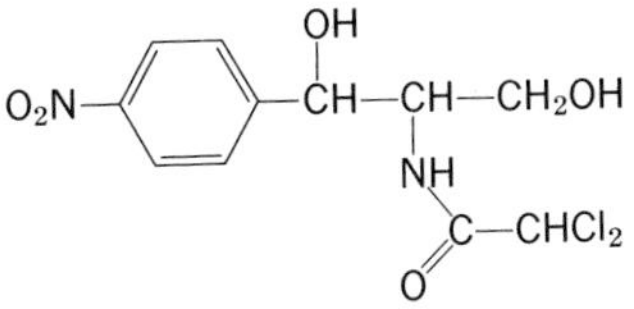

30. 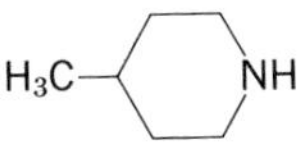 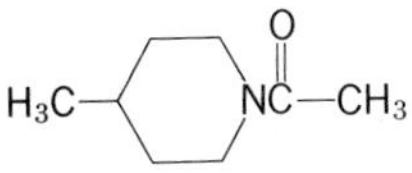 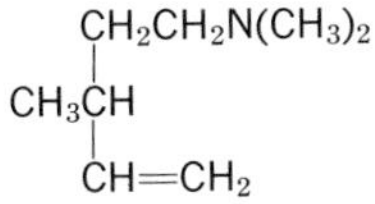

(2,8, 2.9 NH)	(6.0 tert amide)	(6.1 C=C, no NH)
A	B	C

Other structures are possible.

CHAPTER 15

3. *a*. $CH_3CH_2SH + HgO \longrightarrow$ precipitate.
CH_3CH_2SH in the infrared has absorption at about 2575 cm^{-1}.
CH_3CH_2OH would do neither of the above. (There would also be a detectable difference in odor!)

c. $HSCH_2CH_2OCH_3$ would give the mercaptan tests in *a*.
$HOCH_2CH_2SCH_3$ would not give above tests and would not reduce.
$HOCH_2CH_2SSCH_3$ would not give above tests but would easily reduce to two compounds that did.

e. 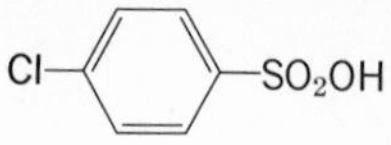

Cl
SO₂OH

is a strong, water-soluble acid.

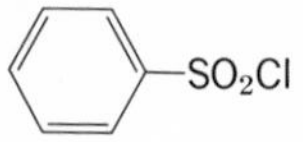

SO₂Cl

is insoluble in water (until it slowly reacts) and is not a strong acid. It liberates Cl^- slowly in H_2O.

5. *a.* *cis*- 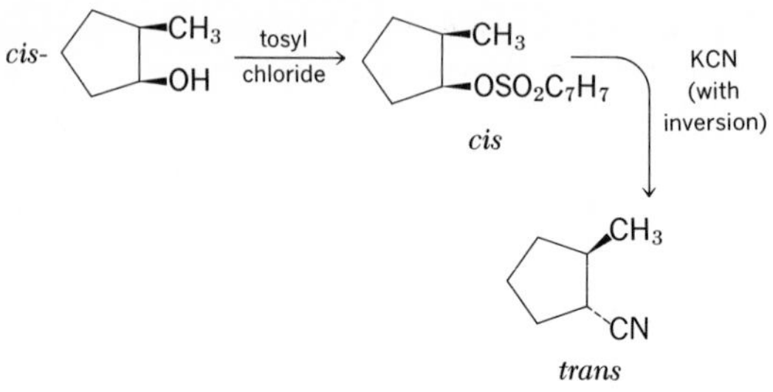

CH₃
OH
tosyl
chloride
CH₃
OSO₂C₇H₇
cis
KCN
(with
inversion)
CH₃
CN
trans

c. 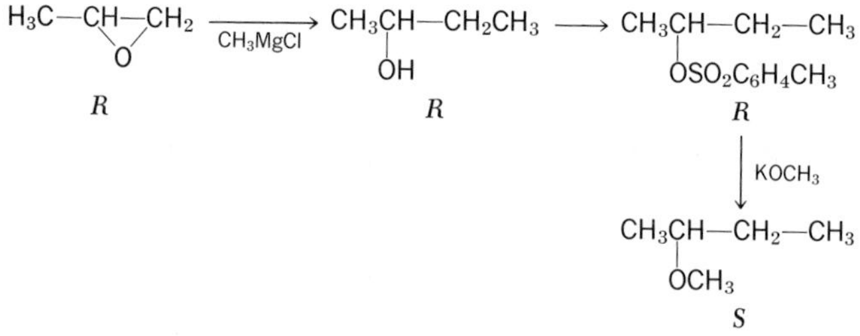

H₃C—CH—CH₂
O
R
CH₃MgCl
CH₃CH—CH₂CH₃
OH
R
CH₃CH—CH₂—CH₃
OSO₂C₆H₄CH₃
R
KOCH₃
CH₃CH—CH₂—CH₃
OCH₃
S

8. *a.* 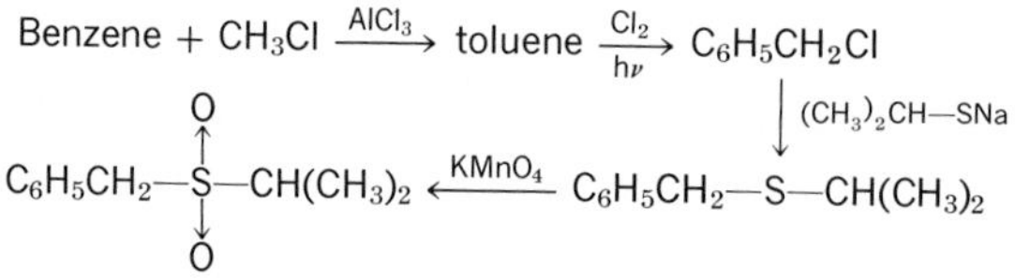

Benzene + CH₃Cl
AlCl₃
toluene
Cl₂
hν
C₆H₅CH₂Cl
(CH₃)₂CH—SNa
C₆H₅CH₂—S—CH(CH₃)₂
KMnO₄
C₆H₅CH₂—S—CH(CH₃)₂
O
O

c. 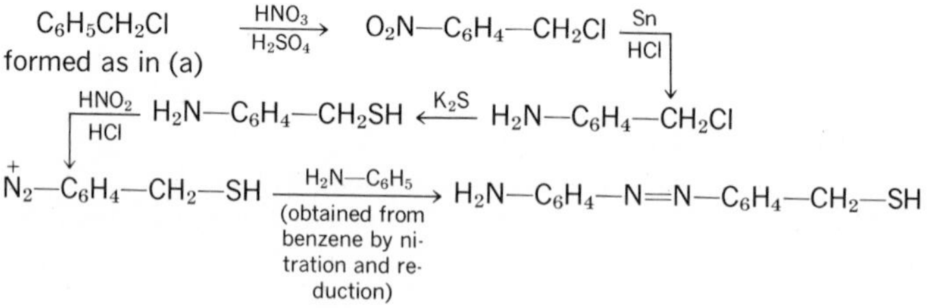

C₆H₅CH₂Cl
formed as in (a)
HNO₃
H₂SO₄
O₂N—C₆H₄—CH₂Cl
Sn
HCl
H₂N—C₆H₄—CH₂Cl
K₂S
H₂N—C₆H₄—CH₂SH
HNO₂
HCl
N₂—C₆H₄—CH₂—SH
H₂N—C₆H₅
(obtained from
benzene by ni-
tration and re-
duction)
H₂N—C₆H₄—N=N—C₆H₄—CH₂—SH

e. 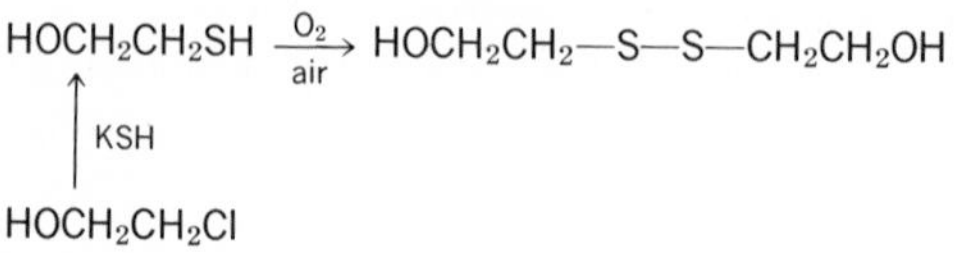

HOCH₂CH₂SH
O₂
air
HOCH₂CH₂—S—S—CH₂CH₂OH
KSH
HOCH₂CH₂Cl

CHAPTER 16

3. *a.* $C_6H_5NH_2 + COCl_2 \longrightarrow C_6H_5N{=}C{=}O \xrightarrow{CH_3NH_2} C_6H_5NHC(=O)NHCH_3$

c.

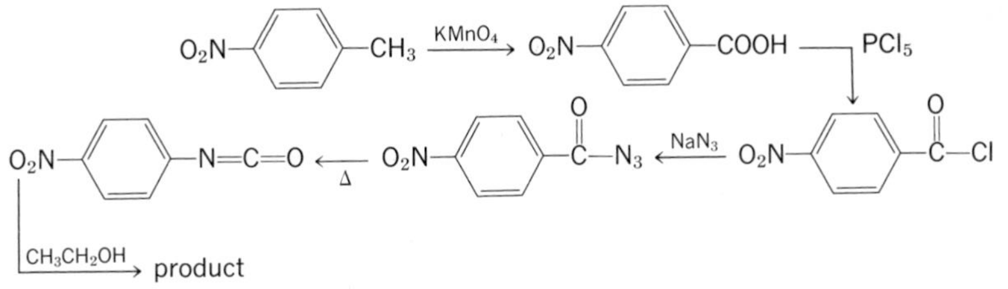

O₂N
CH₃
KMnO₄
COOH
PCl₅
N=C=O
Δ
C—N₃
NaN₃
C—Cl
CH₃CH₂OH
product

5.

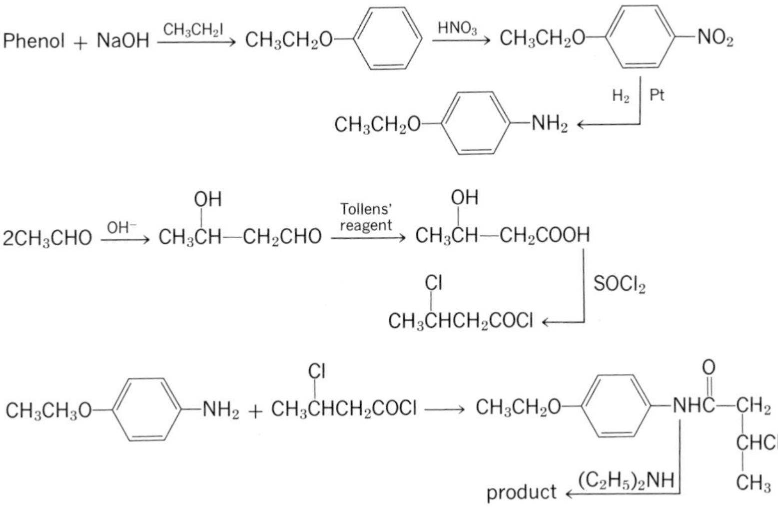

Phenol + NaOH
CH₃CH₂I
CH₃CH₂O
HNO₃
NO₂
H₂
Pt
NH₂
2CH₃CHO
OH⁻
CH₃CH—CH₂CHO
Tollens' reagent
CH₃CH—CH₂COOH
SOCl₂
CH₃CHCH₂COCl
CH₃CH₃O
NH₂ + CH₃CHCH₂COCl
NHC—CH₂
CHCl
CH₃
product
(C₂H₅)₂NH

7. The synthesis of malonic acid from elemental carbon is nontrivial. The following reactions will produce key intermediates (underlined) which should enable you to synthesize malonic ester and, thereby, barbituric acid.

$C + O_2 \xrightarrow[\Delta]{} \underline{CO_2}$ $\qquad CO_2 + 2NaOH \longrightarrow Na_2CO_3$

$CaO + 3C \longrightarrow CaC_2 + CO$ $\qquad CaC_2 + N_2 \longrightarrow CaNCN$

$CaNCN + C \longrightarrow Ca(CN)_2$ $\qquad Ca(CN)_2 + Na_2CO_3 \longrightarrow \underline{2NaCN} + CaCO_3$

$CO_2 + LiAlH_4 \longrightarrow \underline{CH_3OH}$ $\qquad CaNCN \xrightarrow[H_2SO_4]{H_2O} \underline{H_2NCONH_2} + CaSO_4$

9. Both compounds may be prepared from the appropriate alcohols and phosgene, followed by treatment with NH_3.

11. Consider all the equilibria involved:

$$H_2O + 2NH_3 + CO_2 \rightleftharpoons (NH_4)_2CO_3$$

$$H_2NCO_2NH_4 \rightleftharpoons H_2NCONH_2 + H_2O$$

$$(NH_4)_2CO_3 \rightleftharpoons H_2NCO_2NH_4 + H_2O$$

Net equation:

$$\underset{\text{3 moles of gas}}{2NH_3 + CO_2} \rightleftharpoons \underset{\substack{\text{0, 1, or 2 moles of gas}\\ \text{(depending on temperature).}}}{H_2NCONH_2 + H_2O}$$

13. $$\phi\text{—N=C=O} + H_2O \longrightarrow \underset{\text{(unstable)}}{\left[\phi\text{—NH—}\overset{\overset{\displaystyle O}{\|}}{C}\text{—OH}\right]} \longrightarrow \phi\text{—}NH_2 + CO_2$$

more $\phi\text{—N=C=O} + \phi\text{—}NH_2 \longrightarrow \phi\text{—NHCONH—}\phi$

CHAPTER 17

5. *a.*

Cis-isomers

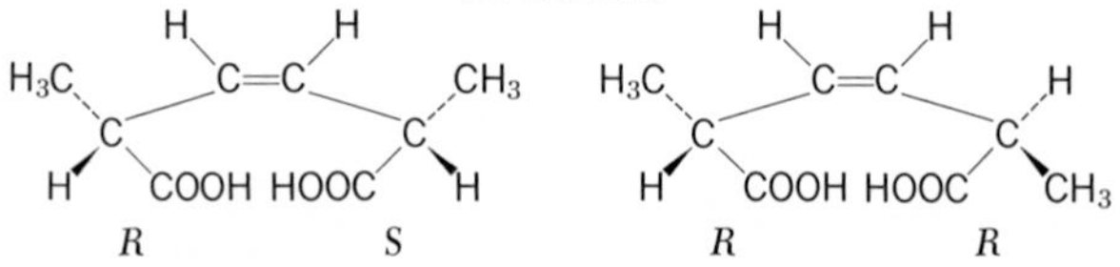

Meso (has a plane of symmetry) | Optically active (same is true for *S,S* isomer)

Trans-isomers

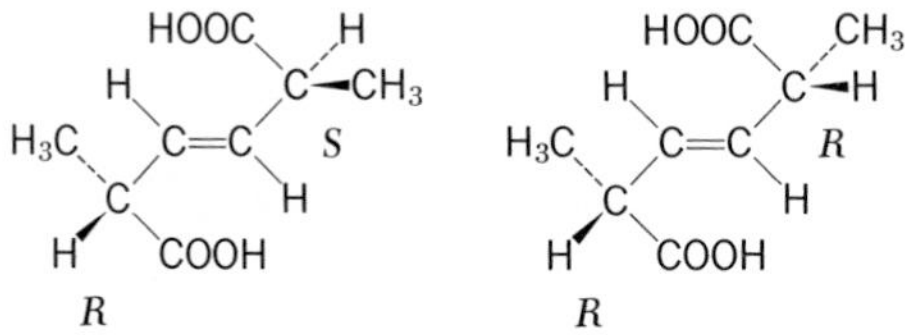

Meso (Trans) | Optically active

Note that the meso-trans isomer does not have a plane of symmetry, and yet the two mirror images are superimposable. This molecule has a center of symmetry—that is, a point through which an imaginary straight line would always connect two identical atoms at equal distances from that point.

c.

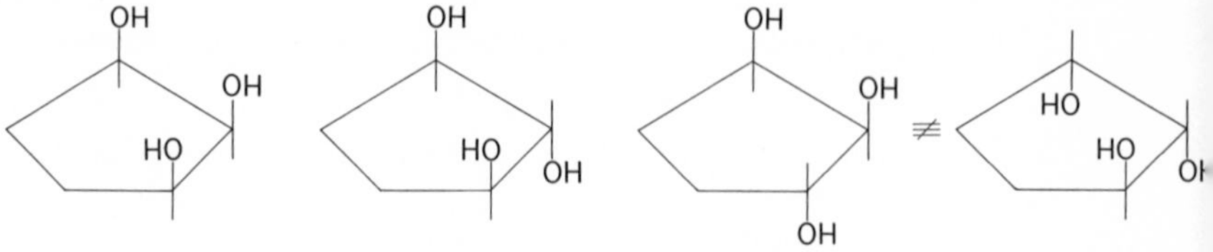

cis,cis (inactive) | *trans,trans* (inactive) | *cis,trans* (active) | *cis,trans* (mirror image)

7. *a.*

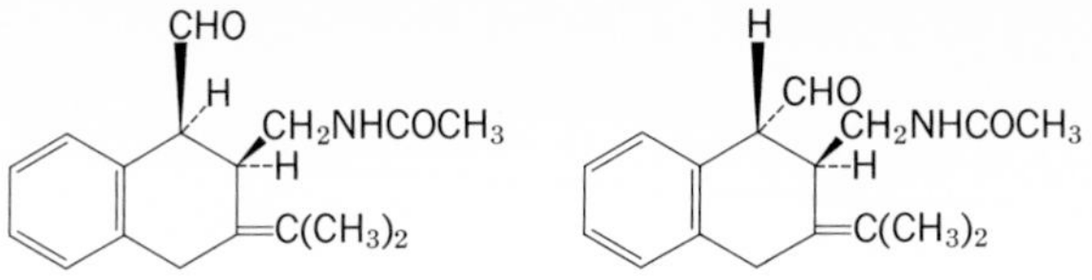

cis *trans*

Oxidation of the aldehyde group to COOH, followed by hydrolysis of the amide and heating would give a cyclic amide in one case (which?) and not in the other.

d. Two cis-trans isomers are possible, each one capable of optical activity.

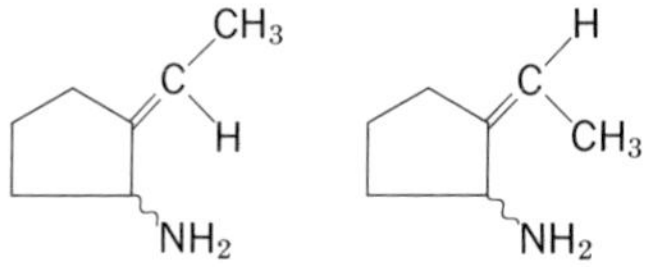

R and *S trans* *R* and *S cis*

If one chlorinated the CH_3 group (free-radical chlorination), the cis isomer would ring close with the NH_2 while the trans would not. If one ozonized the double bond, the resulting ketone would have only an *R* and *S* form so that the two *R* and *S* compounds would become identical.

9.

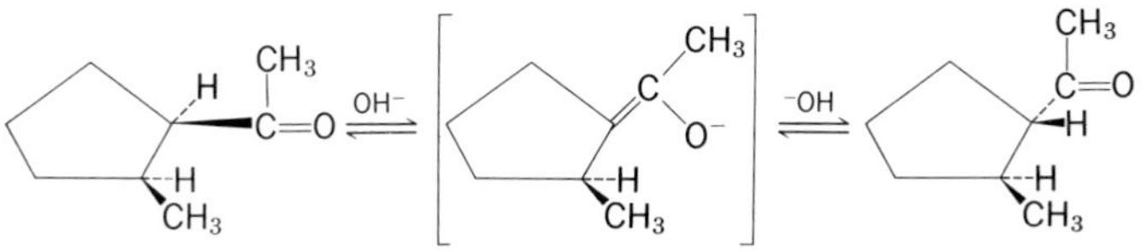

cis Common enolate ion *trans*

11. There are five asymmetrically substituted double bonds in vitamin A so that 32 isomers are theoretically possible (2^5). However, the strain of a trans double bond in the six-membered ring reduces this number to 16. Several isomers, such as the all cis which would curl up on itself, are also highly unlikely.

13.

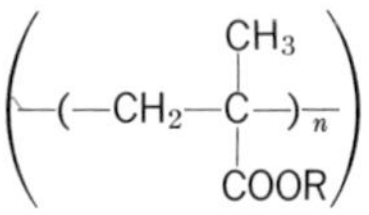

is the repeating unit. Since an asymmetric center is generated every time a new methacrylic unit is added, there is a high probability of a gross mixture of *R* and *S* centers being generated, depending on which side of the molecule the chain adds. This disorder in the chain will tend to disrupt any crystallization process in the polymer molecule.

CHAPTER 18

1. Recall the mixed aldol condensation (Sec. 11-12).
The Claisen condensation with ethyl formate proceeds in analogous fashion as a mixed condensation.

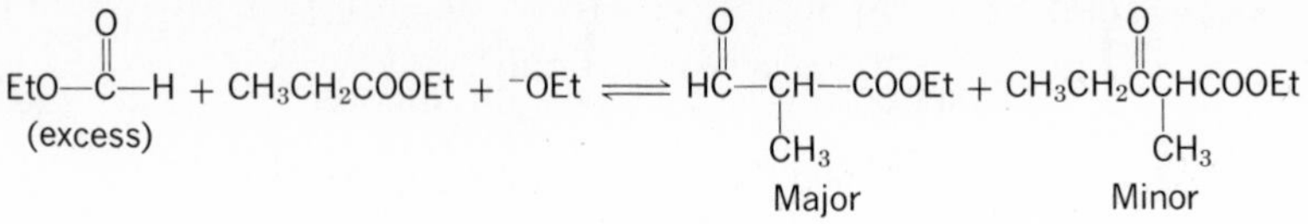

$$EtO-\overset{\overset{\displaystyle O}{\|}}{C}-H + CH_3CH_2COOEt + {}^-OEt \rightleftharpoons HC-\underset{\underset{\displaystyle CH_3}{|}}{CH}-COOEt + CH_3CH_2\overset{\overset{\displaystyle O}{\|}}{C}\underset{\underset{\displaystyle CH_3}{|}}{C}HCOOEt$$

(excess) ⠀⠀⠀⠀⠀⠀⠀⠀⠀⠀⠀⠀⠀⠀⠀⠀⠀⠀⠀⠀⠀Major⠀⠀⠀⠀⠀⠀⠀⠀⠀Minor

3. Both isomers are converted to a common enolate ion.

5. S_N2 displacements on tertiary halides are exceedingly poor.

7. *a.* $CH_3O\overset{\overset{\displaystyle O}{\|}}{C}(CH_2)_4\overset{\overset{\displaystyle O}{\|}}{C}OCH_3 + NaOC_2H_5 \rightleftharpoons CH_3O\overset{\overset{\displaystyle O}{\|}}{C}-\overset{-}{C}H(CH_2)_3\overset{\overset{\displaystyle O}{\|}}{C}OCH_3$

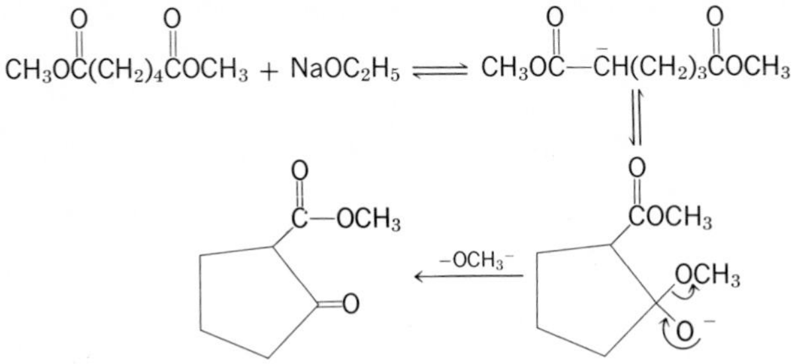

c.

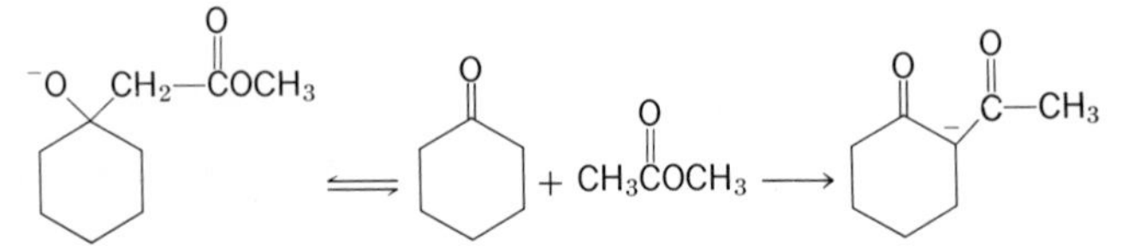

Stabilized as the ion

e. No reaction.

g. $\phi COCH_3 + \phi COOEt \xrightarrow{NaOEt} \phi-\overset{\overset{\displaystyle O}{\|}}{C}-CH_2-\overset{\overset{\displaystyle O}{\|}}{C}-\phi$

11. Use C* for C^{14}.

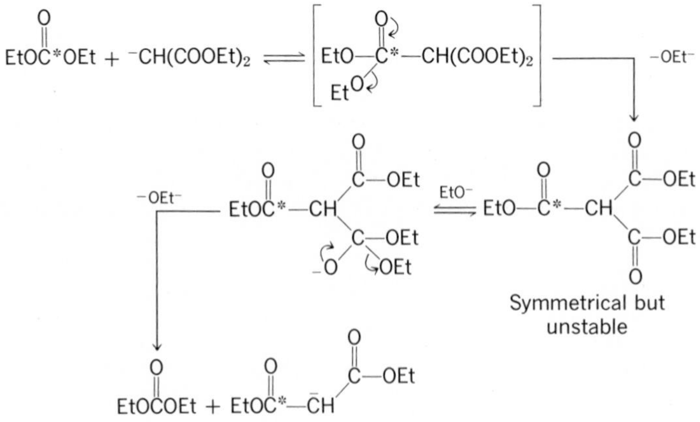

Symmetrical but unstable

13. *a.*

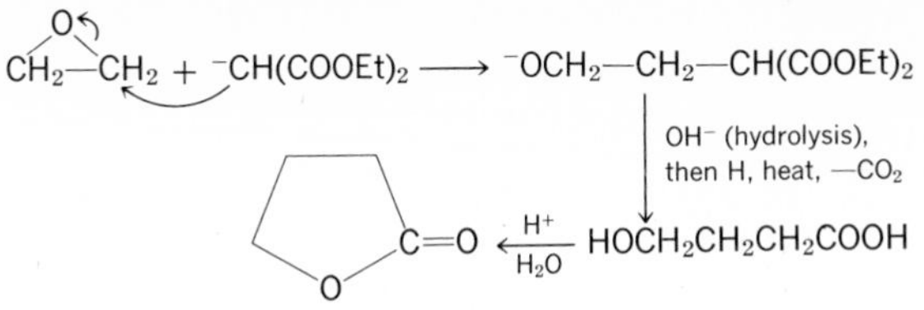

c.

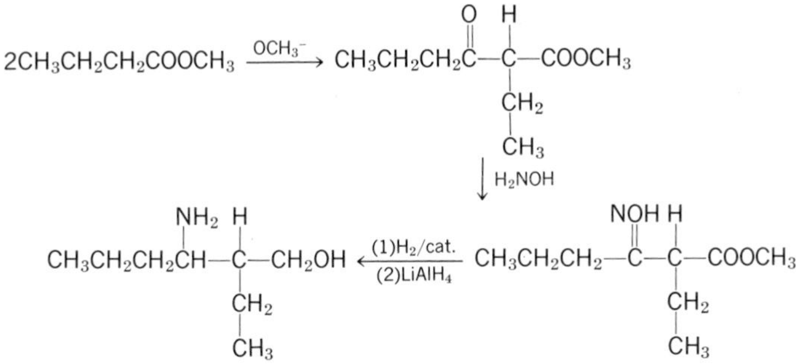

e. Malonic ester ⟶ alkylation with CH_3I, then alkylation with CH_3CH_2Br, hydrolysis and decarboxylation gives

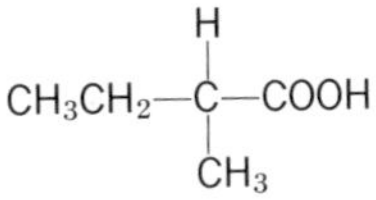

Reduction with $LiAlH_4$, followed by conversion to the *p*-toluenesulfonate ester and displacement with K_2S gives the appropriate thiol. Oxidation with air yields the product.

15. The compound shown has no enolizable hydrogen in the position between the two carbonyl groups. Loss of CO_2 involves the formation of an enol or enolate ion—such a process here would require a double bond to the bridgehead position. These structures are too highly strained to be formed.

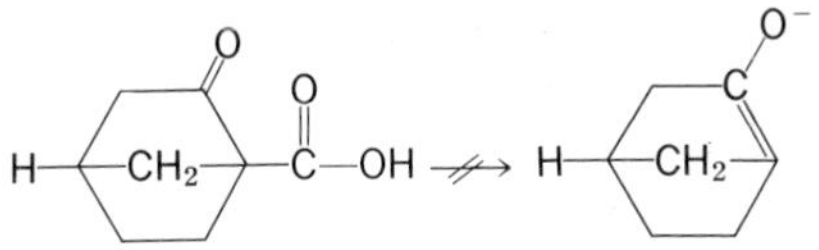

17. Interconversions between keto and enol forms proceed by the enol or enolate ions in which the stereochemistry of the parent molecules is lost (either about double bonds by way of singly bonded intermediates or about rings by way of doubly bonded intermediates).

CHAPTER 19

7. *a.* Cysteine, pH 5.0.

c. Hippuric acid, pH < 3 ($>$99% in the unionized form).

e. $CH_3\underset{\displaystyle COO^-}{\underset{|}{C}}H—N(CH_3)_3$, pH > 7 (>99% of the COOH in COO^- form)

g. $(CH_3)_2CHCOOCH_3$, all pH's.

i. β and γ-aminobutyric acids, pH between 6 and 8 (approaching the limit for completely isolated COOH and NH_2 of about 7.7).

9. *a.* Anode (+ pole). *c.* Neither.
b. Cathode. *d.* Anode.

11. They are amphoteric and have similar solution properties over a range of pH's. Moreover, they are poorly soluble in solvents other than water and do not distill.

13. Tyrosine, tryptophane, histidine, thyroxine.

15. 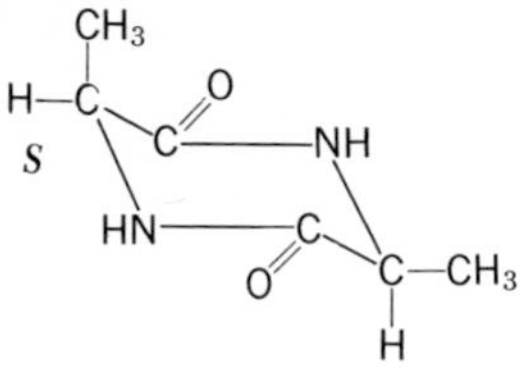

S

Optically active (nonsuperimposable on its mirror image)

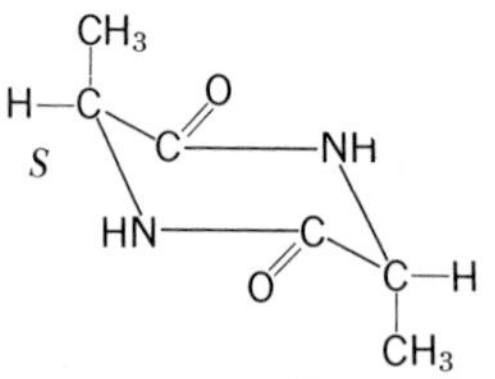

R

Inactive (has a center of symmetry) (see Exercise in Chap. 17)

CHAPTER 20

1. *a.* Unstable. The carbonyl group of one molecule would react rapidly with the Grignard function of a like molecule.

c. Stable. The esterification reaction is reasonably slow and reversible.

e. Reasonably stable. Dehydration requires catalysis.

g. Stable. Formation of this enolate anion actually protects the molecule from carbonyl addition reactions by virtue of the negative charge on the molecule.

i. Unstable. Grignard functions are reactive toward all types of carbonyl functions. The fact that ring size would be poor does not protect it from bimolecular reaction.

3. The Grignard reagent has anionic character and, therefore, can undergo 1,4-addition to conjugated systems in competition with normal (1,2-) addition to the carbonyl. The product from such a reaction is the enolate ion.

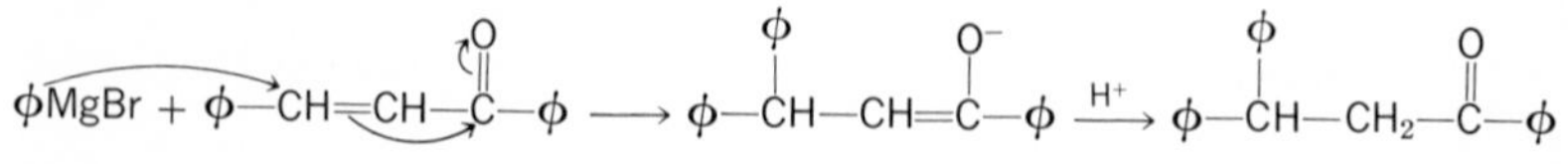

Since there is no source of protons to convert the enolate back to the ketone until workup, and workup destroys the Grignard reagent, addition to the enolate does not occur.

5. *a.*

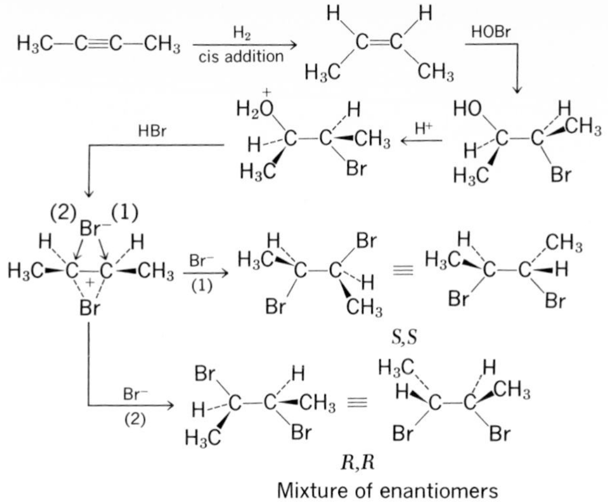

H_2
cis addition
HOBr
H^+
HBr
Br^-
(1)
(2)
S,S
R,R
Mixture of enantiomers

c.

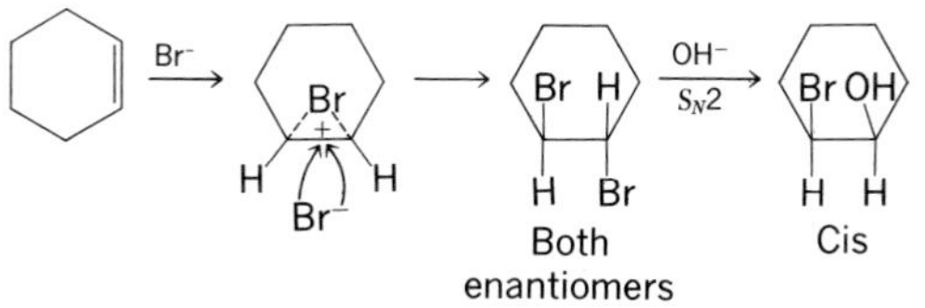

Br^-
OH^-
S_N2
Both enantiomers
Cis

e.

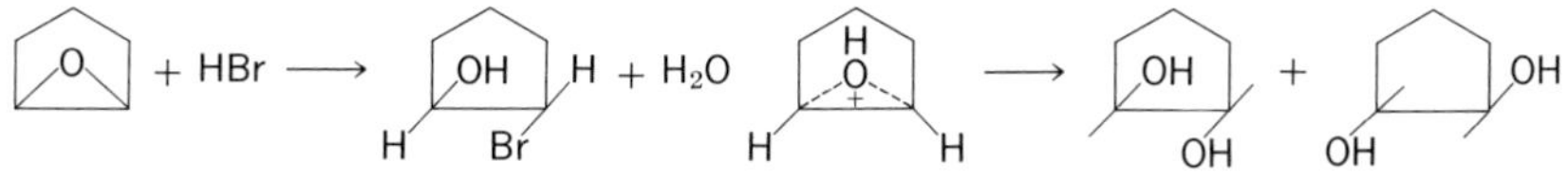

g.

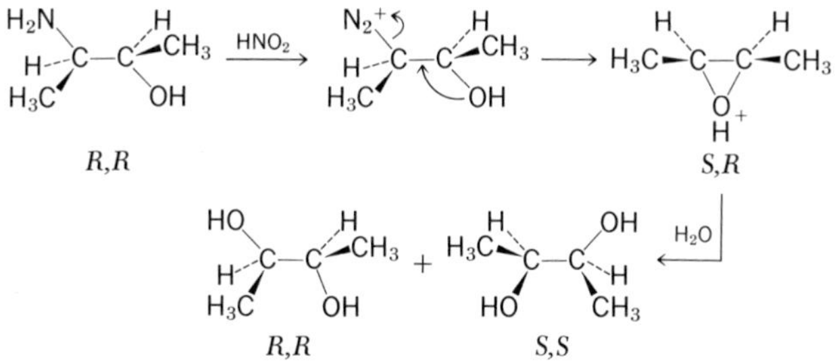

HNO_2
R,R
S,R
H_2O
R,R
S,S

CHAPTER 21

5.

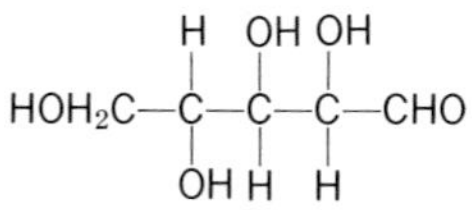

7. A is D-glucose; B is L-glucose; C is L-idose. There are other possible answers.

14. One possibility is

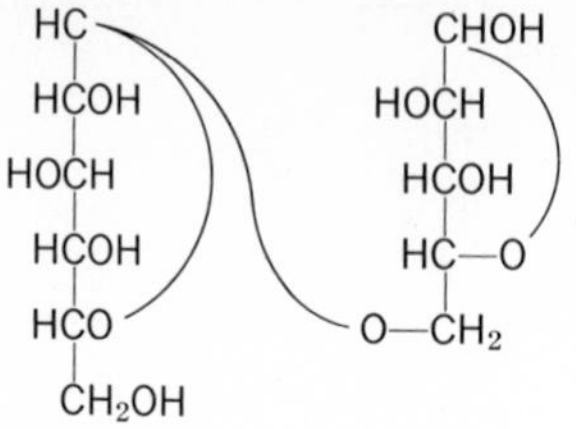

24.

A = 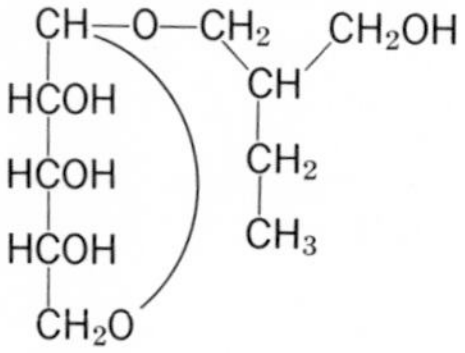

25.

A = 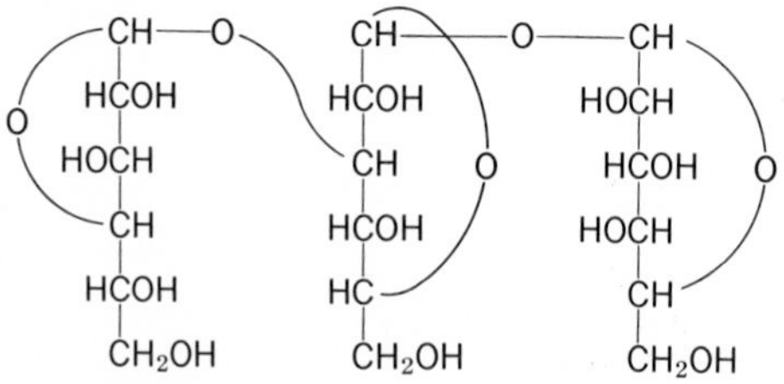

ATOMIC WEIGHTS

Aluminum	Al	26.98
Barium	Ba	137.36
Boron	B	10.82
Bromine	Br	79.92
Calcium	Ca	40.08
Carbon	C	12.01
Chlorine	Cl	35.46
Chromium	Cr	52.01
Copper	Cu	63.54
Fluorine	F	19.00
Hydrogen	H	1.01
Iodine	I	126.91
Magnesium	Mg	24.32
Manganese	Mn	54.93
Nitrogen	N	14.01
Oxygen	O	16.00
Phosphorus	P	30.98
Potassium	K	39.10
Silver	Ag	107.88
Sodium	Na	23.0
Sulfur	S	32.07
Tin	Sn	118.70
Zinc	Zn	65.38

NOTE: The weights given are rounded off to nearest two decimals.

INDEX